Adaptive Signal Processing

OTHER IEEE PRESS BOOKS

System Design for Human Interaction, *Edited by A. P. Sage*
Microcomputer Control of Power Electronics and Drives, *Edited by B. K. Bose*
Advances in Local Area Networks, *Edited by K. Kümmerle, J. O. Limb, and F. A. Tobagi*
Load Management, *Edited by S. Talukdar and C. W. Gellings*
Computers and Manufacturing Productivity, *Edited by R. K. Jurgen*
Being the Boss, *By L. K. Lineback*
Effective Meetings for Busy People, *By W. T. Carnes*
Modern Acoustical Imaging, *Edited by H. Lee and G. Wade*
Low-Temperature Electronics, *Edited by R. K. Kirschman*
Undersea Lightwave Communications, *Edited by P. K. Runge and P. R. Trischitta*
Multidimensional Digital Signal Processing, *Edited by the IEEE Multidimensional Signal Processing Committee*
Adaptive Methods for Control System Design, *Edited by M. M. Gupta*
Residue Number System Arithmetic, *Edited by M. A. Soderstrand, W. K. Jenkins, G. A. Jullien, and F. J. Taylor*
Singular Perturbations in Systems and Control, *Edited by P. V. Kokotovic and H. K. Khalil*
Getting the Picture, *By S. B. Weinstein*
Space Science and Applications, *Edited by J. H. McElroy*
Medical Applications of Microwave Imaging, *Edited by L. Larsen and J. H. Jacobi*
Modern Spectrum Analysis, *Edited by S. B. Kesler*
The Calculus Tutoring Book, *By C. Ash and R. Ash*
Imaging Technology, *Edited by H. Lee and G. Wade*
Phase-Locked Loops, *Edited by W. C. Lindsey and C. M. Chie*
VLSI Circuit Layout: Theory and Design, *Edited by T. C. Hu and E. S. Kuh*
Monolithic Microwave Integrated Circuits, *Edited by R. A. Pucel*
Next-Generation Computers, *Edited by E. A. Torrero*
Kalman Filtering: Theory and Application, *Edited by H. W. Sorenson*
Spectrum Management and Engineering, *Edited by F. Matos*
Digital VLSI Systems, *Edited by M. I. Elmasry*
Introduction to Magnetic Recording, *Edited by R. M. White*
Insights into Personal Computers, *Edited by A. Gupta and H. D. Toong*
Television Technology Today, *Edited by T. S. Rzeszewski*
The Space Station: An Idea Whose Time Has Come, *Edited by T. R. Simpson*
Marketing Technical Ideas and Products Successfully! *Edited by L. K. Moore and D. L. Plung*
The Making of a Profession: A Century of Electrical Engineering in America, *By A. M. McMahon*
Power Transistors: Device Design and Applications, *Edited by B. J. Baliga and D. Y. Chen*
VLSI: Technology and Design, *Edited by O. G. Folberth and W. D. Grobman*
General and Industrial Management, *By H. Fayol; revised by I. Gray*
A Century of Honors, *An IEEE Centennial Directory*
MOS Switched-Capacitor Filters: Analysis and Design, *Edited by G. S. Moschytz*
Distributed Computing: Concepts and Implementations, *Edited by P. L. McEntire, J. G. O'Reilly, and R. E. Larson*
Engineers and Electrons, *By J. D. Ryder and D. G. Fink*
Land-Mobile Communications Engineering, *Edited by D. Bodson, G. F. McClure, and S. R. McConoughey*
Frequency Stability: Fundamentals and Measurement, *Edited by V. F. Kroupa*
Electronic Displays, *Edited by H. I. Refioglu*
Spread-Spectrum Communications, *Edited by C. E. Cook, F. W. Ellersick, L. B. Milstein, and D. L. Schilling*
Color Television, *Edited by T. Rzeszewski*
Advanced Microprocessors, *Edited by A. Gupta and H. D. Toong*
Biological Effects of Electromagnetic Radiation, *Edited by J. M. Osepchuk*
Engineering Contributions to Biophysical Electrocardiography, *Edited by T. C. Pilkington and R. Plonsey*
The World of Large Scale Systems, *Edited by J. D. Palmer and R. Saeks*
Electronic Switching: Central Office Systems of the World, *Edited by A. E. Joel, Jr.*
A Guide for Writing Better Technical Papers, *Edited by C. Harkins and D. L. Plung*
Low-Noise Microwave Transistors and Amplifiers, *Edited by H. Fukui*
Digital MOS Integrated Circuits, *Edited by M. I. Elmasry*
Geometric Theory of Diffraction, *Edited by R. C. Hansen*
Modern Active Filter Design, *Edited by R. Schaumann, M. A. Soderstrand, and K. R. Laker*
Adjustable Speed AC Drive Systems, *Edited by B. K. Bose*
Optical Fiber Technology, II, *Edited by C. K. Kao*
Protective Relaying for Power Systems, *Edited by S. H. Horowitz*

Adaptive Signal Processing

Edited by
Leon H. Sibul
Senior Scientist and Professor of Acoustics
Pennsylvania State University

A volume in the IEEE PRESS Selected Reprint
Series, prepared under the sponsorship of the
IEEE Acoustics, Speech, and Signal Processing Society

The Institute of Electrical and Electronics Engineers, Inc., New York.

PRINTED IN THE UNITED STATES OF AMERICA

IEEE Order Number: PC02147

Library of Congress Cataloging-in-Publication Data

Adaptive signal processing.

(IEEE Press selected reprint series)
Bibliography: p.
Includes indexes.
1. Adaptive signal processing. I. Sibul, Leon H.
TK5102.5.A296 1987 621.38′043 87-2812

ISBN 0-87942-224-6

Contents

Preface

FROM 1980 to 1986, a number of books have been published on adaptive signal processing; however, in teaching my graduate course on adaptive signal processing at Pennsylvania State University, I find it useful to supplement the regular text with a collection of selected reprints. Reading the original papers gives students deeper insights into essential concepts of adaptve signal processing. A collection of classical, tutorial, and selected original research papers helps students to acquire a good perspective of the whole field of adaptive signal processing. A wide community of readers benefit from a set of selected papers that have previously been scattered throughout several journals. An IEEE reprint book is the most appropriate vehicle for such a collection of papers.

Selecting papers for this book has not been easy. Due to the page limitation, many good papers had to be excluded. The initial criteria for the selection of papers was the applicability of thc papcrs to my own learning and teaching of adaptive signal processing. I gave preference to easily readable papers in the tutorial style. The final selection of papers contains both seminal papers and papers of current research interest.

This book is organized around three themes: applications, adaptive algorithms and their properties, and lattice filters and their properties. The collection of papers on the application of adaptive signal processing techniques shows how the same fundamental concepts can be used for diverse applications. These applications include linear prediction, noise and echo cancellers, adaptive equalization, system identification, and adaptive beamforming. The key to successful adaptive signal processing is the understanding of adaptive algorithms and their properties. Some of the fundamental papers on this topic are collected in Part II. Part III contains papers on the latest research topic in adaptive signal processing: lattice filters and their properties.

Due to the space limitations, papers in two areas have not been included: applications of adaptive filters to spectral estimation, and VLSI implementation of adaptive processors. An IEEE reprint book on spectrum analysis is presently available, and the second topic deserves a reprint book in its own right.

I would like to thank the reviewers who made many helpful suggestions and pointed out several important papers that I had overlooked. Their valuable comments helped me to broaden my perspective on adaptive signal processing. Similarly, many of my students and colleagues offered valuable comments on the selection of papers. However, responsibility for the final selection of papers is mine.

I would like to acknowledge the very valuable contributions of John A. Tague, who helped me to compile and update the bibliography, and Curtis I. Caldwell, who reviewed and edited my manuscript.

Introduction

THE literature on adaptive signal processing spans more than twenty years. During this period theoretical problems associated with adaptive signal processing have been pursued with unprecedented intellectual vigor. This activity has been motivated by the practical need for understanding these rather sophisticated signal processing systems. What is adaptive signal processing, what are its applications, how is it done, and what is its future? How did these interesting signal processing concepts develop? These are the questions that this collection of papers attempts to answer.

Adaptive signal processors are learning and self-optimizing systems. In general, adaptive signal processors have a finite number of parameters that are adjusted by adaptive algorithms to optimize some performance criterion. For example, in an adaptive beamformer, array shading coefficients are adjusted to maximize the signal-to-interference ratio. In an adaptive predictor, prediction filter coefficients are adjusted to minimize prediction error in the mean square sense. Adaptive noise and echo cancellers minimize undesired interferences. Adaptive signal processors are learning systems in the sense that they "sense and learn" their unique operating environment and adjust their parameters so that the system will be optimized. For example, an adaptive beamformer adjusts its shading coefficients so that the beam pattern has its nulls in the direction of major noise sources. The information needed to optimize a second order performance criterion is contained in the noise covariance matrix. System optimization requires estimation and inversion of this covariance matrix. This is done by adaptive algorithms. Hence, the analysis and synthesis of adaptive algorithms is a recurring theme in most papers in this collection. Understanding properties of adaptive algorithms is the central issue in adaptive signal processing research.

The adaptive signal processing ideas started along two tracks: adaptive beamforming and adaptive equalization.

IEEE Transactions on Antennas and Propagation published its first special issue on adaptive antennas in March 1964 [1]. Papers in this issue emphasized retrodirective and self-focusing array systems. These arrays were largely based upon phase lock loop and phase-conjugate network ideas. At the same time, ideas for adaptive interference nulling techniques were developing. Based on his earlier work, Paul Howells obtained a patent for an IF sidelobe canceller in 1965 [2]. This was the first successful interference nulling technique. That same year in France, H. Mermoz published his Ph.D. dissertation entitled "Adaptive Filtering and Optimal Utilization of an Antenna." [3] This seminal work started the very productive French adaptive beamformer research. In 1967, B. Widrow, *et al* published their classic paper on adaptive antenna systems [4]. In this paper, minimization of the mean square error was accomplished by the now well-known LMS algorithm. The LMS algorithm is a simple, easily understandable, robust algorithm. Even though it might converge slowly for large eigenvalue spreads, this simple algorithm has contributed much to the development and application of digital adaptive algorithms. After publication of this paper, the majority of adaptive signal processing systems used digital algorithms. A more detailed discussion of the development of various adaptive algorithms is presented in the introduction to Part II.

1965 was also the beginning of a sequence of papers on adaptive equalization of digital transmission systems. Lucky, of Bell Labs, published his classic papers on adaptive equalization [5], [6]. A large number of important papers by his colleagues followed. Adaptive signal processing ideas were also applied to adaptive noise and echo cancellation. At the present time, literally thousands of echo cancellers are being used in the telephone systems throughout the world.

The adaptive signal processing theory and applications continued to develop. In 1976, the second special issue on adaptive antennas was published by the IEEE Antennas and Propagation Society under the able editorship of William F. Gabriel [7]. This issue contained 15 important papers by major contributors to adaptive antenna research. This special issue is a good summary of adaptive antenna research up to the date of its publication. The classic paper by S. P. Applebaum is reprinted from this time. Other papers in this special issue are highly recommended supplementary reading.

The third special issue on adaptive processing antenna systems, again edited by William E. Gabriel, includes a number of papers on the rapidly evolving new technology of high-resolution spatial spectrum estimation [8]. This special issue clearly illustrated enormous growth and progress in adaptive array processing during the past decade.

In parallel to the adaptive array processing work, other applications of adaptive signal processing continued to develop. In June 1981, *IEEE Transactions on Circuits and Systems* and *IEEE Transactions on Acoustics, Speech, and Signal Processing* published a joint special issue on adaptive signal processing [9]. Many papers in this special issue were concerned with the lattice filter realizations of adaptive algorithms. Lattice filters can be viewed as stage-by-stage orthogonalization of input data. This property results in faster convergence than the previous tapped-delay-line implementation of adaptive filters. When a lattice filter is used as a linear predictor, the predictor order can be increased simply by adding another lattice section without changing any of the previous sections. Thus, lattice implementation of a predictor of order N also implements all lower order predictors. The tapped-delay-line implementation of the linear predictor does not have this useful modularity property. Because the fast convergence and modularity of lattice filters have become popular in the adaptive signal processing community, Part III is a compilation of selected papers on lattice filters. Discussion of lattice filters has taken us to the current state of research in adaptive signal processing. Another interesting and important area of current research is matching algorithms to VLSI architectures for efficient, real-time signal processing. This evolving field is not included in this collection, since there is insufficient space to do full justice to this topic.

References

[1] *IEEE Trans. Antennas Propagation*, Special issue on active and adaptive antennas, vol. AP-12, Mar. 1964.

[2] Howells, P. W., ''Intermediate Frequency Sidelobe Canceller,'' *U.S. Patent 3,202,990,* August 24, 1965 (filed May 4, 1959).

[3] Mermoz, H., ''Adaptive filtering and optimal utilization of an antenna,'' Ph.D. Thesis, Institute Polytechnique, Grenoble, France, October 4, 1965.

[4] Widrow, B., *et al*, ''Adaptive antenna systems,'' *Proc. IEEE*, vol. 55, no. 12, pp. 2143–2159, Dec. 1967.

[5] Lucky, R. W., ''Automatic equalization for digital communication,'' *Bell Syst. Tech. J.,* vol. XLIV, no. 4, pp. 547–588, Apr. 1965.

[6] Lucky, R. W., ''Techniques for adaptive equalization of digital communication systems,'' *Bell Syst. Tech. J.,* pp. 255–286, Feb. 1966.

[7] *IEEE Trans. Antennas Propagation*, Special issue on adaptive antennas, vol. AP-24, no. 5, Sept. 1976.

[8] *IEEE Trans. Antennas Propagation*, Special issue on adaptive processing antenna systems, vol. AP-34, no. 3, Mar. 1986.

[9] *IEEE Trans. on Circuits and Systems,* Joint special issue on adaptive signal processing also published in *IEEE Trans. Acous., Speech, Signal Processing,* vol. ASSP-29, no. 3, CAS-28, no. 6, June 1981.

Part I
Applications

MANY adaptive signal processing techniques are based on the solution of the linear prediction problem. For this reason we begin this collection of papers with J. Makhoul's popular paper "Linear Prediction: A Tutorial Review." Linear prediction has applications to noise and echo cancellation, system identification and modeling, time series modeling, speech processing, spectral estimation, and many other adaptive signal processing applications. The basic concepts of linear prediction as well as many applications of linear prediction are discussed in this very readable tutorial review paper. This paper also contains an extensive list of references on applications and theory of linear predictions. Another very highly recommended paper with a different perspective on linear prediction is T. Kailath's paper "A View of Three Decades of Linear Filtering Theory." [1]

One of the simplest and most effective adaptive signal processing techniques is adaptive noise cancelling. The second paper in this chapter is "Adaptive Noise Cancelling: Principles and Applications," by Widrow, *et al.* This paper describes the concept of noise cancellation and presents theoretical performance analyses for different kinds of signals and noise. The second part of the paper discusses interesting applications of adaptive noise cancellation technique. These applications include cancelling interference in electrocardiography, noise in speech signal, and antenna sidelobe interference. A special interference canceller that is widely used is an echo canceller. In the echo canceller, an echo in the telephone network is suppressed by subtracting a predicted echo from the transmission path which contains the actual echo. The paper by D. Messerschmitt is a recent tutorial review paper on echo cancellation techniques as used in voice and data transmission applications. Two outstanding papers on echo cancellation that have not been reprinted here are Sondhi and Berkley's *IEEE Proceedings* invited paper [2] and Sondhi's original *Bell System Technical Journal* paper [3].

In 1965 and 1966, R. W. Lucky published two important papers on adaptive equalization to correct intersymbol interference in digital communication systems [4], [5]. In his first paper, the tap gains of the transversal equalization filter were adjusted during a training period of test pulse transmission before the actual data transmission. His second paper, which is reprinted here, allowed adjustment of tap gains during the data transmission in response to the changes in the channel characteristics. These two papers are considered classic papers on adaptive equalization. The next paper in this collection on adaptive equalization is A. Gersho's "Adaptive Equalization of Highly Dispersive Channels for Data Transmission." This paper describes how a sequence of isolated pulses were used as a test signal in the training mode. An adaptive equalization filter was adjusted to minimize the mean square error between actual and desired pulse shapes at the filter output.

Convergence of the adaptive algorithm was achieved even for highly dispersive channels for which binary data transmission would have been impossible without equalization. This paper also derives important results on convergence properties of stochastic gradient algorithms. The paper by Satorius and Pack discusses applications of lattice algorithms to transmission channel equalization. The self orthogonalization properties of lattice algorithms increase their convergence rate even when the channel correlation matrix has a large eigenvalue spread. In such cases, the simple gradient-based algorithms can converge agonizingly slowly. Other papers that discuss properties of adaptive algorithms and their application to adaptive equalization are the papers by Ungerboeck, Godard, Gitlin and Magee, Falconer and Ljung, and Gitlin, Mazo, and Taylor. These papers have been reprinted in the next part. In addition to the above papers on adaptive linear equalization, the reader's attention is called to a considerable body of literature on decision feedback equalization [6–8]. In this collection we have focussed on linear adaptive equalization.

System identification is an important topic in automatic control system theory. Many problems in adaptive signal processing involve representation of signals in terms of a finite parameter linear filter that is driven by an uncorrelated noise sequence. This point of view reduces the signal representation problem to a parameter estimation problem. B. Friedlander, in his paper "System Identification Techniques for Adaptive Signal Processing," explores the interconnection between system identification and adaptive signal processing. This paper has a good list of references on system identification and parameter estimation.

An important application of adaptive signal processing is adaptive beamforming or adaptive array processing. In many cases adaptive array processing can give dramatic improvements in the signal-to-interference ratio as compared to conventional beamforming. In conventional beamforming, the beam pattern design is based on implicit *a priori* assumptions, such as isotropic background noise. These assumptions may be grossly violated in a given situation. On the other hand, adaptive beamformers adjust their beam patterns to suppress directional noise sources that they encounter in a given situation. Adaptive beamformers are truly learning and self-optimizing systems.

From the large body of literature on adaptive beamforming we have selected only three papers for this collection. The first reason for this is the space limitation. The second reason is that there is already a full collection of papers on array processing [9]. In addition to this there are three special issues of *IEEE Transactions on Antennas and Propagation* which are focused on adaptive arrays [10–12]. Also, there are several books on adaptive arrays [13–16]. Of particular interest is the forthcoming book by Professor Compton, a productive contributor to adaptive array processing literature. A recent book by B. Widrow also has two chapters on adaptive beamforming [17].

Three papers are reprinted here: Widrow's classic paper "Adaptive Antenna Systems," O. L. Frost's "An Algorithm for Linearly Constrained Adaptive Array Processing," and S. Applebaum's "Adaptive Arrays." An additional introductory paper on adaptive array processing is W. F. Gabriel's invited paper in the *IEEE Proceedings* [18]. Unfortunately, Gabriel's paper is too lengthy for this volume. However, it is highly recommended to readers interested in adaptive array processing.

In Widrow's paper, adaptation is based on minimization of the mean square error by the now well-known LMS algorithm. In this paper, the array processing system operates with knowledge of the direction and spectrum of the desired signal. The directional characteristics of the noise field are learned during adaptation. In his paper, Frost presents a linearly constrained algorithm which is able to maintain a chosen frequency response in the look-direction while minimizing output noise power. Frost's algorithm has the unique advantage over other constrained optimization algorithms in that Frost's algorithm has a self-correcting feature which makes the algorithm numerically stable in spite of roundoff and truncation errors. The algorithm can be operated for an arbitrarily long time because error accumulation has been eliminated. S. Applebaum's paper was first published in 1966 as a Syracuse University Research Corporation technical report. This classic report deals with some of the basic concepts of the adaptive arrays, and presents a method for determining the array weights that maximize signal-to-noise ratio for any arbitrary noise environment. The paper shows that this adaptive array processing technique is a generalization of the sidelobe cancellation technique. As discussed in the Introduction, the sidelobe cancellation technique was the first workable adaptive array processing technique. It was a simple but effective array processing technique. Much of the later research in adaptive beamforming was motivated by the elegance and success of these seminal concepts. These three papers are only a small sample of the vast body of adaptive beamforming literature and do not fully represent the scope of this research. Nevertheless, these adaptive beamforming papers illustrate an important application of adaptive signal processing concepts. As such, they form an important complement to temporal applications of adaptive signal processing.

References

[1] Kailath, T., "A view of three decades of linear-filtering theory," *IEEE Trans. Inform. Theory,* vol. IT-20, pp. 145–181, Mar. 1974.

[2] Sondhi, M. M., and Berkley, D. A., "Silencing echoes on the telephone network," *Proc. IEEE,* vol. 68, no. 8, pp. 948–963, Aug. 1980.

[3] Sondhi, M. M., "An adaptive echo canceller," *Bell Syst. Tech. J.,* vol. 46, pp. 497–520, Mar. 1967.

[4] Lucky, R. W., "Automatic equalization for digital communication," *Bell Syst. Tech. J.,* vol. 44, pp. 547–588, Apr. 1965.

[5] Lucky, R. W., "Techniques for adaptive equalization of digital communication systems," *Bell Syst. Tech. J.,* vol. 45, no. 2, pp. 255–286, Feb. 1966.

[6] Monsen, P., "Feedback equalization for fading dispersive channels," *IEEE Trans. Inform. Theory,* vol. IT-17, no. 2, pp. 56–64, Jan. 1971.

[7] Salz, J., "Optimum mean-square decision feedback equalization," *Bell Syst. Tech. J.,* vol. 52, pp. 1341–1373, Oct. 1973.

[8] Dutweiler, D. L., Mazo, J. E., and Messerschmitt, D. G., "Error propagation in decision-feedback equalizer," *IEEE Trans. Inform. Theory,* vol. IT-20, pp. 490–497, July 1974.

[9] Haykin, S., *Array Processing Application to Radar.* Stroudsburg, PA: Dowden, Hutchinson & Ross, Inc., 1980.

[10] *IEEE Trans. on Antennas and Propagation,* Special Issue on Active and Adaptive Antennas, vol. AP-12, no. 2, Mar. 1964.

[11] *IEEE Trans. on Antennas and Propagation,* Special Issue on Adaptive Antennas, vol. AP-24, no. 5, Sept. 1976.

[12] *IEEE Trans. on Antennas and Propagation,* Special Issue on Adaptive Processing Antenna Systems, vol. AP-34, no. 3, Mar. 1986.

[13] Monzingo, R. A., and Miller, T. W., *Introduction to Adaptive Arrays.* New York, NY: John Wiley, 1980.

[14] Hudson, J. E., *Adaptive Array Principles.* Stevenage, U.K.: P. Peregrinus, 1981.

[15] Haykin, S., Ed., *Array Signal Processing.* Englewood Cliffs, NJ: Prentice-Hall, 1985.

[16] Compton, R. T., Jr., *Adaptive Antennas: Concepts and Applications.* Englewood Cliffs, NJ: Prentice-Hall, 1988.

[17] Widrow, B., and Stearns, S. D., *Adaptive Signal Processing.* Englewood Cliffs, NJ: Prentice-Hall, 1985.

[18] Gabriel, W. F., "Adaptive arrays—an introduction," *Proc. IEEE,* vol. 64, no. 2, pp. 239–272, Feb. 1976.

Linear Prediction: A Tutorial Review

JOHN MAKHOUL, MEMBER, IEEE

Invited Paper

Abstract—This paper gives an exposition of linear prediction in the analysis of discrete signals. The signal is modeled as a linear combination of its past values and present and past values of a hypothetical input to a system whose output is the given signal. In the frequency domain, this is equivalent to modeling the signal spectrum by a pole-zero spectrum. The major part of the paper is devoted to all-pole models. The model parameters are obtained by a least squares analysis in the time domain. Two methods result, depending on whether the signal is assumed to be stationary or nonstationary. The same results are then derived in the frequency domain. The resulting spectral matching formulation allows for the modeling of selected portions of a spectrum, for arbitrary spectral shaping in the frequency domain, and for the modeling of continuous as well as discrete spectra. This also leads to a discussion of the advantages and disadvantages of the least squares error criterion. A spectral interpretation is given to the normalized minimum prediction error. Applications of the normalized error are given, including the determination of an "optimal" number of poles. The use of linear prediction in data compression is reviewed. For purposes of transmission, particular attention is given to the quantization and encoding of the reflection (or partial correlation) coefficients. Finally, a brief introduction to pole-zero modeling is given.

I. Introduction

A. Overview

THE MATHEMATICAL analysis of the behavior of general dynamic systems (be they engineering, social, or economic) has been an area of concern since the beginning of this century. The problem has been pursued with accelerated vigor since the advent of electronic digital computers over two decades ago. The analysis of the outputs of dynamic systems was for the most part the concern of "time series analysis," which was developed mainly within the fields of statistics, econometrics, and communications. Most of the work on time series analysis was actually done by statisticians. More recently, advances in the analysis of dynamic systems have been made in the field of control theory based on state-space concepts and time domain analysis.

This paper is a tutorial review of one aspect of time series analysis: linear prediction (defined here). The exposition is based on an intuitive approach, with emphasis on the clarity of ideas rather than mathematical rigor. Although the large body of related literature available on this topic often requires advanced knowledge of statistics and/or control theory concepts, this paper employs no control theory concepts *per se* and only the basic notions of statistics and random processes. For example, the very important statistical concepts of *consistency* and *efficiency* [74], [75] in the estimation of parameters will not be dealt with. It is hoped this paper will serve as a simple introduction to some of the tools used in time series analysis, as well as be a detailed analysis of those aspects of linear prediction of interest to the specialist.

Manuscript received July 21, 1974; revised November, 1974. This work was supported by the Information Processing Techniques Branch of the Advanced Research Projects Agency under Contract DAHC15-17-C-0088.

The author is with Bolt Beranek and Newman, Inc., Cambridge, Mass. 02138.

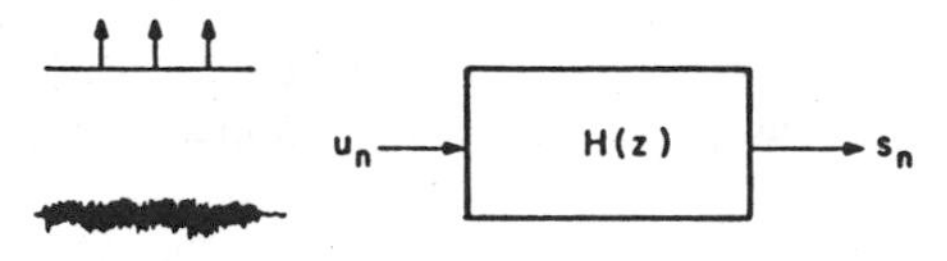

Fig. 1. Discrete speech production model.

B. Current Applications

Before we delve into signal analysis, we shall give three examples of the types of problems that are of current interest.[1] These examples will then serve to illustrate some of the concepts that are developed.

Neurophysics [15], [32], [36], [102]: The spontaneous electrical brain activity is normally measured by means of electrodes placed on the patient's scalp. The recordings, known as electroencephalograms (or EEG signals), show certain periodicities (sharp resonances) accompanied by some randomness. These signals are believed to carry information about the medical status of the brain and are used by physicians as a means of diagnosis. It is of interest to detect the presence, position, and strength of the different resonances, known as rhythms. The three most common rhythms are known as the alpha, beta, and delta rhythms. Therefore, the basic interest here is to describe the spectrum in a simple mathematical manner that would yield the characteristics of the different rhythms.

Geophysics [84]-[87], [114]: In one of the successful methods of oil exploration, a charge of dynamite is exploded in the earth, and the resulting vibrations at various points on the surface of the ground are recorded by a seismograph as seismic traces. The job of the geophysicist is to use these traces in the determination of the structure of the sedimentary rock layers. Such information is then used to decide on the presence of oil in that area. Of interest here are the direct arrival times and strengths of the deep reflections of the explosion, which are then used to determine the layered structure. If somehow one is able to remove (deconvolve) the impulse response of the structure from the seismic trace, the desired arrival times should appear as impulses of different phases and amplitudes.

Speech Communication [10], [33], [47], [50], [51], [62], [68], [89]: In EEG analysis, the spectrum of the recorded signal was of interest. In seismic analysis, the spectral properties of the seismic trace were of interest only to facilitate the deconvolution process in order to obtain the desired impulses. In the analysis of speech, both types of information are of interest.

Fig. 1 shows a rather successful model of speech production. The model consists of a filter that is excited by either a quasi-periodic train of impulses or a random noise source. The periodic source produces voiced sounds such as vowels and

[1] For applications to economic and industrial time series, see for example [17].

Reprinted from *Proc. IEEE,* vol. 63, no. 4, pp. 561–580, Apr. 1975.

nasals, and the noise source produces unvoiced or fricated sounds such as the fricatives (*f, th, s, sh*). The parameters of the filter determine the identity (spectral characteristics) of the particular sound for each of the two types of excitation.

Given a particular speech signal, it is of interest to determine the general type of sound it is, voiced or fricated, and if voiced what the pitch period is (i.e., distance between pitch pulses). In addition, one is interested in the identity of the sound which can be obtained from the spectrum. Such derived information can then be used in an automatic speech recognition system or a speech compression system.

C. Linear Prediction

In applying time series analysis to the aforementioned applications, each continuous-time signal $s(t)$ is sampled to obtain a discrete-time[2] signal $s(nT)$, also known as a time series, where n is an integer variable and T is the sampling interval. The sampling frequency is then $f_s = 1/T$. (Henceforth, we shall abbreviate $s(nT)$ by s_n with no loss in generality.)

A major concern of time series analysis [6], [11], [12], [14], [17], [43], [45], [46], [54], [105], [112] has been the estimation of power spectra, cross-spectra, coherence functions, autocorrelation and cross-correlation functions. A more active concern at this time is that of system modeling. It is clear that if one is successful in developing a parametric model for the behavior of some signal, then that model can be used for different applications, such as prediction or forecasting, control, and data compression.

One of the most powerful models currently in use is that where a signal s_n is considered to be the output of some system with some unknown input u_n such that the following relation holds:

$$s_n = -\sum_{k=1}^{p} a_k s_{n-k} + G \sum_{l=0}^{q} b_l u_{n-l}, \qquad b_0 = 1 \tag{1}$$

where a_k, $1 \leqslant k \leqslant p$, b_l, $1 \leqslant l \leqslant q$, and the gain G are the parameters of the hypothesized system. Equation (1) says that the "output" s_n is a linear function of past outputs and present and past inputs. That is, the signal s_n is *predictable* from *linear* combinations of past outputs and inputs. Hence the name *linear prediction*.

Equation (1) can also be specified in the frequency domain by taking the z transform on both sides of (1). If $H(z)$ is the transfer function of the system, as in Fig. 1, then we have from (1):

$$H(z) = \frac{S(z)}{U(z)} = G \frac{1 + \sum_{l=1}^{q} b_l z^{-l}}{1 + \sum_{k=1}^{p} a_k z^{-k}} \tag{2}$$

where

$$S(z) = \sum_{n=-\infty}^{\infty} s_n z^{-n} \tag{3}$$

is the z transform of s_n, and $U(z)$ is the z transform of u_n. $H(z)$ in (2) is the general *pole-zero model*. The roots of the numerator and denominator polynomials are the zeros and poles of the model, respectively.

There are two special cases of the model that are of interest:

1) all-zero model: $a_k = 0$, $1 \leqslant k \leqslant p$
2) all-pole model: $b_l = 0$, $1 \leqslant l \leqslant q$.

The all-zero model is known in the statistical literature as the *moving average* (MA) model, and the all-pole model is known as the *autoregressive* (AR) model [17]. The pole-zero model is then known as the *autoregressive moving average* (ARMA) model. In this paper we shall use the pole-zero terminology since it is more familiar to engineers.

The major part of this paper will be devoted to the all-pole model. This has been, by far, the most widely used model. Historically, the first use of an all-pole model in the analysis of time series is attributed to Yule [115] in a paper on sunspot analysis. Work on this subject, as well as on time series analysis in general, proceeded vigorously after 1933 when Kolmogorov laid a rigorous foundation for the theory of probability. Later developments by statisticians, such as Cramér and Wold, culminated in the parallel and independent work of Kolmogorov [58] and Norbert Wiener [107] on the prediction and filtering of stationary time series. For a bibliography on time series through the year 1959, see the encyclopedic work edited by Wold [113]. For a discussion of all-pole (autoregressive) models see, for example, [17], [45], [105], [112].

Much of the recent work on system modeling has been done in the area of control theory under the subjects of estimation and system identification. Recent survey papers with extensive references are those of Åström and Eykhoff [8] and Nieman *et al.* [73]. The December 1974 issue of the IEEE TRANSACTIONS ON AUTOMATIC CONTROL is devoted to the subject of system identification. Another relevant survey paper is that of Kailath [55] on linear filtering theory. Related books are those of Lee [60], Sage and Melsa [88], and Eykhoff [30].

D. Paper Outline

Sections II–V deal exclusively with the all-pole model. In Section II, the estimation of model parameters is derived in the time domain by the method of least squares. The resulting normal equations are obtained for deterministic as well as random signals[3] (both stationary and nonstationary). Direct and iterative techniques are presented for the computation of the predictor coefficients, and the stability of the all-pole filter $H(z)$ is discussed. The response of the all-pole filter is then analyzed for two important types of input excitation: a deterministic impulse and statistical white noise.

In Section III, the all-pole modeling of a signal is derived completely in the frequency domain. The method of least squares translates into a spectral matching method where the signal spectrum is to be matched or fitted by a model spectrum. This formulation allows one to perform arbitrary spectral shaping before modeling. This viewpoint has special relevance today with the availability of hardware spectrum analyzers and fast Fourier transform techniques [21]. (We point out that all-pole modeling by linear prediction is identical to the method of maximum entropy spectral estimation [18], [96].)

Section IV gives a detailed discussion of the advantages and disadvantages of the least squares error criterion. The properties of the normalized error are reviewed. Its use is discussed

[2] See [80] for an exposition of the terminology in digital signal processing.

[3] See [12] for a description of deterministic and random signals.

in measuring the ill-conditioning of the normal equations, and in determining an optimal number of poles.

Section V discusses the use of linear prediction in data compression. Alternate representations of the linear predictor are presented and their properties under quantization are discussed. Particular emphasis is given to the quantization and encoding of the reflection (or partial correlation) coefficients.

Finally, in Section VI, a brief discussion of pole-zero modeling is given, with emphasis on methods presented earlier for the all-pole case.

II. Parameter Estimation

A. All-Pole Model

In the all-pole model, we assume that the signal s_n is given as a linear combination of past values and some input u_n:

$$s_n = -\sum_{k=1}^{p} a_k s_{n-k} + G u_n \tag{4}$$

where G is a gain factor. This model is shown in Fig. 2 in the time and frequency domains. The transfer function $H(z)$ in (2) now reduces to an all-pole transfer function

$$H(z) = \frac{G}{1 + \sum_{k=1}^{p} a_k z^{-k}}. \tag{5}$$

Given a particular signal s_n, the problem is to determine the predictor coefficients a_k and the gain G in some manner.

The derivations will be given using an intuitive least squares approach, assuming first that s_n is a deterministic signal and then that s_n is a sample from a random process. The results are identical to those obtained by the method of maximum likelihood [6], [74], [75] with the assumption that the signal is Gaussian [60], [73]. The reader is reminded of the existence of more general least squares methods such as weighted and *a priori* least squares [16], [90].

B. Method of Least Squares

Here we assume that the input u_n is totally unknown, which is the case in many applications, such as EEG analysis. Therefore, the signal s_n can be predicted only approximately from a linearly weighted summation of past samples. Let this approximation of s_n be $\tilde{s}_n$, where

$$\tilde{s}_n = -\sum_{k=1}^{p} a_k s_{n-k}. \tag{6}$$

Then the error between the actual value s_n and the predicted value $\tilde{s}_n$ is given by

$$e_n = s_n - \tilde{s}_n = s_n + \sum_{k=1}^{p} a_k s_{n-k}. \tag{7}$$

e_n is also known as the *residual*. In the method of least squares the parameters a_k are obtained as a result of the minimization of the mean or total squared error with respect to each of the parameters. (Note that this problem is identical to the problem of designing the optimal one-step prediction digital Wiener filter [85].)

The analysis will be developed along two lines. First, we assume that s_n is a deterministic signal, and then we give

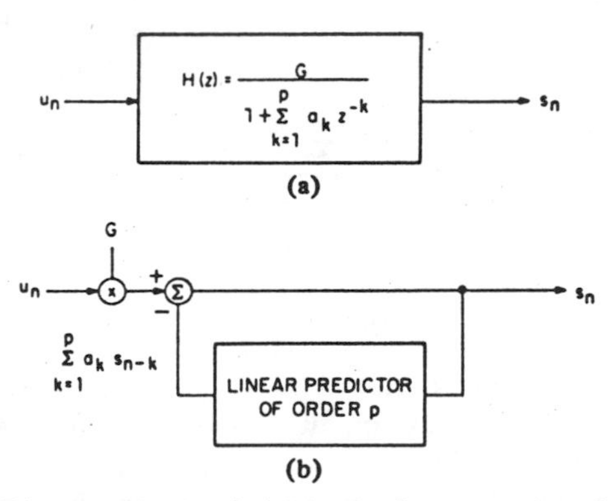

Fig. 2. (a) Discrete all-pole model in the frequency domain. (b) Discrete all-pole model in the time domain.

analogous derivations assuming that s_n is a sample from a random process.

1) Deterministic Signal: Denote the total squared error by E, where

$$E = \sum_n e_n^2 = \sum_n \left(s_n + \sum_{k=1}^{p} a_k s_{n-k}\right)^2. \tag{8}$$

The range of the summation in (8) and the definition of s_n in that range is of importance. However, let us first minimize E without specifying the range of the summation. E is minimized by setting

$$\frac{\partial E}{\partial a_i} = 0, \qquad 1 \leq i \leq p. \tag{9}$$

From (8) and (9) we obtain the set of equations:

$$\sum_{k=1}^{p} a_k \sum_n s_{n-k} s_{n-i} = -\sum_n s_n s_{n-i}, \qquad 1 \leq i \leq p. \tag{10}$$

Equations (10) are known in least squares terminology as the *normal equations.* For any definition of the signal s_n, (10) forms a set of p equations in p unknowns which can be solved for the predictor coefficients $\{a_k, 1 \leq k \leq p\}$ which minimize E in (8).

The minimum total squared error, denoted by E_p, is obtained by expanding (8) and substituting (10). The result can be shown to be

$$E_p = \sum_n s_n^2 + \sum_{k=1}^{p} a_k \sum_n s_n s_{n-k}. \tag{11}$$

We shall now specify the range of summation over n in (8), (10), and (11). There are two cases of interest, which will lead to two distinct methods for the estimation of the parameters.

a) Autocorrelation method: Here we assume that the error in (8) is minimized over the infinite duration $-\infty < n < \infty$. Equations (10) and (11) then reduce to

$$\sum_{k=1}^{p} a_k R(i-k) = -R(i), \qquad 1 \leq i \leq p \tag{12}$$

$$E_p = R(0) + \sum_{k=1}^{p} a_k R(k) \tag{13}$$

where

$$R(i) = \sum_{n=-\infty}^{\infty} s_n s_{n+i} \tag{14}$$

is the autocorrelation function of the signal s_n. Note that $R(i)$ is an even function of i, i.e.,

$$R(-i) = R(i). \tag{15}$$

Since the coefficients $R(i-k)$ form what often is known as an autocorrelation matrix, we shall call this method the *autocorrelation method*. An autocorrelation matrix is a symmetric Toeplitz matrix. (A Toeplitz matrix is one where all the elements along each diagonal are equal [42].)

In practice, the signal s_n is known over only a finite interval, or we are interested in the signal over only a finite interval. One popular method is to multiply the signal s_n by a *window* function w_n to obtain another signal s'_n that is zero outside some interval $0 \leqslant n \leqslant N-1$:

$$s'_n = \begin{cases} s_n w_n, & 0 \leqslant n \leqslant N-1 \\ 0, & \text{otherwise.} \end{cases} \tag{16}$$

The autocorrelation function is then given by

$$R(i) = \sum_{n=0}^{N-1-i} s'_n s'_{n+i}, \quad i \geqslant 0. \tag{17}$$

The shape of the window function w_n can be of importance. The subject is discussed further in Section III.

b) Covariance method: In contrast with the autocorrelation method, here we assume that the error E in (8) is minimized over a finite interval, say, $0 \leqslant n \leqslant N-1$. Equations (10) and (11) then reduce to

$$\sum_{k=1}^{p} a_k \varphi_{ki} = -\varphi_{0i}, \quad 1 \leqslant i \leqslant p \tag{18}$$

$$E_p = \varphi_{00} + \sum_{k=1}^{p} a_k \varphi_{0k} \tag{19}$$

where

$$\varphi_{ik} = \sum_{n=0}^{N-1} s_{n-i} s_{n-k} \tag{20}$$

is the covariance of the signal s_n in the given interval. The coefficients φ_{ki} in (18) form a covariance matrix, and, therefore, we shall call this method the *covariance method*. From (20) it can be easily shown that the covariance matrix φ_{ik} is symmetric, i.e., $\varphi_{ik} = \varphi_{ki}$. However, unlike the autocorrelation matrix, the terms along each diagonal are not equal. This can be seen by writing from (20)

$$\varphi_{i+1,k+1} = \varphi_{ik} + s_{-i-1} s_{-k-1} - s_{N-1-i} s_{N-1-k}. \tag{21}$$

Note from (21) also that values of the signal s_n for $-p \leqslant n \leqslant N-1$ must be known: a total of $p+N$ samples. The covariance method reduces to the autocorrelation method as the interval over which n varies goes to infinity.

We point out here that the covariance method is similar to the method of Prony [49], [71] where a signal is approximated by the summation of a set of damped exponentials.

2) Random Signal: If the signal s_n is assumed to be a sample of a random process, then the error e_n in (7) is also a sample of a random process. In the least squares method, we minimize the expected value of the square of the error. Thus

$$E = \mathcal{E}(e_n^2) = \mathcal{E}\left(s_n + \sum_{k=1}^{p} a_k s_{n-k}\right)^2. \tag{22}$$

Applying (9) to (22), we obtain the normal equations:

$$\sum_{k=1}^{p} a_k \, \mathcal{E}(s_{n-k} s_{n-i}) = -\mathcal{E}(s_n s_{n-i}), \quad 1 \leqslant i \leqslant p. \tag{23}$$

The minimum average error is then given by

$$E_p = \mathcal{E}(s_n^2) + \sum_{k=1}^{p} a_k \, \mathcal{E}(s_n s_{n-k}). \tag{24}$$

Taking the expectations in (23) and (24) depends on whether the process s_n is stationary or nonstationary.

a) Stationary case: For a stationary process s_n, we have

$$\mathcal{E}(s_{n-k} s_{n-i}) = R(i-k) \tag{25}$$

where $R(i)$ is the autocorrelation of the process. Equations (23) and (24) now reduce to equations identical to (12) and (13), respectively. The only difference is that here the autocorrelation is that of a stationary process instead of a deterministic signal. For a stationary (and ergodic) process the autocorrelation can be computed as a time average [12]. Different approximations have been suggested in the literature [54] for estimating $R(i)$ from a finite known signal s_n. One such approximation is given by (17).[4] Using this estimate in the stationary case gives the same solution for the coefficients a_k as the autocorrelation method in the deterministic case.

b) Nonstationary case: For a nonstationary process s_n, we have

$$\mathcal{E}(s_{n-k} s_{n-i}) = R(n-k, n-i) \tag{26}$$

where $R(t, t')$ is the nonstationary autocorrelation between times t and t'. $R(n-k, n-i)$ is a function of the time index n. Without loss of generality, we shall assume that we are interested in estimating the parameters a_k at time $n = 0$. Then, (23) and (24) reduce to

$$\sum_{k=1}^{p} a_k R(-k, -i) = -R(0, -i) \tag{27}$$

$$E'_p = R(0,0) + \sum_{k=1}^{p} a_k R(0,k). \tag{28}$$

In estimating the nonstationary autocorrelation coefficients from the signal s_n, we note that nonstationary processes are not ergodic, and, therefore, one cannot substitute the ensemble average by a time average. However, for a certain class of nonstationary processes known as *locally stationary processes* [12], [92], it is reasonable to estimate the autocorrelation function with respect to a point in time as a short-time average. Examples of nonstationary processes that can be considered to be locally stationary are speech and EEG signals.

[4] Usually the estimate given by (17) is divided by N, but that does not affect the solution for the predictor coefficients.

In a manner analogous to the stationary case, we estimate $R(-k, -i)$ by φ_{ik} in (20). Using this approximation for the nonstationary autocorrelation leads to a solution for the parameters a_k in (27) that is identical to that given by (18) in the covariance method in the deterministic case.

Note that for a stationary signal: $R(t, t') = R(t - t')$, and therefore, the normal equations (27) and (28) reduce to (12) and (13).

3) Gain Computation: Since in the least squares method we assumed that the input was unknown, it does not make much sense to determine a value for the gain G. However, there are certain interesting observations that can be made.

Equation (7) can be rewritten as

$$s_n = -\sum_{k=1}^{p} a_k s_{n-k} + e_n. \tag{29}$$

Comparing (4) and (29) we see that the only input signal u_n that will result in the signal s_n as output is that where $Gu_n = e_n$. That is, the input signal is proportional to the error signal. For any other input u_n, the output from the filter $H(z)$ in Fig. 2 will be different from s_n. However, if we insist that whatever the input u_n, the energy in the output signal must equal that of the original signal s_n, then we can at least specify the total energy in the input signal. Since the filter $H(z)$ is fixed, it is clear from the above that the total energy in the input signal Gu_n must equal the total energy in the error signal, which is given by E_p in (13) or (19), depending on the method used.

Two types of input that are of special interest are: the deterministic impulse and stationary white noise. By examining the response of the filter $H(z)$ to each of these two inputs we shall gain further insight into the time domain properties of the all-pole model. The input gain is then determined as a byproduct of an autocorrelation analysis.

a) Impulse input: Let the input to the all-pole filter $H(z)$ be an impulse or unit sample at $n = 0$, i.e. $u_n = \delta_{n0}$, where δ_{nm} is the Kronecker delta. The output of the filter $H(z)$ is then its impulse response h_n, where

$$h_n = -\sum_{k=1}^{p} a_k h_{n-k} + G\delta_{n0}. \tag{30}$$

The autocorrelation $\hat{R}(i)$ of the impulse response h_n has an interesting relationship to the autocorrelation $R(i)$ of the signal s_n. Multiply (30) by h_{n-i} and sum over all n. The result can be shown to be [10], [62]:

$$\hat{R}(i) = -\sum_{k=1}^{p} a_k \hat{R}(i-k), \qquad 1 \leq |i| \leq \infty \tag{31}$$

$$\hat{R}(0) = -\sum_{k=1}^{p} a_k \hat{R}(k) + G^2. \tag{32}$$

Given our condition that the total energy in h_n must equal that in s_n, we must have

$$\hat{R}(0) = R(0) \tag{33}$$

since the zeroth autocorrelation coefficient is equal to the total energy in the signal. From (33) and the similarity between (12) and (31) we conclude that [62]

$$\hat{R}(i) = R(i), \qquad 0 \leq i \leq p. \tag{34}$$

This says that the first $p + 1$ coefficients of the autocorrelation of the impulse response of $H(z)$ are identical to the corresponding autocorrelation coefficients of the signal. The problem of linear prediction using the autocorrelation method can be stated in a new way as follows. Find a filter of the form $H(z)$ in (5) such that the first $p + 1$ values of the autocorrelation of its impulse response are equal to the first $p + 1$ values of the signal autocorrelation, and such that (31) applies.

From (32), (34), and (13), the gain is equal to

$$G^2 = E_p = R(0) + \sum_{k=1}^{p} a_k R(k) \tag{35}$$

where G^2 is the total energy in the input $G\delta_{n0}$.

b) White noise input: Here the input u_n is assumed to be a sequence of uncorrelated samples (white noise) with zero mean and unit variance, i.e., $\mathcal{E}(u_n) = 0$, all n, and $\mathcal{E}(u_n u_m) = \delta_{nm}$. Denote the output of the filter by $\hat{s}_n$. For a fixed filter $H(z)$, the output $\hat{s}_n$ forms a stationary random process:

$$\hat{s}_n = -\sum_{k=1}^{p} a_k \hat{s}_{n-k} + Gu_n. \tag{36}$$

Multiply (36) by $\hat{s}_{n-i}$ and take expected values. By noting that u_n and $\hat{s}_{n-i}$ are uncorrelated for $i > 0$, the result can be shown [17] to be identical to (31) and (32), where $\hat{R}(i) = \mathcal{E}(\hat{s}_n \hat{s}_{n-i})$ is the autocorrelation of the output $\hat{s}_n$. Therefore, (31) and (32) completely specify an all-pole random process as well. Equations (31) are known in the statistical literature as the *Yule-Walker equations* [17], [98], [115].

For the random case we require that the average energy (or variance) of the output $\hat{s}_n$ be equal to the variance of the original signal s_n, or $\hat{R}(0) = R(0)$, since the zeroth autocorrelation of a zero-mean random process is the variance. By a reasoning similar to that given in the previous section, we conclude that (34) and (35) also apply for the random case.

From the preceding, we see that the relations linking the autocorrelation coefficients of the output of an all-pole filter are the same whether the input is a single impulse or white noise. This is to be expected since both types of input have identical autocorrelations and, of course, identical flat spectra. This dualism between the deterministic impulse and statistical white noise is an intriguing one. Its usefulness surfaces very elegantly in modeling the speech process, as in Fig. 1, where both unit impulses as well as white noise are actually used to synthesize speech.

C. Computation of Predictor Parameters

1) Direct Methods: In each of the two formulations of linear prediction presented in the previous section, the predictor coefficients a_k, $1 \leq k \leq p$, can be computed by solving a set of p equations with p unknowns. These equations are (12) for the autocorrelation (stationary) method and (18) for the covariance (nonstationary) method. There exist several standard methods for performing the necessary computations, e.g., the Gauss reduction or elimination method and the Crout reduction method [49]. These general methods require $p^3/3 + O(p^2)$ operations (multiplications or divisions) and p^2 storage locations. However, we note from (12) and (18) that the matrix of coefficients in each case is a covariance matrix. Covariance matrices are symmetric and in general positive semidefinite, although in practice they are usually positive definite. Therefore, (12) and (18) can be solved more efficiently by the square-root or Cholesky decomposition method [31], [39], [59], [110]. This method requires about half the computation $p^3/6 + O(p^2)$ and about half the storage $p^2/2$ of the general methods. The numerical stability properties of this method are well understood [109], [111]; the method is considered to be quite stable.

Further reduction in storage and computation time is possible in solving the autocorrelation normal equations (12) because of their special form. Equation (12) can be expanded in matrix form as

$$\begin{bmatrix} R_0 & R_1 & R_2 & \cdots & R_{p-1} \\ R_1 & R_0 & R_1 & \cdots & R_{p-2} \\ R_2 & R_1 & R_0 & \cdots & R_{p-3} \\ \vdots & \vdots & \vdots & & \vdots \\ R_{p-1} & R_{p-2} & R_{p-3} & \cdots & R_0 \end{bmatrix} \begin{bmatrix} a_1 \\ a_2 \\ a_3 \\ \vdots \\ a_p \end{bmatrix} = - \begin{bmatrix} R_1 \\ R_2 \\ R_3 \\ \vdots \\ R_p \end{bmatrix}. \tag{37}$$

Note that the $p \times p$ autocorrelation matrix is symmetric and the elements along any diagonal are identical (i.e., a Toeplitz matrix). Levinson [61] derived an elegant recursive procedure for solving this type of equation. The procedure was later reformulated by Robinson [85]. Levinson's method assumes the column vector on the right hand side of (37) to be a general column vector. By making use of the fact that this column vector comprises the same elements found in the autocorrelation matrix, another method attributed to Durbin [25] emerges which is twice as fast as Levinson's. The method requires only $2p$ storage locations and $p^2 + O(p)$ operations: a big saving from the more general methods. Durbin's recursive procedure can be specified as follows:

$$E_0 = R(0) \tag{38a}$$

$$k_i = -\left[R(i) + \sum_{j=1}^{i-1} a_j^{(i-1)} R(i-j)\right] \Big/ E_{i-1} \tag{38b}$$

$$a_i^{(i)} = k_i$$

$$a_j^{(i)} = a_j^{(i-1)} + k_i a_{i-j}^{(i-1)}, \qquad 1 \leq j \leq i-1 \tag{38c}$$

$$E_i = (1 - k_i^2) E_{i-1}. \tag{38d}$$

Equations (38b)–(38d) are solved recursively for $i = 1, 2, \cdots, p$. The final solution is given by

$$a_j = a_j^{(p)}, \qquad 1 \leq j \leq p. \tag{38e}$$

Note that in obtaining the solution for a predictor of order p, one actually computes the solutions for all predictors of order less than p. It has been reported [78] that this solution is numerically relatively unstable. However, most researchers have not found this to be a problem in practice.

It should be emphasized that, for many applications, the solution of the normal equations (12) or (18) does not form the major computational load. The computation of the autocorrelation or covariance coefficients require pN operations, which can dominate the computation time if $N \gg p$, as is often the case.

The solution to (37) is unaffected if all the autocorrelation coefficients are scaled by a constant. In particular, if all $R(i)$ are normalized by dividing by $R(0)$, we have what are known as the *normalized autocorrelation coefficients* $r(i)$:

$$r(i) = \frac{R(i)}{R(0)} \tag{39}$$

which have the property that $|r(i)| \leq 1$. This can be useful in the proper application of scaling to a fixed point solution to (37).

A byproduct of the solution in (38) is the computation of the minimum total error E_i at every step. It can easily be shown that the minimum error E_i decreases (or remains the same) as the order of the predictor increases [61]. E_i is never negative, of course, since it is a squared error. Therefore, we must have

$$0 \leq E_i \leq E_{i-1}, \qquad E_0 = R(0). \tag{40}$$

If the autocorrelation coefficients are normalized as in (39), then the minimum error E_i is also divided by $R(0)$. We shall call the resulting quantity the *normalized error* V_i:

$$V_i = \frac{E_i}{R(0)} = 1 + \sum_{k=1}^{i} a_k r(k). \tag{41}$$

From (40) it is clear that

$$0 \leq V_i \leq 1, \qquad i \geq 0. \tag{42}$$

Also, from (38d) and (41), the final normalized error V_p is

$$V_p = \prod_{i=1}^{p} (1 - k_i^2). \tag{43}$$

The intermediate quantities k_i, $1 \leq i \leq p$, are known as the *reflection coefficients*. In the statistical literature, they are known as *partial correlation coefficients* [6], [17]. k_i can be interpreted as the (negative) partial correlation between s_n and s_{n+i} holding $s_{n+1}, \cdots, s_{n+i-1}$ fixed. The use of the term "reflection coefficient" comes from transmission line theory, where k_i can be considered as the reflection coefficient at the boundary between two sections with impedances Z_i and Z_{i+1}. k_i is then given by

$$k_i = \frac{Z_{i+1} - Z_i}{Z_{i+1} + Z_i}. \tag{44}$$

The transfer function $H(z)$ can then be considered as that of a sequence of these sections with impedance ratios given from (44) by

$$\frac{Z_{i+1}}{Z_i} = \frac{1 + k_i}{1 - k_i}, \qquad 1 \leq i \leq p. \tag{45}$$

The same explanation can be given for any type of situation where there is plane wave transmission with normal incidence in a medium consisting of a sequence of sections or slabs with different impedances. In the case of an acoustic tube with p sections of equal thickness, the impedance ratios reduce to the inverse ratio of the consecutive cross-sectional areas. This fact has been used recently in speech analysis [10], [52], [97]. Because of the more familiar "engineering interpretation" for k_i, we shall refer to them in this paper as reflection coefficients.

2) Iterative Methods: Beside the direct methods for the solution of simultaneous linear equations, there exist a number of iterative methods. In these methods, one begins by an initial guess for the solution. The solution is then updated by adding a correction term that is usually based on the gradient of some error criterion. In general, iterative methods require more computation to achieve a desired degree of convergence than the direct methods. However, in some applications [100] one often has a good initial guess, which might lead to the solution in only a few iterations. This can be a big saving over direct methods if the number of equations is large. Some of the iterative methods are the gradient method, the steepest descent method, Newton's method, conjugate gradient method and the stochastic approximation method [81], [108].

Up till now we have assumed that the whole signal is given all at once. For certain real time applications it is useful to be able to perform the computations as the signal is coming in. Adaptive schemes exist which update the solution based on every new observation of the signal [106]. The update is usually proportional to the difference between the new observation and the predicted value given the present solution. Another application for adaptive procedures is in the processing of very long data records, where the solution might converge long before all the data is analyzed. It is worth noting that Kalman filtering notions [56] are very useful in obtaining adaptive solutions [60].

3) Filter Stability: After the predictor parameters are computed, the question of the stability of the resulting filter $H(z)$ arises. Filter stability is important for many applications. A causal all-pole filter is stable if all its poles lie inside the unit circle (in which case it is also a filter with minimum phase). The poles of $H(z)$ are simply the roots of the denominator polynomial $A(z)$, where

$$A(z) = 1 + \sum_{k=1}^{p} a_k z^{-k} \tag{46}$$

and

$$H(z) = \frac{G}{A(z)}. \tag{47}$$

$A(z)$ is also known as the *inverse filter*.

If the coefficients $R(i)$ in (12) are positive definite [79] (which is assured if $R(i)$ is computed from a nonzero signal using (17) or from a positive definite spectrum[5]), the solution of the autocorrelation equation (12) gives predictor parameters which guarantee that all the roots of $A(z)$ lie inside the unit circle, i.e., a stable $H(z)$ [42], [85], [104]. This result can also be obtained from orthogonal polynomial theory. In fact, if one denotes the inverse filter at step i in iteration (38) by $A_i(z)$, then it can be shown that the polynomials $A_i(z)$ for $i = 0, 1, 2, \cdots$, form an orthogonal set over the unit circle [35], [42], [93]:

$$\frac{1}{2\pi}\int_{-\pi}^{\pi} P(\omega)\, A_n(e^{j\omega})\, A_m(e^{-j\omega})\, d\omega = E_n \delta_{nm}, \qquad n, m = 0, 1, 2, \cdots \tag{48}$$

where E_n is the minimum error for an nth order predictor, and $P(\omega)$ is any positive definite spectrum whose Fourier transform results in the autocorrelation coefficients $R(i)$ that are used in (12). The recurrence relation for these polynomials is as follows:

$$A_i(z) = A_{i-1}(z) + k_i z^{-i} A_{i-1}(z^{-1}) \tag{49}$$

which is the same as the recursion in (38c).

The positive definiteness of $R(i)$ can often be lost if one uses a small word length to represent $R(i)$ in a computer. Also, roundoff errors can cause the autocorrelation matrix to become ill-conditioned. Therefore, it is often necessary to check for the stability of $H(z)$. Checking if the roots of $A(z)$ are inside the unit circle is a costly procedure that is best avoided One method is to check if all the successive errors are positive. In fact, the condition $E_i > 0$, $1 \leq i \leq p$, is a necessary and sufficient condition for the stability of $H(z)$. From (38d) and (40) it is clear that an equivalent condition for the stability of $H(z)$ is that

$$|k_i| < 1, \qquad 1 \leq i \leq p. \tag{50}$$

Therefore, the recursive procedure (38) also facilitates the check for the stability of the filter $H(z)$.

The predictor parameters resulting from a solution to the covariance matrix equation (18) cannot in general be guaranteed to form a stable filter. The computed filter tends to be more stable as the number of signal samples N is increased, i.e., as the covariance matrix approaches an autocorrelation matrix. Given the computed predictor parameters, it is useful to be able to test for the stability of the filter $H(z)$. One method is to compute the reflection coefficients k_i from the predictor parameters by a backward recursion, and then check for stability using (50). The recursion is as follows:

$$k_i = a_i^{(i)}$$
$$a_j^{(i-1)} = \frac{a_j^{(i)} - a_i^{(i)} a_{i-j}^{(i)}}{1 - k_i^2}, \qquad 1 \leq j \leq i-1 \tag{51}$$

where the index i takes values $p, p-1, \cdots, 1$ in that order. Initially $a_j^{(p)} = a_j$, $1 \leq j \leq p$. It is interesting to note that this method for checking the stability of $H(z)$ is essentially the same as the Lehmer–Schur method [81] for testing whether or not the zeros of a polynomial lie inside the unit circle. An unstable filter can be made stable by reflecting the poles outside the unit circle inside [10], such that the magnitude of the system frequency response remains the same. Filter instability can often be avoided by adding a very small number to the diagonal elements in the covariance matrix.

A question always arises as to whether to use the autocorrelation method or covariance method in estimating the predictor parameters. The covariance method is quite general and can be used with no restrictions. The only problem is that of the stability of the resulting filter, which is not a severe problem generally. In the autocorrelation method, on the other hand, the filter is guaranteed to be stable, but problems of parameter accuracy can arise because of the necessity of windowing (truncating) the time signal. This is usually a problem if the signal is a portion of an impulse response. For example, if the impulse response of an all-pole filter is analyzed by the covariance method, the filter parameters can be computed accurately from only a finite number of samples of the signal. Using the autocorrelation method, one cannot obtain the exact parameter values unless the whole infinite impulse response is used in the analysis. However, in practice, very good approximations can be obtained by truncating the impulse response at a point where most of the decay of the response has already occurred.

[5] A spectrum that can be zero at most at a countable set of frequencies.

III. Spectral Estimation

In Section II, the stationary and nonstationary methods of linear prediction were derived from a time domain formulation. In this section we show that the same normal equations can be derived from a frequency domain formulation. It will become clear that linear prediction is basically a correlation type of analysis which can be approached either from the time or frequency domain. The insights gained from the frequency domain analysis will lead to new applications for linear predictive analysis. This section and the following are based mainly on references [62]-[64].

A. Frequency Domain Formulations

1) Stationary Case: The error e_n between the actual signal and the predicted signal is given by (7). Applying the z transform to (7), we obtain

$$E(z) = \left[1 + \sum_{k=1}^{p} a_k z^{-k}\right] S(z) = A(z)\, S(z) \tag{52}$$

where $A(z)$ is the inverse filter defined in (46), and $E(z)$ and $S(z)$ are the z transforms of e_n and s_n, respectively. Therefore, e_n can be viewed as the result of passing s_n through the inverse filter $A(z)$. Assuming a deterministic signal[6] s_n, and applying Parseval's theorem, the total error to be minimized is given by

$$E = \sum_{n=-\infty}^{\infty} e_n^2 = \frac{1}{2\pi}\int_{-\pi}^{\pi} |E(e^{j\omega})|^2 \, d\omega \tag{53}$$

where $E(e^{j\omega})$ is obtained by evaluating $E(z)$ on the unit circle $z = e^{j\omega}$. Denoting the power spectrum of the signal s_n by $P(\omega)$, where

$$P(\omega) = |S(e^{j\omega})|^2 \tag{54}$$

we have from (52)-(54)

$$E = \frac{1}{2\pi}\int_{-\pi}^{\pi} P(\omega)\, A(e^{j\omega}) A(e^{-j\omega})\, d\omega. \tag{55}$$

Following the same procedure as in Section II, E is minimized by applying (9) to (55). The result can be shown [64] to be identical to the autocorrelation normal equations (12), but with the autocorrelation $R(i)$ obtained from the signal spectrum $P(\omega)$ by an inverse Fourier transform

$$R(i) = \frac{1}{2\pi}\int_{-\pi}^{\pi} P(\omega) \cos(i\omega)\, d\omega. \tag{56}$$

Note that in (56) the cosine transform is adequate since $P(\omega)$ is real and even. The minimum squared error E_p can be obtained by substituting (12) and (56) in (55), which results in the same equation as in (13).

2) Nonstationary Case: Here the signal s_n and the error e_n are assumed to be nonstationary. If $R(t, t')$ is the nonstationary autocorrelation of s_n, then we define the nonstationary two-dimensional (2-*D*) spectrum $Q(\omega, \omega')$ of s_n by [12], [64], [67], [79]

$$Q(\omega, \omega') = \sum_{t'=-\infty}^{\infty} \sum_{t=-\infty}^{\infty} R(t, t') \exp\left[-j(\omega t - \omega' t')\right]. \tag{57}$$

[6] A similar development assuming a random signal gives the same results.

$R(t, t')$ can then be recovered from $Q(\omega, \omega')$ by an inverse 2-*D* Fourier transform

$$R(t, t') = \left(\frac{1}{2\pi}\right)^2 \int_{-\pi}^{\pi}\int_{-\pi}^{\pi} Q(\omega, \omega') \exp\left[j(\omega t - \omega' t')\right] d\omega\, d\omega'. \tag{58}$$

As in the time domain formulation, we are interested in minimizing the error variance for time $n = 0$, which is now given by [64]

$$E = \left(\frac{1}{2\pi}\right)^2 \int_{-\pi}^{\pi}\int_{-\pi}^{\pi} Q(\omega, \omega')\, A(e^{j\omega})\, A(e^{-j\omega'})\, d\omega\, d\omega'. \tag{59}$$

Applying (9) to (59) results in equations identical to the nonstationary normal equations (27), where $R(t, t')$ is now defined by (58). The minimum error is then obtained by substituting (27) and (58) in (59). The answer is identical to (28).

B. Linear Predictive Spectral Matching

In this section we shall examine in what manner the signal spectrum $P(\omega)$ is approximated by the all-pole model spectrum, which we shall denote by $\hat{P}(\omega)$. From (5) and (47):

$$\hat{P}(\omega) = |H(e^{j\omega})|^2 = \frac{G^2}{|A(e^{j\omega})|^2} = \frac{G^2}{\left|1 + \sum_{k=1}^{p} a_k e^{-jk\omega}\right|^2}. \tag{60}$$

From (52) and (54) we have

$$P(\omega) = \frac{|E(e^{j\omega})|^2}{|A(e^{j\omega})|^2}. \tag{61}$$

By comparing (60) and (61) we see that if $P(\omega)$ is being modeled by $\hat{P}(\omega)$, then the error power spectrum $|E(e^{j\omega})|^2$ is being modeled by a flat spectrum equal to G^2. This means that the actual error signal e_n is being approximated by another signal that has a flat spectrum, such as a unit impulse, white noise, or any other signal with a flat spectrum. The filter $A(z)$ is sometimes known as a "whitening filter" since it attempts to produce an output signal e_n that is white, i.e., has a flat spectrum.

From (53), (60), and (61), the total error can be written as

$$E = \frac{G^2}{2\pi}\int_{-\pi}^{\pi} \frac{P(\omega)}{\hat{P}(\omega)}\, d\omega. \tag{62}$$

Therefore, minimizing the total error E is equivalent to the minimization of the integrated ratio of the signal spectrum $P(\omega)$ to its approximation $\hat{P}(\omega)$. (This interpretation of the least squares error was proposed in a classic paper by Whittle [103]. An equivalent formulation using maximum likelihood estimation has been given by Itakura [50], [51].) Now, we can back up and restate the problem of linear prediction as follows. Given some spectrum $P(\omega)$, we wish to model it by another spectrum $\hat{P}(\omega)$ such that the integrated ratio between the two spectra as in (62) is minimized. The parameters of the model spectrum are computed from the normal equations (12), where the needed autocorrelation coefficients $R(i)$ are easily computed from $P(\omega)$ by a simple Fourier transform. The gain factor G is obtained by equating the total energy in the two

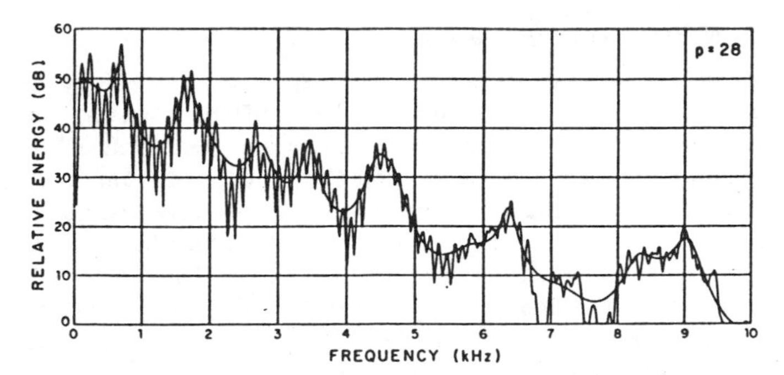

Fig. 3. A 28-pole fit to an FFT-computed signal spectrum. The signal was sampled at 20 kHz.

spectra, i.e., $\hat{R}(0) = R(0)$, where

$$\hat{R}(i) = \frac{1}{2\pi} \int_{-\pi}^{\pi} \hat{P}(\omega) \cos (i\omega)\, d\omega. \tag{63}$$

Note that $\hat{R}(i)$ is the autocorrelation of the impulse response of $H(z)$, which is given by (31) and (32). As then, the gain is computed from (35).

The manner in which the model spectrum $\hat{P}(\omega)$ approximates $P(\omega)$ is largely reflected in the relation between the corresponding autocorrelation functions. From (34), we have $\hat{R}(i) = R(i)$, $0 \leq i \leq p$. Since $P(\omega)$ and $\hat{P}(\omega)$ are Fourier transforms of $R(i)$ and $\hat{R}(i)$, respectively, it follows that increasing the value of the order of the model p increases the range over which $R(i)$ and $\hat{R}(i)$ are equal, resulting in a better fit of $\hat{P}(\omega)$ to $P(\omega)$. In the limit, as $p \to \infty$, $\hat{R}(i)$ becomes identical to $R(i)$ for all i, and hence the two spectra become identical:

$$\hat{P}(\omega) = P(\omega), \qquad \text{as } p \to \infty \tag{64}$$

This statement says that we can approximate any spectrum arbitrarily closely by an all-pole model.

Another important conclusion is that since linear predictive analysis can be viewed as a process of spectrum or autocorrelation matching, one must be careful how to estimate the spectrum $P(\omega)$ or the corresponding autocorrelation that is to be modeled. Since the signal is often weighted or windowed[7] before either the autocorrelation or the spectrum is computed, it can be quite important to properly choose the type and width of the data window to be used. The choice of window depends very much on the type of signal to be analyzed. If the signal can be considered to be stationary for a long period of time (relative to the effective length of the system impulse response), then a rectangular window suffices. However, for signals that result from systems that are varying relatively quickly, the time of analysis must necessarily be limited. For example, in many transient speech sounds, the signal can be considered stationary for a duration of only one or two pitch periods. In that case a window such as Hamming or Hanning [14] is more appropriate. See [13], [14], [26], [54], [64], [99], [101] for more on the issue of windowing and spectral estimation in general.

An example of linear predictive (LP) spectral estimation is shown in Fig. 3, where the original spectrum $P(\omega)$ was obtained by computing the fast Fourier transform (FFT) of a 20-ms, Hamming windowed, 20-kHz sampled speech signal. The speech sound was the vowel [æ] as in the word "bat." The harmonics due to the periodicity of the sound are evident in the FFT spectrum. Fig. 3 also shows a 28-pole fit ($p = 28$) to the signal spectrum. In this case the autocorrelation coefficients needed to solve the normal equations (12) were computed directly from the time signal. The all-pole spectrum $\hat{P}(\omega)$ was computed from (60) by dividing G^2 by the magnitude squared of the FFT of the sequence: $1, a_1, a_2, \cdots, a_p$. Arbitrary frequency resolution in computing $\hat{P}(\omega)$ can be obtained by simply appending an appropriate number of zeros to this sequence before taking the FFT. An alternate method of computing $\hat{P}(\omega)$ is obtained by rewriting (60) as

$$\hat{P}(\omega) = \frac{G^2}{\rho(0) + 2 \sum_{i=1}^{p} \rho(i) \cos (i\omega)} \tag{65}$$

where

$$\rho(i) = \sum_{k=0}^{p-i} a_k a_{k+i}, \qquad a_0 = 1, \quad 0 \leq i \leq p \tag{66}$$

is the autocorrelation of the impulse response of the inverse filter $A(z)$. From (65), $\hat{P}(\omega)$ can be computed by dividing G^2 by the real part of the FFT of the sequence: $\rho(0)$, $2\rho(1)$, $2\rho(2), \cdots, 2\rho(p)$. Note that the slope of $\hat{P}(\omega)$ is always zero at $\omega = 0$ and $\omega = \pi$.

Another property of $\hat{P}(\omega)$ is obtained by noting that the minimum error $E_p = G^2$, and, therefore, from (62) we have

$$\frac{1}{2\pi} \int_{-\pi}^{\pi} \frac{P(\omega)}{\hat{P}(\omega)}\, d\omega = 1. \tag{67}$$

(This relation is a special case of a more general result (48) relating the fact that the polynomials $A_0(z), A_1(z), \cdots, A_p(z), \cdots$, form a complete set of orthogonal polynomials with weight $P(\omega)$.) Equation (67) is true for all values of p. In particular, it is true as $p \to \infty$, in which case from (64) we see that (67) becomes an identity. Another important case where (67) becomes an identity is when $P(\omega)$ is an all-pole spectrum with p_0 poles, then $\hat{P}(\omega)$ will be identical to $P(\omega)$ for all $p \geq p_0$. Relation (67) will be useful in discussing the properties of the error measure in Section IV.

The transfer functions $S(z)$ and $H(z)$ corresponding to $P(\omega)$ and $\hat{P}(\omega)$ are also related. It can be shown [62] that as $p \to \infty$, $H(z)$ is given by

$$H_\infty(z) = \frac{G}{1 + \sum_{k=1}^{\infty} a_k z^{-k}} = \sum_{n=0}^{N-1} h_\infty(n) z^{-n}, \qquad p \to \infty \tag{68}$$

where $h_\infty(n)$, $0 \leq n \leq N-1$, is the minimum phase sequence corresponding to s_n, $0 \leq n \leq N-1$. Note that the minimum phase sequence is of the same length as the original signal. Fig. 4 shows a signal ($N = 256$) and its approximate minimum phase counterpart, obtained by first performing a LP analysis for $p = 250$, and then computing the sequence $h(n)$ by long division.

C. *Selective Linear Prediction*

The major point of the previous section was that LP analysis can be regarded as a method of spectral modeling. We had tacitly assumed that the model spectrum spans the same frequency range as the signal spectrum. We now generalize the

[7] Note that here we are discussing *data* windows which are applied directly to the signal, as opposed to *lag* windows, which statisticians have traditionally applied to the autocorrelation.

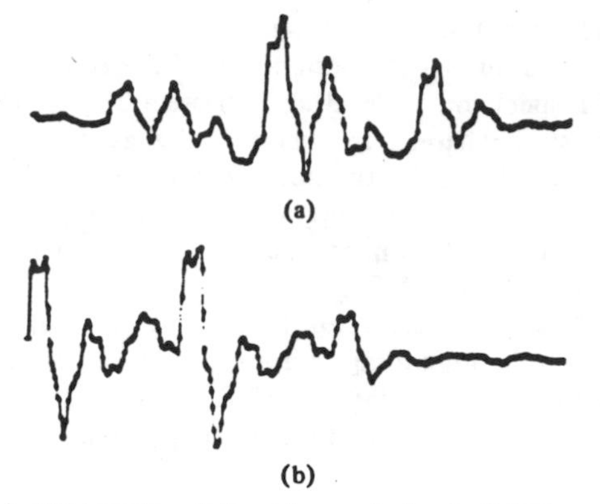

Fig. 4. (a) A 256-sample windowed speech signal. (b) The corresponding approximate minimum phase sequence obtained using a linear predictor of order $p = 250$.

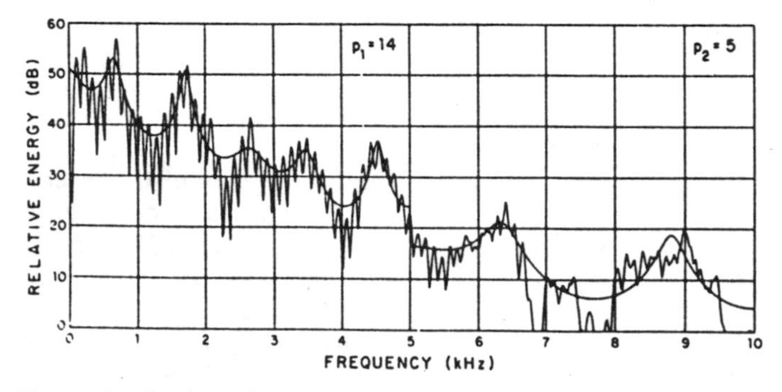

Fig. 5. Application of selective linear prediction to the same signal spectrum as in Fig. 3, with a 14-pole fit to the 0-5 kHz region and a 5-pole fit to the 5-10 kHz region.

LP spectral modeling method to the case where we wish to fit only a selected portion of a given spectrum.

Suppose we wish to model the spectrum $P(\omega)$ only in the region[8] $\omega_\alpha \leq \omega \leq \omega_\beta$ by an all-pole spectrum given by (60). Call the signal spectrum in that region $P'(\omega)$. In order to compute the parameters of the model spectrum $\hat{P}(\omega)$, we first perform a linear mapping of the given region onto the upper half of the unit circle in the z plane. This can be accomplished by the mapping $\omega' = \pi(\omega - \omega_\alpha)/(\omega_\beta - \omega_\alpha)$, so that the given region is mapped onto $0 \leq \omega' \leq \pi$. In addition, let $P'(-\omega') = P'(\omega')$ define the spectrum over the lower half of the unit circle. The model parameters are then computed from the normal equations (12), where the autocorrelation coefficients are obtained by a Fourier transform with $P'(\omega')$ replacing $P(\omega)$ and ω' replacing ω in (56).

Selective linear prediction has had applications in speech recognition and speech compression [63]. An example of its usage is shown in Fig. 5. For speech recognition applications, the 0-5 kHz region is more important than the 5-10 kHz. Even when the 5-10 kHz region is important, only a rough idea of the shape of the spectrum is sufficient. In Fig. 5, the signal spectrum is the same as in Fig. 3. The 0-5 kHz region is modeled by a 14-pole spectrum, while the 5-10 kHz region is modeled independently by only a 5-pole model.

An important point, which should be clear by now, is that since we assume the availability of the signal spectrum $P(\omega)$, any desired frequency shaping or scaling can be performed directly on the signal spectrum before linear predictive modeling is applied.

[8] The remainder of the spectrum is simply neglected.

D. Modeling Discrete Spectra

Thus far we have assumed that the spectrum $P(\omega)$ is a continuous function of frequency. More often, however, the spectrum is known at only a finite number of frequencies. For example, FFT-derived spectra and those obtained from many commercially available spectrum analyzers have values at equally spaced frequency points. On the other hand, filter bank spectra, and, for example, third-octave band spectrum analyzers have values at frequencies that are not necessarily equally spaced. In order to be able to model these discrete spectra, only one change in our analysis need be made. The error measure E in (62) is defined as a summation instead of an integral. The rest of the analysis remains the same except that the autocorrelation coefficients $R(i)$ are now computed from

$$R(i) = \frac{1}{M} \sum_{m=0}^{M-1} P(\omega_m) \cos(i\omega_m) \tag{69}$$

where M is the total number of spectral points on the unit circle. The frequencies ω_m are those for which a spectral value exists, and they need not be equally spaced. Below we demonstrate the application of LP modeling for filter bank and harmonic spectra.

Fig. 6(a) shows a typical 14-pole fit to a spectrum of the fricative [s] that was FFT computed from the time signal. Fig. 6(b) shows a similar fit to a line spectrum that is typical of filter bank spectra. What we have actually done here is to simulate a filter bank where the filters are linearly spaced up to 1.6 kHz and logarithmically spaced thereafter. Note that the all-pole spectrum for the simulated filter bank is remarkably similar to the one in the top figure, even though the number of spectral points is much smaller.

The dashed curve in Fig. 7(a) is a 14-pole spectrum. If one applied LP analysis to this spectrum, the all-pole model for $p = 14$ would be identical to the dashed spectrum. The situation is not so favorable for discrete spectra. Let us assume that the dashed spectrum corresponds to the transfer function of a 14-pole filter. If this filter is excited by a periodic train of impulses with fundamental frequency F_0, the spectrum of the output signal will be a discrete line spectrum with spectral values only at the harmonics (multiples of F_0). The line spectrum for $F_0 = 312$ Hz is shown in Fig. 7(a). Note that the dashed spectrum is an envelope of the harmonic spectrum. The result of applying a 14-pole LP analysis to the harmonic spectrum is shown as the solid curve in Fig. 7(a). The discrepancy between the two all-pole spectra is obvious. In general, the types of discrepancies that can occur between the model and original spectra include merging or splitting of pole peaks, and increasing or decreasing of pole frequencies and bandwidths. Pole movements are generally in the direction of the nearest harmonic. As the fundamental frequency decreases, these discrepancies decrease, as shown in Fig. 7(b) for $F_0 = 156$ Hz.

It is important to note in Fig. 7 that the dashed curve is the *only* possible 14-pole spectrum that coincides with the line spectrum at the harmonics.[9] It is significant that the all-pole spectrum resulting from LP modeling does not yield the spectrum we desire. The immediate reason for this is that the solution for the model parameters from (12) depends on the values of the signal autocorrelation, which for the periodic signal are

[9] In general this is true only if the period between input impulses is greater than twice the number of poles in the filter.

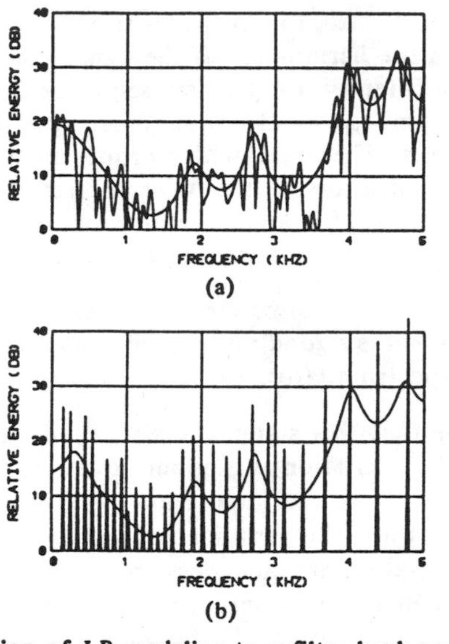

Fig. 6. Application of LP modeling to a filter bank spectrum. (a) A 14-pole fit to the original spectrum. (b) A 14-pole fit to the simulated filter bank spectrum.

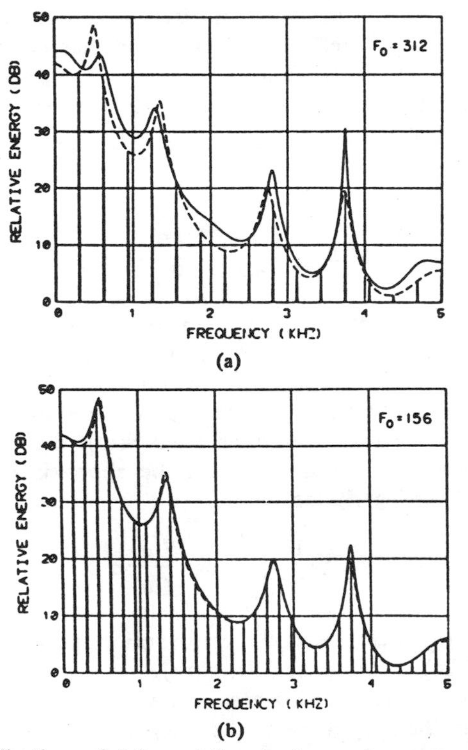

Fig. 7. Application of LP modeling to harmonic spectra. Dashed curve: 14-pole filter spectrum. Vertical lines: Corresponding harmonic spectrum for (a) and (b). (a) $F_0 = 312$ Hz. (b) $F_0 = 156$ Hz. Solid curve: 14-pole fit to the discrete harmonic spectrum. (For display purposes, the energy in the model spectrum (solid curve) was set equal to the energy in the filter spectrum (dashed curve).)

different from that for the single impulse response. However, the major underlying reason lies in the properties of the error measure used. This is the topic of the next section.

IV. Error Analysis

An important aspect of any fitting or matching procedure is the properties of the error measure that is employed, and whether those properties are commensurate with certain objectives. In this section we shall examine the properties of the error measure used in LP analysis and we shall discuss its strengths and weaknesses in order to be able to fully utilize its capabilities. The analysis will be restricted to the stationary (autocorrelation) case, although the conclusions can be extrapolated to the nonstationary (covariance) case.

The error measure used in Section II-B to determine the predictor parameters is the least squares error measure due to Gauss, who first reported on it in the early 1800's. This error measure has been used extensively since then, and is quite well understood. Its major asset is its mathematical tractability. Its main characteristic is that it puts great emphasis on large errors and little emphasis on small errors. Purely from the time domain, it is often difficult to say whether such an error measure is a desirable one or not for the problem at hand. Many would probably agree that it does not really matter which error measure one uses as long as it is a reasonable function of the magnitude of the error at each point. For the linear prediction problem, we are fortunate that the error measure can also be written in the frequency domain and can be interpreted as a goodness of fit between a given signal spectrum and a model spectrum that approximates it. The insights gained in the frequency domain should enhance our understanding of the least squares error criterion.

A. *The Minimum Error*

For each value of p, minimization of the error measure E in (62) leads to the minimum error E_p in (13), which is given in terms of the predictor and autocorrelation coefficients. Here we derive an expression for E_p in the frequency domain, which will help us determine some of its properties. Other properties will be discussed when we discuss the normalized minimum error.

Let

$$\hat{c}_0 = \frac{1}{2\pi} \int_{-\pi}^{\pi} \log \hat{P}(\omega)\, d\omega \tag{70}$$

be the zeroth coefficient (quefrency) of the cepstrum (inverse Fourier transform of log spectrum) [38], [77] corresponding to $\hat{P}(\omega)$. From (60), (70) reduces to

$$\hat{c}_0 = \log G^2 - \frac{1}{2\pi} \int_{-\pi}^{\pi} \log |A(e^{j\omega})|^2\, d\omega. \tag{71}$$

$A(z)$ has all its zeros inside the unit circle. Therefore, the integral in (71) is equal to zero [64], [69], [103]. Since $G^2 = E_p$, we conclude from (71) that

$$E_p = e^{\hat{c}_0}. \tag{72}$$

From (72) and (70), E_p can be interpreted as the geometric mean of the model spectrum $\hat{P}(\omega)$. From (40) we know that E_p decreases as p increases. The minimum occurs as $p \to \infty$, and is equal to

$$E_{\min} = E_\infty = e^{c_0} \tag{73}$$

where c_0 is obtained by substituting $P(\omega)$ for $\hat{P}(\omega)$ in (70).[10] Therefore, the absolute minimum error is a function of $P(\omega)$ only, and is equal to its geometric mean, which is always positive for positive definite spectra.[11] This is a curious result, because it says that the minimum error can be nonzero even

[10] If $P(\omega)$ is a p_0-pole spectrum then $E_p = E_{\min}$ for all $p > p_0$.

[11] $E_{\min}$ is equal to zero only if $P(\omega)$ is zero over a noncountable set of frequencies (i.e., over a line segment). In that case, the signal is perfectly predictable and the prediction error is zero [107].

when the matching spectrum $\hat{P}(\omega)$ is identical to the matched spectrum $P(\omega)$. Therefore, although E_p is a measure of fit of the model spectrum to the signal spectrum, it is not an absolute one. The measure is always relative to $E_{\min}$. The nonzero aspect of $E_{\min}$ can be understood by realizing that, for any p, E_p is equal to that portion of the signal energy that is not predictable by a pth order predictor. For example, the impulse response of an all-pole filter is perfectly predictable *except* for the initial nonzero value. It is the energy in this initial value that shows up in E_p. (Note that in the covariance method one can choose the region of analysis to exclude the initial value, in which case the prediction error would be zero for this example.)

B. Spectral Matching Properties

The LP error measure E in (62) has two major properties:[12] a global property and a local property.

1) Global Property: Because the contributions to the total error are determined by the *ratio* of the two spectra, the matching process should perform uniformly over the whole frequency range, irrespective of the general shaping of the spectrum.

This is an important property for spectral estimation because it makes sure that the spectral match at frequencies with little energy is just as good, on the average, as the match at frequencies with high energy (see Fig. 3). If the error measure had been of the form $\int |P(\omega) - \hat{P}(\omega)|\, d\omega$, the spectral matches would have been best at high energy frequency points.

2) Local Property: This property deals with how the match is done in each small region of the spectrum.

Let the ratio of $P(\omega)$ to $\hat{P}(\omega)$ be given by

$$E(\omega) = \frac{P(\omega)}{\hat{P}(\omega)}. \tag{74}$$

Then from (67) we have

$$\frac{1}{2\pi}\int_{-\pi}^{\pi} E(\omega)\, d\omega = 1, \qquad \text{for all } p. \tag{75}$$

$E(\omega)$ can be interpreted as the "instantaneous error" between $P(\omega)$ and $\hat{P}(\omega)$ at frequency ω. Equation (75) says that the arithmetic mean of $E(\omega)$ is equal to 1, which means that there are values of $E(\omega)$ greater and less than 1 such that the average is equal to 1.[13] In terms of the two spectra, this means that $P(\omega)$ will be greater than $\hat{P}(\omega)$ in some regions and less in others such that (75) applies. However, the contribution to the total error is more significant when $P(\omega)$ is greater than $\hat{P}(\omega)$ than when $P(\omega)$ is smaller, e.g., a ratio of $E(\omega) = 2$ contributes more to the total error than a ratio of 1/2. We conclude that:

> after the minimization of error, we expect a better fit of $\hat{P}(\omega)$ to $P(\omega)$ where $P(\omega)$ is greater than $\hat{P}(\omega)$, than where $P(\omega)$ is smaller (on the average).

For example, if $P(\omega)$ is the power spectrum of a quasi-periodic signal (such as in Fig. 3), then most of the energy in $P(\omega)$ will exist in the harmonics, and very little energy will reside between harmonics. The error measure in (62) insures that the approximation of $\hat{P}(\omega)$ to $P(\omega)$ is far superior at the harmonics than between the harmonics. If the signal had been generated by exciting a filter with a periodic sequence of impulses, then the system response of the filter must pass through all the harmonic peaks. Therefore, with a proper choice of the model order p, minimization of the LP error measure results in a model spectrum that is a good approximation to that system response. This leads to one characteristic of the local property:

> minimization of the error measure E results in a model spectrum $\hat{P}(\omega)$ that is a good estimate of the *spectral envelope* of the signal spectrum $P(\omega)$.

Fig. 6 shows that this statement also applies in a qualitative way when the excitation is random noise. It should be clear from the above that the importance of the local property is not as crucial when the variations of the signal spectrum from the spectral envelope are much less pronounced.

In the modeling of harmonic spectra, we showed an example in Fig. 7(a) where, although the all-pole spectrum resulting from LP modeling was a reasonably good estimate of the harmonic spectral envelope, it did not yield the unique all-pole transfer function that coincides with the line spectrum at the harmonics. This is a significant *disadvantage* of LP modeling, and is an indirect reflection of another characteristic of the local property: *the cancellation of errors.* This is evident from (75) where the instantaneous errors $E(\omega)$ are greater and less than 1 such that the average is 1. To help elucidate this point, let us define a new error measure E' that is the logarithm of E in (62):

$$E' = \log\left[\frac{1}{2\pi}\int_{-\pi}^{\pi} \frac{P(\omega)}{\hat{P}(\omega)}\, d\omega\right] \tag{76}$$

where the gain factor has been omitted since it is not relevant to this discussion. It is simple to show that the minimization of E' is equivalent to the minimization of E. For cases where $P(\omega)$ is smooth relative to $\hat{P}(\omega)$ and the values of $P(\omega)$ are not expected to deviate very much from $\hat{P}(\omega)$, the logarithm of the average of spectral ratios can be approximated by the average of the logarithms, i.e.,

$$E' \cong \frac{1}{2\pi}\int_{-\pi}^{\pi} \log \frac{P(\omega)}{\hat{P}(\omega)}\, d\omega. \tag{77}$$

From (77) it is clear that the contributions to the error when $P(\omega) > \hat{P}(\omega)$ *cancel* those when $P(\omega) < \hat{P}(\omega)$.

The above discussion suggests the use of an error measure that takes the magnitude of the integrand in (77). One such error measure is

$$E'' = \frac{1}{2\pi}\int_{-\pi}^{\pi} \left[\log \frac{P(\omega)}{\hat{P}(\omega)}\right]^2 d\omega$$

$$= \frac{1}{2\pi}\int_{-\pi}^{\pi} [\log P(\omega) - \log \hat{P}(\omega)]^2\, d\omega. \tag{78}$$

E'' is just the mean squared error between the two log spectra. It has the important property that the minimum error of zero occurs if and only if $\hat{P}(\omega)$ is identical to $P(\omega)$. Therefore, if we use (the discrete form of) E'' in modeling the harmonic spectrum in Fig. 7(a), the resulting model spectrum (for $p = 14$) will be identical to the dashed spectrum, since the minimum error of zero is achievable by that spectrum. However, while the error measure E'' solves one problem, it introduces an-

[12] Itakura [50], [51] discusses a maximum likelihood error criterion having the same properties.

[13] Except for the special case when $P(\omega)$ is all-pole, the condition $E(\omega) = 1$ for all ω is true only as $p \to \infty$.

other. Note that the contributions to the total error in (78) are equally important whether $P(\omega) > \hat{P}(\omega)$ or vice versa. This means that if the variations of $P(\omega)$ are large relative to $\hat{P}(\omega)$ (such as in Fig. 3), the resulting model spectrum will *not* be a good estimate of the spectral envelope. In addition, the minimization of E'' in (78) results in a set of nonlinear equations that must be solved iteratively, thus increasing the computational load tremendously.

Our conclusion is that the LP error measure in (62) is to be preferred in general, except for certain special cases (as in Fig. 7(a)) where an error measure such as E'' in (78) can be used, provided one is willing to carry the extra computational burden.

The global and local properties described here are properties of the error measure in (62) and do not depend on the details of $P(\omega)$ and $\hat{P}(\omega)$. These properties apply *on the average* over the whole frequency range. Depending on the detailed shapes of $P(\omega)$ and $\hat{P}(\omega)$, the resulting match can be better in one spectral region than in another. For example, if $\hat{P}(\omega)$ is an all-pole model spectrum and if the signal spectrum $P(\omega)$ contains zeros as well as poles, then one would not expect as good a match at the zeros as at the poles. This is especially true if the zeros have bandwidths of the same order as the poles or less. (Wide bandwidth zeros are usually well approximated by poles.) On the other hand, if $\hat{P}(\omega)$ is an all-zero spectrum then the preceding statement would have to be reversed.

C. The Normalized Error

The normalized error has been a very useful parameter for the determination of the optimal number of parameters to be used in the model spectrum. This subject will be discussed in the following section. Here we shall present some of the properties of the normalized error, especially as they relate to the signal and model spectra.

1) Relation to the Spectral Dynamic Range: The normalized error was defined in Section II as the ratio of the minimum error E_p to the energy in the signal $R(0)$. Keeping in mind that $R(0) = \hat{R}(0)$, and substituting for E_p from (72), we obtain

$$V_p = \frac{E_p}{R(0)} = \frac{e^{\hat{c}_0}}{\hat{R}(0)}. \tag{79}$$

Also, from (73), we have in the limit as $p \to \infty$:

$$V_{\min} = V_\infty = \frac{e^{c_0}}{R(0)}. \tag{80}$$

Therefore, the normalized error is always equal to the normalized zero quefrency of the model spectrum. From (40) and (79) it is clear that V_p is a monotonically decreasing function of p, with $V_0 = 1$ and $V_\infty = V_{\min}$ in (80). Fig. 8 shows plots of V_p as a function of p for two speech sounds (sampled at 10 kHz) whose spectra are similar to those in Figs. 3 and 6.

It is instructive to write V_p as a function of $\hat{P}(\omega)$. From (63) and (70), (79) can be rewritten as

$$V_p = \frac{\exp\left[\frac{1}{2\pi}\int_{-\pi}^{\pi} \log \hat{P}(\omega)\, d\omega\right]}{\frac{1}{2\pi}\int_{-\pi}^{\pi} \hat{P}(\omega)\, d\omega}. \tag{81}$$

It is clear from (81) that V_p depends completely on the shape of the model spectrum, and from (80), $V_{\min}$ is determined solely by the shape of the signal spectrum. An interesting way to view (81) is that V_p is equal to the ratio of the geometric mean of the model spectrum to its arithmetic mean. This ratio has been used in the past as a measure of the spread of the data [22], [48]. When the spread of the data is small, the ratio is close to 1. Indeed, from (81) it is easy to see that if $\hat{P}(\omega)$ is flat, $V_p = 1$. On the other hand, if the data spread is large, then V_p becomes close to zero. Again, from (81) we see that if $\hat{P}(\omega)$ is zero for a portion of the spectrum (hence a large spread), then $V_p = 0$. (Another way of looking at V_p is in terms of the flatness of the spectrum [40].)

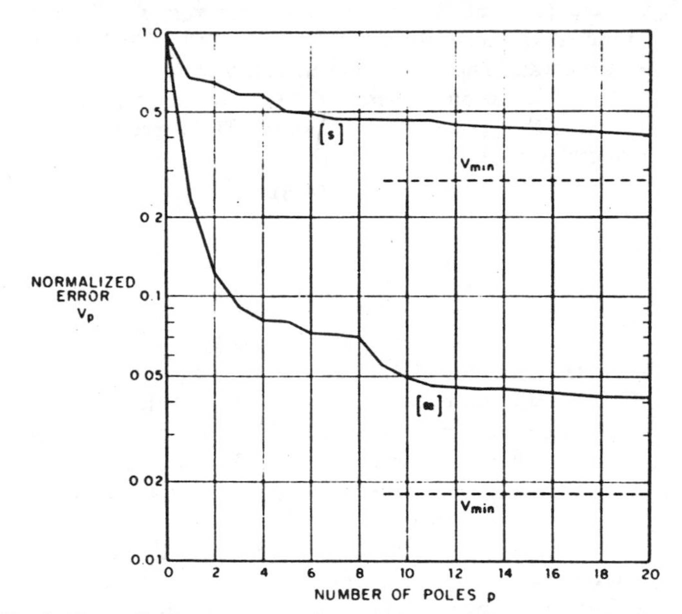

Fig. 8. Normalized error curves for the sounds [s] in the word "list" and [æ] in the word "potassium."

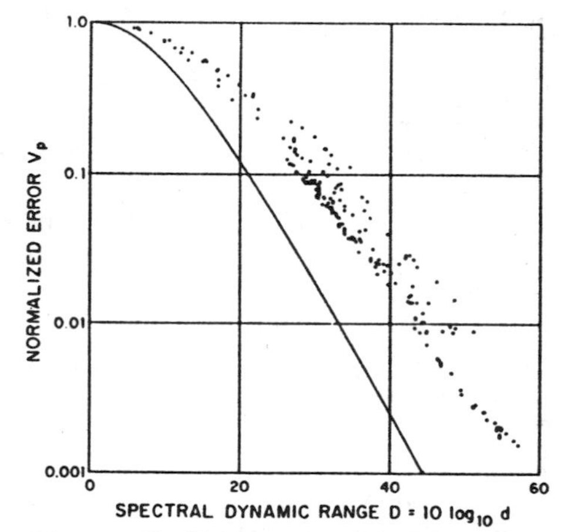

Fig. 9. Two-pole normalized error versus spectral dynamic range for 200 different two-pole models. The solid curve is V_a, the absolute lower bound on the normalized error.

Another measure of data spread is the dynamic range. We define the spectral dynamic range d as the ratio of the highest to the lowest amplitude points on the spectrum:

$$d = H/L$$

where

$$H = \max_\omega \hat{P}(\omega) \qquad L = \min_\omega \hat{P}(\omega). \tag{82}$$

The relation between the normalized error and the spectral dynamic range is illustrated in Fig. 9. The dark dots in the figure are plots of the normalized error versus the spectral dynamic range (in decibels) for 2-pole models of 200 different speech spectra. The solid curve in Fig. 9 is an absolute lower bound on the geometric-to-arithmetic mean ratio for *any* spectrum with a given dynamic range. The curve is a plot of the following relation [22], [64]

$$V_a = \gamma e^{(1-\gamma)} \tag{83}$$

where

$$\gamma = \frac{\log d}{d - 1} \tag{84}$$

and V_a stands for the absolute lower bound on V_p for a given d. The overall impression from Fig. 9 is that the normalized error generally decreases as the dynamic range of the spectrum increases. This is apparent in Fig. 8 where V_p for the vowel [æ] is less than that for the fricative [*s*], and [æ] has a much higher spectral dynamic range than [*s*].

2) A Measure of Ill-Conditioning: In solving the autocorrelation normal equations (12), the condition of the autocorrelation matrix is an important consideration in deciding the accuracy of the computation needed. An ill-conditioned matrix can cause numerical problems in the solution. An accepted measure of ill-conditioning in a matrix is given by the ratio

$$d' = \lambda_{max}/\lambda_{min} \tag{85}$$

where λ_{max} and λ_{min} are the maximum and minimum eigenvalues of the matrix [27], [81]. Grenander and Szegö [41], [42] have shown that all the eigenvalues of an autocorrelation matrix lie in the range $\lambda_i \in [H, L]$, $1 \leqslant i \leqslant p$, where H and L are defined in (82). In addition, as the order of the matrix p increases, the eigenvalues become approximately equal to $\hat{P}(\omega)$ evaluated at equally spaced points with separation $2\pi/(p+1)$. Therefore, the ratio d' given in (85) can be well approximated by the dynamic range of $\hat{P}(\omega)$:

$$d' \cong d. \tag{86}$$

Therefore, the spectral dynamic range is a good measure of the ill-conditioning of the autocorrelation matrix. The larger the dynamic range, the greater is the chance that the matrix is ill-conditioned.

But in the previous section we noted that an increase in d usually results in a decrease in the normalized error V_p. Therefore, V_p can also be used as a measure of ill-conditioning: the ill-conditioning is greater with decreased V_p. The problem becomes more and more serious as $V_p \to 0$, i.e., as the signal becomes highly predictable.

If ill-conditioning occurs sporadically, then one way of patching the problem is to increase the values along the principal diagonal of the matrix by a small fraction of a percent. However, if the problem is a regular one, then it is a good idea if one can reduce the dynamic range of the signal spectrum. For example, if the spectrum has a general slope, then a single-zero filter of the form $1 + az^{-1}$ applied to the signal can be very effective. The new signal is given by

$$s_n' = s_n + as_{n-1}. \tag{87}$$

An optimal value for a is obtained by solving for the filter $A(z)$ that "whitens" (flattens) s_n'. This is, of course, given by the first order predictor, where

$$a = -\frac{R(1)}{R(0)}. \tag{88}$$

$R(1)$ and $R(0)$ are autocorrelation coefficients of the signal s_n. The filtered signal s_n' is then guaranteed to have a smaller spectral dynamic range. The above process is usually referred to as *preemphasis*.

One conclusion from the above concerns the design of the low-pass filter that one uses before sampling the signal to reduce aliasing. In order to ensure against aliasing, it is usually recommended that the cutoff frequency of the filter be lower than half the sampling frequency. However, if the cutoff frequency is appreciably lower than half the sampling frequency, then the spectral dynamic range of the signal spectrum increases, especially if the filter has a sharp cutoff and the stop band is very low relative to the pass band. This increases problems of ill-conditioning. Therefore, if one uses a lowpass filter with a sharp cutoff, the cutoff frequency should be set as close to half the sampling frequency as possible.

D. Optimal Number of Poles

One of the important decisions that usually has to be made in fitting of all-pole models is the determination of an "optimal" number of poles. It is a nontrivial exercise to define the word "optimal" here, for as we have seen, the fit of the model "improves" as the number of poles p increases. The problem is where to stop. Clearly we would like the minimum value of p that is adequate for the problem at hand, both to reduce our computation and to minimize the possibility of ill-conditioning (which increases with p since V_p decreases).

If the signal spectrum is an all-pole spectrum with p_0 poles, then we know that $V_p = V_{p_0}, p \geqslant p_0$, and $k_p = 0, p > p_0$, i.e., the error curve remains flat for $p > p_0$. Therefore, if we expect the signal spectrum to be an all-pole spectrum, a simple test to obtain the optimal p is to check when the error curve becomes flat. But, if the signal is the output of a p_0-pole filter with white noise excitation, then the suggested test will not work, because the estimates of the poles are based on a finite number of data points and the error curve will not be flat for $p > p_0$. In practice, however, the error curve will be almost flat for $p > p_0$. This suggests the use of the following threshold test

$$1 - \frac{V_{p+1}}{V_p} < \delta. \tag{89}$$

This test must succeed for several consecutive values before one is sure that the error curve has actually flattened out.

The use of the ratio V_{p+1}/V_p has been an accepted method in the statistical literature [6], [17], [112] for the determination of the optimal p. The test is based on hypothesis testing procedures using maximum likelihood ratios. A critical review of hypothesis testing procedures has been given recently by Akaike [5]. Akaike's main point is that model fitting is a problem where multiple decision procedures are required rather than hypothesis testing. The fitting problem should be stated as an estimation problem with an associated measure of fit. Akaike suggests the use of an information theoretic criterion that is an estimate of the mean log-likelihood [3], [5].[14] This is given by

$$I(p) = -2 \log(\text{maximum likelihood}) + 2p. \tag{90}$$

The value of p for which $I(p)$ is minimum is taken to be the optimal value. In our problem of all-pole modeling, if we assume that the signal has a Gaussian probability distribution,

[14] An earlier criterion used by Akaike is what he called the "final prediction error" [1], [2], [37].

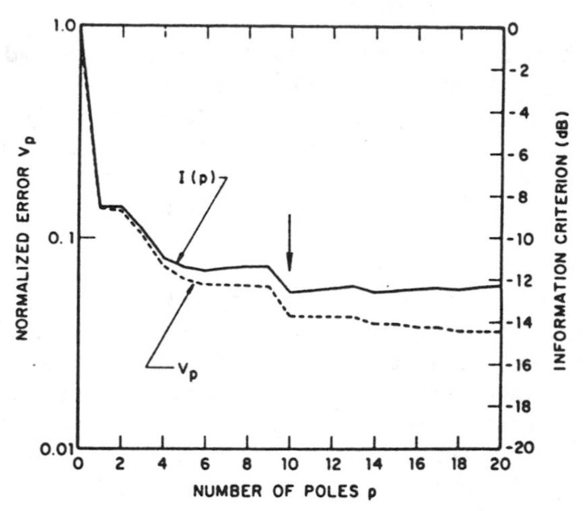

Fig. 10. A plot of Akaike's information criterion versus the order of the predictor p. Here, $I(p) = 10 \log_{10} V_p + (8.686p/0.4N)$, ($N = 200$, Hamming windowed). The "optimal" value of p occurs at the minimum of $I(p)$, shown by the arrow at $p = 10$.

then (90) reduces to (neglecting additive constants and dividing by N) [6], [17], [51]

$$I(p) = \log V_p + \frac{2p}{N_e} \tag{91}$$

where N_e is the "effective" number of data points in the signal. The word "effective" is used to indicate that one must compensate for possible windowing. The effective width of a window can be taken as the ratio of the energy under the window relative to that of a rectangular window. For example, for a Hamming window, $N_e = 0.4N$.

Note that the first term in (91) decreases as a function of p, and the second term increases. Therefore, a minimum can occur. In practice, there are usually several local minima, then the value of p corresponding to the absolute minimum of $I(p)$ is taken as the optimal value. Usually $I(p)$ is computed up to the maximum value of interest,[15] and the minimum of $I(p)$ is found in that region.

Fig. 10 shows an example of the application of Akaike's criterion. The dotted curve is the usual error curve and the solid curve is a plot of $I(p)$ in (91) multiplied by $10 \log_{10} e$ to obtain the results in decibels. In Fig. 10, the optimal predictor order is $p = 10$. Note that $I(p)$ for $p > 10$ slopes upward, but very gently. This indicates that the actual absolute minimum is quite sensitive to the linear term in (91). In practice, the criterion in (91) should not be regarded as an absolute, because it is based on several assumptions which might not apply for the signal of interest. For example, the assumptions of uncorrelated noise excitation and Gaussian distributions might not hold. Therefore, the experimenter should feel free to adjust the criterion to suit one's application. One simple way of "tuning" the criterion is to multiply N_e by an appropriate factor.

V. Data Compression by Linear Prediction

The methods outlined in Section II for the modeling of the behavior of a signal can be very useful in data compression. The process of signal or system modeling is essentially one of redundancy removal, which is the essence of data compression.

[15] Akaike informed me that he usually recommends $p_{max} < 3N^{1/2}$ as the maximum value of p that should be used if one is interested in a reliable estimate.

The idea of attempting to predict the value of a signal from previous sample values has been labeled in communications as "predictive coding" [28]. Adaptive linear prediction has been used extensively in speech and video transmission [7], [9], [23], [44], [50], [57], [83]. For the purposes of transmission one must quantize and transmit the predictor parameters or some transformation thereof. It has been known for some time that the quantization of the predictor parameters themselves is quite inefficient since a large number of bits is required to retain the desired fidelity in the reconstructed signal at the receiver [72]. Below, several equivalent representations of the predictor are presented and their quantization properties are discussed. We shall continue to assume that $H(z)$ is always stable, and hence minimum phase. $A(z)$ is, of course, then also minimum phase.

A. Alternate Representations of Linear Predictor

The following is a list of possible sets of parameters that characterize uniquely the all-pole filter $H(z)$ or its inverse $A(z)$.

1) (a) Impulse response of the inverse filter $A(z)$, i.e., predictor parameters a_k, $1 \leq k \leq p$.

(b) Impulse response of the all-pole model h_n, $0 \leq n \leq p$, which is defined in (30). Note that the first $p + 1$ coefficients uniquely specify the filter.

2) (a) Autocorrelation coefficients of a_k, $\rho(i)$, $0 \leq i \leq p$, as defined in (66).

(b) Autocorrelation coefficients of h_n, $\hat{R}(i)$, $0 \leq i \leq p$, as defined in (31) and (32).

3) Spectral coefficients of $A(z)$, Γ_i, $0 \leq i \leq p$ (or equivalently spectral coefficients of $H(z)$, G^2/Γ_i):

$$\Gamma_i = \rho(0) + 2 \sum_{j=1}^{p} \rho(j) \cos \frac{2\pi i j}{2p + 1}, \qquad 0 \leq i \leq p \tag{92}$$

where $\rho(i)$ are as defined in (66). In other words, $\{\Gamma_i\}$ is obtained from $\{\rho(i)\}$ by a discrete Fourier transform.

4) Cepstral coefficients of $A(z)$, c_n, $1 \leq n \leq p$ (or equivalently cepstral coefficients of $H(z)$, $-c_n$):

$$c_n = \frac{1}{2\pi} \int_{-\pi}^{\pi} \log A(e^{j\omega})\, e^{jn\omega}\, d\omega. \tag{93}$$

Since $A(z)$ is minimum phase, (93) reduces to [38], [77]

$$c_n = a_n - \sum_{m=1}^{n-1} \frac{m}{n} c_m a_{n-m}, \qquad 1 \leq n \leq p. \tag{94}$$

Equation (94) is an iterative method for the computation of the cepstral coefficients directly from the predictor coefficients.

5) Poles of $H(z)$ or zeros of $A(z)$, z_k, $1 \leq k \leq p$, where $\{z_k\}$ are either real or form complex conjugate pairs. Conversion of the roots to the s plane can be achieved by setting each root $z_k = e^{s_k T}$, where $s_k = \sigma_k + j\omega_k$ is the corresponding pole in the s plane, and T is the sampling period. If the root $z_k = z_{kr} + jz_{ki}$, then

$$\omega_k = \frac{1}{T} \arctan \frac{z_{ki}}{z_{kr}}$$

$$\sigma_k = \frac{1}{2T} \log (z_{kr}^2 + z_{ki}^2) \tag{95}$$

where z_{kr} and z_{ki} are the real and imaginary parts of z_k, respectively.

6) Reflection coefficients k_i, $1 \leq i \leq p$, which are obtained as a byproduct of the solution of the autocorrelation normal equations, as in (38), or from the backward recursion (51).

Some of the preceding sets of parameters have $p + 1$ coefficients while others have only p coefficients. However, for the latter sets the gain G needs to be specified as well, thus keeping the total number of parameters as $p + 1$ for all the cases.

For purposes of data transmission, one is usually interested in recovering the predictor coefficients from the parameters that are chosen for transmission. The required transformations are clear for most of the above parameters, except perhaps for the parameters $\rho(i)$ and Γ_i. Through an inverse DFT, the spectral coefficients Γ_i can be converted to autocorrelation coefficients $\rho(i)$. One method of recovering the predictor parameters from $\{\rho(i)\}$ is as follows. Apply a DFT to the sequence $\{\rho(i)\}$ after appending it with an appropriate number of zeros to achieve sufficient resolution in the resulting spectrum of $A(z)$. Divide G^2 by this spectrum to obtain the spectrum of the filter $H(z)$. Inverse Fourier transformation of the spectrum of $H(z)$ yields the autocorrelation coefficients $\hat{R}(i)$. The first $p + 1$ coefficients $\hat{R}(i)$, $0 \leq i \leq p$, are then used to compute the predictor coefficients via the normal equations (12) with $R(i) = \hat{R}(i)$.

B. *Quantization Properties*

Although the sets of parameters given above provide equivalent information about the linear predictor, their properties under quantization are different. For the purpose of quantization, two desirable properties for a parameter set to have are: 1) filter stability upon quantization and 2) a natural ordering of the parameters. Property 1) means that the poles of $H(z)$ continue to be inside the unit circle even after parameter quantization. By 2), we mean that the parameters exhibit an inherent ordering, e.g., the predictor coefficients are ordered as $a_1, a_2, \cdots, a_p$. If a_1 and a_2 are interchanged then $H(z)$ is no longer the same in general, thus illustrating the existence of an ordering. The poles of $H(z)$, on the other hand, are not naturally ordered since interchanging the values of any two poles does not change the filter. When an ordering is present, a statistical study on the distribution of individual parameters can be used to develop better encoding schemes. Only the poles and the reflection coefficients insure stability upon quantization, while all the sets of parameters except the poles possess a natural ordering. Thus only the reflection coefficients possess both of these properties.

In an experimental study [63] of the quantization properties of the different parameters, it was found that the impulse responses $\{a_k\}$ and $\{h_n\}$ and the autocorrelations $\{\rho(i)\}$ and $\{\hat{R}(i)\}$ are highly susceptible to causing instability of the filter upon quantization. Therefore, these sets of parameters can be used only under minimal quantization, in which case the transmission rate would be excessive.

In the experimental investigation of the spectral and cepstral parameters, it was found that the quantization properties of these parameters are generally superior to those of the impulse responses and autocorrelation coefficients. The spectral parameters often yield results comparable to those obtained by quantizing the reflection coefficients. However, for the cases when the spectrum consists of one or more very sharp peaks (narrow bandwidths), the effects of quantizing the spectral coefficients often cause certain regions in the reconstructed spectrum (as described in the previous section) to become negative, which leads to instability of the computed filter. Quantization of the cepstral coefficients can also lead to instabilities. It should be noted here that the quantization properties of these parameters give better results if the spectral dynamic range of the signal is limited by some form of preprocessing.

Filter stability is preserved under quantization of the poles. But poles are expensive to compute, and they do not possess a natural ordering.

The conclusion is that, of the sets of parameters given in the preceding, the reflection coefficients are the best set to use as transmission parameters. In addition to ease of computation, stability under quantization, and natural ordering, the values of the reflection coefficients k_i, $i < p$, do not change as p is increased, unlike any of the other parameters. In the following, we discuss the optimal quantization of the reflection coefficients.

C. *Optimal Quantization [53], [65]*

Optimal quantization of the reflection coefficients depends on the fidelity criterion chosen. For many applications, it is important that the log spectrum of the all-pole model be preserved. In this case, it is reasonable to study the sensitivity of the log spectrum with respect to changes in the reflection coefficients. In a recent study [65], a spectral sensitivity curve was plotted versus each of the reflection coefficients k_i for many different all-pole models obtained by analyzing a large number of speech samples. The results of the study show that each sensitivity curve versus k_i has the same general shape, irrespective of the index i. Each sensitivity curve is U-shaped; it is even-symmetric about $k_i = 0$, with large values when $|k_i| \to 1$, and small values when $|k_i|$ is close to zero. These properties indicate that linear quantization of the reflection coefficients is not desirable, especially if some of them take values very close to 1, which happens when the spectrum contains sharp resonances. Nonlinear quantization of k_i is equivalent to a linear quantization of another parameter, say g_i, which is related to k_i by a nonlinear transformation. The requirement that the spectral sensitivity of the new parameters be flat resulted in the following optimal transformation [65]:

$$g_i = \log \frac{1 + k_i}{1 - k_i}, \qquad \text{all } i. \tag{96}$$

It is interesting to note from (45) that g_i is simply the logarithm of the hypothetical impedance ratios corresponding to k_i.

The optimality of the preceding transformation was based on a specific spectral fidelity criterion. Other transformations would result if other quantization fidelity criteria were adopted.[16]

The transmission rate can be reduced further without affecting the fidelity by proper encoding of each parameter. Variable word length encoding [34] (such as Huffman) can be used for this purpose if the statistical distributions of each of the parameters is known. These distributions can be obtained very simply from a representative sample of signals.

VI. Pole-Zero Modeling

Given the spectrum of some arbitrary signal, it is generally not possible to determine for certain the identity of the system that generated the signal in terms of a set of poles and

[16] Using a log likelihood criterion, Itakura informed me that a transformation proportional to arcsin (k_i) is optimal.

zeros. The problem is inherently nondeterministic, for a zero can be approximated arbitrarily closely by a large number of poles and vice versa. Indeed, we have seen in this paper that the all-pole model spectrum can approximate the signal spectrum arbitrarily closely by simply increasing the number of poles. However, if there are a number of influential zeros in the signal spectrum, the number of model poles can become very large. For data compression applications, this is an undesirable situation. Also, there are applications where the identification of the zeros is important. Therefore, it is useful to be able to model a spectrum in terms of poles and zeros.

Much effort is currently being expended on the problem of pole-zero modeling [8], [17], [19], [20], [29], [30], [88], [91]. Most of these methods are purely in the time domain. However, there seems to be a growing interest in frequency domain methods [4], [46], [94], partly due to the speed offered by the FFT. Add to this the market availability of spectrum analyzers and special hardware FFT processors. Of course, the time and frequency domain approaches should give similar results since LP analysis is actually performed in the autocorrelation domain.

The beauty of all-pole modeling is that it is relatively simple, straightforward, well understood, inexpensive, and "always" works. Unfortunately, none of these properties apply to pole-zero modeling. The main difficulty is that the pole-zero problem is nonlinear. We show this below for the stationary case. Then, we sketch out representative schemes for iterative and noniterative estimation of the pole and zero parameters. No exhaustive analysis is attempted; the reader is referred to the aforementioned references.

A. Normal Equations

The transfer function of the pole-zero model is given by $H(z)$ in (2). The corresponding model spectrum is given by

$$\hat{P}(\omega) = |H(e^{j\omega})|^2 = G^2 \frac{|B(e^{j\omega})|^2}{|A(e^{j\omega})|^2} = G^2 \frac{N(\omega)}{D(\omega)} \tag{97}$$

where $B(z)$ and $A(z)$ are the numerator and denominator polynomials in $H(z)$, and the all-zero spectra $N(\omega)$ and $D(\omega)$ form the numerator and denominator of $\hat{P}(\omega)$ and are given by

$$N(\omega) = \left| 1 + \sum_{l=1}^{q} b_l e^{-jl\omega} \right|^2 \tag{98}$$

and

$$D(\omega) = \left| 1 + \sum_{k=1}^{p} a_k e^{-jk\omega} \right|^2. \tag{99}$$

The matching error between the signal spectrum $P(\omega)$ and $\hat{P}(\omega)$ is given by (62), and from (97) is equal to

$$E = \frac{1}{2\pi} \int_{-\pi}^{\pi} P(\omega) \frac{D(\omega)}{N(\omega)} d\omega. \tag{100}$$

E can be interpreted as the residual energy obtained by passing the signal through the filter $A(z)/B(z)$. The problem is to determine $\{a_k\}$ and $\{b_l\}$ such that E in (100) is minimized.

In the sequel we shall make use of the following two relations:

$$\frac{\partial N(\omega)}{\partial b_i} = 2 \sum_{l=0}^{q} b_l \cos{(i-l)\omega}, \qquad b_0 = 1 \tag{101}$$

$$\frac{\partial D(\omega)}{\partial a_i} = 2 \sum_{k=0}^{p} a_k \cos{(i-k)\omega}, \qquad a_0 = 1. \tag{102}$$

In addition, we shall use the notation $R_{\alpha\beta}(i)$ to represent the autocorrelation defined by

$$R_{\alpha\beta}(i) = \frac{1}{2\pi} \int_{-\pi}^{\pi} P(\omega) \frac{[D(\omega)]^\beta}{[N(\omega)]^\alpha} \cos{(i\omega)}\, d\omega. \tag{103}$$

Thus, $R_{00}(i)$ is simply the Fourier transform of $P(\omega)$.[17] Taking $\partial E/\partial a_i$ in (100) one obtains

$$\frac{\partial E}{\partial a_i} = 2 \sum_{k=0}^{p} a_k R_{10}(i-k), \qquad 1 \leq i \leq p. \tag{104}$$

Similarly, one can show that

$$\frac{\partial E}{\partial b_i} = -2 \sum_{l=0}^{q} b_l R_{21}(i-l), \qquad 1 \leq i \leq q. \tag{105}$$

In order to minimize E, we set $\partial E/\partial a_i = 0$, $1 \leq i \leq p$, and $\partial E/\partial b_i = 0$, $1 \leq i \leq q$, simultaneously. These, then, comprise the normal equations.

From (103), it is clear that $R_{10}(i-k)$ is not a function of a_k. Therefore, setting (104) to zero results in a set of linear equations, identical in form to the autocorrelation normal equations (12). However, $R_{21}(i-l)$ in (105) is a function of b_l, as can be deduced from (103) with $\alpha = 2$ and $\beta = 1$. Therefore, setting (105) to zero results in a set of nonlinear equations in b_l. If one wishes to solve for $\{a_k\}$ and $\{b_l\}$ simultaneously, then one solves a set of $p + q$ *nonlinear* equations.

Note that if the signal is assumed to be nonstationary, the above analysis can be modified accordingly in a manner similar to that in Section III-A for the all-pole case. The resulting equations will be very similar in form to the preceding equations, with nonstationary autocorrelations replacing the stationary autocorrelations.

B. Iterative Solutions

Since the minimization of E in (100) leads to a set of nonlinear equations, the problem of minimizing E must then be solved iteratively. There are many methods in the literature for finding the extrema of a function [30], [108], many of which are applicable in our case. In particular, gradient methods are appropriate here since it is possible to evaluate the error gradient, as in (104) and (105). One such method was used by Tretter and Steiglitz [94] in pole-zero modeling. Other schemes, such as the Newton–Raphson method, require the evaluation of the Hessian (i.e., second derivative). This can be very cumbersome in many problems, but is straightforward in our case. This is illustrated below by giving a Newton–Raphson solution.

Let $x' = [a_1 a_2 \cdots a_p b_1 b_2 \cdots b_q]$ be the transpose of a column vector x whose elements are the coefficients a_k and b_l. If $x(m)$ is the solution at iteration m, then $x(m+1)$ is given by

$$x(m+1) = x(m) - J^{-1} \left. \frac{\partial E}{\partial x} \right|_{x=x(m)} \tag{106}$$

where J is the $(p+q) \times (p+q)$ symmetric Hessian matrix

[17] For a discrete spectrum $P(\omega_n)$ the integrals in (100) and (103) are replaced by summations.

given by $J = \partial^2 E/\partial x \partial x'$. Setting $a' = [a_1 a_2 \cdots a_p]$ and $b' = [b_1 b_2 \cdots b_q]$, (106) can be partitioned, with $x' = [a'b']$, as

$$\begin{bmatrix} a(m+1) \\ b(m+1) \end{bmatrix} = \begin{bmatrix} a(m) \\ b(m) \end{bmatrix} - \begin{bmatrix} \dfrac{\partial^2 E}{\partial a \partial a'} & \dfrac{\partial^2 E}{\partial a \partial b'} \\ \dfrac{\partial^2 E}{\partial b \partial a'} & \dfrac{\partial^2 E}{\partial b \partial b'} \end{bmatrix}^{-1}_{\substack{a=a(m)\\ b=b(m)}} \begin{bmatrix} \dfrac{\partial E}{\partial a} \\ \dfrac{\partial E}{\partial b} \end{bmatrix}_{\substack{a=a(m)\\ b=b(m)}} \tag{107}$$

The elements of the first-order partial derivatives in (107) are given by (104) and (105). The elements of the second partial derivatives can be shown to be equal to

$$\frac{\partial^2 E}{\partial a_i \partial a_j} = 2R_{10}(i-j) \tag{108}$$

$$\frac{\partial^2 E}{\partial a_i \partial b_j} = -2 \sum_{k=0}^{p} \sum_{l=0}^{q} a_k b_l \cdot [R_{20}(j+i-l-k) + R_{20}(j-i-l+k)] \tag{109}$$

$$\frac{\partial^2 E}{\partial b_i \partial b_j} = -2R_{21}(i-j) + 4 \sum_{k=0}^{p} \sum_{l=0}^{q} b_k b_l \cdot [R_{31}(j+i-l-k) + R_{31}(j-i-l+k)]. \tag{110}$$

Given the estimates $a(m)$ and $b(m)$, one can compute $N(\omega)$ and $D(\omega)$ from (98) and (99) using FFT's, and then use (103) to compute the autocorrelations R_{10}, R_{20}, R_{21}, and R_{31}, which can then be used in (107)–(110) to compute the new estimates $a(m+1)$ and $b(m+1)$. The iterations are halted when the error gradient goes below some prespecified threshold. The minimum error is then computed from (100).

The Newton–Raphson method works very well if the initial estimate is close to the optimum. In that case, the Hessian J is positive definite and the convergence is quadratic [30]. In the next section we discuss noniterative methods which can be used to give these good initial estimates.

C. Noniterative Solutions

One property that is common to noniterative methods is that a good estimate of the number of poles and zeros seems to be necessary for a reasonable solution. Indeed, in that case, there is not much need to go to expensive iterative methods. However, in general, such information is unavailable and one is interested in obtaining the best estimate for a given p and q. Then, noniterative methods can be used profitably to give good initial estimates that are necessary in iterative methods.

1) Pole Estimation: Assume that the signal s_n had been generated by exciting the pole-zero filter $H(z)$ in (2) by either an impulse or white noise. Then it is simple to show that the signal autocorrelation obeys the autocorrelation equation (31) for $i > q$. Therefore, the coefficients a_k can be estimated by solving (31) with $q+1 \leq i \leq q+p$.

The effect of the poles can now be removed by applying the inverse filter $A(z)$ to the signal. In the spectral domain this can be done by computing a new spectrum $P_1(\omega) = P(\omega) D(\omega)$. The problem now reduces to the estimation of the zeros in $P_1(\omega)$.

2) Zero Estimation: A promising noniterative method for pole-zero estimation is that of *cepstral prediction* [76], [95]. The basic idea is that the poles of nc_n, where c_n is the complex cepstrum, comprise the poles and zeros of the signal [77]. Therefore, for zero estimation, the problem reduces to finding the poles of nc_n which can be computed by the method just described above, where c_n here is the cepstrum corresponding to $P_1(\omega)$.

Another method for zero estimation is that of *inverse LP modeling* [63]. The idea is quite simple: Invert the spectrum $P_1(\omega)$ and apply a q-pole LP analysis. The resulting predictor coefficients are then good estimates of b_l. This method gives good results if $P_1(\omega)$ is smooth relative to the model spectrum. Problems arise if the variations of the signal spectrum about the model spectrum are large. The reason is that LP modeling attempts to make a good fit to the spectral envelope, and the envelope of the inverted spectrum is usually different from the inverse of the desired spectral envelope. One solution is to smooth the spectrum $P_1(\omega)$ before inversion. Spectral smoothing is usually performed by applying a low-pass filter to the spectrum (autocorrelation smoothing) or to the log spectrum (cepstral smoothing). Another method is all-pole smoothing. Indeed, all-pole modeling can be thought of as just another method of smoothing the spectrum, where the degree of smoothing is controlled by the order of the predictor p, which is usually chosen to be much larger than the number of zeros in the model q. We point out that zero estimation by inverse LP modeling with all-pole smoothing is similar to the method of Durbin [24], [25] in the time domain.

VII. Conclusion

Linear prediction is an autocorrelation-domain analysis. Therefore, it can be approached from either the time or frequency domain. The least squares error criterion in the time domain translates into a spectral matching criterion in the frequency domain. This viewpoint was helpful in exploring the advantages and disadvantages of the least squares error criterion.

The major portion of this paper was devoted to all-pole modeling. This type of modeling is simple, inexpensive and effective; hence its wide applicability and acceptance. In contrast, pole-zero modeling is not simple, generally expensive, and is not yet well understood. Future research should be directed at acquiring a better understanding of the problems in pole-zero modeling and developing appropriate methodologies to deal with these problems.

Acknowledgment

The author wishes to thank the following friends and colleagues who have read and commented on earlier versions of this paper: H. Akaike, R. Barakat, C. Cook, T. Fortmann, F. Itakura, G. Kopec, A. Oppenheim, L. Rabiner, R. Viswanathan, and V. Zue. He is especially grateful to R. Viswanathan for the fruitful discussions they have had throughout the writing of the paper and for his help in supplying many of the references. He also wishes to thank B. Aighes, C. Williams, and R. Schwartz for their assistance in the preparation of the manuscript.

References

[1] H. Akaike, "Power spectrum estimation through autoregressive model fitting," *Ann. Inst. Statist. Math.*, vol. 21, pp. 407–419, 1969.

[2] ——, "Statistical predictor identification," *Ann. Inst. Statist. Math.*, vol. 22, pp. 203–217, 1970.

[3] ——, "Information theory and an extension of the maximum likelihood principle," in *Proc. 2nd Int. Symp. Information Theory* (Supplement to Problems of Control and Information Theory), 1972.

[4] —, "Maximum likelihood identification of Gaussian autoregressive moving average models," *Biometrika*, vol. 60, no. 2, pp. 255–265, 1973.
[5] —, "A new look at statistical model identification," *IEEE Trans. Automat. Contr.*, vol. AC-19, pp. 716–723, Dec. 1974.
[6] T. W. Anderson, *The Statistical Analysis of Time Series.* New York: Wiley, 1971.
[7] C. A. Andrews, J. M. Davies, and G. R. Schwartz, "Adaptive data compression," *Proc. IEEE*, vol. 55, pp. 267–277, Mar. 1967.
[8] K. J. Åström and P. Eykhoff, "System identification—A survey," *Automatica*, vol. 7, pp. 123–162, 1971.
[9] B. S. Atal and M. R. Schroeder, "Adaptive predictive coding of speech signals," *Bell Syst. Tech. J.*, vol. 49, no. 6, pp. 1973–1986, Oct. 1970.
[10] B. S. Atal and S. L. Hanauer, "Speech analysis and synthesis by linear prediction of the speech wave," *J. Acoust. Soc. Amer.*, vol. 50, no. 2, pp. 637–655, 1971.
[11] M. S. Bartlett, *An Introduction to Stochastic Processes with Special Reference to Methods and Applications.* Cambridge, England: Cambridge Univ. Press, 1956.
[12] J. S. Bendat and A. G. Piersol, *Random Data: Analysis and Measurement Procedures.* New York: Wiley, 1971.
[13] C. Bingham, M. D. Godfrey, and J. W. Tukey, "Modern techniques of power spectrum estimation," *IEEE Trans. Audio Electroacoust.*, vol. AU-15, pp. 56–66, June 1967.
[14] R. B. Blackman and J. W. Tukey, *The Measurement of Power Spectra.* New York: Dover, 1958.
[15] T. Bohlin, "Comparison of two methods of modeling stationary EEG signals," *IBM J. Res. Dev.*, pp. 194–205, May 1973.
[16] S. F. Boll, "*A priori* digital speech analysis," Computer Science Div., Univ. Utah, Salt Lake City, UTEC-CSC-73-123, 1973.
[17] G. E. Box and G. M. Jenkins, *Time Series Analysis Forecasting and Control.* San Francisco, Calif.: Holden-Day, 1970.
[18] J. P. Burg, "The relationship between maximum entropy spectra and maximum likelihood spectra," *Geophysics*, vol. 37, no. 2, pp. 375–376, Apr. 1972.
[19] C. S. Burrus and T. W. Parks, "Time domain design of recursive digital filters," *IEEE Trans. Audio Electroacoust.*, vol. AU-18, pp. 137–141, June 1970.
[20] J. C. Chow, "On estimating the orders of an autoregressive moving-average process with uncertain observations," *IEEE Trans. Automat. Contr.*, vol. AC-17, pp. 707–709, Oct. 1972.
[21] W. T. Cochran *et al.*, "What is the fast Fourier transform?," *IEEE Trans. Audio Electroacoust.*, vol. AU-15, pp. 45–55, June 1967.
[22] H. Cox, "Linear versus logarithmic averaging," *J. Acoust. Soc. Amer.*, vol. 39, no. 4, pp. 688–690, 1966.
[23] L. D. Davisson, "The theoretical analysis of data compression systems," *Proc. IEEE*, vol. 56, pp. 176–186, Feb. 1968.
[24] J. Durbin, "Efficient estimation of parameters in moving-average models," *Biometrika*, vol. 46, Parts 1 and 2, pp. 306–316, 1959.
[25] —, "The fitting of time-series models," *Rev. Inst. Int. Statist.*, vol. 28, no. 3, pp. 233–243, 1960.
[26] A. Eberhard, "An optimal discrete window for the calculation of power spectra," *IEEE Trans. Audio Electroacoust.*, vol. AU-21, pp. 37–43, Feb. 1973.
[27] M. P. Ekstrom, "A spectral characterization of the ill-conditioning in numerical deconvolution," *IEEE Trans. Audio Electroacoust.*, vol. AU-21, pp. 344–348, Aug. 1973.
[28] P. Elias, "Predictive coding, Parts I and II," *IRE Trans. Inform. Theory*, vol. IT-1, p. 16, Mar. 1955.
[29] A. G. Evans and R. Fischl, "Optimal least squares time-domain synthesis of recursive digital filters," *IEEE Trans. Audio Electroacoust.*, vol. AU-21, pp. 61–65, Feb. 1973.
[30] P. Eykhoff, *System Identification: Parameter and State Estimation.* New York: Wiley, 1974.
[31] D. K. Faddeev and V. N. Faddeeva, *Computational Methods of Linear Algebra.* San Francisco, Calif.: Freeman, 1963.
[32] P. B. C. Fenwick, P. Michie, J. Dollimore, and G. W. Fenton, "Mathematical simulation of the electroencephalogram using an autoregressive series," *Bio-Med. Comput.*, vol. 2, pp. 281–307, 1971.
[33] J. L. Flanagan, *Speech Analysis Synthesis and Perception*, 2nd Edition. New York: Springer-Verlag, 1972.
[34] R. G. Gallager, *Information Theory and Reliable Communication.* New York: Wiley, 1968.
[35] L. Y. Geronimus, *Orthogonal Polynomials.* New York: Consultants Bureau, 1961.
[36] W. Gersch, "Spectral analysis of EEG's by autoregressive decomposition of time series," *Math. Biosci.*, vol. 7, pp. 205–222, 1970.
[37] W. Gersch and D. R. Sharpe, "Estimation of power spectra with finite-order autoregressive models," *IEEE Trans. Automat. Contr.*, vol. AC-18, pp. 367–369, Aug. 1973.
[38] B. Gold and C. M. Rader, *Digital Processing of Signals.* New York: McGraw-Hill, 1969.
[39] G. H. Golub and C. Reinsch, "Singular value decomposition and least squares solutions," *Numer. Math.*, vol. 14, pp. 403–420, 1970.
[40] A. H. Gray and J. D. Markel, "A spectral-flatness measure for studying the autocorrelation method of linear prediction of speech analysis," *IEEE Trans. Acoust., Speech, Signal Processing*, vol. ASSP-22, pp. 207–216, June 1974.
[41] R. M. Gray, "On the asymptotic eigenvalue distribution of Toeplitz matrices," *IEEE Trans. Inform. Theory*, vol. IT-18, pp. 725–730, Nov. 1972.
[42] U. Grenander and G. Szegö, *Toeplitz Forms and Their Applications.* Berkeley, Calif.: Univ. California Press, 1958.
[43] U. Grenander and M. Rosenblatt, *Statistical Analysis of Stationary Time Series.* New York: Wiley, 1957.
[44] A. Habibi and G. S. Robinson, "A survey of digital picture coding," *Computer*, pp. 22–34, May 1974.
[45] E. J. Hannan, *Time Series Analysis.* London, England: Methuen, 1960.
[46] —, *Multiple Time Series.* New York: Wiley, 1970.
[47] J. R. Haskew, J. M. Kelly, R. M. Kelly, Jr., and T. H. McKinney, "Results of a study of the linear prediction vocoder," *IEEE Trans. Commun.*, vol. COM-21, pp. 1008–1014, Sept. 1973.
[48] R. L. Hershey, "Analysis of the difference between log mean and mean log averaging," *J. Acoust. Soc. Amer.*, vol. 51, pp. 1194–1197, 1972.
[49] F. B. Hildebrand, *Introduction to Numerical Values.* New York: McGraw-Hill, 1956.
[50] F. Itakura and S. Saito, "Analysis synthesis telephony based on the maximum likelihood method," in *Rep. 6th Int. Congr. Acoustics*, Y. Kohasi, Ed., pp. C17–C20, Paper C-5-5, Aug. 1968.
[51] —, "A statistical method for estimation of speech spectral density and formant frequencies," *Electron. Commun. Japan*, vol. 53-A, no. 1, pp. 36–43, 1970.
[52] —, "Digital filtering techniques for speech analysis and synthesis," in *Conf. Rec., 7th Int. Congr. Acoustics*, Paper 25 C 1, 1971.
[53] —, "On the optimum quantization of feature parameters in the Parcor speech synthesizer," in *IEEE Conf. Rec., 1972 Conf. Speech Communication and Processing*, pp. 434–437, Apr. 1972.
[54] G. M. Jenkins and D. G. Watts, *Spectral Analysis and Its Applications.* San Francisco, Calif.: Holden-Day, 1968.
[55] T. Kailath, "A view of three decades of linear filtering theory," *IEEE Trans. Inform. Theory*, vol. IT-20, pp. 146–181, Mar. 1974.
[56] R. E. Kalman, "A new approach to linear filtering and prediction problems," *Trans. ASME, J. Basic Eng.*, Series D82, pp. 35–45, 1960.
[57] H. Kobayashi and L. R. Bahl, "Image data compression by predictive coding I: Prediction algorithms," *IBM J. Res. Dev.*, pp. 164–171, Mar. 1974.
[58] A. Kolmogorov, "Interpolation und Extrapolation von stationären zufälligen Folgen," *Bull. Acad. Sci., U.S.S.R., Ser. Math.*, vol. 5, pp. 3–14, 1941.
[59] K. S. Kunz, *Numerical Analysis.* New York: McGraw-Hill, 1957.
[60] R. C. K. Lee, *Optimal Estimation, Identification, and Control*, Research Monograph no. 28. Cambridge, Mass.: M.I.T. Press, 1964.
[61] N. Levinson, "The Wiener RMS (root mean square) error criterion in filter design and prediction," *J. Math. Phys.*, vol. 25, no. 4, pp. 261–278, 1947. Also Appendix B, in N. Wiener, *Extrapolation, Interpolation and Smoothing of Stationary Time Series.* Cambridge, Mass.: M.I.T. Press, 1949.
[62] J. Makhoul, "Spectral analysis of speech by linear prediction," *IEEE Trans. Audio Electroacoust.*, vol. AU-21, pp. 140–148, June 1973.
[63] —, "Selective linear prediction and analysis-by-synthesis in speech analysis," Bolt Beranek and Newman Inc., Cambridge, Mass., Rep. 2578, Apr. 1974.
[64] J. Makhoul and J. J. Wolf, "Linear prediction and the spectral analysis of speech," Bolt Beranek and Newman Inc., Cambridge, Mass., NTIS AD-749066, Rep. 2304, Aug. 1972.
[65] J. Makhoul and R. Viswanathan, "Quantization properties of transmission parameters in linear predictive systems," Bolt Beranek and Newman Inc., Cambridge, Mass., Rep. 2800, Apr. 1974.
[66] H. B. Mann and A. Wald, "On the statistical treatment of linear stochastic difference equations," *Econometrica*, vol. 11, nos. 3 and 4, pp. 173–220, 1943.
[67] W. D. Mark, "Spectral analysis of the convolution and filtering of non-stationary stochastic processes," *J. Sound Vib.*, no. 11, pp. 19–63, 1970.
[68] J. D. Markel, "Digital inverse filtering—A new tool for formant trajectory estimation," *IEEE Trans. Audio Electroacoust.*, vol. AU-20, pp. 129–137, June 1972.

[69] J. D. Markel and A. H. Gray, "On autocorrelation equations as applied to speech analysis," *IEEE Trans. Audio Electroacoust.*, vol. AU-20, pp. 69-79, Apr. 1973.

[70] E. Matsui *et al.*, "An adaptive method for speech analysis based on Kalman filtering theory," *Bull Electrotech. Lab*, vol. 36, no. 3, pp. 42-51, 1972 (in Japanese).

[71] R. N. McDonough and W. H. Huggins, "Best least-squares representation of signals by exponentials," *IEEE Trans. Automat. Contr.*, vol. AC-13, pp. 408-412, Aug. 1968.

[72] S. K. Mitra and R. J. Sherwood, "Estimation of pole-zero displacements of a digital filter due to coefficient quantization," *IEEE Trans. Circuits Syst.*, vol. CAS-21, pp. 116-124, Jan. 1974.

[73] R. E. Nieman, D. G. Fisher, and D. E. Seborg, "A review of process identification and parameter estimation techniques," *Int. J. Control*, vol. 13, no. 2, pp. 209-264, 1971.

[74] R. H. Norden, "A survey of maximum likelihood estimation," *Int. Statist. Rev.*, vol. 40, no. 3, pp. 329-354, 1972.

[75] ——, "A survey of maximum likelihood estimation, Part 2," *Int. Statist. Rev.*, vol. 41, no. 1, pp. 39-58, 1973.

[76] A. Oppenheim and J. M. Tribolet, "Pole-zero modeling using cepstral prediction," Res. Lab. Electronics, M.I.T., Cambridge, Mass., QPR 111, pp. 157-159, 1973.

[77] A. V. Oppenheim, R. W. Schafer and T. G. Stockham, "Nonlinear filtering of multiplied and convolved signals," *Proc. IEEE*, vol. 56, pp. 1264-1291, Aug. 1968.

[78] M. Pagano, "An algorithm for fitting autoregressive schemes," *J. Royal Statist. Soc., Series C (Applied Statistics)*, vol. 21, no. 3, pp. 274-281, 1972.

[79] A. Papoulis, *Probability, Random Variables, and Stochastic Processes.* New York: McGraw-Hill, 1965.

[80] L. R. Rabiner *et al.*, "Terminology in digital signal processing," *IEEE Trans. Audio Electroacoust.*, vol. AU-20, pp. 322-337, Dec. 1972.

[81] A. Ralston, *A First Course in Numerical Analysis.* New York: McGraw-Hill, 1965.

[82] A. Ralston and H. Wilf, *Mathematical Methods for Digital Computers*, vol. II. New York: Wiley, 1967.

[83] M. P. Ristenbatt, "Alternatives in digital communications," *Proc. IEEE*, vol. 61, pp. 703-721, June 1973.

[84] E. A. Robinson, *Multichannel Time Series Analysis with Digital Computer Programs.* San Francisco, Calif.: Holden-Day, 1967.

[85] ——, *Statistical Communication and Detection.* New York: Hafner, 1967.

[86] ——, "Predictive decomposition of time series with application to seismic exploration," *Geophysics*, vol. 32, no. 3, pp. 418-484, June 1967.

[87] E. A. Robinson and S. Treitel, *The Robinson-Treitel Reader*, 3rd ed. Tulsa, Okla.: Seismograph Service Corp., 1973.

[88] A. P. Sage and J. L. Melsa, *System Identification.* New York: Academic Press, 1971.

[89] R. W. Schafer and L. R. Rabiner, "Digital representations of speech signals," this issue, pp. 662-677.

[90] F. C. Schweppe, *Uncertain Dynamic Systems.* Englewood Cliffs, N.J.: Prentice-Hall, 1973.

[91] J. L. Shanks, "Recursion filters for digital processing," *Geophysics*, vol. XXXII, no. 1, pp. 33-51, 1967.

[92] R. A. Silverman, "Locally stationary random processes," *IRE Trans. Inform. Theory*, vol. IT-3, pp. 182-187, Sept. 1957.

[93] G. Szegö, *Orthogonal Polynomials.* New York: Amer. Math. Soc. Colloquium Publ., vol. XXIII, N.Y., 1959.

[94] S. A. Tretter and K. Steiglitz, "Power-spectrum identification in terms of rational models," *IEEE Trans. Automat. Control*, vol. AC-12, pp. 185-188, Apr. 1967.

[95] J. M. Tribolet, "Identification of linear discrete systems with applications to speech processing," Master's thesis, Dep. Elec. Eng., M.I.T., Cambridge, Mass., Jan. 1974.

[96] A. Van den Bos, "Alternative interpretation of maximum entropy spectral analysis," *IEEE Trans. Inform. Theory*, vol. IT-17, pp. 493-494, July 1971.

[97] H. Wakita, "Direct estimation of the vocal tract shape by inverse filtering of acoustic speech waveforms," *IEEE Trans. Audio Electroacoust.*, vol. AU-21, pp. 417-427, Oct. 1973.

[98] G. Walker, "On periodicity in series of related terms," *Proc. Royal Soc.*, vol. 131-A, p. 518, 1931.

[99] R. J. Wang, "Optimum window length for the measurement of time-varying power spectra," *J. Acoust. Soc. Amer.*, vol. 52, no. 1 (part 1), pp. 33-38, 1971.

[100] R. J. Wang and S. Treitel, "The determination of digital Wiener filters by means of gradient methods," *Geophysics*, vol. 38, no. 2, pp. 310-326, Apr. 1973.

[101] P. D. Welch, "A direct digital method of power spectrum estimation," *IBM J. Res. Dev.*, vol. 5, pp. 141-156, Apr. 1961.

[102] A. Wennberg and L. H. Zetterberg, "Application of a computer-based model for EEG analysis," *Electroencephalogr. Clin. Neurophys.*, vol. 31, no. 5, pp. 457-468, 1971.

[103] P. Whittle, "Some recent contributions to the theory of stationary processes," Appendix 2, in H. Wold, *A Study in the Analysis of Stationary Time Series.* Stockholm, Sweden: Almqvist and Wiksell, 1954.

[104] ——, "On the fitting of multivariate autoregressions, and the approximate canonical factorization of a spectral density matrix," *Biometrika*, vol. 50, nos. 1 and 2, pp. 129-134, 1963.

[105] ——, *Prediction and Regulation by Linear Least Square Methods.* London, England: English Universities Press, 1963.

[106] B. Widrow, P. E. Mantey, L. J. Griffiths, and B. B. Goode, "Adaptive antenna systems," *Proc. IEEE*, vol. 55, pp. 2143-2159, Dec. 1967.

[107] N. Wiener, *Extrapolation, Interpolation and Smoothing of Stationary Time Series With Engineering Applications.* Cambridge, Mass.: M.I.T. Press, 1949.

[108] D. J. Wilde, *Optimum Seeking Methods.* Englewood Cliffs, N.J.: Prentice-Hall, 1964.

[109] J. H. Wilkinson, "Error analysis of direct methods of matrix inversion," *J. Ass. Comput. Mach.*, vol. 8, no. 3, pp. 281-330, 1961.

[110] ——, "The solution of ill-conditioned linear equations," in Ralston and Wilf, *Mathematical Methods for Digital Computers*, pp. 65-93, 1967.

[111] J. H. Wilkinson and C. Reinsch, *Linear Algebra*, vol. II. New York: Springer-Verlag, 1971.

[112] H. Wold, *A Study in the Analysis of Stationary Time Series.* Stockholm, Sweden: Almqvist and Wiksell, 1954.

[113] H. O. A. Wold, *Bibliography on Time Series and Stochastic Processes.* Cambridge, Mass.: M.I.T. Press, 1965.

[114] L. C. Wood and S. Treitel, "Seismic signal processing," this issue, pp. 649-661.

[115] G. U. Yule, "On a method of investigating periodicities in disturbed series, with special reference to Wolfer's sunspot numbers," *Phil. Trans. Roy. Soc.*, vol. 226-A, pp. 267-298, 1927.

Adaptive Noise Cancelling: Principles and Applications

BERNARD WIDROW, SENIOR MEMBER, IEEE, JOHN R. GLOVER, JR., MEMBER, IEEE, JOHN M. McCOOL, SENIOR MEMBER, IEEE, JOHN KAUNITZ, MEMBER, IEEE, CHARLES S. WILLIAMS, STUDENT MEMBER, IEEE, ROBERT H. HEARN, JAMES R. ZEIDLER, EUGENE DONG, JR., AND ROBERT C. GOODLIN

Abstract—This paper describes the concept of adaptive noise cancelling, an alternative method of estimating signals corrupted by additive noise or interference. The method uses a "primary" input containing the corrupted signal and a "reference" input containing noise correlated in some unknown way with the primary noise. The reference input is adaptively filtered and subtracted from the primary input to obtain the signal estimate. Adaptive filtering before subtraction allows the treatment of inputs that are deterministic or stochastic, stationary or time variable. Wiener solutions are developed to describe asymptotic adaptive performance and output signal-to-noise ratio for stationary stochastic inputs, including single and multiple reference inputs. These solutions show that when the reference input is free of signal and certain other conditions are met noise in the primary input can be essentially eliminated without signal distortion. It is further shown that in treating periodic interference the adaptive noise canceller acts as a notch filter with narrow bandwidth, infinite null, and the capability of tracking the exact frequency of the interference; in this case the canceller behaves as a linear, time-invariant system, with the adaptive filter converging on a dynamic rather than a static solution. Experimental results are presented that illustrate the usefulness of the adaptive noise cancelling technique in a variety of practical applications. These applications include the cancelling of various forms of periodic interference in electrocardiography, the cancelling of periodic interference in speech signals, and the cancelling of broad-band interference in the side-lobes of an antenna array. In further experiments it is shown that a sine wave and Gaussian noise can be separated by using a reference input that is a delayed version of the primary input. Suggested applications include the elimination of tape hum or turntable rumble during the playback of recorded broad-band signals and the automatic detection of very-low-level periodic signals masked by broad-band noise.

Manuscript received March 24, 1975; August 7, 1975.

This work was supported in part by the National Science Foundation under Grant ENGR 74-21752, the National Institutes of Health under Grant 1R01HL18307-01CVB, and the Naval Ship Systems Command under Task Assignment SF 11-121-102.

B. Widrow and C. S. Williams are with the Information Systems Laboratory, Department of Electrical Engineering, Stanford University, Stanford, Calif. 94305.

J. R. Glover, Jr., was with the Information Systems Laboratory, Department of Electrical Engineering, Stanford University, Stanford, Calif. He is now with the Department of Electrical Engineering, University of Houston, Houston, Tex.

J. M. McCool, R. H. Hearn, and J. R. Zeidler are with the Fleet Engineering Department, Naval Undersea Center, San Diego, Calif. 92132.

J. Kaunitz was with the Information Systems Laboratory, Department of Electrical Engineering, Stanford University, Stanford, Calif. He is now with Computer Sciences of Australia, St. Leonards, N. S. W., Australia, 2065.

E. Dong, Jr., and R. C. Goodlin are with the School of Medicine, Stanford University, Stanford, Cailf. 94305.

I. Introduction

THE USUAL method of estimating a signal corrupted by additive noise[1] is to pass it through a filter that tends to suppress the noise while leaving the signal relatively unchanged. The design of such filters is the domain of optimal filtering, which originated with the pioneering work of Wiener

[1] For simplicity the term "noise" is used in this paper to signify all forms of interference, deterministic as well as stochastic.

Reprinted from *Proc. IEEE,* vol. 63, no. 12, pp. 1692–1716, Dec. 1975.

and was extended and enhanced by the work of Kalman, Bucy, and others [1]-[5].

Filters used for the above purpose can be fixed or adaptive. The design of fixed filters is based on prior knowledge of both the signal and the noise. Adaptive filters, on the other hand, have the ability to adjust their own parameters automatically, and their design requires little or no *a priori* knowledge of signal or noise characteristics.

Noise cancelling is a variation of optimal filtering that is highly advantageous in many applications. It makes use of an auxiliary or reference input derived from one or more sensors located at points in the noise field where the signal is weak or undetectable. This input is filtered and subtracted from a primary input containing both signal and noise. As a result the primary noise is attenuated or eliminated by cancellation.

At first glance, subtracting noise from a received signal would seem to be a dangerous procedure. If done improperly it could result in an increase in output noise power. If, however, filtering and subtraction are controlled by an appropriate adaptive process, noise reduction can be accomplished with little risk of distorting the signal or increasing the output noise level. In circumstances where adaptive noise cancelling is applicable, levels of noise rejection are often attainable that would be difficult or impossible to achieve by direct filtering.

The purpose of this paper is to describe the concept of adaptive noise cancelling, to provide a theoretical treatment of its advantages and limitations, and to describe some of the applications where it is most useful.

II. Early Work in Adaptive Noise Cancelling

The earliest work in adaptive noise cancelling known to the authors was performed by Howells and Applebaum and their colleagues at the General Electric Company between 1957 and 1960. They designed and built a system for antenna sidelobe cancelling that used a reference input derived from an auxiliary antenna and a simple two-weight adaptive filter [6].

At the time of this work, only a handful of people were interested in adaptive systems, and development of the multiweight adaptive filter was just beginning. In 1959, Widrow and Hoff at Stanford University were devising the least-mean-square (LMS) adaptive algorithm and the pattern recognition scheme known as Adaline (for "adaptive linear threshold logic element") [7], [8]. Rosenblatt had recently built his Perceptron at the Cornell Aeronautical Laboratory [9]-[11].[2] Aizermann and his colleagues at the Institute of Automatics and Telemechanics in Moscow, U.S.S.R., were constructing an automatic gradient searching machine. In Great Britain, D. Gabor and his associates were developing adaptive filters [12]. Each of these efforts was proceeding independently.

In the early and middle 1960's, work on adaptive systems intensified. Hundreds of papers on adaptation, adaptive controls, adaptive filtering, and adaptive signal processing appeared in the literature. The best known commercial application of adaptive filtering grew from the work during this period of Lucky at the Bell Laboratories [13], [14]. His high-speed MODEM's for digital communication are now widely used in connecting remote terminals to computers as well as one computer to another, allowing an increase in the rate and accuracy of data transmission by a reduction of intersymbol interference.

[2] This pioneering equipment now resides at the Smithsonian Institution in Washington, D.C.

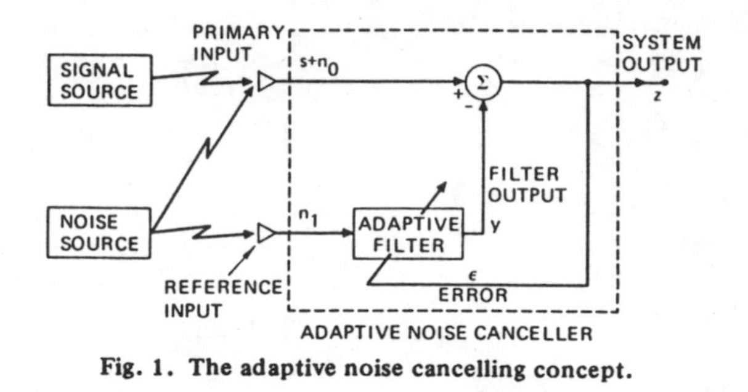

Fig. 1. The adaptive noise cancelling concept.

The first adaptive noise cancelling system at Stanford University was designed and built in 1965 by two students. Their work was undertaken as part of a term paper project for a course in adaptive systems given by the Electrical Engineering Department. The purpose was to cancel 60-Hz interference at the output of an electrocardiographic amplifier and recorder. A description of the system, which made use of a two-weight analog adaptive filter, together with results recently obtained by computer implementation, is presented in Section VIII.

Since 1965, adaptive noise cancelling has been successfully applied to a number of additional problems, including other aspects of electrocardiography, also described in Section VIII, to the elimination of periodic interference in general [15], and to the elimination of echoes on long-distance telephone transmission lines [16], [17]. A recent paper on adaptive antennas by Riegler and Compton [18] generalizes the work originally performed by Howells and Applebaum. Riegler and Compton's approach is based on the LMS algorithm and is an application of the adaptive antenna concepts of Widrow *et al.* [19], [20].

III. The Concept of Adaptive Noise Cancelling

Fig. 1 shows the basic problem and the adaptive noise cancelling solution to it. A signal s is transmitted over a channel to a sensor that also receives a noise n_0 uncorrelated with the signal. The combined signal and noise $s + n_0$ form the primary input to the canceller. A second sensor receives a noise n_1 uncorrelated with the signal but correlated in some unknown way with the noise n_0. This sensor provides the reference input to the canceller. The noise n_1 is filtered to produce an output y that is as close a replica as possible of n_0. This output is subtracted from the primary input $s + n_0$ to produce the system output $z = s + n_0 - y$.

If one knew the characteristics of the channels over which the noise was transmitted to the primary and reference sensors, it would theoretically be possible to design a fixed filter capable of changing n_1 into n_0. The filter output could then be subtracted from the primary input, and the system output would be signal alone. Since, however, the characteristics of the transmission paths are as a rule unknown or known only approximately and are seldom of a fixed nature, the use of a fixed filter is not feasible. Moreover, even if a fixed filter were feasible, its characteristics would have to be adjusted with a precision difficult to attain, and the slightest error could result in an increase in output noise power.

In the system shown in Fig. 1 the reference input is processed by an adaptive filter. An adaptive filter differs from a fixed filter in that it automatically adjusts its own impulse response. Adjustment is accomplished through an algorithm that responds to an error signal dependent, among other things, on the filter's output. Thus with the proper algorithm, the filter can operate under changing conditions and can readjust itself continuously to minimize the error signal.

The error signal used in an adaptive process depends on the nature of the application. In noise cancelling systems the practical objective is to produce a system output $z = s + n_0 - y$ that is a best fit in the least squares sense to the signal s. This objective is accomplished by feeding the system output back to the adaptive filter and adjusting the filter through an LMS adaptive algorithm to minimize total system output power.[3] In an adaptive noise cancelling system, in other words, the system output serves as the error signal for the adaptive process.

It might seem that some prior knowledge of the signal s or of the noises n_0 and n_1 would be necessary before the filter could be designed, or before it could adapt, to produce the noise cancelling signal y. A simple argument will show, however, that little or no prior knowledge of s, n_0, or n_1, or of their interrelationships, either statistical or deterministic, is required.

Assume that s, n_0, n_1, and y are statistically stationary and have zero means. Assume that s is uncorrelated with n_0 and n_1, and suppose that n_1 is correlated with n_0. The output z is

$$z = s + n_0 - y. \tag{1}$$

Squaring, one obtains

$$z^2 = s^2 + (n_0 - y)^2 + 2s(n_0 - y). \tag{2}$$

Taking expectations of both sides of (2), and realizing that s is uncorrelated with n_0 and with y, yields

$$\begin{aligned} E[z^2] &= E[s^2] + E[(n_0 - y)^2] + 2E[s(n_0 - y)] \\ &= E[s^2] + E[(n_0 - y)^2]. \end{aligned} \tag{3}$$

The signal power $E[s^2]$ will be unaffected as the filter is adjusted to minimize $E[z^2]$. Accordingly, the minimum output power is

$$\min E[z^2] = E[s^2] + \min E[(n_0 - y)^2]. \tag{4}$$

When the filter is adjusted so that $E[z^2]$ is minimized, $E[(n_0 - y)^2]$ is, therefore, also minimized. The filter output y is then a best least squares estimate of the primary noise n_0. Moreover, when $E[(n_0 - y)^2]$ is minimized, $E[(z - s)^2]$ is also minimized, since, from (1),

$$(z - s) = (n_0 - y). \tag{5}$$

Adjusting or adapting the filter to minimize the total output power is thus tantamount to causing the output z to be a best least squares estimate of the signal s for the given structure and adjustability of the adaptive filter and for the given reference input.

The output z will contain the signal s plus noise. From (1), the output noise is given by $(n_0 - y)$. Since minimizing $E[z^2]$ minimizes $E[(n_0 - y)^2]$, *minimizing the total output power minimizes the output noise power*. Since the signal in the output remains constant, *minimizing the total output power maximizes the output signal-to-noise ratio*.

It is seen from (3) that the smallest possible output power is $E[z^2] = E[s^2]$. When this is achievable, $E[(n_0 - y)^2] = 0$. Therefore, $y = n_0$, and $z = s$. In this case, minimizing output power causes the output signal to be perfectly noise free.[4]

These arguments can readily be extended to the case where the primary and reference inputs contain, in addition to n_0 and n_1, additive random noises uncorrelated with each other and with s, n_0, and n_1. They can also readily be extended to the case where n_0 and n_1 are deterministic rather than stochastic.

[3] The characteristics and terminology of the LMS adaptive filter used in the noise cancelling systems described in this paper are presented in Appendix A.

IV. Wiener Solutions to Statistical Noise Cancelling Problems

In this section, optimal unconstrained Wiener solutions to certain statistical noise cancelling problems are derived. The purpose is to demonstrate analytically the increase in signal-to-noise ratio and other advantages of the noise cancelling technique. Though the idealized solutions presented do not take into account the issues of finite filter length or causality, which are important in practical applications, means of approximating optimal unconstrained Wiener performance with physically realizable adaptive transversal filters are readily available and are described in Appendix B.

As previously noted, fixed filters are for the most part inapplicable in noise cancelling because the correlation and cross correlation functions of the primary and reference inputs are generally unknown and often variable with time. Adaptive filters are required to "learn" the statistics initially and to track them if they vary slowly. For stationary stochastic inputs, however, the steady-state performance of adaptive filters closely approximates that of fixed Wiener filters, and Wiener filter theory thus provides a convenient method of mathematically analyzing statistical noise cancelling problems.

Fig. 2 shows a classic single-input single-output Wiener filter. The input signal is x_j, the output signal y_j, and the desired response d_j. The input and output signals are assumed to be discrete in time, and the input signal and desired response are assumed to be statistically stationary. The error signal is $\epsilon_j = d_j - y_j$. The filter is linear, discrete, and designed to be optimal in the minimum mean-square-error sense. It is composed of an infinitely long, two-sided tapped delay line.

The optimal impulse response of this filter may be described in the following manner. The discrete autocorrelation function of the input signal x_j is defined as

$$\phi_{xx}(k) \triangleq E[x_j x_{j+k}]. \tag{6}$$

The cross-correlation function between x_j and the desired response d_j is similarly defined as

$$\phi_{xd}(k) \triangleq E[x_j d_{j+k}]. \tag{7}$$

The optimal impulse response $w^*(k)$ can then be obtained from the discrete Wiener–Hopf equation:

$$\sum_{l=-\infty}^{\infty} w^*(l)\, \phi_{xx}(k - l) = \phi_{xd}(k). \tag{8}$$

[4] Note that, on the other hand, when the reference input is completely uncorrelated with the primary input, the filter will "turn itself off" and will not increase output noise. In this case the filter output y will be uncorrelated with the primary input. The output power will be $E[z^2] = E[(s + n_0)^2] + 2E[-y(s + n_0)] + E[y^2] = E[(s + n_0)^2] + E[y^2]$. Minimizing output power requires that $E[y^2]$ be minimized, which is accomplished by making all weights zero, bringing $E[y^2]$ to zero.

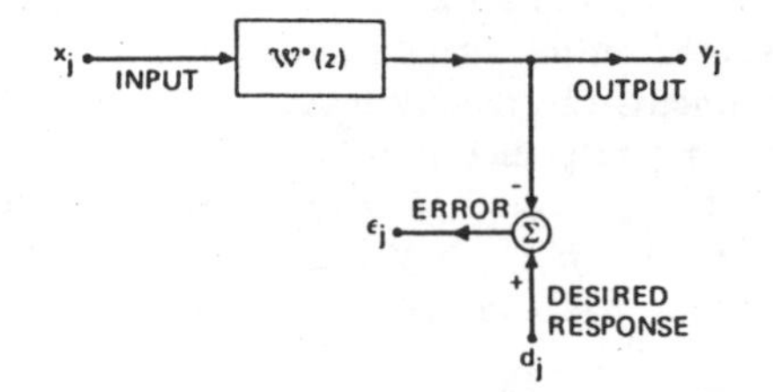

Fig. 2. Single-channel Wiener filter.

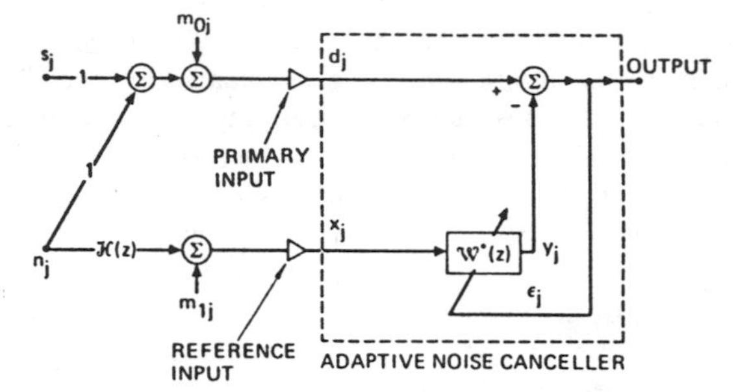

Fig. 3. Single-channel adaptive noise canceller with correlated and uncorrelated noises in the primary and reference inputs.

The convolution can be more simply written as

$$w^*(k) * \phi_{xx}(k) = \phi_{xd}(k). \tag{9}$$

This form of the Wiener solution is unconstrained in that the impulse response $w^*(k)$ may be causal or noncausal and of finite or infinite extent to the left or right of the time origin.[5]

The transfer function of the Wiener filter may now be derived as follows. The power-density spectrum of the input signal is the Z transform of $\phi_{xx}(k)$:

$$\mathcal{S}_{xx}(z) \triangleq \sum_{k=-\infty}^{\infty} \phi_{xx}(k)\, z^{-k}. \tag{10}$$

The cross power spectrum between the input signal and desired response is

$$\mathcal{S}_{xd}(z) \triangleq \sum_{k=-\infty}^{\infty} \phi_{xd}(k)\, z^{-k}. \tag{11}$$

The transfer function of the Wiener filter is

$$\mathcal{W}^*(z) \triangleq \sum w^*(k)\, z^{-k}. \tag{12}$$

Transforming (8) then yields the optimal unconstrained Wiener transfer function:

$$\mathcal{W}^*(z) = \frac{\mathcal{S}_{xd}(z)}{\mathcal{S}_{xx}(z)}. \tag{13}$$

The application of Wiener filter theory to adaptive noise cancelling may now be considered. Fig. 3 shows a single-channel adaptive noise canceller with a typical set of inputs. The primary input consists of a signal s_j plus a sum of two noises m_{0j} and n_j. The reference input consists of a sum of two other noises m_{1j} and $n_j * h(j)$, where $h(j)$ is the impulse response of the channel whose transfer function is $\mathcal{H}(z)$.[6] The noises n_j and $n_j * h(j)$ have a common origin, are correlated with each other, and are uncorrelated with s_j. They further are assumed to have a finite power spectrum at all frequencies. The noises m_{0j} and m_{1j} are uncorrelated with each other, with s_j, and with n_j and $n_j * h(j)$. For the purposes of analysis all noise propagation paths are assumed to be equivalent to linear, time-invariant filters.

The noise canceller of Fig. 3 includes an adaptive filter whose input x_j, the reference input to the canceller, is $m_{1j} + n_j * h(j)$ and whose desired response d_j, the primary input to the canceller, is $s_j + m_{0j} + n_j$. The error signal ϵ_j is the noise canceller's output. If one assumes that the adaptive process has converged and the minimum mean-square-error solution has been found, then the adaptive filter is equivalent to a Wiener filter. The optimal unconstrained transfer function of the adaptive filter is thus given by (13) and may be written as follows.

The spectrum of the filter's input $\mathcal{S}_{xx}(z)$ can be expressed in terms of the spectra of its two mutually uncorrelated additive components. The spectrum of the noise m_1 is $\mathcal{S}_{m_1 m_1}(z)$, and that of the noise n arriving via $\mathcal{H}(z)$ is $\mathcal{S}_{nn}(z)|\mathcal{H}(z)|^2$. The filter's input spectrum is thus

$$\mathcal{S}_{xx}(z) = \mathcal{S}_{m_1 m_1}(z) + \mathcal{S}_{nn}(z)|\mathcal{H}(z)|^2. \tag{14}$$

The cross power spectrum between the filter's input and the desired response depends only on the mutually correlated primary and reference components and is given by

$$\mathcal{S}_{xd}(z) = \mathcal{S}_{nn}(z)\, \mathcal{H}(z^{-1}). \tag{15}$$

The Wiener transfer function is thus

$$\mathcal{W}^*(z) = \frac{\mathcal{S}_{nn}(z)\, \mathcal{H}(z^{-1})}{\mathcal{S}_{m_1 m_1}(z) + \mathcal{S}_{nn}(z)|\mathcal{H}(z)|^2}. \tag{16}$$

Note that $\mathcal{W}^*(z)$ is independent of the primary signal spectrum $\mathcal{S}_{ss}(z)$ and of the primary uncorrelated noise spectrum $\mathcal{S}_{m_0 m_0}(z)$.

An interesting special case occurs when the additive noise m_1 in the reference input is zero. Then $\mathcal{S}_{m_1 m_1}(z)$ is zero and the optimal transfer function (16) becomes

$$\mathcal{W}^*(z) = 1/\mathcal{H}(z). \tag{17}$$

This result is intuitively appealing. The adaptive filter, as in the balancing of a bridge, causes the noise n_j to be perfectly nulled at the noise canceller output. The primary uncorrelated noise m_{0j} remains uncancelled.

The performance of the single-channel noise canceller can be evaluated more generally in terms of the ratio of the signal-to-noise density ratio at the output, $\rho_{out}(z)$ to the signal-to-noise density ratio at the primary input $\rho_{pri}(z)$.[7] Assuming that the signal spectrum is greater than zero at all frequencies and

[5] The Shannon-Bode realization of the Wiener solution, by contrast, is constrained to a causal response. This constraint generally leads to a loss of performance and, as shown in Appendix B, can normally be avoided in adaptive noise cancelling applications.

[6] To simplify the notation the transfer function of the noise path from n_j to the primary input has been set at unity. This procedure does not restrict the analysis, since by a suitable choice of $\mathcal{H}(z)$ and of statistics for n_j any combination of mutually correlated noises can be made to appear at the primary and reference inputs. Though $\mathcal{H}(z)$ may consequently be required to have poles inside and outside the unit circle in the Z-plane, a stable two-sided impulse response $h(j)$ will always exist.

[7] Signal-to-noise density ratio is here defined as the ratio of signal power density to noise power density and is thus a function of frequency.

factoring out the signal power spectrum yields

$$\frac{\rho_{out}(z)}{\rho_{pri}(z)} = \frac{\text{primary noise power spectrum}}{\text{output noise power spectrum}} = \frac{\mathcal{S}_{nn}(z) + \mathcal{S}_{m_0 m_0}(z)}{\mathcal{S}_{\text{output noise}}(z)}. \tag{18}$$

The canceller's output noise power spectrum, as may be seen from Fig. 3, is a sum of three components, one due to the propagation of m_{0j} directly to the output, another due to the propagation of m_{1j} to the output via the transfer function $-\mathcal{W}^*(z)$, and another due to the propagation of n_j to the output via the transfer function $1 - \mathcal{H}(z)\,\mathcal{W}^*(z)$. The output noise power spectrum is thus

$$\mathcal{S}_{\text{output noise}}(z) = \mathcal{S}_{m_0 m_0}(z) + \mathcal{S}_{m_1 m_1}(z)\,|\,\mathcal{W}^*(z)|^2 + \mathcal{S}_{nn}(z)\,|[1 - \mathcal{H}(z)\,\mathcal{W}^*(z)]\,|^2. \tag{19}$$

If one lets the ratios of the spectra of the uncorrelated to the spectra of the correlated noises ("noise-to-noise density ratios") at the primary and reference inputs now be defined as

$$A(z) \triangleq \frac{\mathcal{S}_{m_0 m_0}(z)}{\mathcal{S}_{nn}(z)} \tag{20}$$

and

$$B(z) \triangleq \frac{\mathcal{S}_{m_1 m_1}(z)}{\mathcal{S}_{nn}(z)\,|\,\mathcal{H}(z)|^2} \tag{21}$$

then the transfer function (17) can be written as

$$\mathcal{W}^*(z) = \frac{1}{\mathcal{H}(z)[B(z) + 1]}. \tag{22}$$

The output noise power spectrum (19) can accordingly be rewritten as

$$\mathcal{S}_{\text{output noise}}(z) = \mathcal{S}_{m_0 m_0}(z) + \frac{\mathcal{S}_{m_1 m_1}(z)}{|\,\mathcal{H}(z)|^2\;|B(z) + 1|^2} + \mathcal{S}_{nn}(z)\left|1 - \frac{1}{B(z) + 1}\right|^2 = \mathcal{S}_{nn}(z)\,A(z) + \mathcal{S}_{nn}(z)\,\frac{B(z)}{B(z) + 1}. \tag{23}$$

The ratio of the output to the primary input noise power spectra is

$$\frac{\rho_{out}(z)}{\rho_{pri}(z)} = \frac{\mathcal{S}_{nn}(z)[1 + A(z)]}{\mathcal{S}_{\text{output noise}}(z)} = \frac{1 + A(z)}{A(z) + B(z)/(B(z) + 1)} = \frac{[A(z) + 1]\,[B(z) + 1]}{A(z) + A(z)\,B(z) \pm B(z)}. \tag{24}$$

This expression is a general representation of ideal noise canceller performance with single primary and reference inputs and stationary signals and noises. It allows one to estimate the level of noise reduction to be expected with an ideal noise cancelling system. In such a system the signal propagates to the output in an undistorted fashion (with a transfer function of unity).[8] Classical configurations of Wiener, Kalman, and adaptive filters, in contrast, generally introduce some signal distortion in the process of noise reduction.

It is apparent from (24) that the ability of a noise cancelling system to reduce noise is limited by the uncorrelated-to-correlated noise density ratios at the primary and reference inputs. The smaller are $A(z)$ and $B(z)$, the greater will be $\rho_{out}(z)/\rho_{pri}(z)$ and the more effective the action of the canceller. The desirability of low levels of uncorrelated noise in both inputs is made still more evident by considering the following special cases.

1) Small $A(z)$:

$$\frac{\rho_{out}(z)}{\rho_{pri}(z)} \cong \frac{1 + B(z)}{B(z)}. \tag{25}$$

2) Small $B(z)$:

$$\frac{\rho_{out}(z)}{\rho_{pri}(z)} \cong \frac{1 + A(z)}{A(z)}. \tag{26}$$

3) Small $A(z)$ and $B(z)$:

$$\frac{\rho_{out}(z)}{\rho_{pri}(z)} \cong \frac{1}{A(z) + B(z)}. \tag{27}$$

Infinite improvement is implied by these relations when both $A(z)$ and $B(z)$ are zero. In this case there is complete removal of noise at the system output, resulting in perfect signal reproduction. When $A(z)$ and $B(z)$ are small, however, other factors become important in limiting system performance. These factors include the finite length of the adaptive filter in practical systems, discussed in Appendix B, and "misadjustment" caused by gradient estimation noise in the adaptive process, discussed in [19] and [20]. A third factor, signal components sometimes present in the reference input, is discussed in the following section.

V. Effect of Signal Components in the Reference Input

In certain instances the available reference input to an adaptive noise canceller may contain low-level signal components in addition to the usual correlated and uncorrelated noise components. There is no doubt that these signal components will cause some cancellation of the primary input signal. The question is whether they will cause sufficient cancellation to render the application of noise cancelling useless. An answer is provided in the present section through a quantitative analysis based, like that of the previous section, on unconstrained Wiener filter theory. In this analysis expressions are derived for signal-to-noise density ratio, signal distortion, and noise spectrum at the canceller output.

Fig. 4 shows an adaptive noise canceller whose reference input contains signal components and whose primary and reference inputs contain additive correlated noises. Additive uncorrelated noises have been omitted to simplify the analysis. The signal components in the reference input are assumed to be propagated through a channel with the transfer function $\mathcal{J}(z)$. The other terminology is the same as that of Fig. 3.

[8] Some signal cancellation is possible when adaptation is rapid (that is, when the value of the adaptation constant μ, defined in Appendix A, is large) because of the dynamic response of the weight vector, which approaches but does not equal the Wiener solution. In most cases this effect is negligible; a particular case where it is not negligible is described in Section VI.

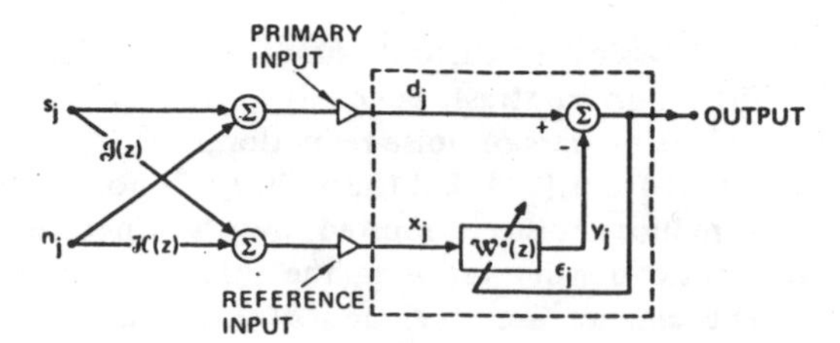

Fig. 4. Adaptive noise canceller with signal components in the reference input.

The spectrum of the signal in Fig. 4 is $\mathcal{S}_{ss}(z)$ and that of the noise $\mathcal{S}_{nn}(z)$. The spectrum of the reference input, which is identical to the spectrum of the input x_j to the adaptive filter, is thus

$$\mathcal{S}_{xx}(z) = \mathcal{S}_{ss}(z)|\mathcal{J}(z)|^2 + \mathcal{S}_{nn}(z)|\mathcal{H}(z)|^2. \tag{28}$$

The cross spectrum between the reference and primary inputs, identical to the cross spectrum between the filter's input x_j and desired response d_j, is similarly

$$\mathcal{S}_{xd}(z) = \mathcal{S}_{ss}(z)\,\mathcal{J}(z^{-1}) + \mathcal{S}_{nn}(z)\,\mathcal{H}(z^{-1}). \tag{29}$$

When the adaptive process has converged, the unconstrained Wiener transfer function of the adaptive filter, given by (13), is thus

$$\mathcal{W}^*(z) = \frac{\mathcal{S}_{ss}(z)\,\mathcal{J}(z^{-1}) + \mathcal{S}_{nn}(z)\,\mathcal{H}(z^{-1})}{\mathcal{S}_{ss}(z)|\mathcal{J}(z)|^2 + \mathcal{S}_{nn}(z)|\mathcal{H}(z)|^2}. \tag{30}$$

The first objective of the analysis is to find the signal-to-noise density ratio $\rho_{out}(z)$ at the noise canceller output. The transfer function of the propagation path from the signal input to the noise canceller output is $1 - \mathcal{J}(z)\,\mathcal{W}^*(z)$ and that of the path from the noise input to the canceller output is $1 - \mathcal{H}(z) \cdot \mathcal{W}^*(z)$. The spectrum of the signal component in the output is thus

$$\begin{aligned}\mathcal{S}_{ss_{out}}(z) &= \mathcal{S}_{ss}(z)|1 - \mathcal{J}(z)\,\mathcal{W}^*(z)|^2 \\ &= \mathcal{S}_{ss}(z)\left|\frac{[\mathcal{H}(z) - \mathcal{J}(z)]\,\mathcal{S}_{nn}(z)\,\mathcal{H}(z^{-1})}{\mathcal{S}_{ss}(z)|\mathcal{J}(z)|^2 + \mathcal{S}_{nn}(z)|\mathcal{H}(z)|^2}\right|^2\end{aligned} \tag{31}$$

and that of the noise component is similarly

$$\begin{aligned}\mathcal{S}_{nn_{out}}(z) &= \mathcal{S}_{nn}(z)|1 - \mathcal{H}(z)\,\mathcal{W}^*(z)|^2 \\ &= \mathcal{S}_{nn}(z)\left|\frac{[\mathcal{J}(z) - \mathcal{H}(z)]\,\mathcal{S}_{ss}(z)\,\mathcal{J}(z^{-1})}{\mathcal{S}_{ss}(z)|\mathcal{J}(z)|^2 + \mathcal{S}_{nn}(z)|\mathcal{H}(z)|^2}\right|^2.\end{aligned} \tag{32}$$

The output signal-to-noise density ratio is thus

$$\begin{aligned}\rho_{out}(z) &= \frac{\mathcal{S}_{ss}(z)}{\mathcal{S}_{nn}(z)}\left|\frac{\mathcal{S}_{nn}(z)\,\mathcal{H}(z^{-1})}{\mathcal{S}_{ss}(z)\,\mathcal{J}(z^{-1})}\right|^2 \\ &= \frac{\mathcal{S}_{nn}(z)|\mathcal{H}(z)|^2}{\mathcal{S}_{ss}(z)|\mathcal{J}(z)|^2}.\end{aligned} \tag{33}$$

The output signal-to-noise density ratio can be conveniently expressed in terms of the signal-to-noise density ratio at the reference input $\rho_{ref}(z)$ as follows. The spectrum of the signal component in the reference input is

$$\mathcal{S}_{ss_{ref}}(z) = \mathcal{S}_{ss}(z)|\mathcal{J}(z)|^2 \tag{34}$$

and that of the noise component is similarly

$$\mathcal{S}_{nn_{ref}}(z) = \mathcal{S}_{nn}(z)|\mathcal{H}(z)|^2. \tag{35}$$

The signal-to-noise density ratio at the reference input is thus

$$\rho_{ref}(z) = \frac{\mathcal{S}_{ss}(z)|\mathcal{J}(z)|^2}{\mathcal{S}_{nn}(z)|\mathcal{H}(z)|^2}. \tag{36}$$

The output signal-to-noise density ratio (33) is, therefore,

$$\rho_{out}(z) = \frac{1}{\rho_{ref}(z)}. \tag{37}$$

This result is exact and somewhat surprising. It shows that, assuming the adaptive solution to be unconstrained and the noises in the primary and reference inputs to be mutually correlated, the signal-to-noise density ratio at the noise canceller output is simply the reciprocal at all frequencies of the signal-to-noise density ratio at the reference input.

The next objective of the analysis is to derive an expression for signal distortion at the noise canceller output. The most useful reference input is one composed almost entirely of noise correlated with the noise in the primary input. When signal components are present some signal distortion will generally occur. The amount will depend on the amount of signal propagated through the adaptive filter, which may be determined as follows. The transfer function of the propagation path through the filter is

$$-\mathcal{J}(z)\,\mathcal{W}^*(z) = -\mathcal{J}(z)\frac{\mathcal{S}_{ss}(z)\,\mathcal{J}(z^{-1}) + \mathcal{S}_{nn}(z)\,\mathcal{H}(z^{-1})}{\mathcal{S}_{ss}(z)|\mathcal{J}(z)|^2 + \mathcal{S}_{nn}(z)|\mathcal{H}(z)|^2}. \tag{38}$$

When $|\mathcal{J}(z)|$ is small, this function can be approximated as

$$-\mathcal{J}(z)\,\mathcal{W}^*(z) \cong -\mathcal{J}(z)/\mathcal{H}(z). \tag{39}$$

The spectrum of the signal component propagated to the noise canceller output through the adaptive filter is thus approximately

$$\mathcal{S}_{ss}(z)|\mathcal{J}(z)/\mathcal{H}(z)|^2. \tag{40}$$

The combining of this component with the signal component in the primary input involves complex addition and is the process that results in signal distortion. The worst case, bounding the distortion to be expected in practice, occurs when the two signal components are of opposite phase.

Let "signal distortion" $\mathcal{D}(z)$ be defined[9] as a dimensionless ratio of the spectrum of the output signal component propagated through the adaptive filter to the spectrum of the signal component at the primary input:

$$\begin{aligned}\mathcal{D}(z) &\triangleq \frac{\mathcal{S}_{ss}(z)|\mathcal{J}(z)\,\mathcal{W}^*(z)|^2}{\mathcal{S}_{ss}(z)} \\ &= |\mathcal{J}(z)\,\mathcal{W}^*(z)|^2.\end{aligned} \tag{41}$$

From (39) it can be seen that, when $\mathcal{J}(z)$ is small, (41) reduces to

$$\mathcal{D}(z) \cong |\mathcal{J}(z)/\mathcal{H}(z)|^2. \tag{42}$$

This expression may be rewritten in a more useful form by combining the expressions for the signal-to-noise density ratio at the primary input:

$$\rho_{pri}(z) \triangleq \mathcal{S}_{ss}(z)/\mathcal{S}_{nn}(z) \tag{43}$$

[9] Note that signal distortion as defined here is a linear phenomenon related to alteration of the signal waveform as it appears at the noise canceller output and is not to be confused with nonlinear harmonic distortion.

and the signal-to-noise density ratio at the reference input (36):

$$\mathfrak{D}(z) \cong \rho_{ref}(z)/\rho_{pri}(z). \tag{44}$$

Equation (44) shows that, with an unconstrained adaptive solution and mutually correlated noises at the primary and reference inputs, low signal distortion results from a high signal-to-noise density ratio at the primary input and a low signal-to-noise density ratio at the reference input. This conclusion is intuitively reasonable.

The final objective of the analysis is to derive an expression for the spectrum of the output noise. The noise n_j propagates to the output with a transfer function

$$1 - \mathcal{H}(z)\, \mathfrak{W}^*(z) = 1 - \mathcal{H}(z) \left[\frac{\mathcal{S}_{ss}(z)\, \mathcal{J}(z^{-1}) + \mathcal{S}_{nn}(z)\, \mathcal{H}(z^{-1})}{\mathcal{S}_{ss}(z)|\mathcal{J}(z)|^2 + \mathcal{S}_{nn}(z)|\mathcal{H}(z)|^2} \right]$$

$$= \frac{\mathcal{S}_{ss}(z)\, \mathcal{J}(z^{-1})[\mathcal{J}(z) - \mathcal{H}(z)]}{\mathcal{S}_{ss}(z)|\mathcal{J}(z)|^2 + \mathcal{S}_{nn}(z)|\mathcal{H}(z)|^2}. \tag{45}$$

When $|\mathcal{J}(z)|$ is small, (45) reduces to

$$1 - \mathcal{H}(z)\, \mathfrak{W}^*(z) \cong \frac{-\mathcal{S}_{ss}(z)\, \mathcal{J}(z^{-1})}{\mathcal{S}_{nn}(z)\, \mathcal{H}(z^{-1})}. \tag{46}$$

The output noise spectrum is

$$\mathcal{S}_{\text{output noise}} = \mathcal{S}_{nn}(z)|1 - \mathcal{H}(z)\, \mathfrak{W}^*(z)|^2. \tag{47}$$

When $|\mathcal{J}(z)|$ is small, (47) similarly reduces to

$$\mathcal{S}_{\text{output noise}}(z) \cong \mathcal{S}_{nn}(z) \left| \frac{\mathcal{S}_{ss}(z)\, \mathcal{J}(z^{-1})}{\mathcal{S}_{nn}(z)\, \mathcal{H}(z^{-1})} \right|^2. \tag{48}$$

This equation can be more conveniently expressed in terms of the signal-to-noise density ratios at the reference input (36) and primary input (43):

$$\mathcal{S}_{\text{output noise}}(z) \cong \mathcal{S}_{nn}(z)|\rho_{ref}(z)||\rho_{pri}(z)|. \tag{49}$$

This result, which may appear strange at first glance, can be understood intuitively as follows. The first factor implies that the output noise spectrum depends on the input noise spectrum and is readily accepted. The second factor implies that, if the signal-to-noise density ratio at the reference input is low, the output noise will be low; that is, the smaller the signal component in the reference input, the more perfectly the noise will be cancelled. The third factor implies that, if the signal-to-noise density ratio in the primary input (the desired response of the adaptive filter) is low, the filter will be trained most effectively to cancel the noise rather than the signal and consequently output noise will be low.

The above analysis shows that signal components of low signal-to-noise ratio in the reference input, though undesirable, do not render the application of adaptive noise cancelling useless.[10] For an illustration of the level of performance attainable in practical circumstances consider the following example. Fig. 5 shows an adaptive noise cancelling system designed to pass a plane-wave signal received in the main beam of an antenna array and to discriminate against strong interference in the near field or in a minor lobe of the array. If one assumes that the signal and interference have overlapping and similar power spectra and that the interference power density is twenty times greater than the signal power density at the individual array element, then the signal-to-noise ratio at the reference input ρ_{ref} is 1/20. If one further assumes that, because of array gain, the signal power equals the interference power at the array output, then the signal-to-noise ratio at the primary input ρ_{pri} is 1. After convergence of the adaptive filter the signal-to-noise ratio at the system output will thus be

$$\rho_{out} = 1/\rho_{ref} = 20.$$

The maximum signal distortion will similarly be

$$\mathfrak{D} = \rho_{ref}/\rho_{pri} = (1/20)/1 = 5 \text{ percent}.$$

In this case, therefore, adaptive noise cancelling improves signal-to-noise ratio twentyfold and introduces only a small amount of signal distortion.

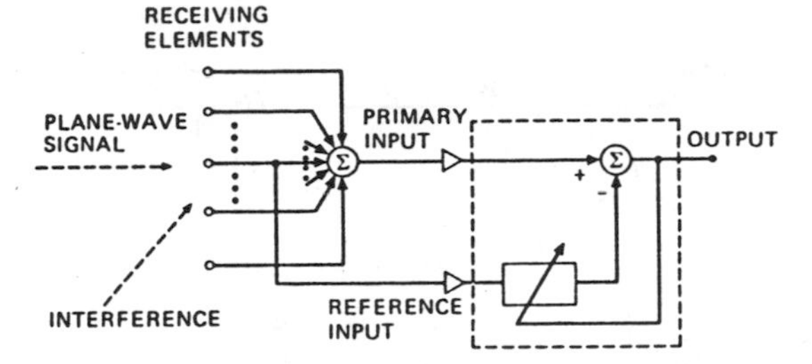

Fig. 5. Adaptive noise cancelling applied to a receiving array.

VI. The Adaptive Noise Canceller as a Notch Filter

In certain situations a primary input is available consisting of a signal component with an additive undesired sinusoidal interference. The conventional method of eliminating such interference is through the use of a notch filter. In this section an unusual form of notch filter, realized by an adaptive noise canceller, is described. The advantages of this form of notch filter are that it offers easy control of bandwidth, an infinite null, and the capability of adaptively tracking the exact frequency of the interference. The analysis presented deals with the formation of a notch at a single frequency. Analytical and experimental results show, however, that if more than one frequency is present in the reference input a notch for each will be formed [21].

Fig. 6 shows a single-frequency noise canceller with two adaptive weights. The primary input is assumed to be any kind of signal—stochastic, deterministic, periodic, transient, etc.—or any combination of signals. The reference input is assumed to be a pure cosine wave $C \cos(\omega_0 t + \phi)$. The primary and reference inputs are sampled at the frequency $\Omega = 2\pi/T$ rad/s. The reference input is sampled directly, giving x_{1j}, and after undergoing a 90° phase shift, giving x_{2j}. The samplers are synchronous and strobe at $t = 0, \pm T, \pm 2T$, etc.

A transfer function for the noise canceller of Fig. 6 may be obtained by analyzing signal propagation from the primary input to the system output.[11] For this purpose the flow diagram of Fig. 7, showing the operation of the LMS algorithm in detail, is constructed. Note that the procedure for updating

[10] It should be noted that if the reference input contained signal components but no noise components, correlated or uncorrelated, then the signal would be completely cancelled. When the reference input is properly derived, however, this condition cannot occur.

[11] It is not obvious, from inspection of Fig. 6, that a transfer function for this propagation path in fact exists. Its existence is shown, however, by the subsequent analysis.

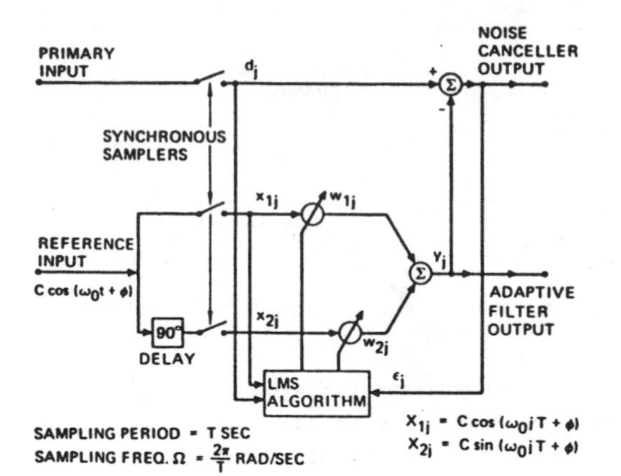

Fig. 6. Single-frequency adaptive noise canceller.

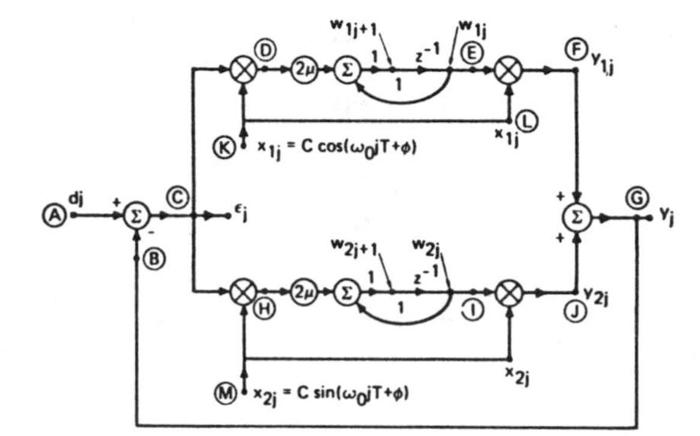

Fig. 7. Flow diagram showing signal propagation in single-frequency adaptive noise canceller.

the weights, as indicated in the diagram, is given by

$$w_{1j+1} = w_{1j} + 2\mu\epsilon_j x_{1j}$$
$$w_{2j+1} = w_{2j} + 2\mu\epsilon_j x_{2j}. \tag{50}$$

The sampled reference inputs are

$$x_{1j} = C\cos(\omega_0 jT + \phi) \tag{51}$$

and

$$x_{2j} = C\sin(\omega_0 jT + \phi). \tag{52}$$

The first step in the analysis is to obtain the isolated impulse response from the error ϵ_j, point C, to the filter output, point G, with the feedback loop from point G to point B broken. Let an impulse of amplitude α be applied at point C at discrete time $j = k$; that is,

$$\epsilon_j = \alpha\delta(j-k) \tag{53}$$

where

$$\delta(j-k) = \begin{cases} 1, & \text{for } j = k \\ 0, & \text{for } j \neq k. \end{cases} \tag{54}$$

The response at point D is then

$$\epsilon_j x_{1j} = \begin{cases} \alpha C\cos(\omega_0 kT + \phi), & \text{for } j = k \\ 0, & \text{for } j \neq k \end{cases} \tag{55}$$

which is the input impulse scaled in amplitude by the instantaneous value of x_{1j} at $j = k$. The signal flow path from point D to point E is that of a digital integrator with transfer function $2\mu/(z-1)$ and impulse response $2\mu u(j-1)$, where $u(j)$ is the discrete unit step function

$$u(j) = \begin{cases} 0, & \text{for } j < 0 \\ 1, & \text{for } j \geq 0. \end{cases} \tag{56}$$

Convolving $2\mu u(j-1)$ with $\epsilon_j x_{1j}$ yields the response at point E:

$$w_{1j} = 2\mu\alpha C\cos(\omega_0 kT + \phi) \tag{57}$$

where $j \geq k+1$. When the scaled and delayed step function is multiplied by x_{1j}, the response at point F is obtained:

$$y_{1j} = 2\mu\alpha C^2\cos(\omega_0 jT + \phi)\cos(\omega_0 kT + \phi) \tag{58}$$

where $j \geq k+1$. The corresponding response at point J, obtained in a similar manner, is

$$y_{2j} = 2\mu\alpha C^2\sin(\omega_0 jT + \phi)\sin(\omega_0 kT + \phi) \tag{59}$$

where $j \geq k+1$. Combining (58) and (59) yields the response at the filter output, point G:

$$y_j = 2\mu\alpha C^2\cos\omega_0 T(j-k)$$
$$= 2\mu\alpha C^2 u(j-k-1)\cos\omega_0 T(j-k). \tag{60}$$

Note that (60) is a function only of $(j-k)$ and is thus a time-invariant impulse response, proportional to the input impulse.

A linear transfer function for the noise canceller may now be derived in the following manner. If the time k is set equal to zero, the unit impulse response of the linear time-invariant signal-flow path from point C to point G is

$$y_j = 2\mu C^2 u(j-1)\cos(\omega_0 jT) \tag{61}$$

and the transfer function of this path is

$$G(z) = 2\mu C^2\left[\frac{z(z-\cos\omega_0 T)}{z^2 - 2z\cos\omega_0 T + 1} - 1\right]$$
$$= \frac{2\mu C^2(z\cos\omega_0 T - 1)}{z^2 - 2z\cos\omega_0 T + 1}. \tag{62}$$

This function can be expressed in terms of a radian sampling frequency $\Omega = 2\pi/T$ as

$$G(z) = \frac{2\mu C^2[z\cos(2\pi\omega_0\Omega^{-1}) - 1]}{z^2 - 2z\cos(2\pi\omega_0\Omega^{-1}) + 1}. \tag{63}$$

If the feedback loop from point G to point B is now closed, the transfer function $H(z)$ from the primary input, point A, to the noise canceller output, point C, can be obtained from the feedback formula:

$$H(z) = \frac{z^2 - 2z\cos(2\pi\omega_0\Omega^{-1}) + 1}{z^2 - 2(1-\mu C^2)z\cos(2\pi\omega_0\Omega^{-1}) + 1 - 2\mu C^2}. \tag{64}$$

Equation (64) shows that the single-frequency noise canceller has the properties of a notch filter at the reference frequency ω_0. The zeros of the transfer function are located in the Z plane at

$$z = \exp(\pm i2\pi\omega_0\Omega^{-1}) \tag{65}$$

and are precisely on the unit circle at angles of $\pm 2\pi\omega_0\Omega^{-1}$ rad. The poles are located at

$$z = (1-\mu C^2)\cos(2\pi\omega_0\Omega^{-1}) \pm i[(1-2\mu C^2)$$
$$- (1-\mu C^2)\cos^2(2\pi\omega_0\Omega^{-1})]^{1/2}. \tag{66}$$

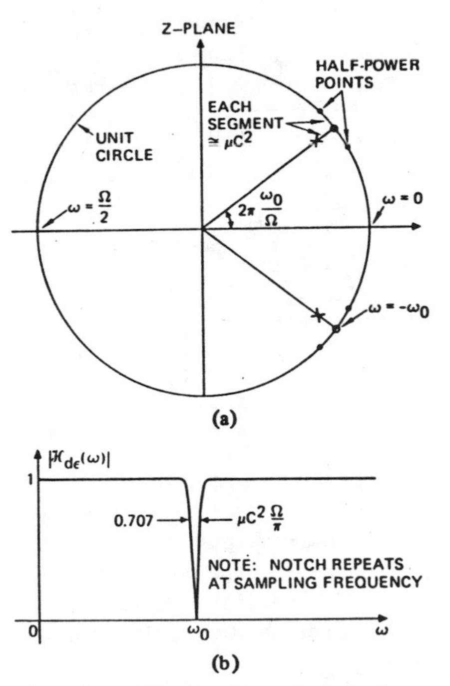

Fig. 8. Properties of transfer function of single-frequency adaptive noise canceller. (a) Location of poles and zeros. (b) Magnitude of transfer function.

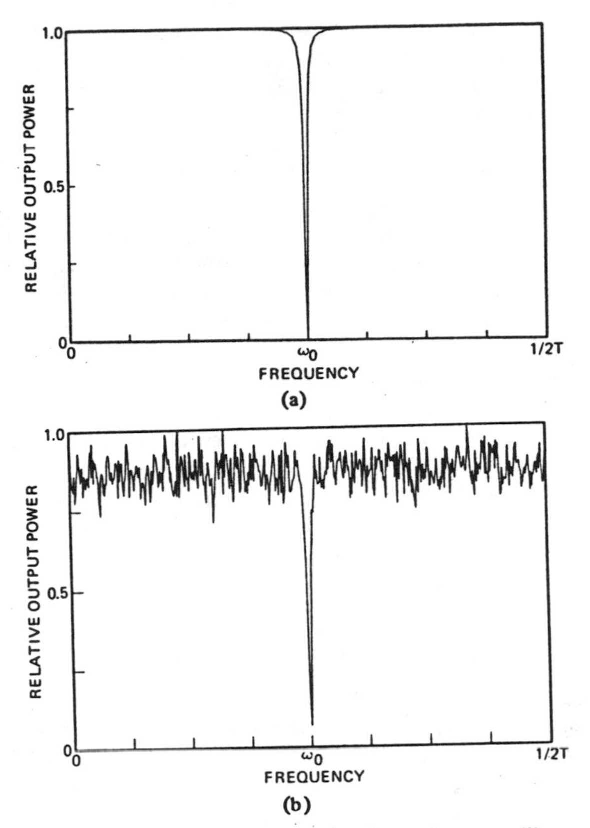

Fig. 9. Results of single-frequency adaptive noise cancelling experiments. (a) Primary input composed of cosine wave at 512 discrete frequencies. (b) Primary input composed of uncorrelated samples of white noise.

The poles are inside the unit circle at a radial distance $(1 - 2\mu C^2)^{1/2}$, approximately equal to $1 - \mu C^2$, from the origin and at angles of

$$\pm \text{arc cos } [(1 - \mu C^2)(1 - 2\mu C^2)^{-1/2} \cos (2\pi\omega_0\Omega^{-1})].$$

For slow adaptation (that is, small values of μC^2) these angles depend on the factor

$$\frac{1 - \mu C^2}{(1 - 2\mu C^2)^{1/2}} = \left(\frac{1 - 2\mu C^2 + \mu^2 C^4}{1 - 2\mu C^2}\right)^{1/2}$$
$$\cong (1 - \mu^2 C^4 + \cdots)^{1/2}$$
$$\cong 1 - \frac{1}{2}\mu^2 C^4 + \cdots \tag{67}$$

which differs only slightly from a value of one. The result is that, in practical instances, the angles of the poles are almost identical to those of the zeros.

The location of the poles and zeros and the magnitude of the transfer function in terms of frequency are shown in Fig. 8. Since the zeros lie on the unit circle, the depth of the notch in the transfer function is infinite at the frequency $\omega = \omega_0$. The sharpness of the notch is determined by the closeness of the poles to the zeros. Corresponding poles and zeros are separated by a distance approximately equal to μC^2. The arc length along the unit circle (centered at the position of a zero) spanning the distance between half-power points is approximately $2\mu C^2$. This length corresponds to a notch bandwidth of

$$\text{BW} = \mu C^2 \Omega/\pi. \tag{68}$$

The Q of the notch is determined by the ratio of the center frequency to the bandwidth:

$$Q \cong \frac{\omega_0 \pi}{\mu C^2 \Omega}. \tag{69}$$

The single-frequency noise canceller is, therefore, equivalent to a stable notch filter when the reference input is a pure cosine wave. The depth of the null achievable is generally superior to that of a fixed digital or analog filter because the adaptive process maintains the null exactly at the reference frequency.

Fig. 9 shows the results of two experiments performed to demonstrate the characteristics of the adaptive notch filter. In the first the primary input was a cosine wave of unit power stepped at 512 discrete frequencies. The reference input was a cosine wave with a frequency ω_0 of $\pi/2T$ rad/s. The value of C was 1, and the value of μ was 1.25×10^{-2}. The frequency resolution of the fast Fourier transform was 512 bins. The output power at each frequency is shown in Fig. 9(a). As the primary frequency approaches the reference frequency, significant cancellation occurs. The weights do not converge to stable values but "tumble" at the difference frequency,[12] and the adaptive filter behaves like a modulator, converting the reference frequency into the primary frequency. The theoretical notch width between half-power points, $1.59 \times 10^{-2}\ \omega_0$, compares closely with the measured notch width of $1.62 \times 10^{-2}\ \omega_0$.

In the second experiment, the primary input was composed of uncorrelated samples of white noise of unit power. The reference input and the processing parameters were the same as in the first experiment. An ensemble average of 4096 power spectra at the noise canceller output is shown in Fig. 9(b). An infinite null was not obtained in this experiment because of the finite frequency resolution of the spectral analysis algorithm.

[12] When the primary and reference frequencies are held at a constant difference, the weights develop a sinusoidal steady state at the difference frequency. In other words, they converge on a dynamic rather than a static solution. This is an unusual form of adaptive behavior.

In these experiments the filtering of a reference cosine wave of a given frequency caused cancellation of primary input components at adjacent frequencies. This result indicates that, under some circumstances, primary input components may be partially cancelled and distorted even though the reference input is uncorrelated with them. In practice this kind of cancellation is of concern only when the adaptive process is rapid; that is, when it is effected with large values of μ. When the adaptive process is slow, the weights converge to values that are nearly stable, and though signal cancellation as described in this section occurs it is generally not significant.

Additional experiments have recently been conducted with reference inputs containing more than one sinusoid. The formation of multiple notches has been achieved by using an adaptive filter with multiple weights (typically an adaptive transversal filter). Two weights are required for each sinusoid to achieve the necessary filter gain and phase. Uncorrelated broad-band noise superposed on the reference input creates a need for additional weights. A full analysis of the multiple notch problem can be found in [21].

VII. The Adaptive Noise Canceller as a High-Pass Filter

The use of a bias weight in an adaptive filter to cancel low-frequency drift in the primary input is a special case of notch filtering with the notch at zero frequency. The method of incorporating the bias weight is shown in Appendix A. Because there is no need to match the phase of the signal, only one weight is needed. The reference input is set to a constant value of one.

The transfer function from the primary input to the noise canceller output is derived as follows. Applying equations (A.3) and (A.15) of Appendix A yields

$$y_j = w_j \cdot 1 = w_j \tag{70}$$

$$w_{j+1} = w_j + 2\mu(\epsilon_j x_j) \tag{71}$$

or

$$\begin{aligned} y_{j+1} &= y_j + 2\mu(d_j - y_j) \\ &= (1 - 2\mu)\, y_j + 2\mu d_j. \end{aligned} \tag{72}$$

Taking the Z transform of (72) yields the steady-state solution:

$$Y(z) = \frac{2\mu}{z - (1 - 2\mu)} D(z). \tag{73}$$

The transfer function is then obtained by substituting $E(z) = D(z) - Y(z)$ in (73):

$$D(z) - E(z) = \frac{2\mu}{z - (1 - 2\mu)} D(z) \tag{74}$$

which reduces to

$$H(z) = \frac{E(z)}{D(z)} = \frac{z - 1}{z - (1 - 2\mu)}. \tag{75}$$

Equation (75) shows that the bias-weight filter is a high-pass filter with a zero on the unit circle at zero frequency and a pole on the real axis at a distance 2μ to the left of the zero. Note that this corresponds to a single-frequency notch filter, described by (64), for the case where $\omega_0 = 0$ and $C = 1$. The half-power frequency of the notch is at $\mu\Omega/\pi$ rad/s.

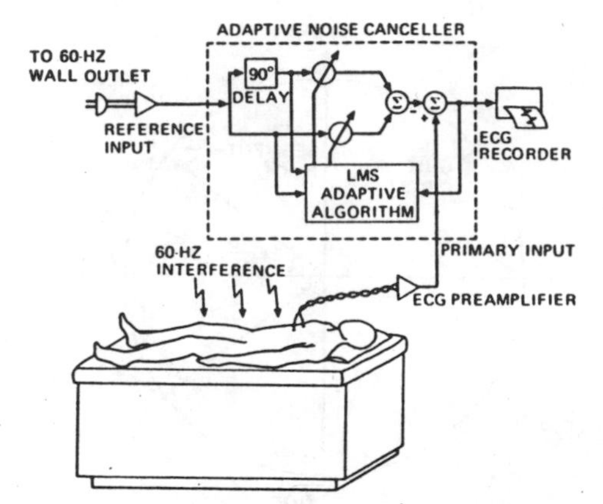

Fig. 10. Cancelling 60-Hz interference in electrocardiography.

The single-weight noise canceller acting as a high-pass filter is capable of removing not only a constant bias but also slowly varying drift in the primary input. Moreover, though it is not demonstrated in this paper, experience has shown that bias or drift removal can be accomplished simultaneously with cancellation of periodic or stochastic interference.

VIII. Applications

The principles of adaptive noise cancelling, including a description of the concept and theoretical analyses of performance with various kinds of signal and noise, have been presented in the preceding pages. This section describes a variety of practical applications of the technique. These applications include the cancelling of several kinds of interference in electrocardiography, of noise in speech signals, of antenna sidelobe interference, and of periodic or broad-band interference for which there is no external reference source. Experimental results are presented that demonstrate the performance of adaptive noise cancelling in these applications and that show its potential value whenever suitable inputs are available.

A. Cancelling 60-Hz Interference in Electrocardiography

In a recent paper [22], the authors point out that a major problem in the recording of electrocardiograms (ECG's) is "the appearance of unwanted 60-Hz interference in the output." They analyze the various causes of such power-line interference, including magnetic induction, displacement currents in leads or in the body of the patient, and equipment interconnections and imperfections. They also describe a number of techniques that are useful for minimizing it and that can be effected in the recording process itself, such as proper grounding and the use of twisted pairs. Another method capable of reducing 60-Hz ECG interference is adaptive noise cancelling, which can be used separately or in conjunction with more conventional approaches.

Fig. 10 shows the application of adaptive noise cancelling in electrocardiography. The primary input is taken from the ECG preamplifier; the 60-Hz reference input is taken from a wall outlet. The adaptive filter contains two variable weights, one applied to the reference input directly and the other to a version of it shifted in phase by 90°. The two weighted versions of the reference are summed to form the filter's output, which is subtracted from the primary input. Selected combinations of the values of the weights allow the reference waveform to be changed in magnitude and phase in any way required for cancellation. The two variable weights, or two

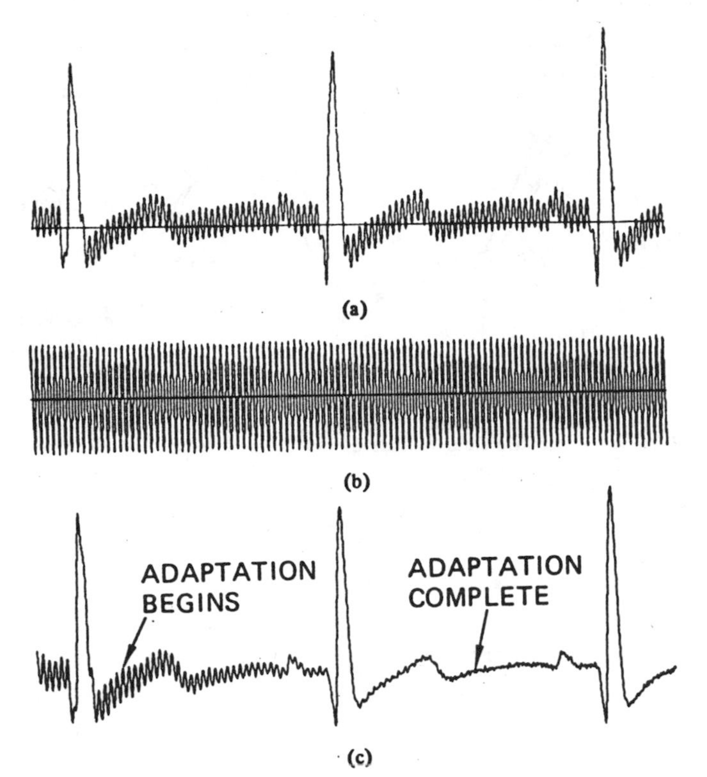

Fig. 11. Result of electrocardiographic noise cancelling experiment. (a) Primary input. (b) Reference input. (c) Noise canceller output.

"degrees of freedom," are required to cancel the single pure sinusoid.

A typical result of a group of experiments performed with a real-time computer system is shown in Fig. 11. Sample size was 10 bits and sampling rate 1000 Hz. Fig. 11(a) shows the primary input, an electrocardiographic waveform with an excessive amount of 60-Hz interference, and Fig. 11(b) shows the reference input from the wall outlet. Fig. 11(c) is the noise canceller output. Note the absence of interference and the clarity of detail once the adaptive process has converged.

B. Cancelling the Donor ECG in Heart-Transplant Electrocardiography

The electrical depolarization of the ventricles of the human heart is triggered by a group of specialized muscle cells known as the atrioventricular (AV) node. Though capable of independent, asynchronous operation, this node is normally controlled by a similar complex, the sinoatrial (SA) node, whose depolarization initiates an electrical impulse transmitted by conduction through the atrial heart muscle to the AV node. The SA node is connected through the vagus and sympathetic nerves to the central nervous system, which by controlling the rate of depolarization controls the frequency of the heartbeat [23], [24].

The cardiac transplantation technique developed by Shumway of the Stanford University Medical Center involves the suturing of the "new" or donor heart to a portion of the atrium of the patient's "old" heart [25]. Scar tissue forms at the suture line and electrically isolates the small remnant of the old heart, containing only the SA node, from the new heart, containing both SA and AV nodes. The SA node of the old heart remains connected to the vagus and sympathetic nerves, and the old heart continues to beat at a rate controlled by the central nervous system. The SA node of the new heart, which is not connected to the central nervous system

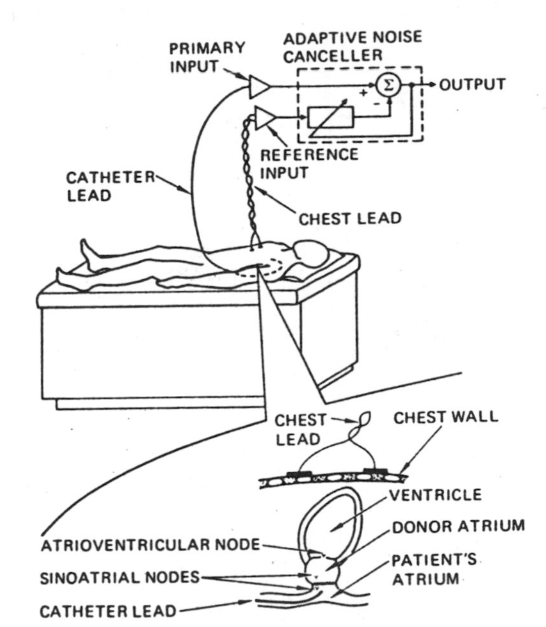

Fig. 12. Deriving and processing ECG signals of a heart-transplant patient.

because the severed vagus nerve cannot be surgically reattached, generates a spontaneous pulse that causes the new heart to beat at a separate self-pacing rate.

It is of interest to cardiac transplant research, and to cardiac research in general, to be able to determine the firing rate of the old heart and, indeed, to be able to see the waveforms of its electrical output. These waveforms, which cannot be obtained by ordinary electrocardiographic means because of interference from the beating of the new heart, are readily obtained with adaptive noise cancelling.

Fig. 12 shows the method of applying adaptive noise cancelling in heart-transplant electrocardiography. The reference input is provided by a pair of ordinary chest leads. These leads receive a signal that comes essentially from the new heart, the source of interference. The primary input is provided by a catheter consisting of a small coaxial cable threaded through the left brachial vein and the vena cava to a position in the atrium of the old heart. The tip of the catheter, a few millimeters long, is an exposed portion of the center conductor that acts as an antenna and is capable of receiving cardiac electrical signals. When it is in the most favorable position, the desired signal from the old heart and the interference from the new heart are received in about equal proportion.

Fig. 13 shows typical reference and primary inputs and the corresponding noise canceller output. The reference input contains the strong QRS waves that, in a normal electrocardiogram, indicate the firing of the ventricles. The primary input contains pulses that are synchronous with the QRS waves of the reference input and indicate the beating of the new heart. The other waves seen in this input are due to the old heart, which is beating at a separate rate. When the reference input is adaptively filtered and subtracted from the primary input, one obtains the waveform shown in Fig. 13(c), which is that of the old heart together with very weak residual pulses originating in the new heart. Note that the pulses of the two hearts are easily separated, even when they occur at the same instant. Note also that the electrical waveform of the new heart is steady and precise, while that of the old heart varies significantly from beat to beat.

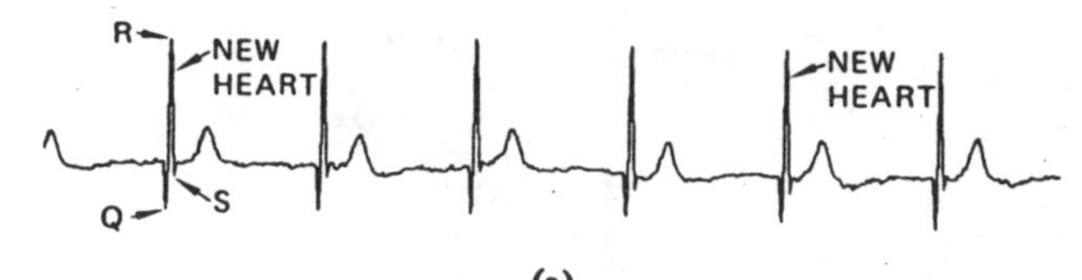

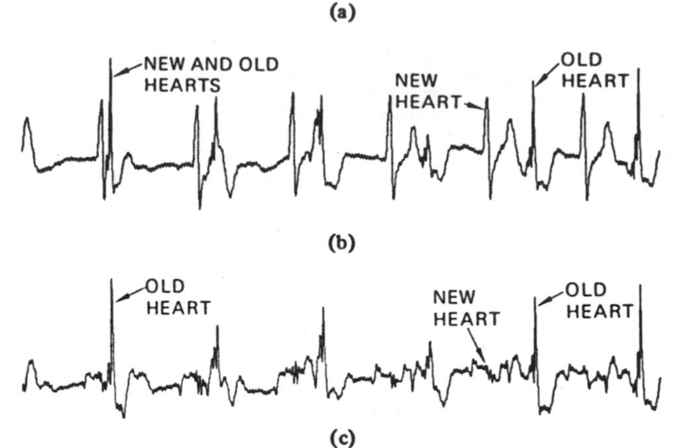

Fig. 13. ECG waveforms of heart-transplant patient. (a) Reference input (new heart). (b) Primary input (new and old heart). (c) Noise canceller output (old heart).

For this experiment the noise canceller was implemented in software with an adaptive transversal filter containing 48 weights. Sampling rate was 500 Hz.

C. Cancelling the Maternal ECG in Fetal Electrocardiography

Abdominal electrocardiograms make it possible to determine fetal heart rate and to detect multiple fetuses and are often used during labor and delivery [26]-[28]. Background noise due to muscle activity and fetal motion, however, often has an amplitude equal to or greater than that of the fetal heartbeat [29]-[31]. A still more serious problem is the mother's heartbeat, which has an amplitude two to ten times greater than that of the fetal heartbeat and often interferes with its recording [32].

In the spring of 1972, a group of experiments was performed to demonstrate the usefulness of adaptive noise cancelling in fetal electrocardiography. The objective was to derive as clear a fetal ECG as possible, so that not only could the heart rate be observed but also the actual waveform of the electrical output. The work was performed by Marie-France Ravat, Dominique Biard, Denys Caraux, and Michel Cotton, at the time students at Stanford University.[13]

Four ordinary chest leads were used to record the mother's heartbeat and provide multiple reference inputs to the canceller.[14] A single abdominal lead was used to record the combined maternal and fetal heartbeats that served as the primary input. Fig. 14 shows the cardiac electric field vectors of mother and fetus and the positions in which the leads were placed. Each lead terminated in a pair of electrodes. The chest and abdominal inputs were prefiltered, digitized, and recorded on tape. A multichannel adaptive noise canceller, shown in Fig. 15 and described theoretically in Appendix C, was used. Each channel had 32 taps with nonuniform (log periodic) spacing and a total delay of 129 ms.

Fig. 16 shows typical reference and primary inputs together with the corresponding noise canceller output. The prefilter-

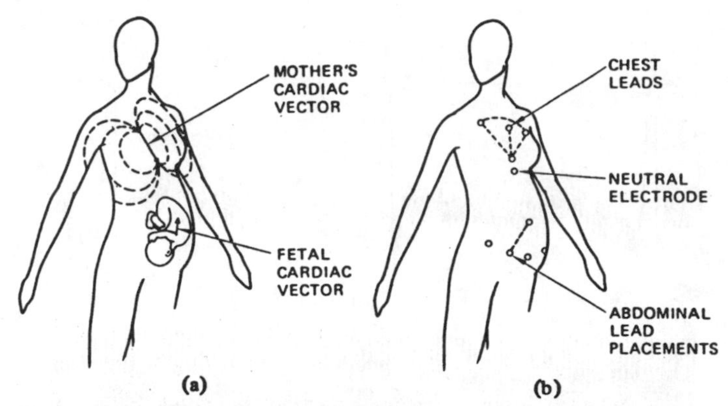

Fig. 14. Cancelling maternal heartbeat in fetal electrocardiography. (a) Cardiac electric field vectors of mother and fetus. (b) Placement of leads.

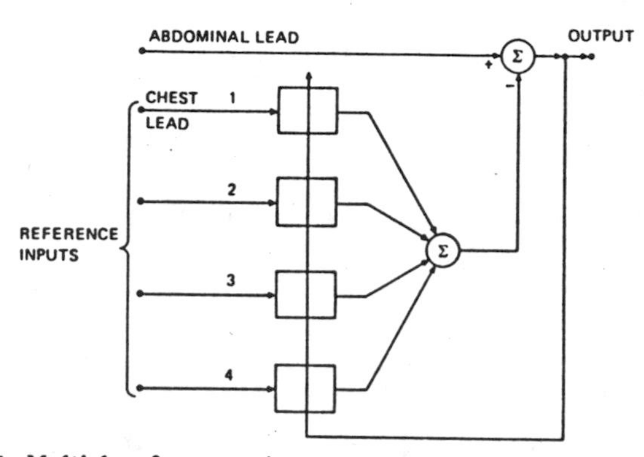

Fig. 15. Multiple-reference noise canceller used in fetal ECG experiment.

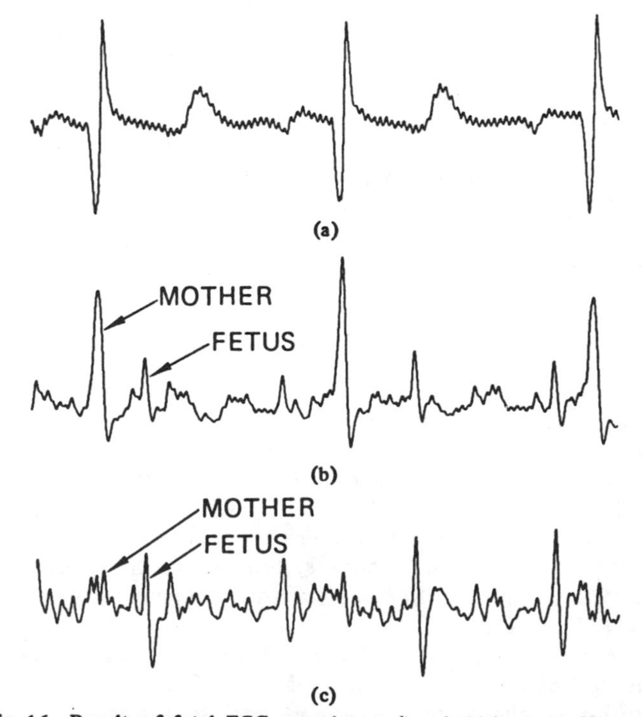

Fig. 16. Result of fetal ECG experiment (bandwidth, 3-35 Hz; sampling rate, 256 Hz). (a) Reference input (chest lead). (b) Primary input (abdominal lead). (c) Noise canceller output.

[13] A similar attempt to cancel the maternal heartbeat had previously been made by Walden and Bimbaum [33] without the use of an adaptive processor. Some reduction of the maternal interference was achieved by the careful placement of leads and adjustment of amplifier gain. It appears that substantially better results can be obtained with adaptive processing.

[14] More than one reference input was used to make the interference filtering task easier. The number of reference inputs required essentially to eliminate the maternal ECG is still under investigation.

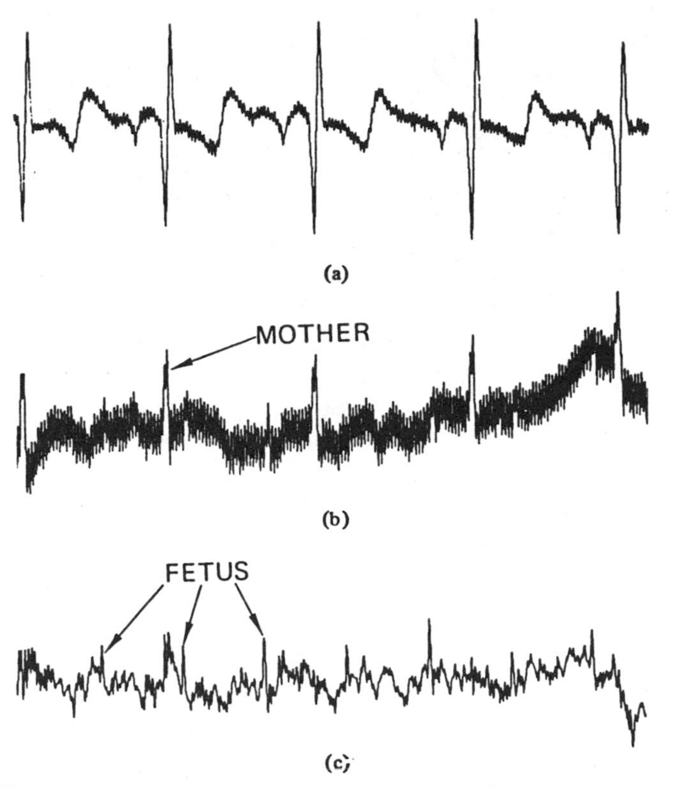

Fig. 17. Result of wide-band fetal ECG experiment (bandwidth, 0.3–75 Hz; sampling rate, 512 Hz). (a) Reference input (chest lead). (b) Primary input (abdominal lead). (c) Noise canceller output.

ing bandwidth was 3 to 35 Hz and the sampling rate 256 Hz. The maternal heartbeat, which dominates the primary input, is almost completely absent in the noise canceller output. Note that the voltage scale of the noise canceller output, Fig. 16(c), is approximately two times greater than that of the primary input, Fig. 16(b).

Fig. 17 shows corresponding results for a prefiltering bandwidth of 0.3 to 75 Hz and a sampling rate of 512 Hz. Baseline drift and 60-Hz interference are clearly present in the primary input, obtained from the abdominal lead. The interference is so strong that it is almost impossible to detect the fetal heartbeat. The inputs obtained from the chest leads contained the maternal heartbeat and a sufficient 60-Hz component to serve as a reference for both of these interferences. In the noise canceller output both interferences have been significantly reduced, and the fetal heartbeat is clearly discernible.

Additional experiments are currently being conducted with the aim of further improving the fetal ECG by reducing the background noise caused by muscle activity. In these experiments various averaging techniques are being investigated together with new adaptive processing methods for signals derived from an array of abdominal leads.

D. Cancelling Noise in Speech Signals

Consider the situation of a pilot communicating by radio from the cockpit of an aircraft where a high level of engine noise is present. The noise contains, among other things, strong periodic components, rich in harmonics, that occupy the same frequency band as speech. These components are picked up by the microphone into which the pilot speaks and severely interfere with the intelligibility of the radio transmission. It would be impractical to process the transmission with a conventional filter because the frequency and intensity of the noise components vary with engine speed and load and position of the pilot's head. By placing a second microphone at a suitable location in the cockpit, however, a sample of the ambient noise field free of the pilot's speech could be obtained. This sample could be filtered and subtracted from the transmission, significantly reducing the interference.

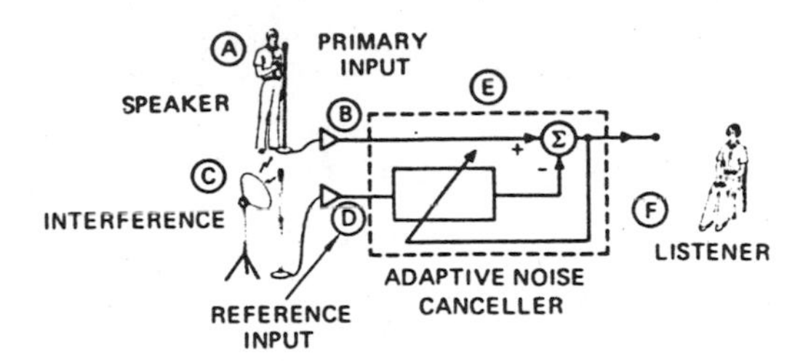

Fig. 18. Cancelling noise in speech signals.

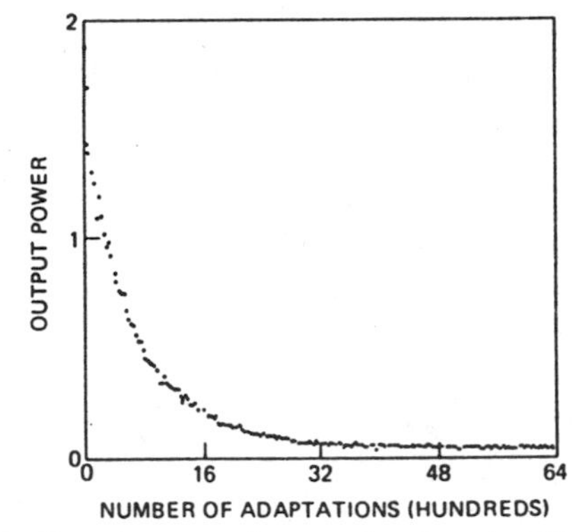

Fig. 19. Typical learning curve for speech noise cancelling experiment.

To demonstrate the feasibility of cancelling noise in speech signals a group of experiments simulating the cockpit noise problem in simplified form was conducted. In these experiments, as shown in Fig. 18, a person (A) spoke into a microphone (B) in a room where strong acoustic interference (C) was present. A second microphone (D) was placed in the room away from the speaker. The output of microphones (B) and (D) formed the primary and reference inputs, respectively, of a noise canceller (E), whose output was monitored by a remote listener (F). The canceller included an adaptive filter with 16 hybrid analog weights whose values were digitally controlled by a computer. The rate of adaptation was approximately 5 kHz. A typical learning curve, showing output power as a function of number of adaptation cycles, is shown in Fig. 19. Convergence was complete after about 5000 adaptations or one second of real time.

In a typical experiment the interference was an audiofrequency triangular wave containing many harmonics that, because of multipath effects, varied in amplitude, phase, and waveform from point to point in the room. The periodic nature of the wave made it possible to ignore the difference in time delay caused by the different transmission paths to the two sensors. The noise canceller was able to reduce the output power of this interference, which otherwise made the speech unintelligible, by 20 to 25 dB, rendering the interference barely perceptible to the remote listener. No noticeable distortion was introduced into the speech signal. Convergence times were on the order of seconds, and the processor was readily able to readapt when the position of the microphones was changed or when the frequency of the interference was varied over the range 100 to 2000 Hz.

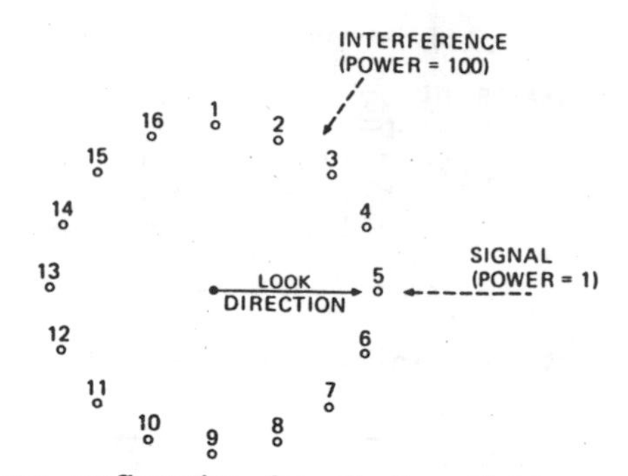

Fig. 20. Array configuration for adaptive sidelobe cancelling experiment.

E. Cancelling Antenna Sidelobe Interference

Strong unwanted signals incident on the sidelobes of an antenna array can severely interfere with the reception of weaker signals in the main beam. The conventional method of reducing such interference, adaptive beamforming [6], [18], [19], [34]-[37], is often complicated and expensive to implement. When the number of spatially discrete interference sources is small, adaptive noise cancelling can provide a simpler and less expensive method of dealing with this problem.

To demonstrate the level of sidelobe reduction achievable with adaptive noise cancelling, a typical interference cancelling problem was simulated on the computer. As shown in Fig. 20, an array consisting of a circular pattern of 16 equally spaced omnidirectional elements was chosen. The outputs of the elements were delayed and summed to form a main beam steered at a relative angle of 0°. A simulated "white" signal consisting of uncorrelated samples of unit power was assumed to be incident on this beam. Simulated interference with the same bandwidth and with a power of 100 was incident on the main beam at a relative angle of 58°. The array was connected to an adaptive noise canceller in the manner shown in Fig. 5. The output of the beamformer served as the canceller's primary input, and the output of element 4 was arbitrarily chosen as the reference input. The canceller included an adaptive filter with 14 weights; the adaptation constant in the LMS algorithm was set at $\mu = 7 \times 10^{-6}$.

Fig. 21 shows two series of computed directivity patterns, one representing a single frequency of $\frac{1}{4}$ the sampling frequency and the other an average of eight frequencies of from $\frac{1}{8}$ to $\frac{3}{8}$ the sampling frequency. These patterns indicate the evolution of the main beam and sidelobes as observed by stopping the adaptive process after the specified number of iterations. Note the deep nulls that develop in the direction of the interference. At the start of adaptation all weights were set at zero, providing a conventional 16-element beam pattern.

The signal-to-noise ratio at the system output, averaged over the eight frequencies, was found after convergence to be +20 dB. The signal-to-noise ratio at the single array element was −20 dB. This result bears out the expectation arising from (37) that the signal-to-noise ratio at the system output would be the reciprocal of the ratio at the reference input, which is derived from a single element.

A small amount of signal cancellation occurred, as evidenced by the changes in sensitivity of the main beam in the steering direction. These changes were not unexpected, since the main-lobe pattern was not constrained by the adaptive process. A method of LMS adaptation with constraints that could have been used to prevent this loss of sensitivity has been developed by Frost [37].

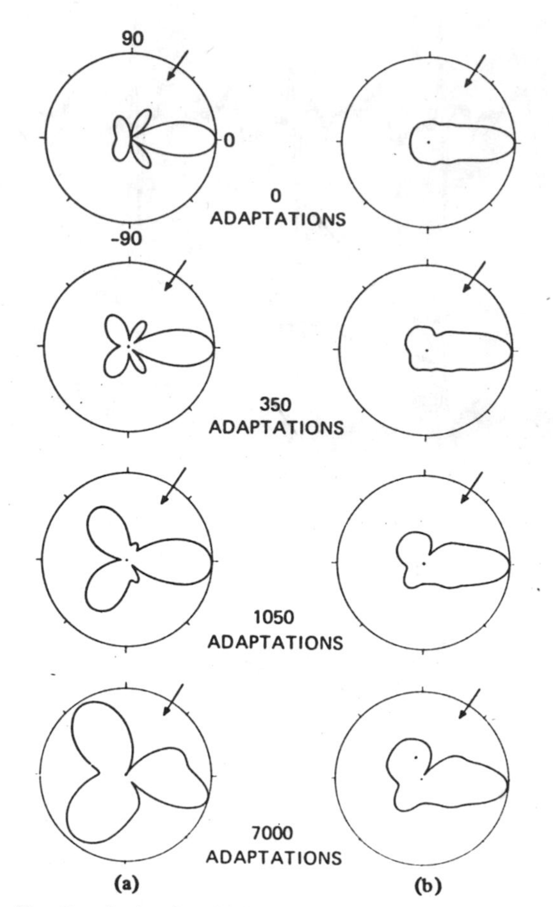

Fig. 21. Results of adaptive sidelobe cancelling experiment. (a) Single frequency (0.5 relative to folding frequency). (b) Average of eight frequencies (0.25 to 0.75 relative to folding frequency).

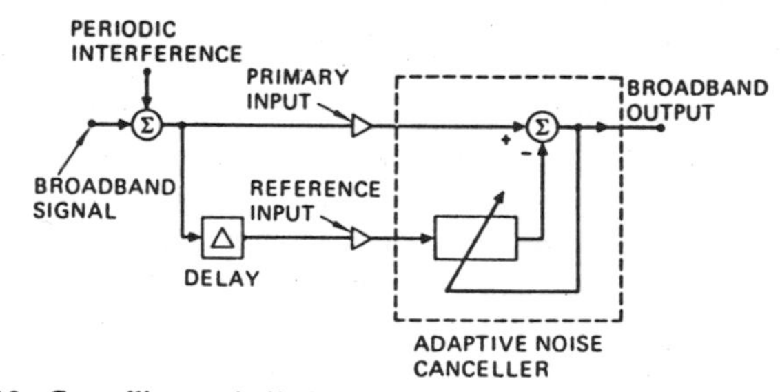

Fig. 22. Cancelling periodic interference without an external reference source.

F. Cancelling Periodic Interference without an External Reference Source

There are a number of circumstances where a broad-band signal is corrupted by periodic interference and no external reference input free of the signal is available. Examples include the playback of speech or music in the presence of tape hum or turntable rumble. It might seem that adaptive noise cancelling could not be applied to reduce or eliminate this kind of interference. If, however, a fixed delay Δ is inserted in a reference input drawn directly from the primary input, as shown in Fig. 22, the periodic interference can in many cases be readily cancelled.[15] The delay chosen must be of sufficient length to cause the broad-band signal components in the reference input to become decorrelated from those in

[15] The delay Δ may be inserted in the primary instead of the reference input if its total length is greater than the total delay of the adaptive filter. Otherwise, the filter will converge to match it and cancel both signal and interference.

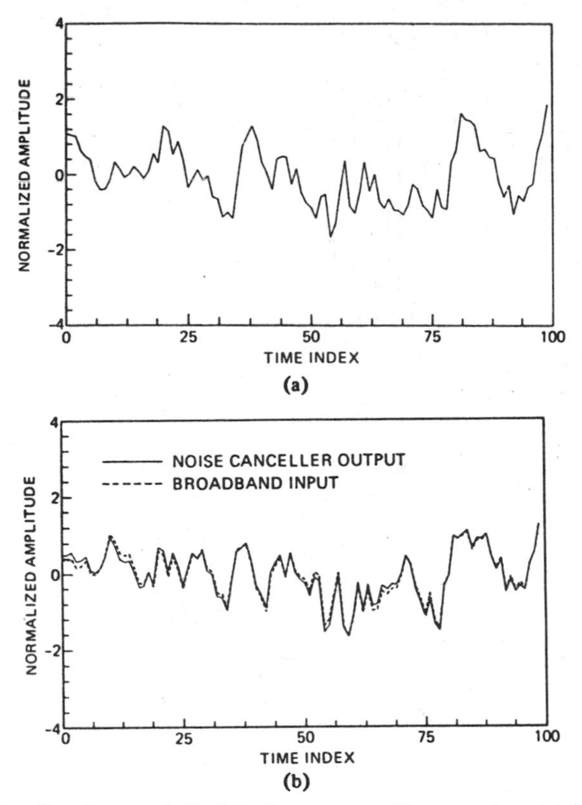

Fig. 23. Result of periodic interference cancelling experiment. (a) Input signal (correlated Gaussian noise and sine wave). (b) Noise canceller output (correlated Gaussian noise).

the primary input. The interference components, because of their periodic nature, will remain correlated with each other.

Fig. 23 presents the results of a computer simulation performed to demonstrate the cancelling of periodic interference without an external reference. Fig. 23(a) shows the primary input to the canceller. This input is composed of colored Gaussian noise representing the signal and a sine wave representing the interference. Fig. 23(b) shows the noise canceller's output. Since the problem was simulated, the exact nature of the broad-band input was known and is plotted together with the output. Note the close correspondence in form and registration. The correspondence is not perfect only because the filter was of finite length and had a finite rate of adaptation.

G. Adaptive Self-Tuning Filter

The previous experiment can also be used to demonstrate another important application of the adaptive noise canceller. In many instances where an input signal consisting of mixed periodic and broad-band components is available, the periodic rather than the broad-band components are of interest. If the system output of the noise canceller of Fig. 22 is taken from the adaptive filter, the result is an adaptive self-tuning filter capable of extracting a periodic signal from broad-band noise.

Fig. 24 shows the adaptive noise canceller as a self-tuning filter. The output of this system was simulated on the computer with the input of sine wave and correlated Gaussian noise used in the previous experiment and shown in Fig. 23(a). The resulting approximation of the input sine wave is shown in Fig. 25 together with the actual input sine wave. Note once again the close agreement in form and registration. The error is a small-amplitude stochastic process.

Fig. 26 shows the impulse response and transfer function of the adaptive filter after convergence. The impulse response,

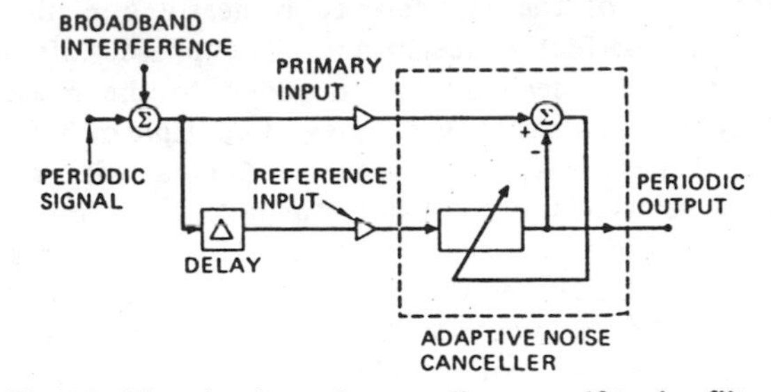

Fig. 24. The adaptive noise canceller as a self-tuning filter.

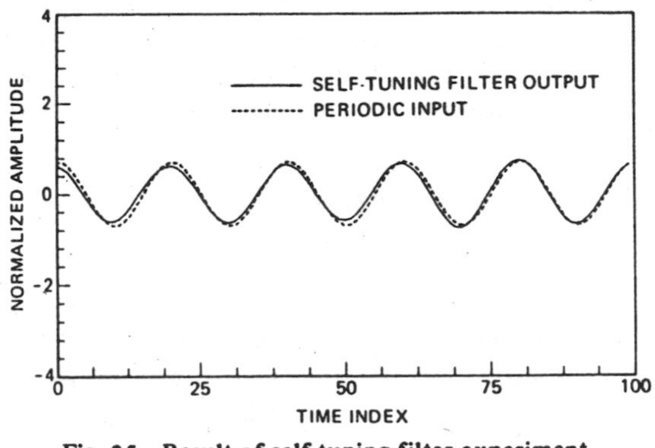

Fig. 25. Result of self-tuning filter experiment.

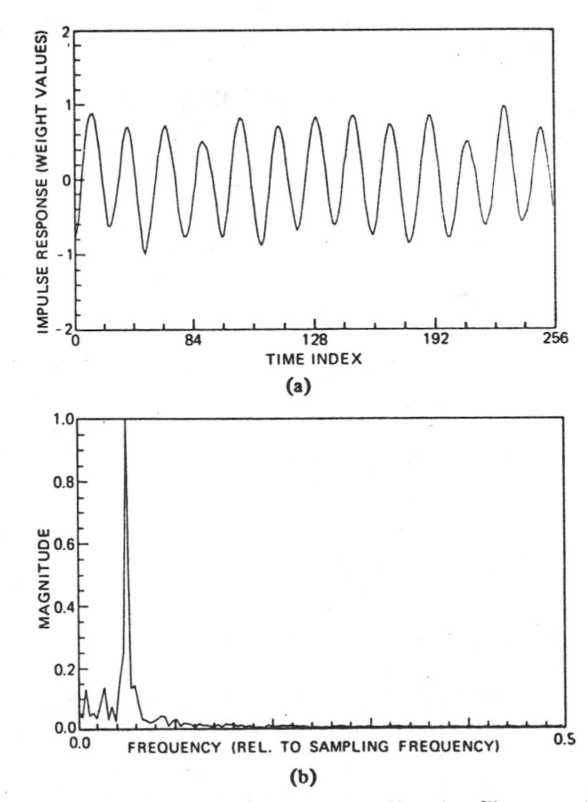

Fig. 26. Adaptive filter characteristics in self-tuning filter experiment. (a) Impulse response of adaptive filter after convergence. (b) Magnitude of transfer function of adaptive filter after convergence.

shown in Fig. 26(a), is somewhat different from but bears a close resemblance to a sine wave. If the broad-band input component had been white noise, the optimal estimator would have been a matched filter, and the impulse response would have been sinusoidal.

The transfer function, shown in Fig. 26(b), is the digital Fourier transform of the impulse response. Its magnitude at the frequency of the interference is nearly one, the value required for perfect cancellation. The phase shift at this frequency is not zero but when added to the phase shift caused by the delay Δ forms an integral multiple of 360°.

Similar experiments have been conducted with sums of sinusoidal signals in broad-band stochastic interference. In these experiments the adaptive filter developed sharp resonance peaks at the frequencies of all the spectral line components of the periodic portion of the primary input. The system thus shows considerable promise as an automatic signal seeker.

Further experiments have shown the ability of the adaptive self-tuning filter to be employed as a line enhancer for the detection of extremely low-level sine waves in noise. An introductory treatment of this application, which promises to be of great importance, is provided in Appendix D.

IX. Conclusion

Adaptive noise cancelling is a method of optimal filtering that can be applied whenever a suitable reference input is available. The principal advantages of the method are its adaptive capability, its low output noise, and its low signal distortion. The adaptive capability allows the processing of inputs whose properties are unknown and in some cases nonstationary. It leads to a stable system that automatically turns itself off when no improvement in signal-to-noise ratio can be achieved. Output noise and signal distortion are generally lower than can be achieved with conventional optimal filter configurations.

The experimental data presented in this paper demonstrate the ability of adaptive noise cancelling greatly to reduce additive periodic or stationary random interference in both periodic and random signals. In each instance cancelling was accomplished with little signal distortion even though the frequencies of the signal and the interference overlapped. The experiments described indicate the wide range of applications in which adaptive noise cancelling has potential usefulness.

Appendix A
The LMS Adaptive Filter

This Appendix provides a brief description of the LMS adaptive filter, the basic element of the adaptive noise cancelling systems described in this paper. For a full description the reader should consult the extensive literature on the subject, including the references cited below.

A. Adaptive Linear Combiner

The principal component of most adaptive systems is the adaptive linear combiner, shown in Fig. 27.[16] The combiner weights and sums a set of input signals to form an output signal. The input signal vector X_j is defined as

$$X_j \triangleq \begin{Bmatrix} x_{0j} \\ x_{1j} \\ \cdot \\ \cdot \\ \cdot \\ x_{nj} \end{Bmatrix}. \tag{A.1}$$

The input signal components are assumed to appear simultaneously on all input lines at discrete times indexed by the subscript j. The component x_{0j} is a constant, normally set to the value +1, used only in cases where biases exist among the inputs (A.1) or in the desired response (defined below). The weighting coefficients or multiplying factors $w_0, w_1, \cdots, w_n$ are adjustable, as symbolized in Fig. 27 by circles with arrows through them. The weight vector is

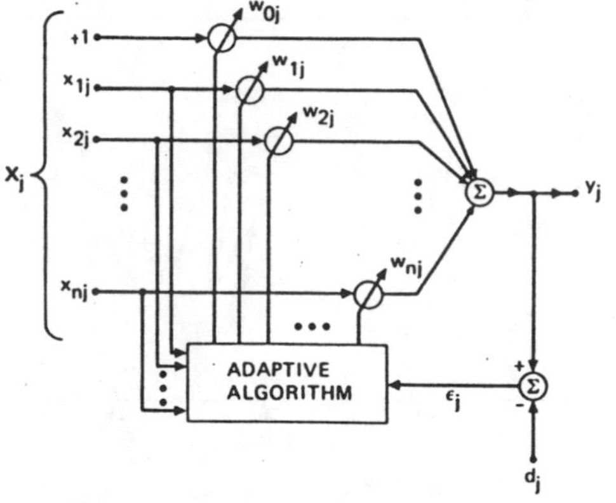

Fig. 27. The adaptive linear combiner.

$$W = \begin{Bmatrix} w_0 \\ w_1 \\ \cdot \\ \cdot \\ \cdot \\ w_n \end{Bmatrix} \tag{A.2}$$

where w_0 is the bias weight.

The output y_j is equal to the inner product of X_j and W:

$$y_j = X_j^T W = W^T X_j. \tag{A.3}$$

The error ϵ_j is defined as the difference between the desired response d_j (an externally supplied input sometimes called the "training signal") and the actual response y_j:

$$\epsilon_j = d_j - X_j^T W = d_j - W^T X_j. \tag{A.4}$$

In most applications some ingenuity is required to obtain a suitable input for d_j. After all, if the actual desired response were known, why would one need an adaptive processor? In noise cancelling systems, however, d_j is simply the primary input.[17]

B. The LMS Adaptive Algorithm

It is the purpose of the adaptive algorithm designated in Fig. 27 to adjust the weights of the adaptive linear combiner to minimize mean-square error. A general expression for mean-square error as a function of the weight values, assuming that the input signals and the desired response are statistically stationary and that the weights are fixed, can be derived in the following manner. Expanding (A.4) one obtains

$$\epsilon_j^2 = d_j^2 - 2d_j X_j^T W + W^T X_j X_j^T W. \tag{A.5}$$

Taking the expected value of both sides yields

$$E[\epsilon_j^2] = E[d_j^2] - 2E[d_j X_j^T]\, W + W^T E[X_j X_j^T]\, W. \tag{A.6}$$

Defining the vector P as the cross correlation between the

[16] This component is linear only when the weighting coefficients are fixed. Adaptive systems, like all systems whose characteristics change with the characteristics of their inputs, are by their very nature nonlinear.

[17] The actual desired response is the primary noise n_0, which is not available apart from the primary input $s + n_0$. The converged weight vector solution is easily shown to be the same when either n_0 or $s + n_0$ serves as the desired response.

desired response (a scalar) and the X vector then yields

$$P \triangleq E[d_j X_j] = E \begin{Bmatrix} d_j x_{0j} \\ d_j x_{1j} \\ \cdot \\ \cdot \\ d_j x_{nj} \end{Bmatrix}. \quad \text{(A.7)}$$

The input correlation matrix R is defined as

$$R \triangleq E[X_j X_j^T] = E \begin{bmatrix} x_{0j}x_{0j} & x_{0j}x_{1j} & x_{0j}x_{2j} & \cdots \\ x_{1j}x_{0j} & x_{1j}x_{1j} & x_{1j}x_{2j} & \cdots \\ x_{2j}x_{0j} & x_{2j}x_{1j} & x_{2j}x_{2j} & \cdots \\ \vdots & \vdots & \vdots & \\ \cdots & \cdots & \cdots & x_{nj}x_{nj} \end{bmatrix}. \quad \text{(A.8)}$$

This matrix is symmetric, positive definite, or in rare cases positive semidefinite. The mean-square error can thus be expressed as

$$E[\epsilon_j^2] = E[d_j^2] - 2P^T W + W^T R W. \quad \text{(A.9)}$$

Note that the error is a quadratic function of the weights that can be pictured as a concave hyperparaboloidal surface, a function that never goes negative. Adjusting the weights to minimize the error involves descending along this surface with the objective of getting to the "bottom of the bowl." Gradient methods are commonly used for this purpose.

The gradient ∇ of the error function is obtained by differentiating (A.9):

$$\nabla \triangleq \begin{Bmatrix} \dfrac{\partial E[\epsilon_j^2]}{\partial w_0} \\ \cdot \\ \cdot \\ \cdot \\ \dfrac{\partial E[\epsilon_j^2]}{\partial w_n} \end{Bmatrix} = -2P + 2RW. \quad \text{(A.10)}$$

The optimal weight vector W^*, generally called the Wiener weight vector, is obtained by setting the gradient of the mean-square error function to zero:

$$W^* = R^{-1} P. \quad \text{(A.11)}$$

This equation is a matrix form of the Wiener-Hopf equation [1], [2].

The LMS adaptive algorithm [7], [8], [19], [20] is a practical method for finding close approximate solutions to (A.11) in real time. The algorithm does not require explicit measurements of correlation functions, nor does it involve matrix inversion. Accuracy is limited by statistical sample size, since the weight values found are based on real-time measurements of input signals.

The LMS algorithm is an implementation of the method of steepest descent. According to this method, the "next" weight vector W_{j+1} is equal to the "present" weight vector W_j plus a change proportional to the negative gradient:

$$W_{j+1} = W_j - \mu \nabla_j. \quad \text{(A.12)}$$

The parameter μ is the factor that controls stability and rate of convergence. Each iteration occupies a unit time period. The true gradient at the jth iteration is represented by ∇_j.

The LMS algorithm estimates an instantaneous gradient in a crude but efficient manner by assuming that ϵ_j^2, the square of a single error sample, is an estimate of the mean-square error and by differentiating ϵ_j^2 with respect to W. The relationships between true and estimated gradients are given by the following expressions:

$$\nabla_j \triangleq \begin{Bmatrix} \dfrac{\partial E[\epsilon_j^2]}{\partial w_0} \\ \cdot \\ \cdot \\ \cdot \\ \dfrac{\partial E[\epsilon_j^2]}{\partial w_n} \end{Bmatrix}_{W=W_j} \qquad \hat{\nabla}_j = \begin{Bmatrix} \dfrac{[\partial \epsilon_j^2]}{\partial w_0} \\ \cdot \\ \cdot \\ \cdot \\ \dfrac{\partial \epsilon_j^2}{\partial w_n} \end{Bmatrix}_{W=W_j} = 2\epsilon_j \begin{Bmatrix} \dfrac{\partial \epsilon_j}{\partial w_0} \\ \cdot \\ \cdot \\ \cdot \\ \dfrac{\partial \epsilon_j}{\partial w_n} \end{Bmatrix}_{W=W_j}. \quad \text{(A.13)}$$

The estimated gradient components are related to the partial derivatives of the instantaneous error with respect to the weight components, which can be obtained by differentiating (A.5). Thus the expression for the gradient estimate can be simplified to

$$\hat{\nabla}_j = -2\epsilon_j X_j. \quad \text{(A.14)}$$

Using this estimate in place of the true gradient in (A.12) yields the Widrow-Hoff LMS algorithm:

$$W_{j+1} = W_j + 2\mu\epsilon_j X_j. \quad \text{(A.15)}$$

This algorithm is simple and generally easy to implement. Although it makes use of gradients of mean-square error functions, it does not require squaring, averaging, or differentiation.

It has been shown [18], [19] that the gradient estimate used in the LMS algorithm is unbiased and that the expected value of the weight vector converges to the Wiener weight vector (A.11) when the input vectors are uncorrelated over time (although they could, of course, be correlated from input component to component).[18] Starting with an arbitrary initial weight vector, the algorithm will converge in the mean and will remain stable as long as the parameter μ is greater than 0 but less than the reciprocal of the largest eigenvalue λ_{max} of the matrix R:

$$1/\lambda_{max} > \mu > 0. \quad \text{(A.16)}$$

Fig. 28 shows a typical individual learning curve resulting from the use of the algorithm. Also shown is an ensemble

[18] Adaptation with correlated input vectors has been analyzed by Senne [38] and Daniell [39]. Extremely high correlation and fast adaptation can cause the weight vector to converge in the mean to something different than the Wiener solution. Practical experience has shown, however, that this effect is generally insignificant. See also Kim and Davisson [40].

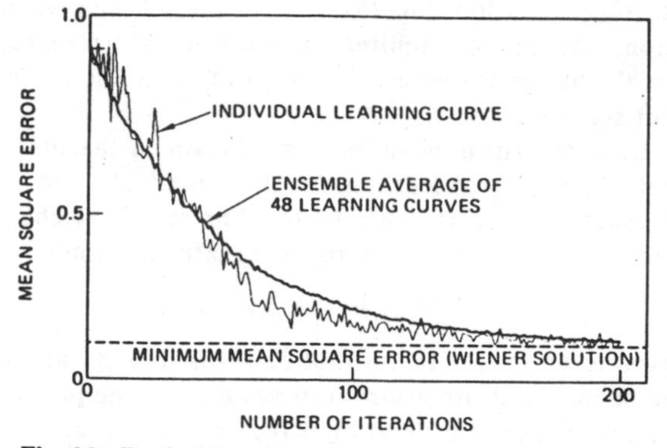

Fig. 28. Typical learning curves for the LMS algorithm.

average of 48 learning curves. The ensemble average reveals the underlying exponential nature of the individual learning curve. The number of natural modes is equal to the number of degrees of freedom (number of weights). The time constant of the pth mode is related to the pth eigenvalue λ_p of the input correlation matrix P and to the parameter μ by

$$\tau_{p_{\text{mse}}} = \frac{1}{4\mu\lambda_p}. \tag{A.17}$$

Although the learning curve consists of a sum of exponentials, it can in many cases be approximated by a single exponential whose time constant is given by (A.17) using the average of the eigenvalues of R:

$$\lambda_{\text{av}} = \frac{\lambda_0 + \lambda_1 + \cdots + \lambda_p + \cdots + \lambda_n}{(n+1)} = \frac{\text{tr}\, R}{(n+1)}. \tag{A.18}$$

Accordingly, the time constant of an exponential roughly approximating the mean-square error learning curve is

$$\tau_{\text{mse}} = \frac{(n+1)}{4\mu\, \text{tr}\, R} = \frac{(\text{number of weights})}{(4\mu)(\text{total input power})}. \tag{A.19}$$

The total input power is the sum of the powers incident to all of the weights.

Proof of these assertions and further discussion of the characteristics and properties of the LMS algorithm are presented in [19], [20], and [41].

C. The LMS Adaptive Filter

The adaptive linear combiner may be implemented in conjunction with a tapped delay line to form the LMS adaptive filter shown in Fig. 29, where the bias weight has been omitted for simplicity. Fig. 29(a) shows the details of the filter, including the adaptive process incorporating the LMS algorithm. Because of the structure of the delay line, the input signal vector is

$$X_j = \begin{Bmatrix} x_j \\ x_{j-1} \\ \cdot \\ \cdot \\ \cdot \\ x_{j-n+1} \end{Bmatrix}. \tag{A.20}$$

The components of this vector are delayed versions of the input signal x_j. Fig. 29(b) is the representation adopted to symbolize the adaptive tapped-delay-line filter.

This kind of filter permits the adjustment of gain and phase at many frequencies simultaneously and is useful in adaptive broad-band signal processing. Simplified design rules, giving

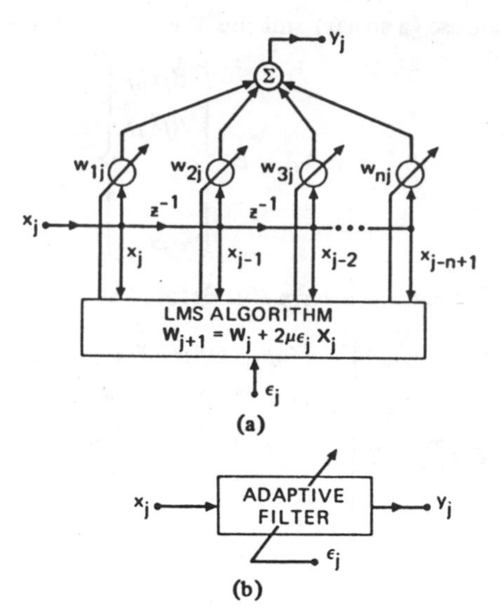

Fig. 29. The LMS adaptive filter. (a) Block diagram. (b) Symbolic representation.

the tap spacings and number of taps (weights), are the following: The tap spacing time must be at least as short as the reciprocal of twice the signal bandwidth (in accord with the sampling theorem). The total real-time length of the delay line is determined by the reciprocal of the desired filter frequency resolution. Thus, the number of weights required is generally equal to twice the ratio of the total signal bandwidth to the frequency resolution of the filter. It may be possible to reduce the number required in some cases by using nonuniform tap spacing, such as log periodic. Whether this is done or not, the means of adaptation remain the same.

Appendix B
Finite-Length, Causal Approximation of the Unconstrained Wiener Noise Canceller

In the analyses of Sections IV and V questions of the physical realizability of Wiener filters were not considered. The expressions derived were ideal, based on the assumption of an infinitely long, two-sided (noncausal) tapped delay line. Though such a delay line cannot in reality be implemented, fortunately its performance, as shown in the following paragraphs, can be closely approximated.

Typical impulse responses of ideal Wiener filters approach amplitudes of zero exponentially over time. Approximate realizations are thus possible with finite-length transversal filters. The more weights used in the transversal filter, the closer its impulse response will be to that of the ideal Wiener filter. Increasing the number of weights, however, also slows the adaptive process and increases the cost of implementation. Performance requirements should thus be carefully considered before a filter is designed for a particular application.

Noncausal filters, of course, are not physically realizable in real-time systems. In many cases, however, they can be realized approximately in delayed form, providing an acceptable delayed real-time response. In practical circumstances excellent performance can be obtained with two-sided filter impulse responses even when they are truncated in time to the left and right. By delaying the truncated response it can be made causal and physically realizable.

Fig. 30 shows an adaptive noise cancelling system with a delay Δ inserted in the primary input. This delay causes an equal delay to develop in the unconstrained optimal filter

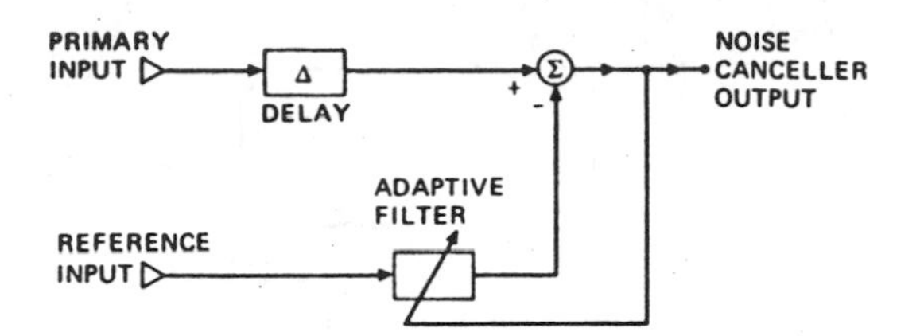

Fig. 30. Adaptive noise canceller with delay in primary input path.

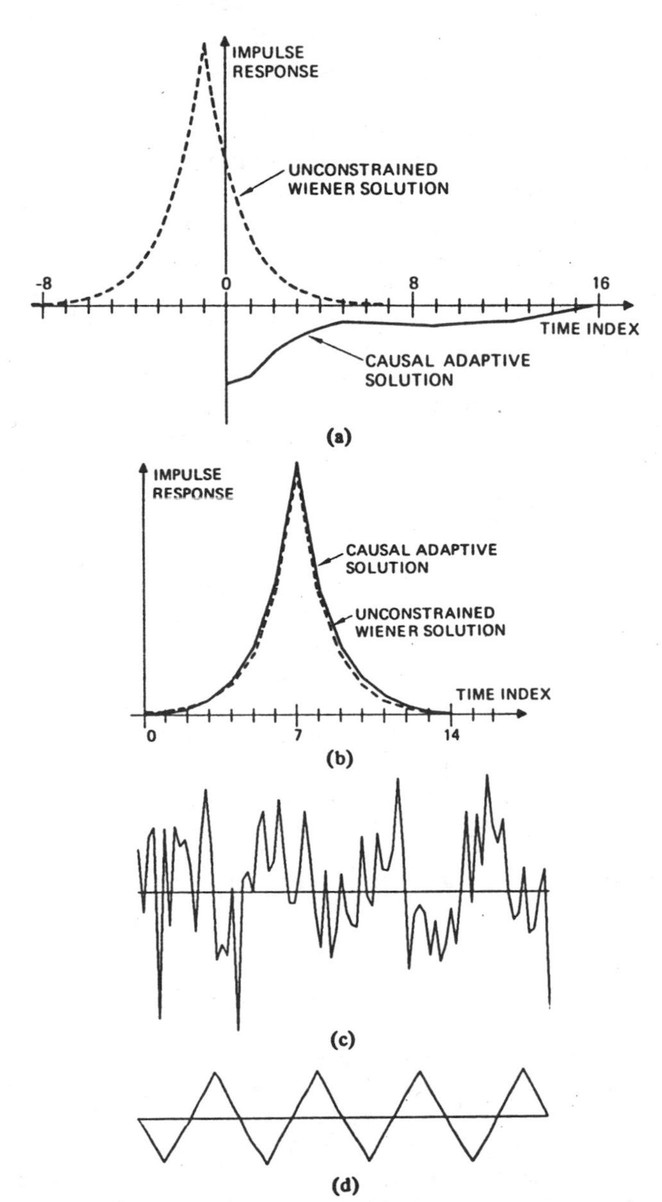

Fig. 31. Results of noise cancelling experiment with delay in primary input path. (a) Optimal solution and adaptive solution found without time delay. (b) Optimal solution and adaptive solution found with delay of eight time units. (c) Noise canceller output without delay. (d) Noise canceller output with delay.

impulse response, which remains otherwise unchanged. In practical, finite-length adaptive transversal filters, on the other hand, the optimal impulse response generally changes shape with changes in the value of Δ, which is chosen to cause the peak of the impulse response to center along the delay line.

Experience has shown that the value of Δ is not critical within a certain optimal range; that is, the curve showing minimum mean-square error as a function of Δ generally has a very broad minimum. A value typically equal to about half the time delay of the adaptive filter produces the least minimum output noise power.

Fig. 31 shows the results of a computer-simulated noise cancelling experiment with an unconstrained optimal filter response that was noncausal. The primary input consisted of a triangular wave and additive colored noise. The reference input consisted of colored noise correlated with the primary noise.[19] The unconstrained Wiener impulse response and the causal, finite time adaptive impulse response obtained without a delay in the primary input are plotted in Fig. 31(a). The large difference in these impulse responses indicates that the noise canceller output will be a poor approximation of the signal. The corresponding Wiener and adaptive impulse responses obtained with a delay of eight time units (half the length of the adaptive filter) are shown in Fig. 31(b). These solutions are similar, indicating that performance of the adaptive filter will be close to optimal. Typical noise canceller outputs with and without delay are shown in Fig. 31(c) and Fig. 31(d). The waveform obtained with the delay is very close to that of the original triangular-wave signal, whereas that obtained with no delay still contains a great amount of noise.

Appendix C
Multiple-Reference Noise Cancelling

When there is more than one noise or interference to be cancelled and a number of linearly independent reference inputs containing mixtures of each can be obtained, it is usually advantageous to use a multiple-reference noise cancelling system. Such a system may be considered a generalization of the single-reference noise cancellers analyzed in this paper. In the model shown in Fig. 32 the ψ_i represent mutually uncorrelated sources of either input signal or noise. The transfer functions $\mathcal{G}_i(z)$ represent the propagation paths from these sources to the primary input. The $\mathcal{F}_i(z)$ similarly represent the propagation paths to the reference inputs and allow for cross-coupling. This model permits treatment not only of multiple noise sources but also of signal components in the reference inputs and uncorrelated noises in the reference and primary inputs. In other words, it is a general representation of an adaptive noise canceller.

The unconstrained Wiener transfer function of the multiple-reference canceller is the matrix equivalent of (13) and is derived in the following manner. The source spectral matrix of ψ_i is defined as

$$[\mathcal{S}_{\psi\psi}(z)] = \begin{bmatrix} \mathcal{S}_{\psi_1\psi_1}(z) & & & & \bigcirc \\ & \mathcal{S}_{\psi_2\psi_2} & & & \\ & & \cdot & & \\ & & & \cdot & \\ \bigcirc & & & & \mathcal{S}_{\psi_m\psi_m}(z) \end{bmatrix}. \qquad \text{(C.1)}$$

The spectral matrix of the k reference inputs to the adaptive

[19] Except for the delay in the primary input, the simulated noise cancelling system was identical with the system shown above in Fig. 3. The transfer function $\mathcal{H}(z)$ was a nonminimum phase, low-pass transversal filter with two zeros and no poles $[\mathcal{H}(z) = 2z^{-1}(1 - \frac{1}{2}z^{-1}) \cdot (1 - \frac{1}{2}z)]$. The optimal unconstrained adaptive filter solution, in this case given by (18), is the reciprocal of $\mathcal{H}(z)$. It has one pole inside and one pole outside the unit circle in the Z plane. A stable realization of $\mathcal{W}^*(z)$ must, therefore, be two sided.

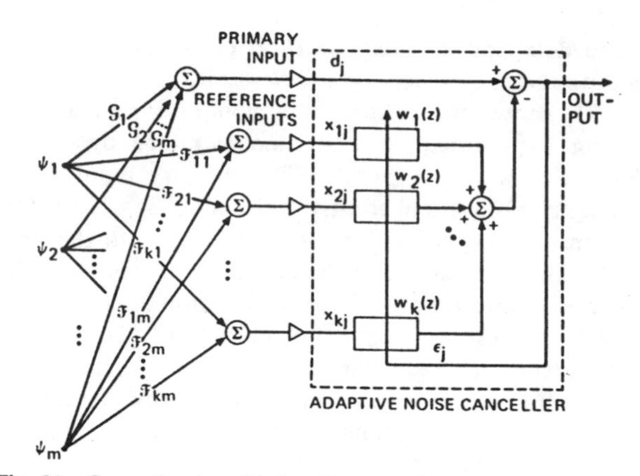

Fig. 32. Generalized multiple-reference adaptive noise canceller.

filters is then

$$[\mathcal{S}_{xx}(z)] = [\mathcal{F}(z^{-1})]\,[\mathcal{S}_{\psi\psi}(z)]\,[\mathcal{F}(z)]^T \tag{C.2}$$

where

$$[\mathcal{F}(z)] = \begin{bmatrix} \mathcal{F}_{11}(z) & \mathcal{F}_{12}(z) & \cdots & \mathcal{F}_{1m}(z) \\ \mathcal{F}_{21}(z) & & & \\ \vdots & & \ddots & \\ \mathcal{F}_{k1}(z) & & \cdots & \mathcal{F}_{km}(z) \end{bmatrix} \tag{C.3}$$

and $\mathcal{F}_{il}(z)$ is the transfer function from input source l to reference input i.

The cross-spectral vector between the reference inputs and the primary input is given by

$$\{\mathcal{S}_{xd}(z)\} = [\mathcal{F}(z^{-1})]\,[\mathcal{S}_{\psi\psi}(z)]\,\{\mathcal{G}(z)\} \tag{C.4}$$

where

$$\mathcal{G}(z) = \begin{bmatrix} \mathcal{G}_1(z) \\ \mathcal{G}_2(z) \\ \vdots \\ \mathcal{G}_m(z) \end{bmatrix} \tag{C.5}$$

and $\mathcal{G}_i(z)$ is the transfer function from input source i to the primary input.

The Wiener optimal weight vector is then

$$\begin{aligned} \mathcal{W}^*(z) &= [\mathcal{S}_{xx}(z)]^{-1}\,\mathcal{S}_{xd}(z) \\ &= [[\mathcal{F}(z^{-1})]\,[\mathcal{S}_{\psi\psi}(z)]\,[\mathcal{F}(z)]^T]^{-1} \\ &\quad \cdot [\mathcal{F}(z^{-1})]\,[\mathcal{S}_{\psi\psi}(z)]\,\{\mathcal{G}(z)\}. \end{aligned} \tag{C.6}$$

If $[\mathcal{F}(z)]$ is square, at those frequencies for which $[\mathcal{F}(z)]$ is invertible (55) simplifies to

$$\{\mathcal{W}^*(z)\} = [\mathcal{F}(z)^T]^{-1}\,\{\mathcal{G}(z)\} \tag{C.7}$$

which is the matrix equivalent of (17).

These expressions can be used to derive steady-state Wiener solutions to multiple-source, multiple-reference noise cancelling problems more general than those of Sections IV and V. An example of a multiple-reference problem is given in Section VIII.

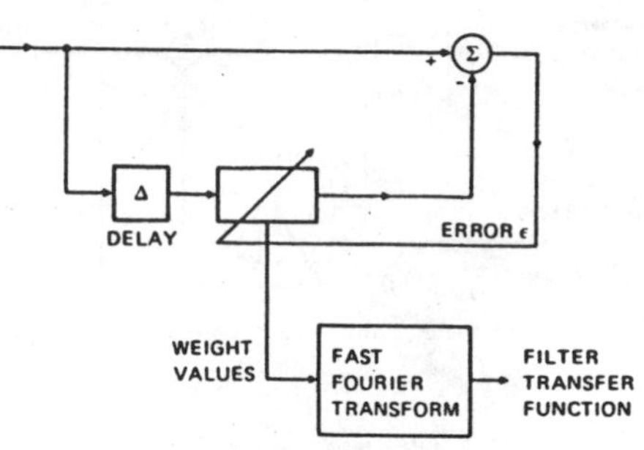

Fig. 33. The adaptive line enhancer.

Appendix D
Adaptive Line Enhancer

A classical detection problem is that of finding a low-level sine wave in noise. The adaptive self-tuning filter, whose capability of separating the periodic and stochastic components of a signal was illustrated above (where these components were of comparable level), is able to serve as an "adaptive line enhancer" for the detection of extremely low-level sine waves in noise. The adaptive line enhancer becomes a competitor of the fast Fourier transform algorithm as a sensitive detector and has capabilities that may exceed those of conventional spectral analyzers when the unknown sine wave has finite bandwidth or is frequency modulated.

The method is illustrated in Fig. 33. The input consists of signal plus noise. The output is the digital Fourier transform of the filter's impulse response. Detection is accomplished when a spectral peak is evident above the background noise. The same method, with minor differences, has been proposed by Griffiths for "maximum entropy spectral estimation" [42], [43].

It should be noted that the filter output signal is also available. This signal could be used directly or as an input to a spectral analyzer or phase-lock loop. The method of Fig. 33 could further be used for the simultaneous detection of multiple sine waves. None of these possibilities is considered here. Only the detection of single low-level sine waves in noise is treated.

A. Optimal Transfer Function

Fig. 34 shows the ideal impulse response and transfer function of the adaptive line enhancer for a given input spectrum. It is assumed that the input noise is white, with a total power of ν^2, and that the input signal has a power of $C^2/2$ at frequency ω_0. The ideal impulse response, equivalent to the matched filter response, is a sampled sinusoid whose frequency is ω_0.[20] The phase shift of this response at frequency ω_0 when added to that of the delay is an integral multiple of 360°. If the peak value of the transfer function is a, the peak value of the weights is to a close approximation $2a/n$, where n is the number of weights.

The adaptive process minimizes the mean square of the error. The error power is the sum of three components, the primary input noise power, the noise power at the output of the adaptive filter, and the sinusoidal signal power. Accordingly, the error power may be expressed as

$$\text{error power} = \nu^2 + (\nu^2/2)\,(2a/n)^2\,n + (C^2/2)\,(1-a)^2. \tag{D.1}$$

[20] This assertion is proved analytically for arbitrary input signal-to-noise ratio in J. R. Zeidler and D. M. Chabries, "An analysis of the LMS adaptive filter used as a spectral line enhancer," Naval Undersea Center, Tech. Note 1476, Feb. 1975.

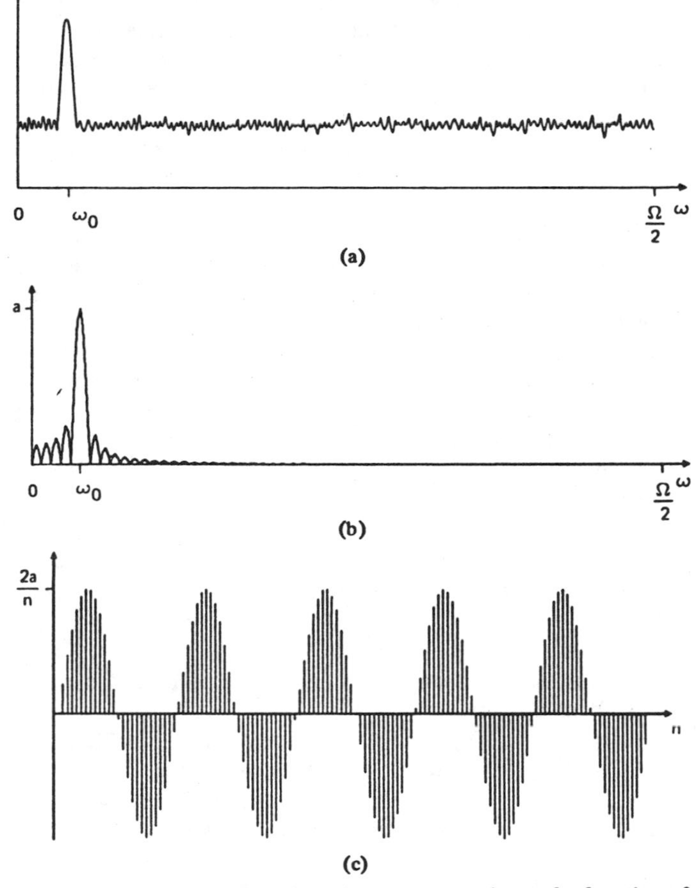

Fig. 34. Ideal adaptive filter impulse response and transfer function of adaptive line enhancer for a given input spectrum. (a) Input spectrum. (b) Transfer function magnitude. (c) Impulse response.

We have used the facts that 1) the output noise power of a digital filter with a white input equals the input power multiplied by the sum of the squares of the impulse values of the impulse response and 2) the primary and filter output sinusoidal components combine coherently at the summing junction. The signal gain from input to error is $(1 - a)$.

The optimal value of a that minimizes error power a^* is obtained by setting the derivative of (D.1) to zero:

$$a^* = \frac{\left(\frac{C^2/2}{\nu^2}\right)\left(\frac{n}{2}\right)}{1+\left(\frac{C^2/2}{\nu^2}\right)\left(\frac{n}{2}\right)} = \frac{(\text{SNR})\,(n/2)}{1+(\text{SNR})\,(n/2)}. \tag{D.2}$$

At high signal-to-noise ratios, $a^* \cong 1$. At low signal-to-noise ratios, $a^* < 1$. Low signal-to-noise conditions can be dealt with by using a large number of adaptive weights, although other problems could result because of weight-vector noise.

B. Noise in the Weight Vector

The ability to detect peaks in the transfer function due to the presence of sinusoidal signals is limited by the presence of spurious peaks caused by noise in the weight vector. One thus needs to know the nature of weight-vector noise and its effects on the transfer function.

The gradient estimate $\hat{\nabla}_j$ used by the LMS algorithm, given by (A.14), may be expressed as

$$\hat{\nabla}_j = -2\epsilon_j X_j = \nabla_j + \mathfrak{N}_j \tag{D.3}$$

where ∇_j is the true gradient and $\mathfrak{N}_j$ is the zero-mean gradient estimation noise. At the minimum point of the quadratic mean-square-error surface the true gradient is zero. The gradient estimate at this point is thus equal to the gradient estimation noise:

$$\hat{\nabla}_j = \mathfrak{N}_j = -2\epsilon_j X_j. \tag{D.4}$$

If one assumes that the input signal vector X_j is uncorrelated over time,[21] then $\mathfrak{N}_j$ is also uncorrelated over time. In addition, when the weight vector W_j is equal to the optimal weight vector W^*, Wiener filter theory shows that the error ϵ_j and the input vector X_j are uncorrelated. If one now assumes that ϵ_j and X_j are Gaussian, then these terms are statistically independent and the covariance of $\mathfrak{N}_j$ is

$$\text{cov}\,[\mathfrak{N}_j] = E[\mathfrak{N}_j\mathfrak{N}_j^T] = 4E[\epsilon_j^2 X_j X_j^T] = 4E[\epsilon_j^2] \cdot E[X_j X_j^T] = 4E[\epsilon_j^2]\,R \tag{D.5}$$

where R is the input correlation matrix. Since at the minimum point of the mean-square-error surface $E[\epsilon_j^2] = \xi_{\min}$, (D.5) can be expressed as

$$\text{cov}\,[\mathfrak{N}_j] = 4\xi_{\min} R. \tag{D.6}$$

In the vicinity of the minimum point the covariance of the gradient noise is closely approximated by (D.6), and the gradient noise is statistically stationary and uncorrelated over time.

For the purpose of the following analysis it is more convenient to work in "primed coordinates." The correlation matrix R may be expressed in normal form as

$$R = Q\Lambda Q^{-1} \tag{D.7}$$

where Q is an orthonormal modal matrix, Λ is a diagonal matrix of eigenvalues, and Q^{-1} is equal to Q^T. The gradient noise in the primed coordinates is then

$$\mathfrak{N}'_j = Q^{-1}\mathfrak{N}_j \tag{D.8}$$

and the covariance of the gradient noise is

$$\begin{aligned}\text{cov}\,[\mathfrak{N}'_j] &= E[\mathfrak{N}'_j\mathfrak{N}'^T_j] = E[Q^{-1}\mathfrak{N}_j\mathfrak{N}_j^T Q] = Q^{-1}E[\mathfrak{N}_j\mathfrak{N}_j^T]\,Q \\ &= Q^{-1}\,\text{cov}\,[\mathfrak{N}_j]\,Q = 4\xi_{\min}Q^{-1}RQ = 4\xi_{\min}\Lambda.\end{aligned} \tag{D.9}$$

It should be noted that the components of $\mathfrak{N}'_j$ are mutually uncorrelated and proportional to the respective eigenvalues.

The effect of gradient noise on the weight vector can now be determined as follows. The LMS algorithm with a noisy gradient estimate can be expressed in accordance with (A.12) as

$$W_{j+1} = W_j + \mu(-\hat{\nabla}_j) = W_j + \mu(-\nabla_j + \mathfrak{N}_j). \tag{D.10}$$

Reexpressing (D.10) in terms of V_j, where V_j is defined as $W_j - W^*$, yields

$$V_{j+1} = V_j + \mu(-2RV_j + \mathfrak{N}_j). \tag{D.11}$$

Projecting into the primed coordinates by premultiplying both sides by Q^{-1} yields

$$V'_{j+1} = V'_j - 2\mu\Lambda V'_j + \mu\mathfrak{N}'_j = (I - 2\mu\Lambda)\,V'_j + \mu\mathfrak{N}'_j. \tag{D.12}$$

Note once again that, since the components of $\mathfrak{N}'_j$ are mutually uncorrelated and since (D.12) is diagonalized, the components of noise in V'_j are mutually uncorrelated.

[21] This common assumption is not strictly correct in this case but greatly simplifies the analysis and yields results that work well in practice.

Near the minimum point of the error surface, in steady state after adaptive transients have died out, the mean of V'_j is zero, and the covariance of the weight-vector noise may be obtained as follows. Postmultiplying both sides of (D.12) by their transposes and taking expected values yields

$$E[V'_{j+1} V'^T_{j+1}] = E[(I - 2\mu\Lambda) V'_j V'^T_j (I - 2\mu\Lambda)] + \mu^2 E[\mathfrak{N}'_j \mathfrak{N}'^T_j] + \mu E[\mathfrak{N}'_j V'^T_j (I - 2\mu\Lambda)] + \mu E[(I - 2\mu\Lambda) V'_j \mathfrak{N}'^T_j]. \quad (D.13)$$

It has been assumed that the input vector X_j is uncorrelated over time; the gradient noise $\mathfrak{N}_j$ is accordingly uncorrelated with the weight vector W_j, and therefore $\mathfrak{N}'_j$ and V'_j are uncorrelated. Equation (D.13) can thus be expressed as

$$E[V'_{j+1} V'^T_{j+1}] = (I - 2\mu\Lambda) E[V'_j V'^T_j] (I - 2\mu\Lambda) + \mu^2 E[\mathfrak{N}'_j \mathfrak{N}'^T_j]. \quad (D.14)$$

Furthermore, if V'_j is stationary, the covariance of V'_{j+1} is equal to the covariance of V'_j, which may be expressed as

$$\text{cov}[V'_j] = (I - 2\mu\Lambda) \text{cov}[V'_j] (I - 2\mu\Lambda) + \mu^2 \text{cov}[\mathfrak{N}'_j]. \quad (D.15)$$

Since the noise components of V'_j are mutually uncorrelated, (D.15) is diagonal. It can thus be rewritten as

$$\text{cov}[V'_j] = (I - 2\mu\Lambda)^2 \text{cov}[V'_j] + \mu^2 (4\xi_{\min}\Lambda) \quad (D.16)$$

or

$$(I - \mu\Lambda) \text{cov}[V'_j] = \mu\xi_{\min}. \quad (D.17)$$

When the value of the adaptive constant μ is small (as is consistent with a converged solution near the minimum point of the error surface), it is implied that

$$\mu\Lambda \ll I. \quad (D.18)$$

Equation (D.17) thus becomes

$$\text{cov}[V'_j] = \mu\xi_{\min} I. \quad (D.19)$$

The covariance of V_j can now be expressed as follows:

$$\text{cov}[V_j] = E[V_j V_j^T] = E[Q V'_j V'^T_j Q^{-1}] = Q \text{cov}[V'_j] Q^{-1} = \mu\xi_{\min} I \quad (D.20)$$

where the components of the weight-vector noise are all of the same variance and are mutually uncorrelated. This derivation of the covariance depends on the assumptions made above. It has been found by experience, however, that (D.20) closely approximates the exact covariance of the weight-vector noise under a considerably wider range of conditions than these assumptions imply. A derivation of bounds on the covariance based on fewer assumptions has been made by Kim and Davisson [40].

C. Noise in the Transfer Function

The filter weights, comprising the impulse response, undergo digital Fourier transformation to yield the transfer function. The noise in each of the weights is uncorrelated over time, uncorrelated from weight to weight, and of variance $\mu\xi_{\min}$. At the jth instant the impulse response has n samples, $w_{0j}, w_{1j}, \cdots, w_{kj}, \cdots, w_{n-1j}$, and their transform is

$$H_j(l) = \sum_{k=0}^{n-1} w_{kj} \exp(-i2\pi kl/n) \quad (D.21)$$

where l is the frequency index. For a single value of l, $H_j(l)$ is a linear combination of all the weights, each weighted by a phasor of unit magnitude. Since $H_j(l)$ is complex, the power of this noise is the sum of its "real" and "imaginary" power and equals the sum of the noise power in the weights themselves. Thus at each frequency l, the noise power in $H_j(l)$ is

$$n\mu\xi_{\min}. \quad (D.22)$$

In spectral analysis, "ensemble averaging" techniques are commonly used. The same approach could be used here, averaging the weights over time before transforming. Although the gradient noise is essentially uncorrelated over j, the weight-vector noise is generally highly correlated over time. Averaging with each adaptive iteration could be done but is not necessary; averaging the weight vector at intervals corresponding to about four adaptive time constants ($4\tau_{\text{mse}}$) would assure noise independence and would be appropriate in gathering the information contained in the time history of the weights. On this basis, averaging N weight vectors would produce, at the lth frequency, a noise power in $\overline{H(l)}$, the averaged transfer function, with the following value:

$$(n/N)\mu\xi_{\min}. \quad (D.23)$$

This expression for weight-vector noise can be put in more usable form by relating $\xi_{\min}$ to the physical line enhancing process shown in Fig. 33. The noise power at the filter output will always be negligible compared to the input noise power, since the optimized filter transfer function will be small in magnitude except at the peaks whose value is a^*. When signal power is low compared to noise power, which is the case of interest in the present context, the error power is essentially equal to the input noise power. Thus

$$\xi_{\min} = \nu^2. \quad (D.24)$$

The noise power in $\overline{H(l)}$ at the lth frequency is accordingly

$$(n/N)\mu\nu^2. \quad (D.25)$$

D. Detectability of Sine Waves by Adaptive Line Enhancing

Detection of a signal is dependent on identification of its adaptive filter transfer function peak (of value a^*) as distinct from other peaks due to weight-vector noise. For this purpose one could compare the value of a^* with the standard deviation of the noise in $\overline{H(l)}$. A still better procedure is to work with signal and noise power by comparing the squares of these quantities. "Detectability" for the adaptive line enhancer (ALE) is accordingly defined as follows:

$$D_{\text{ALE}} \triangleq \frac{(a^*)^2}{\text{noise power in } \overline{H(l)}}. \quad (D.26)$$

This measure must typically be one or greater to achieve signal detection. Using (D.2) and (D.22), equation (D.26) can be re-expressed as

$$D_{\text{ALE}} = \frac{(\text{SNR})^2 (n^2/4)}{(n/N)\mu\nu^2 [1 + (\text{SNR})(n/2)]^2}. \quad (D.27)$$

The power of the adaptive filter input is essentially that of the noise, equal to ν^2. Since the filter input is essentially white, the input correlation matrix can be well represented by

$$R = \nu^2 I. \quad (D.28)$$

All eigenvalues are equal to ν^2. The trace of R is equal to $n\nu^2$.

Using (A.19) of Appendix A, one may thus write

$$\tau_{mse} = \frac{n}{4\mu \operatorname{tr} R} = \frac{n}{4\mu n \nu^2} = \frac{1}{4\mu\nu^2}. \tag{D.29}$$

This is the time constant of the mean-square error learning curve. Note that the line enhancer does not have a bias weight and that the number of weights is thus n rather than $n + 1$. Equation (D.27) may now be expressed in more useful form as follows:

$$D_{ALE} = \left(\frac{4N\tau_{mse}}{n}\right)\left[\frac{(SNR)(n/2)}{1 + (SNR)(n/2)}\right]^2. \tag{D.30}$$

For high signal-to-noise ratios—that is, for $(SNR)(n/2) \gg 1$—equation (D.30) becomes

$$D_{ALE} \cong \frac{4N\tau_{mse}}{n}. \tag{D.31}$$

For low signal-to-noise ratios—that is, for $(SNR)(n/2) \ll 1$—equation (D.30) shows that

$$D_{ALE} \cong (N\tau_{mse})(SNR)^2 n. \tag{D.32}$$

Intermediate values must be independently calculated.

Choice of the number of weights has an influence on the value of D_{ALE} for a given input signal-to-noise ratio. Differentiating (D.30) with respect to n and setting the derivative to zero yields the following expression for the optimal value of n:

$$n^* = 2/SNR. \tag{D.33}$$

Substituting (D.33) into (D.30) then yields the optimal value[22] of D_{ALE}:

$$D^*_{ALE} = (N\tau_{mse})(SNR/2). \tag{D.34}$$

It is interesting to note that, when n is so optimized,

$$a^* = 1/2. \tag{D.35}$$

E. Detectability of Sine Waves by Spectral Analysis

Let the power spectrum of a signal in white Gaussian noise be derived from an L-point digital Fourier transform. The frequency of the signal is assumed to be at the center of a spectral bin. Input signal power is assumed to be $C^2/2$ and noise power to be ν^2. At the signal frequency, the component of the power spectrum due to the signal will have the value $C^2L^2/4$. Each spectral bin will have an identical average noise power of $\nu^2 L$.

For the signal to be detected its spectral peak must be distinguishable from noise peaks that are deviations about the mean noise power. The variance of the noise power about the mean can be reduced by ensemble averaging; that is, by averaging N power spectra, each derived from L data points. With Gaussian noise the variance of the noise power about its mean in any spectral bin can be shown [23] to be $(2/N)$ (average noise power)^2, which is equivalent to

$$(2/N)(\nu^2 L)^2. \tag{D.36}$$

[22] The exact value of n is not critical; it may be as much as 8 times larger or smaller than n^* and D_{ALE} will remain within approximately 50 percent of D^*_{ALE}.

[23] The variance in the estimate of variance from N samples of a zero-mean process equals (mean fourth − [mean square]^2)/N.

It is reasonable to compare the average signal power in the selected spectral bin with the standard deviation of the noise power fluctuations that occur in each spectral bin; that is, with the square root of (D.36). We thus define "detectability" for spectral analysis as

$$D_{DFT} \triangleq \frac{C^2L^2/4}{(2/N)^{1/2}\nu^2 L} = \frac{1}{2}(SNR)\,L(N/2)^{1/2} = (SNR)\,L(N/8)^{1/2}. \tag{D.37}$$

The motive for this definition is derived from the early work of Woodward [44], Skolnick [45], Swerling [46], Marcum [47], and others.

F. Comparison of Adaptive Line Enhancing and Spectral Analysis

Fig. 35 illustrates the definitions of the detectability of a sine wave by adaptive line enhancing and spectral analysis given in (D.30) and (D.37). It is useful to compare Fig. 35(a) with Fig. 35(b). Note that in the former case the measure of detectability is based on the magnitude of the adaptive filter transfer function, whereas in the latter it is based on the digital power spectrum. Since the measure of detectability is different for the two techniques, in a sense one is comparing "apples and oranges." Yet both D_{ALE} and D_{DFT} are ratios of signal power to noise power.

Fig. 36 presents experimental results, obtained by computer simulation, showing the performance of the adaptive line enhancing and spectral analysis techniques for three values of D_{ALE} and D_{DFT}. Visual examination indicates that D_{ALE} and D_{DFT} do provide a reasonable basis for comparing the performance of the two techniques.

Equation (D.34) describes the detectability of a sine wave by the adaptive technique when n is optimized. This equation can be rewritten as

$$D^*_{ALE} = (4N\tau_{mse})(SNR/8). \tag{D.38}$$

Since weight vectors are taken for ensemble averaging at $4\tau_{mse}$ intervals and N vectors are averaged, $4N\tau_{mse}$ represents the total number of input data samples. Note that the time constant τ_{mse} is not expressed in seconds but in number of adaptive iterations, which is equivalent to number of input data samples. Thus (D.38) can be rewritten as

$$D^*_{ALE} = (\text{number of data samples})(SNR/8). \tag{D.39}$$

The detectability of a sine wave by spectral analysis is given by (D.37). Since N sample spectra are ensemble-averaged, and since each requires L data points, the number of data samples required is the product of N and L. Equation (D.37) can thus be rewritten as

$$D_{DFT} = (\text{number of data samples})(SNR/[8N]^{1/2}). \tag{D.40}$$

The ratio of detectabilities is, therefore,

$$\frac{D^*_{ALE}}{D_{DFT}} = \left(\frac{N}{8}\right)^{1/2}. \tag{D.41}$$

Accordingly, spectral analysis is advantageous as long as the number of ensemble members is less than eight. Adaptive line enhancing would be advantageous when the number of ensemble members required for spectral analysis is greater than eight.

For the comparative experiment represented by Fig. 36, input signal-to-noise ratio in each case was 0.01562. The number of data samples used with spectral analysis was the same as

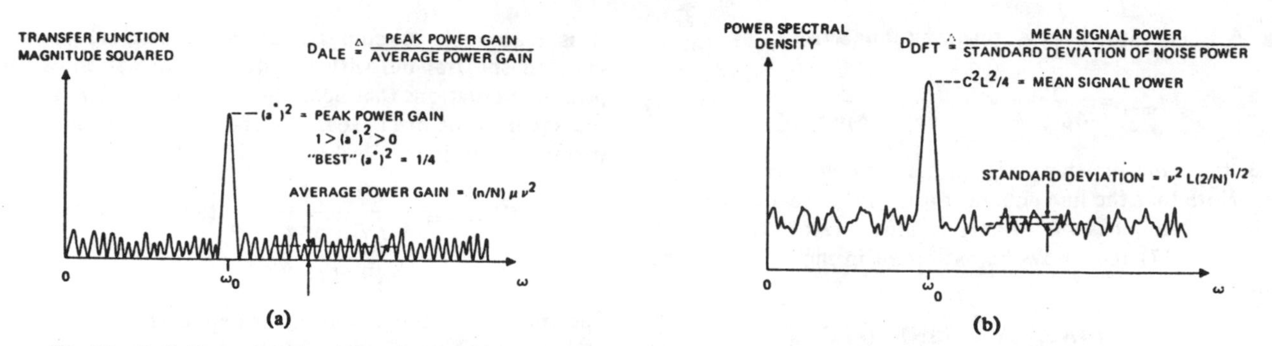

Fig. 35. Definition of detectability D of a sine wave in noise. (a) With adaptive line enhancing. (b) With spectral analysis.

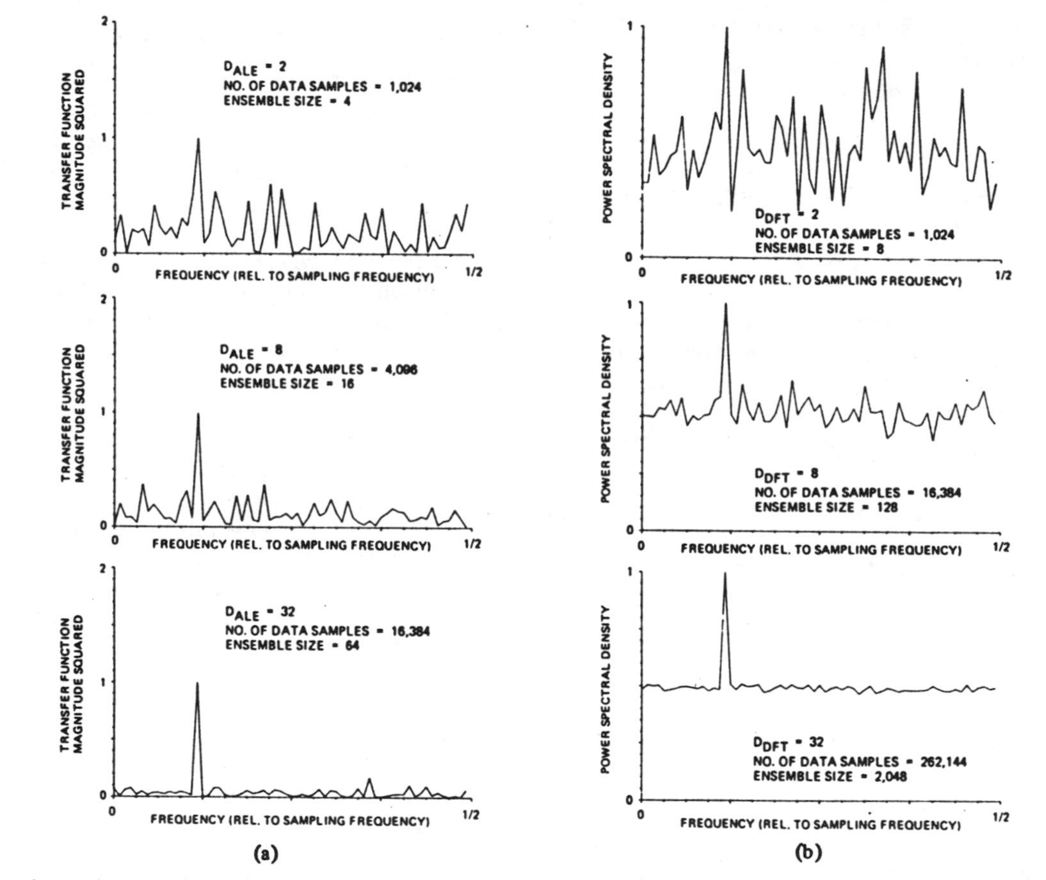

Fig. 36. Experimental comparison of adaptive line enhancing and spectral analysis for three values of detectability D; input signal-to-noise ratio, 0.01562; number of weights and transform points, 128. (a) Adaptive line enhancing. (b) Spectral analysis.

the number used with adaptive line enhancing when the value of D_{ALE} and D_{DFT} was 2 but became 16 times greater when the value of D_{ALE} and D_{DFT} was 32.

With adaptive line enhancing one could freely trade N for τ_{mse}. Their product is all that is important. Ensemble averaging may not even be required, since τ_{mse} can be made large by making μ small (although this may cause one to go to "double precision" arithmetic). With spectral analysis, on the other hand, ensemble averaging cannot be avoided in most cases. The size of L may be limited by cost considerations, computer speed, or in the case where the signal is an imperfect or modulated sine wave by signal bandwidth. Large values of N are required when input signal-to-noise ratio is low, and values in the thousands are not uncommon.

The reason that adaptive methods may be superior to spectral methods in certain cases, especially those of low signal-to-noise ratio, can be stated as follows. Averaging within the digital Fourier transform itself provides coherent signal enhancement. Thus the detectability D_{DFT} of the signal is proportional to L. Since ensemble averaging is incoherent ("postdetection averaging"), however, the detectability D_{DFT} is proportional only to the square root of N. The adaptive process, on the other hand, provides coherent signal averaging, making the detectability D_{ALE} proportional to τ_{mse}. It is equally coherent in averaging the weight vector ensemble, making D_{ALE} proportional also to N.

An analytical comparison of the computational requirements of the two techniques has not yet been made, but it appears that the adaptive process will provide a simpler implementation when spectral analysis involves large values of L. The adaptive process has the advantage of being a smooth, steadily flowing process, whereas spectral analysis is performed with consecutive time segments of data.

The subject of signal detection by adaptive filtering is relatively new, and the analysis presented here should be regarded as preliminary. The formulas derived have been verified by simulation and experiment, but the concepts they describe have not been in existence long enough to provide an adequate perspective. It is hoped that this work can be extended in the future.

Acknowledgment

Many people have contributed support, assistance, and ideas to the work described in this paper. The authors especially wish to acknowledge the contributions of Prof. T. Kailath, Prof. M. Hellman, Dr. H. Garland, J. Treichler, and M. Larimore of Stanford University; Prof. L. Griffiths of the University of Colorado; Dr. D. Chabries and M. Ball of the Naval Undersea Center; Dr. O. Frost of Argo Systems, Inc.; Dr. M. Hoff of the Intel Corp.; and the students in two classes in the Department of Electrical Engineering at Stanford University: EE 280, Computer Applications Laboratory, and EE 373, Adaptive Systems. Special thanks are also due to R. Fraser of the Naval Undersea Center, who assisted in editing the paper; his efforts led to significant improvements in its organization and clarity.

References

[1] N. Wiener, *Extrapolation, Interpolation and Smoothing of Stationary Time Series, with Engineering Applications.* New York: Wiley, 1949.

[2] H. Bode and C. Shannon, "A simplified derivation of linear least squares smoothing and prediction theory," *Proc. IRE*, vol. 38, pp. 417-425, Apr. 1950.

[3] R. Kalman, "On the general theory of control," in *Proc. 1st IFAC Congress.* London: Butterworth, 1960.

[4] R. Kalman and R. Bucy, "New results in linear filtering and prediction theory," *Trans. ASME, ser. D, J. Basic Eng.*, vol. 83, pp. 95-107, Dec. 1961.

[5] T. Kailath, "A view of three decades of linear filtering theory," *IEEE Trans. Inform. Theory*, vol. IT-20, pp. 145-181, Mar. 1974.

[6] P. Howells, "Intermediate frequency side-lobe canceller," U.S. Patent 3 202 990, Aug. 24, 1965.

[7] B. Widrow and M. Hoff, Jr., "Adaptive switching circuits," in *IRE WESCON Conv. Rec.*, pt. 4, pp. 96-104, 1960.

[8] J. Koford and G. Groner, "The use of an adaptive threshold element to design a linear optimal pattern classifier," *IEEE Trans. Inform. Theory*, vol. IT-12, pp. 42-50, Jan. 1966.

[9] F. Rosenblatt, "The Perceptron: A perceiving and recognizing automaton, Project PARA," Cornell Aeronaut. Lab., Rep. 85-460-1, Jan. 1957.

[10] ——, *Principles of Neurodynamics: Perceptrons and the Theory of Brain Mechanisms.* Washington, D.C.: Spartan Books, 1961.

[11] N. Nilsson, *Learning Machines.* New York: McGraw-Hill, 1965.

[12] D. Gabor, W. P. L. Wilby, and R. Woodcock, "A universal nonlinear filter predictor and simulator which optimizes itself by a learning process," *Proc. Inst. Elec. Eng.*, vol. 108B, July 1960.

[13] R. Lucky, "Automatic equalization for digital communication," *Bell Syst. Tech. J.*, vol. 44, pp. 547-588, Apr. 1965.

[14] R. Lucky *et al.*, *Principles of Data Communication.* New York: McGraw-Hill, 1968.

[15] J. Kaunitz, "Adaptive filtering of broadband signals as applied to noise cancelling," Stanford Electronics Lab., Stanford Univ., Stanford, Calif., Rep. SU-SEL-72-038, Aug. 1972 (Ph.D. dissertation).

[16] M. Sondhi, "An adaptive echo canceller," *Bell Syst. Tech. J.*, vol. 46, pp. 497-511, Mar. 1967.

[17] J. Rosenberger and E. Thomas, "Performance of an adaptive echo canceller operating in a noisy, linear, time-invariant environment," *Bell Syst. Tech. J.*, vol. 50, pp. 785-813, Mar. 1971.

[18] R. Riegler and R. Compton, Jr., "An adaptive array for interference rejection," *Proc. IEEE*, vol. 61, pp. 748-758, June 1973.

[19] B. Widrow, P. Mantey, L. Griffiths, and B. Goode, "Adaptive antenna systems," *Proc. IEEE*, vol. 55, pp. 2143-2159, Dec. 1967.

[20] ——, "Adaptive filters," in *Aspects of Network and System Theory*, R. Kalman and N. DeClaris, Eds. New York: Holt, Rinehart, and Winston, 1971, pp. 563-587.

[21] J. Glover, "Adaptive noise cancelling of sinusoidal interferences," Ph.D. dissertation, Stanford Univ., Stanford, Calif., May 1975.

[22] J. C. Huhta and J. G. Webster, "60-Hz interference in electrocardiography," *IEEE Trans. Biomed. Eng.*, vol. BME-20, pp. 91-101, Mar. 1973.

[23] W. Adams and P. Moulder, "Anatomy of heart," in *Encycl. Britannica*, vol. 11, pp. 219-229, 1971.

[24] G. von Anrep and L. Arey, "Circulation of blood," in *Encycl. Britannica*, vol. 5, pp. 783-797, 1971.

[25] R. R. Lower, R. C. Stofer, and N. E. Shumway, "Homovital transplantation of the heart," *J. Thoracic and Cardiovascular Surgery*, vol. 41, p. 196, 1961.

[26] T. Buxton, I. Hsu, and R. Barter, "Fetal electrocardiography," *J.A.M.A.*, vol. 185, pp. 441-444, Aug. 10, 1963.

[27] J. Roche and E. Hon, "The fetal electrocardiogram," *Amer. J. Obst. and Gynecol.*, vol. 92, pp. 1149-1159, Aug. 15, 1965.

[28] S. Yeh, L. Betyar, and E. Hon, "Computer diagnosis of fetal heart rate patterns," *Amer. J. Obst. and Gynecol.*, vol. 114, pp. 890-897, Dec. 1, 1972.

[29] E. Hon and S. Lee, "Noise reduction in fetal electrocardiography," *Amer. J. Obst. and Gynecol.*, vol. 87, pp. 1087-1096, Dec. 15, 1963.

[30] J. Van Bemmel, "Detection of weak foetal electrocardiograms by autocorrelation and crosscorrelation of envelopes," *IEEE Trans. Biomed. Eng.*, vol. BME-15, pp. 17-23, Jan. 1968.

[31] J. R. Cox, Jr., and L. N. Medgyesi-Mitschang, "An algorithmic approach to signal estimation useful in fetal electrocardiography," *IEEE Trans. Biomed. Eng.*, vol. BME-16, pp. 215-219, July 1969.

[32] J. Van Bemmel, L. Peeters, and S. Hengeveld, "Influence of the maternal ECG on the abdominal fetal ECG complex," *Amer. J. Obst. and Gynecol.*, vol. 102, pp. 556-562, Oct. 15, 1968.

[33] W. Walden and S. Birnbaum, "Fetal electrocardiography with cancellation of maternal complexes," *Amer. J. Obst. and Gynecol.*, vol. 94, pp. 596-598, Feb. 15, 1966.

[34] J. Capon, R. J. Greenfield, and R. J. Kolker, "Multidimensional maximum likelihood processing of a large aperture seismic array," *Proc. IEEE*, vol. 55, pp. 192-211, Feb. 1967.

[35] S. P. Applebaum, "Adaptive arrays," Special Projects Lab., Syracuse Univ. Res. Corp., Rep. SPL 769.

[36] L. J. Griffiths, "A simple adaptive algorithm for real-time processing in antenna arrays," *Proc. IEEE*, vol. 57, pp. 1696-1704, Oct. 1969.

[37] O. L. Frost, III, "An algorithm for linearly constrained adaptive array processing," *Proc. IEEE*, vol. 60, pp. 926-935, Aug. 1972.

[38] K. Senne, "Adaptive linear discrete-time estimation," Stanford Electronics Lab., Stanford Univ., Rep. SEL-68-090, June 1968 (Ph.D. dissertation).

[39] T. Daniell, "Adaptive estimation with mutually correlated training samples," Stanford Electronics Lab., Stanford Univ., Rep. SEL-68-083, Aug. 1968 (Ph.D. dissertation).

[40] J. K. Kim and L. D. Davisson, "Adaptive linear estimation for stationary M-dependent processes," *IEEE Trans. Inform. Theory*, vol. IT-21, pp. 23-31, Jan. 1975.

[41] B. Widrow, "Adaptive filters 1: Fundamentals," Stanford Electronics Lab., Stanford Univ., Rep. SU-SEL-66-126, Dec. 1966.

[42] L. J. Griffiths, "Rapid measurement of instantaneous frequency," *IEEE Trans. Acoustics, Speech, and Signal Processing*, vol. ASSP-23, pp. 209-222, Apr. 1975.

[43] J. P. Burg, "Maximum entropy spectral analysis," presented at the 37th Annual Meeting, Soc. Exploration Geophysicists, Oklahoma City, Okla., 1967.

[44] P. M. Woodward, *Probability and Information Theory with Applications to Radar*, 2nd ed. London: Pergamon Press, 1964.

[45] M. I. Skolnik, *Introduction to Radar Systems.* New York: McGraw-Hill, 1962.

[46] P. Swerling, "Probability of detection for fluctuating targets," *IRE Trans. Inform. Theory*, vol. IT-6, pp. 269-308, Apr. 1960.

[47] J. I. Marcum, "A statistical theory of target detection by pulsed radar: Mathematical appendix," *IRE Trans. Inform. Theory*, vol. IT-6, pp. 145-267, Apr. 1960.

Echo Cancellation in Speech and Data Transmission

DAVID G. MESSERSCHMITT, FELLOW, IEEE

***Abstract*—This tutorial paper reviews echo cancellation techniques as applied to both voice and data transmission applications. The echo control problem in telephone voice transmission is described, and the several measures taken to counteract echo are outlined. The problem of full-duplex data transmission over two-wire lines is described, with application to both digital subscriber loop and voiceband data transmission. The unique characteristics of each of these applications, more advanced methods recently proposed, and approaches to implementation are described.**

I. Introduction

ECHO cancellation is an application of adaptive filtering technology to the control of echo in the telephone network. It has been studied in the research lab for two decades, and due to the advances in microelectronics technology is now finding widespread practical application. This tutorial paper describes the echo control problems addressed by echo cancellation in the context of both voice and data transmission, the echo cancellation algorithms which are used in these applications, and the implementation of echo cancellers.

In Section II, the application of echo cancellation to speech transmission is described, and in Section III applications in full-duplex data transmission are discussed. Section IV outlines the algorithms used for adaptation of echo cancellers in all these applications, and evaluates their performance. Finally, in Section V more recently proposed advanced techniques in echo cancellation are summarized.

II. Echo Cancellation in Speech Transmission

The telephone network generates echos at points internal to and near the end of a telephone connection. Echo cancellation is one means used to combat this echo for both speech and data transmission. The requirements for speech and data are quite different, so this section will concentrate on speech and Section III will deal with data. See [1] for another excellent reference on this topic.

A starting point on echo terminology is given In Fig. 1. In Fig. 1(a), a simplified telephone connection is shown.

Manuscript received September 20, 1983; revised October 20, 1983. This research was supported by Racal-Vadic, Advanced Micro Devices, Harris Semiconductor, National Semiconductor, Fairchild Semiconductor, Harris Semiconductor and Digital Telephone Systems, and National Semiconductor, with a matching grant from the University of California Microelectronics and Computer Research Opportunities Program.

The author is with the Department of Electrical Engineering and Computer Science, University of California, Berkeley, CA 94720.

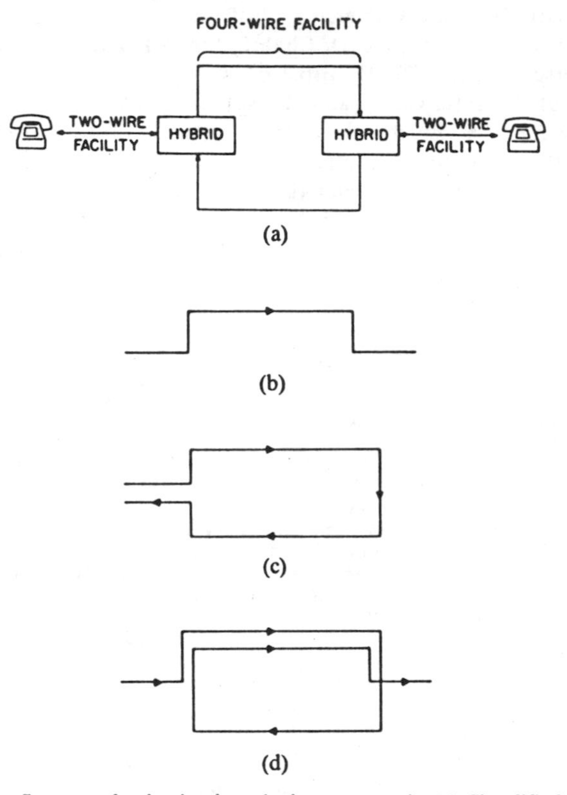

Fig. 1. Sources of echo in the telephone network. (a) Simplified telephone connection. (b) Talker speech path. (c) Talker echo. (d) Listener echo.

This connection is typical in that it contains two-wire segments on the ends (the subscriber loops and possibly some portion of the local network), in which both directions of transmission are carried on a single wire pair. The center of the connection is four wire, in which the two directions of transmission are segregated on physically different facilities (typical of carrier systems).

There is a potential feedback loop around the four-wire portion of the connection, and without sufficient loss in this path there is degradation of the transmission or in extreme cases oscillation (called singing). The hybrid is a device which provides a large loss around this loop without affecting the loss in the two talker speech paths. The remainder of Fig. 1 illustrates more graphically the function of this hybrid. One of the two talker speech paths is shown in Fig. 1(b). In order that this path not have a large attenuation, it is necessary for the hybrid not to have an appreciable attenuation between its two-wire and either

Reprinted from *IEEE J. Selected Areas in Comm.*, vol. SAC-2, no. 2, pp. 283–297, Mar. 1982.

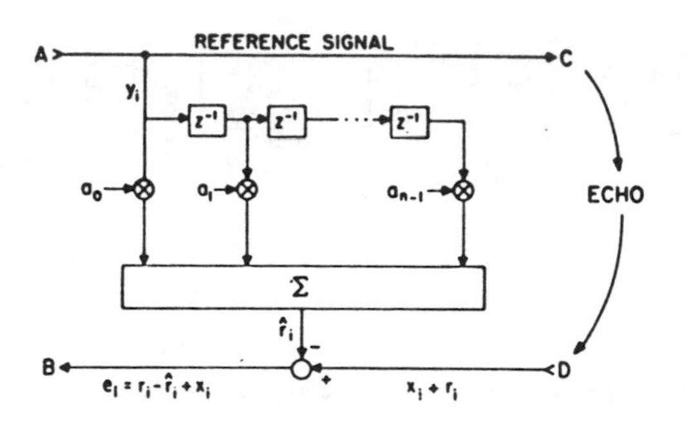

Fig. 2. Principle of the echo canceller for one direction of transmission.

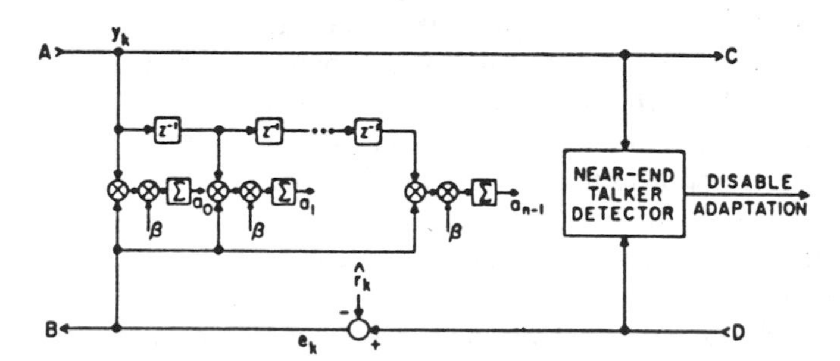

Fig. 3. Adaptation mechanism for the echo canceller.

four-wire port. There are two distinct echo mechanisms shown in Fig. 1(c) and (d), "Talker echo" results in the talker hearing a delayed version of his or her own speech, while in "listener echo" the listener hears a delayed version of the talker's speech.

On connections with small propagation delay, the subjective impairment from this echo can be controlled by inserting loss in the connection since the echo suffers this loss twice or three times whereas the talker path suffers this loss only once. For roundtrip delays over about 40 ms, it is necessary to use echo suppressors [1]. For very long delays, such as would be experienced on a satellite connection, the echo canceller is used [2].

The principle of the echo canceller for only one direction of transmission is shown in Fig. 2. The canceller is placed in the four-wire path near the origin of the echo. One direction of transmission is from port A to C (the "reference signal" y_i for the canceller), and the other direction from port D to B. The reference signal passing through the echo channel results in the echo signal r_i, which together with a near-end talker signal x_i appears on port D. Included in x_i is any thermal or quantization noise. A replica of the echo r_i is generated by applying the reference signal to a transversal filter (tapped delay line), and this replica is subtracted from the signal on port D to yield the cancellation error signal e_i (which also contains the near-end talker signal).

The $n-1$ filter coefficients $a_0, \cdots, a_{n-1}$ are caused to adapt to the echo transfer function (minimize the cancellation error) using the circuitry of Fig. 3. While there are more sophisticated adaptation algorithms, as discussed in Section V, this simple algorithm (described in Section IV-A) is adequate for most echo cancellation applications. The residual cancellation error e_k is used to drive the canceller adaptation by inferring from this error the appropriate correction to the transversal filter coefficients so as to reduce this error. Specifically, this error is cross correlated with the successive delays of the reference signal. The summation box is an accumulator, the output of which is the corresponding filter coefficient. The correction to the accumulated value is scaled by a step size β that controls the speed of adaptation and the asymptotic error in a manner discussed in Section IV-A.

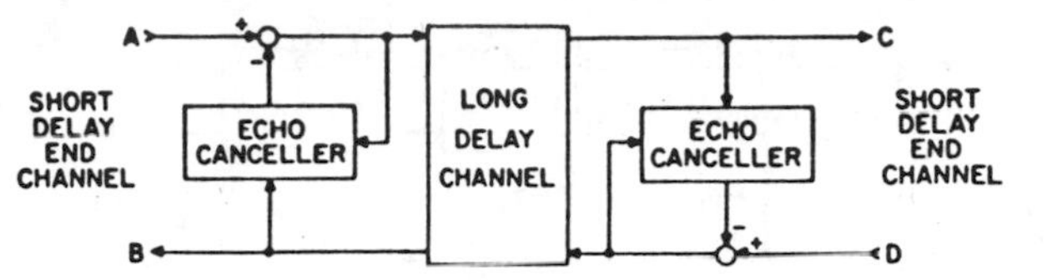

Fig. 4. Split echo canceller configuration for two directions.

The error signal e_k contains a component of the near-end talker signal x_k in addition to the residual echo cancellation error. The effect of this on the cancellation is naturally of some concern. As long as y_k is uncorrelated with x_k, which should be the case, this near-end talker will not affect the asymptotic mean value of the filter coefficients. However, the asymptotic variation in the filter coefficients about this mean will be increased substantially in the presence of a near-end talker due to the introduction of another (large) stochastic component in the adaptation. This effect can be compensated by choosing a very small speed of adaptation (small β), but it is generally preferable to employ a near-end talker detector as shown. The adaptation proceeds only when a there is no detectable near-end talker signal.

Of course, in practice it is desirable to cancel the echos in both directions of the connection. For this purpose two adaptive cancellers are necessary, as shown in Fig. 4, where one cancels the echo from each end of the connection. The near-end talker for one of the cancellers is the far-end talker for the other. In each case, the near-end talker is the "closest" talker, and the far-end talker is the talker generating the echo which is being cancelled. It is desirable to position these two "halves" of the canceller in a split configuration, as shown in the figure where the bulk of the delay in the four-wire portion of the connection is in the middle. The reason is that number of coefficients in the echo cancellation filter is directly related to the delay of the channel between where the echo canceller is located and the hybrid which generates the echo. In the split configuration the largest delay is not in the echo path of either half of the canceller, and hence the number of coefficients is minimized. Typically cancellers in this configuration require only 128 or 256 coefficients, whereas the number of coefficients required to accommodate a satellite connection in the end link would be impractically large.

Mathematically, the transversal filter of Fig. 2 is a direct realization of a finite impulse response (FIR)

$$r_k = \sum_{m=0}^{n-1} a_m y_{k-m} \tag{2.1}$$

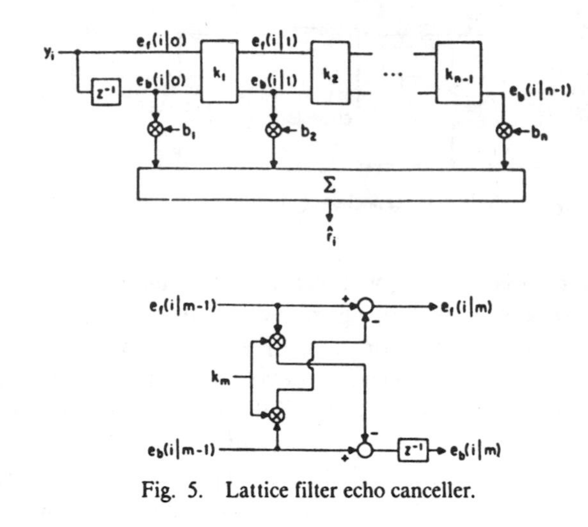

Fig. 5. Lattice filter echo canceller.

where there are n filter coefficients $a_0,\cdots,a_{n-1}$. Of course there are many filter realizations other than the transversal filter with the same response. One of these alternatives of particular importance is the lattice structure shown in Fig. 5 [3], [4]. The lattice filter structure consists first of all of a set of $n-1$ stages with internal coefficients k_j, $1 < j \leqslant n-1$, which are commonly called the *reflection* or *PARCOR* coefficients. The signals at the output of the mth stage, $e_f(1|m)$ and $e_b(i|m)$, are called, respectively, the forward and backward prediction error of order m. For purposes of realization of the echo canceller, the salient property of these prediction errors is that when the reflection coefficients are chosen appropriately, the successive backward prediction errors are uncorrelated. The so-called order update equations for the prediction errors are given by

$$e_f(i|m) = e_f(i|m-1) - k_m e_b(i|m-1) \tag{2.2}$$

$$e_f(i|0) = y_i \tag{2.3}$$

$$e_b(i|m) = e_b(i-1|m-1) - k_m e_f(i-1|m-1) \tag{2.4}$$

$$e_b(i|0) = y_{i-1}. \tag{2.5}$$

The echo replica is generated by forming a weighted linear combination of the backward prediction errors of successive orders with weights $b_1,\cdots,b_n$, given by

$$r_i = \sum_{m=1}^{n} b_m e_b(i|m-1). \tag{2.6}$$

One way to think of this structure is as a transversal filter where the delay elements have been replaced by the more complicated lattice stages. The purpose of the lattice stages are to decorrelate the input signal, so that the final replica is formed from a sum of uncorrelated signals. The reflection coefficients are therefore chosen irrespective of the desired echo replica on the basis of the reference signal input statistics alone. The transfer function of the overall filter is controlled by the choice of the $b_1,\cdots,b_n$.

The advantage of the lattice structure arises from its greater speed of adaptation, as discussed in Section V-B.

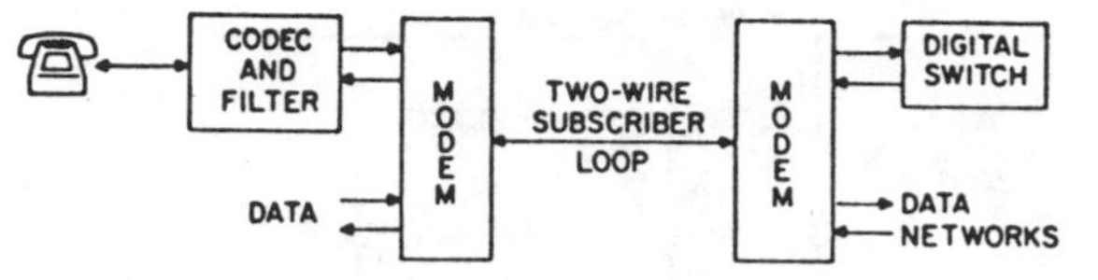

Fig. 6. Application of digital subscriber loop.

III. Echo Cancellation in Data Transmission

When it is attempted to transmit data signals through the network, the same echoes are encountered as in speech transmission. When half-duplex data transmission is used, echoes present no problem since there is no receiver on the transmitting end to be affected by the echo. In full-duplex transmission, where the data signals are transmitted in both directions simultaneously, echoes from the data signal transmitted in one direction can interfere with the data signal flowing in the opposite direction, unless these two data signals are in nonoverlapping frequency bands.

Full-duplex transmission over a common media has arisen in two important applications. In both these applications, the need for a common media arises because the public telephone network typically only provides a two-wire connection to each customer premise. (This is because of the high cost of copper wire, and the large percentage of the telephone network investment in this facility.)

The first application, Fig. 6, is digital transmission on the subscriber loop, in which the basic voice service as well as enhanced data services are provided over the two-wire subscriber loop. Total bit rates for this application that have been proposed are 80 and 144 kbits/s in each direction where the latter rate includes provision for two voice/data channels at 64 kbits/s each plus a data channel at 16 kbits/s, and the first alternative allows only a single 64 kbits/s voice/data channel. This digital subscriber loop capability is an important element of the emerging integrated services digital network (ISDN), in which integrated voice and data services will be provided to the customer over a common facility. As shown in Fig. 6, voice transmission requires a codec and filter to perform the analog-to-digital and digital-to-analog conversion on the customer premises, together with the modem for transmitting the full-duplex data stream over the two-wire subscriber loop. Any data signals to be accommodated are connected directly to the modem. The central office end of the loop has another full-duplex modem, with connections to the digital central office switch for voice or circuit-switched data transmission, and to data networks for packet switched data transport capability.

The second application for full-duplex data transmission is in voiceband data transmission where the basic customer interface to the network is often the same two-wire subscriber loop as shown in Fig. 7. In this case, however, the transmission link is usually more complicated, due to the possible presence of four-wire trunk facilities in the middle of the connection. The situation can be further complicated by the presence of two-wire toll switches, allowing inter-

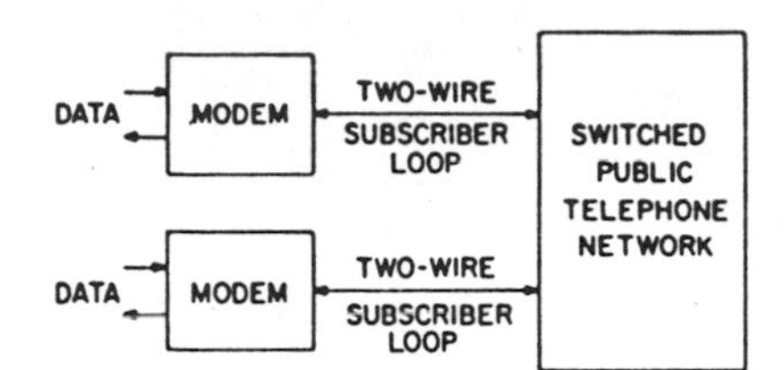

Fig. 7. Application of full-duplex voiceband data modems.

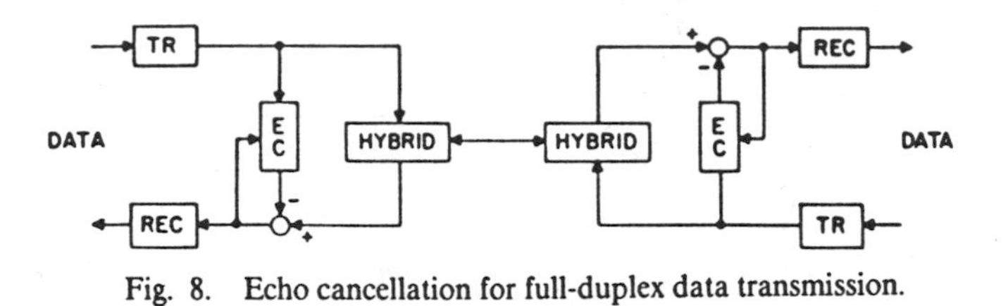

Fig. 8. Echo cancellation for full-duplex data transmission.

mediate four-two-four wire conversions internal to the network.

The two applications differ substantially in the types of problems which must be overcome. For the digital subscriber loop, the transmission medium is fairly ideal, consisting of cable pairs with a relatively wide bandwidth. The biggest complication is the presence in some countries of bridged taps—open-circuited wire pairs bridged onto the main line. The voiceband data modem, while requiring a lower speed of transmission, encounters many more impairments. In addition to the severe bandlimiting when carrier facilities are used, there are problems with noise, nonlinearities, and sometimes even frequency offset. Another difference is that the subscriber loop can use baseband transmission, while the voiceband data set always uses passband transmission; that is, modulates a carrier with the data stream.

One approach to full-duplex data transmission is the use of an echo canceller to isolate the two directions of transmission as shown in Fig. 8. There is a transmitter and receiver on each end of the connection, and a hybrid is used to provide a virtual four-wire connection between the transmitter on each end and the receiver on the opposite end. Unfortunately, there is some feedthrough the hybrid of the high-level signal from the transmitter into the local receiver. The hybrid depends on knowledge of the two-wire impedance for complete isolation, and in practice no better than 10 dB or so of attenuation through the hybrid can be guaranteed with a single compromise termination.

In both subscriber loop and voiceband data transmission, the loss of the channel from one transmitter to the receiver on the other end can be as high as 40–50 dB. This implies that in the worst case, the undesired transmitter local feedthrough (echo) signal can be 30–40 dB higher in level than the data signal from the far end (assuming both transmitted signals are at the same level). Since signal to interference ratios on the order of 20 dB or more are required for reliable data transmission, the echo canceller is added to each end to give an additional attenuation of the undesired feedthrough signal of about 50–60 dB.

A. Subscriber Loop Digital Transmission

There are two competing methods of providing full-duplex data capability on the subscriber loop, frequency-division multiplexing (FDM) and time-compression multiplexing (TCM) [5]. In spite of this competition, echo cancellation has achieved significant attention because of its ability to provide greater range (distance between subscriber and central office) in some circumstances. This greater range is due to the approximate halving of the transmitted signal bandwidth relative to the alternatives.

The implementation of a data echo canceller differs from the speech echo canceller discussed in Section II in several respects. One difference is that the reference signal for a data echo canceller is the sequence of transmitted data symbols. The fact that these symbols assume only a small number of values simplifies the implementation of the canceller [6]. Another difference is that signals in both directions are present at all times (except perhaps during a training period), making the effect of the near-end signal on adaptation much more significant an issue. A third difference is that the desired degree of cancellation is much larger, making design and particularly implementation much more challenging. Finally, there are significant issues of timing recovery, synchronization, and equalization which interact with the echo cancellation and do not exist in speech cancellation [7].

A particularly important constraint imposed by timing recovery considerations is that the sampling rate at the output of the echo canceller must usually be chosen to be a integral multiple of the data symbol rate. The data symbol rate is usually not adequate because the received data signal bandwidth exceeds half the data symbol rate. This implies that the echo canceller has different sampling rates at input and output since the input rate is equal to the symbol rate. To see how this is done, assume that the transmitted signal is

$$s(t)=\sum_m C_m g(t-mT) \tag{3.1}$$

where C_m is the sequence of transmitted data symbols, $g(t)$ is the transmitted pulse shape, and T is the interval between transmitted data symbols (the baud interval). Suppose this signal is passed through a filter with transfer function $H_e(w)$ representing the echo response. If we denote the response of this filter to $g(t)$ as $h(t)$ and the echo response as $r(t)$, then the latter becomes

$$r(t)=\sum_m C_m h(t-mT). \tag{3.2}$$

Assuming that this signal is to be reconstructed by the echo canceller with a sampling rate equal to RT, where R is an appropriately chosen integer (typically 2 or 4), define

$$r_i(l)=r\left(\left(i+\frac{l}{R}\right)T\right), \qquad 0\leqslant l\leqslant R-1 \tag{3.3}$$

where the index i represents the data symbol epoch and l represents the sample from among R samples uniformly

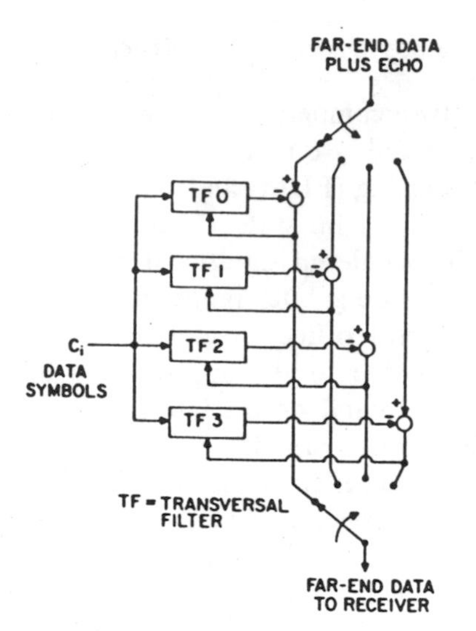

Fig. 9. Interleaved echo cancellers to achieve higher output sampling rate ($R = 4$).

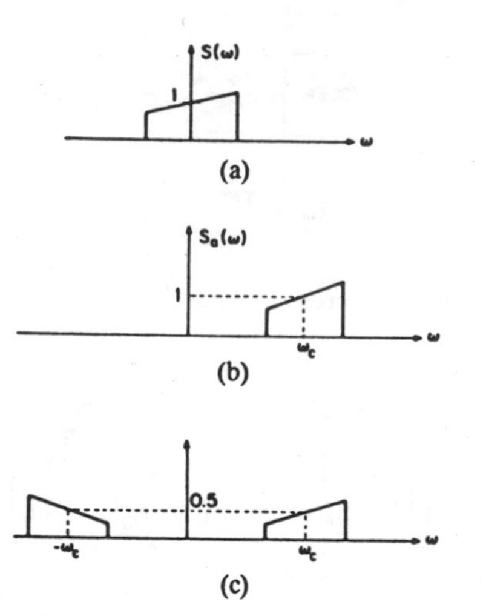

Fig. 10. Spectra of signals for quadrature amplitude modulation (QAM). (a) Baseband signal with complex-valued data symbols. (b) After modulation by complex exponential. (c) After taking real part.

spaced in this epoch. Similarly, define a notation for the samples of the echo pulse response

$$h_i(l) = h\left(\left(i + \frac{l}{R}\right)T\right), \qquad 0 \leqslant l \leqslant R - 1. \tag{3.4}$$

Using this convenient notation, from (3.2) the samples of the received echo become

$$r_i(l) = \sum_m h_m(l) C_{i-m}. \tag{3.5}$$

This relation shows that the samples of the echo can be thought of as R independent echo channels, each channel being driven by the same sequence of data symbols. The impulse response of the lth echo channel is $h_i(l)$.

Since the R echo channels are independent, the index l can be dropped. The transversal filter has a finite impulse response, so consider building an echo canceller to generate the replica

$$r_i = \sum_{m=0}^{n-1} C_{i-m} a_m \tag{3.6}$$

where a_i, $0 \leqslant i < n-1$, are the n filter coefficients of one of the R interleaved transversal filters. This transversal filter generates an FIR approximation to the echo response. Note the analogy to (2.1) where the far-end talker samples of the voice canceller have been replaced by the transmitted data symbols at the canceller reference input.

The R independent echo cancellers can be used in the manner of Fig. 9 (shown for $R = 4$). Each canceller has the same input data sequence at its reference input. The far-end data plus echo signal from the hybrid is decimated to four independent signals at the data symbol rate, and these are independently cancelled and recombined into a single sample stream representing the far-end data signal alone. Each of the cancellation error signals is used to drive the adaptation of the corresponding canceller.

Each canceller can be thought of as adapting to the impulse response of the echo channel sampled at a rate equal to the symbol rate, but with a particular phase out of R possible phases. These cancellers independently converge, although they do have in common the same input sequence of data symbols, and the presence of multiple interleaved cancellers does not affect the speed of adaptation.

B. Voiceband Data Transmission

The preceding section described the application of echo cancellation to full-duplex baseband data transmission (the digital subscriber loop); the present section extends this to the passband case typical of voiceband data transmission. Full-duplex voiceband data transmission has been provided for some years at 1.2 kbits/s and below (and more recently at 2.4 kbits/s) using FDM. However, at 4.8 kbits/s and above, frequency separation becomes impractical due to inadequate total bandwidth, and the use of echo cancellation must be considered.

The usual method of voiceband data transmission is quadrature amplitude modulation (QAM), which is briefly reviewed here. In a mathematical formulation of this passband modulation, it is convenient to use complex notation, so consider a baseband data signal identical to (3.1) except that the data symbols are complex

$$C_i = a_i + jb_i \tag{3.7}$$

and the transmitted pulse $g(t)$ is still real-valued. Since this signal is complex-valued, it has an asymmetrical spectrum as illustrated in Fig. 10(a). If this complex data signal is modulated up to passband by multiplying by a complex exponential with frequency ω_c, the carrier frequency, the spectrum is shifted by ω_c radians as illustrated in Fig. 10(b). If the time domain equivalent is called $s_a(t)$, this complex valued signal is given by

$$s_a(t) = \sum_m C_m g(t - mT) e^{j\omega_c t}. \tag{3.8}$$

Finally, to make the transmitted signal real-valued, simply

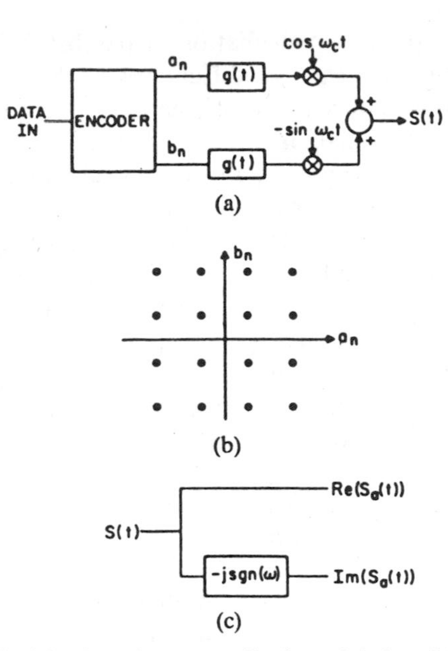

Fig. 11. Principle of quadrature amplitude modulation (QAM). (a) QAM modulator. (b) 16 point signal constellation. (c) Phase splitter in receiver.

take the real part of $s_a(t)$,

$$s(t) = \text{Re}(s_a(t)). \tag{3.9}$$

Since the real part can be written alternatively as

$$s(t) = \frac{s_s(t) + s_a^*(t)}{2} \tag{3.10}$$

and since the Fourier transform of $s_a^*(t)$ is $S_a^*(-\omega)$, the Fourier transform of $s(t)$ is

$$S(\omega) = \frac{S_a(\omega) + S_a^*(-\omega)}{2} \tag{3.11}$$

as illustrated in Fig. 10(c).

For purposes of implementation of the transmitter, it is convenient to expand in terms of (3.7), so that (3.9) becomes

$$s(t) = \sum_m a_m g(t - mT)\cos(\omega t) - \sum_m b_m g(t - mT)\sin(\omega t) \tag{3.12}$$

where the amplitude modulation of two independent data signals by quadrature carriers is evident. A method of generating this modulated signal is illustrated in Fig. 11(a). The input data stream is encoded to yield the a_k and b_k values. There are many ways of doing this, and Fig. 11(b) illustrates one method. In this case, the complex-valued data symbol C_k assumes one of 16 distinct values, with the real part a_k having four values and the imaginary part b_k having four values (each equally spaced). The encoder in this case maps four input bits into the complex-valued data symbol, two of those bits into the real part and two into the imaginary part. The remainder of the transmitter in Fig. 11(a) modulates a pulse $g(t)$ by the real and imaginary parts, modulates them up to passband by multiplying by quadrature carriers, and sums the results.

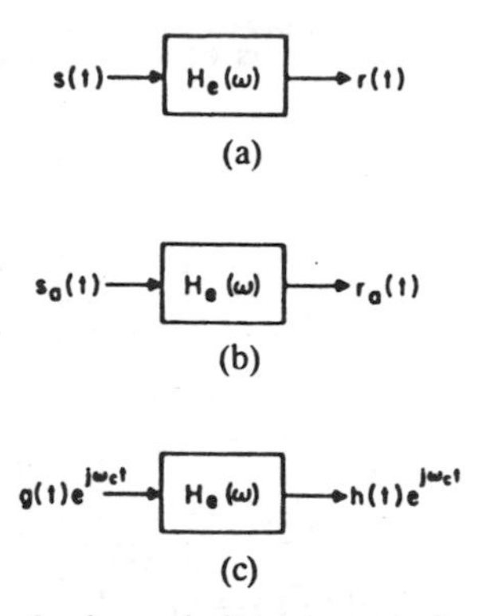

Fig. 12. Response of echo path filter to a single modulated pulse. (a) Transmitted signal and echo. (b) Analytic signals. (c) Isolated pulse response.

The time domain signal with only positive frequency components corresponding to Fig. 10(b) is called the "analytic signal." In the receiver processing, it is convenient to recover this complex valued signal prior to further processing. Given the passband QAM modulated signal, the analytic signal can be recovered in the manner illustrated in Fig. 11(c) utilizing a phase splitter. The imaginary part of the analytic signal is simply the output of a 90° phase shift network [transfer function-$j\text{sgn}(\omega)$], also called a Hilbert transform filter. To see this, note that the Fourier transform of the complex-valued signal generated in Fig. 11(c) is

$$S(\omega) + \text{sgn}(\omega)S(\omega) = \begin{cases} 2S(\omega) & x \geqslant 0 \\ 0 & x < 0 \end{cases} \tag{3.13}$$

and this signal has only positive frequency components.

In practice, the phase splitter consists of two filters which are all-pass (have unity magnitude response) and a phase difference of 90°. In addition, it would be common to implement the phase splitter in discrete time.

The passband echo canceller is similar to the baseband case except that a complex-valued data sequence and the complex-valued output of the phase splitter are input to the canceller rather than a real-valued sequence, and the filter coefficients are also complex-valued [8]. Consider the response of the echo transfer function [denoted by $H_e(\omega)$ as in the baseband case] to the transmitted data signal given by (3.9). In fact, since the analytic signal of this echo response will be generated by the phase splitter at the receiver input, what is really desired is the analytic signal of this echo signal. Examining Fig. 12, if the response of the echo filter to the transmitted signal is denoted $r(t)$ as shown in Fig. 12(a), and the corresponding analytic signal is $r_a(t)$, then as shown in Fig. 12(b) $r_a(t)$ is the response of the echo filter to the transmitted analytic signal $s_a(t)$. This is because the impulse response of the echo filter is real-valued. Determining the response to the analytic transmitted signal is simple if the response to a single isolated analytic pulse can be determined, as illustrated in Fig. 12(c). The output is expressed in terms of another analytic signal where $h(t)$ is the equivalent baseband pulse which returns through the echo path. Taking the Fourier transform of the input and output of the echo path filter in Fig. 12(c),

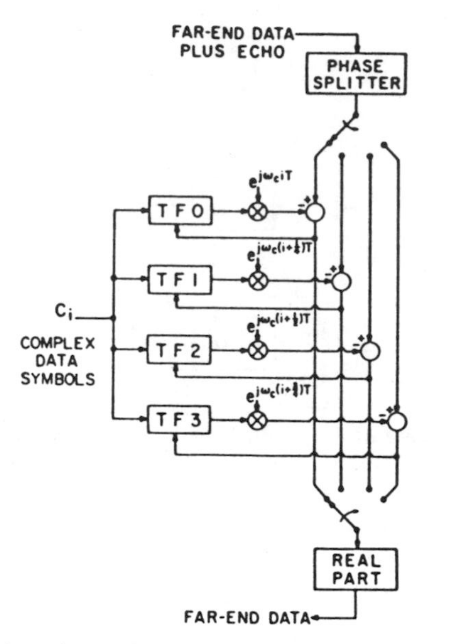

Fig. 13. Interleaved complex baseband echo cancellers in a passband data transmission system ($R = 4$).

$$G(\omega - \omega_c) H_e(\omega) = H(\omega - \omega_c) \tag{3.14}$$

or equivalently,

$$H(\omega) = G(\omega) H_e(\omega + \omega_c). \tag{3.15}$$

This demonstrates that the baseband output pulse can be obtained by shifting the echo transfer function in the vicinity of the carrier frequency down to dc, as might be expected. It also follows from (3.15) that $h(t)$ is complex-valued in general, even though $g(t)$ is real-valued.

The analytic echo signal follows from superposition as

$$r(t) = \sum_m C_m h(t - mT) e^{j\omega_c t} \tag{3.16}$$

very much in analogy to (3.2). As in the baseband case, the echo canceller must be implemented in discrete time, and in order to be able to reconstruct the echo signal the sampling rate must be an integer R times as high as the data symbol rate, where R is an appropriate integer on the order of 2 or 4. Defining $r_{a,i}(l)$ and $h_i(l)$ as in (3.3) and (3.4), a relation similar to (3.6) is obtained,

$$r_{a,i}(l) = \left(\sum_m C_m h_{i-m}(l) \right) e^{j\omega_c(i+(l/R))T}. \tag{3.17}$$

A transversal filter FIR approximation to this echo becomes

$$r_{a,i} = \left(\sum_{m=0}^{n-1} a_m C_{i-m} \right) e^{j\omega_c(i+(l/R))T} \tag{3.18}$$

where the l has again been suppressed and the filter coefficients a_m are complex-valued.

An illustration of this approach for $R = 4$ is shown in Fig. 13. The same complex data symbols C_i are applied to each of the transversal filters. These transversal filters are operating at baseband, have complex-valued filter coefficients, and the outputs are modulated up to the carrier frequency ω_c prior to cancellation at passband. The phase of the modulation carriers differs for each of the interleaved cancellers. The output of the hybrid, containing the undesired echo signal, is first converted to the analytic signal with a phase splitter, and then decimated by 4. The four interleaved echo cancellations are performed, and then the signals are recombined into a single canceller error signal containing the far-end data signal. The real value of this signal is taken to yield a passband representation of the far-end data signal. Typically, this signal would be applied to an interpolation filter and then resampled synchronously with the far-end data since it is desired that the data signals in the two directions be asynchronous.

IV. Design Considerations in Echo Cancellation

The design of an echo canceller involves many considerations, such as the speed of adaptation, the effect of near- and far-end signals, the impact of signal levels and spectra, and the impact of nonlinearity. Many of these considerations are outlined in this section [8]–[11].

A. Adaptation Algorithm

There are two important measures of performance for an echo canceller adaptation algorithm: the speed of adaptation and the accuracy of the cancellation after adaptation. Generally there is a tradeoff between these two measures: for a particular class of adaptation algorithm, as the speed of adaptation is increased the accuracy of the transfer function after adaptation becomes poorer. This tradeoff is fundamental since a longer averaging time is necessary to increase asymptotic accuracy, but slows the rate of convergence.

Most often the motivation for adapting an echo canceller is that the transfer function of the echo is not known in advance. It is also probable that the echo transfer function is changing with time, although in most cases the change will be quite slow (say in response to temperature changes of the transmission facilities). In most instances the accuracy of the final cancellation of the echo is a critical design factor. Although the ability of the canceller to rapidly track a changing echo response is usually not important, the speed of initial adaptation to the echo response from an arbitrary initial condition is often important. For a voice echo canceller, where the canceller is not dedicated to a particular subscriber, the adaptation must occur on each new call, and during the adaptation period the voice quality is degraded. For a data echo canceller, the adaptation is a part of the training or initialization period during which useful data transmission cannot occur, and there is, therefore, motivation to speed adaptation. However, speed of adaptation in echo cancellers is usually not so critical that the simple stochastic gradient algorithm cannot give adequate performance. Nevertheless, some approaches to speeding up the adaptation are discussed in Section V.

It is convenient to define a vector notation for the filter coefficients

$$a^T \equiv [a_0, a_1, \cdots, a_{n-1}], \tag{4.1}$$

where T denotes transpose, and the reference input samples,

$$y_i^T \equiv [y_i, y_{i-1}, \cdots, y_{i-n+1}]. \tag{4.2}$$

If the impulse response of the echo channel is h_i, $0 \leqslant k < \infty$, then

$$h^T = [h_0, h_1, \cdots, h_{n-1}] \tag{4.3}$$

is defined as a vector of the first n of these samples. The error signal for a voice echo canceller as shown in Fig. 2 can be written as

$$\begin{aligned} e_k &= \sum_{m=0}^{\infty} h_m y_{k-m} - \sum_{m=0}^{n-1} a_m y_{k-m} + x_k \\ &= (h - a) y_k + v_k \end{aligned} \tag{4.4}$$

where

$$v_k = \sum_{m=n}^{\infty} h_m y_{k-m} + x_k \tag{4.5}$$

is the residual uncancellable echo corresponding to echo delays which exceed the number of coefficients in the transversal filter, plus the noise and near-end talker signal. For the baseband data echo canceller, (4.4) is still valid where the reference data symbols C_{i-m} replace y_{i-m}. It should be noted that the two terms in (4.4) are correlated when the reference signal samples are correlated.

For purposes of analyzing a given adaptation algorithm, it is often necessary to assume that the reference signal y_i and near-end talker v_i are jointly wide-sense stationary. For this case, define

$$p = E[v_i y_i] \tag{4.6}$$

$$\Phi = E[y_i y_i^T] = \begin{bmatrix} R_0 & R_1 & \cdots & R_{n-1} \\ R_1 & & \ddots & \vdots \\ \vdots & & & \\ R_{n-1} & & \cdots & R_0 \end{bmatrix} \tag{4.7}$$

where

$$R_j = E[y_i y_{i+j}] \tag{4.8}$$

are the autocorrelation coefficients of the reference process. It is readily shown that the filter coefficient vector which minimizes the mean-square error $E[e_i^2]$ under this assumption is

$$a_{\text{opt}} = h + \Phi^{-1} p. \tag{4.9}$$

Note that in general the optimum filter coefficient vector is not equal to the first n samples of the echo impulse response. However, if we assume that the reference samples are mutually uncorrelated, and that x_k is uncorrelated with the reference samples, then we get

$$p = 0 \tag{4.10}$$

$$\Phi = R_0 I \tag{4.11}$$

where I is the identity matrix. In this case the optimum filter coefficient vector is equal to the echo impulse response

$$a_{\text{opt}} = h. \tag{4.12}$$

Another condition under which this is true is when the echo impulse response is truncated to n coefficients or less. In practice, n will be chosen sufficiently large for this latter condition to be nearly valid.

The most widely used practical algorithm for adaptation of an echo canceller is the stochastic gradient algorithm (sometimes known as the LMS algorithm). It is based on the following idea. If the ensemble statistics of the reference and near-end talker processes were known, the matrix inversion of (4.9) could be avoided by using an iterative gradient algorithm. In this approach, successive corrections to a are made proportional to the negative of the gradient of $E[e_i^2]$ with respect to a. Since this gradient is in the direction of maximum rate of increase of $E[e_i^2]$, subtracting this gradient from a should reduce the error. It can be shown that this algorithm does converge to the minimum of (4.9) as long as the constant multiplying the gradient is small enough.

An expression for the gradient is

$$\nabla_a\{E[e_i^2]\} = -2E[e_i y_i]. \tag{4.13}$$

The troublesome part of this expression is the expectation operator since the underlying statistics are not known in practice. The principle behind the stochastic gradient algorithm is to ignore the expectation operator. The quantity which is left, while it is random, has an expected value equal to the desired gradient. Thus, it is an unbiased estimate of the gradient. This "noisy" or "stochastic" gradient is substituted for the actual gradient, resulting in the so-called stochastic gradient algorithm

$$a_i = a_{i-1} - \frac{\beta}{2} \nabla_a [e_i^2] \tag{4.14}$$

or

$$\begin{aligned} a_i &= a_i + \beta e_i y_i \\ &= (I - \beta y_i y_i^T) a_i + \beta (y_i y_i^T h + v_i y_i). \end{aligned} \tag{4.15}$$

The first form of (4.15) is suitable for implementation since all the quantities are known, while the second is used for analysis. The constant β is used to control the convergence of the algorithm. In general, making β larger speeds convergence, while making β smaller reduces the asymptotic error. (It will be found later that there is an optimum β from the point of view of convergence rate.)

The nature of adaptation algorithm (4.5) was illustrated in Fig. 3, which is based on (4.15) rewritten for the jth component as

$$[a_i]_j = [a_{i-1}]_j + \beta e_i y_{i-j}. \tag{4.17}$$

The convergence properties of this algorithm will be analyzed in two steps in the following two subsections: First the average trajectory of the filter coefficient vector will be considered, and then the fluctuation of the trajectories about this average trajectory will be discussed. In both cases, wide-sense stationarity will be assumed. This is unrealistic for a speech canceller, but nevertheless useful for developing insight.

1) Average Trajectory of a Coefficient Vector: If the expectation of the filter coefficient vector in (4.15) is taken,

$$E[\boldsymbol{a}_i] = E\left[\left(\boldsymbol{I} - \beta \boldsymbol{y}_i \boldsymbol{y}_i^T\right)\boldsymbol{a}_{i-1}\right] + \beta E\left[\boldsymbol{y}_i \boldsymbol{y}_i^T \boldsymbol{h} + v_i \boldsymbol{y}_i\right] \approx (\boldsymbol{I} - \beta\Phi) E\boldsymbol{a}_{i-1} + \beta(\Phi\boldsymbol{h} + \boldsymbol{p}). \tag{4.18}$$

The key approximation that has been made is that the reference signal is uncorrelated with the filter coefficient vector. This approximation, which is necessary to make tractable the stochastic analysis of the gradient algorithm, is fortunately valid for small β because of the slow trajectory of $\boldsymbol{a}_i$ which results.

If $\boldsymbol{a}_{\text{opt}}$ from (4.9) is subtracted from both sides of (4.18) and if

$$\epsilon_i = E[\boldsymbol{a}_i] - \boldsymbol{h} - \Phi^{-1}\boldsymbol{p} \tag{4.19}$$

is defined as the error between the actual and optimal coefficient vector, then an iterative equation for the error results,

$$\epsilon_{i+1} = (\boldsymbol{I} - \beta\Phi)\epsilon_i. \tag{4.20}$$

Iterating this equation,

$$\epsilon_i = (\boldsymbol{I} - \beta\Phi)^i \epsilon_0. \tag{4.21}$$

In addressing the question of whether this error approaches zero, several important properties of the Φ matrix are critical. Specifically, it is symmetric, Toeplitz, and positive definite. The latter implies that it has positive real eigenvalues and is invertible. This inverse Φ^{-1} is also a symmetric matrix. Φ can be written in the form

$$\Phi = V\Lambda V^T \tag{4.22}$$

where Λ is a diagonal matrix of eigenvalues of Φ

$$\Lambda = \text{diag}[\lambda_1, \cdots, \lambda_n] \tag{4.23}$$

and V is an orthonormal matrix,

$$VV^T = I \tag{4.24}$$

whose jth column is the eigenvector of Φ associated with the jth eigenvalue. Since Φ is assumed to be positive definite, the eigenvalues are positive real-valued.

The matrix in (4.21) can be written in the form

$$(\boldsymbol{I} - \beta\Phi)^i = (VV^T - \beta V\Lambda V^T)^i = (V(I - \beta\Lambda)V^T)^i = V(I - \beta\Lambda)^i V^T. \tag{4.25}$$

Thus, the vector ϵ_i obeys a trajectory which is the sum of n modes, the jth of which is proportional to $(1 - \beta\lambda_j)^i$. Assume for simplicity also that all the eigenvalues are distinct, and order them for smallest λ_{min} to largest λ_{max}. It follows that $\epsilon(j)$ decays exponentially to zero as long as

$$0 < \beta < \frac{2}{\lambda_{\text{max}}} \tag{4.26}$$

which is the condition for each of the factors to be less than unity. Convergence of the average trajectory is thus assured if β is sufficiently small. Conversely, an excessively large value for β leads to instability in the form of an exponentially growing error.

For a fixed β, the speed of convergence of the algorithm is determined by the smallest eigenvalue. Specifically, the slowest converging mode is proportional to $(1 - \beta\lambda_{\text{min}})^i$. If β is chosen to be near its maximum value, say

$$\beta = \frac{1}{\lambda_{\text{max}}} \tag{4.27}$$

then the slowest converging mode is proportional to

$$\left(1 - \frac{\lambda_{\text{min}}}{\lambda_{\text{max}}}\right)^i. \tag{4.28}$$

The ratio of largest to smallest eigenvalue is thus seen to be of fundamental importance; it is called the *eigenvalue spread*. The eigenvalue spread has a minimum value of unity, and can be arbitrarily large. The larger the eigenvalue spread of the autocorrelation matrix, the slower the convergence of the filter coefficients.

It is instructive to relate the eigenvalue spread to the power spectral density of the reference random process. It is a classical result of Toeplitz form theory that

$$\min_\omega S(\omega) < \lambda_j < \max_\omega S(\omega) \tag{4.29}$$

where $S(\omega)$ is the power spectral density. While the eigenvalues depend on the order of the matrix n as $n \to \infty$ the maximum eigenvalue

$$\lambda_{\text{max}} \to \max_\omega S(\omega) \tag{4.30}$$

and the minimum eigenvalue

$$\lambda_{\text{min}} \to \min_\omega S(\omega). \tag{4.31}$$

It follows that spectra of the reference signal which result in slow convergence of the average trajectory are those for which the ratio of the maximum to minimum spectrum is large, and spectra which are almost flat (have an eigenvalue spread near unity) result in fast convergence.

Intuitively, the reason for slow convergence of the average trajectory for some spectra is the interaction among the adaptation of the different coefficients. As the successive reference samples become more correlated, there is more undesirable interaction among the coefficient adaptations.

Since the modes of convergence of the stochastic gradient algorithm are all of the form of γ^i where γ is a positive real number less than unity and i is the iteration number, the error in decibels can be determined by taking the logarithm of the square,

$$10\log_{10}(\gamma^{2i}) = (10\log_{10}(\gamma^2))i \tag{4.32}$$

and thus the error expressed in decibels decreases linearly with iteration number (the constant factors multiplying these exponentially decaying terms give a constant factor in decibels). The convergence of a gradient algorithm is thus often expressed in units of dB/s., which is the number of decibels of decrease in the error power per second.

Unfortunately, the convergence of the mean filter coefficient vector does not mean that any particular coefficient vector trajectory itself converges to the optimum, but only that the average of all trajectories converges to the optimum. In fact, the coefficient vector does not converge to the optimum. To see this, observe that even after convergence of the coefficient vector in the mean-value sense, difference equation (4.15) still has a stochastic driving term, and therefore the coefficient vector continues to fluctuate about the optimum coefficient vector randomly. The larger the value of the step size β, the larger this fluctuation.

It is therefore appropriate to study the fluctuation of the filter coefficient vector about its mean after convergence. This is done in the next section.

2) Fluctuation of Trajectories About the Average: The last section studied the convergence of the average filter coefficient trajectory; this section studies the fluctuation of a given trajectory about this average. To greatly simplify this analysis it will be assumed that the reference signal samples are uncorrelated (the reference signal is white). This assumption is not valid for speech signals, but the analysis nevertheless gives useful insight. This assumption is usually valid for a data echo canceller.

Calculating the norm of the difference between the filter coefficient vector and the optimum vector (which is $\boldsymbol{h}$ for this case), taking the expectation, and assuming that v_i is zero-mean and uncorrelated with y_i, and that the reference signal and the filter coefficient vector are uncorrelated,

$$E|\boldsymbol{a}_i - \boldsymbol{h}|^2 = E\left[(\boldsymbol{a}_{i-1} - \boldsymbol{h})^T E(\boldsymbol{I} - \beta y_i y_i^T)^2 (\boldsymbol{a}_{i-1} - \boldsymbol{h})\right] + \beta^2 \sigma_v^2 n R_0. \tag{4.33}$$

The final quantity to calculate,

$$E(\boldsymbol{I} - \beta y_i y_i^T)^2 = \boldsymbol{I} - 2\beta R_0 \boldsymbol{I} + \beta^2 E\left[(y_i^T y_i)(y_i y_i^T)\right] \tag{4.34}$$

depends on fourth-order statistics of the reference signal. This fourth moment can be evaluated simply for two cases: the reference signal white and Gaussian, and the reference signal a binary-valued independent identically distributed sequence of data symbols. For each case, the answer is

$$E\left[(y_i^T y_i)(y_i y_i^T)\right] = nR_0^2 \boldsymbol{I} \tag{4.35}$$

(for the Gaussian case n is replaced by $n+2$, which is nearly the same for large n). Substituting (4.35) back into (4.33),

$$E|\boldsymbol{a}_i - \boldsymbol{h}|^2 = (1 - 2\beta R_0 + \beta^2 n R_0^2) E|\boldsymbol{a}_{i-1} - \boldsymbol{h}|^2 + \beta^2 n R_0 \sigma_v^2. \tag{4.36}$$

This simple recursion shows that for small β the mean-square filter coefficient error approaches an asymptotic value of

$$E|\boldsymbol{a}_i - \boldsymbol{h}|^2 \to \frac{\beta n}{2 - \beta n R_0} \sigma_v^2 \tag{4.37}$$

or the rms error per coefficient is

$$\frac{E|\boldsymbol{a}_i - \boldsymbol{h}|^2}{n} \approx \frac{\beta \sigma_v^2}{2}. \tag{4.38}$$

This illustrates that the filter coefficient vector can be made more accurate asymptotically by reducing β, which also slows adaptation, or by keeping σ_v^2 small during periods of adaptation. The latter is accomplished for a voice canceller by disabling adaptation during periods of significant near-end talker energy.

The error approaches the asymptote of (4.37) as

$$(1 - 2\beta R_0 + \beta^2 n R_0^2)^i \tag{4.39}$$

and hence only approaches this asymptote if the quantity in parentheses is less than unity in magnitude. This in turn requires that

$$0 < \beta < \frac{2}{nR_0}. \tag{4.40}$$

It is instructive to define the time constant of the convergence of the mean-square filter coefficient vector error as the τ such that

$$(1 - 2\beta R_0 + \beta^2 n R_0^2)^\tau = \frac{1}{e}. \tag{4.41}$$

When β is very small, this can be solved for τ as

$$\tau \approx \frac{1}{2\beta R_0} \tag{4.42}$$

iterations.

In comparing (4.40) to (4.26), note that for the simpler white reference signal case considered here, all the eigenvalues of Φ are equal to R_0. Hence, (4.40) is considerably more stringent than (4.26), particularly for large n. Further, while (4.25) implies that β should be chosen as large as possible for fastest convergence (subject to stability constraints), (4.39) has an optimum β for fastest convergence given by

$$\beta = \frac{1}{nR_0} \tag{4.43}$$

(half the largest β), with a resultant convergence as

$$\left(1-\frac{1}{n}\right)^i \tag{4.44}$$

or a time constant of

$$\tau \approx \frac{n}{2}. \tag{4.45}$$

The preceding gives the mean-square error in the filter coefficient. This quantity can also be related to the mean-square echo residual. The latter is given by

$$E\left[e_i^2\right] = R_0 E|\boldsymbol{a}_i - \boldsymbol{h}|^2 + \sigma_v^2 \tag{4.46}$$

assuming again that $\boldsymbol{p} = 0$ and $\Phi = R_0 \boldsymbol{I}$. The second term is the minimum mean-square error which would be possible if the correct constant filter coefficient vector was used. The first term therefore represents an "excess" mean-square error which is due to the adaptation algorithm. When the step size of (4.43) which results in maximum convergence rate is used, (4.46) becomes

$$E\left[e_i^2\right] = 2\sigma_v^2 \tag{4.47}$$

and the excess mean-square error is equal to the variance of the uncancellable signal.

Several properties of this convergence should be noted. First, the β which should be chosen to keep the excess mean-square error small is considerably smaller than would be predicted by (4.26), which relates only to the convergence of the mean-value of the filter coefficient vector. Second, the fastest convergence of the mean-square error depends on the number of coefficients in the echo canceller, as the number of coefficients n increases, the fastest convergence slows down. Finally, the optimum step size for fastest convergence in (4.43) is inversely proportional to the reference signal power, R_0. Thus, to properly choose β, R_0 should be known, and if R_0 is actually varying slowly with time (as in speech cancellers), best performance requires that β track this variation. This is considered in Section IV-C.

B. Deterministic Theory of Convergence

There are several shortcomings of the convergence analysis just performed. The assumption of wide-sense stationarity is invalid for speech, and the approximation of uncorrelated coefficient vector and reference signal is of doubtful validity as β gets large.

There is also a fundamental question of what happens when the input signal does not cover the whole band up to half the sampling rate (an extreme case would occur when the reference signal is a sinusoid). Then from (4.31) the minimum eigenvalue approaches zero and the eigenvalue spread approaches infinity. Does this imply that the average coefficient trajectory does not approach the echo impulse response? Indeed it does since there are many coefficient vectors which will result in complete cancellation under this condition, each of these vectors having a different transfer function in the frequency band where there is no reference signal power.

There is an important deterministic theory of canceller adaptation which is able to rigorously derive upper and lower bounds on coefficient vector error for a given input reference signal given waveform [1], [12]. This theory gives qualitatively similar results to the stochastic analysis previously described, but it is very comforting to be able to state conditions under which convergence is guaranteed. In particular, this theory assumes a "mixing condition" to be satisfied in order for convergence of the coefficient vector to the region of the actual echo impulse response. This mixing condition is analogous to the reference signal having power over the entire band up to half the sampling rate.

C. Modifications to the Adaptation Algorithm

There are several useful modifications which can be made to the adaptation algorithm to improve performance or simplify implementation which are described in this section.

A useful modification which is often made is to replace the error signal e_i by the sign of this signal $\mathrm{sgn}(e_i)$. The motivation for doing this is simplification of the hardware: the adaptation equation of (4.17) becomes

$$[\boldsymbol{a}_i]_j = [\boldsymbol{a}_{i-1}]_j + \beta \, \mathrm{sgn}(e_i) y_{i-j} \tag{4.48}$$

where the only multiplication is by β or $-\beta$. This limited multiplication is particularly simple to implement if β is chosen to be a power of 2^{-1} in a digital implementation.

One would expect intuitively that the value of the cross correlation would not be affected in a major way by using the sign of the error. It has been shown [13] that the convergence of the algorithm is not compromised by this simplification, but that the speed of adaptation is reduced somewhat. In fact, it is shown more generally that any nonlinearity in the multiplication adversely affects convergence to some extent.

Another common modification to the stochastic gradient algorithm is to normalize the step size to eliminate an undesirable dependence of speed of convergence on input signal power. This dependence can be seen from (4.42) where the time constant of adaptation is inversely proportional to R_0. Further, if β is kept constant and the signal power is increased, eventually from (4.40) the algorithm becomes unstable. This is particularly a problem in a speech canceller where the input signal power varies considerably. From (4.43), it is desirable to normalize β by an estimate of the reference signal power, as in

$$\beta_i = \frac{a}{\sigma_i^2 + b} \tag{4.49}$$

where β_i is the step size at sample i, a, and b are some appropriately chosen constants, and σ_i^2 is an estimate of the reference signal power at time i. The purpose of the b in the denominator is to prevent β_i from getting too large (causing instability) when the input signal power gets very small, and is obviated if adaptation is disabled under this condition.

Another modification to the stochastic gradient algorithm is required for the adaptation of a passband canceller [8]. For this case, an algorithm analogous to (4.15) is

$$a_i = a_{i-1} + \beta e_i C_i^* e^{-j\omega_c(i+(l/R))T} \tag{4.50}$$

where e_i is the analytic cancellation error and C_i^* is the conjugate of a vector of complex data symbols.

V. Recent Advances in Echo Cancellation

The major portion of recent work in echo cancellation has focused on three areas: the extension to nonlinear echo models, speedup of the adaptation, and implementation. This section reviews this recent activity.

A. Nonlinear Echo Cancellation

The basic echo cancellation model presented earlier was capable, within the constraints of a finite impulse response, of exactly cancelling only an echo which is a linear function of the reference signal. Some older work in speech cancellers and some more recent work in data cancellers has extended the adaptive echo canceller to nonlinear echo generation phenomena [14], [15]. This extension is not critical to speech echo cancellers because the actual echo generation mechanisms, while not necessarily precisely linear, are close enough to linear that a linear canceller is able to meet the cancellation objectives. In data transmission, on the other hand, the objectives for degree of cancellation are sufficiently ambitious that nonlinear echo generation phenomena are of importance. The primary sources of nonlinearity in data cancellers are the data converters in the implementation of the canceller itself, as well as transmitted pulse asymmetry and saturation of transformer magnetic media. While these mechanisms result in a very small degree of nonlinearity, they are nevertheless of importance when the objective is 50–60 dB of cancellation.

The method of extending echo cancellation to nonlinear echo mechanisms that has been proposed is the use of the Volterra expansion. This expansion is capable of representing any echo mechanism which is time invariant. That is, the nonlinearity can have memory, such as in the hysteresis in a magnetic medium, but the nature of the nonlinearity cannot change with time. Of course, since the canceller is adaptive, the nonlinearity can change slowly in time, as long as the adaptation mechanism can keep up. For a data canceller, the number of coefficients required for a Volterra expansion is finite even for perfect cancellation because of the finite number of possible transmitted signals [15]. This assumes as in the linear case that the echo can be represented as a (nonlinear) function of a finite number of past transmitted data symbols.

B. Speedup of Adaptation

Where faster adaptation than is afforded by the stochastic gradient algorithm is desired, there are several techniques for speeding adaptation.

In almost all echo cancellation applications, high speed of convergence is desirable for initial acquisition, but is not necessary for subsequent tracking since the echo transfer function changes very slowly. Thus, a large constant of adaptation β is desirable during acquisition, but is not necessary following reasonable convergence. A natural algorithm to consider is to use a relatively large β initially, and then "gear-shift" to a smaller β after convergence is achieved.

A more radical means for increasing adaptation rate is to use a lattice filter in place of the transversal filter. As was detailed in Section IV-A-2), the transversal filter algorithm convergence suffers when the reference samples are highly correlated. In effect the lattice filter does an adaptive prefiltering of the reference signal prior to the adaptation to the echo transfer function. This prefilter serves the purpose of whitening the reference signal, thereby speeding the convergence of the canceller.

The extension of the stochastic gradient algorithm to the lattice filter is straightforward. The simplest, but not necessarily best performing, approach will be described here (see [16], [17] for more details). Referring to (2.6), if the derivative of e_i^2 with respect to the coefficient b_m is taken, a stochastic gradient algorithm becomes

$$[b_i]_m = [b_{i-1}]_m + \beta e_i e_b(i|m-1) \tag{5.1}$$

which is identical to the transversal filter algorithm of (4.17) except for the substitution of $e_b(i|m-1)$ for the reference signal y_{i-m}. The adaptation of the reflection coefficients can proceed by taking the derivative of $e_f^2(i|m)$ with respect to k_m, yielding the stochastic gradient algorithm

$$k_m(i) = \left(1 - \beta e_b^2(i|m-1)\right) k_m(i-1) + \beta e_f(i|m-1) e_b(i|m). \tag{5.2}$$

Note that the adaptation of each lattice stage is based on the input signals to that stage. Therefore, the algorithm converges sequentially by stage: the adaptation of the first stage is necessary before the adaptation of the second state, etc. Since each stage requires only a scalar adaptation, the speed of adaptation is essentially independent of eigenvalue spread. The adaptation serves to minimize the forward prediction error, which it turns out is equivalent to making the successive orders of backward prediction error uncorrelated. The latter in turn speeds the adaptation of the b_m coefficients since it eliminates the interaction among those coefficient adaptations. Experience has shown that for input signals with a large eigenvalue spread, the overall adaptation of the lattice filter is considerably faster than the adaptation of the transversal filter.

For the data canceller, the data symbols are usually approximately uncorrelated. This is particularly true in the presence of scrambling, which "breaks up" probable highly correlated patterns such as marking sequences. While the primary purpose of scrambling is to aid timing recovery, it is also of benefit to canceller adaptation. Thus, for data cancellers, there is little benefit to using the lattice filter.

For speech cancellers, on the other hand, the signal samples are usually highly correlated. Unfortunately, however, the exact nature of the correlation changes fairly rapidly with time. The reflection coefficients of the lattice filter therefore have to continually readapt to the spectrum of the reference speech signal. As the reflection coefficients adapt, the b_m coefficients also have to readapt to maintain the same overall transfer function. This illustrates a disadvantage of the lattice filter for echo cancellation applications: while the transversal filter coefficients are largely independent of the reference signal statistics, and depend mostly on the impulse response of the echo channel, the lattice filter coefficients depend strongly on the reference signal statistics and must continually readapt as those statistics change.

In cases where the lattice filter is of little benefit or even faster convergence is desired, another class of least-square (LS) adaptation algorithms closely related to the Kalman filter in control theory can be used. These algorithms are based on the minimization of the least-square cancellation error over the choice of adaptive filter parameters, with a weighting function decreasing exponentially into the past to give the algorithm finite memory. A simple example is appropriate for illustration. Define a squared error at time i as

$$\epsilon_i = \sum_{j=0}^{i} \alpha^j \left(r_{i-j} - \boldsymbol{a}_i^T y_{i-j} \right)^2. \tag{5.3}$$

Note that the filter coefficient vector at iteration i, $\boldsymbol{a}_i$, is used in forming the error for the past, and the error is minimized by the choice of this vector. The adaptive nature of the solution is enabled by the exponentially decreasing weighting into the past, resulting in greater weight for more recent reference samples. Defining

$$\begin{aligned} \Phi_i &= \sum_{j=0}^{i} \alpha^j y_{i-j} y_{i-j}^T \\ &= \alpha \Phi_{i-1} + y_i y_i^T \end{aligned} \tag{5.4}$$

which is an estimate of the reference signal correlation matrix (4.7), and also defining the cancellation error

$$e_i = r_i - \boldsymbol{a}_{i-1}^T y_i \tag{5.5}$$

then it is readily shown that the vector $\boldsymbol{a}_i$ which minimizes (5.3) can be defined iteratively as

$$\boldsymbol{a}_i = \boldsymbol{a}_{i-1} + \Phi_i^{-1} e_i y_i. \tag{5.6}$$

Comparing this to the stochastic gradient algorithm of (4.15), the only difference is the replacement of β with Φ_i^{-1}. This replacement, which substantially increases the complexity of the algorithm, also speeds up adaptation substantially for reference signals with a large eigenvalue spread. This matrix adaptation constant can be thought of again as uncoupling the adaptation of the individual components of the filter coefficient vector.

There are many versions of the LS algorithms, including those based on both a transversal filter [18] and lattice filter realization [19]. Neglecting finite precision effects, the performance is identical since the quantity being minimized in each case is identical. There are also versions of the LS algorithms which have much reduced computational requirements relative to the simple algorithm just displayed [19].

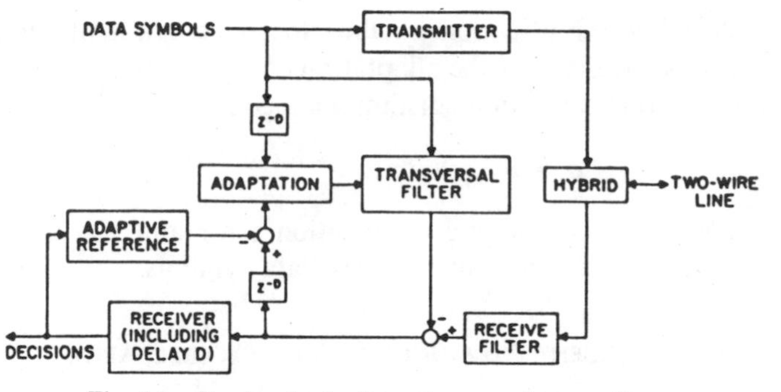

Fig. 14. Baseband adaptive reference echo canceller.

The LS algorithms are particularly effective in speeding adaptation for the data canceller case where the data symbols can be chosen during a training period to assist in the canceller adaptation. In this case, the reference signal algebraic properties become much more important than the stochastic properties, and it has been shown that the mean values of the filter coefficients of a canceller based on least squares can adapt in n data symbols for an n-coefficient canceller [20]. Furthermore, it has been shown that the least-squares algorithm can be virtually as simple as the stochastic gradient algorithm for a reference signal which is chosen to be a pseudorandom sequence [21]. This sequence is also particularly simple to generate.

In data echo cancellation, a significant factor slowing adaptation is the far-end data signal. This suggests yet another way of speeding adaptation: the data signal can be adaptively removed from the cancellation error in a decision-directed fashion [22] in an approach called an "adaptive reference" canceller. This is illustrated in Fig. 14 where the adaptive reference is attached to the output of the receiver. It forms a linear combination of the current and past receiver decisions to form a replica of the far-end data signal appearing in the cancellation error signal. This replica is subtracted to yield a new error signal which drives the echo canceller adaptation algorithm. Because the far-end data signal has been removed from the error, σ_v^2 in (4.37) is made smaller and the adaptation constant β can be chosen to be larger for a given excess mean-square cancellation error. This in turn speeds adaptation.

Because the receiver decision circuit typically includes delay, a compensating delay is inserted in the other inputs to the canceller adaptation algorithm. In effect the entire adaptation operates on a delayed basis.

Of course there is a problem getting started, when the cancellation may not be adequate to support a reasonable receiver error rate. This problem is solved by either starting out with a known training sequence (not requiring receiver decisions), or by disabling the adaptive reference and using

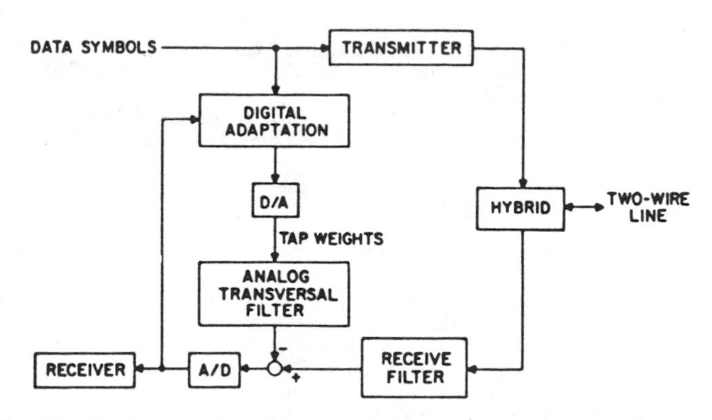

Fig. 15. Implementation of baseband data echo canceller insensitive to data converter nonlinearity.

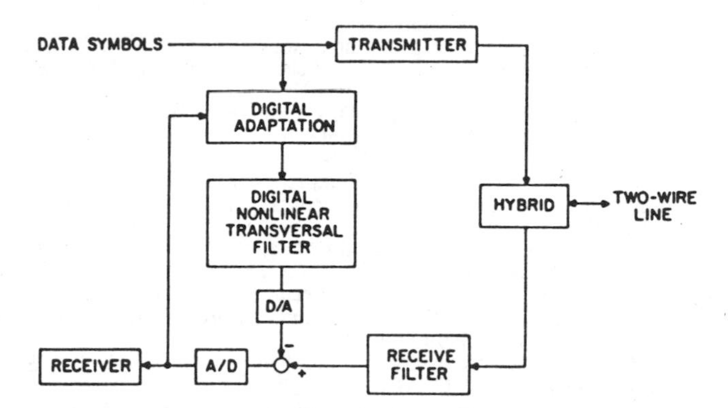

Fig. 16. Baseband data nonlinear echo canceller.

a larger β initially until the echo is cancelled sufficiently for a reasonable error rate.

C. Implementation of Echo Cancellation

The widespread application of echo cancellation has recently been stimulated by advances in microelectronics, making the computational requirements of the canceller within the reach of an inexpensive single-ship implementation. Because the potential applications of echo cancellation are numerous, there has been considerable activity in the design of devices for echo cancellation.

The first special-purpose chip for speech cancellation was developed by Bell Laboratories [23]. This device implemented a single 128 coefficient canceller using digital techniques, with an interface to the standard μ-255 format widely used in the digital transmission network. Cleverly, advantage was taken of the nearly logarithmic nature of this companding law to simplify the multiplications required in the implementation of the transversal filter. A later version of this device added the capability to cascade chips to achieve up to four times the number of coefficients [24].

The implementation of a data echo canceller in monolithic form represents special challenges [7], [10]. The major difficulty is in achieving the required accuracy without incorporating expensive off-chip precision components or trimming during manufacture. The source of the problem is that the adaptation must be implemented digitally to achieve the long time constants required for high asymptotic accuracy, but the interface to the transmission medium is inherently analog. This implies the need for high-speed analog-to-digital conversion somewhere between the medium and the adaptation, and this conversion must have an accuracy on the order of 12 bits with perfect linearity. While this accuracy can be achieved with trimming, without trimming special measures must be taken to overcome the nonlinearity of the data converters.

Two solutions have been proposed to this problem. One solution [7] is shown in Fig. 15, in which the transversal filter is implemented in analog circuitry where good linearity can be insured. The adaptation circuitry is digital, ensuring that the long time constants which are necessary can be achieved. The D/A conversion occurs at the output of the adaptation circuitry where the filter coefficients are converted to analog. In this case, a mild monotonic nonlinearity can be compensated by the adaptation algorithm.

The second solution [15] is to use the nonlinear cancellation algorithm described earlier. This approach is illustrated in Fig. 16 where the transversal filter is implemented digitally. The nonlinearity in the D/A converter in this case is compensated by extra nonlinear taps in the canceller. This method has the advantage that sources of nonlinearity external to the canceller, such as transmitted pulse asymmetry and transformer saturation, can also be compensated. Finally, the two solutions can be combined: extra nonlinear taps can be added to the analog transversal filter in Fig. 15 for the compensation of any external nonlinearities.

VI. Conclusions

This paper has described the applications, algorithms, and technology of echo cancellation. Due to advances in microelectronics, which make echo cancellation more economical, broad application of these techniques can be expected. New applications of echo cancellation, such as the elimination of acoustic reverberation, can also be expected.

Acknowledgment

The author is indebted to many colleagues who have contributed to his understanding of echo cancellation, and particularly Dr. D. Duttweiler, Dr. O. Agazzi, D. Falconer, S. Weinstein, and D. Hodges.

References

[1] M. Sondhi and D. A. Berkley, "Silencing echos on the telephone network," *Proc. IEEE*, vol. 68, Aug. 1980.
[2] H. G. Suyderhoud, S. J. Campanella, and M. Onufry, "Results and analysis of a worldwide echo cancellation field trial," *COMSAT Tech. Rev.*, vol. 5, p. 253, Fall 1975.
[3] A. H. Gray and J. D. Markel, "Digital lattice and ladder filter synthesis," *IEEE Trans. Audio Electroacoust.*, vol. AU-21, p. 491, 1973.
[4] J. Makhoul, "A class of all-zero lattice digital filters," *IEEE Trans. Acoust., Speech, Signal Processing*, vol. ASSP-26, p. 304, Aug. 1978.
[5] S. V. Ahamed, P. P. Bohn, and N. L. Gottfried, "A tutorial on two-wire digital transmission in the loop plant," *IEEE Trans. Commun.*, vol. COM-29, Nov. 1981.

[6] K. H. Mueller, "A new digital echo canceler for two-wire full-duplex data transmission," *IEEE Trans. Commun.*, vol. COM-24, Sept. 1976.
[7] O. Agazzi, D. A. Hodges, and D. G. Messerschmitt, "Large-scale integration of hybrid-method digital subscriber loops," *IEEE Trans. Commun.*, vol. COM-30, p. 2095, Sept. 1982.
[8] S. B. Weinstein, "A passband data-driven echo canceller for full-duplex transmission on two-wire circuits," *IEEE Trans. Commun.*, vol. COM-25, July 1977.
[9] B. Widrow *et al.*, "Stationary and nonstationary learning characteristics of the LMS adaptive filter," *Proc. IEEE*, vol. 64, Aug. 1976.
[10] N. A. M. Verhoeckx *et al.*, "Digital echo cancellation for baseband data transmission," *IEEE Trans. Acoust., Speech, Signal Processing*, vol. ASSP-27, Dec. 1979.
[11] D. L. Duttweiler, "A twelve-channel digital echo canceller," *IEEE Trans. Commun.*, vol. COM-26, May 1978.
[12] A. Weiss and D. Mitra, "Digital adaptive filters: conditions for convergence, rates of convergence, effects of noise and errors arising from the implementation," *IEEE Trans. Inform. Theory*, vol. IT-25, pp. 637–652, Nov. 1979.
[13] D. L. Duttweiler, "Adaptive filter performance with nonlinearities in the correlation multiplier," *IEEE Trans. Acoust., Speech, Signal Processing*, vol. ASSP-30, p. 578, Aug. 1982.
[14] E. J. Thomas, "An adaptive echo canceller in a nonideal environment (nonlinear or time variant)," *Bell Syst. Tech. J.*, vol. 50, pp. 2779–2795, Oct. 1971.
[15] O. Agazzi, D. G. Messerschmitt, and D. A. Hodges, "Nonlinear echo cancellation of data signals," *IEEE Trans. Commun.*, vol. COM-30, p. 2421, Nov. 1982.
[16] J. Makhoul, "Stable and efficient lattice methods for linear prediction," *IEEE Trans. Acoust., Speech, Signal Processing*, vol. ASSP-25, p. 423, Oct. 1977.
[17] E. H. Satorius and S. T. Alexander, "Channel equalization using adaptive lattice algorithms," *IEEE Trans. Commun.*, vol. COM-27, p. 899, June 1979.
[18] D. D. Falconer and L. Ljung, "Application of fast Kalman estimation to adaptive equalization," *IEEE Trans. Commun.*, vol. COM-26, p. 1439, Oct. 1976.
[19] B. Friedlander, "Lattice filters for adaptive processing," *Proc. IEEE*, vol. 70, p. 829, Aug. 1982.
[20] M. S. Mueller, "On the rapid initial convergence of least-squares equalizer adjustment algorithms," *Bell Syst. Tech. J.*, vol. 60, pp. 2345–2358, Dec. 1981.
[21] J. Salz, "On the start-up problem in digital echo cancelers," *Bell Syst. Tech. J.*, vol. 62, p. 1353, July–Aug. 1983.
[22] D. D. Falconer, "Adaptive reference echo cancellation," *IEEE Trans. Commun.*, vol. COM-30, p. 2083, Sept. 1982.
[23] D. L. Duttweiler and Y. S. Chen, "A single-chip VLSI echo canceler," *Bell Syst. Tech. J.*, vol. 59, pp. 149–160, Feb. 1980.
[24] Y. G. Tao, K. D. Kolwicz, C. W. K. Gritton, and D. L. Duttweiler, "A cascadable VLSI echo canceller," this issue, pp. 297–303.

Techniques for Adaptive Equalization of Digital Communication Systems

By R. W. LUCKY

(Manuscript received October 14, 1965)

A previous paper described a simple adjustment algorithm which could be employed to set the tap gains of a transversal filter for the equalization of data transmission systems. An automatic equalizer was shown which used this algorithm during a training period of test pulse transmission prior to actual data transmission. The present paper extends the utility of this automatic equalization system by permitting it to change settings during the data transmission period in response to changes in transmission channel characteristics. Three schemes for accomplishing this adaptive equalization without the use of test signals are described and evaluated analytically. The first such scheme uses periodic estimates of channel response based on the received data signal to adjust or update the transversal filter settings. The second system is entirely digital and employs a sequential testing procedure to make adjustments aperiodically as they are required by changing conditions. The third system uses information obtained from a forward-acting error correction system for the purposes of adaptive equalization. Of the three systems described, the second is not only theoretically superior, but is practically the simplest. Experimental results for this second system are described.

I. INTRODUCTION

A previous paper[1] has dealt with the problem of automatic equalization for data transmission systems. In that paper, it was assumed that a finite-length transversal filter was to be used to correct the pulse response of a baseband (VSB) system at the sampling instants. A simple control system was shown which could be used to adjust the tap gains of the transversal filter to optimum positions using a series of test pulses transmitted prior to actual data transmission. After this training period the control system is disconnected, the tap gains remain fixed, and normal data transmission ensues. This automatic equalization system has been used in conjunction with multilevel vestigial sideband modulation to achieve a rate of 9600 bits-per-second on private line voice facilities.[2,3,4]

Two limitations of this automatic equalization system are immediately apparent — it requires that test pulses be transmitted and it must be reset in another training period whenever the channel characteristics change. Other disadvantages of the present mode of operation include the long training period required to establish accurate final settings and the possibility of a nonlinear channel causing the transmission characteristics for data transmission to be slightly different from those for isolated pulse transmission.

For these reasons, it has been found advantageous to develop an equalizer capable of deriving its control signals directly from the transmitted data signal itself. Such an equalizer would be capable of tracking a time varying channel and would also circumvent the other difficulties associated with preset equalizer operation. To distinguish this equalizer from the previously described preset automatic equalizer, we shall call the tracking equalizer an *adaptive* equalizer. The purpose of the adaptive equalizer is to continually monitor channel conditions and to readjust itself when required so as to provide optimum equalization. To conserve signal power and bandwidth, the channel monitoring (or system identification) of the adaptive equalizer must be done using only the normal received data signal and without the benefit of added test information.

In this paper, we will present a few techniques which may be used to achieve adaptive equalization. Since these techniques are based on the use of a particular tap gain adjustment algorithm used in the preset equalizer, we will begin with a brief description of this equalizer and the algorithm.

II. PRINCIPLES OF TRANSVERSAL FILTER EQUALIZATION

In the preset automatic equalizer, a sequence of isolated test pulses is transmitted through the channel and demodulator. At the output of the demodulator, the pulse waveform is designated $x(t)$ as shown in Fig. 1. The pulse then passes through a $(2N + 1)$-tap transversal filter whose tap gain settings are $c_{-N}, \cdots, c_N$. The output pulse of the equalizer is

$$h(t) = \sum_{j=-N}^{N} c_j x(t - jT). \tag{1}$$

Reprinted with permission from *The Bell System Technical Journal,* vol. 45, no. 2, pp. 255–286, Feb. 1966.

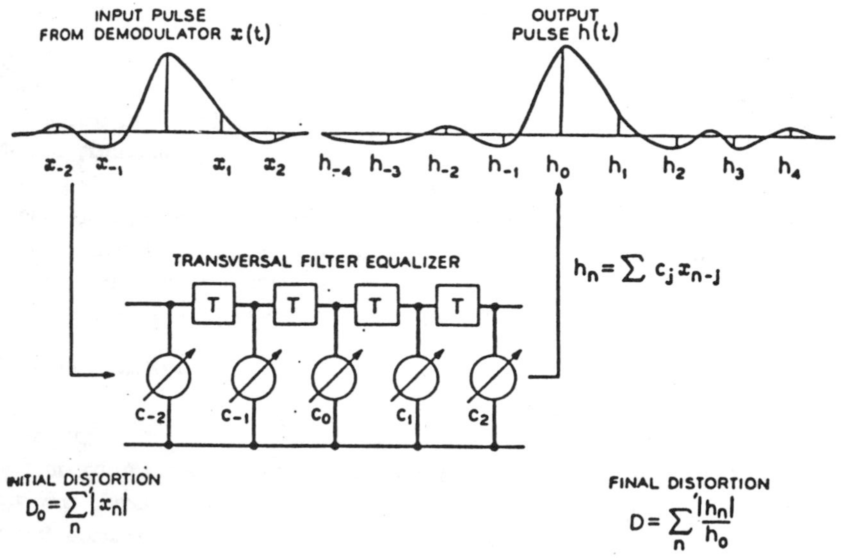

Fig. 1 — Transversal filter equalizer.

Since the output of the equalizer will be sampled at T second intervals during data transmission, we are only interested in the samples

$$h_n = h(nT)$$

of the output pulse. In terms of the input samples x_n we can write

$$h_n = \sum_{j=-N}^{N} c_j x_{n-j}. \tag{2}$$

The objective of the equalizer control circuitry is to set the tap gains such that the pulse distortion D is minimized, where we define*

$$D = \frac{1}{h_0} \sum_{n=-\infty}^{\infty}{}' \, | h_n |. \tag{3}$$

This criterion is equivalent to requiring that the equalizer maximize the eye opening.

We assume that the input is normalized so that $x_0 = 1$ and that the center tap c_0 is used to satisfy the practical constraint $h_0 = 1$. In Ref. 1 it was proved that if the initial distortion $D_0 < 1$, where

$$D_0 = \sum_{n=-\infty}^{\infty}{}' \, | x_n |, \tag{4}$$

then the output distortion D is at a minimum when the $2N$ tap gains c_j for $| j | \leqq N, j \neq 0$ are adjusted so that $h_n = 0$ for $| n | \leqq N$, $n \neq 0$. In other words, if a binary eye is open before equalization, then using the tap gains to force zeros in the output response is optimum.

Also, if the initial distortion is less than unity, it was shown that a simple iterative procedure could be used to obtain optimum tap gain settings. In this procedure each tap gain c_j is adjusted an amount $-\Delta$ sgn h_j after each test pulse. (The center tap gain c_0 is adjusted by $-\Delta$ sgn $(h_0 - 1)$.) Thus one simply inspects the polarities of the output pulse $h(t)$ at the sampling times and uses this polarity information to advance or retard counter-controlled attenuators along the tapped delay line.

In constructing an adaptive equalizer we shall use this same simple adjustment algorithm. The problem now becomes one of finding the polarities of the channel pulse response $h(t)$ without the benefit of test pulses. Once the decisions have been made as to the most likely polarities for the impulse response samples h_j, we make discrete adjustments in the corresponding tap gains c_j to correct the equalizer. (This is shown diagrammatically in Fig. 2.) The adjustments in tap gains can be made periodically or only as required by changes in transmission characteristics. Since the normal telephone channel's characteristics change very slowly, the decisions regarding the channel response polarities can be

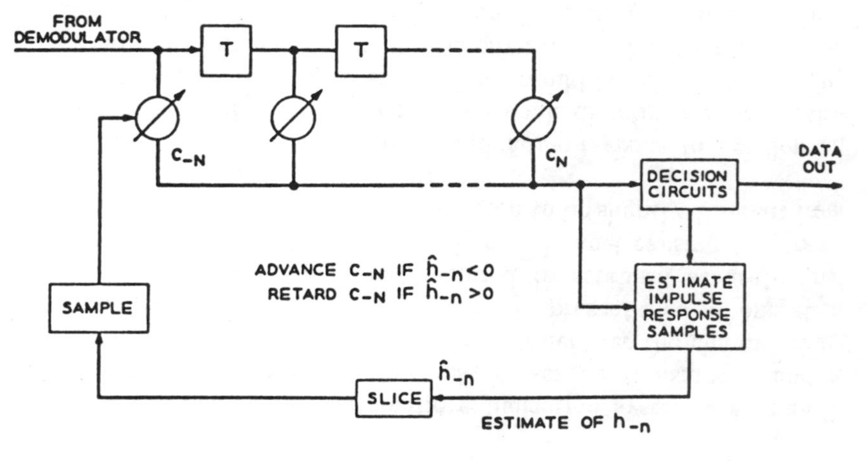

Fig. 2 — Adaptive equalizer.

* Primes on summations indicate deletion of the zeroth term.

made extremely accurate. In general the more accurate these decisions, the more precise the equalization which can be obtained, but with a corresponding increase in adaptive equalizer response, or settling time. We shall evaluate the accuracies and response times of the adaptive equalizers described in subsequent sections.

III. PERIODIC EQUALIZER ADJUSTMENT

3.1 *Maximum Likelihood Estimation of Response Values*

Let us assume that at the end of every KT seconds we wish to make a decision as to the current most probable impulse response polarities and to effect an incremental adjustment of the equalizer. The demodulated and equalized voltage at time t_k is

$$y_k = \sum_{n=-\infty}^{\infty} a_n h_{k-n} + \eta_k , \tag{5}$$

where the a_n's are the input symbols chosen from an M-symbol alphabet, the h_n's are the samples of the overall (equalized) system impulse response, and the η_k's are noise samples.* We also make the following key assumptions.

(*i*) The noise samples η_k are independent, identically distributed Gaussian variables with variance σ^2.

(*ii*) The input data symbols are uncorrelated.

(*iii*) The probability of error is relatively small, so that for practical purposes the sequence $\{a_n\}$ is available at the output of the detector.

(*iv*) The channel response samples h_n are essentially constant over the observation interval of KT seconds.

It is important to note that these assumptions are not made because they are true in a practical situation, but are made at this time to enable us to derive a reasonable system configuration and to assess its probable performance. Assumption (*ii*) is a particularly important one which we shall have to consider more carefully in predicting actual system performance.

Since we are using a decision-directed system — i.e., the detected symbols a_n are assumed correct — we can regard the signal samples y_k as being determined by noise and the set of parameters h_n. By an appropriate statistical technique we can make estimates of the response samples h_n from the signal samples y_k. Using the set of assumptions (*i*) through (*iv*), the probability of receiving the sequence $\{y_k\}$, $k = 1, \cdots, K$ for a particular choice of the parameter set h_n is

$$p(\mathbf{y}|\mathbf{h}) = \prod_{k=1}^{K} \frac{\exp - \frac{1}{2\sigma^2}\left(y_k - \sum_{n=-\infty}^{\infty} a_n h_{k-n}\right)^2}{\sigma\sqrt{2\pi}} . \tag{6}$$

The likelihood function $L(\mathbf{y}|\mathbf{h})$ is the logarithm of $p(\mathbf{y}|\mathbf{h})$. Apart from a constant, this is

$$L(\mathbf{y}|\mathbf{h}) = \sum_{k=1}^{K} - \frac{1}{2\sigma^2}\left(y_k - \sum_{n=-\infty}^{\infty} a_n h_{k-n}\right)^2 . \tag{7}$$

The maximum likelihood estimates of the $(2N+1)$ response values of h_j, $j = -N, \cdots, +N$, needed to adjust the transversal filter tap gains are determined by the $(2N+1)$ simultaneous equations $\partial L/\partial h_j = 0$. Thus

$$\sum_{k=1}^{K} a_{k-j}\left(y_k - \sum_{n=-\infty}^{\infty} a_n \hat{h}_{k-n}\right) = 0, \qquad \text{for } j = -N, \cdots, +N. \tag{8}$$

These equations are more conveniently rewritten in the form

$$\frac{1}{K}\sum_{k=1}^{K} a_{k-j} y_k - \sum_{n=-\infty}^{\infty} \hat{h}_n A_{nj} = 0, \tag{9}$$

where

$$A_{nj} = \frac{1}{K}\sum_{k=1}^{K} a_{k-n} a_{k-j} . \tag{10}$$

Because of assumption (*ii*) that the input symbols are uncorrelated, for a reasonably long averaging period K we can use $A_{nj} = S\delta_{nj}$ where S is the average signal power. This simplification yields the estimators

$$\hat{h}_j = \frac{1}{KS}\sum_{k=1}^{K} a_{k-j} y_k . \tag{11}$$

We could easily now find the variance of the estimates (11) under the assumption of random data, and we would discover that they are poorly converging estimates with typically about 50,000 samples y_k required before we could move the equalizer taps with any degree of confidence. This exorbitant settling time is caused by the presence of a large parameter, $h_0 \approx 1$, among a set of typically very small parameters.

* It is true that the noise sequence $\{\eta_k\}$ must pass through the equalizer. However, since in practice, all the tap gains c_j of the equilizer except for c_0 are very small we assume that the statistics of the noise sequence are unaffected by the equalizer.

Bear in mind that in normal operation the equalization will be very close to perfect and the samples $\hat{h}_j$ for $j \neq 0$ will be generally much less than 0.01 in magnitude.

This difficulty is circumvented by estimating $(\hat{h}_0 - 1)$ which is comparable in size to $\hat{h}_j$ for $j \neq 0$ instead of directly estimating $\hat{h}_0$. Therefore, we define

$$h_j = \begin{cases} \hat{h}_j & \text{for } j \neq 0, \\ \hat{h}_0 - 1 & \text{for } j = 0. \end{cases} \tag{12}$$

The samples h_j represent equalization error in the output pulse response. Following estimation of these values each of the taps c_j is advanced if h_j is negative and retarded if h_j is positive — the center tap being handled the same as any other tap.

Substitution of (12) in (7) gives the likelihood function

$$L(\mathbf{y} \mid \mathbf{h}) = \sum_{k=1}^{K} -\frac{1}{2\sigma^2}\left(y_k - a_k - \sum_{n=-\infty}^{\infty} a_n h_{k-n}\right)^2. \tag{13}$$

The quantities $(y_k - a_k)$ will be used frequently so we designate these e_k since they represent the error between the received sample y_k and the detected level a_k. The maximum likelihood estimates of the equalization error become

$$\hat{h}_j = \frac{1}{KS}\sum_{k=1}^{K} a_{k-j} e_k . \tag{14}$$

Fig. 3 shows a block diagram of an adaptive equalizer employing the estimates (14). In this system, the detected levels a_k are converted to analog form and subtracted from the received samples y_k to form the error samples e_k. The error samples e_k are then correlated simultaneously with each of the detected symbols a_{k-j} for $j = -N, \cdots, +N$. To accomplish this, the error samples e_k must be delayed NT seconds while the detected samples a_k are passed along a $(2N + 1)$-tap delay line. The delay line may consist of parallel shift registers since the samples a_k are in digital form. The outputs of the correlators (multipliers — low-pass filters) are sampled at K symbol intervals and appropriate actions are taken on the transversal filter attenuators.

3.2 *Performance of the Adaptive Equalizer Under Ideal Conditions*

In assessing the performance of an adaptive equalizer, we are primarily concerned with accuracy and settling time. The action of the equalizer

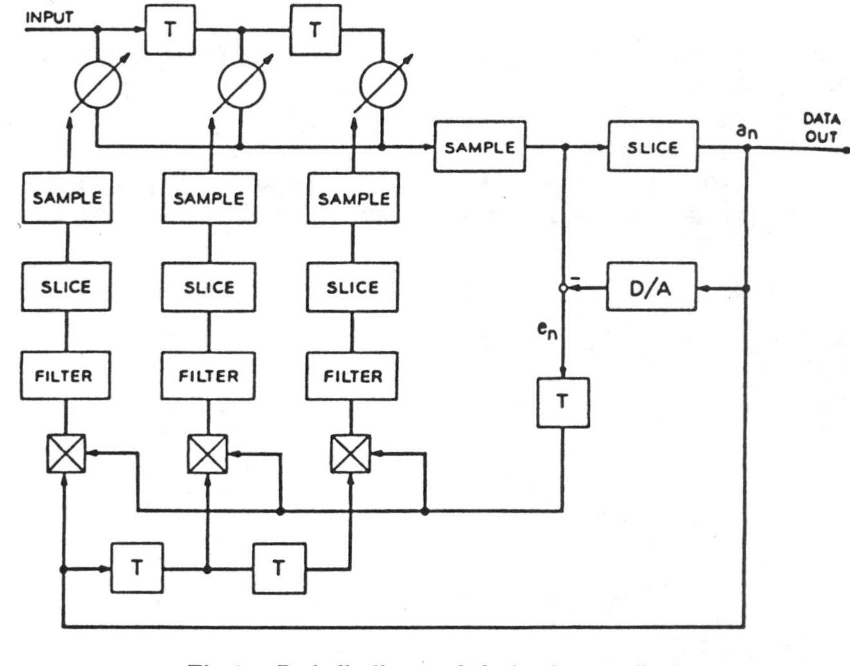

Fig. 3 — Periodically sampled adaptive equalizer.

is that of a multidimensional random walk with a bias toward correct equalization being supplied by the estimation circuitry. We shall make the assumption of small tap interaction and consider that each of the tap gains exhibits a one-dimensional walk independent of the action of other taps. The fundamental quantity involved in questions of accuracy and settling time is the probability $P(\text{sgn } \hat{h}_j = \text{sgn } h_j) = p_c$. For given noise and data statistics this probability is a function of the value of h_j and represents the probability of making a correct adjustment of tap c_j. When $|h_j|$ is large, i.e., equalization is poor, we expect that p_c will be close to unity, while for small $|h_j|$ the probability p_c approaches 0.5 and the tap gain tends to wander.

The adaptive equalizer must be designed to keep the inevitable wander of the tap gain within bounds imposed by accuracy requirements. Generally, we would wish to take full advantage of the inherent accuracy of the attenuator setting apparatus, i.e., each attenuator is adjusted in steps of Δ. Thus, when the equalizer is in perfect adjustment each tap gain will have an error of about 0.5Δ, which means that each of the samples h_j will be about 0.5Δ in magnitude. At the end of K symbol durations, each of the gains will be increased or decreased by Δ. If $|h_j| = 0.5\Delta$ and a mistake is made in the polarity of $\hat{h}_j$, then the next value of h_j will be about 1.5Δ in magnitude and distortion will be considerably increased. We should generally design the system so that each tap gain spends a great majority of the time in the state where $|h_j| = 0.5\Delta$ and a small amount at 1.5Δ. Thus, for example, we may wish p_c to be 0.99 when $h_j = 0.5\Delta$.

We now evaluate $p_c(h_j)$ under the ideal conditions given in assumptions (*i*) through (*iv*). When the averaging period K is large the estimates $\hat{h}_j$ become normally distributed. The mean value of $\hat{h}_j$ is

$$\overline{\hat{h}_j} = \frac{1}{KS} \sum_{k=1}^{K} \overline{a_{k-j} e_k}. \tag{15}$$

The error sample e_k may be written

$$e_k = y_k - a_k = \sum_{n=-\infty}^{\infty} a_n h_{k-n} + \eta_k \tag{16}$$

so that (15) becomes

$$\overline{\hat{h}_j} = \frac{1}{KS} \sum_{k=1}^{K} \sum_{n=-\infty}^{\infty} \overline{a_{k-j} a_n}\, h_{k-n} + \frac{1}{KS} \sum_{k=1}^{K} \overline{a_{k-j} \eta_k}. \tag{17}$$

Since $\overline{a_k a_j} = S\delta_{kj}$ we have the necessary result

$$\overline{\hat{h}_j} = h_j. \tag{18}$$

The variance, σ_j^2, of $\hat{h}_j$ may be evaluated in straightforward fashion.

$$\sigma_j^2 = \frac{1}{K^2S^2} \sum_{k=1}^{K} \sum_{e=1}^{K} \sum_{n=-\infty}^{\infty} \sum_{m=-\infty}^{\infty} \overline{a_{k-j} a_n a_{e-j} a_m}\, h_{k-n} h_{e-m} + \frac{1}{K^2S^2} \sum_{k=1}^{K} \sum_{e=1}^{K} \overline{a_{k-j} a_{e-j} \eta_k \eta_e} - h_j^2. \tag{19}$$

The four-fold sum in (19) may be partitioned into three sums involving pairwise equality in subscripts and an overlap term. A little manipulation yields

$$\sigma_j^2 = \frac{1}{K} \sum_{\substack{n=-\infty \\ n \neq j}}^{\infty} h_n^2 + \frac{2}{K^2} \sum_{n=1}^{K-1} (K-n) h_{j+n} h_{j-n} + \frac{\sigma^2}{KS}. \tag{20}$$

To get some sort of feel for the values involved, let's assume that $h_j = 0.5\Delta$ or -0.5Δ with equal probability when $|j| \leq N$ and is zero otherwise. Then the expected value of σ_j^2 becomes

$$\overline{\sigma_j^2} = \frac{1}{K}\left(\frac{N\Delta^2}{2} + \frac{\sigma^2}{S}\right). \tag{21}$$

For p_c to be 0.99 when $h_j = 0.5\Delta$, we require that $0.5\Delta = 2.33\sigma_j$, so that for the zero noise condition

$$K \approx 10.8N. \tag{22}$$

For example, a 13-tap equalizer in the absence of noise requires about 140 symbols (0.058 sec. at a baud rate of 2400) to make a sufficiently accurate estimate. However, for a typical phone line application we require extremely accurate equalization so that Δ might typically be on the order of 0.0025 while the signal-to-noise ratio (S/σ^2) is about 30 db. In this case, the noise term in (21) completely swamps the "intersymbol clutter" term $N\Delta^2/2$ and the 0.99 accuracy at 0.5Δ condition means that

$$K \approx \frac{21.7\sigma^2}{\Delta^2 S}. \tag{23}$$

For this example, 3470 samples (1.45 sec) are required per step of equalization.

3.3 *Performance of the Adaptive Equalizer Under Adverse Conditions*

Now that the equalization system configuration has been established under the assumption of ideal conditions, it becomes necessary to judge deterioration when these conditions are not met in practice. Throughout this paper, the specific application is assumed to be high-speed, voice telephone channel, data transmission.

Assumption (*i*) regarding Gaussian noise is usually justified in practice (the impulsive noise is not a determining factor for high-speed transmission) and in any event is not a very crucial assumption. Assumption (*iii*) that the received symbols are detected correctly is generally amply satisfied during normal data transmission with error rates of 0.01 or less. Remember that the information from a thousand or more symbols may be averaged to make one decision regarding equalization. Thus, we only require that a large majority of decisions are correct. Obviously the system performance deteriorates and finally "breaks" as the error rate becomes higher and higher, but the analytical evaluation of the effect appears difficult. Experimental results will be mentioned in a later section of this paper, but it should be said here that the system will work well with error rates on the order of 0.1.

Once the equalizer begins to work, the error rate quickly drops back to a more normal value.

The time variation rate of the channel must be matched with the accuracy requirement to arrive at an averaging time KT seconds during which the channel does not vary greatly compared to the size step Δ being taken on the equalizer taps. In the phone line application, this appears simple even for very high accuracies since the transmission characteristics are not usually observed to continually change at any great rate.

The most troublesome of the four ideal assumptions is the one involving uncorrelated input symbols. During normal data transmission one would expect that the sequence $\{a_n\}$ would appear random over an interval of a thousand symbols, but unfortunately this is frequently not the case. Long steady sequences of ones or of zeros may be used to hold the line, or the dotting pattern of alternate ones and zeros may be employed for some such purpose. In any event where a short, repetitive pattern is transmitted the spectrum of the transmitted signal consists of a number of discrete lines. Obviously it is impossible at the receiver to extract any information about the channel's transmission characteristics except at a few discrete points. Any adaptive equalizer must be prepared to weather this period and await new random data upon which meaningful decisions can be made. Fortunately, it can be shown that the adaptive equalizer of Fig. 3 acts on whatever information is available in the received data and retains its settings through periods of bad sequences.

We return now to (17) and remove the assumption concerning uncorrelated data. Then

$$\bar{h}_j = \sum_{n=-\infty}^{\infty} h_n r_{j-n} \tag{24}$$

where r_j is the normalized autocorrelation function of the input data sequence (which is assumed to be stationary).

$$r_j = \lim_{K\to\infty} \frac{1}{2KS} \sum_{k=-K}^{K} a_k a_{k-j} \,. \tag{25}$$

For ideal equalization, we require that the action of the equalizer cause $h_n = 0$ for $|n| \leqq N$. The adaptive equalizer tries to accomplish this by forcing $\bar{h}_j = 0$ for $|j| \leqq N$. As can be seen from (24), a slight error may be introduced if $r_j \neq 0$ for $j \neq 0$. The only error involved, however, is to cause some influence of the samples h_n for $|n| > N$ on the tap settings. Generally, these samples which are outside the range of equalization are quite small (otherwise the number of equalizer taps needs to be increased) and their effect is only multiplied by the tails of the input autocorrelation function. Suppose, for example, that $h_n = 0$ for $|n| > N$. Then after equalization

$$\bar{h}_j = \sum_{n=-N}^{N} h_n r_{j-n} = 0 \quad \text{for} \quad |j| \leqq N \tag{26}$$

and the only solution to this set of $(2N + 1)$ equations is the perfect state $h_n = 0$ for $|n| \leqq N$ provided the matrix

$$R = \begin{matrix} r_0 & r_1 \cdots r_{2N} \\ r_{-1} & r_0 & \\ \vdots & & \\ r_{-2N} & & r_0 \end{matrix} \tag{27}$$

is nonsingular.

If the data sequence consists of a repetitive pattern of period L symbols, then $r_j = r_{j+L}$ and the rank of R cannot be greater than L. Thus, if the period of pattern repetition L is less than the number of taps on the equalizer $(2N + 1)$, then the equalizer does not reach optimum settings. Consider then what settings the equalizer does reach. In view of $r_j = r_{j+L}$ we can rewrite (24) in the form

$$\bar{h}_j = \sum_{n=0}^{L-1} r_n \left\{ \sum_{m=-\infty}^{\infty} h_{j-n-mL} \right\} \tag{28}$$

and

$$\bar{h}_j = \bar{h}_{j+L} \,. \tag{29}$$

Thus, only L taps of the equalizer represent independent feedback loops, while the other $(2N + 1 - L)$ tap gains are slaves to the L gains considered independent. In fact, they receive the identical error signals and are thus incremented identical amounts. The equalizer solves the L simultaneous equations to arrive at

$$\sum_{m=-\infty}^{\infty} h_{j-mL} = 0 \tag{30}$$

for $j = 0, \cdots, L - 1$.

Actually, this solution minimizes the data distortion for the particular sequence being transmitted. The received samples y_k may be written in terms of the equalization error samples h_n using (5) and (12).

$$y_k = \sum_{n=-\infty}^{\infty} a_n h_{k-n} + a_k + \eta_k . \tag{31}$$

But, since $a_k = a_{k+L}$, we have

$$y_k = \sum_{n=0}^{L-1} a_n \left\{ \sum_{m=-\infty}^{\infty} h_{k-n-mL} \right\} + a_k + \eta_k \tag{32}$$

and we see that an equivalent channel response could be defined as $\{g_n\}$ where

$$g_n = \sum_{m=-\infty}^{\infty} h_{n-mL} \qquad 0 \leq n \leq L-1$$
$$g_n = 0 \qquad \text{elsewhere.} \tag{33}$$

The distortion for the equivalent channel is minimized by zeroing the samples g_n within the range of the equalizer, but this is precisely what (30) indicates is done.

If the equalizer is at a nominally perfect setting ($h_n = 0$; $| n | \leq N$) when the repetitive sequence is begun, then the equalizer holds its settings over the time of periodic transmission. There is no possibility of a drift in settings since taps c_k and c_{k+L} are "locked" together and yet must maintain the solutions $g_n = 0$. There are L free taps and L independent equations for which the taps are initially at a solution. Thus, the equalizer can hold its settings over an indefinite time while periodic sequences are transmitted so long as the channel characteristics remain fixed.

If the channel response changes while periodic sequences are being transmitted then the equalizer will move to a new solution of (30) to minimize the distortion for the particular periodic sequence. However, this solution will not infer that $h_n = 0$ for $| n | \leq N$ and the equalization will not be perfect when random data starts again.

The action of the equalizer for unfavorable data sequences can be summarized by saying that the system always attempts to do the best equalization possible for the data being transmitted. In no case is the equalization deteriorated because of the adaptive loop over the performance of a fixed equalizer.

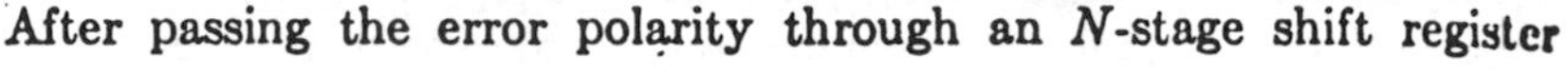

IV. AN IMPROVED DIGITALIZED ADAPTIVE EQUALIZER

4.1 *Description*

There is a great premium attached to the use of digital circuits where possible for reasons of equipment cost and size. It is possible to considerably simplify the implementation of the adaptive equalizer of Fig. 3 by discarding linear concepts and using only polarity information throughout. Thus, instead of correlating the error signal e_k with the detected symbols a_{k-j} we add mod 2 the binary symbols corresponding to the polarities of e_k and a_{k-j}. The resulting simplification can be seen in Fig. 4. The symbol polarity sgn a_{k-j} is obtained by passing the most significant digit of the detected symbol (in binary format) through a shift register. In the Gray code commonly used for binary-to-multilevel conversion the first bit indicates polarity of the symbol.

The polarity of error sgn e_k can be produced by the simple expedient of adding an additional stage of slicing to the $\log_2 M$ slicers required for M-level transmission. Each stage of the $\log_2 M$ slicers "folds" the signal value about the last threshold, so that an extra stage simply produces automatically the polarity of the error sgn e_k. Fig. 4 illustrates a 16-level transmission system. Five stages of slicing are employed. The first four stages deliver the four detected bits indicating the received symbol a_k while the first and fifth bits are used for equalization purposes.

After passing the error polarity through an N-stage shift register

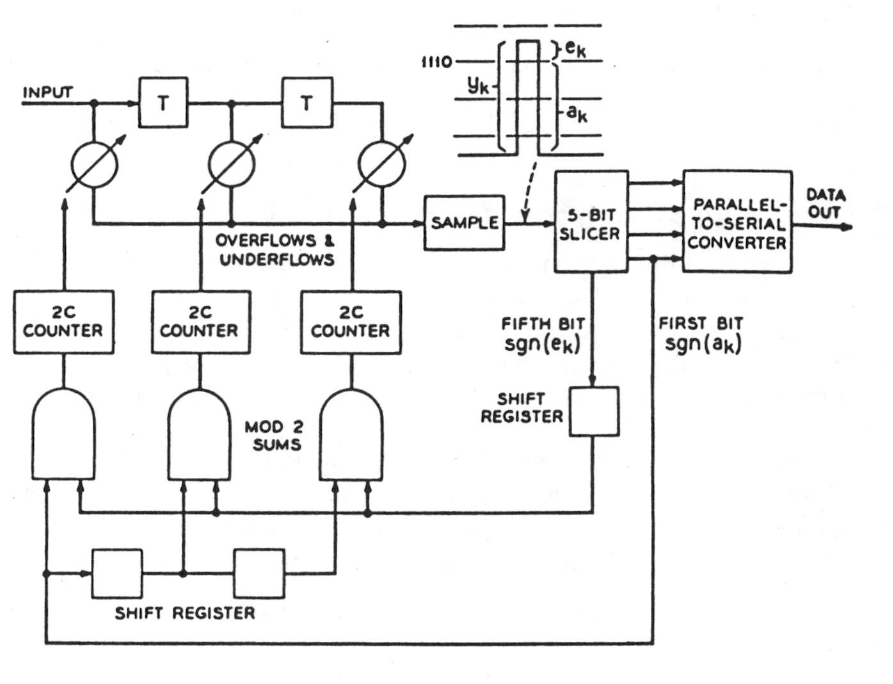

Fig. 4 — 3-tap digital adaptive equalizer.

(in order to be able to correlate with N future and N past symbol polarities), the error polarities and symbol polarities are summed using exclusive-OR circuits. At this point we are able to easily introduce an improvement over the low-pass filter and periodic slicer used in the equalizer of Fig. 3. The problem with periodic slicing is that a compromise time interval KT must be chosen for averaging which is based on the most critical situation of near perfect equalization. The equalizer moves just as slowly when equalization is poor and the correlator output samples need much less averaging for a given accuracy.

A sequential testing procedure is clearly called for in this application. In a sequential test, the interval between decisions is determined by the input data itself. Instead of averaging data over a KT second interval and then sampling to determine polarity, running sums of the exclusive-OR outputs are kept. Positive and negative thresholds are set and the tests are terminated whenever these thresholds are crossed. This procedure is most easily implemented in digital form using up-down binary counters whose capacities of $2C$ counts determine the decision threshold value. Whenever a one is emitted from the exclusive-OR the counter advances one count, while a zero retards the counter one count. When the counter overflows we decide the polarity of h_k is positive and reduce the gain of on tap c_k. The counter is then reset to the center position of C counts. Similarly, an underflow adjusts c_k one step higher and resets the counters. The $2C$ storage counters are of course tied directly to the up-down counters which control the tap gain to accomplish this task in a most simple manner.

Thus, the equalizer of Fig. 4 is surprisingly simple. It requires only an N-stage shift register, a slicer, and $(2N + 1)$ binary counters of capacity $2C$ in order to convert a preset equalizer to the adaptive mode. Since the storage counters are used for averaging during stepup for the preset equalizer we finally arrive at an adaptive equalizer which costs almost nothing more than a preset equalizer.

The question arises as to why the preset mode (test pulses before transmission) is needed at all. In many cases it may not be needed provided a period of initial equalization is alloted during which data is transmitted, but not used due to its unreliability. As we shall find, the adaptive equalizer can accomplish a given degree of accuracy in equalization in less time than a preset equalizer providing the error rate is not too high. However, during initial setup the error rate is generally so high that the adaptive equalizer operates very slowly or not at all. Thus, a short period of test pulses can be profitably used to bring the error rate down to manageable values before adaptive equalization is begun.

4.2 *Analytical Evaluation*

In this section we will evaluate the probability of correct adjustment of tap c_j and the average time required for an adjustment. The probability of correct adjustment p_c depends on the size $(2C)$ of the storage counters and on the probability of an up-count p, and of a down-count q, on the jth storage counter.

The kth count of the jth counter is obtained by multiplying the polarities of e_k and a_{k-j}. Let us assume for convenience that h_j is positive (for negative h_j the situation is, of course, entirely similar). The probability of a correct adjustment is then the probability of an overflow occurring before an underflow. The probability of an up-count is

$$p = P(e_k > 0, a_{k-j} > 0) + P(e_k < 0, a_{k-j} < 0). \tag{34}$$

These probabilities are identical so we use

$$p = 2P(e_k > 0, a_{k-j} > 0). \tag{35}$$

Equation (35) can be rewritten in terms of the conditional probability

$$p = 2P(e_k > 0 \mid a_{k-j} > 0)P(a_{k-j} > 0). \tag{36}$$

The symbols a_j will be taken as independent and equally likely to assume any of the M values. Take $2d$ as the distance between adjacent levels, so that d is the distance from any level to the nearest slicing (decision) threshold. The amplitudes a_j can assume are then $d(2i - 1)$ for $i = -M/2 + 1, \cdots, M/2$. Since a_{k-j} can be positive or negative with equal likelihood, (36) becomes

$$p = P(e_k > 0 \mid a_{k-j} > 0) \tag{37}$$

$$p = \frac{2}{M} \sum_{i=1}^{M/2} P[e_k > 0 \mid a_{k-j} = d(2i - 1)]. \tag{38}$$

Now we need to evaluate the conditional probabilities in (38). With equalization error samples h_n at times nT and noise samples η_n, the received voltage at time kT is

$$y_k = \sum_n a_n h_{k-n} + a_k + \eta_k . \tag{39}$$

The error voltage e_k is

$$e_k = \sum_{n=-\infty}^{\infty} a_n h_{k-n} + \eta_k . \tag{40}$$

We remove the term involving a_{k-j} in (40) to obtain

$$e_k = a_{k-j}h_j + [\sum_{n \neq k-j} a_n h_{k-n} + \eta_k]. \tag{41}$$

The assumption is made that the sum of the intersymbol interference and noise (the terms in brackets) is Gaussian distributed, with mean zero and variance σ^2. The error e_k is then Gaussian with mean $a_{k-j}h_j$, and variance σ^2.

The probability density of e_k is sketched in Fig. 5. The conditional probabilities in (38) should be interpreted as the probability that, given given a_{k-j} was at level i, we *decide* that e_k is positive. If e_k crosses the decision threshold on the right-hand side of Fig. 5, it will appear to the receiver that e_k was negative since a_k will be incorrectly received as the next higher symbol. If e_k crosses the decision threshold on the left, e_k is interpreted as being positive. Thus, the conditional probability may be written (refer to Fig. 5)

$$P(e_k > 0 \mid a_{k-j}) = \tfrac{1}{2} + p_1 - (p_2 - p_3) \tag{42}$$

where

$$p_1 = \frac{1}{\sqrt{2\pi}\sigma} \int_{-a_{k-j}h_j}^{0} \exp\left(-\frac{x^2}{2\sigma^2}\right) dx \tag{43}$$

$$(p_2 - p_3) = \frac{1}{\sqrt{2\pi}\sigma} \int_{d-a_{k-j}h_j}^{d+a_{k-j}h_j} \exp\left(-\frac{x^2}{2\sigma^2}\right) dx. \tag{44}$$

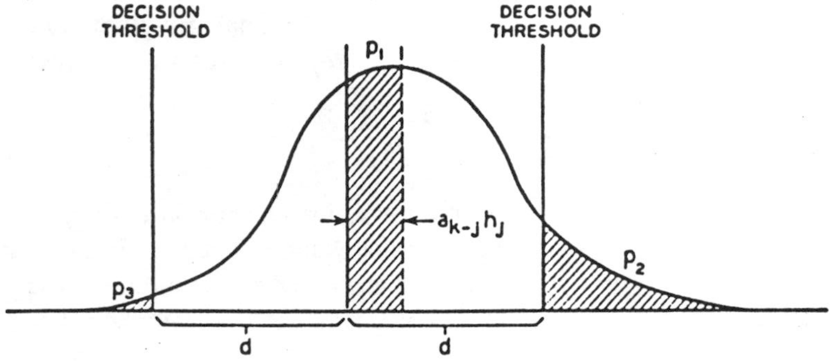

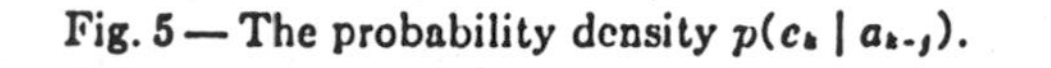

Fig. 5 — The probability density $p(e_k \mid a_{k-j})$.

With the equalization even fairly good (as it must be if meaningful data are being transmitted), $a_{k-j}h_j$ must be a small number so that the ranges of integration in (43) and (44) are small. Thus, we make the approximations

$$p_1 \cong \frac{a_{k-j}h_j}{\sqrt{2\pi}\sigma} \tag{45}$$

$$(p_2 - p_3) \cong \frac{2a_{k-j}h_j \exp\left(-\dfrac{d^2}{2\sigma^2}\right)}{\sqrt{2\pi}\sigma}. \tag{46}$$

Using these approximations we can write

$$P[e_k > 0 \mid a_{k-j} = d(2i-1)] = \tfrac{1}{2} + \frac{d(2i-1)h_j}{\sqrt{2\pi}\sigma}\left[1 - 2\exp\left(-\frac{d^2}{2\sigma^2}\right)\right] \tag{47}$$

This expression is substituted into (38) and the summation is easily performed over i to give

$$p = \tfrac{1}{2} + \left[1 - 2\exp\left(-\frac{d^2}{2\sigma^2}\right)\right]\frac{h_j dM}{2\sigma\sqrt{2\pi}}. \tag{48}$$

For normal operation the exponential term in the brackets is small in comparison with unity and may be neglected. This term gives the contribution due to slightly more errors being made in the right-hand region (p_2) than in the left-hand region (p_3). An examination of the relative probabilities p and $(p_2 - p_3)$ without the approximations (45) and (46) shows that $(p_2 - p_3)$ is generally small in comparison with p, even for error rates if 0.01 and higher. Thus, the system is able to estimate the polarity of h_j even when the eye is completely closed.

The final approximation for the probability p of a correct step in the jth counter becomes

$$p = \tfrac{1}{2} + \frac{h_j dM}{2\sigma\sqrt{2\pi}}. \tag{49}$$

Generally, this probability is only a little larger than 0.5, hence many counts must be averaged before a decision can be made as to whether $p > \tfrac{1}{2}$, in which case $h_j > 0$, or whether $p < \tfrac{1}{2}$ and consequently $h_j < 0$. This averaging is best done using the $2C$-count storage devices, since these devices effect a sequential test of the two hypotheses $p > \tfrac{1}{2}$ and $p < \tfrac{1}{2}$. This sequential test will require less time on the average for a given accuracy than straightforward averaging with a fixed sample size as in the adaptive equalizer of Fig. 3. In addition, it is more easily implemented than the former technique.

The probability p_e of an overflow before an underflow when the probability of an up-count is p and the probability of a down-count is $q = 1 - p$ is taken from Feller's analysis of the problem of the gambler's ruin.[5] For a $2C$ counter initially set to its midpoint of C, Feller gives

$$p_e = 1 - \frac{(q/p)^{2C} - (q/p)^C}{(q/p)^{2C} - 1}. \tag{50}$$

If p_e is to be close to unity, $(q/p)^{2C}$ must be small compared with $(q/p)^C$, so we approximate (50) as

$$p_e \approx 1 - (q/p)^C. \tag{51}$$

We can further simplify this expression since q and p are both very close to 0.5. Writing

$$\left.\begin{aligned} p &= 0.5 + \varepsilon \\ q &= 0.5 - \varepsilon \end{aligned}\right\} \tag{52}$$

we obtain

$$(q/p) \approx 1 - 4\varepsilon \tag{53}$$

and we use

$$(1 - 4\varepsilon)^C = \exp\,[C \log\,(1 - 4\varepsilon)] \tag{54}$$

to obtain the approximation

$$p_e = 1 - \exp\,(-4C\varepsilon). \tag{55}$$

Finally, we substitute the value of ε from (49).

$$p_e = 1 - \exp\left(-\frac{2Ch_j dM}{\sigma\sqrt{2\pi}}\right). \tag{56}$$

The average number of counts required for an overflow (or underflow) is also given by Feller.

$$\bar{n} = \frac{C}{(q - p)} - \frac{2C}{(q - p)}\left[\frac{1 - (q/p)^C}{1 - (q/p)^{2C}}\right]. \tag{57}$$

Using the same approximations in (57) as we used in (51) results in the approximation

$$\bar{n} \cong \frac{C\sigma\sqrt{2\pi}}{dMh_j}. \tag{58}$$

In order to be able to compare this system with the previous adaptive equalization system which uses linear techniques and averages over a fixed interval of K symbols, we need to find p_e in terms of h_j, $\bar{n}$ and S/σ^2. The signal power S for an M-level system with separation $2d$ is

$$S = \frac{2}{M}\sum_{i=1}^{M/2} d^2(2i - 1)^2 = \frac{d^2}{3}(M^2 - 1). \tag{59}$$

Now combining (59), (58), and (56) we arrive at (for $M \gg 1$)

$$p_e = 1 - \exp\left(-\frac{3}{\pi}\bar{n}h_j^2\frac{S}{\sigma^2}\right) \quad \text{(digitalized)} \tag{60}$$

whereas an equivalent approximation for the previous equalizer is

$$p_e = 1 - \frac{\sigma}{h_j\sqrt{2\pi KS}}\exp\left(-\tfrac{1}{2}Kh_j^2\frac{S}{\sigma^2}\right). \tag{61}$$

As expected, this comparison shows the previous system requires about twice as much time to achieve a given degree of accuracy as does the digitalized system. For the specific example used earlier 3090 symbols (1.29 seconds) are required to achieve an accuracy $p_e = 0.99$ when $h_j = 0.5\Delta$, $\Delta = 0.0025$, and $S/\sigma^2 = 1000$.

This does not mean that the digitalized equalizer operates twice as fast as the previous equalizer. Actually, it operates much faster than that. The average time of 3090 symbols is required only when $h_j = 0.5\Delta$ and since this represents perfect equalization we don't care how long the equalizer takes to move to the state -0.5Δ. When the gain c_j is out of equalization by a single step h_j is approximately 1.5Δ and according to (58) the time $\bar{n}$ required is only a third as much — or 1030 symbols. Similarly, when c_j is 2 steps out $\bar{n}$ becomes 618 symbols. If the equalizer is turned on when equalization is relatively poor the steps are taken in nearly the minimum time of C symbols. The counter capacity may be calculated from the accuracy requirement using (56). For our example we obtain $C = 85$. (Of course, either a 7-stage ($2C = 128$) or an 8-stage ($2C = 256$) would have to be used in practice.)

The longest average time before an equalizer change is required when $h_j = 0$ and, consequently, $p_e = 0.5$. Here a long average time is desirable since disturbing the equalizer is detrimental. The time in such case is C^2 symbols — in our example about 7200 symbols. Thus, the average time between equalizer adjustments varies between C (85) and C^2 (7200) symbols with short times used when urgency of movement is greatest and longer times used when leisurely adjustment is possible and, in fact, necessary because of stringent accuracy requirements.

V. ADAPTIVE EQUALIZATION USING ERROR CONTROL INFORMATION

5.1 *Description*

In many applications for adaptive equalization a forward-acting error correction system will be associated with the data transmission system. When the objective of system design is high-speed transmission, the modem is generally operated at an unacceptably high error rate. A detection-retransmission system cannot be solely relied upon, since the high error rate would necessitate constant requests for retransmission.

In the exploratory VSB system described in Refs. 2, 3, and 4, a (200, 175) Bose-Chaudhuri code with a minimum distance of 8 was used for triple error correction. In the event of a detectable error pattern containing more than three errors, a retransmission request was made. Using triple error correction with a modem error rate of roughly 2×10^{-3}, the frequency of requests for retransmission was low enough to not have appreciably affected the throughput of the system.

When the equalization is imperfect the error rate is naturally increased, but moreover the data system becomes pattern sensitive. Some patterns of input data are more likely to result in errors than other patterns because of the memory of the system. Given that an error has occurred, it is quite likely that such a bad pattern was transmitted. Since the bad patterns are simply related to the system impulse response we have the interesting possibility of using the information available in the error correction system for the purpose of adjusting the equalizer.

Briefly, a scheme based on this principle works as follows. Whenever an error is corrected by the error control unit, the direction of the error and the polarities of the surrounding symbols are observed. By "direction" of the error is meant whether a symbol has been changed to a higher amplitude level (+) or to a lower level (−). If the direction of the error is positive the polarities of the surrounding symbols are taken directly to the equalizer. If the direction of the error is negative all symbol polarities are inverted. These polarities are used to either advance or retard counters attached to the variable attenuators of the equalizer. The attenuators are incremented positively or negatively whenever the corresponding counters underflow or overflow. A block diagram of this system is shown in Fig. 6.

For a specific example, suppose that 16-level transmission is used. The incoming binary data train is converted 4 bits at a time into the 16 symbols using a Gray code. Suppose that at the receiver the following sequence is received

$$0011 \quad 1011 \quad 1\text{X}10 \quad 0110 \quad 0001.$$

The error control locates and corrects an error in the bit marked with an X. As part of the error correction procedure the entire 200-bit word has been stored, so the polarities of the symbols surrounding the error are readily available. In fact, with a Gray code the polarity of the amplitude level corresponding to a given 4-bit symbol is determined by the first bit. Therefore, the polarities of the two symbols either side of the error are, in order,

$$- + \text{(error)} - -.$$

The symbol which was in error was 1010 changed to 1110. From a table of the Gray code we can determine that this error carried amplitude level 13 into amplitude level 12. Thus, the direction of the error is nega-

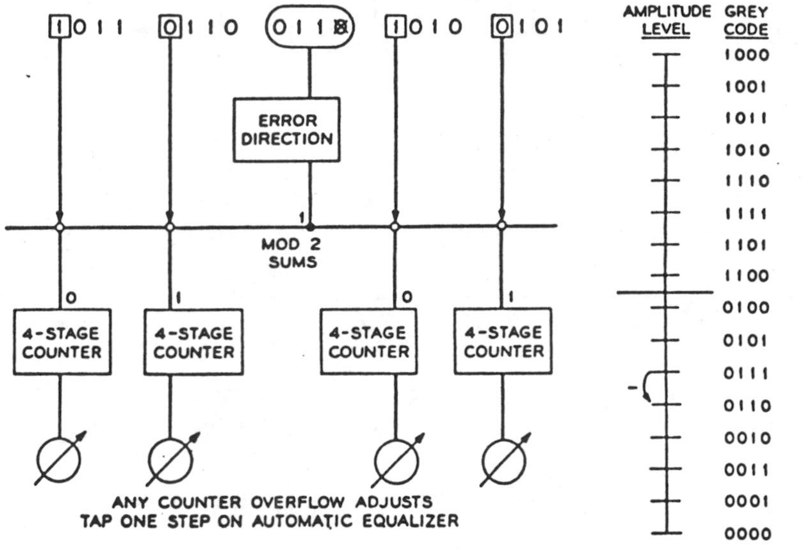

Fig. 6 — Example error control — adaptive equalizer coordination ($2C = 16$).

tive and the symbol polarities are reversed and used to increment counters.

Notice that in this system the storage counters are only incremented when errors occur. If no errors are being made, then the equalizer settings are not changed. Therefore, the counters are changed much less frequently in this system than in the previous adaptive equalizer. However, when the counters are changed, we shall find that the changes are more reliable and that smaller counters may be used to effect a comparably reliable statistical test.

5.2 *Analytical Evaluation*

Again we are going to evaluate the probability p of counting in the correct direction on the storage counter attached to attenuator c_j. We suppose that h_j is positive and with a spacing $2d$ between levels we have

$$p = P(a_{k-j} > 0 \mid e_k > d) + P(a_{k-j} < 0 \mid e_k < -d) \quad (62)$$

$$p = 2P(a_{k-j} > 0 \mid e_k > d). \quad (63)$$

With the M possible symbols equally likely we can write (63) in the form

$$p = \frac{\sum_{i=1}^{m/2} P[e_k > d \mid a_{k-j} = d(2i-1)]}{MP(e_k > d)}. \quad (64)$$

As in Section 4.2, we write the error voltage e_k

$$e_k = a_{k-j}h_j + [\sum_{n \neq k-j} a_n h_{k-n} + \eta_k] \quad (65)$$

and assume the bracketed term is Gaussian, mean zero, variance σ^2.

Thus,

$$P[e_k > d \mid a_{k-j} = d(2i-1)] = \frac{1}{\sqrt{2\pi}\sigma} \int_{d-d(2i-1)h_j}^{\infty} \exp\left(-\frac{x^2}{2\sigma^2}\right) dx \quad (66)$$

and for reasonably small error rates we make the approximation

$$P[e_k > d \mid a_{k-j} = d(2i-1)] \cong \frac{\sigma}{\sqrt{2\pi}\, d[1-(2i-1)h_j]} \exp\left(-\frac{d^2[1-(2i-1)h_j]^2}{2\sigma^2}\right). \quad (67)$$

If we also assume that when a_{k-j} is a random variable, e_k is Gaussian distributed and make the same approximation involved in (67), we arrive at

$$P(e_k > d) \cong \frac{\sqrt{\sigma^2 + Sh_j}}{\sqrt{2\pi}\, d} \exp\left(-\frac{d^2}{2(\sigma^2 + Sh_j)}\right). \quad (68)$$

Presumably $\sigma^2 \gg Sh_j$ when the system is near perfect equalization, so we drop the Sh_j terms. Equations (68) and (67) are then inserted into (64) to get

$$p = \frac{1}{M} \sum_{i=1}^{M/2} \frac{1}{[1-(2i-1)h_j]} \exp\left(-\frac{d^2[1-(2i-1)h_j]^2 + d^2}{2\sigma^2}\right). \quad (69)$$

Our aim in evaluating (69) is an approximation which is accurate to terms linear in h_j, a small number. The denominator in (69) does not contribute terms on this order, so we are able to sum the geometric series giving

$$p = \frac{1}{M} \exp\left(\frac{d^2 h_j}{\sigma^2}\right) \left[\frac{1 - \exp\left(\frac{Md^2h_j}{\sigma^2}\right)}{1 - \exp\left(\frac{2d^2h_j}{\sigma^2}\right)}\right]. \quad (70)$$

Finally, we retain only terms linear in h_j to obtain the result

$$p = \tfrac{1}{2} + \frac{Md^2h_j}{4\sigma^2}. \quad (71)$$

Equation (71) is similar in form to (49) for the probability of a correct count in the digitalized equalizer. The principal difference is that (71) involves the threshold-to-noise ratio squared (d^2/σ^2) whereas (49) uses the ratio (d/σ). Thus, p in (71) is considerably more reliable than the probability for a correct count in the digitalized adaptive equalizer. However, the counts in the error control system occur at a much slower rate, namely $2P(e_k > d)$.

A counter of capacity $2C$ is used to store the counts from the error correction circuitry. The equations of Feller may again be used with suitable approximations to find the probability p_c of a correct equalizer adjustment and the average number of counts $\bar{n}$ required before a correction.

$$p_c = 1 - \exp\left(-\frac{Ch_jd^2M}{\sigma^2}\right) \quad (72)$$

$$\bar{n} = \frac{2C\sigma^2}{Md^2h_j}. \quad (73)$$

In order to get the average number of symbols required before an equalizer adjustment, we must multiply $\bar{n}$ by the average number of symbols per error, which is approximately [from (68)]

$$n_0 = \frac{\sqrt{2\pi}\, d}{2\sigma e^{-d^2/2\sigma^2}}. \qquad (74)$$

The comparison of this system with the previous system becomes quite complicated because of the dependence of the error control system upon the number of levels M and on the error rate of the system (for which our approximation is only valid when equalization is exact). In general, it seems that for a given accuracy of equalization, the previous digitalized equalizer will require less time per adjustment. To follow through with our example we assume $h_j = 0.5\Delta$, $\Delta = 0.0025$, $S/\sigma^2 = 10^3$, and now $M = 16$. Since

$$S/\sigma^2 = \frac{d^2}{3\sigma^2}(M^2 - 1) \qquad (75)$$

the threshold-to-noise ratio is

$$d^2/\sigma^2 = 11.76. \qquad (76)$$

The probability of error for the system is about 6.5×10^{-4} and $n_0 = 1540$ symbols per error. From (72), we find that a counter capacity $2C = 40$ is required to ensure $p_c = 0.99$. This may be compared with $2C = 170$ for the previous system. The number of counts per adjustment $\bar{n} = 170$, but $n_0\bar{n} = 2.62 \times 10^5$ symbols per adjustment.

This comparison is somewhat unfair to the error control system since it must be pointed out that the speed of movement at $h_j = 0.5\Delta$ is immaterial. This condition merely determines the counter sizes necessary to meet accuracy requirements. We are much more concerned with the equalizer response when the equalization is imperfect. In this case, not only is $\bar{n}$ inversely proportional to h_j, but the error rate also increases so that counts are made more frequently. Even if one is willing to stretch the point quite a bit, it does seem that the error control equalizer coordination is unattractive in comparison with the previous adaptive system. Nevertheless the system implementation is quite simple and the concept sufficiently intriguing that perhaps a use can be found for such a coordination.

VI. EXPERIMENTAL RESULTS

Three systems for adaptive equalization of digital data systems have been described and analyzed. One of these three systems, the digitalized adaptive equalizer described in Section IV, appears to be much more attractive than the other two, both from the standpoint of instrumentation ease and of performance. Therefore, although the error control coordinated system has also been constructed, only the digitalized adaptive system has been subjected to extensive testing.

The system constructed used a 13-tap delay line and a tap increment Δ of 0.0025, although this increment is tapered to considerably smaller values near the outside taps of the delay line. In line with the example values computed in Section IV, eight stages were used in the storage counters, resulting in a capacity of $2C = 256$ counts for each tap.

The system was tested in conjunction with the 9600 bit-per-second, 16-level VSB system described by F. K. Becker in Ref. 2. A good pictorial demonstration of the adaptive equalizer is shown in the sequence of photos in Fig. 7 where the VSB system is transmitting at 4800 bits-per-second, using only four levels. The reduced speed here is to enable us to easily discern the distinct levels in the eye pictures of Fig. 7. The first photo shows the normal, equalized eye pattern. The intensified dots on this picture indicate the position of the sampling time which in this first photo has been artificially moved to the right, completely out of the eye opening. This is equivalent to the sudden introduction of a constant delay into the transmission channel. It can be observed in this first photo that the error rate would be relatively high due to the mistiming.

In the subsequent pictures the timing position has been left fixed while the adaptive equalizer changes its settings to move the entire eye pattern to the right effecting a reequalization of the system. At first the pattern moves quite rapidly since decisions are made quickly by the testing counters. As the eye approaches the timing position again decisions are made at a slower rate and the movement slows. The entire process takes only about a few seconds in spite of the quite abnormally large disturbance of the transmission characteristic.

Fig. 8 shows a sequence of photographs of a 16-level folded eye picture following a sudden change in transmission characteristics. If the 16-level eye were shown in the same format as the 4-level eyes of Fig. 4, the 15 "holes" in the eye diagram would be too small to distinguish. Therefore, all 15 "holes" have been superimposed by folding the 16-level eye diagram over and over until the picture resembles a binary eye diagram. In the first second of operation, from 0 to 1 second, the eye is recaptured by the adaptive equalizer.

At the positions shown in the first photos of Figs. 7 and 8 the equalizer

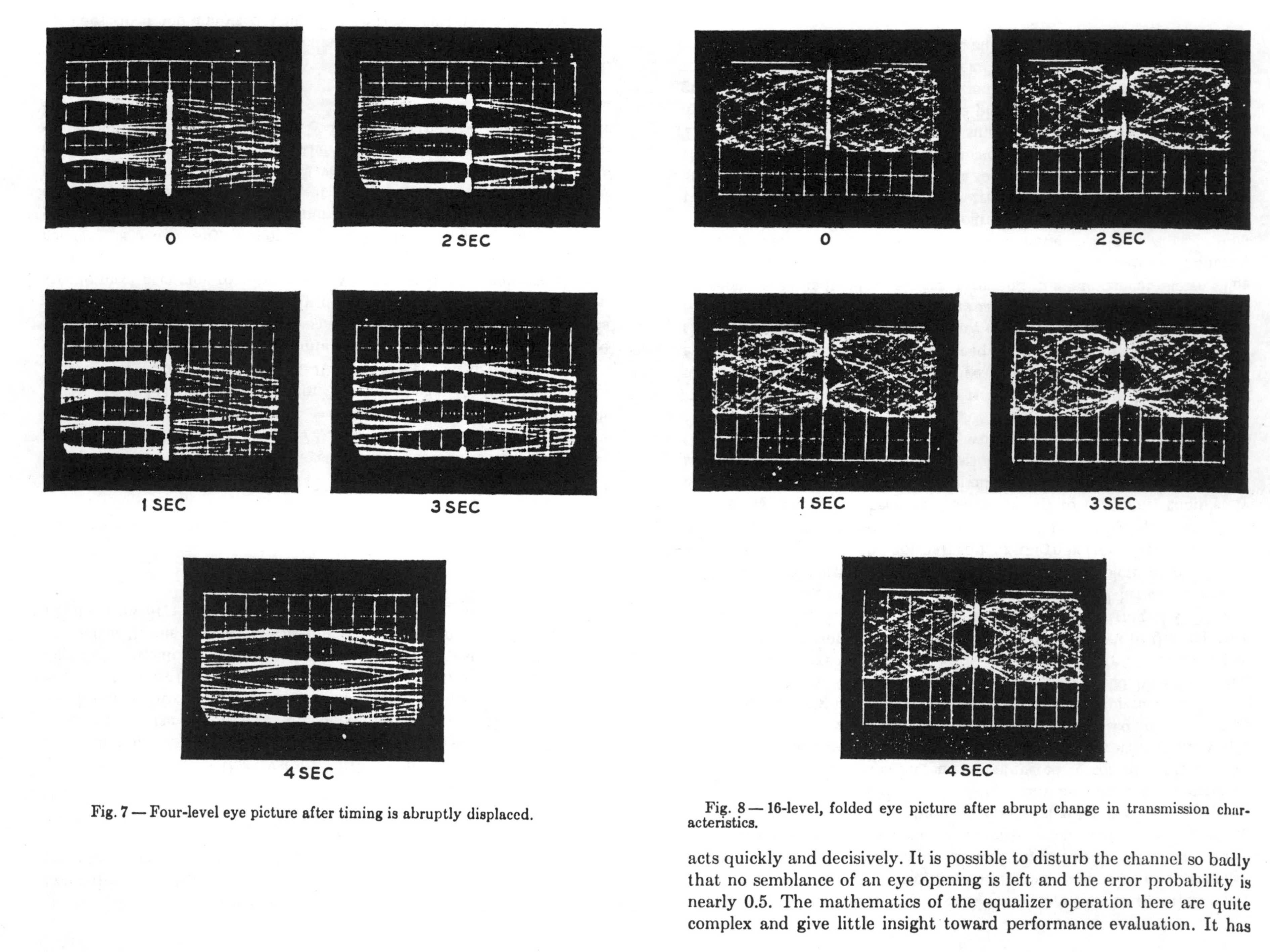

Fig. 7 — Four-level eye picture after timing is abruptly displaced.

Fig. 8 — 16-level, folded eye picture after abrupt change in transmission characteristics.

acts quickly and decisively. It is possible to disturb the channel so badly that no semblance of an eye opening is left and the error probability is nearly 0.5. The mathematics of the equalizer operation here are quite complex and give little insight toward performance evaluation. It has been observed experimentally that for binary (2-level) operation it has been impossible to find a setting or disturbance from which the equalizer will not eventually converge. Sometimes the eventuality takes as long as a minute as the equalizer makes slow (on the order of C^7) decisions — a sure indication of inaccuracy. After a period of what seems to be random hunting the equalizer reaches a position which it suddenly recognizes. The decisions come very quickly (on the order of C) and the eye appears seemingly from nowhere. Finally, perfect equalization is approached and the decisions once more come very slowly.

For higher level systems, 8- and 16-level, positions can be found from which the equalizer will not in all probability converge in a time that one is willing to wait around and watch. There seem to exist certain stable states where, for example, a 16-level eye pattern exists where 8-level transmission is being used — each level being split into two balanced levels. It should be emphasized that these conditions never occur during normal data transmission when the equalizer has an open eye to begin with and only has to track this eye through changing transmission characteristics. They are relevant, however, for the acquisition period if the equalizer is turned on without any setup period. Such procedure is not recommended for the higher level systems. (It will be obvious to the reader that it is possible to send a pattern of "outside" levels — i.e., a 2-level signal to start a higher level transmission. Also, it is possible to use quasi-random pattern generators, say maximal-length shift registers, at both transmitter and receiver to start transmission without worrying about transmission errors.)

A series of error performance runs was made on two test facilities — one looped via K-carrier to Boston from Holmdel, New Jersey and the other looped to Chicago via LMX-1 carrier from Holmdel. A number of 2 minute runs were made at 9600 bits-per-second at various times of day on each facility. In every case a control run of 2 minutes (1.152×10^6 bits) was made using preset equalization as described in Ref. 3. The results of these runs are plotted in Fig. 9 with preset error rate as the abscissa and adaptive error rate as the ordinate. In all cases, the error rate was diminished by the use of adaptive equalization. Even in the worst case the improvement factor was over three, while the best case represented an improvement factor of 50 and the average factor was approximately 10.

The importance of this improvement factor in the performance of this system cannot be overemphasized. While the order-of-magnitude improvement may not seem significant, it must be pointed out that this is raw error rate previous to error correction. If the curves relating customer error rate to raw error rate for the error control system used in

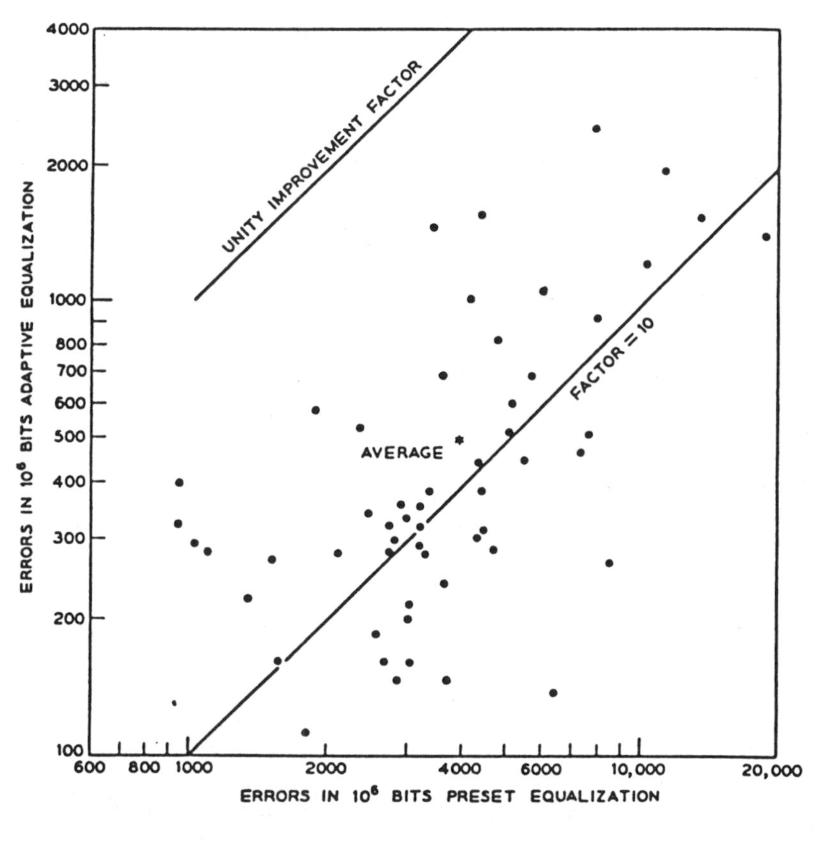

Fig. 9 — Error rates for adaptive and preset equalization on test runs.

conjunction with the VSB system which are shown in Ref. 4 are examined, it will be seen that the customer error rate is an extremely sensitive function of the raw error rate. An improvement of an order of magnitude in raw error rate results in an improvement of 2 or 3 orders of magnitude in the customer error rate. The customer error rates are not shown in Fig. 9 since in many cases the error rate for preset equalization was so high as to preclude synchronization, while frequently the adaptive error rate was too low to get any customer errors at all in the two-minute runs.

The question arises as to where the improvement comes from. First the improvement factor does *not* come from tracking the changing channel characteristics over the two-minute period. The error rate is virtually unchanged if the adaptive equalizer is turned off after a few second's operation. The improvement comes from a number of other factors, chief of which is the improved setting accuracy. The preset equalizer operates for a period of 7 seconds during which 100 test pulses per second are transmitted. After this period an equilibrium distribution of equalizer positions has been reached in which the average tap error is about Δ, whereas with the adaptive equalizer 2400 symbols per second are used to extract information and the average tap error has been designed to be close to the minimum value of 0.5Δ.

The other factors, whose importance has not been quantitatively assessed, are the possibility of a nonlinearity in the transmission channel so that different characteristics are presented to the data signals than to test pulses, the possibility that the test pulses have some overlap, and the possibility of some bias in the examination of test pulse sample polarities. There seems to be no question that adaptive equalization using received data is superior to test pulse equalization.

Finally, an attempt was made to measure the time variation of some of the test lines. A critical measure of this variation can be achieved by averaging in low-pass filters the outputs of the exclusive-OR circuits in Fig. 4. The filter outputs are extremely sensitive indicators of the parameters h_j, although they are difficult to calibrate since the proportionality constant is a function of the noise variance [see (49)]. The recorder outputs from these filters are purposely not reproduced here lest the reader grant too much significance to the time variation records obtained. Generally, about a peak variation of one percent was found on the center 3 taps (h_{-1}, h_0, and h_1) and a negligible amount on other taps. The period of variation varied from about 10 seconds to a few minutes and then grew short again periodically. It was easily demonstrated that the variation was due to phase wander in the carrier system. Such slow phase wander was subsequently tracked using more sophisticated phase control apparatus.

A record of the exact setting reached by each tap gain control using adaptive equalizer setup was made periodically over a period of several weeks. Within statistical error the values reached were constant with the exception of the three central, phase-sensitive taps. (The reason these taps are phase sensitive is that the quadrature pulse ideally has nonzero values only at these points.)

The transmission facilities have been designed through the years to be relatively insensitive to temperature and humidity changes. It would seem that even for the most critical current data transmission usage, the telephone channel's phase and amplitude characteristics can be assumed to be time invariant. Thus, it seemed that adaptive equalization was more useful as a refining device after test pulse equalization and as an insurance system rather than as a tracking equalizer for a significantly time varying channel. Nevertheless, the potentialities of the system considerably exceeded the demands of the environment and even so its benefits were quite strikingly apparent.

REFERENCES

1. Lucky, R. W., Automatic Equalization for Digital Communication, B.S.T.J., *44*, April, 1965, pp. 547–588.
2. Becker, F. K., An Exploratory, Multi-Level, Vestigial Sideband Data Terminal for Use on High Grade Voice Facilities, IEEE Annual Communications Convention, Boulder, Colorado, June 7–9, 1965.
3. Lucky, R. W., An Automatic Equalization System for Use with a High Speed, Voiceband Data Set, IEEE Annual Communications Convention, Boulder, Colorado, June 7–9, 1965.
4. Burton, H. O. and Weldon, E. J., An Error Control System for Use with a High Speed, Voiceband Data Set, IEEE Annual Communications Convention, Boulder, Colorado, June 7–9, 1965.
5. Feller, W., *An Introduction to Probability Theory and Its Applications*, John Wiley & Sons, New York, 1950.

Adaptive Equalization of Highly Dispersive Channels for Data Transmission

By ALLEN GERSHO

(Manuscript received May 29, 1968)

This paper analyzes an adaptive training algorithm for adjusting the tap weights of a tapped delay line filter to minimize mean-square intersymbol interference for synchronous data transmission. The significant feature of the adjustment procedure is that convergence is guaranteed for all channel response pulses, even for very severe amplitude and phase distortion.

The author examines convergence, rate of convergence, and the effect of noisy observations of the received pulses, and he shows that the noisy observations result in a random sequence of tap weight settings whose mean value converges to a suboptimal setting. The mean-square deviation of the tap weights from the suboptimal values is asymptotically bounded with a bound that can be made as small as desired by sufficiently reducing the speed of convergence.

The suboptimality arising here results from the use of isolated test pulses for the training signal. However, a training scheme using pseudorandom sequences or the actual data signal does not suffer from the suboptimality effect. Hence, although of possible utility in other pulse shaping applications, the technique presented here appears to be primarily of value in providing a conceptual framework for the closely related but more practical techniques to be examined in the sequel to this paper to be published shortly.

I. INTRODUCTION

A common approach to data transmission is to code the amplitudes of successive pulses in a periodic pulse train with a discrete set of possible amplitude levels. The coded pulse train is then linearly modulated, transmitted through the channel, demodulated, equalized, and synchronously sampled and quantized. As a result of dispersion of the pulse shape by the channel, the number of detectable amplitude levels has very often been limited by intersymbol interference rather than by additive noise.

In principle, if the channel is known precisely it is virtually always possible to design an equalizer that will make the intersymbol interference (at the sampling instants) arbitrarily small. However, in practice a channel is random in the sense of being one of an ensemble of possible channels. Consequently, a fixed equalizer designed on average channel characteristics may not adequately reduce intersymbol interference. An adaptive equalizer is then needed which can be "trained," with the guidance of a suitable training signal transmitted through the channel, to adjust its parameters to optimal values. If the channel is also time-varying, an adaptive equalizer operating in a tracking mode is needed which can update its parameter values by tracking the changing channel characteristics during the course of normal data transmission. In both cases the adaptation may be achieved by observing or estimating the error between actual and desired equalizer responses and using this error to estimate the direction in which the parameters should be changed to approach the optimal values.

A simple and effective technique for adaptive equalization was developed by Lucky using the tapped delay line filter structure for the equalizer.[1, 2] The main limitation of this technique is that convergence of the tap weight adjustment algorithm is assured only for relatively low dispersion channels. The convergence condition requires that the dispersed pulse shape have adequate quality so that, in the absence of noise, error-free binary data transmission would be possible without equalization. In other words the dispersed pulse must have an open binary "eye."

Using an approach to adaptation[3, 4] with virtually unrestricted convergence properties, Lucky and Rudin subsequently proposed and implemented an adaptive equalizer for minimizing the mean square error in frequency response of an analog channel.[5, 6] This approach was applied to synchronous data transmission by the author and independently by Lytle and by Niessen.[7–9] An implementation of the technique was described by Niessen and Drouilhet.[10] It has also been implemented for data communication at Bell Laboratories.

In this paper the approach is used for synchronous data transmission in a training mode where a sequence of isolated pulses is used as a test signal. The technique may be viewed equally as an adaptive design procedure for a sampled-data pulse shaping filter where the

Reprinted with permission from *The Bell System Technical Journal,* vol. 48, no 1, pp. 55–70, Jan. 1969.

error criterion is to minimize the mean square error between actual and desired pulse shapes at the filter output. The important feature of the technique is that convergence is achieved for any channel pulse response whatever, thereby including highly dispersed pulses for which even binary data transmission would be impossible without equalization. Of particular interest are: (*i*) the analogous optimality condition to Lucky's zero forcing condition resulting with the change from a summed absolute error to a summed squared error criterion,[1] (*ii*) the manner in which noisy observations introduce randomness in the iterative corrections to the weights and the resulting stochastic convergence properties, (*iii*) the possibility of applying the technique where isolated pulses applied to a filter must be used to adaptively adjust the filter for optimum pulse shaping (unrelated to equalization), and (*iv*) the conceptual framework for the more practical adaptation techniques to be described in a sequel to this paper, planned for publication soon.

Perhaps the earliest application of the tapped delay line or "transversal" filter to pulse shaping for data transmission was made by W. P. Boothroyd and E. M. Creamer.[11] Tufts and George have shown that under a mean-square error criterion the optimal receiver structure includes a tapped-delay line filter with delay between taps equal to the symbol period.[12, 13] Aaron and Tufts have also shown that the same receiver structure is needed to minimize the average error probability for binary data transmission.[14]

The basic approach to adaptive adjustment of a set of weights where a mean-square error criterion is used with a gradient search procedure was considered by Widrow and Hoff who noticed that no derivative computation is needed.[3] Narendra and McBride proposed a self-optimizing Wiener filter using a continuous-time gradient algorithm and a filter structure whose transfer function is a weighted sum of fixed functions.[4] Koford and Groner used a mean-square error criterion and a gradient learning algorithm to find an optimum set of weights for pattern classifying.[15] Widrow described a general adaptive filtering problem with the tapped delay line filter.[16] Coll and George discussed the performance of George's optimum equalizer and indicated a possible adaptive adjustment technique.[17] Lucky and Rudin were the first to apply the mean square error criterion with the gradient search procedure to the field of adaptive equalization.[5, 6] This paper expands on a short presentation given at an international symposium on information theory.[18]

II. PERFORMANCE OBJECTIVES FOR EQUALIZATION

The objective of equalization, viewed as a pulse shaping problem, is to adjust the parameters of the equalizer to a setting which minimizes a suitable measure of the error between actual and desired pulse shapes. For the usual synchronous data transmission application, the desired pulse shape is one with the Nyquist property that the sample values y_k at the sampling instant kT are given by $y_k = \delta_{kr}$ where δ_{kr} is unity for $k = r$ and zero for all other integers k. The criterion used by Lucky[1] is peak distortion, D, given by

$$D = \sum_{k \neq r} | y_k | / | y_r | .$$

An alternate criterion of interest is the mean square distortion E, defined by

$$E = \sum_{k \neq r} y_k^2 / y_r^2 .$$

The physical interpretation of the peak distortion is that it is directly related to eye opening and determines the error probability for a worst case message pattern. The mean square distortion has a different interpretation. If the message pattern is such that the transmitted level for each time slot is statistically independent of the levels for other time slots, then the variance of the intersymbol interference in a given time slot is proportional to the mean square distortion. If the pulse shape has a large number of small sidelobes so that the intersymbol interference is normally distributed, then minimizing mean square distortion is equivalent to minimizing error rate.

Closely related to the mean square distortion is the mean square error

$$\mathcal{E} = \sum_k (y_k - d_k)^2 \qquad (1)$$

where d_k is the desired pulse sample value at time instant kT. For the usual equalization problem where $d_k = \delta_{kr}$, the measure $\mathcal{E}$ has virtually the same interpretation as E; however, E is a normalized measure independent of pulse amplitude while $\mathcal{E}$ depends on both shape and amplitude. Optimization of the tapped delay line equalizer with respect to either criterion leads to equivalent results.

III. FORMULATION

Consider the transversal equalizer with N taps and tap spacing T equal to the symbol period. Let c_k be the weight at the kth tap for

$k = 0, 1, \cdots, N-1$ so that the input output relation of the transversal filter at the sample times is

$$y_n = \sum_{k=0}^{N-1} c_k x_{n-k} = \mathbf{c}'\mathbf{x}_n \tag{2}$$

where x_k and y_k denote the input and output pulse samples, respectively, at time instants kT, $\mathbf{c} = (c_0, c_1, \cdots, c_{N-1})$ is the tap weight vector, and $\mathbf{x}_n = (x_n, x_{n-1}, \cdots, x_{n-N+1})$ is the sample memory state of the delay line at the time instant nT; the vectors $\mathbf{c}$ and $\mathbf{x}_n$ are to be regarded as column matrices, and the prime denotes the transpose. We assume that the input sequence x_k has finite energy. Let $\epsilon_n = y_n - d_n$. Then from equation (1), using (2), the gradient of the error with respect to $\mathbf{c}$ may be written as

$$\nabla \mathcal{E} = 2 \sum_k \epsilon_k \mathbf{x}_k . \tag{3}$$

Therefore the optimality condition for minimum error $\nabla \mathcal{E} = 0$ is equivalent to the requirement that the (deterministic) corss-correlation between the input sequence x_k and output error sequence ϵ_k must have zeros for the N components with index values corresponding to the index values of the available tap weights. That is,

$$\varphi_{x\epsilon}(k) = \sum_n \epsilon_n x_{n-k} = 0 \qquad \text{for} \quad k = 0, 1, 2, \cdots, N-1.$$

This condition has an interesting similarity to Lucky's condition which states that the peak distortion, D, is minimized when the error sequence ϵ_n has zeros for the N components with index values corresponding to the index values of the available tap weights.[1] An important distinction is that Lucky's condition is generally not valid when the input pulse distortion D exceeds unity, while the mean square optimizing condition is valid for any input pulse with finite mean square distortion.

Using equation (2), the gradient (3) can be expressed explicitly as a function of the tap weight vector $\mathbf{c}$, namely:

$$\nabla \mathcal{E} = 2(\mathbf{Ac} - \mathbf{g}) \tag{4}$$

where

$$\mathbf{A} = \sum_n \mathbf{x}_n \mathbf{x}_n', \qquad \text{and} \qquad \mathbf{g} = \sum_n d_n \mathbf{x}_n .$$

Notice that $\mathbf{A}$ is symmetric and positive definite (see Appendix A). Setting equation (4) equal to zero yields the solution for the optimum tap weight vector $\mathbf{c}^*$,

$$\mathbf{c}^* = \mathbf{A}^{-1}\mathbf{g}.$$

Using equation (2), the error expression given by equation (1) may be expressed in the convenient form:

$$\mathcal{E}(\mathbf{c}) = \mathcal{E}(\mathbf{c}^*) + (\mathbf{c} - \mathbf{c}^*)'\mathbf{A}(\mathbf{c} - \mathbf{c}^*) \tag{5}$$

which shows explicitly the simple quadratic nature of the error surface and the unique optimality of the minimizing weight vector $\mathbf{c}^*$. It can be shown that the residual error $\mathcal{E}(\mathbf{c}^*)$ can be made as small as desired for all channels of practical interest by using a sufficiently large number, N, of taps.[19]

It is intuitively reasonable that successive corrections to the tap weight vector in the direction of steepest descent of the error surface should lead to the minimum error where $\mathbf{c} = \mathbf{c}^*$. This is the idea of the well-known[20] gradient algorithm:

$$\mathbf{c}_{i+1} = \mathbf{c}_i - \tfrac{1}{2}\alpha \nabla \mathcal{E}(\mathbf{c}_i) \tag{6}$$

where α is a suitably small positive proportionality constant, $\mathbf{c}_0$ is arbitrary, and $\mathbf{c}_i$ is the tap weight vector after the ith iteration.

The significant feature of the gradient algorithm for our quadratic error surface (5) is that the gradient can be conveniently evaluated without knowledge of the error surface itself. We have seen from equation (3) that the components of the gradient vector are values of the crosscorrelation between the input sequence and the output error sequence. This suggests the conceptually simple implementation where an isolated test pulse is transmitted through the channel and the requisite crosscorrelation values are formed by multiplying the delayed input pulse with the error pulse, sampling, and summing (or averaging). The tap weights are then incremented according to (6), the old crosscorrelation values "dumped" and a new iteration is begun with the transmission of a new test pulse.

The error pulse is formed by subtracting from the equalizer output pulse an "ideal" pulse whose sample values are the desired values d_k; the ideal pulse is locally generated at the appropriate time. The basic scheme is shown in Fig. 1. Naturally, the summation given by equation (3) cannot be performed over an infinite time interval. Suppose κT is a practical upper bound on the possible time duration of the input pulse, ξT is the time interval between successive test pulses with $\xi T > \kappa T$, ξ and κ as positive integers. Then if we include the effect of perturbing

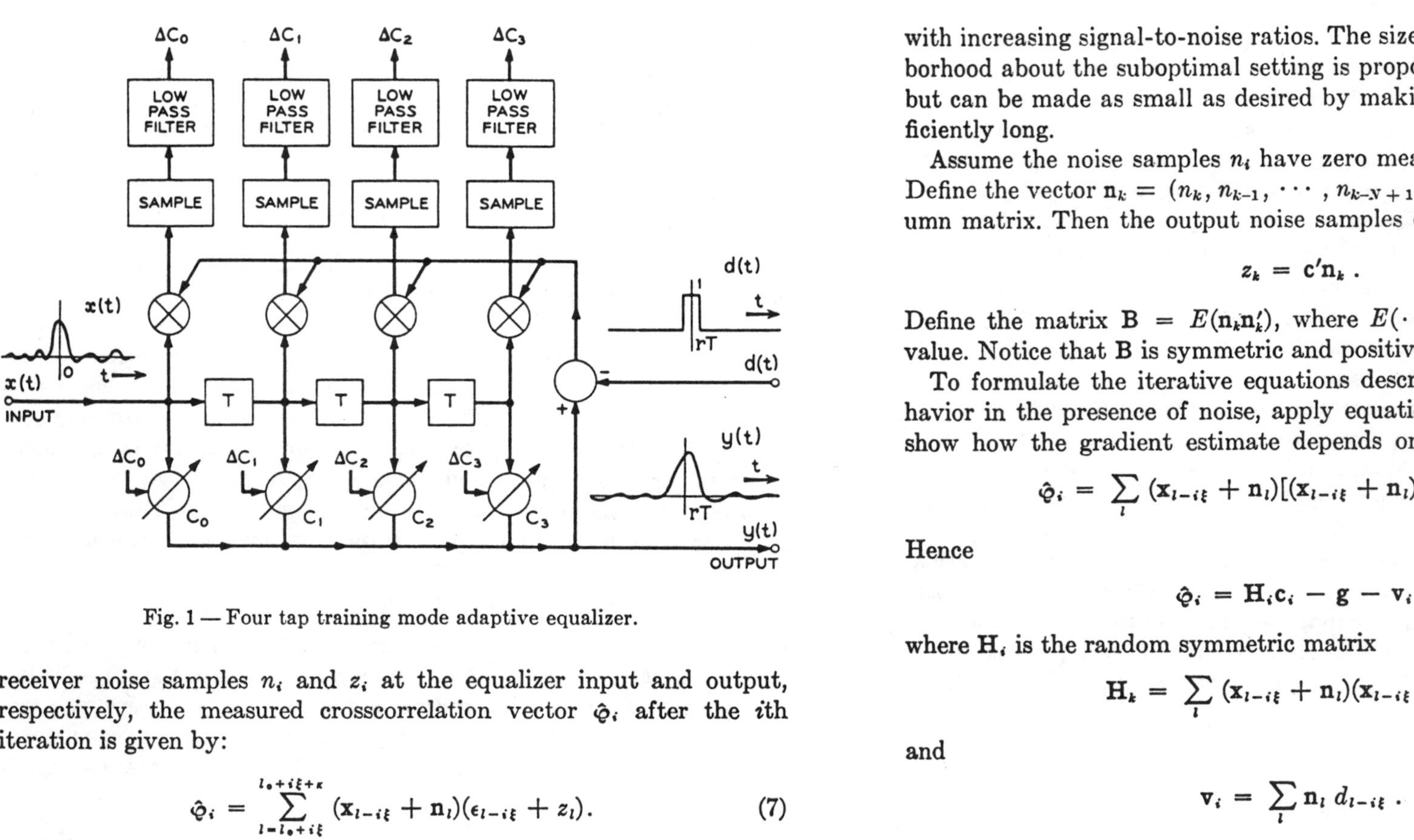

Fig. 1 — Four tap training mode adaptive equalizer.

receiver noise samples n_i and z_i at the equalizer input and output, respectively, the measured crosscorrelation vector $\hat{\varphi}_i$ after the ith iteration is given by:

$$\hat{\varphi}_i = \sum_{l=l_0+i\xi}^{l_0+i\xi+\kappa} (\mathbf{x}_{l-i\xi} + \mathbf{n}_l)(\epsilon_{l-i\xi} + z_l). \tag{7}$$

In the noiseless case the estimate $\hat{\varphi}_i$ reduces to one-half the deterministic gradient, that is, $\frac{1}{2}\nabla\mathcal{E}(\mathbf{c}_i)$ under the assumption that the pulse sequence x_l and desired sequence d_l are virtually zero outside of the interval $l_0 \leqq l \leqq l_0 + \kappa - N + 1$.

IV. CONVERGENCE PROPERTIES

In the presence of noise the tap weight corrections contain undesired random components consisting of products of input and output noise samples and products of pulse and noise sample. As a result, the random tap weights no longer converge to the optimal values but instead approach some neighborhood of a suboptimal setting and then fluctuate randomly about this setting. The error between the optimal and suboptimal settings is small for low noise levels and decreases with increasing signal-to-noise ratios. The size of the fluctuation neighborhood about the suboptimal setting is proportional to the noise level but can be made as small as desired by making the training time sufficiently long.

Assume the noise samples n_i have zero mean and finite variance σ^2. Define the vector $\mathbf{n}_k = (n_k, n_{k-1}, \cdots, n_{k-N+1})$ to be regarded as a column matrix. Then the output noise samples of the equalizer are:

$$z_k = \mathbf{c}'\mathbf{n}_k . \tag{8}$$

Define the matrix $\mathbf{B} = E(\mathbf{n}_k\mathbf{n}_k')$, where $E(\cdots)$ denotes the expected value. Notice that $\mathbf{B}$ is symmetric and positive semidefinite.

To formulate the iterative equations describing the tap weight behavior in the presence of noise, apply equations (2) and (8) to (7) to show how the gradient estimate depends on the tap weight vector:

$$\hat{\varphi}_i = \sum_l (\mathbf{x}_{l-i\xi} + \mathbf{n}_l)[(\mathbf{x}_{l-i\xi} + \mathbf{n}_l)'\mathbf{c}_i - d_{l-i\xi}].$$

Hence

$$\hat{\varphi}_i = \mathbf{H}_i\mathbf{c}_i - \mathbf{g} - \mathbf{v}_i , \tag{9}$$

where $\mathbf{H}_i$ is the random symmetric matrix

$$\mathbf{H}_k = \sum_l (\mathbf{x}_{l-i\xi} + \mathbf{n}_l)(\mathbf{x}_{l-i\xi} + \mathbf{n}_l)' \tag{10}$$

and

$$\mathbf{v}_i = \sum_l \mathbf{n}_l\, d_{l-i\xi} . \tag{11}$$

Let $\mathbf{\alpha} = E(\mathbf{H}_i)$, the expected value of $\mathbf{H}_i$. Then equation (10) yields

$$\mathbf{\alpha} = \mathbf{A} + \kappa\mathbf{B}. \tag{12}$$

which is positive definite since $\mathbf{A}$ is positive definite and $\mathbf{B}$ is positive semidefinite.

It is convenient to examine the random variation of the tap weight vector c_k about the suboptimal setting defined by

$$\bar{\mathbf{c}} = \mathbf{\alpha}^{-1}\mathbf{g}, \tag{13}$$

and let $\mathbf{q}_i = \mathbf{c}_i - \bar{\mathbf{c}}$. From equation (12) it is evident that the suboptimal setting $\bar{\mathbf{c}}$ approaches the optimal setting $\mathbf{c}^*$ as the ratio of noise variance to input pulse sequence energy approaches zero. The iterative algorithm may be expressed in the form

$$\mathbf{q}_{i+1} = \mathbf{q}_i - \alpha\hat{\varphi}_i , \tag{14}$$

$$\hat{\varphi}_i = \mathbf{H}_i \mathbf{q}_i + \mathbf{h}_i \tag{15}$$

where

$$\mathbf{h}_i = \mathbf{H}_i \bar{\mathbf{c}} - \mathbf{g} - \mathbf{v}_i \, . \tag{16}$$

Equations (14) and (15) constitute a system of first-order stochastic difference equations with a forcing function $\mathbf{h}_i$ which is statistically dependent on the stochastic state matrix $\mathbf{H}_i$. We assume that the perturbing noise samples in different iterations are uncorrelated, so that $\mathbf{H}_i$ and $\mathbf{h}_i$ are independent of $\mathbf{H}_j$ and $\mathbf{h}_j$ for $i \neq j$. Notice that the expected value of any function of $\mathbf{H}_i$ and $\mathbf{h}_i$ is independent of i. Under these conditions it is proved in Appendix C that for suitably small values of α the mean value of the solution vector $\mathbf{q}_i$ approaches zero as $i \to \infty$ and the sum of the variances of the components of $\mathbf{q}_i$ is bounded with a bound that approaches zero as α approaches zero. Consequently the mean value of the tap weight vector converges to the suboptimal setting $\bar{\mathbf{c}}$ while the actual tap weights fluctuate randomly about the converging mean values with a variability that can be made arbitrarily small.

Notice from Appendix C that the norm of the mean solution vector $\langle \mathbf{q} \rangle_i$ is reduced at least by the factor ζ, the spectral norm[20] of $I - \alpha\mathfrak{A}$. Let ρ_1 and ρ_N denote the minimum and maximum eigenvalues, respectively, of $\mathfrak{A}$. Then

$$\zeta = \min |\, 1 - \alpha\rho_1 \,| , \quad |\, 1 - \alpha\rho_N \,| \, . \tag{17}$$

(For proof see p. 24 of Ref. 20.)

Then for $0 < \alpha < 2/(\rho_1 + \rho_N)$, we obtain $\zeta = 1 - \alpha\rho_1$. Consequently, while decreasing α offers a smaller bound on variability of the tap weight vector, increasing α assures a stronger bound on convergence rate. For the training mode it is likely that speed of adaptation will be relatively unimportant so that a very small value of α could be used to approach a tap weight setting that is very close to the suboptimal setting.

It is useful to obtain bounds on the eigenvalues of $\mathfrak{A}$ which can be determined without specific knowledge of the channel characteristics. If $x(t)$ denotes the channel pulse response and $n(t)$ the additive receiver noise so that the sampled values used earlier are given by $x_k = x(kT)$ and $n_k = n(kT)$, then the sampled spectrum $X^*(\omega)$ of x_k is

$$X^*(\omega) = \sum_k x_k e^{-i\omega kT} = \frac{1}{T} \sum_k X(\omega - 2\pi/T)$$

and the sampled spectral density $S^*(\omega)$ of n_k is

$$S^*(\omega) = \sum_k E(n_i n_{i+k}) e^{-i\omega kT} = \frac{1}{T} \sum_k S(\omega - 2\pi/T)$$

where $X(\omega)$ is the Fourier transform of $x(t)$ and $S(\omega)$ is the spectral density of $n(t)$. Let m and M denote the infimum and supremum, respectively, of $|\, X^*(\omega) \,|^2 + \kappa S^*(\omega)$ so that

$$m \leqq |\, X^*(\omega) \,|^2 + \kappa S^*(\omega) \leqq M. \tag{18}$$

In all cases of practical interest M will be finite; furthermore generally m will be greater than zero. It is shown in Appendix B that each eigenvalue, ρ_i , of $\mathfrak{A}$ will be bounded according to

$$m \leqq \rho_i \leqq M. \tag{19}$$

To illustrate the use of this bound, notice from Appendix C that the condition for convergence of the mean tap weight vector to the suboptimal solution is that $\alpha < 2/\rho_N$. Thus a sufficient condition is that

$$\alpha < 2/M. \tag{20}$$

Furthermore, the mean tap setting converges exponentially with the convergence factor ζ, given by equation (17). Hence it can be inferred that the choice of α which provides the strongest bound (least value of ζ) is $\alpha = 2/(\rho_1 + \rho_N)$ yielding

$$\zeta = \frac{p - 1}{p + 1}$$

where $p = \rho_N/\rho_1$. Using the bounds given in (19) we obtain $p \leqq M/m$, and so

$$\zeta = \frac{M - m}{M + m}. \tag{21}$$

Therefore, for the best choice of α, convergence of the mean proceeds at least at a rate given by the geometric factor $(M - m)/(M + m)$. Thus useful information regarding the convergence speed can be determined without knowledge of the channel characteristics.

V. CONCLUSION

The degree of suboptimality of the tap weight setting reached by the training algorithm may or may not be consequential, depending on the application. In applications where multilevel pulse transmission with a large number of levels could be achieved with adequate equalization, the signal-to-noise ratio is necessarily very high and therefore the degree of suboptimality is not large. Even when the noise level is fairly substantial the suboptimal setting may still be adequate if the error surface given by $\mathcal{E}(\mathbf{c})$ is "shallow" in a large neighborhood of the minimum. Then a fairly large departure of $\tilde{\mathbf{c}}$ from $\mathbf{c}^*$ may correspond to a relatively small increase in mean-square error. Also, if training mode adaptation is used as a prelude to tracking mode adaptation, a fairly large degree of suboptimality may be a tolerable starting point for a tracking mode operation such as the one we plan to describe in a future paper.

When the noise level is substantial the criterion for optimality used here becomes inadequate because it does not consider the effect of the equalizer on the receiver noise. The price of reducing intersymbol interference may be a sizable increase in noise level at the equalizer output. In our future paper the error criterion is modified to include noise with the result that the problem of suboptimality does not arise.

The random fluctuation of the tap weights which prevents true convergence to the suboptimal setting can be eliminated by reducing the proportionality constant α in each iteration using a sequence of step sizes α_k with the properties

$$\sum \alpha_k = \infty \qquad \text{and} \qquad \sum \alpha_k^2 < \infty .$$

It may then be shown that the tap weight vector converges to the suboptimal solution with probability 1. The proof uses stochastic approximation theory and follows the lines taken by Tong and Liu who considered a training mode algorithm for low dispersion channels.[21] However, this modification complicates the implementation somewhat and cannot be applied to the tracking mode adaptation problem.

APPENDIX A

Proof that A *is Positive Definite*

The matrix $\mathbf{A}$ is defined by

$$\mathbf{A} = \sum_{k=-\infty}^{\infty} \mathbf{x}_k \mathbf{x}_k' . \tag{22}$$

Consequently

$$\mathbf{c}'\mathbf{A}\mathbf{c} = \sum_{-\infty}^{\infty} \mathbf{c}'\mathbf{x}_k\mathbf{x}_k'\mathbf{c} = \sum_{-\infty}^{\infty} y_k^2 .$$

But the sequence y_k is the convolution of the x_k sequence with the finite tap weight sequence c_k. Hence, using Parseval's equality,

$$\mathbf{c}'\mathbf{A}\mathbf{c} = \frac{1}{2\pi} \int_{-\pi/T}^{\pi/T} | X^*(\omega) |^2 \, | C(\omega) |^2 \, d\omega, \tag{23}$$

where

$$C(\omega) = \sum_{k=0}^{N-1} c_k e^{-ik\omega T} .$$

Equation (23) shows immediately that $\mathbf{c}'\mathbf{A}\mathbf{c}$ is nonnegative for all vectors $\mathbf{c}$. Also, $C(\omega)$ can have only isolated zeros and $| X^*(\omega) |$ is square integrable since the input pulse has finite mean square distortion. It may then be inferred that $\mathbf{c}'\mathbf{A}\mathbf{c} > 0$ unless $\mathbf{c} = 0$, which proves that $\mathbf{A}$ is positive definite.

APPENDIX B

Bounds on the Eigenvalues of $\mathfrak{A}$

Since $\mathbf{B} = E(\mathbf{n}_k\mathbf{n}_k')$ the quadratic form $\mathbf{c}'\mathbf{B}\mathbf{c}$ is the mean squared value of $\mathbf{y}_k^2$ of the response of the equalizer with weight vector $\mathbf{c}$ to the input noise n_l . Consequently

$$\mathbf{c}'\mathbf{B}\mathbf{c} = \frac{1}{2\pi} \int_{-\pi/T}^{\pi/T} S^*(\omega) \, | C(\omega) |^2 \, d\omega. \tag{24}$$

Combining equations (23) and (24) yields

$$\mathbf{c}'\mathfrak{A}\mathbf{c} = \frac{1}{2\pi} \int_{-\pi/T}^{\pi/T} \{ | X^*(\omega) |^2 + \kappa S^*(\omega) \} \, | C(\omega) |^2 \, d\omega. \tag{25}$$

Applying to equation (25) the bounds m and M given by equation (18) yields

$$m \, \mathbf{c}'\mathbf{c} \leqq \mathbf{c}'\mathfrak{A}\mathbf{c} \leqq M \mathbf{c}'\mathbf{c}. \tag{26}$$

Let $\mathbf{c}$ be the eigenvector of $\mathfrak{A}$ corresponding to eigenvalue ρ. Then $\mathfrak{A}\mathbf{c} = \rho_i\mathbf{c}$ and equation (26) yields

$$m \leqq \rho_i \leqq M \tag{27}$$

which provides a convenient bound for the largest and smallest eigenvalues of $\mathfrak{A}$.

APPENDIX C

Convergence Proof

To examine the convergence properties of the tap weight adjustment algorithm, it is convenient to define the norm of a random vector $\mathbf{u}$ as

$$|| \mathbf{u} || = [E(\mathbf{u}'\mathbf{u})]^{1/2}, \tag{28}$$

so that the squared norm of $\mathbf{u}$ is the sum of the second moments of the components of $\mathbf{u}$. For a deterministic vector the norm reduces to the usual Euclidian norm. The norm of a deterministic matrix will denote the usual spectral norm.[20]

Theorem: Let $\mathbf{H}_k$ *be a sequence of random symmetric* $N \times N$ *matrices and* $\mathbf{h}_k$ *a sequence of random N-tuple column vectors. Suppose* $\mathbf{H}_k$ *and* $\mathbf{h}_k$ *are stationary in k with* $\mathbf{H}_k$ *and* $\mathbf{h}_k$ *independent of* $\mathbf{H}_j$ *and* $\mathbf{h}_j$ *for* $k \neq j$. *Assume* $\mathbf{h}_k$ *has zero mean, and the elements of* $\mathbf{H}_k$ *and* $\mathbf{h}_k$ *have finite variance,* $E\mathbf{H}_k = \mathfrak{A}$, *independent of k with* $\mathfrak{A}$ *positive definite. Define the random vector sequence* $\mathbf{q}_k$ *according to:*

$$\mathbf{q}_{k+1} = \mathbf{q}_k - \alpha \varphi_k \tag{29}$$

where

$$\varphi_k = \mathbf{H}_k \mathbf{q}_k + \mathbf{h}_k \tag{30}$$

for $k = 0, 1, 2, \cdots$ *and* $\mathbf{q}_0$ *is an arbitrary deterministic vector. Then for* α *positive and sufficiently small,*

$$\lim_{k\to\infty} \| E\, \mathbf{q}_k \| = 0 \tag{31a}$$

and

$$\limsup_{k\to\infty} \| \mathbf{q}_k \| \leqq V(\alpha) \tag{31b}$$

with $V(\alpha)$, *given in* (47), *satisfying:*

$$\lim_{\alpha\to 0} V(\alpha) = 0. \tag{32}$$

Proof: Combining equations (29) and (30) yields

$$\mathbf{q}_{k+1} = (\mathbf{I} - \alpha \mathbf{H}_k)\mathbf{q}_k - \alpha \mathbf{h}_k\,. \tag{33}$$

Noting that $\mathbf{q}_k$ is independent of $\mathbf{H}_k$, taking the expected value in equation (33), we find

$$E(\mathbf{q}_{k+1}) = (\mathbf{I} - \alpha \mathfrak{A})E(\mathbf{q}_k). \tag{34}$$

It follows then that

$$\| E(\mathbf{q}_k) \| \leqq \zeta^k \| E\, \mathbf{q}_0 \| \tag{35}$$

where

$$\zeta = \| I - \alpha \mathfrak{A} \|\,. \tag{36}$$

Hence equation (31a) follows when $\zeta < 1$, or equivalently, for

$$0 < \alpha < 2/\rho_N \tag{37}$$

where ρ_N is the largest eigenvalue of $\mathfrak{A}$.

To prove equation (31b), observe that

$$E(\mathbf{q}_{k+1}'\mathbf{q}_{k+1}) = E[\mathbf{q}_k'(I - \alpha H_k)^2\mathbf{q}_k] - E[2\alpha \mathbf{q}_k'(I - \alpha H_k)\mathbf{h}_k] + \alpha^2 \| \mathbf{h}_k \|^2 \tag{38}$$

from equation (33). Noting again that $\mathbf{q}_k$ is independent of H_k, we have

$$E[\mathbf{q}_k'(I - \alpha \mathbf{H}_k)^2\mathbf{q}_k] = E\{\mathbf{q}_k' E[(I - \alpha \mathbf{H}_k)^2]\mathbf{q}_k\} \leqq \mu \| \mathbf{q}_k \|^2, \tag{39}$$

where

$$\mu = \| E[(I - \alpha \mathbf{H}_k)^2] \|\,. \tag{40}$$

Also, using the Schwarz inequality,

$$E[-\mathbf{q}_k'(I - \alpha \mathbf{H}_k)\mathbf{h}_k] = \alpha \mathbf{q}_k' E(\mathbf{H}_k \mathbf{h}_k) \leqq \alpha \| \mathbf{q}_k' \| f$$

where $f = \| E(\mathbf{H}_k \mathbf{h}_k) \|$. Using equation (35) we obtain

$$-E[\mathbf{q}_k(I - \alpha \mathbf{H}_k)\mathbf{h}_k] \leqq \alpha \zeta^k \| E(\mathbf{q}_0) \| f. \tag{41}$$

The bounds (39) and (41) may be applied to equation (38), yielding

$$\| \mathbf{q}_{k+1} \|^2 \leqq \mu \| \mathbf{q}_k \|^2 + \alpha^2 f \| E\, \mathbf{q}_0 \| \zeta^k + \alpha^2 \| \mathbf{h}_k \|^2\,. \tag{42}$$

If we now define the bounding sequence of positive numbers Q_k according to

$$Q_0 = \| E\, \mathbf{q}_0 \|^2$$

and

$$Q_{k+1} = \mu Q_k + \alpha^2 f \| E(\mathbf{q}_0) \| \zeta^k + \alpha^2 \|\mathbf{h}_k \|^2\,, \tag{43}$$

then it follows from (42) that

$$\| \mathbf{q}_k \|^2 \leqq Q_k\,.$$

But the difference equation given by (43) has the asymptotic solution

$$\lim_{k\to\infty} Q_k = \frac{\alpha^2 \| \mathbf{h}_k \|^2}{1 - \mu}\,,$$

for $\zeta < 1$ and $\mu < 1$. Then

$$\limsup_{k\to\infty} \| q_k \|^2 \leqq \frac{\alpha^2 \| \mathbf{h}_k \|^2}{1 - \mu}. \tag{44}$$

Notice that $\| h_k \|$ is independent of k by the hypothesis of stationarity.

Since

$$(I - \alpha \mathbf{H}_k)^2 = (I - \alpha \mathfrak{A})^2 + \alpha^2 E(\mathbf{G}_k^2).$$

where

$$\mathbf{G}_k = \mathbf{H}_k - \mathfrak{A}, \tag{45}$$

we find that

$$\begin{aligned} \mu &\leqq \| I - \alpha \mathfrak{A} \|^2 + \alpha^2 \| E(\mathbf{G}_k^2) \| \\ \mu &\leqq \zeta^2 + \alpha^2 \gamma \end{aligned} \tag{46}$$

where $\gamma = \| \mathbf{G}_k^2 \|$. Furthermore for $\alpha < 2/(\rho_1 + \rho_N)$, we have $\zeta = 1 - \alpha\rho_1$. Then, using (46), we see that

$$\frac{\alpha^2}{1 - \mu} \leqq \frac{\alpha^2}{2\alpha\rho_1 + \alpha^2(\rho_1^2 + \gamma)}.$$

We have therefore shown that for positive and sufficiently small α, equations (31b) and (32) are valid where

$$V(\alpha) = \frac{\alpha}{2\rho_1 + \alpha^2(\rho_1^2 + \gamma)}. \tag{47}$$

REFERENCES

1. Lucky, R. W., "Automatic Equalization for Digital Communication," B.S.T.J., *44,* No. 4 (April 1965), pp. 547–588.
2. Lucky, R. W., "Techniques for Adaptive Equalization of Digital Communication Systems," B.S.T.J., *45,* No. 2 (February 1966), pp. 255–286.
3. Widrow, B., and Hoff, M. E., Jr., "Adaptive Switching Circuits," IRE Wescon Conv. Record, Pt. 4 (August 1960), pp. 96–104.
4. Narenda, K. S. and McBride, L. E., "Multiparameter Self-Optimizing Systems Using Correlation Techniques," IEEE Trans. Automatic Control, *AC-9,* No. 1 (January 1964), pp. 31–38.
5. Lucky, R. W. and Rudin, H. R., Generalized Automatic Equalization for Communication Channels", Proc. IEEE (Letters), *54,* No. 3, pp. 439–40, March 1966.
6. Lucky, R. W. and Rudin, H. R. An Automatic Equalizer for General-Purpose Communication Channels, B.S.T.J., *46,* No. 9 (November 1967), pp. 2179–2208.
7. Gersho, A., unpublished work.
8. Lytle, D. W., unpublished work.
9. Niessen, C. W., "Automatic Channel Equalization Algorithm," Proc. IEEE, *55,* No. 5 (May 1967), p. 689.
10. Niessen, C. W., and Drouilhet, P. R., "Adaptive Equalizer for Pulse Transmission," Conf. Digest, IEEE Int. Conf. on Commun. (June 1967), p. 117.
11. Boothroyd, W. P. and Creamer, E. M., Jr., "A Time Division Multiplexing System," Trans. AIEE, *68,* Pt. I, (1949) pp. 92–97.
12. Tufts, D. W., "Nyquist's Problem—the Joint Optimization of Transmitter and Receiver in Pulse Amplitude Modulation," Proc. IEEE, *53,* No. 3 (March 1965), pp. 248–259.
13. George, D. A., "Matched Filters for Interfering Signals," IEEE Trans. Inform. Theory *IT-11,* No. 1 (January 1965), pp. 153–154.
14. Aaron, M. R. and Tufts, D. W., "Intersymbol Interference and Error Probability," IEEE Trans. Inform. Theory *IT-12,* No. 1 (January 1966), pp. 26–34.
15. Koford, J. S., and Groner, G. F., "The Use of an Adaptive Threshold Element to Design a Linear Optimal Pattern Classifier," IEEE Trans. Inform. Theory, *IT-12,* No. 1 (January 1966), pp. 42–50.
16. Widrow, B., "Adaptive Filters I: Fundamentals," Stanford Electronics Lab., Stanford, Calif., Report SEL-60-126 (Technical Report 6764-6), December 1966.
17. Coll, D. C. and George, D. A., "A Receiver for Time-Dispersed Pulses, Conf. Record, 195, IEEE Annual Commun. Conv., pp. 753–757.
18. Gersho, A., "Automatic Equalization of Highly Dispersive Channels for Data Transmission," Int. Symp. on Inform. Theory, San Remo, Italy, September 1967.
19. Gersho, A. and Freeny, S. L., "Performance Capabilities of Transversal Equalizers for Digital Communication," Conf. Digest, IEEE Int. Conf. on Commun. 1967, p. 88.
20. Goldstein, A. A., *Constructive Real Analysis,* New York: Harper and Row, 1967, Chap. 1.
21. Tong, P. S. and Liu, B., "Automatic Time Domain Equalization in the Presence of Noise," Proc. Nat. Elec. Conf., *23,* (October 1967), pp. 262–266.
22. Varga, R. S., *Matrix Iterative Analysis,* New York: Prentice-Hall, 1962.

Application of Least Squares Lattice Algorithms to Adaptive Equalization

E. H. SATORIUS AND J. D. PACK

Abstract—**In many applications of adaptive data equalization, rapid initial convergence of the adaptive equalizer is of paramount importance. Apparently, the fastest known equalizer adaptation algorithm is based on a recursive least squares estimation algorithm. In this paper we show how the least squares lattice algorithms, recently introduced by Morf and Lee, can be adapted to the equalizer adjustment algorithm. The resulting algorithm, although computationally more complex than certain other equalizer algorithms (including the fast Kalman algorithm), has a number of desirable features which should prove useful in many applications.**

I. INTRODUCTION

Traditionally, equalization of a baseband pulse amplitude modulated (PAM) system [1] has been accomplished with a tapped delay line (TDL) filter. Recently, however, the application of lattice structures to channel equalization has been presented in [2]. Lattice structures offer a number of potential advantages over TDL filters. First, an N-stage lattice filter automatically generates all of the outputs which could be generated by N different TDL filters with lengths ranging from 1 to N filter taps. This property of lattice structures allows the dynamic assignment of the particular length of the lattice equalizer which proves most effective at any instant of equalization. A second advantage of lattice structures is that longer lattice filters may be built up from shorter ones by simply adding on more lattice stages. This property should prove useful in designing large scale systems which employ lattice equalizers. Finally, an important property of lattice structures in general is their high insensitivity to roundoff noise. This aspect of fixed coefficient (nonadaptive) digital lattice filters has already been investigated by a number of authors (see, e.g., [3]). However, the effects of quantization errors on digital adaptive lattice filters has not been as thoroughly studied, although preliminary results discussed in [17] suggest that, at least for a particular adaptive lattice and adaptive TDL filter, the lattice filter was less sensitive to quantization errors than the TDL filter. Certainly, this is an important area for future investigation.

In this paper we will be primarily concerned with the adaptation algorithm that is used to update the filter coefficients in the lattice equalizer. In [2] a gradient estimation algorithm was used to update the lattice filter coefficients. Although the resulting adaptive equalizer algorithm was found to have a faster initial startup than a conventional TDL gradient adaptive equalizer, an even faster startup can be achieved with a least squares adaptive equalizer algorithm. The least squares algorithm, which is a special case of the Kalman estimation algorithm, was first applied to channel equalization by Godard in a seminal paper [8]. Simulations presented by Godard, as well as others (e.g., [9]), have confirmed the extremely rapid startup properties of the least squares adaptive equalizer. One disadvantage with the Godard algorithm, however, is that the complexity (number of additions and multiplications) of the algorithm grows quadratically with the number of filter coefficients. An important recent development is the work of Morf, Ljung, Lee, and others [4]–[7], [10], who have shown how the complexity of the conventionally implemented least squares algorithms (e.g., Godard's algorithm) can be made to grow only linearly with the number of filter coefficients. Furthermore, these computationally simpler least squares algorithms may be implemented either in TDL or lattice form. The application of the TDL form (i.e., the fast Kalman algorithm) to channel equalization has already been considered by Falconer and Ljung [10]. In this paper we show how the least squares lattice algorithms, originally introduced by Morf and Lee in [4]–[7], can be adapted to the equalizer adjustment algorithm. The extremely rapid startup properties of the least squares lattice equalizer are confirmed by computer simulation.

The basic outline of this paper will consist of first briefly reviewing the general least squares problem from which the least squares lattice equalizer arises. Then in Section III the least squares lattice equalizer algorithm will be presented, and finally in Section IV simulation results will be presented and discussed.

II. LEAST SQUARES CHANNEL EQUALIZATION

As noted in [10], apparently the fastest converging equalizer algorithm (which is a special case of the Kalman estimation algorithm [8]–[10]) arises in the context of the classical least squares problem; at each sampling instant, find the set of equalizer tap coefficients that minimize the accumulation of the squared errors between the filter output and a desired output up to that time. In particular, suppose $a(0), a(1) \cdots, a(n)$ is a training sequence of data symbols which are known by the equalizer adjustment algorithm. Further, assume that the $a(n)$ sequence is transmitted over a channel, resulting in a sequence of equalizer inputs, $x(0), x(1), \cdots, x(n)$ and define $X_N(n)$ to be an $(N+1)$-dimensional vector consisting of time-delayed samples of $x(n)$, i.e.,

$$X_N(n) = (x(n), x(n-1), \cdots, x(n-N))^T \tag{1}$$

where a superscript T denotes transpose. The least squares problem is equivalent to finding the $(N+1)$-dimensional vector of equalizer tap values $-F_N(n)$, which minimizes the following sum of squared errors:[1]

$$\sum_{p=0}^{n} w^{n-p}\{a(p) + F_N{}^T(n)X_N(p)\}^2.$$

The parameter w is a real constant $0 \leqslant w \leqslant 1$ and is included in the sum of squared errors in order to allow the equalizer to track slow time variations in the channel. Typically, w is close

[1] In this paper a negative sign convention is used in representing the equalizer tap vector $-F_N(n)$.

Paper approved by the Editor for Communication Theory of the IEEE Communications Society for publication after presentation at the 13th Annual Asilomar Conference on Circuits, Systems, and Computers, Pacific Grove, CA, November 1979. Manuscript received May 30, 1979; revised July 20, 1980. This work was supported by the Naval Ocean Systems Center Internal Research Project ZR94.

E. H. Satorius is with Dynamics Technology, Inc., Torrance, CA 90505.

J. D. Pack is with the Naval Ocean Systems Center, San Diego, CA 92152.

Reprinted from *IEEE Trans. Comm.*, vol. COM-29, no. 2, pp. 136–142, Feb. 1981.

to 1. The inverse of $(1 - w)$ is, approximately, the memory of the equalizer.

Differentiating the above sum with respect to the components of $F_N(n)$ and setting the result to zero leads to the following equation for the $F_N(n)$ vector:

$$R_N(n)F_N(n) = -\sum_{p=0}^{n} w^{n-p} a(p) X_N(p). \tag{2}$$

In (2) the $(N+1) \times (N+1)$ matrix $R_N(n)$ is given by

$$R_N(n) = \sum_{p=0}^{n} w^{n-p} X_N(p) X_N^T(p). \tag{3}$$

The solution of (2) and (3) provides the least squares equalizer tap vector $-F_N(n)$ at the nth data sample. Notice that in the above equations the limits on all the summation signs extend from $p = 0$ to $p = n$. The lower limit, therefore, imposes the assumption on the data that $x(p) = 0$ for $p = -1, \cdots, -N$. These limits lead to the so-called "pre-windowed" least squares algorithm (which we will consider in this paper). If the limits are $p = N$ and $p = n$, the unwindowed or "convariance" algorithm is obtained. A more complete discussion of the different windowing methods may be found in [5].

At this point it should be noted that the solution of (2) and (3) could proceed basically along one of two possible lines of development. The first method of development arises by noting that $F_N(n)$ obeys the following time update relation [8], [10]:

$$F_N(n) = F_N(n-1) - \beta_N(n)\xi(n) \tag{4a}$$

where

$$\beta_N(n) = R_N(n)^{-1} X_N(n) \tag{4b}$$

and

$$\xi(n) = a(n) + F_N(n-1)^T X_N(n). \tag{4c}$$

As is seen, the major computational burden in this formulation lies in computing the $(N+1)$-dimensional vector $\beta_N(n)$ at each iteration. Conventional methods of computing $\beta_N(n)$ (i.e., the Godard algorithm [8]) require on the order of N^2 operations (additions and multiplications) per update. However, by exploiting certain shift properties of $X_N(n)$, it is possible to compute $\beta_N(n)$ at each iteration using a number of operations which depends only linearly on N (i.e., the fast Kalman algorithm [10]). In any case, it is seen that this formulation, i.e., (4) of the solution to (2) leads to a TDL equalizer structure for which $-F_N(n)$ represents the vector of equalizer taps and $-F_N(n-1)^T X_N(n)$ represents the equalizer output.

A second method of solution of (2) is the lattice method which is the subject of this paper. In this method, certain properties of $R_N(n)$ (which are discussed in the Appendix) are exploited to derive *order* update equations for the outputs $y_m(n) = -F_m(n-1)^T X_m(n) (m = 0, 1, \cdots, N)$. In particular, lattice recursion relations are derived which relate $y_m(n)$ to $y_{m-1}(n)$ as well as certain other auxiliary variables. In this way all of the least squares outputs $y_m(n) (m = 0, \cdots, N)$ are automatically generated. This is in contrast to the TDL formulation where only $y_N(n)$ is computed every iteration. The least squares lattice equalizer algorithm is presented in the next section.

III. THE LEAST SQUARES LATTICE ALGORITHM

In this section we present the lattice, least squares equalizer algorithm. This algorithm is a modified version of the least squares algorithm originally presented in [4]-[7]. In the Appendix we present a brief summary of the algorithm's development. A complete derivation is given in [12]. Also, a more complete discussion of least squares lattice algorithms in general and their properties may be found in [4]-[7], as well as [11]-[16].

In the least squares adaptive lattice equalizer (LSALE) algorithm, the following scalar quantities are stored and updated every iteration:

1) $\epsilon_m^e(n)$, $\epsilon_m^r(n)$, $r_m(n)$, $e_m(n)$, $\bar{e}_m(n)$, $\bar{k}_m(n)$, $y_m(n)$ $(m = 0, 1, \cdots, N)$.
2) $\gamma_m(n)$, $k_m(n) (m = 0, 1, \cdots, N-1)$.

To initialize the algorithm we set

$$\epsilon_m^e(-1) = \epsilon_m^r(-2) = \epsilon_m^r(-1) = \delta \tag{5a}$$

$$r_m(-1) = k_m(-1) = \bar{k}_m(-1) = \gamma_m(-1) = 0. \tag{5b}$$

Then, the following computations are performed starting with $n = 0$:

$$e_0(n) = r_0(n) = x(n) \tag{6a}$$

$$\epsilon_0^e(n) = \epsilon_0^r(n) = w\, \epsilon_0^e(n-1) + x^2(n) \tag{6b}$$

$$y_{-1}(n) = \gamma_{-1}(n) = \gamma_{-1}(n-1) = 0 \tag{6c}$$

$$\bar{e}_{-1}(n) = a(n), \tag{6d}$$

$$k_m(n) = wk_m(n-1) + (1 - \gamma_{m-1}(n-1)) e_m(n) r_m(n-1) \tag{7a}$$

$$e_{m+1}(n) = e_m(n) - \frac{k_m(n-1) r_m(n-1)}{\epsilon_m^r(n-2)} \tag{7b}$$

$$r_{m+1}(n) = r_m(n-1) - \frac{k_m(n-1)}{\epsilon_m^e(n-1)} e_m(n) \tag{7c}$$

$$\epsilon_{m+1}^e(n) = \epsilon_m^e(n) - \frac{k_m^2(n)}{\epsilon_m^r(n-1)} \tag{7d}$$

$$\epsilon_{m+1}^r(n) = \epsilon_m^r(n-1) - \frac{k_m^2(n)}{\epsilon_m^e(n)} \tag{7e}$$

$$\gamma_m(n) = \gamma_{m-1}(n) + \frac{\{(1 - \gamma_{m-1}(n)) r_m(n)\}^2}{\epsilon_m^r(n)} \tag{7f}$$

$$y_m(n) = y_{m-1}(n) + \frac{\bar{k}_m(n-1)}{\epsilon_m^r(n-1)} r_m(n) \tag{7g}$$

$$\bar{e}_m(n) = a(n) - y_m(n). \tag{7h}$$

The recursions in (7a)-(7f) are computed for $m = 0, 1, \cdots, N-1$. The recursions in (7g)-(7h) are computed for $m = 0, 1, \cdots, N$. Finally, the coefficients $\bar{k}_m(n)$ are updated by the

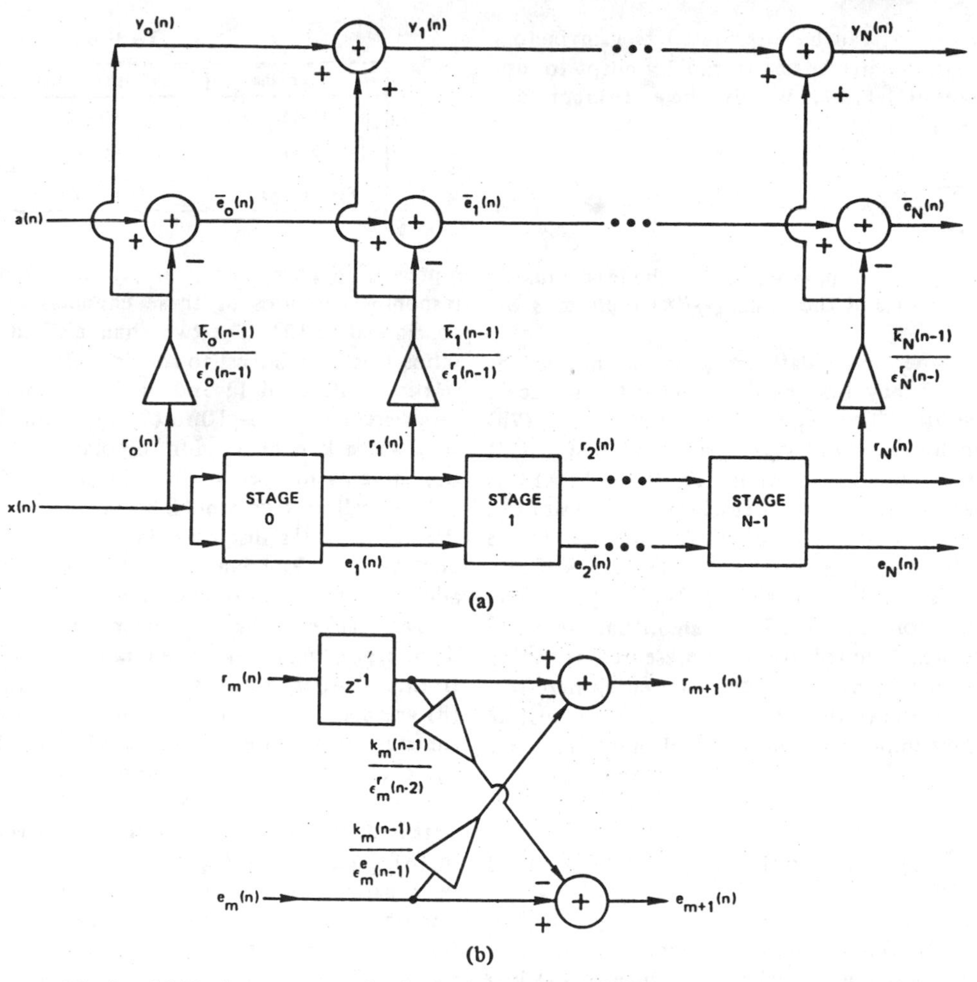

Fig. 1. (a) The least squares, adaptive lattice equalizer. (b) The mth stage of the lattice.

recursion

$$\bar{k}_m(n) = w\bar{k}_m(n-1) + (1-\gamma_{m-1}(n))\bar{e}_{m-1}(n)r_m(n) \quad (8)$$

$$m = 0, \cdots, N.$$

It should be noted that δ in (5a) is a small positive constant which is used to ensure nonsingularity of the $R_N(n)$ matrix.

Equations (5)–(8) represent the LSALE algorithm. A schematic diagram of the lattice equalizer is presented in Fig. 1(a) and (b). The $\epsilon_m{}^r$ are the least squares backward prediction errors of order m for $m = 0, \cdots, N-1$; i.e.,

$$\epsilon_m{}^r(n) = \min_b \sum_{p=0}^{n} w^{n-p} \cdot \left\{x(p-m) + \sum_{i=0}^{m-1} b_i{}^m(n)x(p-i)\right\}^2. \quad (9)$$

Let $\hat{b}_i{}^m(n)$ be the minimizing coefficients. Then $r_m(n)$, the backward error residual at time n, is given by

$$r_m(n) = x(n-m) + \sum_{i=0}^{m-1} \hat{b}_i{}^m(n-1)x(n-i). \quad (10)$$

(Note the presence of $\hat{b}_i{}^m(n-1)$ in (10) and not $\hat{b}_i{}^m(n)$, i.e., the current backward error residuals $r_m(n)$ are formed from the previous backward prediction error filter coefficients calculated at time $n-1$.) Similarly, the $\epsilon_m{}^e$ are the least squares forward prediction errors

$$\epsilon_m{}^e(n) = \min_a \sum_{p=0}^{n} w^{n-p} \cdot \left\{x(p) + \sum_{i=0}^{m-1} a_i{}^m(n)x(p-i-1)\right\}^2 \quad (11)$$

and the $e_m(n)$ are the forward error residuals at time n

$$e_m(n) = x(n) + \sum_{i=0}^{m-1} \hat{a}_i{}^m(n-1)x(n-i-1). \quad (12)$$

As discussed in [13], the constants $k_m{}^e \equiv k_m/\epsilon_m{}^e$ and $k_m{}^r \equiv k_m/\epsilon_m{}^r$ are equivalent to the so-called reflection or partial correlation (PAR COR) coefficients. In many lattice algorithms, $k_m{}^e$ and $k_m{}^r$ are constrained to equal each other (see [2]). However, in the least squares version of the lattice equalizer, $k_m{}^e \neq k_m{}^r$ in general. The constants $\bar{k}_m/\epsilon_m{}^r$ are the lattice equalizer taps which multiply the orthogonal signals $r_m(n)$[2] to form the equalizer output $y_N(n)$.

[2] The $r_m(n)$ are orthogonal in a least squares sense for all n greater than the maximum filter order (see [15]). This is in contrast to the gradient lattice equalizer where the input signals to the equalizer taps are only asymptotically orthogonal [2].

The constants $1 - \gamma_m(n)$ can be interpreted as gain factors which enable the least squares lattice to rapidly adapt to sudden changes in the input data [7]. Finally, the $\bar{e}_m(n)$ represent least squares error residuals.

$$\bar{e}_m(n) = a(n) - \sum_{i=0}^{m} \hat{f}_i^m(n-1)x(n-i) \tag{13}$$

where the $\hat{f}_i^m(n-1)$ are the components of the least squares tap vector $-F_m(n-1)$ and the scalar $y_N(n)$ represents the LSALE output.

As noted in [2], when actual data are being transmitted, the LSALE output $y_N(n)$ may first be thresholded to produce a reference sequence $\hat{a}(n)$. Then $\bar{e}_m(n)$ is computed as in (7h) with $a(n)$ replaced by $\hat{a}(n)$, and $\bar{k}_m(n)$ is updated by (8). It should be noted that the above formulation of the LSALE is for the scalar input case (i.e., a linear equalizer). Extensions to the vector input case (which is required, for instance, in the case of decision feedback equalization [10]) are straightforward and are discussed further in [4]-[7], [14], and [16]. In [14] extensions of the LSALE algorithm presented here to decision feedback equalization are presented. It should also be noted that the LSALE algorithm presented here represents a modified version of the least squares lattice algorithm presented in [7]. One important feature of this modification is our use of the residuals $\bar{e}_m(n)$ as opposed to the residuals

$$\hat{e}_m(n) = a(n) - \sum_{i=0}^{m} \hat{f}_i^m(n)x(n-i) \tag{14}$$

which were used in [7]. Our version of the algorithm is directly applicable to decision-directed equalization, since $\bar{e}_m(n)$ is a function of the previous filter coefficients $\hat{f}_i^m(n-1)$ [or equivalently, the $\bar{k}_m(n-1)$]. This permits one to calculate the current lattice equalizer coefficients $\bar{k}_m(n)$ via thresholding as described above. In contrast, the version of the algorithm which employs the $\hat{e}_m(n)$ residuals requires a current value of the reference input $a(n)$ in order to update its coefficients ([7], [11]). This, of course, would only be applicable to reference-directed equalization where $a(n)$ is known at the receiver.

A count of the number of operations needed to calculate each equalizer output and update the LSALE algorithm [(5)-(8)] indicates that approximately $19L$-12 multiplications and $11L$-7 additions are required where $L = N + 1$ is the total number of equalizer taps. It is interesting to compare these figures with those corresponding to the fast Kalman [10] as well as the gradient lattice equalizer algorithm (referred to as ALCE in [2]). The results are summarized in Table I. For all of these algorithms, divisions are counted as multiplications. As is seen, both of the lattice algorithms have more multiplications than the TDL fast Kalman algorithm. This is basically due to the power normalizations which are required at each stage of the lattice. It is interesting to note that LSALE is more complex than ALCE. This increased complexity is the price which must be paid for the faster convergence rate of the LSALE algorithm. The improved convergence performance of LSALE over ALCE will be observed in the next section.

TABLE I

Algorithm	Multiplications	Additions
LSALE	19L-12	11L-7
ALCE	13L-8	10L-7
Fast Kalman	10L+4	12L+5

IV. SIMULATION RESULTS

For purposes of examining the convergence performance of LSALE, we have simulated this algorithm for two channels representing pure, heavy amplitude distortion. The impulse response structures of these channels are identical to those considered in [2]. The two channels had eigenvalue disparity ratios (ratio of largest-to-smallest eigenvalues of channel correlation matrix) of 11 and 21. In all simulations 11-tap equalizers were used [$N = 10$ in (7)-(8)] and the symbol sequence $a(n)$ was a known, random bipolar training sequence ($a(n) = \pm 1$), suitably delayed so that the minimum mean square error TDL equalizer taps would be symmetric about the center of the equalizer (as discussed in [2]). Also, a small amount of uncorrelated Gaussian noise (noise variance = 0.001) was added to the output of the channel.

The results of the simulations are presented in Figs. 2 and 3 corresponding to eigenvalue ratios of 11 and 21, respectively. In these figures, δ in (5a) was set equal to 0.001 and w in (6)-(8) was set equal to one. (Additional simulation revealed that the startup performance of LSALE was highly insensitive to the choice of δ.) It was found in our simulation studies that the best startup performance for the LSALE algorithm was obtained by suppressing the update of the $\bar{k}_m(n)$ coefficients in (8), (i.e., holding $\bar{k}_m(n) = 0$, $m = 0, \cdots, N$) until after the main part of the first data pulse was received. The number of iterations in Figs. 2 and 3, however, are reckoned from the initial updating of (6)-(7) and not from the initial updating of the $\bar{k}_m(n)$ coefficients. It should be noted that the improved performance of LSALE which results from suppressing the $\bar{k}_m(n)$ update is analogous to the improved startup performance of the fast Kalman equalizer, which results from initially suppressing the tap update $F_N(n)$ in (4a), as discussed in [10].

For the sake of comparison, we have also included in Figs. 2 and 3 corresponding learning curves for the gradient lattice algorithm (ALCE) described in [2], as well as the simple gradient TDL equalizer algorithm (also described in [2]). In face, the parameters of the ALCE algorithm and the simple TDL gradient algorithm are identical to those parameters used in [2], which provided the best startup performance (i.e., the $P_0 = 0.075$ curves in Figs. 5 and 6 of [2]). As is clearly seen in Figs. 2 and 3, the convergence rate of both lattice algorithms is highly insensitive to the channel eigenvalue disparity. However, the LSALE algorithm converges in approximately 40 iterations for both channels, whereas ALCE required approximately 120 iterations to converge. The improved convergence performance of both lattice algorithms over the TDL gradient algorithm is clearly observed.

V. CONCLUSIONS

In this paper we have shown how the new least squares, lattice algorithms may be applied to data equalization. In particular, a new least squares adaptive lattice equalizer (LSALE) algorithm was presented and the improved initial convergence performance of LSALE over gradient based algorithms (implemented in either TDL or lattice structure) was clearly observed in Section IV. Although LSALE is computationally more complex than certain other equalizer algorithms (including the fast Kalman algorithm), its modular structure may potentially pro-

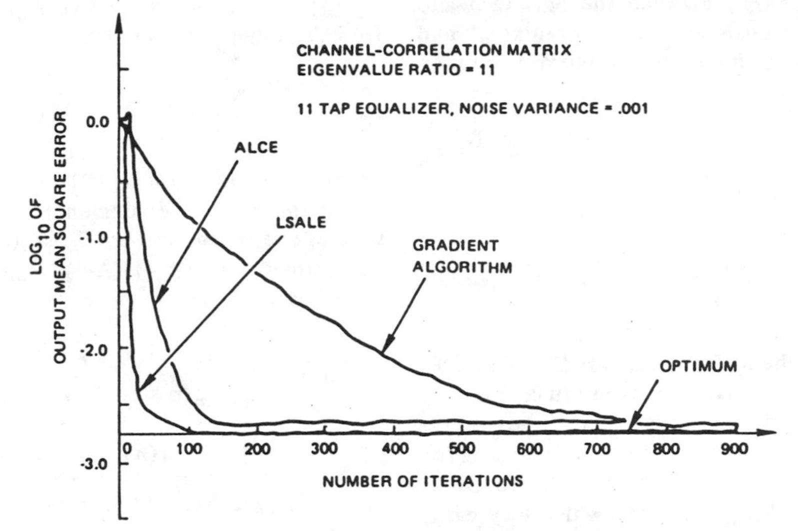

Fig. 2. Comparison by simulation of convergence properties for eigenvalue ratio = 11.

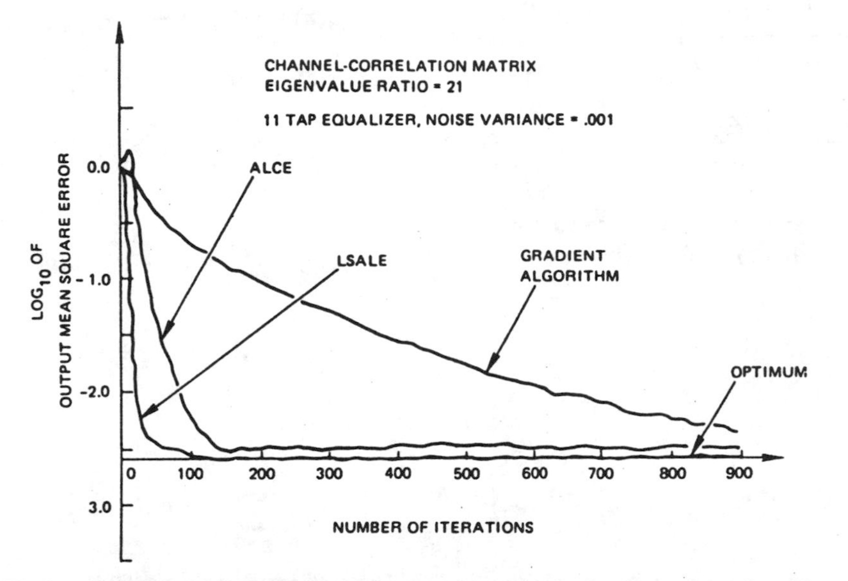

Fig. 3. Comparison by simulation of convergence properties for eigenvalue ratio = 21.

vide certain advantages over conventional TDL equalizers (such as greater insensitivity to quantization errors). Also, it should be pointed out that recent work [16] has shown how the computational complexity of the least squares lattice algorithm may be reduced by using square-root normalized forms. As noted in [16], the normalized lattice forms may have even better numerical properties (e.g., insensitivity to quantization errors) than the unnormalized forms such as that considered in this paper.

APPENDIX

SUMMARY OF THE LEAST SQUARES ADAPTIVE LATTICE EQUALIZER ALGORITHM

A complete development of the least squares lattice algorithm used in this paper may be found in [12] (see also [11], [13], [15], and [16]). In this Appendix we will only present a brief summary of its derivation.

As with the case of the gradient lattice equalizer [2], the basic idea behind the least squares lattice equalizer is to expand the equalizer output $y_m(n)$(m = order of the equalizer), in terms of orthogonal (uncorrelated) signals $r_m(n)$, which are derived from the received data $x(n)$, i.e.,

$$y_m(n) = \sum_{i=0}^{m} G_i(n-1) r_i(n). \tag{A-1}$$

The $G_i(n)$ represent lattice equalizer tap gains (their computation will be discussed subsequently).

The orthogonal signals $r_m(n)$ are generated through the recursions

$$r_{m+1}(n) = r_m(n-1) - k_m{}^e(n-1) e_m(n) \tag{A-2a}$$

and

$$e_{m+1}(n) = e_m(n) - k_m{}^r(n-1) r_m(n-1) \tag{A-2b}$$

$$(m = 0, \cdots, N-1)$$

where

$$e_0(n) = r_0(n) = x(n). \tag{A-2c}$$

The $r_m(n)$ and $e_m(n)$ in (A-2) are called the backward and forward prediction error residuals and the constants k_m^e and k_m^r (which are termed reflection coefficients) have a common factor, i.e.,

$$k_m^e(n) = \frac{k_m(n)}{\epsilon_m^e(n)} \tag{A-3a}$$

$$k_m^r(n) = \frac{k_m(n)}{\epsilon_m^r(n-1)}. \tag{A-3b}$$

As discussed in Section III, the ϵ_m^e and ϵ_m^r are the minimized sum of forward and backward squared error residuals and obey the recursions (7d)–(7e). The common factor $k_m(n)$ obeys the time recursion (7a) along with (7f). These recursions [(7a), (7d)–(7f)], as well as (A-2), are all derived from the following important properties of the $R_m(n)$ matrix [which may easily be derived from its definition (3)]:

$$R_m(n) = \left[\begin{array}{c:c} q_m(n) & Q_m^T(n) \\ \hdashline Q_m(n) & R_{m-1}(n-1) \end{array}\right] \tag{A-4a}$$

$$= \left[\begin{array}{c:c} R_{m-1}(n) & V_m(n) \\ \hdashline V_m^T(n) & v_m(n) \end{array}\right]. \tag{A-4b}$$

The dashed lines in (A-4a) and (A-4b) are used to denote partitioning. The quantities $q_m(n)$ and $v_m(n)$ denote scalars which are given by

$$q_m(n) = \sum_{p=0}^{n} w^{n-p} x^2(p) \tag{A-5a}$$

and

$$v_m(n) = \sum_{p=0}^{n} w^{n-p} x^2(p-m). \tag{A-5b}$$

Also, $Q_m(n)$ and $V_m(n)$ in (A-4a), (A-4b) denote m-dimensional vectors given by

$$Q_m(n) = \sum_{p=0}^{n} w^{n-p} x(p) X_{m-1}(p-1) \tag{A-5c}$$

and

$$V_m(n) = \sum_{p=0}^{n} w^{n-p} x(p-m) X_{m-1}(p). \tag{A-5d}$$

A third property of the $R_m(n)$ matrix is the following time shift relation:

$$R_m(n) = wR_m(n-1) + X_m(n)X_m^T(n). \tag{A-6}$$

Note that the generation of the signals $r_m(n)$ involve only the received data $x(n)$ and not the reference signal $a(n)$. To include this signal, consider the extended data vector

$$\bar{X}_m(p) = (a(p), x(p), x(p-1), \cdots, x(p-m))^T \tag{A-7}$$

which is generated by appending $a(p)$ onto the data vector $X_m(p)$. In analogy with (3), $\bar{X}_m(p)$ defines the $(m+2) \times (m+2)$ dimensional matrix

$$\bar{R}_m(n) = \sum_{p=0}^{n} w^{n-p} \bar{X}_m(p) \bar{X}_m^T(p). \tag{A-8}$$

As can be easily shown [12], $R_m(n)$ is the coefficient matrix which appears in the augmented matrix equation for the least squares coefficient vectors $F_m(n)$ (defined by (2) with $N = m$). In analogy with (A-4)–(A-6), $\bar{R}_m(n)$ possesses the following properties:

$$\bar{R}_m(n) = \left[\begin{array}{c:c} \bar{q}_m(n) & \bar{Q}_m^T(n) \\ \hdashline \bar{Q}_m(n) & R_m(n) \end{array}\right] \tag{A-9a}$$

$$= \left[\begin{array}{c:c} \bar{R}_{m-1}(n) & \bar{V}_m(n) \\ \hdashline \bar{V}_m^T(n) & \bar{v}_m(n) \end{array}\right] \tag{A-9b}$$

and

$$\bar{R}_m(n) = w\bar{R}_m(n-1) + \bar{X}_m(n)\bar{X}_m^T(n). \tag{A-9c}$$

In (A-9a)–(A-9c),

$$\bar{q}_m(n) = \sum_{p=0}^{n} w^{n-p} a^2(p) \tag{A-10a}$$

$$\bar{Q}_m(n) = \sum_{p=0}^{n} w^{n-p} a(p) X_m(p) \tag{A-10b}$$

$$\bar{V}_m(n) = \sum_{p=0}^{n} w^{n-p} x(p-m) \bar{X}_{m-1}(p) \tag{A-10c}$$

and

$$\bar{v}_m(n) = \sum_{p=0}^{n} w^{n-p} x^2(p-m) = v_m(n). \tag{A-10d}$$

The last equality in (A-10d) follows directly from the definition of $v_m(n)$ (A-5b). The recursions (7g), (7h) for $y_m(n)$ and $\bar{e}_m(n)$ follow directly from (A-9) and (A-10). Finally, the equalizer tap gains $G_m(n)$ in (A-1) can be factored as

$$G_m(n) = \frac{\bar{k}_m(n)}{\epsilon_m^r(n)} \tag{A-11}$$

and the $\bar{k}_m(n)$ obeys the time recursion (8). Equation (8) also follows from (A-9) and (A-10).

The development summarized here is based upon a matrix formulation. A completely different development which is based upon a projection operator (geometric) formulation is presented in [15] and [16].

ACKNOWLEDGMENT

The authors wish to thank M. J. Shensa of the Naval Ocean Systems Center and W. S. Hodgkiss of the Scripps Institution of Oceanography (Marine Physical Laboratory) for their helpful discussions relating to this paper.

REFERENCES

[1] R. W. Lucky, J. Salz, and E. J. Weldon, Jr., *Principles of Data Communication*. New York: McGraw-Hill, 1968.

[2] E. H. Satorius and S. T. Alexander, "Channel equalization using

adaptive lattice algorithms," *IEEE Trans. Commun.*, vol. COM-27, pp. 899–905, June 1979.

[3] J. D. Markel and A. H. Gray, Jr., "Roundoff noise characteristics of a class of orthogonal polynomial structures," *IEEE Trans. Acoust., Speech, and Signal Processing*, vol. ASSP-23, pp. 473–486, Oct. 1975.

[4] M. Morf, D. Lee, J. Nickolls, and A. Vieira, "A classification of algorithms for ARMA models and ladder realizations," in *Proc. IEEE Int. Conf. on Acoust., Speech, and Signal Processing*, Hartford, CT, May 1977, pp. 13–19.

[5] M. Morf, A. Vieira, and D. Lee, "Ladder forms for identification and speech processing," in *Proc. 1977 IEEE Conf. Decision and Contr.*, New Orleans, LA, Dec. 1977, pp. 1074–1078.

[6] M. Morf, "Ladder forms in estimation and system identification," presented at 11th Annu. Asilomar Conf. on Circuits, Syst., and Comput., Monterey, CA, Nov. 7–9, 1977.

[7] M. Morf and D. Lee, "Recursive least squares ladder forms for fast parameter tracking," in *Proc. 1978 IEEE Conf. on Decision and Contr.*, San Diego, CA, Jan. 10–12, 1979, pp. 1362–1367.

[8] D. Godard, "Channel equalization using a Kalman filter for fast data transmission," *IBM J. Res. Develop.*, pp. 267–273, May 1974.

[9] R. D. Gitlin and F. R. Magee, "Self-orthogonalizing adaptive equalization algorithms," *IEEE Trans. Commun.*, vol. COM-25, July 1977.

[10] D. D. Falconer and L. Ljung, "Application of fast Kalman estimation to adaptive equalization," *IEEE Trans. Commun.*, vol. COM-26, pp. 1439–1446, Oct. 1978.

[11] J. D. Pack and E. H. Satorius, "Least squares, adaptive lattice algorithms," Naval Ocean Syst. Center, San Diego, CA, Tech. Rep. TR423, Apr. 1979.

[12] E. H. Satorius and J. D. Pack, "A least squares adaptive lattice equalizer algorithm," Naval Ocean Syst. Center, San Diego, CA, Tech. Rep. TR575, Sept. 1980.

[13] E. H. Satorius and M. J. Shensa, "On the application of recursive least squares methods to adaptive processing," presented at the Int. Workshop on Applications of Adaptive Contr., Yale Univ., New Haven, CT, Aug. 23–25, 1979.

[14] M. J. Shensa, "A least squares lattice decision feedback equalizer," presented at the Int. Conf. on Commun., Seattle, WA, June, 8–11 1980.

[15] ——, "Recursive least squares lattice algorithms—A geometrical approach," Naval Ocean Syst. Center, San Diego, CA, Tech. Rep. TR552, Dec. 1979.

[16] D. Lee and M. Morf, "Recursive square-root ladder estimation algorithms," in *Proc. IEEE Int. Conf. on Acoust., Speech, Signal Processing*, Denver, CO, April 9–11, 1980, pp. 1005–1017.

[17] T. L. Lim and M. S. Mueller, "Rapid equalizer start-up using least-squares algorithms," presented at the Int. Conf. on Commun., Seattle, WA, June 8–11, 1980.

System Identification Techniques for Adaptive Noise Cancelling

BENJAMIN FRIEDLANDER, SENIOR MEMBER, IEEE

Abstract—**A general form of a noise canceller is derived, using a signal modeling approach. The proposed structure combines two infinite impulse response filters: a noise canceller and a line enhancer. A recursive prediction error algorithm is used to adaptively estimate the filter coefficients. Preliminary simulation results on the performance of the generalized adaptive noise canceller are presented.**

I. Introduction

The problem of estimating a signal corrupted by additive noise from measurements by multiple sensors has found many applications. A special form of this problem was discussed in the literature under the name of noise cancelling [1]-[6]. This technique makes use of an auxiliary or reference input which provides some information about the noise component that is corrupting the signal. From this "side information," an estimate of the noise is obtained and then subtracted from the primary input to "cancel out" the additive noise component. This technique is very sensitive to modeling errors since it is essentially equivalent to "bridge balancing." Therefore, the noise cancelling is typically performed in the context of an adaptive process, which properly adjusts the filter parameters. A typical model of an adaptive noise canceller (ANC) is depicted in Fig. 1.

The signal x_t, the common noise of interference z_t, and the measurement noises n_t, m_t are all assumed to be statistically uncorrelated. The objective of the ANC is to adjust the filter $W(z)$ so as to make its output the best estimate of the common noise process z_t. This will minimize the noise power at the output s_t of the ANC. To see this, note that

$$y_t = x_t + z_t + n_t = x_t + \hat{z}_t + (z_t - \hat{z}_t) + n_t \tag{1}$$

or

$$s_t \triangleq (y_t - \hat{z}_t) = x_t + (z_t - \hat{z}_t) + n_t. \tag{2}$$

The effective noise corrupting the signal x_t is the sum of $(z_t - \hat{z}_t)$ and an uncorrelated noise process n_t. The variance of this sum will be minimized if the variance of $(z_t - \hat{z}_t)$ is minimized, i.e., if $\hat{z}_t$ is the least-squares estimate of z_t.

The form of the optimal filter $W(z)$ is well known [7]. Let $H(z)$ denote the transfer function relating the noise z_t to the reference input u_t, i.e.,

$$u_t = H(z)\, z_t + m_t. \tag{3}$$

Manuscript received November 22, 1980; revised November 17, 1981. This work was supported by the Office of Naval Research under Contract N00014-79-C-0743.

The author is with Systems Control Technology, Inc., Palo Alto, CA 94304.

Then

$$W(z) = \frac{S_{zz}(z)\, H(z^{-1})}{S_{mm}(z) + S_{zz}(z)\, H(z)\, H(z^{-1})} \tag{4}$$

where S_{zz} and S_{mm} are the spectra of the noise processes z_t and m_t. Note that the optimal filter $W(z)$ has generally an infinite impulse response (IIR).[1]

In typical implementations of the ANC, the filter $W(z)$ is constrained to have a finite impulse response (FIR). The filter coefficients are then adjusted using a gradient search algorithm of the Widrow-Hoff LMS type [8]-[10]. These implementations of the ANC perform well in many situations. Their performance is, however, suboptimal due to two principal factors. 1) The FIR filter structure is only an approximation of the optimal filter. The quality of the approximation depends on the filter order as well as on channel characteristics $H(z)$ and noise spectra $S_{zz}(z)$, $S_{mm}(z)$. 2) The gradient search algorithm sometimes converges very slowly (especially when the input covariance matrix has a large eigenvalue spread). In this paper, we explore some alternative implementations which attempt to circumvent these difficulties. We propose an IIR filter structure and a recursive prediction error algorithm for estimating its parameters. Our approach is based on some ideas from the area of system identification, as was further elaborated in [11].

II. The Modeling Approach to ANC

A large class of stationary random processes can be represented as the output of a linear filter driven by white noise. The transfer function of the filter is obtained by factoring the power spectrum of the process. Consider, for example, the output of the ANC as given by (2). Let $S_{xx}(z)$, $S_{zz}(z)$, and $S_{nn}(z)$ denote the power spectra of x, $\tilde{z} = z - \hat{z}$, and n, respectively. Define $C(z)/D(z)$ as the (causal) spectral factors of the output process, i.e.,

$$\frac{C(z)\, C(z^{-1})}{D(z)\, D(z^{-1})}\sigma_e^2 = S_{xx}(z) + S_{\tilde{z}\tilde{z}}(z) + S_{nn}(z) \tag{5}$$

where

$$C(z) = 1 + c_1 z^{-1} + \cdots + c_{n_c} z^{-n_c}$$

$$D(z) = 1 + d_1 z^{-1} + \cdots d_{n_d} z^{-n_d}$$

σ_e^2 = variance of e_t, a white noise process.

Then

$$s_t = y_t - \hat{z}_t = x_t + (z_t - \hat{z}_t) + n_t = \frac{C(z)}{D(z)} e_t. \tag{6}$$

[1] In the form it is presented here, $W(z)$ may be noncausal. This can be fixed by properly delaying the input.

Reprinted from *IEEE Trans. Acoust., Speech, Signal Processing,* vol. 1, no. 1, pp. 699–709, Oct. 1982.

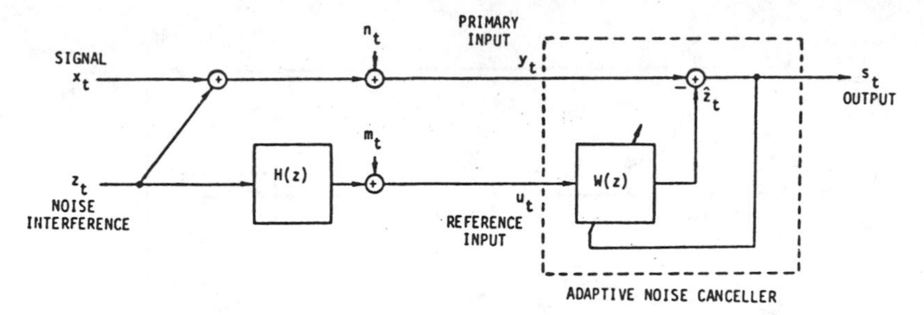

Fig. 1. The single-channel ANC.

Next, recall that the estimate $\hat{z}_t$ is computed by $\hat{z}_t = W(z)\,u_t$. We will find it convenient to write out explicitly the numerator and denominator polynomials of the IIR filter $W(z)$, i.e.,

$$\hat{z}_t = W(z)\,u_t = \frac{B(z)}{F(z)}\,u_t \tag{7}$$

where

$$B(z) = b_0 + b_1 z^{-1} + \cdots + b_{n_b} z^{-n_b}$$

$$F(z) = 1 + f_1 z^{-1} + \cdots + f_{n_f} z^{-n_f}.$$

Thus, we obtain the following equation which relates the primary and reference inputs:

$$y_t = \frac{B(z)}{F(z)}\,u_t + \frac{C(z)}{D(z)}\,e_t. \tag{8}$$

Note that e_t is uncorrelated with u_t since x_t, n_t, and $z_t - \hat{z}_t$ are all uncorrelated with u_t. Models of this type have been studied extensively in the context of system identification [12] and time-series analysis [13]. Various techniques have been developed for estimating the model parameters $\{B(z), F(z), C(z), D(z)\}$ from observed data $\{y_t, u_t\}$. Both off-line and on-line (recursive) techniques are available. One recursive parameter estimation algorithm is described in detail in Section III, and it is proposed as a candidate for adaptive noise cancelling. However, the main point we want to make here is that casting the ANC problem as a model fitting or system identification problem opens the way for using algorithms that were developed in these areas.

The structure of the ANC corresponding to the model given (8) is depicted in Fig. 2. The parameters $\{b_i, f_i\}$ will be updated at each time step by the parameter estimation algorithm. The feedback of $\hat{z}_t$ makes this an IIR filter.

The filter $B(z)/F(z)$ provides the optimal estimate of the noise component z_t. Note, however, that the model relating y_t and u_t also involves another filter: $C(z)/D(z)$. It is interesting to observe that this filter can be interpreted as an adaptive line enhancer (ALE) for the signal x_t. In other words, $C(z)/D(z)$ is, under certain conditions, the prediction filter for x_t. To see this, we rewrite (6) as

$$s_t = -\sum_{i=1}^{n_d} d_i s_{t-i} + \sum_{i=1}^{n_c} c_i e_{t-i} + e_t \tag{9}$$

where e_t is a white noise sequence. Thus, the one-step ahead predictor for s_t is given by

$$\hat{s}_{t|t-1} = -\sum_{i=1}^{n_d} d_i s_{t-i} + \sum_{i=1}^{n_c} c_i e_{t-i}. \tag{10}$$

Next, recall that $s_t = x_t + \tilde{z}_t + n_t$. Therefore,

$$\hat{s}_{t|t-1} = \hat{x}_{t|t-1} + (\hat{\tilde{z}}_t)_{t|t-1} + \hat{n}_{t|t-1} \tag{11}$$

where $(\hat{\cdot})_{t|t-1}$ denotes the least-squares estimate of $(\cdot)$, given past data $\{s_i, e_i\}_{i=0}^{t-1}$. The measurement noise n_t is assumed to be white and, therefore $\hat{n}_{t|t-1} = 0$. The prediction error z_t of the common noise process z_t is *not* necessarily white, and thus $(\hat{\tilde{z}})_{t|t-1} \neq 0$ in general. However, in many situations, $\tilde{z}_t$ can be shown to be nearly white (see Appendix A for details). Furthermore, if the ANC is working effectively, $\tilde{z}_t$ has a small variance compared to $x_t + n_t$. Thus,

$$\hat{x}_{t|t-1} \cong \hat{s}_{t|t-1} \tag{12}$$

which means that the filter $\{c_i, d_i\}$ in (10) is (approximately) the least-squares predictor of the signal x_t from the ANC output s_t. For narrow-band signals in noise, such a predictor is known under the name of the adaptive line enhancer (ALE) [1], [14], [15]. The ALE provides adaptive filtering of the narrow-band process x_t from the broad-band noise $\tilde{z}_t + n_t$, and it can improve the signal-to-noise ratio beyond the improvement provided by the ANC alone. The combined ANC–ALE is depicted in Fig. 3. It is interesting to note that all of the coefficients of the ANC–ALE can be computed by a single parameter estimation algorithm, as will be shown in Section III. This is different from the ANC followed by an ALE, each operating independently, which was discussed in [16]. Note also that the resulting ALE has an infinite impulse response. This type of ALE is discussed in much more detail in [17] and [18]. Here we shall only state that it is relatively straightforward to develop a version of the ALE for the colored noise case, i.e., for the case where $n_t + \tilde{z}_t$ cannot be assumed to be approximately white. The coefficients of this ALE can be computed from $C(z)$ and $D(z)$ or they can be estimated directly by a slight modification of the parameter estimation algorithm [17]. The IIR-ALE contains as a special case the more familiar FIR-ALE, which has the structure depicted in Fig. 3, with $c_i = 0$.

To get a better insight at the operation of the generalized ANC, consider Fig. 3 where

$$\epsilon_t = s_t - \hat{x}_t = y_t - \hat{z}_t - \hat{x}_t. \tag{13}$$

Inserting y_t from (1) gives

$$\epsilon_t = (x_t - \hat{x}_t) + (z_t - \hat{z}_t) + n_t. \tag{14}$$

Under the assumption that $\tilde{z}_t$ is approximately white, it can be shown that $E\{(x_t - \hat{x}_t)(z_t - \hat{z}_t)\} \cong 0$ (see Appendix A). The prediction error ϵ_t will thus be minimized if $E\{(x_t - \hat{x}_t)^2\}$ and $E\{(z_t - \hat{z}_t)^2\}$ are both minimized. In other words, the adaptive algorithm which minimizes $E\{\epsilon_t^2\}$ does indeed provide the least-squares estimates of the signal (x_t) and noise (z_t) processes.

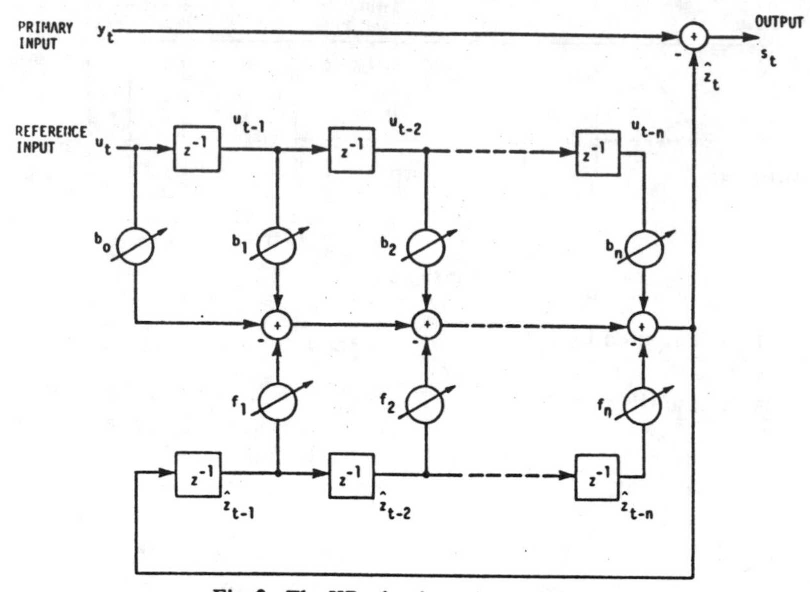

Fig. 2. The IIR adaptive noise canceller.

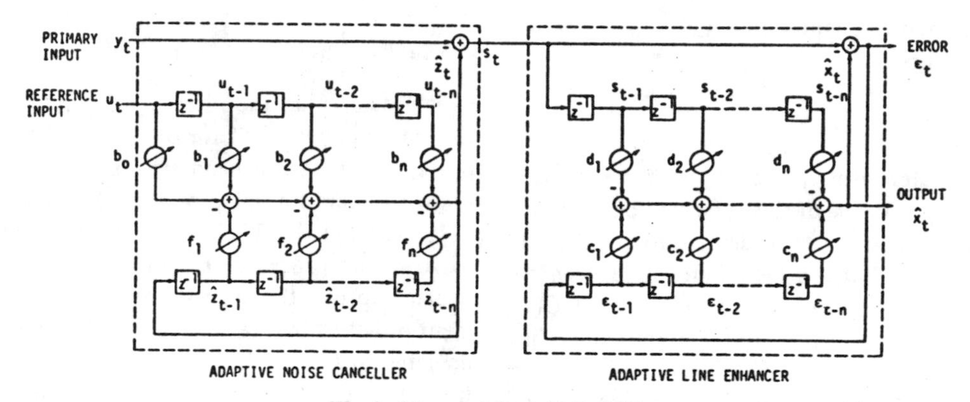

Fig. 3. The combined ALE-ANC.

The ANC discussed so far was assumed to have a tapped-delay line implementation. An alternative implementation using a multichannel least-squares lattice structure is discussed in [19]. A gradient adaptive lattice noise canceller is presented in [33].

III. A Recursive Identification Algorithm

A perusal of the vast literature on system identification [12], [20]-[24] shows a large number of algorithms which have been developed for estimating parameters of "black box" models from input-output data. Recently, a substantial body of theory was developed for a very general class of recursive identification techniques under the name of prediction error algorithms [24], [25]. The study of such algorithms has reached a certain level of maturity: their asymptotic properties (bias, efficiency) are well understood [26], and tools are available for convergence analysis [27]-[30]. Some effort has been made to present the various algorithms in a unified framework to help compare their performance. There is, however, no general agreement on the relative merits of the various algorithms. Based on our own experience, we have chosen a variant of the prediction error (or approximate maximum likelihood) algorithm [24], [25]. It is not clear at this stage whether this algorithm is the best for ANC applications. We believe, however, that its performance is typical of recursive identification algorithms. The basic algorithm is described below. The full derivation of this algorithm is beyond the scope of this paper; see [24] and [25] for details. Instead, we present a heuristic explanation of the algorithm based on its similarity to recursive least-squares techniques.

A Prediction Model

Consider the problem of estimating the parameters $\{b_i, f_i, c_i, d_i\}$ of the model in (8) from observations of the data sequence $\{y_t, u_t\}$. Note that the data seem to be a complicated nonlinear function of the parameters. Estimation problems involving nonlinear functions are usually more difficult than linear estimation problems. We will, therefore, attempt to rewrite the model (8) in a "linear-in-the-parameters" form. To do this, we rewrite the model (8) as

$$\hat{y}_{t|t-1} = y_t - e_t = \left(1 - \frac{D(z)}{C(z)}\right) y_t + \frac{D(z)B(z)}{C(z)F(z)} u_t \tag{15}$$

where $\hat{y}_{t|t-1}$ denotes the one-step ahead predictor of y_t. It is convenient to introduce the auxiliary variables

$$w_t = \frac{B(z)}{F(z)} u_t \tag{16a}$$

$$v_t = y_t - w_t. \tag{16b}$$

Inserting these definitions in (15), we get

$$e_t = y_t - \hat{y}_{t|t-1} = \frac{D(z)}{C(z)} v_t. \tag{17}$$

Next, rewrite (17) and (16a) in difference equation form:

$$e_t = -\sum_{i=1}^{n_c} c_i e_{t-i} + \sum_{i=1}^{n_d} d_i v_{t-i} + v_t \tag{18a}$$

$$w_t = -\sum_{i=1}^{n_f} f_i w_{t-i} + \sum_{i=0}^{n_b} b_i u_{t-i}. \tag{18b}$$

From (16b), it follows that

$$\hat{y}_{t|t-1} = y_t - e_t = w_t + (v_t - e_t). \tag{19}$$

Using (18a), (18b), and (19), we can now write a prediction model for y_t:

$$\hat{y}_{t|t-1} = \Theta^T \phi_t \tag{20a}$$

where

$$\Theta^T = [b_0, \cdots, b_{n_b}, f_1, \cdots, f_{n_f}, c_1, \cdots, c_{n_c}, d_1, \cdots, d_{n_b}] = \text{parameter vector} \tag{20b}$$

$$\phi_t = [u_t, \cdots, u_{t-n_b}, -w_{t-1}, \cdots, -w_{t-n_f}, e_{t-1}, \cdots, e_{t-n_c}, \cdot -v_{t-1}, \cdots, -v_{t-n_d}]^T = \text{data vector}. \tag{20c}$$

The Extended Least-Squares Algorithm

Note that if we could measure the variables $\{u_t, w_t, e_t, v_t\}$, the predictor will be linear in the parameters. Many different techniques are available for estimating the parameters of linear predictors, including the following recursive least-squares (RLS) algorithm:

$$e_t = y_t - \hat{\Theta}_{t-1}^T \phi_t \quad \text{(prediction error)} \tag{21a}$$

$$P_t = [P_{t-1} - P_{t-1}\phi_t \phi_t^T P_{t-1}/(\lambda + \phi_t^T P_{t-1}\phi_t)]/\lambda = \text{error covariance matrix} \tag{21b}$$

$$\hat{\Theta}_t = \hat{\Theta}_{t-1} + P_t \phi_t e_t \tag{21c}$$

where $0 \leqslant \lambda \leqslant 1$ is an exponential "forgetting factor" and P_t is an $n \times n$ matrix, $n = n_b + n_f + n_c + n_d + 1$. This algorithm is widely used for system identification [12], [24], [30], time series analysis, and linear regression. It is closely related to the Kalman filter for the "state vector" $\hat{\Theta}$ (e.g., compare to [7]).

In the problem under consideration, the variables $\{w_t, e_t, v_t\}$ are not directly measurable. However, at each step, it is possible to compute estimates of these variables $\{\hat{w}_t, \hat{e}_t, \hat{v}_t\}$ using (16a), (16b), and (17) with the unknown model parameters $B(z)$, $F(z)$, $C(z)$, $D(z)$ replaced by their most recent estimates $\{\hat{B}_t(z), \hat{F}_t(z), \hat{C}_t(z), \hat{D}_t(z)\}$. The subscript t denotes that the coefficients of these filters were taken from the parameter vector $\hat{\Theta}_{t-1}$. More precisely, let

$$\hat{w}_t = \frac{\hat{B}_t(z)}{\hat{F}_t(z)} u_t \tag{22a}$$

$$\hat{v}_t = y_t - \hat{w}_t \tag{22b}$$

$$\hat{e}_t = \frac{\hat{D}_t(z)}{\hat{C}_t(z)} \hat{v}_t. \tag{22c}$$

These variables will define an estimated data vector

$$\hat{\phi}_t = [u_t, \cdots, u_{t-n_b}, -\hat{w}_{t-1}, \cdots, -\hat{w}_{t-n_f}, \hat{e}_{t-1}, \cdots, \hat{e}_{t-n_c}, -\hat{v}_{t-1}, \cdots, -\hat{v}_{t-n_d}]^T. \tag{23}$$

Using this data vector in (21a)-(21c) gives an algorithm for estimating the model parameters (8) from the measured data $\{y_t, u_t\}$. The question arises, of course, whether this algorithm will converge. Observe that if the parameter estimate vector $\hat{\Theta}_t$ is close to its true value Θ, then the variables $\hat{w}_t$, $\hat{v}_t$, $\hat{e}_t$ will be close to w_t, v_t, e_t. Therefore, the algorithm is expected to behave like a true least-squares algorithm. In fact, Ljung *et al.* [28]-[30] and others have shown that the algorithm described above will converge, provided that the polynomials $C(z)$, $F(z)$ obey a certain positive real condition. This algorithm appears in the literature under different names. Here we shall call it the extended least-squares (ELS) algorithm.

The Recursive Maximum Likelihood Algorithm

While the ELS algorithm can be used in many practical situations, it is preferable to have an estimation technique that is guaranteed to converge for all possible signal models. Such an algorithm can be obtained by a slight modification of the data vectors that appear in (21b) and (21c). More precisely, define the following filtered quantities:

$$\tilde{u}_t = \frac{\hat{D}_t(z)}{\hat{C}_t(z)\hat{F}_t(z)} u_t \tag{24a}$$

$$\tilde{w}_t = \frac{\hat{D}_t(z)}{\hat{C}_t(z)\hat{F}_t(z)} \hat{w}_t \tag{24b}$$

$$\tilde{e}_t = \frac{1}{\hat{C}_t(z)} \hat{e}_t \tag{24c}$$

$$\tilde{v}_t = \frac{1}{\hat{C}_t(z)} \hat{v}_t \tag{24d}$$

and a corresponding filtered data vector ψ_t:

$$\psi_t = [\tilde{u}_t, \cdots, \tilde{u}_{t-n_b}, -\tilde{w}_{t-1}, \cdots, -\tilde{w}_{t-n_f}, \tilde{e}_{t-1}, \cdots, \tilde{e}_{t-n_c}, -\tilde{v}_{t-1}, \cdots, \tilde{v}_{t-n_d}]^T. \tag{24e}$$

Replace ϕ_t in (21b) and (21c) by ψ_t. The resulting algorithm is called the recursive maximum likelihood (RML) algorithm [24], [25], [28]-[30]. The properties of this algorithm were studied extensively in recent years, and it was shown to be asymptotically unbiased and asymptotically efficient, i.e., the variance of the estimates approaches the Cramer-Rao lower bound [26]. The RML algorithm is guaranteed to converge, provided that the filters used in (22) and (24) remain stable during its operation. Note that the stability of all the filters appearing in the algorithm depends only on the roots of $\hat{C}_t(z)$ and $\hat{F}_t(z)$. A straightforward approach is to test the stability of $\hat{C}_t(z)$ and $\hat{F}_t(z)$ at each time step. In Appendix C, we present a simple algorithm for doing this. (A set of reflection coefficients is computed using the Levinson algorithm [31] "in reverse"; if all of these coefficients have magnitudes less than one, the filter is stable.) If either of the polynomials turns out to be unstable, we project the parameter estimates back into a region of stability [27]. This can be done by setting $\hat{\Theta}_t$ back to its value before the last update which caused $\hat{C}_t(z)$ or $\hat{F}_t(z)$ to become unstable. In many situations, it is possible to avoid stability monitoring altogether. (Once the algorithm has reached the neighborhood of the true parameter values, the probability of the filters becoming unstable is very small. In our experience, stability monitoring seemed to be needed only during the start up of the algorithm.) However, performing the stability test does not significantly increase the computational complexity of the algorithm, and we found it simpler to always use it. The complete RML algorithm is summarized in Table I.

TABLE I
THE RML ALGORITHM

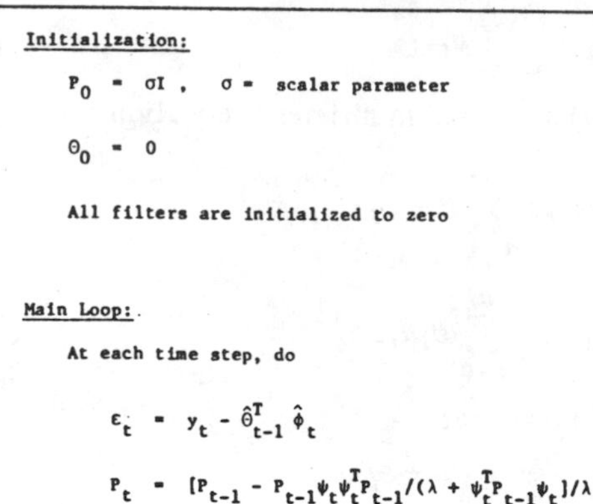

Initialization:

$P_0 = \sigma I$, σ = scalar parameter

$\Theta_0 = 0$

All filters are initialized to zero

Main Loop:

At each time step, do

$\epsilon_t = y_t - \hat{\Theta}^T_{t-1} \hat{\phi}_t$

$P_t = [P_{t-1} - P_{t-1}\psi_t\psi_t^T P_{t-1}/(\lambda + \psi_t^T P_{t-1}\psi_t)]/\lambda$

$\hat{\Theta}_t = \hat{\Theta}_{t-1} + P_t\psi_t\epsilon_t$

Test stability of F(z),C(z) . If unstable, set $\hat{\Theta}_t = \hat{\Theta}_{t-1}$

Update $\hat{\phi}_t$ using Eqs. (22),(23)

Update ψ_t using Eqs. (24),(25)

Computational Requirements

It is difficult to evaluate the computational complexity of the proposed algorithm without considering a specific implementation on a particular computer. However, we can get a rough idea of its complexity by counting the number of operations (multiplies and adds) involved in the update equations in Table I. An approximate count gives $3n^2 + 6n$ operations per time step, where n is the number of parameters. (To get this count, note the following: 1) P_t is symmetric, and only half of its entries need to be computed; 2) the quantity $P_{t-1}\psi_t$ can be computed once and used several times; 3) the gain vector in the $\hat{\Theta}_t$ update can be computed by $P_t\psi_t = P_{t-1}\psi_t/(\lambda + \psi_t^T P_t \psi_t)$; 4) the stability test for a polynomial of order m requires approximately m^2 operations.)

For comparison, we note that the gradient LMS algorithm for an FIR-ANC with n coefficients requires only n operations per time step. Thus, the RML algorithm appears to require substantially more computations than currently used gradient techniques. There are, however, a number of factors which need to be taken into account when making this comparison.

1) Efficient Versions of the RML: Most of the computations required by the RML algorithm are related to updating the gain vector $K_t = P_t\psi_t$. Morf and others [32] developed more efficient versions of this algorithm in which the gain vector K_t is computed directly (without first evaluating P_t). These "fast" algorithms require on the order of n, rather than n^2, operations per time step. Lattice implementations of the ANC and the combined ANC-ALE were recently developed which also require on the order of n operations [19].

2) The Number of Estimated Parameters: The FIR adaptive filters developed by Widrow *et al.* [1], [8]-[10] often require the use of high-order filters to achieve good performance, especially at low signal-to-noise ratios. Adaptive filters with several hundred and even several thousand weights are not uncommon. An adaptive IIR filter can achieve the same performance with a smaller number of parameters. For example, if the input has p spectral lines, the orders of the various filters (n_c, n_d, n_f, n_b) will be $\leqslant 2p$. Many applications involve only one or two spectral lines, in which case $n \leqslant 16$. For a more detailed comparison of FIR versus IIR adaptive filters, see [17]. It was shown there that to achieve comparable performance at low SNR, we need to have a much smaller number of RML parameters than gradient LMS parameters. Thus, the increased complexity of the proposed algorithm is compensated in part by the reduction in the number of estimated parameters. Consider, for example, an RML algorithm with $n = 12$ and an LMS adaptive filter with $n = 512$. In this case, the operation count for the two algorithms is about the same.

3) Performance: The IIR ANC/ALE described in Section II has the potential for achieving better performance than the conventional FIR ANC. There are some applications (e.g., surveillance systems) in which even small performance improvements are worth the added computational complexity.

Finally, it should be noted that the advent of LSI and VLSI technology and the availability of relatively cheap computing power are changing traditional tradeoffs between performance and computational requirements. In the near future, it will be possible to implement adaptive processing algorithms of greater complexity. We believe that it is, therefore, worthwhile to study the RML and similar algorithms, even though they lack the attractive simplicity of gradient LMS techniques.

The Recursive Instrumental Variable Algorithm

The RML algorithm provides parameter estimates for both the ANC and ALE. In some situations (e.g., narrow-band interference and good signal-to-noise ratio in the primary input), the ALE part of the processor will be ineffective. The question then arises of how to avoid estimating the unnecessary parameters $\{c_i, d_i\}$. To see the problem more clearly, we rewrite (8) as

$$y_t = -\sum_{i=1}^{n_f} f_i y_{t-i} + \sum_{i=0}^{n_b} b_i u_{t-i} + v_t \tag{25a}$$

where

$$v_t = \frac{F(z)\,C(z)}{D(z)} e_t. \tag{25b}$$

Using any standard least-squares technique to estimate the parameters, $\{f_i, b_i\}$ will give biased estimates, unless v_t is a white noise process (i.e., if $F(z)\,C(z)/D(z) = 1$) [12], [20]

The instrumental variable method of system identification [35], [38] was developed to provide unbiased estimates of the model parameters in (25) without estimating noise dynamics. A recursive version of this algorithm can, therefore, be used to estimate the ANC coefficients without estimating the ALE coefficients. This means a substantial reduction in the problem dimension ($n = n_b + n_f + 1$) and the computational requirements. One version of the recursive instrumental variable (RIV) algorithm [36], [37] is presented in Appendix B.

In [24] it was shown that the RML algorithm can be specialized to a symmetric form of the instrument variable method (different from the nonsymmetric version presented in Appendix B). This is done by setting $n_c = n_d = 0$ in the RML algorithm and eliminating the prefiltering step in (24). Thus, the adaptive algorithm presented in this section can be easily modified to eliminate the ALE part of the filter.

IV. Performance of the ANC

In this section, we present some preliminary simulation results which illustrate the expected performance of a combined ANC-ALE. The algorithm used in these simulations was a simplified version of the algorithm presented in Section III.[2] The behavior of the adaptive algorithm was tested for different types of signal y_t, interference z_t, and signal-to-noise ratio (SNR). Three typical test cases are described below. The following definitions of SNR at the primary input (ρ_p) and SNR at the reference input (ρ_r) will be used:

$$\rho_p = \frac{\text{signal power}}{\text{measurement noise power}} = \frac{\sigma_x^2}{\sigma_n^2} \tag{26a}$$

$$\rho_r = \frac{\text{interference power}}{\text{measurement noise power}} = \frac{\sigma_r^2}{\sigma_m^2} \tag{26b}$$

where σ_r^2 is the variance of $H(z)\, z_t$, i.e., of the interference as it appears at the reference input.

Case 1

The signal and interference are both sinusoids of equal power and different frequencies. The adaptive algorithm was used to estimate a total of six parameters ($n_b = n_f = n_c = 2$), on the basis of 1024 data points. The spectra of $y_t, s_t, \hat{x}_t$ for the last 512 points are depicted in Figs. 4-6. When the signal-to-noise ratio at the reference input is good, excellent cancellation is achieved (Fig. 4). When the reference input is fairly noisy, the interference is only partially cancelled, as expected (Figs. 5 and 6). An interesting possibility for improving the performance of the ANC for narrow-band interference is to add an adaptive filter (e.g., a separate ALE) at the reference input to improve its SNR.

Figs. 4(c), 5(c), and 6(c) indicate the additional noise suppression provided by the ALE portion of the generalized ANC. It should be noted, however, that the adaptive filter is not fully converged after 1024 data points. Running the filter longer gave significant improvements in the SNR of the output $\hat{x}_t$. For a more detailed discussion of the ALE, see [17].

[2]This algorithm was capable of estimating only the parameters of $B(z)$, $C(z)$ and $F(z)$, and it assumed that $D(z) = F(z)$.

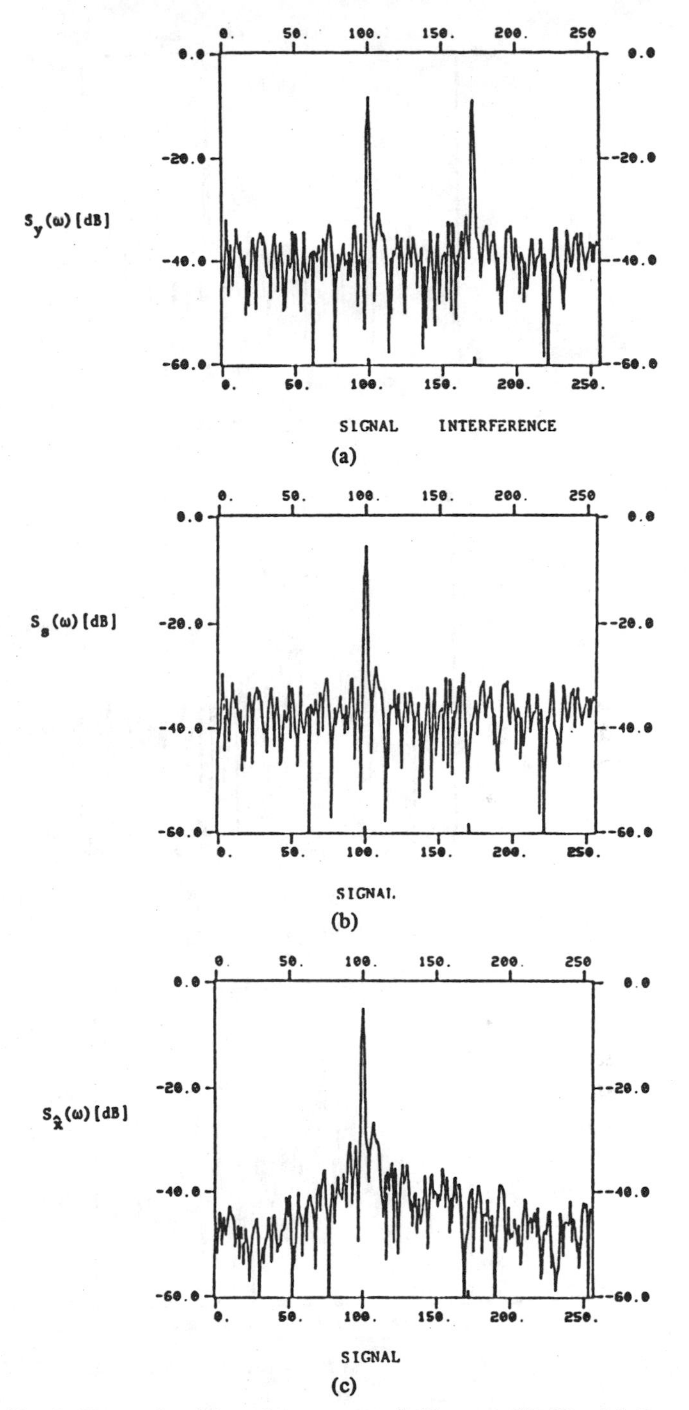

Fig. 4. Narrow-band interference, $\rho_p = 7$ dB, $\rho_R = 20$ dB. (a) Spectrum of primary input y_t. (b) Spectrum of ANC output s_t. (c) Spectrum of ANC-ALE output $\hat{x}_t$.

Case 2

This case is the same as Case 1, except that the interference was 10 dB stronger than the signal. The observed performance was very similar. A typical test result is depicted in Fig. 7.

Case 3

The signal was sinusoid, and the interference was broad-band noise with total power equal to the signal power. The adaptive algorithm was used to estimate eight parameters ($n_b = n_c = 3$, $n_f = 2$) on the basis of 1024 data points. Figs. 8 and 9 depict some typical results. Note the significant improvement provided

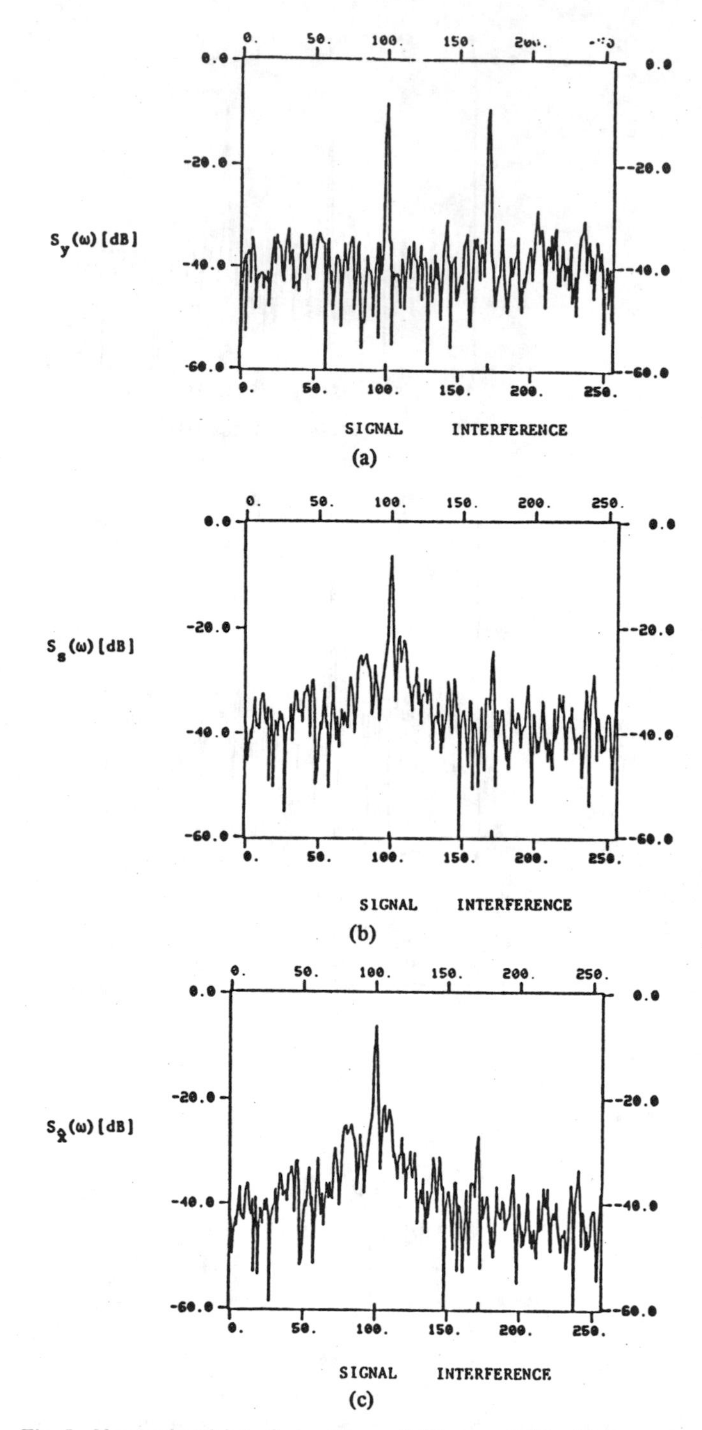

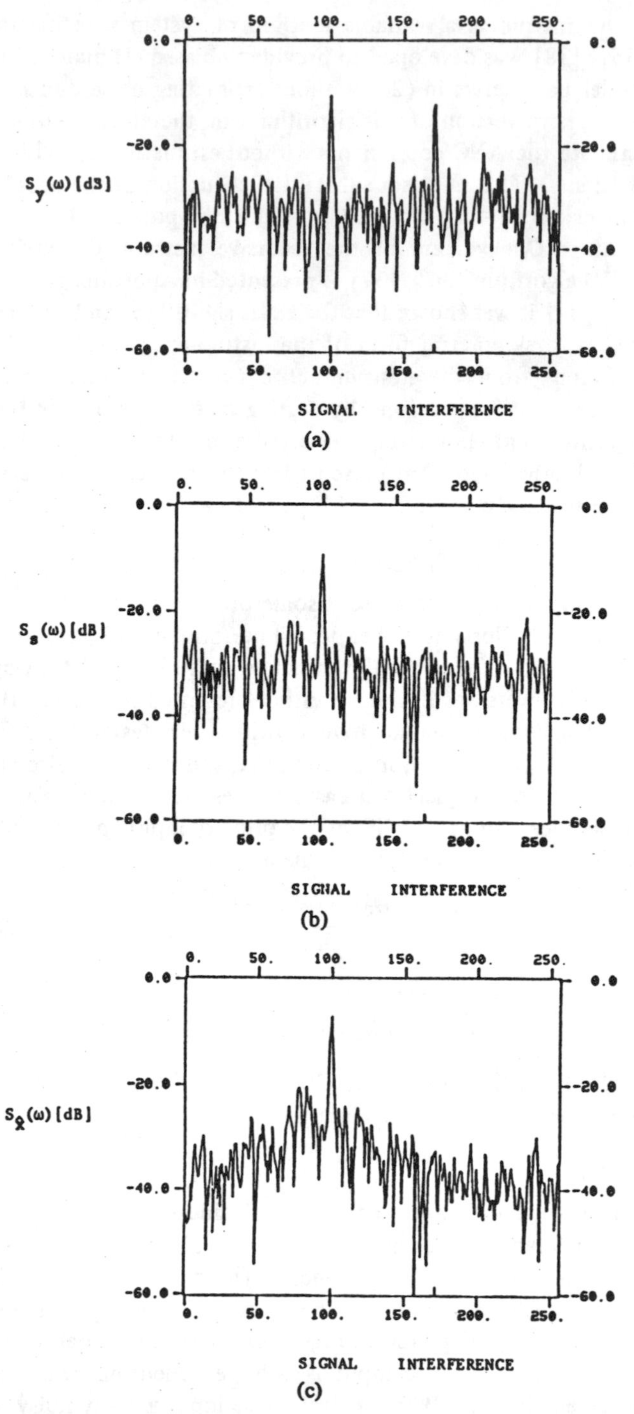

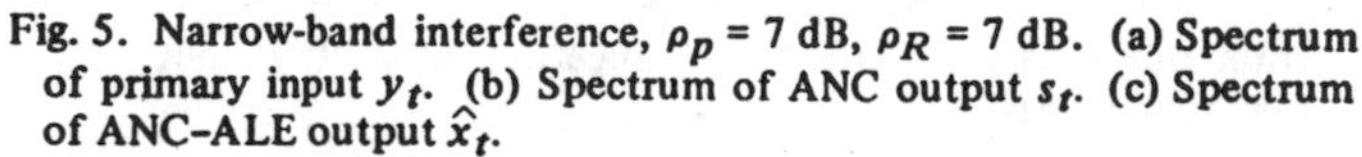

Fig. 5. Narrow-band interference, $\rho_P = 7$ dB, $\rho_R = 7$ dB. (a) Spectrum of primary input y_t. (b) Spectrum of ANC output s_t. (c) Spectrum of ANC-ALE output $\hat{x}_t$.

Fig. 6. Narrow-band interference $\rho_P = 3$ dB, $\rho_R = 7$ dB. (a) Spectrum of primary input y_t. (b) Spectrum of ANC output s_t. (c) Spectrum of ANC-ALE output $\hat{x}_t$.

by the ALE portion of the filter. In both of these examples, the measurement noise n_t has the same power as the interference z_t. Thus, noise cancelling is not too effective, and most of the noise suppression is provided by the ALE.

V. Conclusions

We presented a generalized ANC structure, which follows naturally from the system identification point of view. This filter combines an IIR-ANC with an IIR-ALE. It contains as special cases all the other forms of noise cancellers and line enhancers currently in use (e.g., set $f_i = d_i = c_i = 0$ to get FIR-ANC or set $b_i = f_i = c_i = 0$ to get the FIR-ALE). Initial experimentation with the proposed algorithm was encouraging. However, considerably more testing and performance analysis is required to fully evaluate the potential of this technique.

The proposed approach can be extended to handle the presence of significant signal components in the reference input. In some related work [34], we presented an algorithm based on multichannel system identification. The basic idea is to perform a true two-channel processing in which the noise and signal processes (z_t, x_t) are estimated simultaneously from both inputs (y_t, u_t). (In the current approach, the estimation of

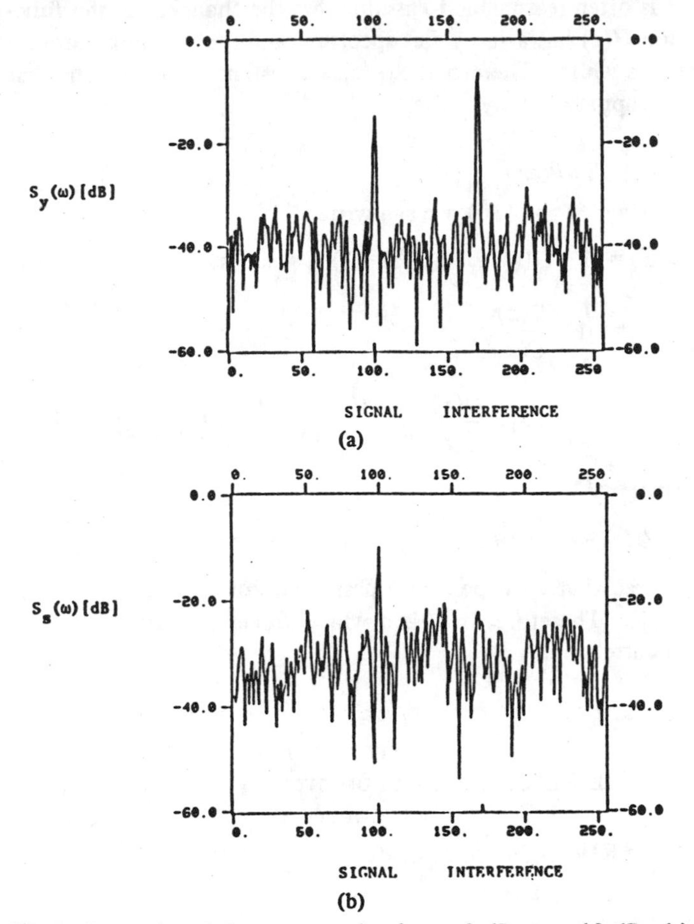

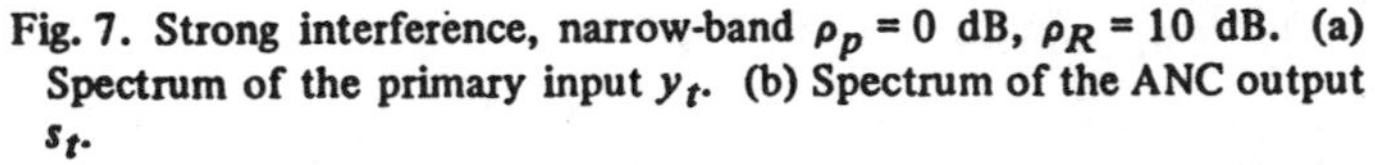

Fig. 7. Strong interference, narrow-band $\rho_P = 0$ dB, $\rho_R = 10$ dB. (a) Spectrum of the primary input y_t. (b) Spectrum of the ANC output s_t.

the noise component is based only on the information in the reference input.) This approach is more complex because it involves a two-input, two-output system, but it is more general than the traditional ANC which involves only single-channel prediction.

Appendix A
Evaluation of the Spectra of Various Processes in the ANC-ALE

The Optimal Filter for $\hat{z}_t$

Using (3) and (7), we can write

$$\tilde{z}_t = z_t - \hat{z}_t = z_t - W(z)\, u_t = (1 - H(z)\, W(z))\, z_t - W(z)\, m_t. \tag{A1}$$

Thus,

$$\begin{aligned} S_{\tilde{z}\tilde{z}}(z) &= (1 - H(z)\, W(z))\,(1 - H(z^{-1})\, W(z^{-1}))\, S_{zz}(z) \\ &\quad + W(z)\, W(z^{-1})\, S_{mm}(z) \\ &= (S_{mm}(z) + H(z)\, H(z^{-1})\, S_{zz}(z))\, W(z)\, W(z^{-1}) \\ &\quad - H(z)\, S_{zz}(z)\, W(z) - H(z^{-1})\, S_{zz}(z)\, W(z^{-1}) \\ &\quad + S_{zz}(z). \end{aligned} \tag{A2}$$

Let

$$X(z) = S_{mm}(z) + H(z)\, H(z^{-1})\, S_{zz}(z); \tag{A3}$$

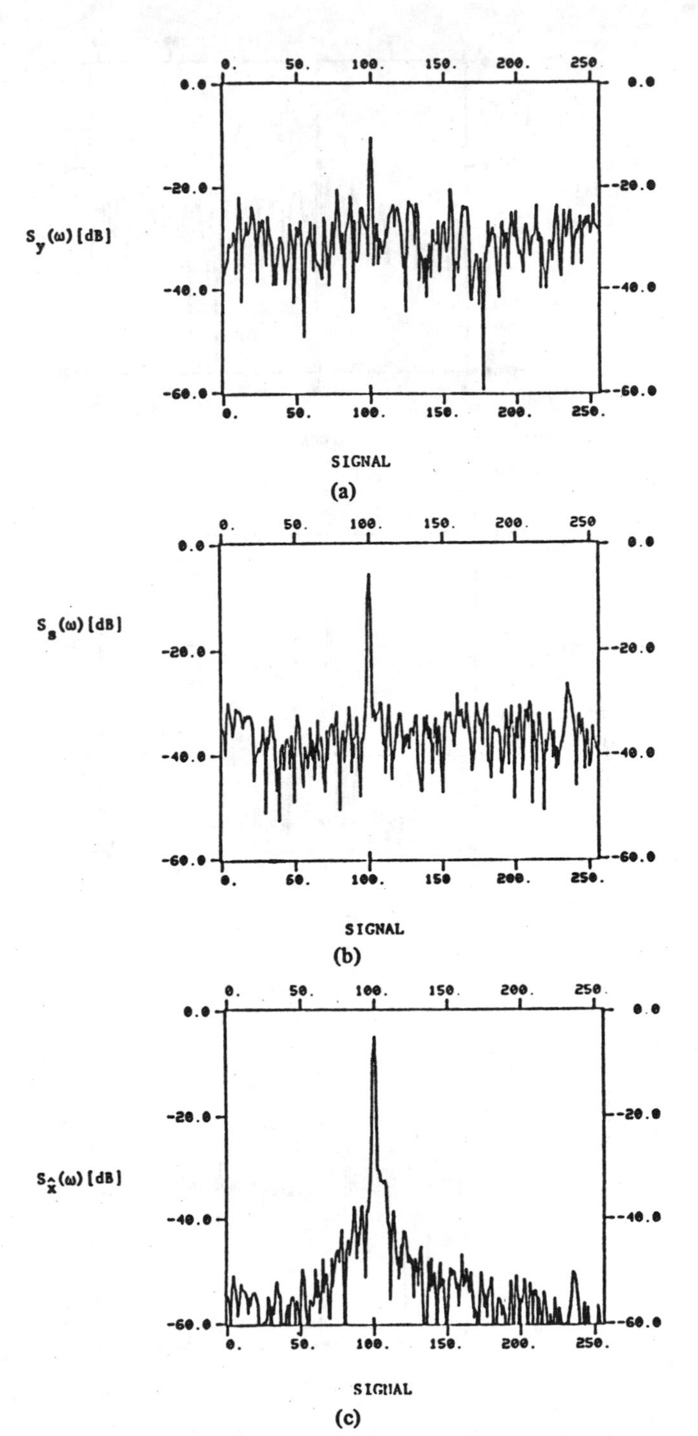

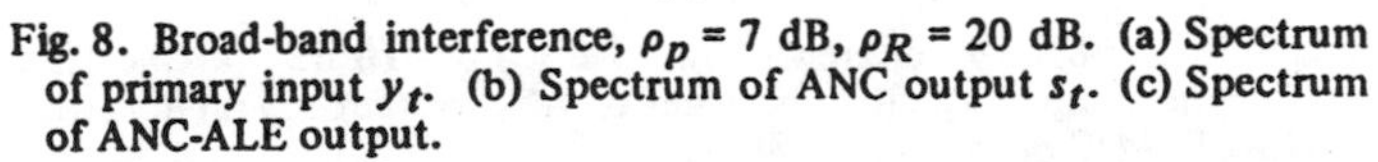

Fig. 8. Broad-band interference, $\rho_P = 7$ dB, $\rho_R = 20$ dB. (a) Spectrum of primary input y_t. (b) Spectrum of ANC output s_t. (c) Spectrum of ANC-ALE output.

then "complete the square" to get

$$\begin{aligned} S_{\tilde{z}\tilde{z}}(z) &= X(z)\, [(W(z) - H(z^{-1})\, S_{zz}(z)/X(z))\, (W(z^{-1}) \\ &\quad - H(z)\, S_{zz}(z)/X(z))] + S_{mm}(z)\, S_{zz}(z)/X(z). \end{aligned} \tag{A4}$$

The choice of $W(z)$ that will minimize the power of $\tilde{z}_t$ is clearly given by setting the first term in the equation to zero, i.e.,

$$W(z) = H(z^{-1})\, S_{zz}(z)/X(z) \tag{A5}$$

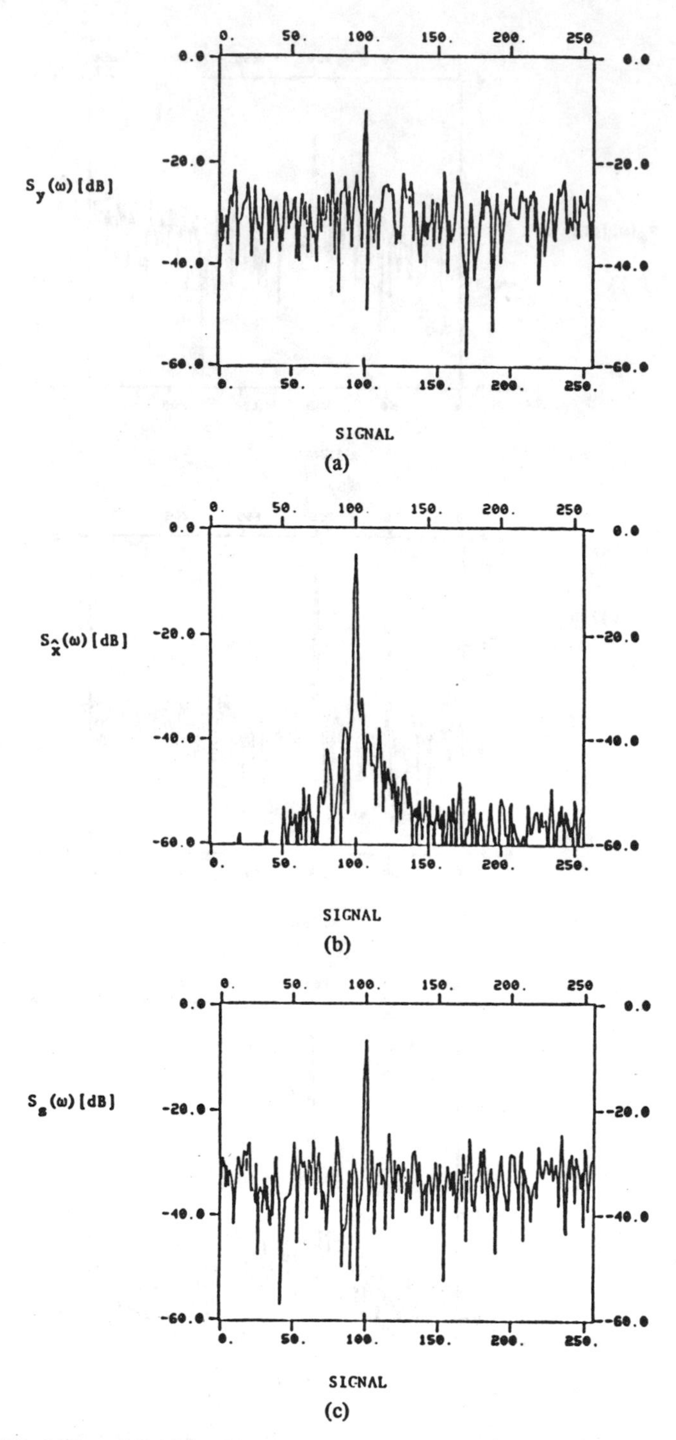

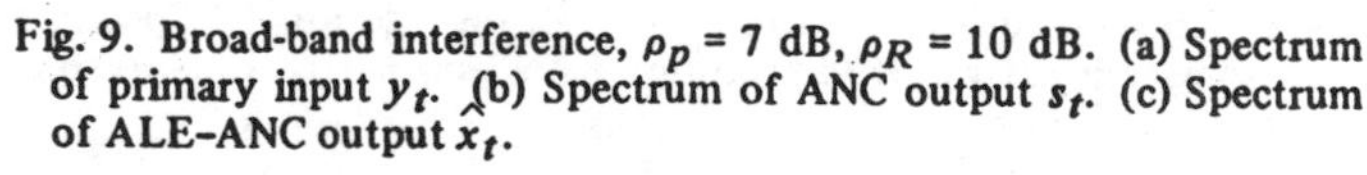

Fig. 9. Broad-band interference, $\rho_P = 7$ dB, $\rho_R = 10$ dB. (a) Spectrum of primary input y_t. (b) Spectrum of ANC output s_t. (c) Spectrum of ALE-ANC output $\hat{x}_t$.

and the corresponding power spectrum of the error $\tilde{z}_t$ is

$$S_{\tilde{z}\tilde{z}} = S_{mm}(z)\, S_{zz}(z)/X(z). \tag{A6}$$

If the signal-to-noise ratio at the reference input is high, then

$$X(z) \cong H(z)\, H(z^{-1})\, S_{zz}(z) \tag{A7}$$

and

$$S_{\tilde{z}\tilde{z}}(z) \cong \frac{S_{mm}(z)}{H(z)\, H(z^{-1})}\,. \tag{8}$$

It is often reasonable to assume that the channel transfer function $H(z)$ has a fairly flat spectrum and that the noise process m_t is white. Therefore, $S_{\tilde{z}\tilde{z}}(z) \cong$ constant, which means that $\tilde{z}_t$ is approximately white.

The ALE Filter

Using (6) and (10), we can write

$$\hat{x}_t = (1 - D(z))\, s_t + (C(z) - 1)\, e_t = s_t - e_t$$

$$= \left(1 - \frac{D(z)}{C(z)}\right) s_t \tag{A9}$$

$$\tilde{x}_t = (x_t - \hat{x}_t) = \frac{D(z)}{C(z)}\, x_t + \left(\frac{D(z) - C(z)}{C(z)}\right) [\tilde{z}_t + n_t]. \tag{A10}$$

Since

$$D(z) - C(z) = (d_1 - c_1)\, z^{-1} + (d_2 - c_2)\, z^{-2} + \cdots, \tag{A11}$$

we see that $\tilde{x}_t$ depends on past values of $\tilde{z}_t$, i.e., $\{\tilde{z}_{t-1}, \tilde{z}_{t-2}, \cdots\}$. Therefore, if $\tilde{z}_t$ is a white process, $\tilde{x}_t$ and $\tilde{z}_t$ will be uncorrelated.

Appendix B
The Recursive Instrumental Variable (RIV) Algorithm

The RIV algorithm estimates the parameters of the model

$$y_t = -\sum_{i=1}^{n_f} f_i y_{t-i} + \sum_{i=0}^{n_b} b_i u_{t-i} + v_t \tag{B1}$$

where v_t is an unmeasurable noise process and y_t, u_t are the observed input/output processes. This equation can be written more compactly as

$$y_t = \beta_t^T\, \theta$$

where

$$\theta = [b_0, \cdots, b_{n_b}, f_1, \cdots, f_{n_f}]^T = \text{parameter vector}$$

$$\beta_t = [u_t, \cdots, u_{t-n_b}, -y_{t-1}, \cdots, -y_{t-n_f}]^T = \text{data vector}.$$

The RIV algorithm is described in [12], [24], [29], [30], [36]-[38]. A summary of the algorithm is presented next.

Initialization

$P_0 = \sigma I$, σ = scalar parameter

$\theta_0 = 0$.

All filters are initialized to zero.

Data Vectors

$$\beta_t = [u_t, \cdots, u_{t-n_b}, -y_{t-1}, \cdots, -y_{t-n_f}]^T$$

$$\eta_t = [u_t, \cdots, u_{t-n_b}, -z_{t-1}, \cdots, -z_{t-n_f}]^T$$

z_t = the instrumental variable.

Main Loop

At each time step, do

$$\epsilon_t = y_t - \hat{\theta}_{t-1}^T \beta_t$$

$$\hat{\theta}_t = \hat{\theta}_{t-1} + P_t \eta_t \epsilon_t$$

$$P_t = [P_{t-1} - P_{t-1} \eta_t \beta_t^T P_{t-1} / (\lambda + \beta_t^T P_{t-1} \eta_t)] / \lambda$$

update β_t

compute z_t (see below)

update η_t.

The instrumental variable z_t is chosen to fulfill certain technical conditions, as described in [35] - [38]. The most common choices seem to be

$$z_t = \frac{\hat{B}_t(z)}{\hat{F}_t(z)} u_t \tag{B2}$$

$$z_t = y_{t-N}, \quad N = \text{delay parameter.} \tag{B3}$$

The delay parameter is chosen to be larger than the correlation time of v_t. For a discussion of these and other choices of the instrumental variable, see [38]. A symmetric version of the RIV is sometimes used in which β_t in the P_t update equation is replaced by η_t. This form of the RIV can be interpreted as a special case of the RML, as mentioned earlier.

Appendix C
Stability Test

FUNCTION TSTAB (A, NA) performs a test whether the polynomial $A(z)$ has all roots inside the unit circle or not.

$$A(z) = 1 + A(1) z^{-1} + \cdots + A(NA) z^{-NA}.$$

The vector A contains the coefficients of the polynomial. NA is the order of the polynomial (MAX = 30). The FUNCTION is returned 0 if all roots are inside the unit circle (i.e., $A(z)$ is stable); otherwise, FUNCTION is returned as 1.

```
      DIMENSION A(1)
      DIMENSION AA(30), D(30)
      TSTAB=0
      DO 10 I=1,NA
10    AA(I)=A(I)
      DO 50 I=NA,1,-1
      K=AA(I)
      IF(ABS(K).GE.1.) GO TO 60
      KC=1.-K*K
      I1=I-1
      IF(I1) 50,50,20
20    DO 30 J=1,I1
30    D(J)=(AA(J)-K*AA(I-J))/KC
      DO 40 J=1,I1
40    AA(J)=D(J)
50    CONTINUE
      GO TO 70
60    TSTAB=1
70    RETURN
      END
```

References

[1] B. Widrow *et al.*, "Adaptive noise cancelling: Principles and applications," *Proc. IEEE*, vol. 63, pp. 1692-1716, Dec. 1975.

[2] J. R. Glover, Jr., "Adaptive noise cancelling applied to sinusoidal interferences," *IEEE Trans. Acoust. Speech, Signal Processing*, vol. ASSP-25, Dec. 1977.

[3] —, "Adaptive noise cancelling of sinusoidal interferences," Ph.D. dissertation, Stanford Univ., Stanford, CA, 1975.

[4] —, "Adaptive noise cancelling of sinusoidal interference," Naval Undersea Cen., San Diego, CA, Tech. Note 1617, Dec. 1975.

[5] J. Kaunitz, "Adaptive filtering of broadband signals as applied to noise cancelling," SEL, Stanford Univ., Stanford, CA, Rep. SEL-72-038, Aug. 1972.

[6] M. Shensa, "Performance of the adaptive noise canceller with a noisy reference: Non-Weiner solutions," NOSC, Tech. Rep. TR 381, Mar. 1979.

[7] T Kailath, *Lectures on Linear Least-Squares Estimation*. New York: Springer-Verlag, 1976.

[8] B. Widrow, J. M. McCool, M. G. Larimore, and C. R. Johnson, Jr., "Stationary and nonstationary learning characteristics of the LMS adaptive filter," *Proc. IEEE*, vol. 64, pp. 1151-1162, Aug. 1976.

[9] B. Widrow and J. M. McCool, "A comparison of adaptive algorithms based on the methods of steepest descent and random search," *IEEE Trans. Antennas Propagat.*, vol. AP-24, pp. 615-637, Sept. 1976.

[10] B. Widrow, "Adaptive filters," in *Aspects of Network and System Theory*, R. Kalman and N. DeClaris, Eds. New York: Holt, Rinehart and Winston, 1971. .

[11] B. Friedlander, "System identification techniques for adaptive signal processing," *IEEE Trans. Acoust., Speech, Signal Processing*, vol. ASSP-30, pp. 240-246, Apr. 1982.

[12] G. C. Goodwin and R. L. Payne, *Dynamic System Identification: Experiment Design and Data Analysis*. New York: Academic, 1977.

[13] G.E.P. Box and G. M. Jenkins, *Time Series Analysis, Forecasting and Control*. San Francisco, CA: Holden-Day, 1967.

[14] J. R. Zeidler, E. H. Satorius, D. M. Chabries, and H. T. Wexler, "Adaptive enhancement of multiple sinusoids in uncorrelated noise," *IEEE Trans. Acoust., Speech, Signal Processing*, vol. ASSP-26, pp. 240-254, June 1978.

[15] J. R. Treichler, "Transient and convergent behavior of the adaptive line enhancer," *IEEE Trans. Acoust., Speech, Signal Processing*, vol. ASSP-27, pp. 53-62, Feb. 1979.

[16] E. Satorius, "Adaptive noise cancelling and enhancement of a sinusoid in uncorrelated noise," Naval Ocean Syst. Cen., San Diego, CA, Tech. Rep. 184, Dec. 1977.

[17] B. Friedlander, "A recursive maximum likelihood algorithm for ARMA line enhancement," in *Proc. IEEE Int. Conf. Acoust., Speech, Signal Processing*, Atlanta, GA, Mar. 30-Apr. 1, 1981, pp. 488-491. Also, *IEEE Trans. Acoust., Speech, Signal Processing*, vol. ASSP-30, pp. 651-657, Aug. 1982.

[18] B. Friedlander, "A pole-zero lattice form for adaptive line enhancement," in *Proc. 14th Asilomar Conf. Circuits, Syst., Comput.*, Pacific Grove, CA, Nov. 1980, pp. 380-384.

[19] B. Friedlander, "Lattice filters for adaptive processing," *Proc. IEEE*, Aug. 1982.

[20] K. J. Aström and P. Eykoff, "System identification–A survey," *Automatica*, vol. 7, pp. 123-162, 1971.

[21] *IEEE Trans. Automat. Contr.* (*Special Issue on System Identification and Time-Series Analysis*), Dec. 1974.

[22] *Automatica* (*Special Section on System Identification Tutorials*), vol. 16, pp. 501-587, Sept. 1980.

[23] *Automatica* (*Special Issue on Identification and System Parameter Estimation*), vol. 17, Jan. 1981.

[24] L. Ljung and T. Söderström, *Theory and Practice of Recursive Identification*. Cambridge, MA: M.I.T. Press, 1982.

[25] L. Ljung, "Analysis of a general recursive prediction error identification algorithm," *Automatica*, vol. 17, no. 1, pp. 89-99, 1981.

[26] P. E. Caines and L. Ljung, "Asymptotic normality and accuracy of prediction error methods," Dep. Elec. Eng., Univ. Toronto, Toronto, Ont., Canada, Res. Rep. 7602, 1976.

[27] L. Ljung, "Analysis of recursive stochastic algorithms," *IEEE Trans. Automat. Contr.*, vol. AC-22, pp. 551-575, 1977.

[28] —, "On positive real functions and the convergence of some recursive schemes," *IEEE Trans. Automat. Contr.*, vol. AC-22, pp. 539-551, 1977.

[29] T. Söderström, L. Ljung, and I. Gustavsson, "A comparative study of recursive identification methods," Dep. Automat. Contr., Lund Inst. Technol., Lund, Sweden, Rep. 7427, 1974.

[30] —, "A theoretical analysis of recursive identification methods," *Automatica*, vol. 14, pp. 231-244, 1978.

[31] J. D. Markel and A. H. Gray, Jr., *Linear Prediction of Speech*. New York: Springer-Verlag, 1976.

[32] L. Ljung, M. Morf, and D. Falconer, "Fast calculations of gain matrices for recursive estimation schemes," *Int. J. Contr.*, vol. 27, pp. 1-19, 1978.

[33] L. J. Griffiths, "An adaptive lattice structure for noise cancelling applications," in *Proc. IEEE Int. Conf. Acoust., Speech, Signal Processing*, Tulsa, OK, Apr. 1978, pp. 87-90.

[34] B. Friedlander and B. Porat, "Localization of multiple targets by sensor arrays: A modeling approach" in *Proc. 20th IEEE Conf. Decision Control*, San Diego, CA, Dec. 1981, pp. 1425-1430.

[35] K. Y. Wong and E. Polak, "Identification of linear discrete time systems using the instrumental variable method," *IEEE Trans. Automat. Contr.*, vol. AC-12, pp. 707-718, Dec. 1967.

[36] P. C. Young, "An instrumental variable method for real-time identification of a noisy process," *Automatica*, vol. 6, pp. 271-287, 1970.

[37] P. C. Young and A. J. Jakeman, "Refined instrumental variable methods of recursive time-series analysis. Part I: Single input, single output systems," *Int. J. Contr.*, vol. 29, no. 1, pp. 1-30, 1979.

[38] T. Söderström and P. Stoica, "Comparison of some instrumental variable methods–Consistency and accuracy aspects," *Automatica*, vol. 17, pp. 101-115, Jan. 1981.

Adaptive Antenna Systems

B. WIDROW, MEMBER, IEEE, P. E. MANTEY, MEMBER, IEEE, L. J. GRIFFITHS, STUDENT MEMBER, IEEE, AND B. B. GOODE, STUDENT MEMBER, IEEE

***Abstract*—A system consisting of an antenna array and an adaptive processor can perform filtering in both the space and the frequency domains, thus reducing the sensitivity of the signal-receiving system to interfering directional noise sources.**

Variable weights of a signal processor can be automatically adjusted by a simple adaptive technique based on the least-mean-squares (LMS) algorithm. During the adaptive process an injected pilot signal simulates a received signal from a desired "look" direction. This allows the array to be "trained" so that its directivity pattern has a main lobe in the previously specified look direction. At the same time, the array processing system can reject any incident noises, whose directions of propagation are different from the desired look direction, by forming appropriate nulls in the antenna directivity pattern. The array adapts itself to form a main lobe, with its direction and bandwidth determined by the pilot signal, and to reject signals or noises occurring outside the main lobe as well as possible in the minimum mean-square error sense.

Several examples illustrate the convergence of the LMS adaptation procedure toward the corresponding Wiener optimum solutions. Rates of adaptation and misadjustments of the solutions are predicted theoretically and checked experimentally. Substantial reductions in noise reception are demonstrated in computer-simulated experiments. The techniques described are applicable to signal-receiving arrays for use over a wide range of frequencies.

INTRODUCTION

THE SENSITIVITY of a signal-receiving array to interfering noise sources can be reduced by suitable processing of the outputs of the individual array elements. The combination of array and processing acts as a filter in both space and frequency. This paper describes a method of applying the techniques of adaptive filtering[1] to the design of a receiving antenna system which can extract directional signals from the medium with minimum distortion due to noise. This system will be called an *adaptive array*. The adaptation process is based on minimization of mean-square error by the LMS algorithm.[2]–[4] The system operates with knowledge of the direction of arrival and spectrum of the signal, but with no knowledge of the noise field. The adaptive array promises to be useful whenever there is interference that possesses some degree of spatial correlation; such conditions manifest themselves over the entire spectrum, from seismic to radar frequencies.

Manuscript received May 29, 1967; revised September 5, 1967.

B. Widrow and L. J. Griffiths are with the Department of Electrical Engineering, Stanford University, Stanford, Calif.

P. E. Mantey was formerly with the Department of Electrical Engineering, Stanford University. He is presently with the Control and Dynamical Systems Group, IBM Research Laboratories, San Jose, Calif.

B. B. Goode is with the Department of Electrical Engineering, Stanford University, Stanford, Calif., and the Navy Electronics Laboratory, San Diego, Calif.

The term "adaptive antenna" has previously been used by Van Atta[5] and others[6] to describe a self-phasing antenna system which reradiates a signal in the direction from which it was received. This type of system is called adaptive because it performs without any prior knowledge of the direction in which it is to transmit. For clarity, such a system might be called an adaptive *transmitting* array; whereas the system described in this paper might be called an adaptive *receiving* array.

The term "adaptive filter" has been used by Jakowatz, Shuey, and White[7] to describe a system which extracts an unknown signal from noise, where the signal waveform recurs frequently at random intervals. Davisson[8] has described a method for estimating an unknown signal waveform in the presence of white noise of unknown variance. Glaser[9] has described an adaptive system suitable for the detection of a pulse signal of fixed but unknown waveform.

Previous work on array signal processing directly related to the present paper was done by Bryn, Mermoz, and Shor. The problem of detecting Gaussian signals in additive Gaussian noise fields was studied by Bryn,[10] who showed that, assuming K antenna elements in the array, the Bayes optimum detector could be implemented by either K^2 linear filters followed by "conventional" beam-forming for each possible signal direction, or by K linear filters for each possible signal direction. In either case, the measurement and inversion of a $2K$ by $2K$ correlation matrix was required at a large number of frequencies in the band of the signal. Mermoz[11] proposed a similar scheme for narrowband known signals, using the signal-to-noise ratio as a performance criterion. Shor[12] also used a signal-to-noise-ratio criterion to detect narrowband pulse signals. He proposed that the sensors be switched off when the signal was known to be absent, and a pilot signal injected as if it were a noise-free signal impinging on the array from a specified direction. The need for specific matrix inversion was circumvented by calculating the gradient of the ratio between the output power due to pilot signal and the output power due to noise, and using the method of steepest descent. At the same time, the number of correlation measurements required was reduced, by Shor's procedure, to $4K$ at each step in the adjustment of the processor. Both Mermoz and Shor have suggested the possibility of real-time adaptation.

This paper presents a potentially simpler scheme for obtaining the desired array processing improvement in real time. The performance criterion used is minimum mean-square error. The statistics of the signal are assumed

Reprinted from *Proc. IEEE,* vol. 55, no. 12, pp. 2143–2159, Dec. 1967.

to be known, but no prior knowledge or direct measurements of the noise field are required in this scheme. The adaptive array processor considered in the study may be automatically adjusted (adapted) according to a simple iterative algorithm, and the procedure does not directly involve the computation of any correlation coefficients or the inversion of matrices. The input signals are used only once, as they occur, in the adaptation process. There is no need to store past input data; but there is a need to store the processor adjustment values, i.e., the processor weighting coefficients ("weights"). Methods of adaptation are presented here, which may be implemented with either analog or digital adaptive circuits, or by digital-computer realization.

Directional and Spatial Filtering

An example of a linear-array receiving antenna is shown in Fig. 1(a) and (b). The antenna of Fig. 1(a) consists of seven isotropic elements spaced $\lambda_0/2$ apart along a straight line, where λ_0 is the wavelength of the center frequency f_0 of the array. The received signals are summed to produce an array output signal. The directivity pattern, i.e., the relative sensitivity of response to signals from various directions, is plotted in this figure in a plane over an angular range of $-\pi/2<\theta<\pi/2$ for frequency f_0. This pattern is symmetric about the vertical line $\theta=0$. The main lobe is centered at $\theta=0$. The largest-amplitude side lobe, at $\theta=24°$, has a maximum sensitivity which is 12.5 dB below the maximum main-lobe sensitivity. This pattern would be different if it were plotted at frequencies other than f_0.

The same array configuration is shown in Fig. 1(b); however, in this case the output of each element is delayed in time before being summed. The resulting directivity pattern now has its main lobe at an angle of ψ radians, where

$$\psi = \sin^{-1}\left(\frac{\lambda_0 \delta f_0}{d}\right) = \sin^{-1}\left(\frac{c\delta}{d}\right) \tag{1}$$

in which

f_0 = frequency of received signal
λ_0 = wavelength at frequency f_0
δ = time-delay difference between neighboring-element outputs
d = spacing between antenna elements
c = signal propagation velocity $=\lambda_0 f_0$.

The sensitivity is maximum at angle ψ because signals received from a plane wave source incident at this angle, and delayed as in Fig. 1(b), are in phase with one another and produce the maximum output signal. For the example illustrated in the figure, $d=\lambda_0/2$, $\delta=(0.12941/f_0)$, and therefore $\psi=\sin^{-1}(2\delta f_0)=15°$.

There are many possible configurations for phased arrays. Fig. 2(a) shows one such configuration where each of the antenna-element outputs is weighted by two weights in parallel, one being preceded by a time delay of a quarter of a cycle at frequency f_0 (i.e., a 90° phase shift), denoted by $1/(4f_0)$. The output signal is the sum of all the weighted signals, and since all weights are set to unit values, the directivity pattern at frequency f_0 is by symmetry the same as that of Fig. 1(a). For purposes of illustration, an interfering directional sinusoidal "noise" of frequency f_0 incident on the array is shown in Fig. 2(a), indicated by the dotted arrow. The angle of incidence (45.5°) of this noise is such that it would be received on one of the side lobes of the directivity pattern with a sensitivity only 17 dB less than that of the main lobe at $\theta=0°$..

If the weights are now set as indicated in Fig. 2(b), the directivity pattern at frequency f_0 becomes as shown in that figure. In this case, the main lobe is almost unchanged from that shown in Figs. 1(a) and 2(a), while the particular side lobe that previously intercepted the sinusoidal noise in Fig. 2(a) has been shifted so that a null is now placed in the direction of that noise. The sensitivity in the noise direction is 77 dB below the main lobe sensitivity, improving the noise rejection by 60 dB.

A simple example follows which illustrates the existence and calculation of a set of weights which will cause a signal from a desired direction to be accepted while a "noise" from a different direction is rejected. Such an example is illustrated in Fig. 3. Let the signal arriving from the desired direction $\theta=0°$ be called the "pilot" signal $p(t)=P \sin \omega_0 t$, where $\omega_0 \triangleq 2\pi f_0$, and let the other signal, the noise, be chosen as $n(t)=N \sin \omega_0 t$, incident to the receiving array at an angle $\theta=\pi/6$ radians. Both the pilot signal and the noise signal are assumed for this example to be at exactly the same frequency f_0. At a point in space midway between the antenna array elements, the signal and the noise are assumed to be in phase. In the example shown, there are two identical omnidirectional array elements, spaced $\lambda_0/2$ apart. The signals received by each element are fed to two variable weights, one weight being preceded by a quarter-wave time delay of $1/(4f_0)$. The four weighted signals are then summed to form the array output.

The problem of obtaining a set of weights to accept $p(t)$ and reject $n(t)$ can now be studied. Note that with any set of nonzero weights, the output is of the form $A \sin(\omega_0 t+\phi)$, and a number of solutions exist which will make the output be $p(t)$. However, the output of the array must be independent of the amplitude and phase of the noise signal if the array is to be regarded as rejecting the noise. Satisfaction of this constraint leads to a unique set of weights determined as follows.

The array output due to the pilot signal is

$$P[(w_1 + w_3)\sin \omega_0 t + (w_2 + w_4)\sin(\omega_0 t - \pi/2)]. \tag{2}$$

For this output to be equal to the desired output of $p(t)=P \sin \omega_0 t$ (which is the pilot signal itself), it is necessary that

$$\left.\begin{array}{l} w_1 + w_3 = 1 \\ w_2 + w_4 = 0 \end{array}\right\}. \tag{3}$$

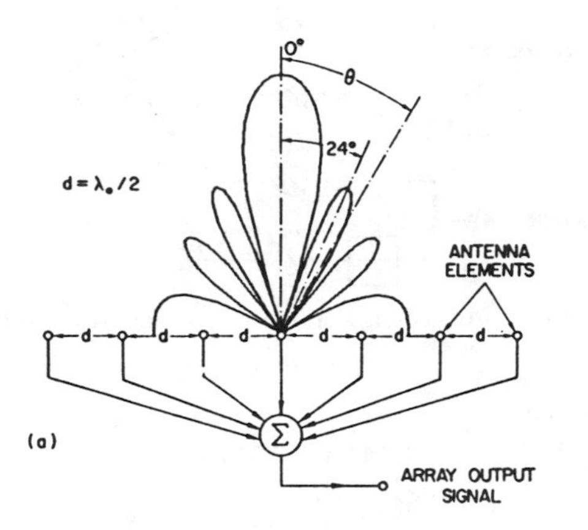

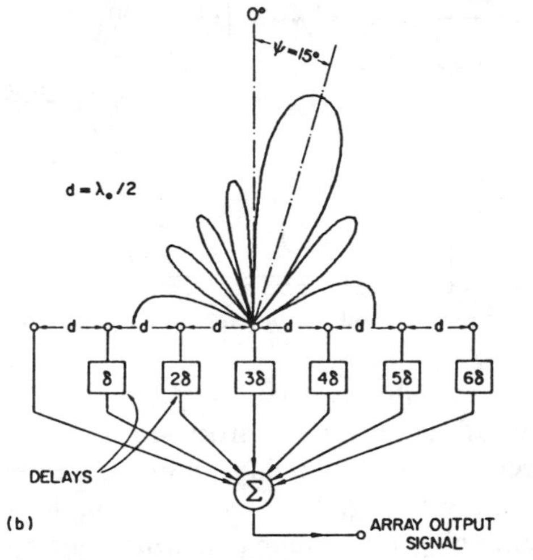

Fig. 1. Directivity pattern for a linear array. (a) Simple array. (b) Delays added.

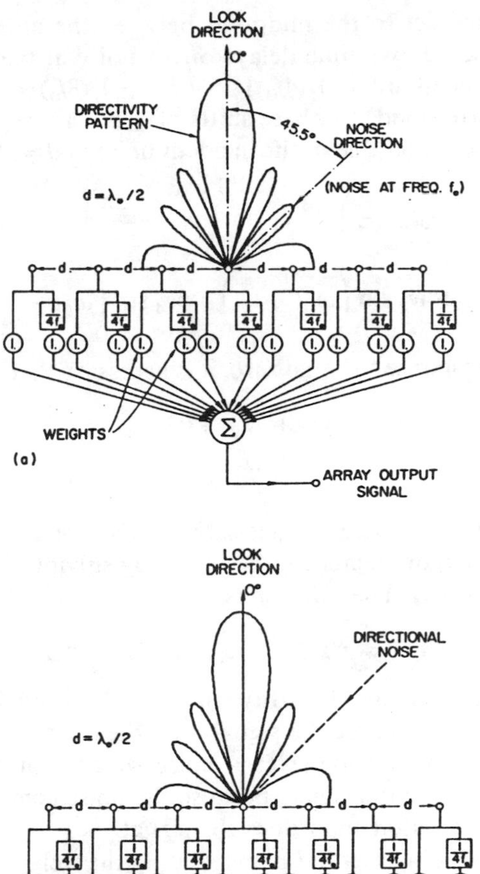

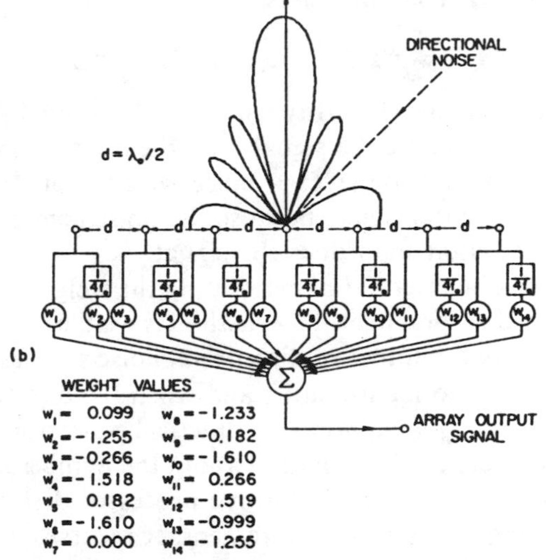

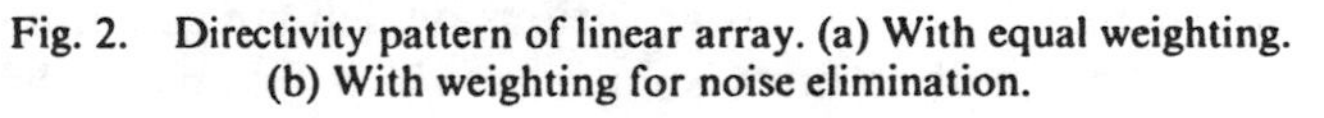

Fig. 2. Directivity pattern of linear array. (a) With equal weighting. (b) With weighting for noise elimination.

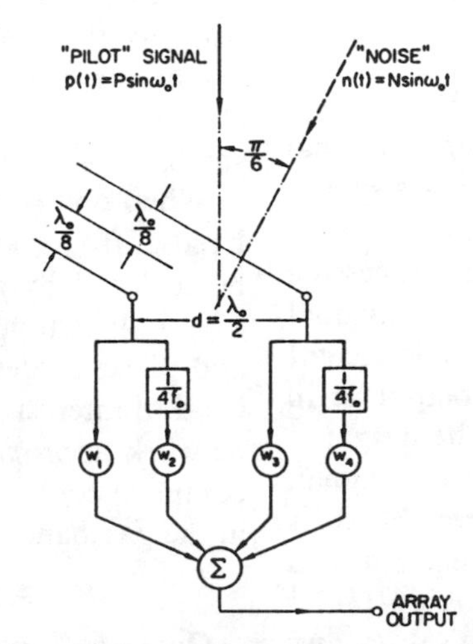

Fig. 3. Array configuration for noise elimination example.

With respect to the midpoint between the antenna elements, the relative time delays of the noise at the two antenna elements are $\pm[1/(4f_0)]\sin\pi/6=\pm1/(8f_0)=\pm\lambda_0/(8c)$, which corresponds to phase shifts of $\pm\pi/4$ at frequency f_0. The array output due to the incident noise at $\theta=\pi/6$ is then

$$N\left[w_1\sin\left(\omega_0 t-\frac{\pi}{4}\right)+w_2\sin\left(\omega_0 t-\frac{3\pi}{4}\right)\right.$$
$$\left.+w_3\sin\left(\omega_0 t+\frac{\pi}{4}\right)+w_4\sin\left(\omega_0 t-\frac{\pi}{4}\right)\right]. \quad (4)$$

For this response to equal zero, it is necessary that

$$\left.\begin{aligned} w_1+w_4&=0\\ w_2-w_3&=0\end{aligned}\right\}. \quad (5)$$

Thus the set of weights that satisfies the signal and noise response requirements can be found by solving (3) and (5) simultaneously. The solution is

$$w_1=\tfrac{1}{2},\ w_2=\tfrac{1}{2},\ w_3=\tfrac{1}{2},\ w_4=-\tfrac{1}{2}. \quad (6)$$

With these weights, the array will have the desired properties in that it will accept a signal from the desired direction, while rejecting a noise, even a noise which is at the same frequency f_0 as the signal, because the noise comes from a different direction than does the signal.

The foregoing method of calculating the weights is more illustrative than practical. This method is usable when there are only a small number of directional noise sources, when the noises are monochromatic, and when the directions of the noises are known *a priori*. A practical processor should not require detailed information about the number and the nature of the noises. The adaptive processor described in the following meets this requirement. It recursively solves a sequence of simultaneous equations, which are generally overspecified, and it finds solutions which minimize the mean-square error between the pilot signal and the total array output.

Configurations of Adaptive Arrays

Before discussing methods of adaptive filtering and signal processing to be used in the adaptive array, various spatial and electrical configurations of antenna arrays will be considered. An adaptive array configuration for processing narrowband signals is shown in Fig. 4. Each individual antenna element is shown connected to a variable weight and to a quarter-period time delay whose output is in turn connected to another variable weight. The weighted signals are summed, as shown in the figure. The signal, assumed to be either monochromatic or narrowband, is received by the antenna element and is thus weighted by a complex gain factor $Ae^{j\phi}$. Any phase angle $\phi=-\tan^{-1}(w_2/w_1)$ can be chosen by setting the two weight values, and the magnitude of this complex gain factor $A=\sqrt{w_1^2+w_2^2}$ can take on a wide range of values limited only by the range limitations of the two individual weights. The latter can assume a continuum of both positive and negative values.

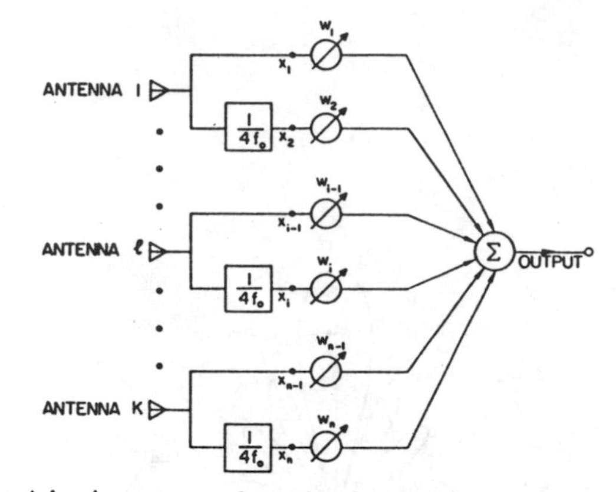

Fig. 4. Adaptive array configuration for receiving narrowband signals.

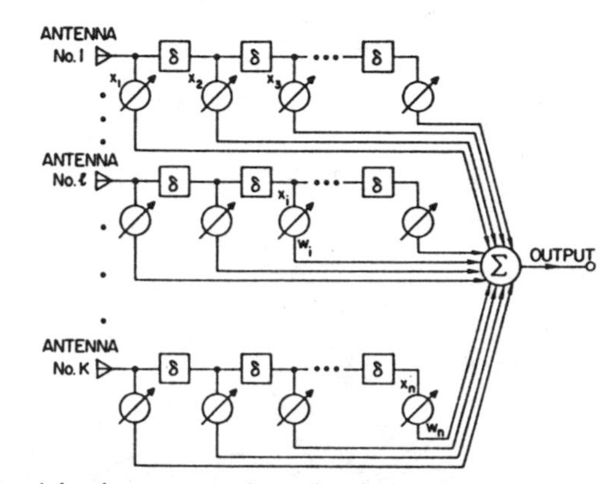

Fig. 5. Adaptive array configuration for receiving broadband signals.

Thus the two weights and the $1/(4f_0)$ time delay provide completely adjustable linear processing for narrowband signals received by each individual antenna element.

The full array of Fig. 4 represents a completely general way of combining the antenna-element signals in an adjustable linear structure when the received signals and noises are narrowband. It should be realized that the same generality (for narrowband signals) can be achieved even when the time delays do not result in a phase shift of exactly $\pi/2$ at the center frequency f_0. Keeping the phase shifts close to $\pi/2$ is desirable for keeping required weight values small, but is not necessary in principle.

When one is interested in receiving signals over a wide band of frequencies, each of the phase shifters in Fig. 4 can be replaced by a tapped-delay-line network as shown in Fig. 5. This tapped delay line permits adjustment of gain and phase as desired at a number of frequencies over the band of interest. If the tap spacing is sufficiently close, this network approximates the ideal filter which would allow complete control of the gain and phase at each frequency in the passband.

Adaptive Signal Processors

Once the form of network connected to each antenna element has been chosen, as shown for example in Fig. 4 or Fig. 5, the next step is to develop an adaptation procedure which can be used to adjust automatically the multiplying weights to achieve the desired spatial and frequency filtering.

The procedure should produce a given array gain in the specified look direction while simultaneously nulling out interfering noise sources.

Fig. 6 shows an adaptive signal-processing element. If this element were combined with an output-signal quantizer, it would then comprise an adaptive threshold logic unit. Such an element has been called an "Adaline"[13] or a threshold logic unit (TLU).[14] Applications of the adaptive threshold element have been made in pattern-recognition systems and in experimental adaptive control systems.[2],[3],[14]–[17]

In Fig. 6 the input signals $x_1(t), \cdots, x_i(t), \cdots, x_n(t)$ are the same signals that are applied to the multiplying weights $w_1, \cdots, w_i, \cdots, w_n$ shown in Fig. 4 or Fig. 5. The heavy lines show the paths of signal flow; the lighter lines show functions related to weight-changing or adaptation processes.

The output signal $s(t)$ in Fig. 6 is the weighted sum

$$s(t) = \sum_{i=1}^{n} x_i(t)w_i \tag{7}$$

where n is the number of weights; or, using vector notation

$$s(t) = W^T X(t) \tag{8}$$

where W^T is the transpose of the weight vector

$$W \triangleq \begin{bmatrix} w_1 \\ \vdots \\ w_i \\ \vdots \\ w_n \end{bmatrix}$$

and the signal-input vector is

$$X(t) \triangleq \begin{bmatrix} x_1(t) \\ \vdots \\ x_i(t) \\ \vdots \\ x_n(t) \end{bmatrix}.$$

For digital systems, the input signals are in discrete-time sampled-data form and the output is written

$$s(j) = W^T X(j) \tag{9}$$

where the index j indicates the jth sampling instant.

In order that adaptation take place, a "desired response" signal, $d(t)$ when continuous or $d(j)$ when sampled, must be supplied to the adaptive element. A method for obtaining this signal for adaptive antenna array processing will be discussed in a following section.

The difference between the desired response and the output response forms the error signal $\varepsilon(j)$:

$$\varepsilon(j) = d(j) - W^T X(j). \tag{10}$$

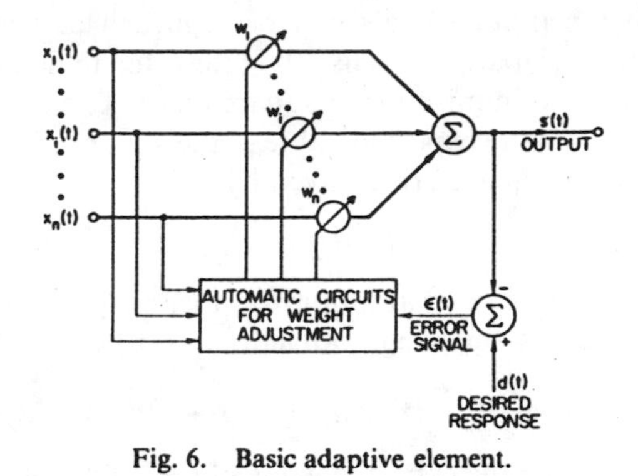

Fig. 6. Basic adaptive element.

This signal is used as a control signal for the "weight adjustment circuits" of Fig. 6.

Solving Simultaneous Equations

The purpose of the adaptation or weight-changing processes is to find a set of weights that will permit the output response of the adaptive element at each instant of time to be equal to or as close as possible to the desired response. For each input-signal vector $X(j)$, the error $\varepsilon(j)$ of (10) should be made as small as possible.

Consider the finite set of linear simultaneous equations

$$\begin{cases} W^T X(1) = d(1) \\ W^T X(2) = d(2) \\ \quad\vdots \qquad\quad \vdots \\ W^T X(j) = d(j) \\ \quad\vdots \qquad\quad \vdots \\ W^T X(N) = d(N) \end{cases} \tag{11}$$

where N is the total number of input-signal vectors; each vector is a measurement of an underlying n-dimensional random process. There are N equations, corresponding to N instants of time at which the output response values are of concern; there are n "unknowns," the n weight values which form the components of W. The set of equations (11) will usually be overspecified and inconsistent, since in the present application, with an ample supply of input data, it is usual that $N \gg n$. [These equations did have a solution in the simple example represented in Fig. 3. The solution is given in (6). Although the simultaneous equations (3) in that example appear to be different from (11), they are really the same, since those in (3) are in a specialized form for the case when all inputs are deterministic sinusoids which can be easily specified over all time in terms of amplitudes, phases, and frequencies.]

When N is very large compared to n, one is generally interested in obtaining a solution of a set of N equations [each equation in the form of (10)] which minimizes the sum of the squares of the errors. That is, a set of weights W is found to minimize

$$\sum_{j=1}^{N} \varepsilon^2(j). \tag{12}$$

When the input signals can be regarded as stationary stochastic variables, one is usually interested in finding a set of weights to minimize mean-square error. The quantity of interest is then the expected value of the square of the error, i.e., the mean-square error, given by

$$E[\varepsilon^2(j)] \triangleq \overline{\varepsilon^2}. \tag{13}$$

The set of weights that minimizes mean-square error can be calculated by squaring both sides of (10) which yields

$$\varepsilon^2(j) = d^2(j) + W^T X(j) X(j)^T W - 2d(j) W^T X(j) \tag{14}$$

and then taking the expected value of both sides of (14)

$$\begin{aligned} E[\varepsilon^2(j)] &= E[d^2 + W^T X(j) X^T(j) W - 2W^T d(j) X(j)] \\ &= E[d^2] + W^T \Phi(x, x) W - 2W^T \Phi(x, d) \end{aligned} \tag{15}$$

where

$$\Phi(x, x) \triangleq E[X(j)X^T(j)] \triangleq E \begin{bmatrix} x_1x_1 & x_1x_2 \cdots x_1x_n \\ x_2x_1 & \cdots x_2x_n \\ \vdots & \\ x_nx_1 & x_nx_n \end{bmatrix} \tag{16}$$

and

$$\Phi(x, d) \triangleq E[X(j)d(j)] \triangleq E \begin{bmatrix} x_1d \\ x_2d \\ \vdots \\ x_id \\ \vdots \\ x_nd \end{bmatrix} \tag{17}$$

The symmetric matrix $\Phi(x, x)$ is a matrix of cross correlations and autocorrelations of the input signals to the adaptive element, and the column matrix $\Phi(x, d)$ is the set of cross correlations betweeen the n input signals and the desired response signal.

The mean-square error defined in (15) is a quadratic function of the weight values. The components of the gradient of the mean-square-error function are the partial derivatives of the mean-square error with respect to the weight values. Differentiating (15) with respect to W yields the gradient $\nabla E[\varepsilon^2]$, a linear function of the weights,

$$\nabla E[\varepsilon^2] = 2\Phi(x, x)W - 2\Phi(x, d). \tag{18}$$

When the choice of the weights is optimized, the gradient is zero. Then

$$\begin{aligned} \Phi(x, x) W_{\text{LMS}} &= \Phi(x, d) \\ W_{\text{LMS}} &= \Phi^{-1}(x, x)\Phi(x, d). \end{aligned} \tag{19}$$

The optimum weight vector W_{LMS} is the one that gives the least mean-square error. Equation (19) is the Wiener-Hopf equation, and is the equation for the multichannel least-squares filter used by Burg[18] and Claerbout[19] in the processing of digital seismic array data.

One way of finding the optimum set of weight values is to solve (19). This solution is generally straightforward, but presents serious computational problems when the number of weights n is large and when data rates are high. In addition to the necessity of inverting an $n \times n$ matrix, this method may require as many as $n(n+1)/2$ autocorrelation and cross-correlation measurements to obtain the elements of $\Phi(x, x)$. Furthermore, this process generally needs to be continually repeated in most practical situations where the input signal statistics change slowly. No perfect solution of (19) is possible in practice because of the fact that an infinite statistical sample would be required to estimate perfectly the elements of the correlation matrices.

Two methods for finding approximate solutions to (19) will be presented in the following. Their accuracy is limited by statistical sample size, since they find weight values based on finite-time measurements of input-data signals. These methods do not require explicit measurements of correlation functions or matrix inversion. They are based on gradient-search techniques applied to mean-square-error functions. One of these methods, the LMS algorithm, does not even require squaring, averaging, or differentiation in order to make use of gradients of mean-square-error functions. The second method, a relaxation method, will be discussed later.

The LMS Algorithm

A number of weight-adjustment procedures or algorithms exist which minimize the mean-square error. Minimization is usually accomplished by gradient-search techniques. One method that has proven to be very useful is the LMS algorithm.[1]–[3],[17] This algorithm is based on the method of steepest descent. Changes in the weight vector are made along the direction of the estimated gradient vector. Accordingly,

$$W(j+1) = W(j) + k_s \hat{\nabla}(j) \tag{20}$$

where

$W(j) \triangleq$ weight vector before adaptation
$W(j+1) \triangleq$ weight vector after adaptation
$k_s \triangleq$ scalar constant controlling rate of convergence and stability ($k_s < 0$)
$\hat{\nabla}(j) \triangleq$ estimated gradient vector of $\overline{\varepsilon^2}$ with respect to W.

One method for obtaining the estimated gradient of the mean-square-error function is to take the gradient of a single time sample of the squared error

$$\hat{\nabla}(j) = \nabla[\varepsilon^2(j)] = 2\varepsilon(j)\nabla[\varepsilon(j)].$$

From (10)

$$\begin{aligned} \nabla[\varepsilon(j)] &= \nabla[d(j) - W^T(j)X(j)] \\ &= -X(j). \end{aligned}$$

Thus

$$\hat{\nabla}(j) = -2\varepsilon(j)X(j). \tag{21}$$

The gradient estimate of (21) is unbiased, as will be shown by the following argument. For a given weight vector $W(j)$, the expected value of the gradient estimate is

$$E[\hat{\nabla}(j)] = -2E[\{d(j) - W^T(j)X(j)\}X(j)]$$
$$= -2[\Phi(x, d) - W^T(j)\Phi(x, x)]. \quad (22)$$

Comparing (18) and (22), we see that

$$E[\hat{\nabla}(j)] = \nabla E[\varepsilon^2]$$

and therefore, for a given weight vector, the expected value of the estimate equals the true value.

Using the gradient estimation formula given in (21), the weight iteration rule (20) becomes

$$W(j + 1) = W(j) - 2k_s\varepsilon(j)X(j) \quad (23)$$

and the next weight vector is obtained by adding to the present weight vector the input vector scaled by the value of the error.

The LMS algorithm is given by (23). It is directly usable as a weight-adaptation formula for digital systems. Fig. 7(a) shows a block-diagram representation of this equation in terms of one component w_i of the weight vector W. An equivalent differential equation which can be used in analog implementation of continuous systems (see Fig. 7(b)) is given by

$$\frac{d}{dt} W(t) = -2k_s\varepsilon(t)X(t).$$

This equation can also be written as

$$W(t) = -2k_s \int_0^t \varepsilon(\xi)X(\xi)\, d\xi.$$

Fig. 8 shows how circuitry of the type indicated in Fig. 7(a) or (b) might be incorporated into the implementation of the basic adaptive element of Fig. 6.

Convergence of the Mean of the Weight Vector

For the purpose of the following discussion, we assume that the time between successive iterations of the LMS algorithm is sufficiently long so that the sample input vectors $X(j)$ and $X(j+1)$ are uncorrelated. This assumption is common in the field of stochastic approximation.[20]-[22]

Because the weight vector $W(j)$ is a function *only* of the input vectors $X(j-1), X(j-2), \cdots, X(0)$ [see (23)] and because the successive input vectors are uncorrelated, $W(j)$ is independent of $X(j)$. For stationary input processes meeting this condition, the expected value $E[W(j)]$ of the weight vector after a large number of iterations can then be shown to converge to the Wiener solution given by (19). Taking the expected value of both sides of (23), we obtain a difference equation in the expected value of the weight vector

$$E[W(j + 1)] = E[W(j)] - 2k_sE[\{d(j) - W^T(j)X(j)\}X(j)]$$
$$= [I + 2k_s\Phi(x, x)]E[W(j)] - 2k_s\Phi(x, d) \quad (24)$$

where I is the identity matrix. With an initial weight vector $W(0)$, $j+1$ iterations of (24) yield

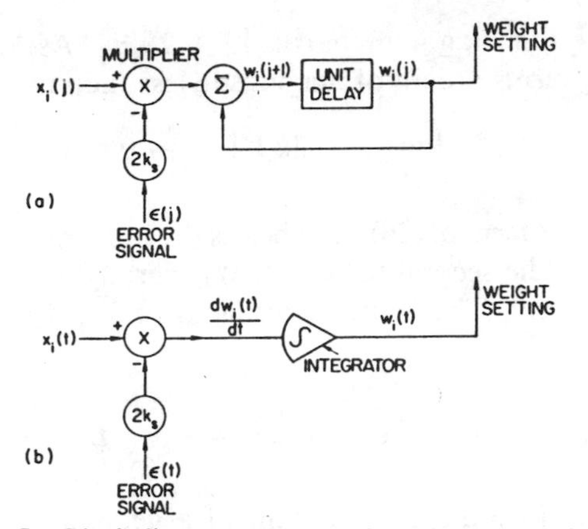

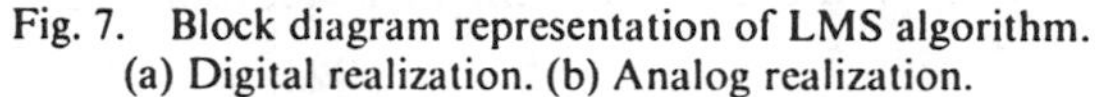

Fig. 7. Block diagram representation of LMS algorithm. (a) Digital realization. (b) Analog realization.

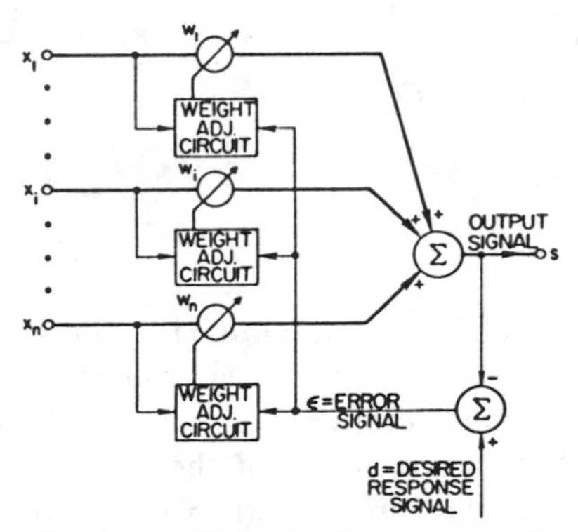

Fig. 8. Analog/digital implementation of LMS weight-adjustment algorithm.

$$E[W(j + 1)] = [I + 2k_s\Phi(x, x)]^{j+1}W(0)$$
$$- 2k_s \sum_{i=0}^{j} [I + 2k_s\Phi(x, x)]^i\Phi(x, d). \quad (25)$$

Equation (25) may be put in diagonal form by using the appropriate similarity transformation Q for the matrix $\Phi(x, x)$, that is,

$$\Phi(x, x) = Q^{-1}EQ$$

where

$$E \triangleq \begin{bmatrix} e_1 & 0 & \cdots & 0 \\ 0 & e_2 & \cdots & 0 \\ \vdots & & \ddots & \vdots \\ 0 & 0 & \cdots & e_n \end{bmatrix}$$

is the diagonal matrix of eigenvalues. The eigenvalues are all positive, since $\Phi(x, x)$ is positive definite [see (16)]. Equation (25) may now be expressed as

$$E[W(j + 1)] = [I + 2k_sQ^{-1}EQ]^{j+1}W(0)$$
$$- 2k_s \sum_{i=0}^{j} [I + 2k_sQ^{-1}EQ]^i\Phi(x, d)$$
$$= Q^{-1}[I + 2k_sE]^{j+1}QW(0)$$
$$- 2k_sQ^{-1} \sum_{i=0}^{j} [I + 2k_sE]^iQ\Phi(x, d). \quad (26)$$

Consider the diagonal matrix $[I + 2k_s E]$. As long as its diagonal terms are all of magnitude less than unity

$$\lim_{j\to\infty} [I + 2k_s E]^{j+1} \to 0$$

and the first term of (26) vanishes as the number of iterations increases. The second term in (26) generally converges to a nonzero limit. The summation factor $\sum_{i=0}^{j}[I+2k_s E]^i$ becomes

$$\lim_{j\to\infty} \sum_{i=0}^{j} [I + 2k_s E]^i = -\frac{1}{2k_s} E^{-1}$$

where the formula for the sum of a geometric series has been used, that is,

$$\sum_{i=0}^{\infty} (1 + 2k_s e_p)^i = \frac{1}{1-(1+2k_s e_p)} = \frac{-1}{2k_s e_p}.$$

Thus, in the limit, (26) becomes

$$\lim_{j\to\infty} E[W(j+1)] = Q^{-1}E^{-1}Q\Phi(x,d)$$
$$= \Phi^{-1}(x,x)\Phi(x,d).$$

Comparison of this result with (19) shows that as the number of iterations increases without limit, the expected value of the weight vector converges to the Wiener solution.

Convergence of the mean of the weight vector to the Wiener solution is insured if and only if the proportionality constant k_s is set within certain bounds. Since the diagonal terms of $[I+2k_s E]$ must all have magnitude less than unity, and since all eigenvalues in E are positive, the bounds on k_s are given by

$$|1 + 2k_s e_{\max}| < 1$$

or

$$\frac{-1}{e_{\max}} < k_s < 0 \tag{27}$$

where $e_{\max}$ is the maximum eigenvalue of $\Phi(x,x)$. This convergence condition on k_s can be related to the total input power as follows.

Since

$$e_{\max} \leqq \text{trace}\,[\Phi(x,x)] \tag{28}$$

where

$$\text{trace}\,[\Phi(x,x)] \triangleq E[X^T(j)X(j)]$$
$$= \sum_{i=1}^{n} E[x_i^2] \triangleq \text{total input power},$$

it follows that satisfactory convergence can be obtained with

$$\frac{-1}{\sum_{i=1}^{n} E[x_i^2]} < k_s < 0.$$

In practice, when slow, precise adaptation is desired, k_s is usually chosen such that

$$\frac{-1}{\sum_{i=1}^{n} E[x_i^2]} \ll k_s < 0. \tag{29}$$

It is the opinion of the authors that the assumption of independent successive input samples used in the foregoing convergence proof is overly restrictive. That is, convergence of the mean of the weight vector to the LMS solution can be achieved under conditions of highly correlated input samples. In fact, the computer-simulation experiments described in this paper *do not* satisfy the condition of independence.

Time Constants and Learning Curve with LMS Adaptation

State-variable methods, which are widely used in modern control theory, have been applied by Widrow[1] and Koford and Groner[2] to the analysis of stability and time constants (related to rate of convergence) of the LMS algorithm. Considerable simplifications in the analysis have been realized by expressing transient phenomena of the system adjustments (which take place during the adaptation process) in terms of the normal coordinates of the system. As shown by Widrow,[1] the weight values undergo transients during adaptation. The transients consist of sums of exponentials with time constants given

$$\tau_p = \frac{1}{2(-k_s)e_p}, \quad p = 1, 2, \cdots, n \tag{30}$$

where e_p is the pth eigenvalue of the input-signal correlation matrix $\Phi(x,x)$.

In the special case when all eigenvalues are equal, all time constants are equal. Accordingly,

$$\tau = \frac{1}{2(-k_s)e}.$$

One very useful way to monitor the progress of an adaptive process is to plot or display its "learning curve." When mean-square error is the performance criterion being used, one can plot the expected mean-square error at each stage of the learning process as a function of the number of adaptation cycles. Since the underlying relaxation phenomenon which takes place in the weight values is of exponential nature, and since from (15) the mean-square error is a quadratic form in the weight values, the transients in the mean-square-error function must also be exponential in nature.

When all the time constants are equal, the mean-square-error learning curve is a pure exponential with a time constant

$$\tau_{\text{mse}} = \frac{\tau}{2} = \frac{1}{4(-k_s)e}.$$

The basic reason for this is that the square of an exponential function is an exponential with half the time constant.

Estimation of the rate of adaptation is more complex when the eigenvalues are unequal.

When actual experimental learning curves are plotted, they are generally of the form of noisy exponentials because of the inherent noise in the adaptation process. The slower the adaptation, the smaller will be the amplitude of the noise apparent in the learning curve.

Misadjustment with LMS Adaptation

All adaptive or learning systems capable of adapting at real-time rates experience losses in performance because their system adjustments are based on statistical averages taken with limited sample sizes. The faster a system adapts, in general, the poorer will be its expected performance.

When the LMS algorithm is used with the basic adaptive element of Fig. 8, the expected level of mean-square error will be greater than that of the Wiener optimum system whose weights are set in accordance with (19). The longer the time constants of adaptation, however, the closer the expected performance comes to the Wiener optimum performance. To get the Wiener performance, i.e., to achieve the minimum mean-square error, one would have to know the input statistics *a priori*, or, if (as is usual) these statistics are unknown, they would have to be measured with an arbitrarily large statistical sample.

When the LMS adaptation algorithm is used, an excess mean-square error therefore develops. A measure of the extent to which the adaptive system is misadjusted as compared to the Wiener optimum system is determined in a performance sense by the ratio of the excess mean-square error to the minimum mean-square error. This dimensionless measure of the loss in performance is defined as the "misadjustment" M. For LMS adaptation of the basic adaptive element, it is shown by Widrow[1] that

$$\text{Misadjustment } M = \frac{1}{2} \sum_{p=1}^{n} \frac{1}{\tau_p}. \tag{31}$$

The value of the misadjustment depends on the time constants (settling times) of the filter adjustment weights. Again, in the special case when all the time constants are equal, *M is proportional to the number of weights and inversely proportional to the time constant.* That is,

$$\begin{aligned} M &= \frac{n}{2\tau} \\ &= \frac{n}{4\tau_{\text{mse}}}. \end{aligned} \tag{32}$$

Although the foregoing results specifically apply to statistically stationary processes, the LMS algorithm can also be used with nonstationary processes. It is shown by Widrow[23] that, under certain assumed conditions, the rate of adaptation is optimized when the loss of performance resulting from adapting too rapidly equals twice the loss in performance resulting from adapting too slowly.

Adaptive Spatial Filtering

If the radiated signals received by the elements of an adaptive antenna array were to consist of signal components plus undesired noise, the signal would be reproduced (and noise eliminated) as best possible in the least-squares sense if the desired response of the adaptive processor were made to be the signal itself. This signal is not generally available for adaptation purposes, however. If it were available, there would be no need for a receiver and a receiving array.

In the adaptive antenna systems to be described here, the desired response signal is provided through the use of an artificially injected signal, the "pilot signal", which is completely known at the receiver and usually generated there. The pilot signal is constructed to have spectral and directional characteristics similar to those of the incoming signal of interest. These characteristics may, in some cases, be known *a priori* but, in general, represent estimates of the parameters of the signal of interest.

Adaptation with the pilot signal causes the array to form a beam in the pilot-signal direction having essentially flat spectral response and linear phase shift within the passband of the pilot signal. Moreover, directional noises impinging on the antenna array will cause reduced array response (nulling) in their directions within their passbands. These notions are demonstrated by experiments which will be described in the following.

Injection of the pilot signal could block the receiver and render useless its output. To circumvent this difficulty, two adaptation algorithms have been devised, the "one-mode" and the "two-mode." The two-mode process alternately adapts on the pilot signal to form the beam and then adapts on the natural inputs with the pilot signal off to eliminate noise. The array output is usable during the second mode, while the pilot signal is off. The one-mode algorithm permits listening at all times, but requires more equipment for its implementation.

The Two-Mode Adaptation Algorithm

Fig. 9 illustrates a method for providing the pilot signal wherein the latter is actually transmitted by an antenna located some distance from the array in the desired look direction. Fig. 10 shows a more practical method for providing the pilot signal. The inputs to the processor are connected either to the actual antenna element outputs (during "mode II"), or to a set of delayed signals derived from the pilot-signal generator (during "mode I"). The filters $\delta_1, \cdots, \delta_K$ (ideal time-delays if the array elements are identical) are chosen to result in a set of input signals identical with those that would appear if the array were actually receiving a radiated plane-wave pilot signal from the desired "look" direction, the direction intended for the main lobe of the antenna directivity pattern.

During adaptation in mode I, the input signals to the adaptive processor derive from the pilot signal, and the desired response of the adaptive processor is the pilot signal

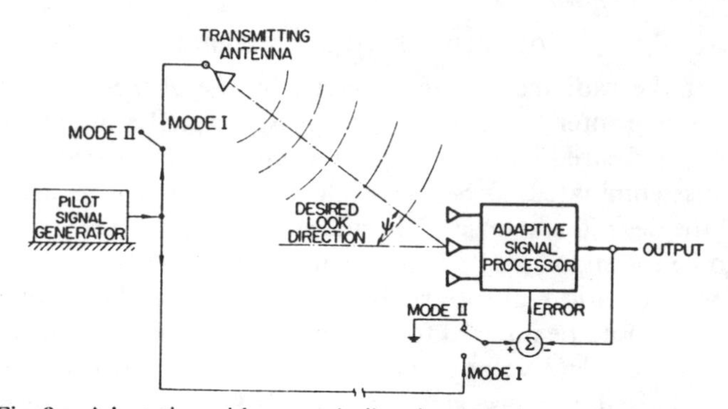

Fig. 9. Adaptation with external pilot-signal generator. Mode I: adaptation with pilot signal present; Mode II: adaptation with pilot signal absent.

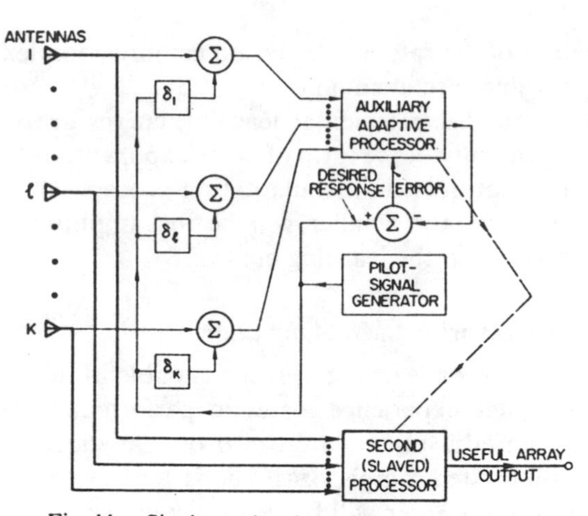

Fig. 11. Single-mode adaptation with pilot signal.

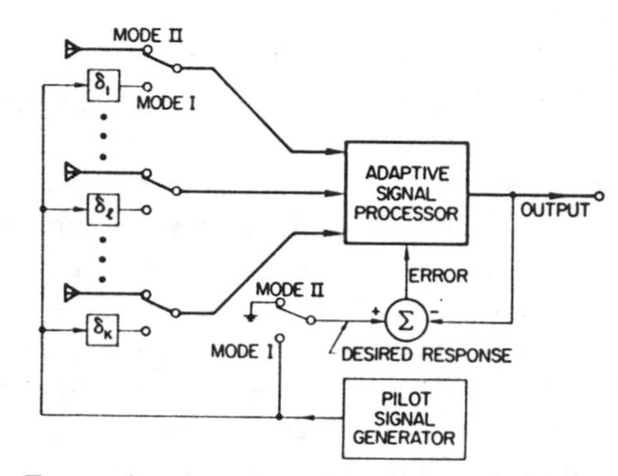

Fig. 10. Two-mode adaptation with internal pilot-signal generator. Mode I: adaptation with pilot signal present; Mode II: adaptation with pilot signal absent.

itself. If a sinusoidal pilot signal at frequency f_0 is used, for example, adapting the weights to minimize mean-square error will force the gain of the antenna array in the look direction to have a specific amplitude and a specific phase shift at frequency f_0.

During adaptation in mode II, all signals applied to the adaptive processor are received by the antenna elements from the actual noise field. In this mode, the adaptation process proceeds to eliminate all received signals, since the desired response is set to zero. Continuous operation in mode II would cause all the weight values to tend to zero, and the system would shut itself off. However, by alternating frequently between mode I and mode II and causing only small changes in the weight vector during each mode of adaptation, it is possible to maintain a beam in the desired look direction and, in addition, to minimize the reception of incident-noise power.

The pilot signal can be chosen as the sum of several sinusoids of differing frequencies. Then adaptation in mode I will constrain the antenna gain and phase in the look direction to have specific values at each of the pilot-signal frequencies. Furthermore, if several pilot signals of different simulated directions are added together, it will be possible to constrain the array gain simultaneously at various frequencies and angles when adapting in mode I. This feature affords some control of the bandwidth and beamwidth in the look direction. The two-mode adaptive process essentially minimizes the mean-square value (the total power) of all signals received by the antenna elements which are uncorrelated with the pilot signals, subject to the constraint that the gain and phase in the beam approximate predetermined values at the frequencies and angles dictated by the pilot-signal components.

The One-Mode Adaptation Algorithm

In the two-mode adaptation algorithm the beam is formed during mode I, and the noises are eliminated in the least-squares sense (subject to the pilot-signal constraints) in mode II. Signal reception during mode I is impossible because the processor is connected to the pilot-signal generator. Reception can therefore take place only during mode II. This difficulty is eliminated in the system of Fig. 11, in which the actions of both mode I and mode II can be accomplished simultaneously. The pilot signals and the received signals enter into an auxiliary, adaptive processor, just as described previously. For this processor, the desired response is the pilot signal $p(t)$. A second weighted processor (linear element) generates the actual array output signal, but it performs no adaptation. Its input signals do not contain the pilot signal. It is slaved to the adaptive processor in such a way that its weights track the corresponding weights of the adapting system, so that it never needs to receive the pilot signal.

In the single-mode system of Fig. 11, the pilot signal is on continuously. Adaptation to minimize mean-square error will force the adaptive processor to reproduce the pilot signal as closely as possible, and, at the same time, to reject as well as possible (in the mean-square sense) all signals received by the antenna elements which are uncorrelated with the pilot signal. Thus the adaptive process forces a directivity pattern having the proper main lobe in the look direction in the passband of the pilot signal (satisfying the pilot signal constraints), and it forces nulls in the directions of the noises and in their frequency bands. Usually, the stronger the noises, the deeper are the corresponding nulls.

Computer Simulation of Adaptive Antenna Systems

To demonstrate the performance characteristics of adaptive antenna systems, many simulation experiments, involving a wide variety of array geometries and signal-

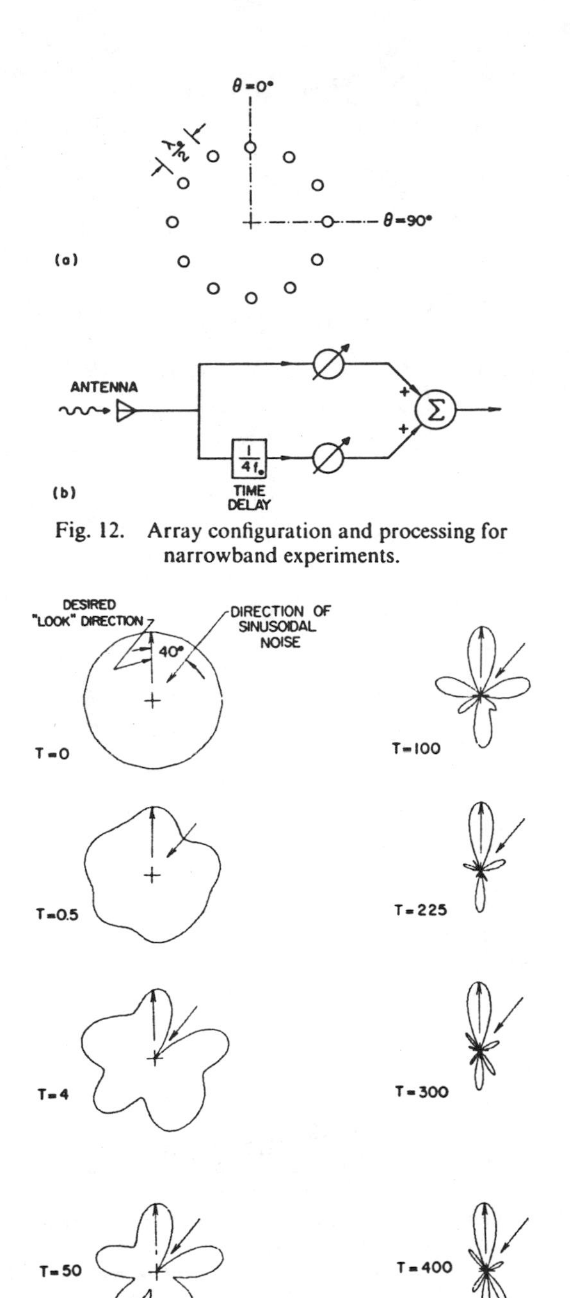

Fig. 12. Array configuration and processing for narrowband experiments.

Fig. 13. Evolution of the directivity pattern while learning to eliminate a directional noise and uncorrelated noises. (Array configuration of Fig. 12.) T = number of elapsed cycles of frequency f_0 (total number of adaptations = $20T$).

and noise-field configurations, have been carried out using an IBM 1620-II computer equipped with a digital output plotter.

For simplicity of presentation, the examples outlined in the following are restricted to planar arrays composed of ideal isotropic radiators. In every case, the LMS adaptation algorithm was used. All experiments were begun with the initial condition that all weight values were equal.

Narrowband Processor Experiments

Fig. 12 shows a twelve-element circular array and signal processor which was used to demonstrate the performance of the narrowband system shown in Fig. 4. In the first computer simulation, the two-mode adaptation algorithm was used. The pilot signal was a unit-amplitude sine wave (power = 0.5, frequency f_0) which was used to train the array to look in the $\theta = 0°$ direction. The noise field consisted of a sinusoidal noise signal (of the same frequency and power as the pilot signal) incident at angle $\theta = 40°$, and a small amount of random, uncorrelated, zero-mean, "white" Gaussian noise of variance (power) = 0.1 at each antenna element. In this simulation, the weights were adapted using the LMS two-mode algorithm.

Fig. 13 shows the sequence of directivity patterns which evolved during the "learning" process. These computer-plotted patterns represent the decibel sensitivity of the array at frequency f_0. Each directivity pattern is computed from the set of weights resulting at various stages of adaptation. The solid arrow indicates the direction of arrival of the interfering sine-wave noise source. Notice that the initial directivity pattern is essentially circular. This is due to the symmetry of the antenna array elements and of the initial weight values. A timing indicator T, the number of elapsed cycles of frequency f_0, is presented with each directivity pattern. The total number of adaptations equals $20T$ in these experiments. Note that if $f_0 = 1$ kHz, $T = 1$ corresponds to 1 ms real time; if $f_0 = 1$ MHz, $T = 1$ corresponds to 1 μs, etc.

Several observations can be made from the series of directivity patterns of Fig. 13. Notice that the sensitivity of the array in the look direction is essentially constant during the adaptation process. Also notice that the array sensitivity drops very rapidly in the direction of the sinusoidal noise source; a deep notch in the directivity pattern forms in the noise direction as the adaptation process progresses. After the adaptive transients died out, the array sensitivity in the noise direction was 27 dB below that of the array in the desired look direction.

The total noise power in the array output is the sum of the sinusoidal noise power due to the directional noise source plus the power due to the "white" Gaussian, mutually uncorrelated noise-input signals. The total noise power generally drops as the adaptation process commences, until it reaches an irreducible level.

A plot of the total received noise power as a function of T is shown in Fig. 14. This curve may be called a "learning curve." Starting with the initial weights, the total output noise power was 0.65, as shown in the figure. After adaptation, the total output noise power was 0.01. In this noise field, the signal-to-noise ratio of the array[1] after adaptation was better than that of a single isotropic receiving element by a factor of about 60.

A second experiment using the same array configuration and the two-mode adaptive process was performed to investigate adaptive array performance in the presence of several interfering directional noise sources. In this example, the noise field was composed of five directional sinus-

[1] Signal-to-noise ratio is defined as

$$\mathrm{SNR} = \frac{\text{array output power due to signal}}{\text{array output power due to noise}}.$$

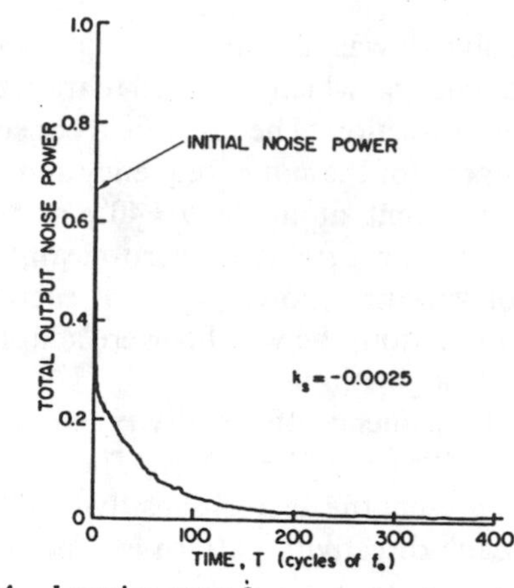

Fig. 14. Learning curve for narrowband system of Fig. 12, with noise from one direction only.

TABLE I

SENSITIVITIES OF ARRAY IN DIRECTIONS OF THE FIVE NOISE SOURCES OF FIG. 15, AFTER ADAPTATION

Noise Direction (degrees)	Noise Frequency (times f_0)	Array Sensitivity in Noise Direction, Relative to Sensitivity in Desired Look Direction (dB)
67	1.10	−26
134	0.95	−30
191	1.00	−28
236	0.90	−30
338	1.05	−38

oidal noises, each of amplitude 0.5 and power 0.125, acting simultaneously, and, in addition, superposed uncorrelated "white" Gaussian noises of power 0.5 at each of the antenna elements. The frequencies of the five directional noises are shown in Table I.

Fig. 15(a) shows the evolution of the directivity pattern, plotted at frequency f_0, from the initial conditions to the finally converged (adapted) state. The latter was achieved after 682 cycles of the frequency f_0. The learning curve for this experiment is shown in Fig. 15(b). The final array sensitivities in the five noise directions relative to the array sensitivity in the desired look direction are shown in Table I. The signal-to-noise ratio was improved by a factor of about 15 over that of a single isotropic radiator. In Fig. 15(b), one can roughly discern a time constant approximately equal to 70 cycles of the frequency f_0. Since there were 20 adaptations per cycle of f_0, the learning curve time constant was approximately $\tau_{mse} = 1400$ adaptations. Within about 400 cycles of f_0, the adaptive process virtually converges to steady state. If f_0 were 1 MHz, 400 μs would be the real-time settling time. The misadjustment for this process can be roughly estimated by using (32), although actually all eigenvalues were not equal as required by this equation:

$$M = \frac{n}{4\tau_{mse}} = \frac{24}{4\tau_{mse}} = \frac{6}{1400} = 0.43 \text{ percent.}$$

This is a very low value of misadjustment, indicating a very slow, precise adaptive process. This is evidenced by the

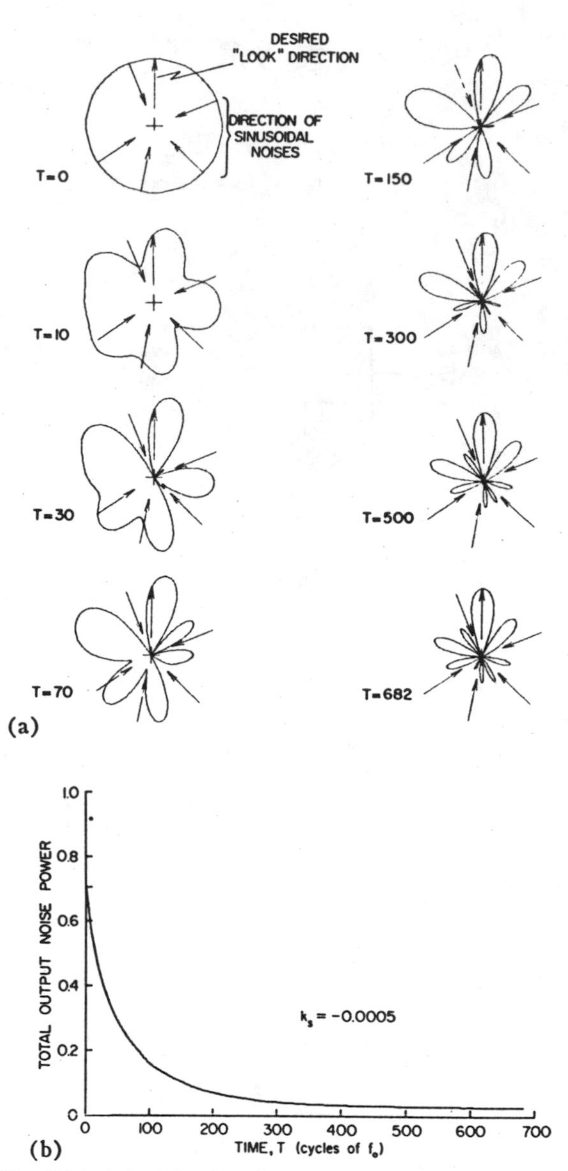

Fig. 15. Evolution of the directivity pattern while learning to eliminate five directional noises and uncorrelated noises. (Array configuration of Fig. 12.) (a) Sequence of directivity patterns during adaptation. (b) Learning curve (total number of adaptations=20T).

learning curve Fig. 15(b) for this experiment, which is very smooth and noise-free.

Broadband Processor Experiments

Fig. 16 shows the antenna array configuration and signal processor used in a series of computer-simulated broadband experiments. In these experiments, the one-mode or simultaneous adaptation process was used to adjust the weights. Each antenna or element in a five-element circular array was connected to a tapped delay line having five variable weights, as shown in the figure. A broadband pilot signal was used, and the desired look direction was chosen (arbitrarily, for purposes of example) to be $\theta = -13°$. The frequency spectrum of the pilot signal is shown in Fig. 17(a). This spectrum is approximately one octave wide and is centered at frequency f_0. A time-delay increment of $1/(4f_0)$ was used in the tapped delay line, thus providing a delay between adjacent weights of a quarter cycle at fre-

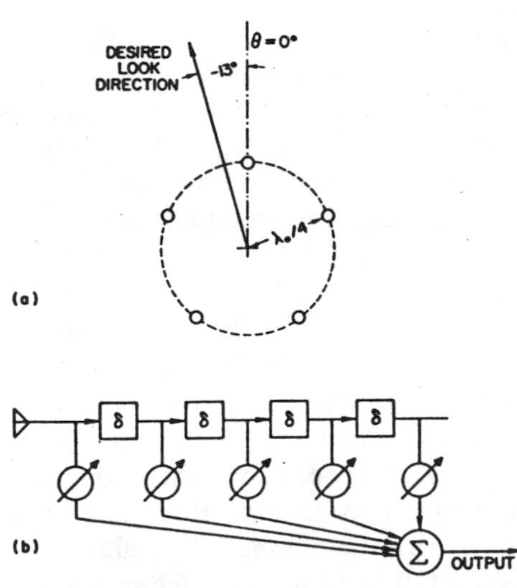

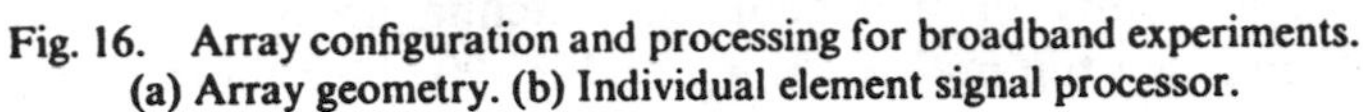
Fig. 16. Array configuration and processing for broadband experiments. (a) Array geometry. (b) Individual element signal processor.

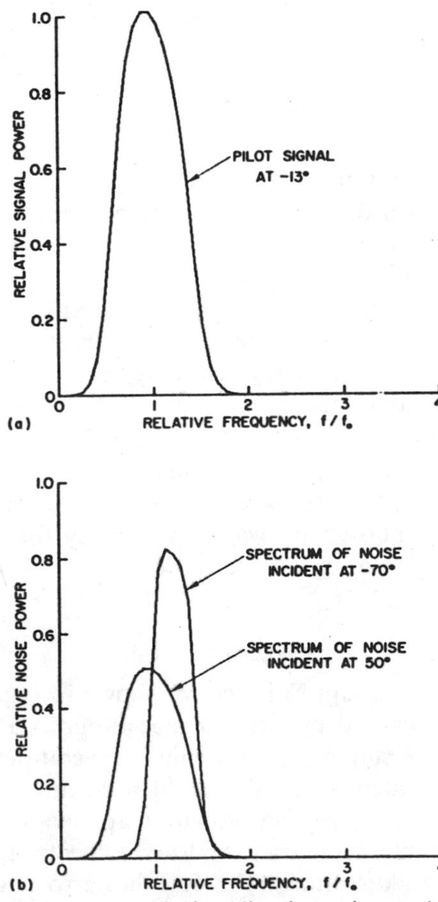

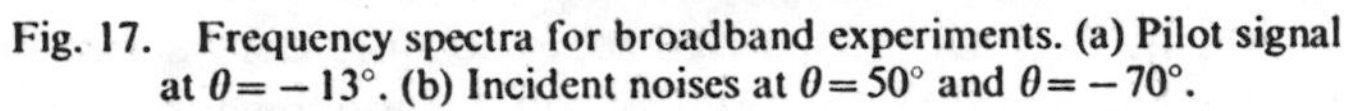
Fig. 17. Frequency spectra for broadband experiments. (a) Pilot signal at $\theta = -13°$. (b) Incident noises at $\theta = 50°$ and $\theta = -70°$.

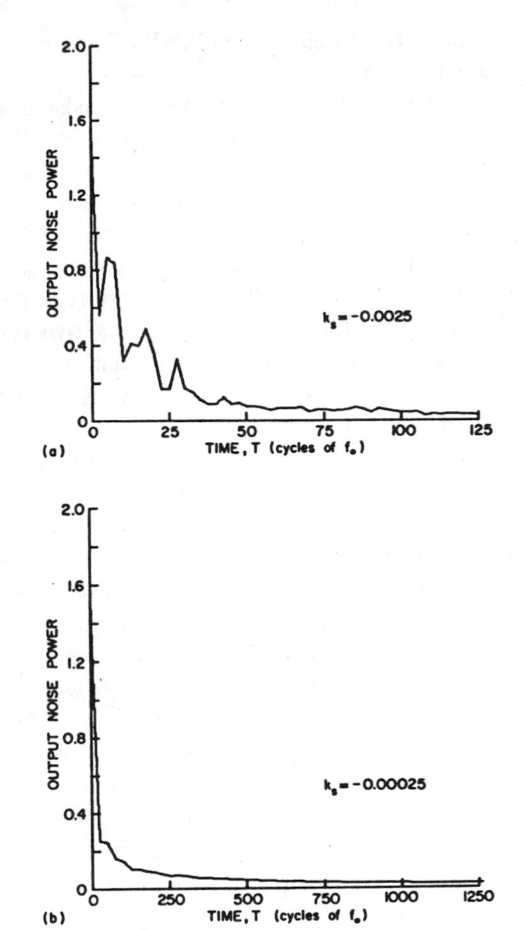

Fig. 18. Learning curves for broadband experiments. (a) Rapid learning ($M = 13$ percent). (b) Slow learning ($M = 1.3$ percent).

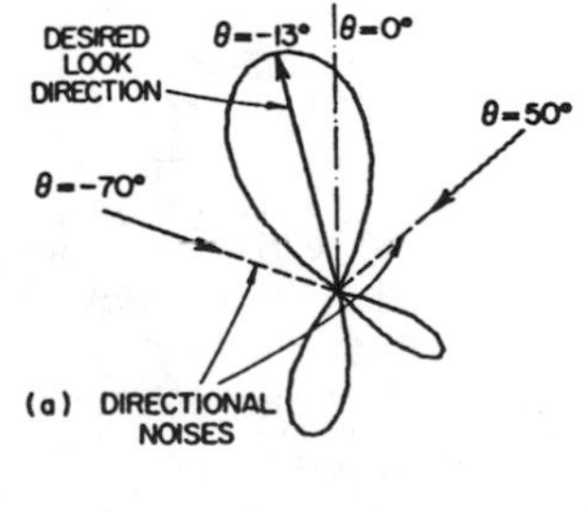

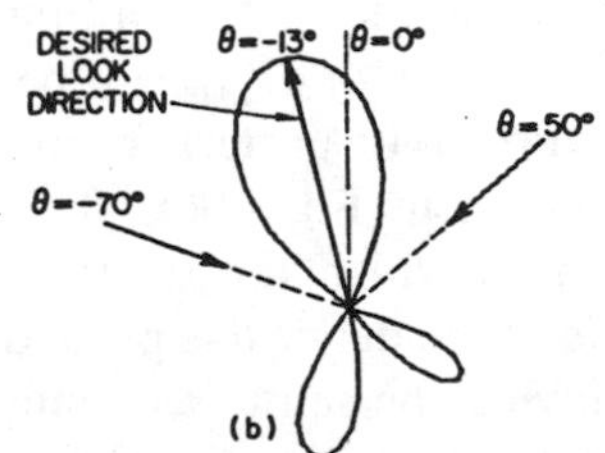

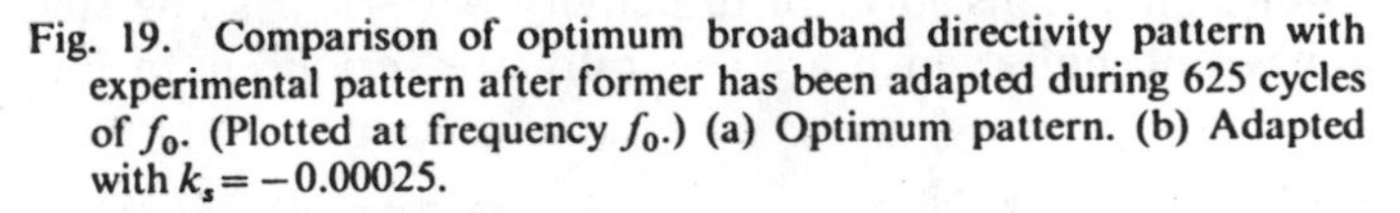
Fig. 19. Comparison of optimum broadband directivity pattern with experimental pattern after former has been adapted during 625 cycles of f_0. (Plotted at frequency f_0.) (a) Optimum pattern. (b) Adapted with $k_s = -0.00025$.

quency f_0, and a total delay-line length of one wavelength at this frequency.

The computer-simulated noise field consisted of two wideband directional noise sources[2] incident on the array at angles $\theta=50°$ and $\theta=-70°$. Each source of noise had power 0.5. The noise at $\theta=50°$ had the same frequency spectrum as the pilot signal (though with reduced power); while the noise at $\theta=-70°$ was narrower and centered at a slightly higher frequency. The noise sources were uncorrelated with the pilot signal. Fig. 17(b) shows these frequency spectra. Additive "white" Gaussian noises (mutually uncorrelated) of power 0.0625 were also present in each of the antenna-element signals.

To demonstrate the effects of adaptation rate, the experiments were performed twice, using two different values (-0.0025 and -0.00025) for k_s, the scalar constant in (23). Fig. 18(a) and (b) shows the learning curves obtained under these conditions. The abscissa of each curve is expressed in cycles of f_0, the array center frequency; and, as before, the array was adapted at a rate of twenty times per cycle of f_0. Note that the faster learning curve is a much more noisy one.

Since the statistics of the pilot signal and directional noises in this example are known (having been generated in the computer simulation), it is possible to check measured values of misadjustment against theoretical values. Thus the $\mathbf{\Phi}(x, x)$ matrix is known, and its eigenvalues have been computed.[3]

Using (30) and (31) and the known eigenvalues, the misadjustment for the two values of k_s is calculated to give the following values:

k_s	Theoretical Value of M	Experimental Value of M
-0.0025	0.1288	0.134
-0.00025	0.0129	0.0170

The theoretical values of misadjustment check quite well with corresponding measured values.

From the known statistics the optimum (in the least-squares sense) weight vector W_{LMS} can be computed, using (19). The antenna directivity pattern for this optimum weight vector W_{LMS} is shown in Fig. 19(a). This is a broadband directivity pattern, in which the relative sensitivity of the array versus angle of incidence θ is plotted for a broadband received signal having the same frequency spectrum as the pilot signal. This form of directivity pattern has few side lobes, and nulls which are generally not very deep. In Fig. 19(b), the broadband directivity pattern which resulted from adaptation (after 625 cycles of f_0, with $k_s=-0.0025$) is plotted for comparison with the optimum broadband pattern. Note that the patterns are almost indistinguishable from each other.

The learning curves of Fig. 18(a) and (b) are composed of decaying exponentials of various time constants. When k_s is set to -0.00025, in Fig. 18(b), the misadjustment is about 1.3 percent, which is a quite small, but practical value. With this rate of adaptation, it can be seen from Fig. 18(b) that adapting transients are essentially finished after about 500 cycles of f_0. If f_0 is 1 MHz, for example, adaptation could be completed (if the adaptation circuitry is fast enough) in about 500 μs. If f_0 is 1 kHz, adaptation could be completed in about one-half second. Faster adaptation is possible, but there will be more misadjustment. These figures are typical for an adaptive antenna with broadband noise inputs with 25 adaptive weights. For the same level of misadjustment, convergence times increase approximately linearly with the number of weights.[1]

The ability of this adaptive antenna array to obtain "frequency tuning" is shown in Fig. 20. This figure gives the sensitivities of the adapted array (after 1250 cycles of f_0 at $k_s=-0.00025$) as a function of frequency for the desired look direction, Fig. 20(a), and for the two noise directions, Fig. 20(b) and (c). The spectra of the pilot signal and noises are also shown in the figures.

In Fig. 20(a), the adaptive process tends to make the sensitivity of this simple array configuration as close as possible to unity over the band of frequencies where the pilot signal has finite power density. Improved performance might be attained by adding antenna elements and by adding more taps to each delay line; or, more simply, by band-limiting the output to the passband of the pilot signal. Fig. 20(b) and (c) shows the sensitivities of the array in the directions of the noises. Illustrated in this figure is the very striking reduction of the array sensitivity in the directions of the noises, within their specific passbands. The same idea is illustrated by the nulls in the broadband directivity patterns which occur in the noise directions, as shown in Fig. 19. After the adaptive transients subsided in this experiment, the signal-to-noise ratio was improved by the array over that of a single isotropic sensor by a factor of 56.

Implementation

The discrete adaptive processor shown in Figs. 7(a) and 8 could be realized by either a special-purpose digital apparatus or a suitably programmed general-purpose machine. The antenna signals would need analog-to-digital conversion, and then they would be applied to shift registers or computer memory to realize the effects of the tapped delay lines as illustrated in Fig. 5. If the narrowband scheme shown in Fig. 4 is to be realized, the time delays can be implemented either digitally or by analog means (phase shifters) before the analog-to-digital conversion process.

The analog adaptive processor shown in Figs. 7(b) and 8 could be realized by using conventional analog-computer

[2] Broadband directional noises were computer-simulated by first generating a series of uncorrelated ("white") pseudorandom numbers, applying them to an appropriate sampled-data (discrete, digital) filter to achieve the proper spectral characteristics, and then applying the resulting correlated noise waveform to each of the simulated antenna elements with the appropriate delays to simulate the effect of a propagating wavefront.

[3] They are: 10.65, 9.83, 5.65, 5.43, 3.59, 3.44, 2.68, 2.13, 1.45, 1.35, 1.20, 0.99, 0.66, 0.60, 0.46, 0.29, 0.24, 0.20, 0.16, 0.12, 0.01, 0.087, 0.083, 0.075, 0.069.

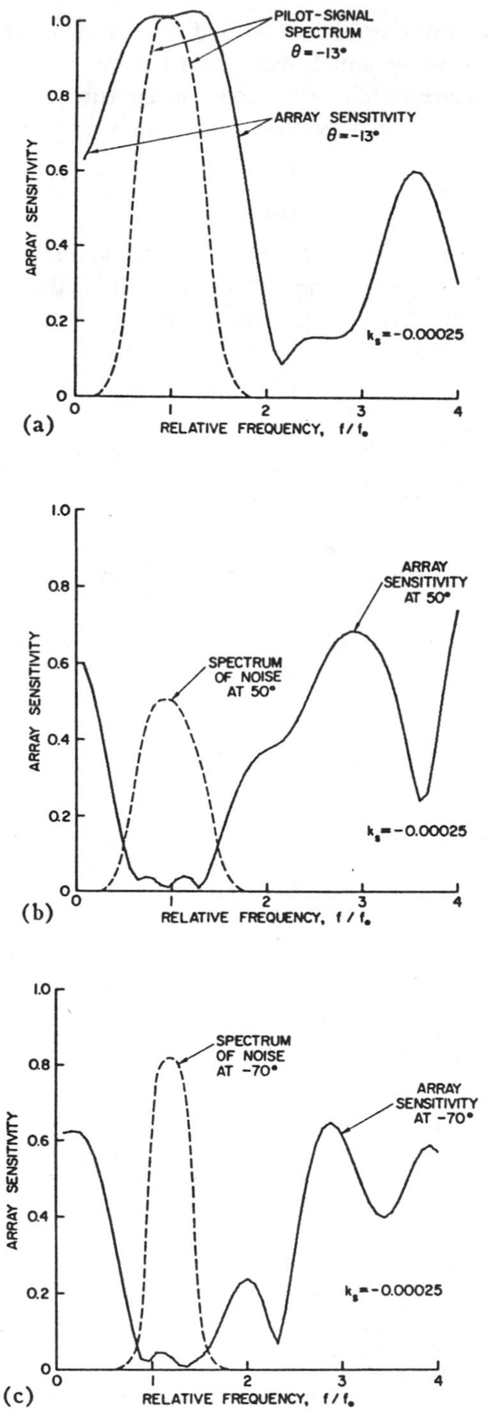

Fig. 20. Array sensitivity versus frequency, for broadband experiment of Fig. 19. (a) Desired look direction, $\theta=-13°$. (b) Sensitivity in one noise direction, $\theta=50°$. (c) Sensitivity in the other noise direction, $\theta=-70°$.

apparatus, such as multipliers, integrators, summers, etc. More economical realizations that would, in addition, be more suitable for high-frequency operation might use field-effect transistors as the variable-gain multipliers, whose control (gate) signals could come from capacitors used as integrators to form and store the weight values. On the other hand, instead of using a variable resistance structure to form the vector dot products, the same function could be achieved using variable-voltage capacitors, with ordinary capacitors again storing the weight values. The resulting structure would be a capacitive voltage divider rather than a resistive one. Other possible realizations of analog weights include the use of a Hall-effect multiplier combiner with magnetic storage[24] and also the electrochemical memistor of Widrow and Hoff.[25]

Further efforts will be required to improve existing weighting elements and to develop new ones which are simple, cheap, and adaptable according to the requirements of the various adaptation algorithms. The realization of the processor ultimately found to be useful in certain applications may be composed of a combination of analog and digital techniques.

Relaxation Algorithms and Their Implementation

Algorithms other than the LMS procedure described in the foregoing exist that may permit considerable decrease in complexity with specific adaptive circuit implementations. One method of adaptation which may be easy to implement electronically is based on a relaxation algorithm described by Southwell.[26] This algorithm uses the same error signal as used in the LMS technique. An estimated mean-square error formed by squaring and averaging this error signal over a finite time interval is used in determining the proper weight adjustment. The relaxation algorithm adjusts one weight at a time in a cyclic sequence. Each weight in its turn is adjusted to minimize the measured mean-square error. This method is in contrast to the simultaneous adjustment procedure of the LMS steepest-descent algorithm. The relaxation procedure can be shown to produce a misadjustment that increases with the *square* of the number of weights, as opposed to the LMS algorithm whose misadjustment increases only linearly with the number of weights. For a given level of misadjustment, the adaptation settling time of the relaxation process increases with the square of the number of weights.

For implementation of the Southwell relaxation algorithm, the configurations of the array and adaptive processor remain the same, as does the use of the pilot signal. The relaxation algorithm will work with either the two-mode or the one-mode adaptation process. Savings in circuitry may result, in that changes in the adjustments of the weight values depend only upon error measurements and not upon configurations of error measurements and simultaneous input-signal measurements. Circuitry for implementing the LMS systems as shown in Fig. 7(a) and (b) may be more complicated.

The relaxation method may be applicable in cases where the adjustments are not obvious "weight" settings. For example, in a microwave system, the adjustments might be a system of motor-driven apertures or tuning stubs in a waveguide or a network of waveguides feeding an antenna. Or the adjustments may be in the antenna geometry itself. In such cases, the mean-square error can still be measured, but it is likely that it would not be a simple quadratic function of the adjustment parameters. In any event, some very interesting possibilities in automatic optimization are presented by relaxation adaptation methods.

Other Applications and Further Work on Adaptive Antennas

Work is continuing on the proper choice of pilot signals to achieve the best trade-off between response in the desired look direction and rejection of noises. The subject of "null-steering," where the adaptive algorithm causes the nulls of the directivity pattern to track moving noise sources, is also being studied.

The LMS criterion used as the performance measure in this paper minimizes the mean-square error between the array output and the pilot signal waveform. It is a useful performance measure for signal *extraction* purposes. For signal *detection*, however, maximization of array output signal-to-noise ratio is desirable. Algorithms which achieve the maximum SNR solution are also being studied. Goode[27] has described a method for synthesizing the optimal Bayes detector for continuous waveforms using Wiener (LMS) filters. A third criterion under investigation has been discussed by Kelley and Levin[28] and, more recently, applied by Capon *et al.*[29] to the processing of large aperture seismic array (LASA) data. This filter, the maximum-likelihood array processor, is constrained to provide a *distortionless* signal estimate and simultaneously minimize output noise power. Griffiths[30] has discussed the relationship between the maximum likelihood array processor and the Wiener filter for discrete systems.

The examples given have illustrated the ability of the adaptive antenna system to counteract directional interfering noises, whether they are monochromatic, narrowband, or broadband. Although adaptation processes have been applied here exclusively to receiving arrays, they may also be applied to transmitting systems. Consider, for example, an application to aid a low-power transmitter. If a fixed amplitude and frequency pilot signal is transmitted from the receiving site on a slightly different frequency than that of the carrier of the low-power information transmitter, the transmitter array could be adapted (in a receiving mode) to place a beam in the direction of this pilot signal, and, therefore, by reciprocity the transmitting beam would be directed toward the receiving site. The performance of such a system would be very similar to that of the retrodirective antenna systems,[5],[6] although the methods of achieving such performance would be quite different. These systems may be useful in satellite communications.

An additional application of interest is that of "signal seeking." The problem is to find a coherent signal of unknown direction in space, and to find this signal by adapting the weights so that the array directivity pattern receives this signal while rejecting all other noise sources. The desired response or pilot signal for this application is the received signal itself processed through a narrowband filter. The use of the output signal of the adaptive processor to provide its own desired response is a form of unsupervised learning that has been referred to as "bootstrap learning."[31] Use of this adaptation algorithm yields a set of weights which accepts all correlated signals (in the desired passband) and rejects all other received signals. This system has been computer simulated and shown to operate as expected. However, much work of a theoretical and experimental nature needs to be done on capture and rejection phenomena in such systems before they can be reported in detail.

Conclusion

It has been shown that the techniques of adaptive filtering can be applied to processing the output of the individual elements in a receiving antenna array. This processing results in reduced sensitivity of the array to interfering noise sources whose characteristics may be unknown *a priori*. The combination of array and processor has been shown to act as an automatically tunable filter in both space and frequency.

Acknowledgment

The authors are indebted to Dr. M. E. Hoff, Jr., for a number of useful discussions in the early development of these ideas, and to Mrs. Mabel Rockwell who edited the manuscript.

References

[1] B. Widrow, "Adaptive filters I: Fundamentals," Stanford Electronics Labs., Stanford, Calif., Rept. SEL-66-126 (Tech. Rept. 6764-6), December 1966.

[2] J. S. Koford and G. F. Groner, "The use of an adaptive threshold element to design a linear optimal pattern classifier," *IEEE Trans. Information Theory*, vol. IT-12, pp. 42–50, January 1966.

[3] K. Steinbuch and B. Widrow, "A critical comparison of two kinds of adaptive classification networks," *IEEE Trans. Electronic Computers (Short Notes)*, vol. EC-14, pp. 737–740, October 1965.

[4] C. H. Mays, "The relationship of algorithms used with adjustable threshold elements to differential equations," *IEEE Trans. Electronic Computers (Short Notes)*, vol. EC-14, pp. 62–63, February 1965.

[5] L. C. Van Atta, "Electromagnetic reflection," U.S. Patent 2 908 002, October 6, 1959.

[6] "Special Issue on Active and Adaptive Antennas," *IEEE Trans. Antennas and Propagation*, vol. AP-12, March 1964.

[7] C. V. Jakowatz, R. L. Shuey, and G. M. White, "Adaptive waveform recognition," *4th London Symp. on Information Theory*. London: Butterworths, September 1960, pp. 317–326.

[8] L. D. Davisson, "A theory of adaptive filtering," *IEEE Trans. Information Theory*, vol. IT-12, pp. 97–102, April 1966.

[9] E. M. Glaser, "Signal detection by adaptive filters," *IRE Trans. Information Theory*, vol. IT-7, pp. 87–98, April 1961.

[10] F. Bryn, "Optimum signal processing of three-dimensional arrays operating on gaussian signals and noise," *J. Acoust. Soc. Am.*, vol. 34, pp. 289–297, March 1962.

[11] H. Mermoz, "Adaptive filtering and optimal utilization of an antenna," U. S. Navy Bureau of Ships (translation 903 of Ph.D. thesis, Institut Polytechnique, Grenoble, France), October 4, 1965.

[12] S. W. W. Shor, "Adaptive technique to discriminate against coherent noise in a narrow-band system," *J. Acoust. Soc. Am.*, vol. 39, pp. 74–78, January 1966.

[13] B. Widrow and M. E. Hoff, Jr., "Adaptive switching circuits," *IRE WESCON Conv. Rec.*, pt. 4, pp. 96–104, 1960.

[14] N. G. Nilsson, *Learning Machines*. New York: McGraw-Hill, 1965.

[15] B. Widrow and F. W. Smith, "Pattern-recognizing control systems," *1963 Computer and Information Sciences (COINS) Symp. Proc.* Washington, D.C.: Spartan, 1964.

[16] L. R. Talbert *et al.*, "A real-time adaptive speech-recognition system," Stanford Electronics Labs., Stanford University, Stanford, Calif., Rept. SEL 63-064 (Tech. Rept. 6760-1), May 1963.

[17] F. W. Smith, "Design of quasi-optimal minimum time controllers," *IEEE Trans. Automatic Control*, vol. AC-11, pp. 71–77, January 1966.

[18] J. P. Burg, "Three-dimensional filtering with an array of seismome-

ters," *Geophysics*, vol. 29, pp. 693–713, October 1964.

[19] J. F. Claerbout, "Detection of *P* waves from weak sources at great distances," *Geophysics*, vol. 29, pp. 197–211, April 1964.

[20] H. Robbins and S. Monro, "A stochastic approximation method," *Ann. Math. Stat.*, vol. 22, pp. 400–407, March 1951.

[21] J. Kiefer and J. Wolfowitz, "Stochastic estimation of the maximum of a regression function," *Ann. Math. Stat.*, vol. 23, pp. 462–466, March 1952.

[22] A. Dvoretzky, "On stochastic approximation," *Proc. 3rd Berkeley Symp. on Math. Stat. and Prob.*, J. Neyman, Ed. Berkeley, Calif.: University of California Press, 1956, pp. 39–55.

[23] B. Widrow, "Adaptive sampled-data systems," *Proc. 1st Internat'l Congress of the Internat'l Federation of Automatic Control* (Moscow, 1960). London: Butterworths, 1960.

[24] D. Gabor, W. P. L. Wilby, and R. Woodcock, "A universal nonlinear filter predictor and simulator which optimizes itself by a learning process," *Proc. IEE* (London), vol. 108 B, July 1960.

[25] B. Widrow and M. E. Hoff, Jr., "Generalization and information storage in networks of adaline 'neurons'," in *Self Organizing Systems 1962*, M. C. Yovits, G. T. Jacobi, and G. D. Goldstein, Eds. Washington, D. C.: Spartan, 1962, pp. 435–461.

[26] R. V. Southwell, *Relaxation Methods in Engineering Science*. London: Oxford University Press, 1940.

[27] B. B. Goode, "Synthesis of a nonlinear Bayes detector for Gaussian signal and noise fields using Wiener filters," *IEEE Trans. Information Theory* (*Correspondence*), vol. IT-13, pp. 116–118, January 1967.

[28] E. J. Kelley and M. J. Levin, "Signal parameter estimation for seismometer arrays," M.I.T. Lincoln Lab., Lexington, Mass., Tech. Rept. 339, January 8, 1964.

[29] J. Capon, R. J. Greenfield, and R. J. Kolker, "Multidimensional maximum-likelihood processing of a large aperture seismic array," *Proc. IEEE*, vol. 55, pp. 192–211, February 1967.

[30] L. J. Griffiths, "A comparison of multidimensional Wiener and maximum-likelihood filters for antenna arrays," *Proc. IEEE* (*Letters*), vol. 55, pp. 2045–2047, November 1967.

[31] B. Widrow, "Bootstrap learning in threshold logic systems," presented at the American Automatic Control Council (Theory Committee), IFAC Meeting, London, England, June 1966.

An Algorithm for Linearly Constrained Adaptive Array Processing

OTIS LAMONT FROST, III, MEMBER, IEEE

Abstract—A constrained least mean-squares algorithm has been derived which is capable of adjusting an array of sensors in real time to respond to a signal coming from a desired direction while discriminating against noises coming from other directions. Analysis and computer simulations confirm that the algorithm is able to iteratively adapt variable weights on the taps of the sensor array to minimize noise power in the array output. A set of linear equality constraints on the weights maintains a chosen frequency characteristic for the array in the direction of interest.

The array problem would be a classical constrained least-mean-squares problem except that the signal and noise statistics are assumed unknown *a priori*.

A geometrical presentation shows that the algorithm is able to maintain the constraints and prevent the accumulation of quantization errors in a digital implementation.

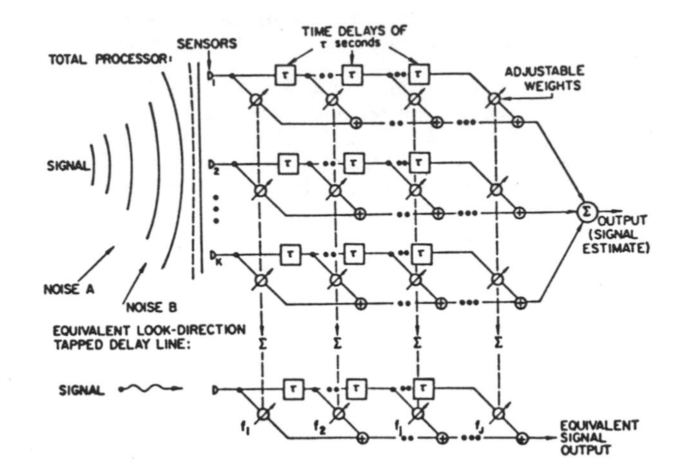

Fig. 1. Broad-band antenna array and equivalent processor for signals coming from the look direction.

I. Introduction

THIS PAPER describes a simple algorithm for adjusting an array of sensors in real time to respond to a desired signal while discriminating against noises. A "signal" is here defined as a waveform of interest which arrives in plane waves from a chosen direction (called the "look direction"). The algorithm iteratively adapts the weights of a broad-band sensor array (Fig. 1) to minimize noise power at at the array output while maintaining a chosen frequency response in the look direction.

The algorithm, called the "Constrained Least Mean-Squares" or "Constrained LMS" algorithm, is a simple stochastic gradient-descent algorithm which requires only that the direction of arrival and a frequency band of interest be specified *a priori*. In the adaptive process, the algorithm progressively learns statistics of noise arriving from directions other than the look direction. Noise arriving from the look direction may be filtered out by a suitable choice of the frequency response characteristic in that direction, or by external means. Subsequent processing of the array output may be done for detection or classification.

A major advantage of the constrained LMS algorithm is that it has a self-correcting feature permitting it to operate for arbitrarily long periods of time in a digital computer implementation without deviating from its constraints because of cumulative roundoff or truncation errors.

The algorithm is applicable to array processing problems in geoscience, sonar, and electromagnetic antenna arrays in which a simple method is required for adjusting an array in real time to discriminate against noises impinging on the array sidelobes.

Manuscript received December 23, 1971; revised May 4, 1972. This research is based on a Ph.D. dissertation in the Department of Electrical Engineering, Stanford University, Stanford, Calif.

The author is with ARGOSystems, Inc., Palo Alto, Calif. 94303.

Previous Work

Previous work on iterative least squares array processing was done by Griffiths [1]; his method uses an unconstrained minimum-mean-square-error optimization criterion which requires *a priori* knowledge of second-order signal statistics. Widrow, Mantey, Griffiths, and Goode [2] proposed a variable-criterion [3] optimization procedure involving the use of a known training signal; this was an application and extension of the original work on adaptive filters done by Widrow and Hoff [4]. Griffiths also proposed a constrained least mean-squares processor not requiring *a priori* knowledge of the signal statistics [5]; a new derivation of this processor, given in [6], shows that it may be considered as putting "soft" constraints on the processor via the quadratic penalty-function method.

"Hard" (i.e., exactly)-constrained iterative optimization was studied by Rosen [7] for the deterministic case. Lacoss [8], Booker *et al.* [9], and Kobayashi [10] studied "hard"-constrained optimization in the array processing context for filtering short lengths of data. All four authors used gradient-projection techniques [11]; Rosen and Booker correctly indicated that gradient-projection methods are susceptible to cumulative roundoff errors and are not suitable for long runs without an additional error-correction procedure. The constrained LMS algorithm developed in the present work is designed to avoid error accumulation while maintaining a "hard" constraint; as a result, it is able to provide continual filtering for arbitrarily large numbers of iterations.

Basic Principle of the Constraints

The algorithm is able to maintain a chosen frequency response in the look direction while minimizing output noise

Reprinted from *Proc. IEEE,* vol. 60, no. 8, pp. 926–935, Aug. 1972.

power because of a simple relation between the look direction frequency response and the weights in the array of Fig. 1. Assume that the look direction is chosen as the direction perpendicular to the line of sensors. Then identical signal components arriving on a plane wavefront parallel to the line of sensors appear at the first taps simultaneously and parade in parallel down the tapped delay lines following each sensor; however, noise waveforms arriving from other than the look direction will not, in general, produce equal voltage components on any given vertical column of taps. The voltages (signal plus noise) at each tap are multiplied by the tap weights and added to form the array output. Thus as far as the signal is concerned, the array processor is equivalent to a single tapped delay line in which each weight is equal to the sum of the weights in the corresponding vertical column of the processor, as indicated in Fig. 1. These summation weights in the equivalent tapped delay line must be selected so as to give the desired frequency response characteristic in the look direction.

If the look direction is chosen to be other than that perpendicular to the line of sensors, then the array can be steered either mechanically or electrically by the addition of steering time delays (not shown) placed immediately after each sensor.

A processor having K sensors and J taps per sensor has KJ weights and requires J constraints to determine its look-direction frequency response. The remaining $KJ-J$ degrees of freedom in choosing the weights may be used to minimize the total power in the array output. Since the look-direction frequency response is fixed by the J constraints, minimization of the total output power is equivalent to minimizing the non-look-direction noise power, so long as the set of signal voltages at the taps is uncorrelated with the set of noise voltages at these taps. The latter assumption has commonly been made in previous work on iterative array processing [1], [5], [8]–[10]. The effect of signal-correlated noise in the array may be to cancel out all or part of the desired signal component in the array output. Sources of signal-correlated noise may be multiple signal-propagation paths, and coherent radar or sonar "clutter."

It is permissible, and in fact desirable for proper noise cancellation that the voltages produced by the noises on the taps of the array be correlated among themselves, although uncorrelated with the signal voltages. Examples of such noises include waveforms from point sources in other than the look direction (e.g., lightning, "jammers," noise from nearby vehicles), spatially localized incoherent clutter, and self-noise from the structure carrying the array.

Noise voltages which are uncorrelated between taps (e.g., amplifier thermal noise) may be partially rejected by the adaptive array in two ways. As in a conventional nonadaptive array, such noises are eliminated to the extent that signal voltages on the taps are added coherently at the array output, while uncorrelated noise voltages are added incoherently. Second, an adaptive array can reduce the weighting on any tap that may have a disportionately large uncorrelated noise power.

II. Optimum-Constrained LMS Weight Vector

The first step in developing the constrained LMS algorithm is to find the optimum weight vector.

Notation

Notation will be as follows (see Fig. 2):

Every Δ seconds, where Δ may be a multiple of the delay τ between taps, the voltages at the array taps are sampled. The vector of tap voltages at the kth sample is written $X(k)$, where

$$X^T(k) \triangleq [x_1(k\Delta), x_2(k\Delta), \cdots, x_{KJ}(k\Delta)].$$

The superscript T denotes transpose. The tap voltages are the sums of voltages due to look-direction waveforms l and non-look-direction noises n, so that

$$X(k) = L(k) + N(k) \tag{1}$$

where the KJ-dimensional vector of look-direction waveforms at the kth sample is

$$L(k) \triangleq \begin{bmatrix} l(k\Delta) \\ \vdots \\ l(k\Delta) \\ l(k\Delta-\tau) \\ \vdots \\ l(k\Delta-\tau) \\ \vdots \\ l(k\Delta-(J-1)\tau) \\ \vdots \\ l(k\Delta-(J-1)\tau) \end{bmatrix} \begin{matrix} \left.\vphantom{\begin{matrix}a\\a\\a\end{matrix}}\right\} K \text{ taps} \\ \left.\vphantom{\begin{matrix}a\\a\\a\end{matrix}}\right\} K \text{ taps} \\ \\ \left.\vphantom{\begin{matrix}a\\a\\a\end{matrix}}\right\} K \text{ taps} \end{matrix}$$

and the vector of non-look-direction noises is

$$N^T(k) \triangleq [n_1(k\Delta), n_2(k\Delta), \cdots, n_{KJ}(k\Delta)].$$

The vector of weights at each tap is W, where

$$W^T \triangleq [w_1, w_2, \cdots, w_{KJ}].$$

It is assumed for this derivation that the signals and noises are adequately modeled as zero-mean random processes with (unknown) second-order statistics:

$$E[X(k)X^T(k)] \triangleq R_{XX} \tag{2a}$$

$$E[N(k)N^T(k)] \triangleq R_{NN} \tag{2b}$$

$$E[L(k)L^T(k)] \triangleq R_{LL}. \tag{2c}$$

As previously stated, it is assumed that the vector of look-direction waveforms is uncorrelated with the vector of non-look-direction noises, i.e.,

$$E[N(k)L^T(k)] = 0. \tag{3}$$

It is assumed that the noise environment is distributed so that R_{XX} and R_{NN} are positive definite [12].

The output of the array (signal estimate) at the time of kth sample is

$$y(k) = W^T X(k) = X^T(k) W. \tag{4}$$

Using (4) the expected output power of the array is

$$E[y^2(k)] = E[W^T X(k) X^T(k) W] = W^T R_{XX} W. \tag{5}$$

The constraint that the weights on the jth vertical column of taps sum to a chosen number f_j (see Fig. 1) is expressed by

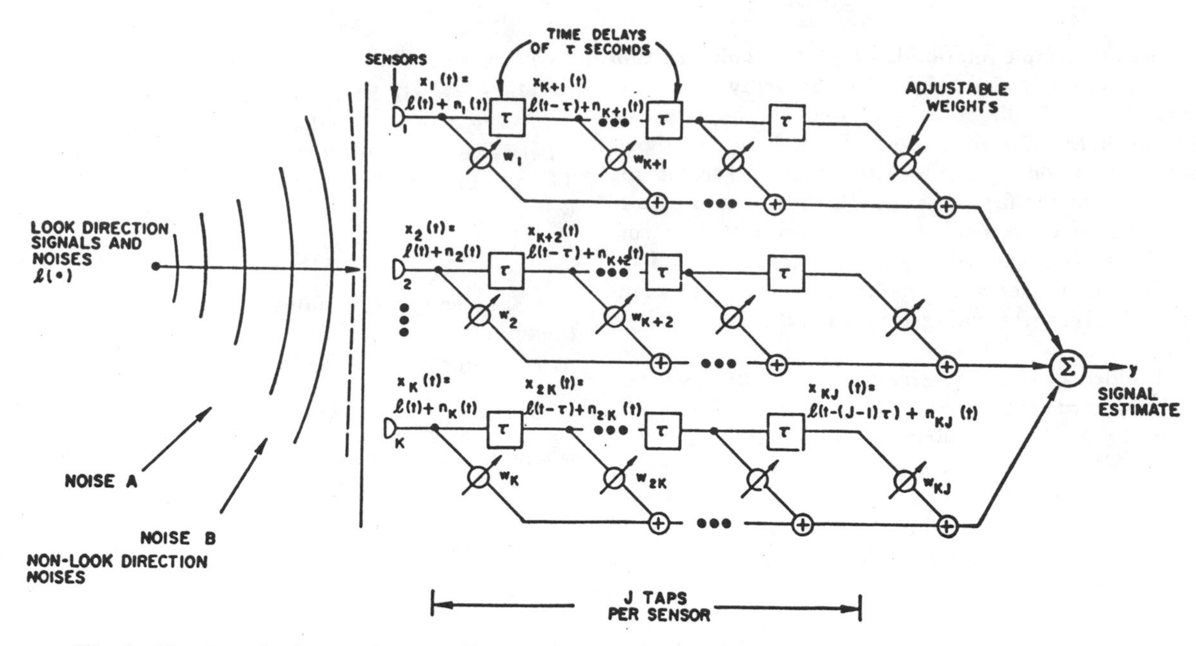

Fig. 2. Signals and noises on the array. Because the array is steered toward the look direction, all beam signal components on any given column of filter taps are identical.

the requirement

$$c_j^T W = f_j, \qquad j = 1, 2, \cdots, J \tag{6}$$

where the KJ-dimensional vector c_j has the form

$$c_j = \begin{bmatrix} 0 \\ \vdots \\ 0 \\ \vdots \\ 0 \\ \vdots \\ 0 \\ 1 \\ \vdots \\ 1 \\ 0 \\ \vdots \\ 0 \\ \vdots \\ 0 \\ \vdots \\ 0 \end{bmatrix} \begin{matrix} \left.\right\} K \\ \vdots \\ \left.\right\} K \\ \left.\right\} j\text{th group of } K \text{ elements.} \\ \left.\right\} K \\ \vdots \\ \left.\right\} K \end{matrix} \tag{7}$$

Constraining the weight vector to satisfy the J equations of (6) restricts W to a $(KJ-J)$-dimensional plane.

Define the constraint matrix C as

$$C \triangleq \underbrace{\left[c_1 \cdots c_j \cdots c_J \right]}_{J} \; \updownarrow KJ \tag{8}$$

and define $\mathfrak{F}$ as the J-dimensional vector of weights of the look-direction-equivalent tapped delay line shown in Fig. 1:

$$\mathfrak{F} \triangleq \begin{bmatrix} f_1 \\ \vdots \\ f_j \\ \vdots \\ f_J \end{bmatrix}. \tag{9}$$

By inspection the constraint vectors c_j are linearly independent, hence, C has full rank equal to J. The constraints (6) are now written

$$C^T W = \mathfrak{F}. \tag{10}$$

Optimum Weight Vector

Since the look-direction-frequency response is fixed by the J constraints, minimization of the non-look-direction noise power is the same as minimization of the total output power. The cost criterion used in this paper will be minimization of total array output power $W^T R_{XX} W$. The problem of finding the optimum set of filter weights W_{opt} is summarized by (5) and (10) as

$$\underset{W}{\text{minimize}} \quad W^T R_{XX} W \tag{11a}$$

$$\text{subject to } C^T W = \mathfrak{F}. \tag{11b}$$

This is the constrained LMS problem.

W_{opt} is found by the method of Lagrange multipliers, which is discussed in [13]. Including a factor of $\frac{1}{2}$ to simplify later arithmetic, the constraint function is adjoined to the cost function by a J-dimensional vector of undetermined Lagrange multipliers λ:

$$H(W) = \tfrac{1}{2} W^T R_{XX} W + \lambda^T (C^T W - \mathfrak{F}). \tag{12}$$

Taking the gradient of (12) with respect to W

$$\nabla_W H(W) \; R_{XX} W + C\lambda. \tag{13}$$

The first term in (13) is a vector proportional to the gradient of the cost function (11a), and the second term is a vector normal to the $(KJ-J)$-dimensional constraint plane defined by $C^TW-\mathcal{F}=0$ [14]. For optimality these vectors must be antiparallel [13], which is achieved by setting the sum of the vectors (13) equal to zero

$$\nabla_W H(W) = R_{XX}W + C\lambda = 0.$$

In terms of the Lagrange multipliers, the optimal weight vector is then

$$W_{opt} = -R_{XX}^{-1}C\lambda \tag{14}$$

where R_{XX}^{-1} exists because R_{XX} was assumed positive definite. Since W_{opt} must satisfy the constraint (11b)

$$C^TW_{opt} = \mathcal{F} = -C^TR_{XX}^{-1}C\lambda$$

and the Lagrange multipliers are found to be

$$\lambda = -[C^TR_{XX}^{-1}C]^{-1}\mathcal{F} \tag{15}$$

where the existence of $[C^TR_{XX}^{-1}C]^{-1}$ follows from the facts that R_{XX} is positive definite and C has full rank [6]. From (14) and (15) the optimum-constrained LMS weight vector solving (11) is

$$W_{opt} = R_{XX}^{-1}C[C^TR_{XX}^{-1}C]^{-1}\mathcal{F}. \tag{16}$$

Using the set of weights W_{opt} in the array processor of Fig. 2 forms the optimum constrained LMS processor, which is a filter in space and frequency. Substituting W_{opt} in (4), the constrained least squares estimate of the look-direction waveform is

$$y_{opt}(k) = W_{opt}^TX(k). \tag{17}$$

Discussion

The constrained LMS filter is sometimes known by other names. If the frequency characteristic in the look-direction is chosen to be all-pass and linear phase (distortionless), the output of the constrained LMS filter is the maximum likelihood estimate of a stationary process in Gaussian noise if the angle of arrival is known [15]. The distortionless form of the constrained LMS filter is called by some authors the "Minimum Variance Distortionless Look" estimator, "Maximum Likelihood Distortionless Estimator," and "Least Squares Unbiased Estimator." By suitable choice of $\mathcal{F}$ a variety of other optimal processors can be obtained [16].

III. The Adaptive Algorithm

In this paper it is assumed that the input correlation matrix R_{XX} is unknown *a priori* and must be learned by an adaptive technique. In stationary environments during learning, and in time-varying environments, an estimate of the optimum filter weights must be recomputed periodically. Direct substitution of a correlation matrix estimate into the optimal-weight equation (16) requires a number of multiplications at each iteration proportional to the cube of the number of weights. The complexity is primarily caused by the required inversion of the input correlation matrix. Recently Saradis *et al.* [17] and Mantey and Griffiths [18] have shown how to iteratively update matrix inversions, requiring only a number of multiplications and storage locations proportional to the square of the number of weights. The gradient-descent constrained LMS algorithm presented here requires only a number of multiplications and storage locations directly proportional to the number of weights. It is therefore simple to implement and, for a given computational cost, is applicable to arrays in which the number of weights is on the order of the square of the number that could be handled by the iterative matrix inversion method and the cube of the number that could be handled by the direct substitution method.

Derivation

For motivation of the algorithm derivation temporarily suppose that the correlation matrix R_{XX} is known. In constrained gradient-descent optimization, the weight vector is initialized at a vector satisfying the constraint (11b), say $W(0)=C(C^TC)^{-1}\mathcal{F}$, and at each iteration the weight vector is moved in the negative direction of the constrained gradient (13). The length of the step is proportional to the magnitude of the constrained gradient and is scaled by a constant μ. After the kth iteration the next weight vector is

$$\begin{aligned} W(k+1) &= W(k) - \mu\nabla_W H[W(k)] \\ &= W(k) - \mu[R_{XX}W(k) + C\lambda(k)] \end{aligned} \tag{18}$$

where the second step is from (13). The Lagrange multipliers are chosen by requiring $W(k+1)$ to satisfy the constraint (11b):

$$\mathcal{F} = C^TW(k+1) = C^TW(k) - \mu C^TR_{XX}W(k) - \mu C^TC\lambda(k).$$

Solving for the Lagrange multipliers $\lambda(k)$ and substituting into the weight-iteration equation (18) we have

$$\begin{aligned} W(k+1) = W(k) &- \mu[I - C(C^TC)^{-1}C^T]R_{XX}W(k) \\ &+ C(C^TC)^{-1}[\mathcal{F} - C^TW(k)]. \end{aligned} \tag{19}$$

The deterministic algorithm (19) is shown in this form to emphasize that the last factor $\mathcal{F}-C^TW(k)$ is not assumed to be zero, as it would be if the weight vector precisely satisfied the constraint at the kth iteration. It will be shown in Section VI that this term permits the algorithm to correct any small deviations from the constraint due to arithmetic inaccuracy and prevents their eventual accumulation and growth.

Defining the KJ-dimensional vector

$$F \triangleq C(C^TC)^{-1}\mathcal{F} \tag{20a}$$

and the $KJ \times KJ$ matrix

$$P \triangleq I - C(C^TC)^{-1}C^T \tag{20b}$$

the algorithm may be rewritten as

$$W(k+1) = P[W(k) - \mu R_{XX}W(k)] + F. \tag{21}$$

Equation (21) is a deterministic constrained gradient descent algorithm requiring knowledge of the input correlation matrix R_{XX}, which, however, in the array problem is unavailable *a priori*. An available and simple approximation for R_{XX} at the kth iteration is the outer product of the tap voltage vector with itself: $X(k)X^T(k)$. Substitution of this estimate into (21) gives the stochastic constrained LMS algorithm

$$\boxed{\begin{aligned} W(0) &= F \\ W(k+1) &= P[W(k) - \mu y(k)X(k)] + F \end{aligned}} \tag{22}$$

where $y(k)$ is the array output (signal estimate) defined by (4).

Discussion

The constrained LMS algorithm (22) satisfies the constraint $C^T W(k+1) = \mathfrak{F}$ at each iteration, as can be verified by premultiplying (22) by C^T and using (20). At each iteration the algorithm requires only the tap voltages $X(k)$ and the array output $y(k)$; no *a priori* knowledge of the input correlation matrix is needed. F is a constant vector that can be precomputed. One of the two most complex operations required by (22) is the multiplication of each of the KJ components of the vector $X(k)$ by the scalar $\mu y(k)$; the other significant operation is indicated by the matrix $P = I - C(C^T C)^{-1} C^T$. Because of the simple form of C [refer to (7)], multiplication of a vector by P as indicated in (22) amounts to little more than a few additions. Expressed in summation notation the iterative equations for the weight vector components are

$$
\begin{aligned}
w_1(k+1) &= w_1(k) - \mu y(k) x_1(k) - \frac{1}{K} \sum_{j=1}^{K} [w_j(k) - \mu y(k) x_j(k)] + \frac{f_1}{K} \\
&\vdots \\
w_K(k+1) &= w_K(k) - \mu y(k) x_K(k) - \frac{1}{K} \sum_{j=1}^{K} [w_j(k) - \mu y(k) x_j(k)] + \frac{f_1}{K} \\
w_{K+1}(k+1) &= w_{K+1}(k) - \mu y(k) x_{K+1}(k) - \frac{1}{K} \sum_{j=K+1}^{2K} [w_j(k) - \mu y(k) x_j(k)] + \frac{f_2}{K} \\
&\vdots \\
w_{2K}(k+1) &= w_{2K}(k) - \mu y(k) x_{2K}(k) - \frac{1}{K} \sum_{j=K+1}^{2K} [w_j(k) - \mu y(k) x_j(k)] + \frac{f_2}{K} \\
&\vdots \\
w_{JK}(k+1) &= w_{JK}(k) - \mu y(k) x_{JK}(k) - \frac{1}{K} \sum_{j=(J-1)K+1}^{JK} [w_j(k) - \mu y(k) x_j(k)] + \frac{f_J}{K}.
\end{aligned}
$$

These equations can readily be implemented on a digital computer.

IV. Performance

Convergence to the Optimum

The weight vector $W(k)$ obtained by the use of the stochastic algorithm (22) is a random vector. Convergence of the mean weight vector to the optimum is demonstrated by showing that the length of the difference vector between the mean weight vector and the optimum (16) asymptotically approaches zero.

Proof of convergence of the mean is greatly simplified by the assumption (used in [2]) that successive samples of the input vector taken Δ seconds apart are statistically independent. This condition can usually be approximated in practice by sampling the input vector at intervals large compared to the correlation time of the input process plus the length of time it takes an input waveform to propagate down the array. The assumption is more restrictive than necessary, since Daniell [19] has shown that the much weaker assumption of asymptotic independence of the input vectors is sufficient to demonstrate convergence in the related unconstrained least squares problem.

Taking the expected value of both sides of the algorithm (22), using (4), (2a), and the independence assumption yields an iterative equation in the mean value of the constrained LMS weight vector

$$E[W(k+1)] = P\{E[W(k)] - \mu R_{XX} E[W(k)]\} + F. \quad (23)$$

Define the vector $V(k+1)$ to be the difference between the mean adaptive weight vector at iteration $k+1$ and the optimal weight vector (16)

$$V(k+1) \triangleq E[W(k+1)] - W_{\text{opt}}.$$

Using (23) and the relations $F = (I-P) W_{\text{opt}}$ and $P R_{XX} W_{\text{opt}} = 0$, which may be verified by direct substitution of (16) and (20b), an equation for the difference process may be constructed

$$V(k+1) = PV(k) - \mu P R_{XX} V(k). \quad (24)$$

The idempotence of P (i.e., $P^2 = P$), which can be verified by carrying out the multiplication using (20b) and premultiplication of equation (24) by P shows that $PV(k) = V(k)$ for all k, so (24) can be written

$$
\begin{aligned}
V(k+1) &= [I - \mu P R_{XX} P] V(k) \\
&= [I - \mu P R_{XX} P]^{k+1} V(0).
\end{aligned}
$$

The matrix $P R_{XX} P$ determines both the rate of convergence of the mean weight vector to the optimum and the steady-state variance of the weight vector about the optimum. It is shown in [6] that $P R_{XX} P$ has precisely J zero eigenvalues, corresponding to the column vectors of the constraint matrix C; this is a result of the fact that during adaption no movement is permitted away from $(KJ-J)$-dimensional constraint plane. It is also shown in [6, appendix C] that $P R_{XX} P$ has $KJ-J$ nonzero eigenvalues σ_i, $i = 1, 2, \cdots, KJ-J$, with values bounded between the smallest and largest eigenvalues of R_{XX}

$$\lambda_{\min} \leq \sigma_{\min} \leq \sigma_i \leq \sigma_{\max} \leq \lambda_{\max}, \qquad i = 1, 2, \cdots, KJ - J$$

where $\lambda_{\min}$ and $\lambda_{\max}$ are the smallest and largest eigenvalues of R_{XX} and $\sigma_{\min}$ and $\sigma_{\max}$ are the smallest and largest nonzero eigenvalues of $P R_{XX} P$.

Examination of $V(0)=F-W_{opt}$ shows that it can be expressed as a linear combination of the eigenvectors of $PR_{XX}P$ corresponding to nonzero eigenvalues. If $V(0)$ is equal to an eigenvector of $PR_{XX}P$, say e_i with eigenvalue $\sigma_i \neq 0$ then

$$V(k+1) = [I - \mu PR_{XX}P]^{k+1} e_i = [1 - \mu\sigma_i]^{k+1} e_i.$$

The convergence of the mean weight vector to the optimum weight vector along any eigenvector of $PR_{XX}P$ is therefore geometric with geometric ratio $(1-\mu\sigma_i)$. The time required for the euclidean length of the difference vector to decrease to e^{-1} of its initial value (time constant) is

$$\tau_i = \Delta/\ln(1 - \mu\sigma_i) \cong \Delta/\mu\sigma_i \quad (25)$$

where the approximation is valid for $\mu\sigma_i \ll 1$.

If μ is chosen so that

$$0 < \mu < 1/\sigma_{max}$$

then the length (norm) of any difference vector is bounded between two ever-decreasing geometric progressions

$$(1 - \mu\sigma_{max})^{k+1} \|V(0)\| \le \|V(k+1)\| \le (1 - \mu\sigma_{min})^{k+1} \|V(0)\|$$

and so if the initial difference is finite the mean weight vector converges to the optimum, i.e.,

$$\lim_{k\to\infty} \|E[W(k)] - W_{opt}\| = 0$$

with time constants given by (25).

Steady-State Performance—Stationary Environment

The algorithm is designed to continually adapt for coping with nonstationary noise environments. In stationary environments this adaptation causes the weight vector to have a variance about the optimum and produces an additional component of noise (above the optimum) to appear at the output of the adaptive processor.

The output power of the optimum processor with a fixed weight vector (17) is

$$E[y_{opt}^2(k)] = W_{opt}^T R_{XX} W_{opt} = \mathfrak{F}^T (C^T R_{XX}^{-1} C)^{-1} \mathfrak{F}.$$

A measure of the fraction of additional noise caused by the adaptive algorithm operating in steady state in a stationary environment is termed "misadjustment" $M(\mu)$ by Widrow

$$M(\mu) \triangleq \lim_{k\to\infty} \frac{E[y^2(k)] - E[y_{opt}^2(k)]}{E[y_{opt}^2(k)]}.$$

By assuming that successive observation vectors [vectors $X(k)$ of tap voltages] are independent and have components $x_1(k), \cdots, x_{KL}(k)$ that are jointly Gaussian distributed, Moschner [20] calculated very tight bounds on the misadjustment, using a method due to Senne [21], [22]. For a convergence constant μ satisfying

$$0 < \mu < \frac{1}{\sigma_{max} + (1/2)\,\mathrm{tr}\,(PR_{XX}P)} \quad (26)$$

the steady-state misadjustment may be bounded by

$$\frac{\mu}{2} \frac{\mathrm{tr}\,(PR_{XX}P)}{1 - \frac{\mu}{2}[\mathrm{tr}\,(PR_{XX}P) + 2\sigma_{min}]} \le M(\mu) \le \frac{\mu}{2} \frac{\mathrm{tr}\,(PR_{XX}P)}{1 - \frac{\mu}{2}[\mathrm{tr}\,(PR_{XX}P) + 2\sigma_{max}]} \quad (27)$$

where tr denotes trace.

$M(\mu)$ can be made arbitrarily close to zero by suitably small choice of μ; this means that the steady-state performance of the constrained LMS algorithm can be made arbitrarily close to the optimum. From (25) it seen that such performance is obtained at the expense of increased convergence time.

If μ is chosen to satisfy

$$0 < \mu < \frac{2}{3\,\mathrm{tr}\,(R_{XX})} \quad (28)$$

then it is guaranteed to satisfy (26). Griffiths [1] shows that the upper bound in (28) can be calculated directly and easily from observations since $\mathrm{tr}\,(R_{XX}) = E[X^T(k)X(k)]$, the sum of the powers of the tap voltages.

Steady-State Performance—Nonstationary Environment

A model of the effect of a nonstationary noise environment proposed by Brown [23] is that the steady-state rms change of the optimal weight vector $W_{opt}(k)$ between iterations has magnitude δ, i.e.,

$$\limsup_{k\to\infty} E\|W_{opt}(k+1) - W_{opt}(k)\|^2 = \delta^2.$$

Brown's general results may be applied to the constrained LMS algorithm by restricting the optimal weight vector to have magnitude less than some number $\|W_{max}\|$ and again assuming the successive input vectors are independent with Gaussian-distributed components. For μ small it can be shown [23, p. 47] that the steady-state rms distance of the weight vector from the optimum is bounded by

$$\left(\limsup_{k\to\infty} E\|W(k) - W_{opt}(k)\|^2\right)^{1/2} \le \frac{\delta + \mu 2^{1/2}[\lambda^* \mathrm{tr}^*\,(PR_{XX}P) + \sigma^{*2}]\|W_{max}\|}{1 - \{1 - 2\mu\sigma_* + \mu^2[3\sigma^{*2} + 2\lambda^* \mathrm{tr}^*\,(PR_{XX}P)]\}^{1/2}} \quad (29)$$

where any starred quantities q_* or q^* are taken to bound the corresponding time-varying quantity $q(k)$, i.e., $q_* \le q(k) \le q^*$ for all k. In general, the optimum convergence constant μ that minimizes the upper bound (29) for a nonstationary environment is nonzero. This contrasts with the stationary case, in which the best steady-state performance is obtained by making μ as small as possible.

V. Geometrical Interpretation

The constrained LMS algorithm (22) has a simple geometrical interpretation that is useful for visualizing the error-correcting property which keeps the weight vector from deviating from its constraints.

In an error-free implementation of the algorithm, the KJ-dimensional weight vectors satisfy the constraint equation (11b) and therefore terminate on a constraint plane Λ defined

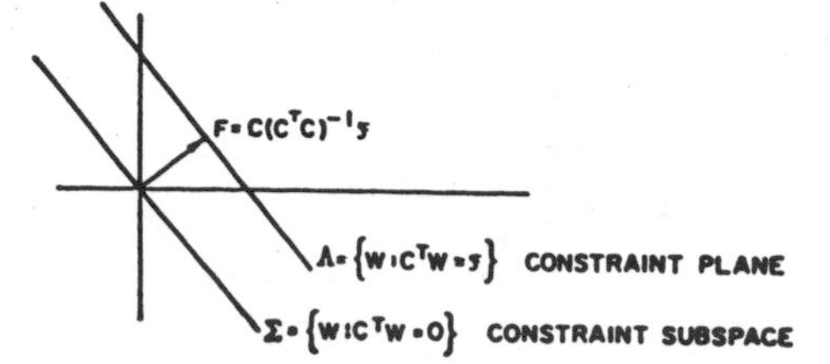

Fig. 3. The $(KJ-J)$-plane Λ and subspace Σ defined by the constraint.

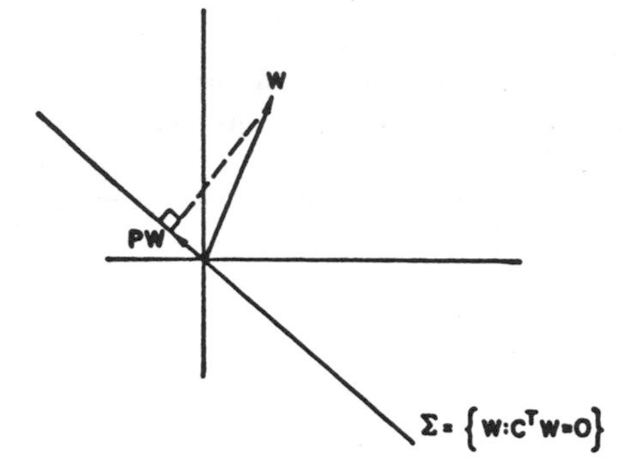

Fig. 4. P projects vectors onto the constraint subspace.

by

$$\Lambda = \{W : C^T W = \mathfrak{F}\}.$$

This $(KJ-J)$-dimensional constraint plane is indicated diagrammatically in Fig. 3.

It is well known [14] that vectors pointing in a direction normal to the constraint plane (but not necessarily normal to the vectors that terminate on that plane) are linear combinations of the constraint matrix column vectors. These vectors have the form CA, where A is a J-dimensional vector determining the linear combination. Thus the vector $F = C(C^TC)^{-1}\mathfrak{F}$, appearing in the algorithm (22) and used as the initial weight vector, points in a direction normal to the constraint plane. F also terminates on the constraint plane since $C^TF = \mathfrak{F}$. Thus F is the shortest vector terminating on the constraint plane (see Fig. 3).

The homogeneous form of the constraint equation

$$C^T W = 0 \tag{30}$$

defines a second $(KJ-J)$-dimensional plane, which includes the zero vector and thus passes through the origin. Such a plane is called a subspace [11] (see Fig. 3).

The matrix P in the algorithm (22) is a projection operator [24]. Premultiplication of any vector by P will annihilate any components perpendicular to Σ, projecting the vector into the constraint subspace (see Fig. 4).

The vector $y(k)X(k)$ in the algorithm is an estimate of the unconstrained gradient. Referring to (12) the unconstrained cost function is $\frac{1}{2}W^T R_{XX} W$. The unconstrained gradient [refer to (13)] is $R_{XX}W$. The estimate of $R_{XX}W$ at the kth iteration, used in deriving (22), is $y(k)X(k)$.

Contours of constant output power (cost) and the optimum constrained weight vector W_{opt} that minimizes the output power are shown in Fig. 5.

The operation of the constrained LMS algorithm is shown in Fig. 6. In this example, the unconstrained negative gradient estimate $-y(k)X(k)$ is scaled by μ and added to the current weight vector $W(k)$. This is an attempt to change the weight vector in a direction that minimizes output power. In

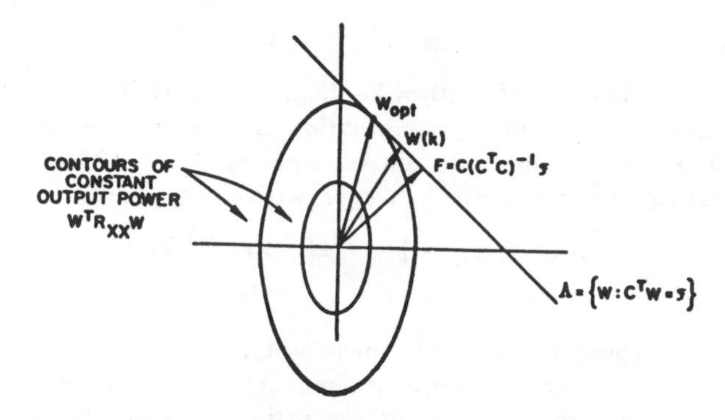

Fig. 5. Example showing contours of constant output power and the constrained weight vector that minimizes output power. $W_{\text{opt}} = R_{XX}^{-1}C(C^TR_{XX}^{-1}C)^{-1}\mathfrak{F}$.

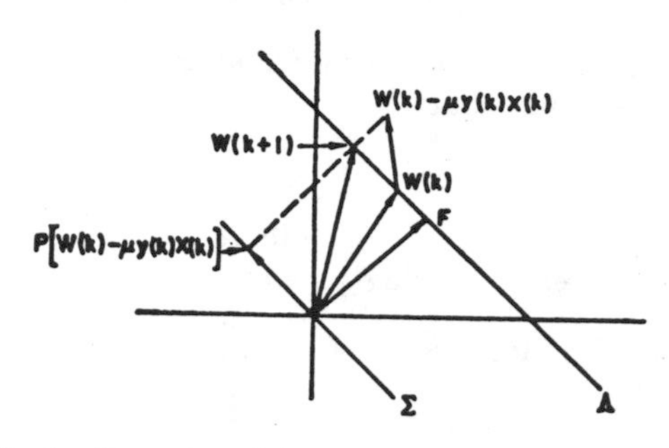

Fig. 6. Operation of the constrained LMS algorithm: $W(k+1) = P[W(k) - \mu y(k)X(k)] + F$.

general, this change moves the resulting vector off the constraint plane. The resulting vector is projected onto the constraint subspace and then returned to the constraint plane by adding F. The new weight vector $W(k+1)$ satisfies the constraint to within the accuracy of the arithmetic used in implementing the algorithm.

VI. Error-Correcting Feature

In a digital-computer implementation of any algorithm, it is likely that small computational errors will occur at each iteration because of truncation, roundoff, or quantization errors. A difficulty in applying the well-known gradient-projection algorithm to the real time array-processing problem is that computational errors causing deviations of the weight vector from the constraint are not corrected [7], [9]. Without additional error-correcting procedures, application of the gradient-projection algorithm is limited to problems requiring few enough iterations that significant deviations from the constraint do not occur. The constrained LMS algorithm, on the other hand, was specifically designed to continuously correct for such errors and prevent them from accumulating. The reason for this characteristic is shown by a geometrical comparison of the two algorithms.

The gradient-projection algorithm may be derived by following the derivation of the constrained LMS algorithm to (19) and dropping the last factor, $\mathfrak{F} - C^TW(k)$. This factor would be equal to zero in a perfect implementation in which the weight vector satisfied the constraint $C^TW(k) = \mathfrak{F}$ at each iteration. The algorithm that results when the term is dropped is

$$\begin{aligned} W(0) &= C(C^TC)^{-1}\mathfrak{F} \\ W(k+1) &= W(k) - \mu P y(k) X(k). \end{aligned} \tag{31}$$

This is a gradient-projection algorithm [11]. It is so named

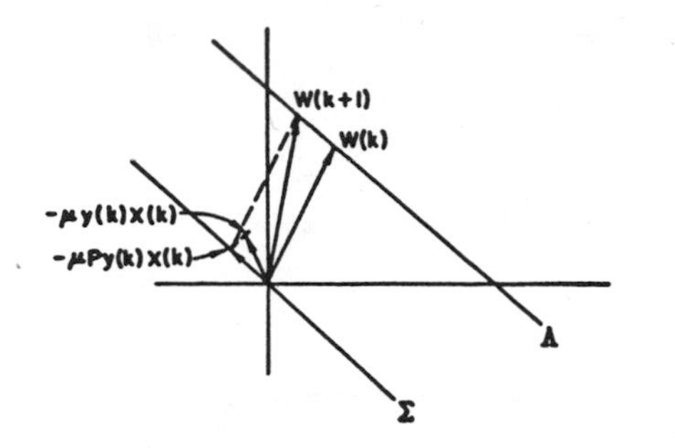

Fig. 7. Operation of the gradient-projection algorithm (31).

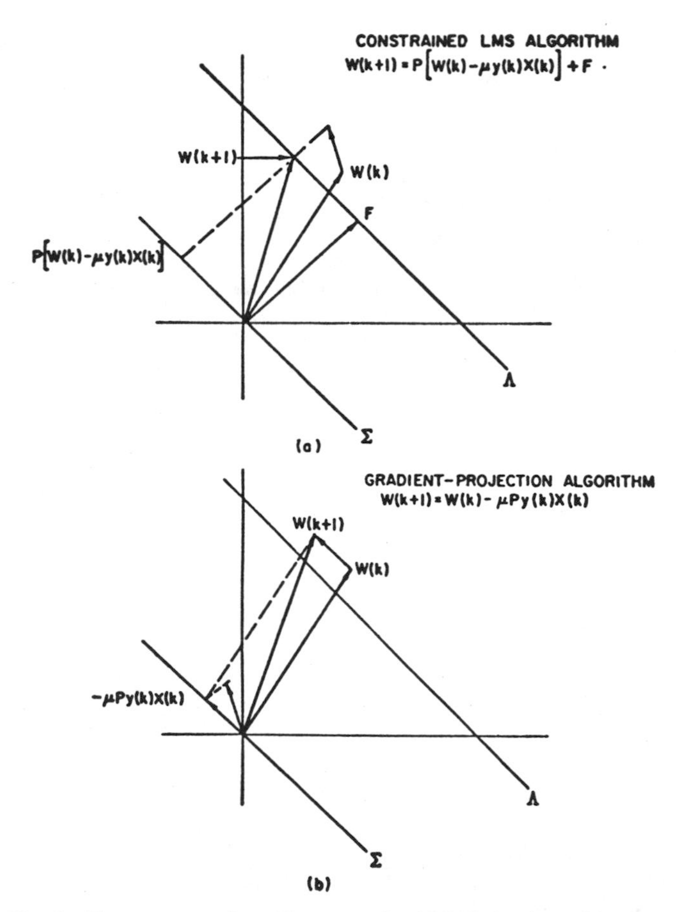

Fig. 8. Error propagation. The constrained LMS algorithm (a) corrects deviations from the constraints while the gradient-projection algorithm (b) allows them to accumulate.

because the unconstrained gradient estimate $y(k)X(k)$ is projected onto the constraint subspace and then added to the current weight vector. Its operation is shown in Fig. 7 (compare with Fig. 6).

A comparison between the effect of computational errors on the gradient-projection algorithm and on the constrained LMS algorithm is shown in Fig. 8. The weight vector is assumed to be off the constraint at the kth iteration because of a quantization error occurring in the previous iteration. It is shown in Fig. 8(a) that the constrained LMS algorithm makes the unconstrained step, projects onto the subspace, and then adds F, producing a new weight vector $W(k+1)$ that satisfies the constraint. The gradient-projection algorithm [Fig. 8(b)], however, projects the gradient estimate onto the subspace and adds the projected vector to the past weight vector, moving parallel to the constraint plane but continuing the error. Note the implicit (incorrect) assumption that $W(k)$ satisfied the constraint, corresponding to the same assumption made in the derivation of the gradient-projection algorithm.

Accumulating errors in the gradient-projection algorithm can be expected to cause the weight vector to do a random walk away from the constraint plane with variance (expected squared distance from the plane) increasing linearly with the number of iterations. By contrast, the expected deviation of the constrained LMS algorithm from the constraint does not grow, remaining at its original value.

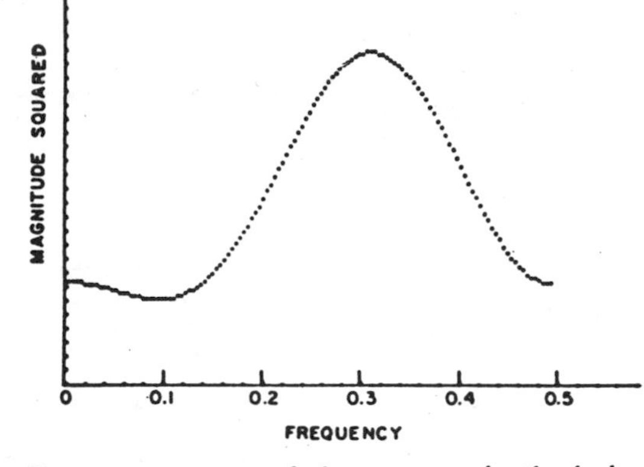

Fig. 9. Frequency response of the processor in the look direction.

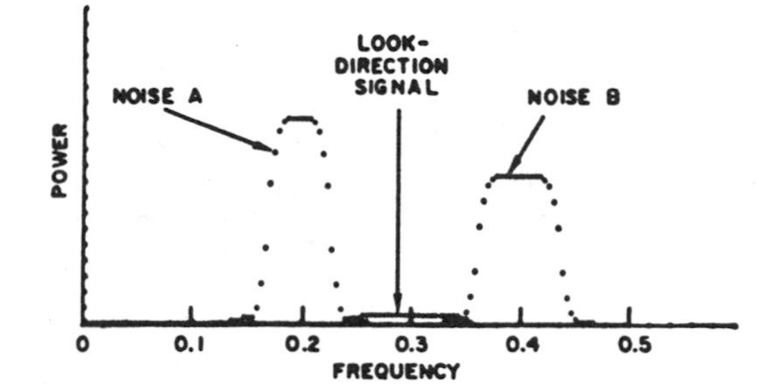

Fig. 10. Power spectral density of incoming signals. See Fig. 2 and Table I for spatial position of noises.

TABLE I

SIGNALS AND NOISES IN THE SIMULATION (SEE FIG. 2)

Source	Power	Direction (0° is normal to array)	Center Frequency (1.0 is $1/\tau$)	Bandwidth
Look-direction signal	0.1	0°	0.3	0.1
Noise A	1.0	45°	0.2	0.05
Noise B	1.0	60°	0.4	0.07
White noise (per tap)	0.1			

VII. SIMULATION

A computer simulation of the processor was made using 6-digit floating point arithmetic on a small computer (the HP-2116). The processor had four sensors on a line spaced at τ-second intervals and had four taps per sensor (thus $KJ=16$). The environment had three point-noise sources, and white noise added to each sensor. Power of the look direction signal was quite small in comparison to the power of interfering noises (see Table I). The tap spacing defined a frequency of 1.0 (i.e., $f=1.0$ is a frequency of $1/\tau$ Hz). In the look-direction, foldover frequency for the processor response was $\frac{1}{2}\tau$, or 0.5. All signals were generated by a pseudo-Gaussian gen-

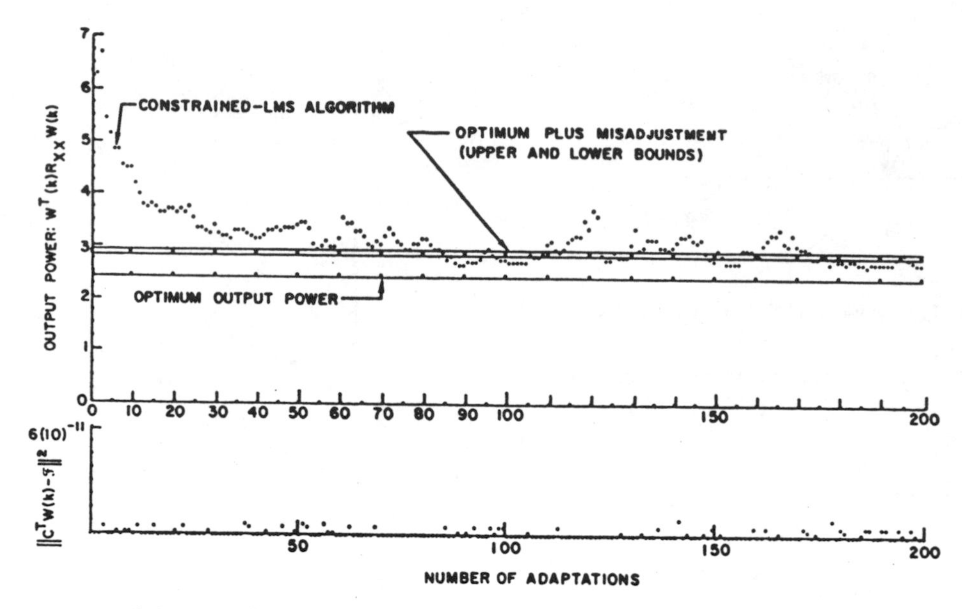

Fig. 11. The output power of the constrained LMS filter (upper graph) decreases as it adapts to discriminate against unwanted noise. Lower curve shows small deviations from the constraint due to quantization.

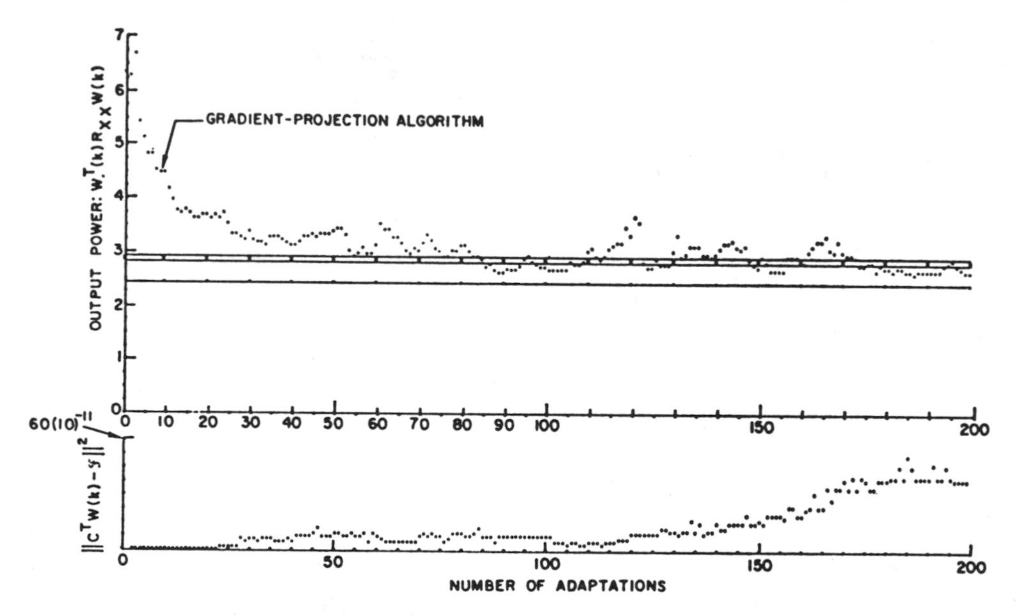

Fig. 12. Output power of the gradient-projection algorithm (upper graph) operated on the same data as the constrained LMS algorithm (c.f., Fig. 11). Lower curve shows that deviations from the constraint tend to increase with time. Note scale.

erator and filtered to give them the proper spatial and temporal correlations. All temporal correlations were arranged to be identically zero for time differences greater than 25τ. The time between adaptations Δ was assumed greater than 58τ, so successive samples of $X(k)$ were uncorrelated.

The look direction filter was specified by the vector $\mathfrak{F}^T=[1, -2, 1.5, 2]$ which resulted in the frequency characteristic shown in Fig. 9. The signal and noise spectra are shown in Fig. 10 and their spatial position in Fig. 2.

In this problem, the eigenvalues of R_{XX} ranged from 0.111 to 8.355. The upper permissible bound on the convergence constant μ calculated by (26) was 0.074; a value of $\mu=0.01$ was selected, which, by (27), would lead to a misadjustment of between 15.2 and 17.0 percent.

The processor was initialized with $W(0)=F=C(C^TC)^{-1}\mathfrak{F}$, and Fig. 11 shows performance as a function of time. The upper graph has three horizontal lines. The lower line is the output power of the optimum weight vector. The closely spaced upper two lines are upper and lower bounds for the adaptive processor output power, which is the optimum output power plus misadjustment. The mean steady-state value of the processor's output power falls somewhere between the upper and lower bounds (but may, at any instant fall above or below these bounds). The difference between the initial and steady-state power levels is the amount of undesirable noise power the processor has been able to remove from the output.

A simulation of the gradient projection algorithm (31) on the array problem was made using exactly the same data as used by the constrained LMS algorithm. The results are shown in Fig. 12. The lower part of Fig. 12 shows how the gradient-projection algorithm walks away from the constraint. Note the change in scale. If the errors of the constrained LMS algorithm (Fig. 11) were plotted on the same scale they would not be discernible. The errors of the gradient-projection method are expected to continue to grow.

The fact that the output power of the gradient-projection processor (upper curve, Fig. 12) is virtually identical to the output power of the constrained LMS processor is a result of the fact that the errors have not yet accumulated to the point of moving the constraint a significant radial distance from the origin.

VIII. Limitations and Extension

Application of the constrained LMS algorithm in some array processing problems is limited by the requirement that the non-look-direction noise voltages on the taps be uncorrelated with the look-direction signal voltages.

This restriction is a result of the fact that if the noise voltages are correlated with the signal then the processor may cancel out portions of the signal with them in spite of the constraints. If the source of correlated noise is known, its effect may be reduced by placing additional constraints to minimize the array response in its direction.

Implementation errors, i.e., deviations from the assumed electrical and spatial properties of the array (such as incorrect amplifier gains, incorrect sensor placements, or unpredicted mutual coupling between sensors) may also limit the effectiveness of the processor by permitting it to discriminate against look-direction signals while still satisfying the letter of the constraints. Injection of known test signals into the array may provide information about the signal paths that can be used to compensate, in part, for the errors.

The algorithm may be extended to a more general stochastic constrained least squares problem

$$\min E\{[d(k) - W^T X(k)]^2\} \text{ subject to } C^T W = \mathfrak{F} \quad (32)$$

where $d(k)$ is a scalar variable related to the observation vector $X(k)$ and C is a general constraint matrix. The scalar $d(k)$ may be a random variable correlated with $X(k)$ or it may be a known test signal used to compensate for array errors. This would be a classical least squares problem except that the statistics of $X(k)$ and $d(k)$ are assumed unknown *a priori*. The general constrained LMS algorithm solving (32) may be derived similarly to (22) and is

$$W(0) = C(C^T C)^{-1}\mathfrak{F}$$

$$W(k+1) = P\{W(k) - \mu[y(k) - d(k)]X(k)\} + F.$$

The general algorithm is applicable to constrained modeling, prediction, estimation, and control. It is discussed in [6].

IX. Conclusion

Analysis and computer simulations have confirmed the ability of the constrained LMS algorithm to adjust an array of sensors in real time to respond to a desired signal while discriminating against noise. Because of a system of constraints on weights in the array, the algorithm is shown to require no prior knowledge of the signal or noise statistics.

A geometrical presentation has shown why the constrained LMS algorithm has an ability to maintain the constraints and prevent the accumulation of quantization errors in a digital implementation. The simulation tests have confirmed the effectiveness of this error-correcting feature, in contrast with the usual uncorrected gradient-projection algorithm. The error-correcting feature and the simplicity of the algorithm make it appropriate for continuous real-time signal estimation and discriminating against noises in a possibly time-varying environment where little *a priori* information is available about the signals or noises. Time constants, steady-state performance, and a proof of convergence are derived for operation of the algorithm in a stationary environment; convergence and steady-state performance in a nonstationary environment are also shown.

A simple extension of the algorithm may be used to solve a general constrained LMS problem, which is to minimize the expected squared difference between a multidimensional filter output and a known desired signal under a set of linear equality constraints.

References

[1] L. J. Griffiths, "A simple adaptive algorithm for real-time processing in antenna arrays," *Proc. IEEE*, vol. 57, pp. 1696–1704, Oct. 1969.

[2] B. Widrow, P. E. Mantey, L. J. Griffiths, and B. B. Goode, "Adaptive antenna systems," *Proc. IEEE*, vol. 55, pp. 2143–2158, Dec. 1967.

[3] L. J. Griffiths, "Comments on 'A simple adaptive algorithm for real-time processing in antenna arrays' (Author's reply)," *Proc. IEEE* (Lett.), vol. 58, p. 798, May 1970.

[4] B. Widrow and M. E. Hoff, Jr., "Adaptive switching circuits," *IRE WESCON Conv. Rec.*, pt. 4, pp. 96–104, 1960.

[5] L. J. Griffiths, "Signal extraction using real-time adaptation of a linear multichannel filter," Stanford Electron. Lab., Stanford, Calif., Doc. SEL-60-017, Tech. Rep. TR 67881-1, Feb. 1968.

[6] O. L. Frost, III, "Adaptive least squares optimization subject to linear equality constraints," Stanford Electron. Lab., Stanford, Calif., Doc. SEL-70-055, Tech. Rep. TR 6796-2, Aug. 1970.

[7] J. B. Rosen, "The gradient projection method for nonlinear programming, pt. 1: Linear constraints," *J. Soc. Indust. Appl. Math.*, vol. 8, p. 181, Mar. 1960.

[8] R. T. Lacoss, "Adaptive combining of wideband array data for optimal reception," *IEEE Trans. Geosci. Electron.*, vol. GE-6, pp. 78–86, May 1968.

[9] A. H. Booker, C. Y. Ong, J. P. Burg, and G. D. Hair, "Multiple-constraint adaptive filtering," Texas Instruments, Sci. Services Div., Dallas, Tex., Apr. 1969.

[10] H. Kobayashi, "Iterative synthesis methods for a seismic array processor," *IEEE Trans. Geosci. Electron*; vol. GE-8, pp. 169–178, July 1970.

[11] D. G. Luenberger, *Optimization by Vector Space Methods.* New York: Wiley, 1969.

[12] I. J. Good and K. Koog, "A paradox concerning rate of information," *Informat. Contr.*, vol. 1, pp. 113–116, May 1958.

[13] A. E. Bryson, Jr., and Y. C. Ho, *Applied Optimal Control.* Waltham, Mass.: Blaisdell, 1969.

[14] W. H. Fleming, *Functions of Several Variables.* Reading, Mass.: Addison-Wesley, 1965.

[15] E. J. Kelly, Jr., and M. J. Levin, "Signal parameter estimation for seismometer arrays," Mass. Inst. Technol. Lincoln Lab. Tech. Rept. 339, Jan. 1964.

[16] A. H. Nuttall and D. W. Hyde, "A unified approach to optimum and suboptimum processing for arrays," U. S. Navy Underwater Sound Lab., New London, Conn., USL Rep. 992, Apr. 1969.

[17] G. N. Saradis, Z. J. Nikolic, and K. S. Fu, "Stochastic approximation algorithms for system identification, estimation, and decomposition of mixtures," *IEEE Trans. Sys. Sci. Cybern.*, Vol. SSC-5, pp. 8–15, Jan. 1969.

[18] P. E. Mantey and L. J. Griffiths, "Iterative least-squares algorithms for signal extraction," in *Proc. 2nd Hawaii Int. Conf. on Syst. Sci.*, pp. 767–770.

[19] T. P. Daniell, "Adaptive estimation with mutually correlated training samples," Stanford Electron. Labs., Stanford, Calif., Doc. SEL-68-083, Tech. Rep. TR 6778-4, Aug. 1968.

[20] J. L. Moschner, "Adaptive filtering with clipped input data," Stanford Electron. Labs., Stanford, Calif., Doc. SEL-70-053, Tech. Rep. TR 6796-1, June 1970.

[21] K. D. Senne, "Adaptive linear discrete-time estimation," Stanford Electron. Labs., Stanford, Calif., Doc. SEL-68-090, Tech. Rep. TR 6778-5, June 1968.

[22] ——, "New results in adaptive estimation theory," Frank J. Seiler Res. Lab., USAF Academy, Colo., Tech. Rep. SRL-TR-70-0013, Apr. 1970.

[23] J. E. Brown, III, "Adaptive estimation in nonstationary, environments," Stanford Electron. Labs., Stanford, Calif., Doc. SEL-70-056, Tech. Rep. TR 6795-1, Aug. 1970.

[24] D. T. Finkbeiner, II, *Introduction to Matrices and Linear Transformations.* San Francisco, Calif.: Freeman, 1966.

Adaptive Arrays

SIDNEY P. APPLEBAUM, FELLOW, IEEE

***Abstract*—A method for adaptively optimizing the signal-to-noise ratio of an array antenna is presented. Optimum element weights are derived for a prescribed environment and a given signal direction. The derivation is extended to the optimization of a "generalized" signal-to-noise ratio which permits specification of preferred weights for the normal quiescent environment. The relation of the adaptive array to sidelobe cancellation is shown, and a real-time adaptive implementation is discussed. For illustration, the performance of an adaptive linear array is presented for various jammer configurations.**

I. Introduction

ARRAY ANTENNAS consisting of many controllable radiating elements are very versatile sensors. The pattern of the array can be steered by applying linear phase weighting across the array and can be shaped by amplitude and phase weighting the outputs of the array elements. Most arrays are built with fixed weights designed to produce a pattern that is a compromise between resolution, gain, and low sidelobes. The versatility of the array antenna, however, invites the use of more sophisticated techniques for array weighting. Particularly attractive are adaptive schemes that can sense and respond to a time-varying environment. In this report, we show how adaptive techniques can be applied to an antenna array to reduce its susceptibility to jamming or interference of any kind. We begin with a derivation of the "control law" for the array weights that will maximize the signal-to-noise ratio (SNR) of the array output in the presence of any spatial configuration of noise sources.

II. Signal-to-Noise Optimization

It is well known that a uniformly weighted array gives the maximum SNR when the noise contributions from the element channels have equal power and are uncorrelated. These conditions are approximately valid when receiver noise and uniformly distributed sky noise are the predominant noise contributions. (They pertain exactly in linear halfwave space array antennas.) However, when there is directional interference from other in-band transmitters, jammers, or natural phenomena, the noise out of the element channels will be correlated, and uniform weighting will not optimize the SNR.

Manuscript received April 20, 1976. This paper is reprinted from Syracuse University Research Corporation, Syracuse, NY, Technical Report SURC TR 66-001, August 1966 (revised March 1975). This study was performed under the sponsorship of the Advanced Projects Agency, Ballistic Missile Defense Office (Project Defender), on Contract AF 30 (602)-3523, ARPA Order 561, Program Code 4720.

The author was with the Syracuse University Research Corporation, Syracuse, NY. He is now with the General Electric Company, Syracuse, NY 13201.

Editor's Note: This paper was first published as a technical report. Its circulation was somewhat limited, consequently, this classic report is being reprinted in this issue with the kind permission of the Syracuse University Research Corporation.

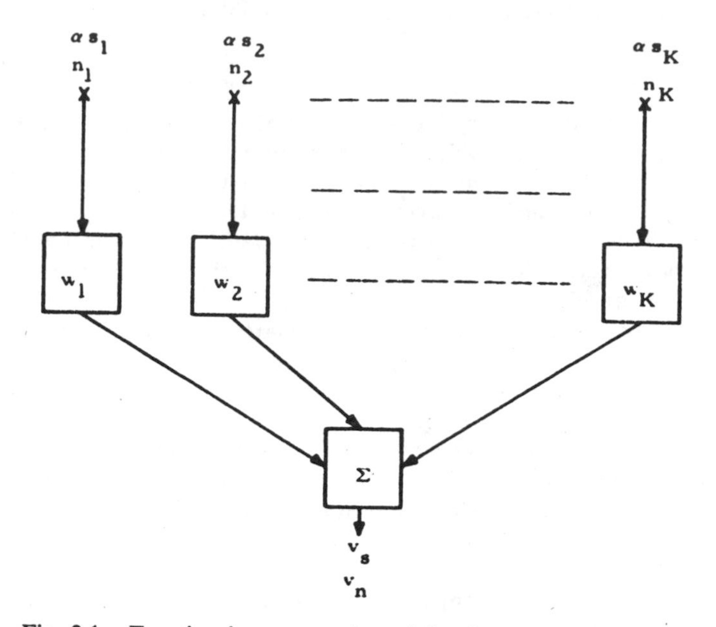

Fig. 2-1. Functional representation of "optimum" linear coherent combiner.

In this report, we consider the problem of determining the array weights that will maximize the SNR for any type of noise environment. The problem may be viewed as that of finding an optimum coherent combiner for K channels as shown in Fig. 2-1. We assume in this discussion that all signals have "bandpass" frequency spectra. All signals are represented by their complex envelopes. These are assumed to modulate a common carrier reference that never appears explicitly.

Each of the K channels contains a noise component whose complex envelope is denoted by n_k. The envelope power in the kth channel is denoted by μ_{kk} and the covariance of n_k and n_l by

$$\mu_{kl} = E(n_k^* n_l) \tag{2-1}$$

where the asterisk (*) denotes the complex conjugate. We note that

$$\mu_{lk} = E(n_l^* n_k) = \mu_{kl}^*. \tag{2-2}$$

When the K channels represent the output of the element of an array antenna, the covariance terms μ_{kl} are determined by receiver noise and the spatial distribution of all noise sources "seen" by the antenna. Here we assume that the covariances are known. The desired signal, when it occurs, is assumed to be present in the K channels in proportion to the known complex numbers s_k. The signal in channel k is represented as αs_k, where α defines the level and time variation of the signal. In a linear array antenna with equally spaced elements, the s_k are determined by the direction of the desired signal. Thus, if the desired direction

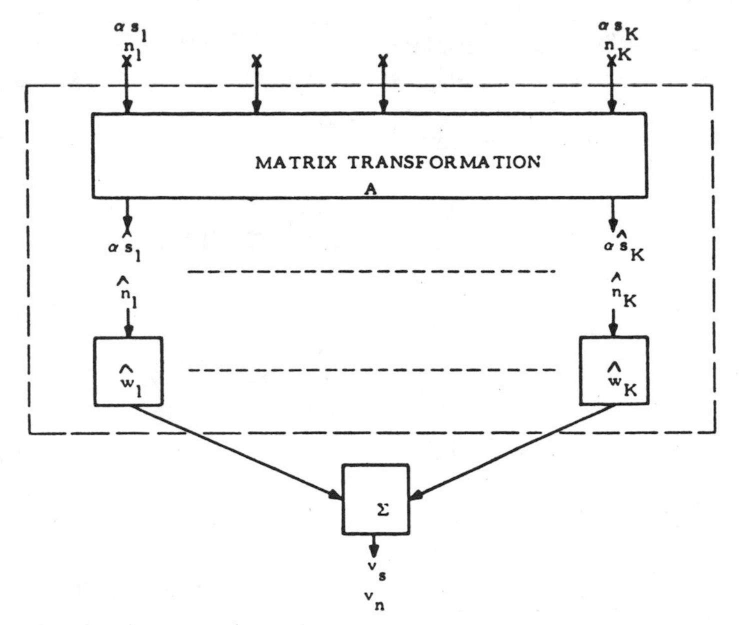

Fig. 2-2. Functional representation of combiner that is equivalent to one shown in Fig. 2-1.

is θ rad from mechanical boresight, we have

$$s_k = \exp\left(j\frac{2\pi kd}{\lambda}\sin\theta\right) \tag{2-3}$$

where d is the element spacing and λ is the wavelength.

The problem is to choose the weights w_k (see Fig. 2-1), so that the SNR at the output of the combiner is maximized. We will show that the optimum weights are determined by

$$MW = \mu S^* \tag{2-4}$$

where

$$M = [\mu_{kl}] = \text{covariance matrix of the noise outputs} \tag{2-5}$$

$$W = \begin{bmatrix} w_1 \\ \vdots \\ w_K \end{bmatrix} \tag{2-6}$$

$$S = \begin{bmatrix} s_1 \\ \vdots \\ s_K \end{bmatrix} \tag{2-7}$$

are column vectors and μ is an arbitrary constant.

To derive (2-4), we start by writing expressions for the signal and noise outputs of the combiner. The signal output is

$$v_s = \alpha \sum_{k=1}^{K} w_k s_k. \tag{2-8}$$

This may be conveniently expressed in matrix notation as

$$v_s = \alpha W_t S = \alpha S_t W \tag{2-9}$$

where the subscript t means transposed. Similarly, the noise output may be expressed as

$$v_n = W_t N = N_t W \tag{2-10}$$

where

$$N = \begin{bmatrix} n_1 \\ \vdots \\ n_K \end{bmatrix}. \tag{2-11}$$

The expected output noise power will be

$$\begin{aligned} P_n &= E\{|v_n|^2\} = E\{|W_t N|^2\} \\ &= E\{(W_t N)^*(N_t W)\} \\ &= E\{W_t^* N^* N_t W\}. \end{aligned} \tag{2-12}$$

The expectation operator E will affect only the noise terms; hence,

$$P_n = W_t^* E\{N^* N_t\} W \tag{2-13}$$

$$= W_t^* M W \tag{2-14}$$

where

$$M = E\{N^* N_t\} = [\mu_{kl}]. \tag{2-15}$$

As noted earlier, M is the covariance matrix of the noise components. If the noise components are uncorrelated, the matrix M will be a diagonal matrix. In general, however, M may have nonzero entries in any position. Since $\mu_{kl} = \mu_{lk}^*$, the matrix M is Hermitian, that is, $M_t = M^*$. It is also positive definite since the output noise power P_n is greater than zero whenever $W \neq 0$.

Since M is a positive definite Hermitian matrix, it can be "diagonalized" by a nonsingular coordinate transformation. This means that a transformation exists that will transform the problem into one in which all channels have equal power noise components that are uncorrelated.

Assume that the transformation matrix is A and consider the block diagram shown in Fig. 2-2. After the transformation shown there, the signal and noise components, respectively, become

$$\hat{S} = AS \tag{2-16}$$

and

$$\hat{N} = AN. \tag{2-17}$$

A caret (^) has been used to denote quantities after the transformation. The transformation matrix A followed by a combiner, as shown in Fig. 2-2, may be made equivalent to the combiner of Fig. 2-1 by properly relating the weights used in each case. If the outputs of the transformation matrix A are combined with the weights $\hat{W}_k$, the output signal will be

$$v_s = \alpha \hat{W}_t \hat{S} \tag{2-18}$$

$$= \alpha \hat{W}_t A S. \tag{2-19}$$

Similarly, the output noise will be

$$v_n = \hat{W}_t \hat{N} \tag{2-20}$$

$$= \hat{W}_t A N. \tag{2-21}$$

Comparing (2-19) and (2-21) with (2-9) and (2-10), we note that combining the channels after the transformation matrix A with the weight vector $\hat{W}$ is equivalent to using the weight vector $A_t \hat{W}$ without the transformation matrix. Thus, for equivalent outputs we have

$$W = A_t \hat{W}. \tag{2-22}$$

The output noise power expressed in terms of the quantities in Fig. 2-2 is

$$P_n = E\{|\hat{W}_t \hat{N}|^2\} \tag{2-23}$$

$$= E\{\hat{W}_t^* \hat{N}^* \hat{N}_t \hat{W}\} \tag{2-24}$$

$$= \hat{W}_t^* E\{\hat{N}^* \hat{N}_t\} \hat{W}. \tag{2-25}$$

Since the transformation matrix A decorrelates the noise components and equalizes their powers, the covariance matrix of the noise components after A is simply the identity matrix of order K; thus,

$$E\{\hat{N}^* \hat{N}_t\} = 1_K. \tag{2-26}$$

Using (2-26) in (2-25), we obtain

$$P_n = \hat{W}_t^* \hat{W} = \|\hat{W}\|^2. \tag{2-27}$$

If the configurations of Figs. 2-1 and 2-2 are equivalent, we may write, from (2-14) and (2-22),

$$P_n = W_t^* M W$$

$$= \hat{W}_t^* A^* M A_t \hat{W}. \tag{2-28}$$

Comparing this with (2-27), we see that

$$A^* M A_t = 1_K \tag{2-29}$$

and, therefore,

$$M = (A_t A^*)^{-1}. \tag{2-30}$$

Equation (2-29) expresses the fact that the transformation matrix A diagonalizes the matrix M. It is well known that the optimum choice for the weighting vector $\hat{W}$ in Fig. 2-2, where the noise components $\hat{n}_k$ have equal power and are uncorrelated, is given by

$$\hat{W}_{\text{opt}} = \mu \hat{S}^* \tag{2-31}$$

where μ is an arbitrary constant.

To show that (2-31) is the optimum, apply the Cauchy–Schwartz inequality to (2-18). This yields

$$|v_s|^2 \leq |\alpha|^2 \|\hat{W}\|^2 \|\hat{S}\|^2 \tag{2-32}$$

where

$$\|\hat{S}\|^2 = \hat{S}_t^* \hat{S} \tag{2-33}$$

and

$$\|\hat{W}\|^2 = \hat{W}_t^* \hat{W}. \tag{2-34}$$

From (2-27) we have $P_n = \|\hat{W}\|^2$; thus, dividing both sides of (2-32) by P_n, we obtain

$$\frac{|v_s|^2}{P_n} \leq |\alpha|^2 \|\hat{S}\|^2. \tag{2-35}$$

This puts an upper bound on the SNR. However, if we substitute $\hat{W} = \mu \hat{S}^*$ into (2-18) and (2-27), we obtain

$$v_s = \alpha \mu \hat{S}_t^* \hat{S} = \alpha \mu \|\hat{S}\|^2 \tag{2-36}$$

and

$$P_n = |\mu|^2 \|\hat{S}\|^2. \tag{2-37}$$

Dividing the magnitude square of (2-36) by (2-37), we have

$$\frac{|v_s|^2}{P_n} = |\alpha|^2 \|\hat{S}\|^2. \tag{2-38}$$

By (2-35), this is the maximum possible value of the SNR. Thus, we have shown that $\mu \hat{S}^*$ is the optimum value for $\hat{W}$. The optimum value of W may now be obtained from (2-22):

$$W_{\text{opt}} = A_t \hat{W}_{\text{opt}} = A_t \mu \hat{S}^* \tag{2-39}$$

or using (2-16),

$$W_{\text{opt}} = \mu A_t A^* S^* \tag{2-40}$$

and, finally, from (2-30),

$$W_{\text{opt}} = \mu M^{-1} S^*. \tag{2-41}$$

Thus the optimum weight vector W_{opt} for the combiner in Fig. 2-1 is the value of W that satisfies the equation

$$MW = \mu S^*. \tag{2-42}$$

The SNR corresponding to the optimum weighting can be obtained by substituting (2-16) into (2-38). The result is

$$\left(\frac{S}{N}\right)_{\text{opt}} = |\alpha|^2 S_t M^{-1} S^*. \tag{2-43}$$

It is important to emphasize that the solution, or control-law, expressed by (2-42) maximizes the SNR. In Section 6, we show how the optimum weights may be determined adaptively in real time.

III. Generalized Signal-to-Noise Optimization

If an array is designed so that the weights are adaptively controlled to satisfy (2-42), then the array designer is forced to accept the SNR as the governing criterion for all noise environments, including the normal "quiescent" environment (no jamming). In most applications, however, array designers are willing to compromise on SNR in order to obtain some control of the pattern shape, particularly side-lobe levels. For example, Dolph–Chebyshev weights can be used to obtain 30 dB sidelobes with less than 2 dB loss in SNR. It is apparent, then, that we need a criterion for the control of array weights that will give more flexibility in beam shaping.

A tractable approach can be developed as follows. Suppose that, in the normal quiescent environment, the most desirable array weights are given by the weight vector W_q. We assume that W_q represents an optimum compromise between gain, sidelobes, etc. Let the covariance matrix in the normal quiescent environment be M_q, and define the column vector T by the equation

$$M_q W_q = \mu T^* \tag{3-1}$$

where μ is a normalizing constant.

The point of view we wish to present is that in choosing W_q as being optimum, we have, in effect, decided to optimize to an equivalent signal vector T instead of the actual signal vector S. If the environment changes so that the covariance matrix becomes M, then from the results of the previous section, to continue optimizing on T, the weights should be adjusted so that

$$MW = \mu T^* = M_q W_q. \tag{3-2}$$

From the results of the previous section, we also know that controlling the array so as to always satisfy (3-2) is equivalent to maximizing the ratio

$$\frac{|W_t T|^2}{W_t^* M W}. \tag{3-3}$$

This is a more general criterion than maximizing the SNR, which it includes as a special case when $T = \alpha S$. We refer to T as a generalized signal vector and the ratio, (3-3), as the generalized signal-to-noise ratio (GSN). The GSN is a measure of how close the weights are to the generalized signal vector T in the "dot" on "inner" product sense, relative to output noise. Since the "level" of T is arbitrary, the numerical value of the GSN by itself has no significance. It is only significant in comparisons in which the vector T is held fixed.

Comparison of the GSN of a fixed array with that of an adaptive array is of interest. Suppose, then, that the weight vector is held fixed at W_q for all noise environments. Then the GSN will be

$$(\mathrm{GSN})_f = \frac{|(W_q)_t T|^2}{(W_q)_t^* M W_q}. \tag{3-4}$$

Using (3-1), this may be written as

$$(\mathrm{GSN})_f = \frac{1}{|\mu|^2}\left(\frac{|(W_q)_t M_q^* W_q^*|^2}{(W_q)_t^* M W_q}\right). \tag{3-5}$$

For an adaptive array controlled by (3-2), we have

$$(\mathrm{GSN})_a = \frac{|W_t T|^2}{W_t^* M W} = \frac{1}{|\mu|^2}\left(\frac{|W_t M^* W^*|^2}{W_t^* M W}\right) \tag{3-6}$$

$$= \frac{1}{|\mu|^2}(W_t^* M W) \tag{3-7}$$

$$= \frac{1}{|\mu|^2}((W_q)_t^* M_q M^{-1} M_q W_q). \tag{3-8}$$

In the normal quiescent environment where $W = W_q$ and $M = M_q$, (3-5) and (3-8) both reduce to

$$(\mathrm{GSN})_q = \frac{1}{|\mu|^2}(W_q)_t^* M_q W_q. \tag{3-9}$$

In order to put (3-6) and (3-8) into more meaningful forms, we note that the covariance matrix M differs from M_q because of the presence of additional noise sources (jamming). If these are statistically independent of the noise sources in the quiescent environment, we may write

$$M = M_q + M_j \tag{3-10}$$

where M_j is the covariance matrix of the noise components due to the additional noise sources.

Using (3-10) and (3-11) in (3-6) and (3-9), we can obtain the following expressions:

$$(\mathrm{GSN})_f = \frac{(\mathrm{GSN})_q}{1 + \dfrac{(W_q)_t^* M_j W_q}{(W_q)_t^* M_q W_q}} \tag{3-11}$$

and

$$(\mathrm{GSN})_a = \frac{(\mathrm{GSN})_q}{1 + \dfrac{(W_q)_t^* M_q M^{-1} M_j W_q}{(W_q)_t^* M_q M^{-1} M_q W_q}}. \tag{3-12}$$

The ratio in the denominator of (3-11) is the ratio of the jammer noise output to the quiescent noise output when the weights are fixed at W_q. Designating this ratio by $(J/N)_q$, we may write (3-11) and (3-12) as

$$(\mathrm{GSN})_f = \frac{(\mathrm{GSN})_q}{1 + (J/N)_q} \tag{3-13}$$

and

$$(\mathrm{GSN})_a = \frac{(\mathrm{GSN})_q}{1 + \Gamma(J/N)_q} \tag{3-14}$$

$$= (\mathrm{GSN})_q\{1 - \gamma(J/N)_q\} \tag{3-15}$$

where

$$\Gamma = \frac{\gamma}{1 - \gamma(J/N)_q} \tag{3-16}$$

and

$$\gamma = \frac{(W_q)_t^* M_q M^{-1} M_j W_q}{(W_q)_t^* M_j W_q}. \tag{3-17}$$

When the weights are fixed, we note from (3-13) that the GSN varies almost as the inverse of the jammer-to-noise ratio. From (3-14), however, we see that the effect of the jammer on the GSN with adaptive control is reduced by the factor Γ. This factor Γ is analogous to the jammer cancellation ratio that appears in analyses of sidelobe cancellation; it is a figure-of-merit that describes the performance of an adaptive array in a specific noise environment.

The output noise power of an adaptive array controlled according to (3-2) will not remain constant as the noise environment varies. This may require using an AGC following the array combiner to maintain a constant noise level.

Without the AGC, the change in the noise level will be

$$\Delta P_n = W_t{}^*MW - (W_q)_t{}^*M_qW_q. \quad (3\text{-}18)$$

Substituting $W = M^{-1}M_qW_q$ from (3-3), we get

$$\Delta P_n = (M^{-1}M_qW_q)_t{}^*MM^{-1}M_qW_q - (W_q)_t{}^*M_qW_q \quad (3\text{-}19)$$

$$= (W_q)_t{}^*M_qM^{-1}M_qW_q \quad (3\text{-}20)$$

$$= -(W_q)_t{}^*M_qM^{-1}M_jW_q. \quad (3\text{-}21)$$

Since M_q, M_j, and M are positive definite Hermitian matrices, M^{-1} and, hence, $M_qM^{-1}M_j$ are also positive definite Hermitian matrices. This implies that, because of the minus sign in (3-21), the output noise level will decrease when additional noise sources are added to the environment.

It is interesting to observe that the control law given by (3-2) can also be developed from another point of view. Assume, again, that the weight vector W_q represents an optimum compromise between gain, sidelobes, and other factors in the quiescent environment. When the environment changes due to the introduction of jamming noise sources, we recognize that the weight vector W_q may no longer be optimum. How should the weights be changed so that the effects of the jamming are reduced? We want a criterion which reduces the jamming effects and still recognizes the optimality of W_q in the quiescent environment. A suitable criterion is to choose a weight vector W that minimizes the quantity η, where

$$\eta = W_t{}^*M_jW + (W - W_q)_t{}^*M_q(W - W_q). \quad (3\text{-}22)$$

The first term in (3-22) represents the output noise power due to the jamming sources. The second term is a measure of the deviation of the weight vector W from the quiescent optimum W_q. The deviation is measured with respect to a metric based on the quiescent covariance matrix M_q. In so doing, the importance of minimizing the deviation, $\|W - W_q\|$, is emphasized when the noise introduced by the jamming is small compared to the quiescent noise; conversely, when the jamming power is large compared to the quiescent noise, minimizing the jamming power will receive greater weight.

The expression for η, (3-22), can be manipulated into the form

$$\eta = (W_q)_t{}^*(M_qM^{-1}M_j)W_q + (MW - M_qW_q)_t{}^*M^{-1}(MW - M_qW_q). \quad (3\text{-}23)$$

It is easy to verify that (3-23) does reduce to (3-22). Now, as has already been remarked, M^{-1} and $M_qM^{-1}M_j$ are positive definite Hermitian matrices. This implies that both terms in (3-23) are nonnegative. Since the first term is fixed, η will have its minimum value when the second term is minimized. The minimum value of the second term is 0, and this occurs if, and only if,

$$MW = M_qW_q. \quad (3\text{-}24)$$

This is the same as (3-2). Note that in the quiescent environment M reduces to M_q and W is then equal to W_q as it should be. It is also interesting to observe that the value of η is the negative of ΔP_n (see (3-21)) when (3-24) is satisfied.

IV. Linear Adaptive Array

To illustrate the concepts and results of the previous section, we consider here a linear, uniformly spaced array that is weighted in accordance with (3-2). We assume first that in the quiescent environment the noise output of the element channels have equal powers and are uncorrelated. The covariance matrix in the quiescent environment would then be a diagonal matrix of the form

$$M_q = \begin{bmatrix} p_q & & & 0 \\ & p_q & & \\ & & \ddots & \\ 0 & & & p_q \end{bmatrix} \quad (4\text{-}1)$$

$$= p_q 1_K \quad (4\text{-}2)$$

where

- p_q noise power output of each element,
- K number of array elements,
- 1_K identity matrix of order K.

Now let the desired weight vector in the quiescent environment for a signal in the direction θ_s from mechanical boresight be

$$W_q = \begin{bmatrix} a_1 \\ a_2 \exp -j\beta_s \\ a_3 \exp -j2\beta_s \\ \vdots \\ a_K \exp -j(K-1)\beta_s \end{bmatrix}. \quad (4\text{-}3)$$

In (4-3), the a_l are amplitude weights (real numbers) and

$$\beta_s = \frac{2\pi d}{\lambda} \sin \theta_s. \quad (4\text{-}4)$$

If the amplitude weights are all equal, the resultant patterns will be of the form $(\sin Kx)/(\sin x)$. In general, however, they will not be equal. The pattern obtained with any weight vector W_q will be

$$G_q(\beta) = \sum_{l=1}^{K} a_l \exp j(l-1)(\beta - \beta_s) \quad (4\text{-}5)$$

where

$$\beta = \frac{2\pi d}{\lambda} \sin \theta. \quad (4\text{-}6)$$

This can be expressed in matrix notation as

$$G_q(\beta) = B_tW_q \quad (4\text{-}7)$$

where

$$B = \begin{bmatrix} 1 \\ \exp j\beta \\ \exp j2\beta \\ \vdots \\ \exp j(K-1)\beta \end{bmatrix}. \quad (4\text{-}8)$$

Since the weights on the array are controlled in accordance with (3-2), the control law for the weights W in any noise environment will be

$$MW = M_q W_q = p_q W_q \tag{4-9}$$

or

$$W = p_q M^{-1} W_q. \tag{4-10}$$

The noise environment we wish to study is that caused by a single jammer added to the quiescent environment. Let the jammer be located at the angle θ_j from mechanical boresight. Then if the jamming signal in the first element channel is $J(t)$, the jamming signal in the lth channel, assuming narrow bandwidth, will be $J(t) \exp j(l - 1)\beta_j$, where

$$\beta_j = \frac{2\pi d}{\lambda} \sin \theta_j. \tag{4-11}$$

The covariance of the jamming signals in the kth and lth channels, therefore, will be $p_j \exp -j(k - l)\beta_j$, where p_j is the envelope jamming power in each channel. The covariance matrix due to the jamming signal, therefore, will be

$$M_j = p_j(\exp [-j(k - l)\beta_j]). \tag{4-12}$$

This is a Hermitian matrix of order K in which all terms on the same diagonal are equal. Because of its simple structure, it can be conveniently expressed as

$$M_j = p_j H^* U H \tag{4-13}$$

where H is the diagonal matrix

$$H = \begin{bmatrix} 1 & & & 0 \\ & \exp j\beta_j & & \\ & & \exp j2\beta_j & \\ 0 & & & \exp j(K - 1)\beta_j \end{bmatrix} \tag{4-14}$$

and U is a $K \times K$ matrix of ones.

The covariance matrix for the total noise environment, quiescent plus jammer, is just the sum of the two covariance matrices

$$M = M_q + M_j \tag{4-15}$$

$$= p_q 1_K + p_j H^* U H. \tag{4-16}$$

To obtain the weight vector W from (4-10), we require the inverse of M. Using the property $H^* = H^{-1}$, it is easy to verify that

$$M^{-1} = \frac{1}{p_q} \left\{ 1_K - \left(\frac{p_j}{p_j + Kp_j} \right) H^* U H \right\}. \tag{4-17}$$

Using (4-17) in (4-10),

$$W = \left\{ 1_K - \left(\frac{p_j}{p_q + Kp_j} \right) H^* U H \right\} W_q \tag{4-18}$$

$$= W_q - \left(\frac{p_j}{p_q + Kp_j} \right) H^* U H W_q. \tag{4-19}$$

From the definitions of W_q and H we have

$$HW_q = \begin{bmatrix} a_1 \\ a_2 \exp j(\beta_j - \beta_s) \\ \vdots \\ a_K \exp j(K - 1)(\beta_j - \beta_s) \end{bmatrix} \tag{4-20}$$

therefore,

$$UHW_q = G_q(\beta_j) \begin{bmatrix} 1 \\ 1 \\ 1 \\ \vdots \\ 1 \end{bmatrix} \tag{4-21}$$

and finally,

$$H^* U H W_q = G_q(\beta_j) B_j^* \tag{4-22}$$

where

$$B_j = \begin{bmatrix} 1 \\ \exp j\beta_j \\ \vdots \\ \exp j(K - 1)\beta_j \end{bmatrix}. \tag{4-23}$$

Substituting (4-22) in (4-19), we get

$$W = W_q - \left(\frac{p_j}{p_q + Kp_j} \right) G_q(\beta_j) B_j^*. \tag{4-24}$$

The pattern obtained with the weight vector W may now be expressed as

$$G(\beta) = B_t W \tag{4-25}$$

$$= B_t W_q - \left(\frac{p_j}{p_q + Kp_j} \right) G_q(\beta_j) B_t B_j^*. \tag{4-26}$$

From (4-7), however,

$$B_t W_q = G_q(\beta) \tag{4-27}$$

and it is easy to show that

$$B_t B_j^* = C(\beta - \beta_j) \tag{4-28}$$

where

$$C(x) = \exp j \left\{ \frac{(K - 1)x}{2} \right\} \frac{\sin Kx/2}{\sin x/2} \tag{4-29}$$

thus, substituting (4-27) and (4-28) into (4-26) gives

$$G(\beta) = G_q(\beta) - \left(\frac{p_j}{p_q + Kp_j} \right) G_q(\beta_j) C(\beta - \beta_j). \tag{4-30}$$

This says that the pattern of the adaptively controlled linear array in the presence of a jammer consists of two parts. The first is the quiescent pattern $G_q(\beta)$, and the second, which is subtracted from the first, is a $(\sin Kx)/(\sin x)$ shaped beam centered on the jammer. This is shown in Fig. 4-1. The gain of the array in the direction of the jammer will be

$$G(\beta_j) = G_q(\beta_j) - \left(\frac{p_j}{p_q + Kp_j} \right) G_q(\beta_j) C(0). \tag{4-31}$$

Since $C(0) = K$, (4-31) reduces to

$$G(\beta_j) = \left(\frac{p_q}{p_q + Kp_j} \right) G_q(\beta_j). \tag{4-32}$$

If the array weights remained fixed at W_q in the presence of the jammer, the gain in the direction of the jammer would be $G_q(\beta_j)$. Hence, the adaptive control reduces the gain in

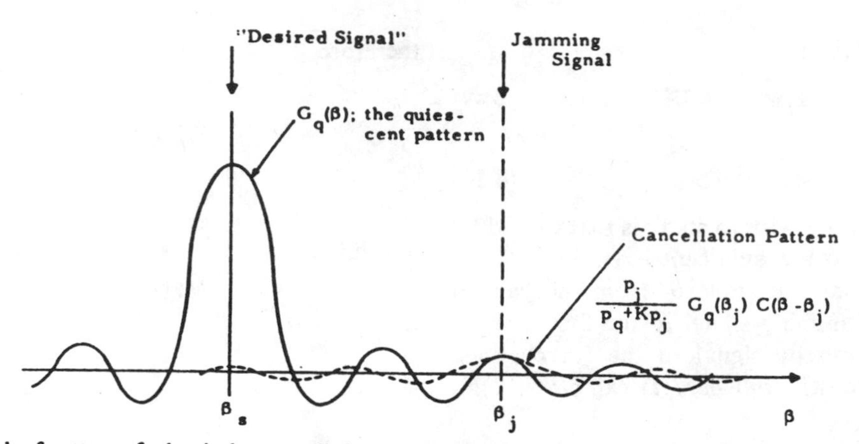

Fig. 4-1. Sketch of pattern of adaptively controlled array. (Resultant pattern is difference between quiescent and cancellation pattern.)

the direction of the jammer by the factor

$$\frac{G(\beta_j)}{G_q(\beta_j)} = \frac{p_j}{p_q + Kp_j} = \frac{1}{1 + \dfrac{Kp_j}{p_q}}. \tag{4-33}$$

The ratio in the denominator of (4-33), Kp_j/p_q, is the jammer-to-noise ratio in the "cancellation" beam, $C(\beta - \beta_j)$. Since $G(\beta)$ is a voltage gain pattern, it is tempting to conclude that the performance of the array against the jammer power had been improved by the square of the inverse of Equation (4-33). This, however, would not be correct. The proper measure of the improvement against the jammer is the cancellation ratio Γ defined by (3-16) and (3-17), which are repeated for convenience:

$$\Gamma = \frac{\gamma}{1 - \gamma(J/N)_q} \tag{3-16}$$

and

$$\gamma = \frac{(W_q)t^* M_q M^{-1} M_j W_q}{(W_q)t^* M_j W_q}. \tag{3-17}$$

To evaluate (3-17) in the present context, we note that

$$M_q M^{-1} = 1_K - \left(\frac{1}{p_q + Kp_j}\right) M_j \tag{4-34}$$

and

$$M_j M_j = Kp_j M_j \tag{4-35}$$

therefore,

$$M_q M^{-1} M_j = M_j - \left(\frac{Kp_j}{p_q + Kp_j}\right) M_j \tag{4-36}$$

$$= \left(\frac{p_q}{p_q + Kp_j}\right) M_j. \tag{4-37}$$

Using (4-37) in (3-17), we see that

$$\gamma = \frac{1}{1 + \dfrac{Kp_j}{p_j}}. \tag{4-38}$$

Now the maximum possible value of $(J/N)_q$ is Kp_j/p_q, and this will occur only if the jammer is at the peak of the main beam of a $(\sin Kx)/(\sin x)$ quiescent beam. In that case, we obtain $\Gamma = 1$, indicating no improvement of performance against the jammer, which is to be expected. The adaptive control will be most effective against jammers in the sidelobe region of the quiescent pattern. In that case, we will have $(J/N)_q \ll Kp_j/p_q$, so that $\gamma(J/N)_q \ll 1$ and, hence, $\Gamma \approx \gamma$. Thus, for a jammer in the sidelobe region, the adaptive control will "cancel" the jammer power by approximately the jammer-to-noise ratio in the cancellation beam.

It is interesting to note that this result is the same as the reduction in voltage gain of the pattern in the direction of the jammer and that it is almost independent of the choice of W_q and, hence, of the shape of the quiescent beam.

V. Sidelobe Cancellation

The concepts presented in this report are actually a generalization of the coherent sidelobe cancellation techniques. For the sake of completeness and as a second illustration of the application of the concepts, we shall show that sidelobe cancellation may be viewed as a special case of an adaptive array.

A sidelobe cancellation system is shown in Fig. 5-1. It consists of a main, high gain antenna whose output is designated as channel "o" and K auxiliary antennas. The auxiliary antenna gains are designed to approximate the average sidelobe level of the main antenna gain pattern. The amount of desired target signal received by the auxiliaries is negligible compared to the target signal in the main channel. The purpose of the auxiliaries is to provide independent replicas of jamming signals in the sidelobes of the main pattern for cancellation.

To apply the concepts of the adaptive array we assume that the outputs of the channels are weighted and summed. The problem is to find an appropriate control law for the weights. The first step is to select a T vector. In this case, since the signal collected by the auxiliaries is negligible and the main antenna has a carefully designed pattern, we choose the $K + 1$ column vector

$$T = \begin{bmatrix} 1 \\ 0 \\ 0 \\ \vdots \\ 0 \end{bmatrix}. \tag{5-1}$$

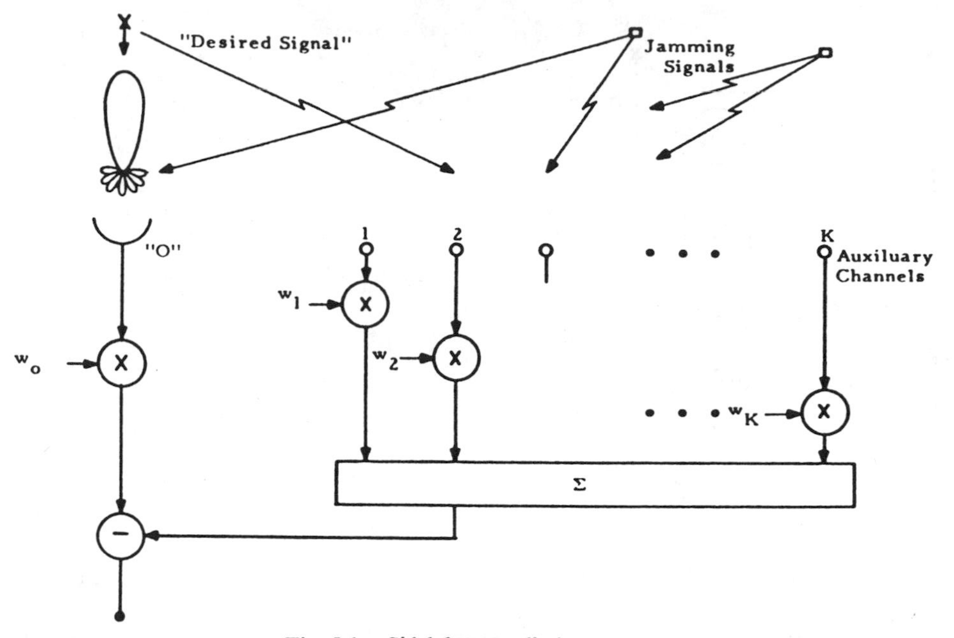

Fig. 5-1. Sidelobe cancellation system.

The optimum control law would then be

$$M'W' = \mu T \tag{5-2}$$

where M' is the $K + 1$ by $K + 1$ covariance matrix of all the channels and W' is the $K + 1$ column vector of all the weights.

Now let M be the K by K covariance matrix of the auxiliary channels only and W be the K column vector of the auxiliary channel weights. Equation (5-2) now may be partitioned as follows:

$$\begin{bmatrix} p_0 & \Lambda_t^* \\ \Lambda & M \end{bmatrix} \begin{bmatrix} w_0 \\ W \end{bmatrix} = \begin{bmatrix} \mu \\ 0 \end{bmatrix}. \tag{5-3}$$

In (5-2) we have set

$$p_0 = \mu_{00}' \text{ of } M' = \text{noise power output of the main channel} \tag{5-4}$$

and

$$\Lambda = \begin{bmatrix} \mu_{1o}' \\ \mu_{2o}' \\ \vdots \\ \mu_{Ko}' \end{bmatrix}. \tag{5-5}$$

Note that μ_{lo}' is the cross-correlation of the output of the lth auxiliary with the output of the main channel. Equation (5-3) may be written as two separate equations: a scalar equation,

$$p_o w_o + \Lambda_t^* W = \mu \tag{5-6}$$

and a matrix equation,

$$MW = -w_0 \Lambda. \tag{5-7}$$

The control law expressed by (5-2) can be implemented using $K + 1$ control loops to control the $K + 1$ weights $w_o, w_1, \cdots, w_k$. However, because of the unique form of the T vector, it is possible to achieve an optimum combiner with only K control loops. To do this, we observe that if the weight vector W' is optimum for a given noise environment (a given M'), then any multiple of W' will also be optimum since the figure of merit, GSN, is a ratio in which the level of W' does not matter. This means, in effect, that we may let the parameter μ in (5-2) vary freely. Thus, instead of fixing μ and trying to control w_o to satisfy (5-6), we can fix w_o at a nonzero value if we are sure that no solution to (5-2) will require w_o to be 0. However, since the inner product $W_t'T = w_o$, the GSN will never be optimized with $w_o = 0$. With w_o fixed at $\hat{w}_o$, we only need control loops for the auxiliaries, and the control law is given by (5-7) with $w_o = \hat{w}_o$,

$$MW = -\hat{w}_o \Lambda. \tag{5-8}$$

Since the inner product $W_t'T = \hat{w}_o$ is fixed, and the control law, (5-8), optimizes the ratio of the inner product to the output noise power, it must clearly be minimizing the output power. Thus, (5-8) is the sidelobe cancellation control law for minimizing the output noise power. Equation (5-8) has been obtained using this criterion in previous studies of sidelobe cancellation.

VI. Implementation

The control loops for an adaptive array can be implemented using the same circuitry as is used for coherent sidelobe cancellation. The basic idea is shown in the functional block diagram in Fig. 6-1. The notation used there gives the complex envelope of signals on phase coherent carriers. The signal in each channel u_k is multiplied by the weight w_k, and then the weighted signals are summed. The multiplication and summation occur on carrier frequencies at IF.

The weights w_k are derived by correlating u_k with the sum signal $\sum$, substracting the correlation from the desired vector component t_k^*, and then using a high gain amplifier.

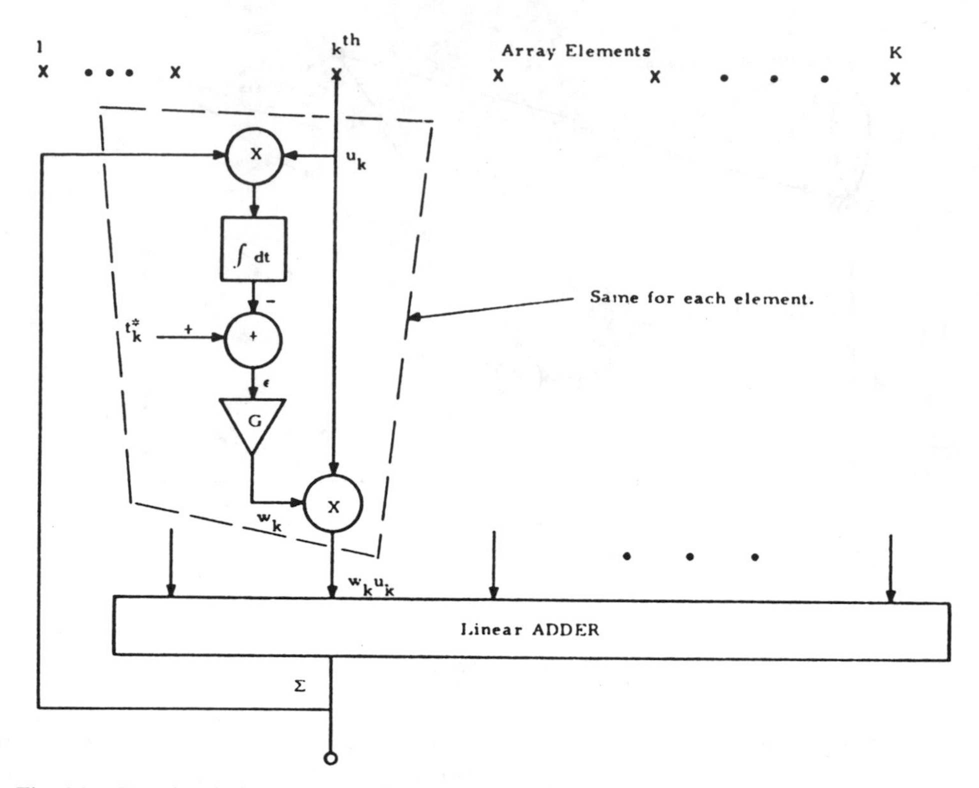

Fig. 6-1. Functional block diagram of implementation of element control loop for adaptive array.

For each w_k, we have

$$w_k = G\left\{t_k^* - u_k^* \sum_{l=1}^{K} w_l u_l\right\} \quad (6\text{-}1)$$

or

$$\sum_{l=1}^{K} w_l\left(u_k^* u_l + \frac{\delta_{lk}}{G}\right) = t_k^* \qquad \text{for } k = 1,\cdots,K \quad (6\text{-}2)$$

where

$$\delta_{lk} = \begin{cases} 1, & \text{when } k = l \\ 0, & \text{all other } k. \end{cases}$$

Recalling that $u_k^* u_l$, is an element of the covariance matrix M, the K equations represented by (6-2) may be written as

$$\left\{M + \frac{1_K}{G}\right\} W = T^*. \quad (6\text{-}3)$$

This differs from the optimum control law, (3-2), by the addition of a term inversely proportional to gain. This term introduces an error analogous to the "servo error" of a type 0 servo. Its effect can be made negligible with sufficiently high gain.

The preceding discussion demonstrates that the steady-state solution of the control loops is essentially the desired one. The dynamic behavior of the control loops is determined by the "integrators" shown in Fig. 6-1. In practice these are high Q, single-pole circuits. The differential equations describing the dynamic behavior of the loops can be shown to be

$$\tau\frac{dw_k}{dt} + w_k = G\left\{t_k^* - u_k^* \sum_{l=1}^{K} w_l u_l\right\} \quad (6\text{-}4)$$

where τ is the time constant of the "integrator" circuits.

The loops are designed so that the weights w_k will vary slowly compared to the bandwidth of the signals u_l. Hence, the weights will be uncorrelated with u_l. Thus, if we apply the expectation operator to both sides of (6-4) we get

$$\tau\frac{d\overline{w}_k}{dt} + \overline{w}_k = G\{t_k^* - \sum \overline{w}_l\overline{u_k^* u_l}\}. \quad (6\text{-}5)$$

This equation determines the dynamic behavior of the expected value of w_k. The K equations obtained from (6-5) may be represented in matrix form as

$$\tau\frac{d\overline{W}}{dt} = -(GM + 1_K)\overline{W} + GT^*. \quad (6\text{-}6)$$

It can be shown easily that matrix equations of this form are stable if the matrix $GM + 1_K$ has only positive eigenvalues. Since M is a positive definite Hermitian matrix, $GM + 1_K$ is also a positive definite Hermitian matrix. It, therefore, has only positive eigenvalues. Thus, the implementation shown in Fig. 6-1 is stable under all conditions for the assumptions made. In practice, second-order effects, neglected in the preceding analysis, may make the loops unstable if the loop gains are allowed to become excessive.

VII. Computer Simulation

Figs. 7-1 through 7-9 show the results of the computer simulation of an adaptive array. The array used in these studies was a 21-element linear array with half-wavelength spacing. The servo error discussed previously was neglected, and it was assumed that the array weights were determined by the control law

$$MW = T^*. \quad (7\text{-}1)$$

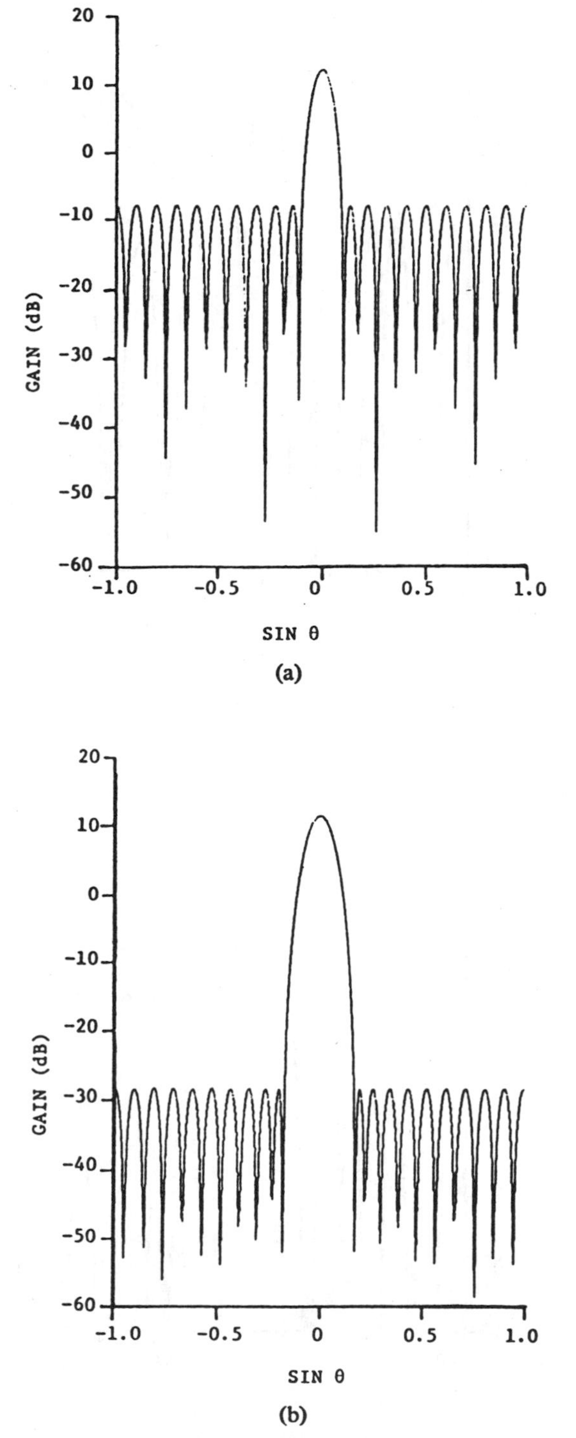

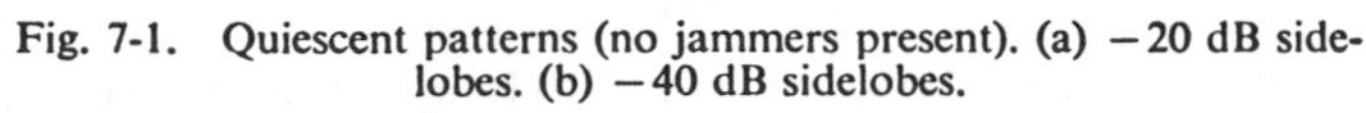
Fig. 7-1. Quiescent patterns (no jammers present). (a) −20 dB sidelobes. (b) −40 dB sidelobes.

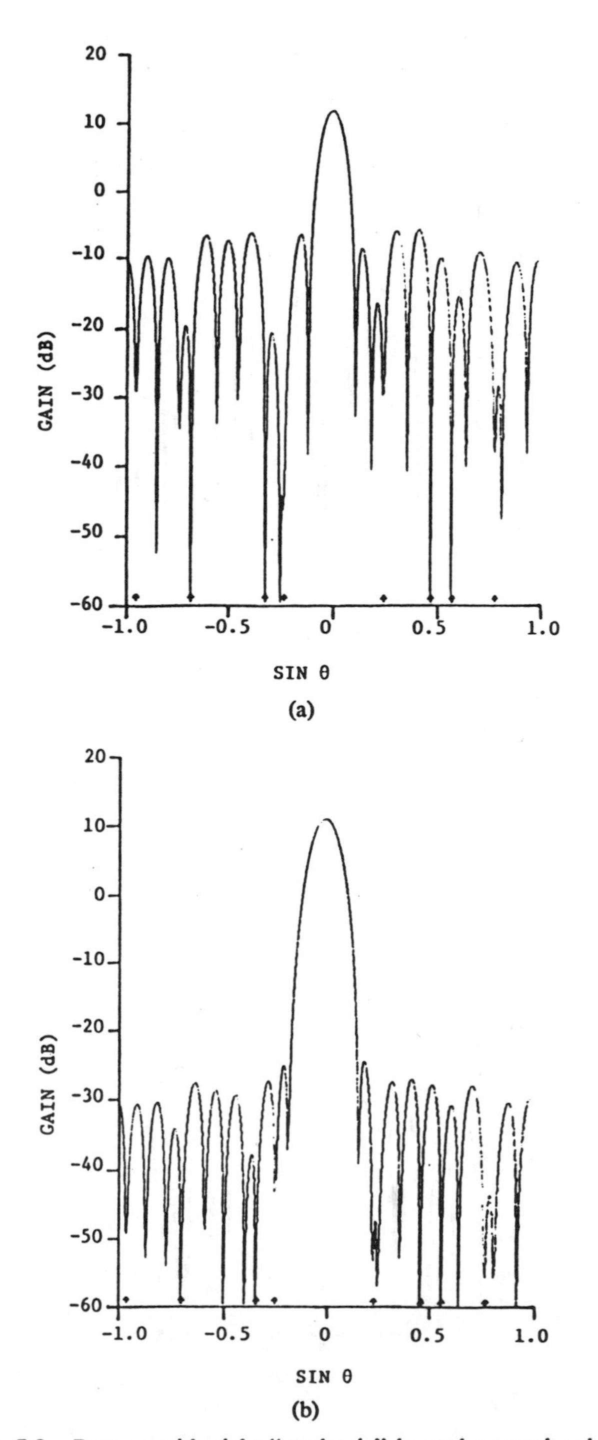

Fig. 7-2. Patterns with eight "randomly" located narrowband jammers. Nulls are produced at locations of jammers. Nulls below −60 dB are plotted as 60 dB. (a) −20 dB sidelobes (quiescent). (b) −40 dB sidelobes (quiescent).

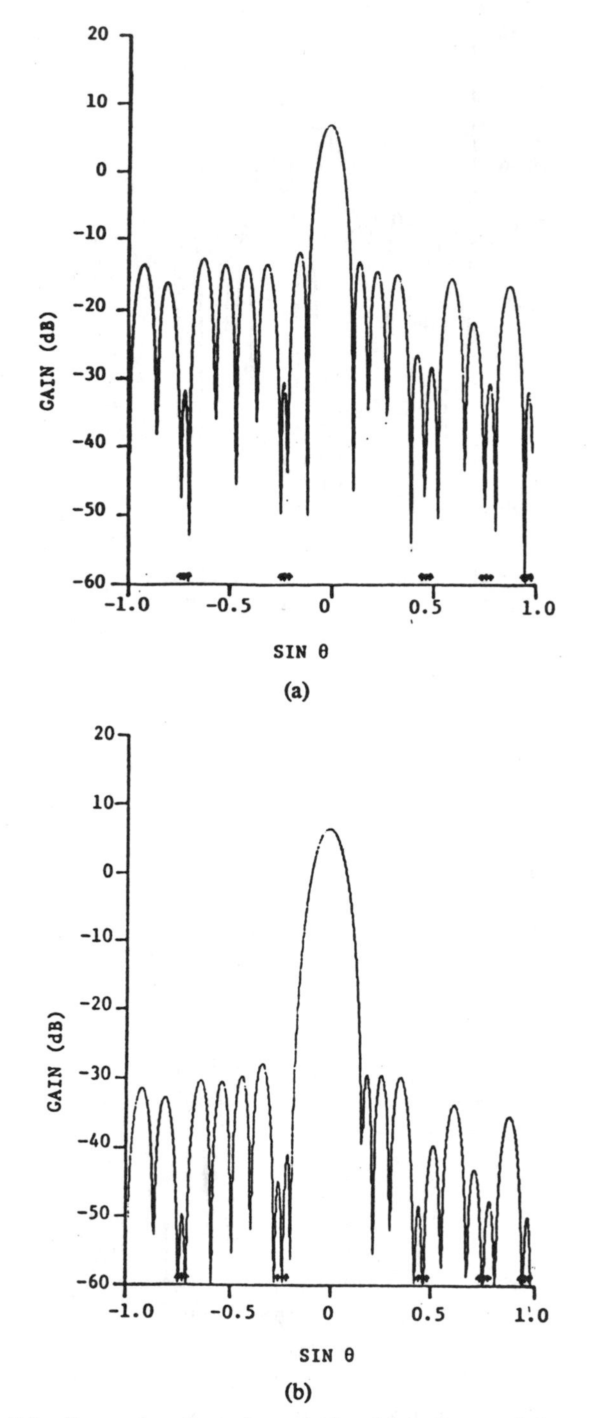

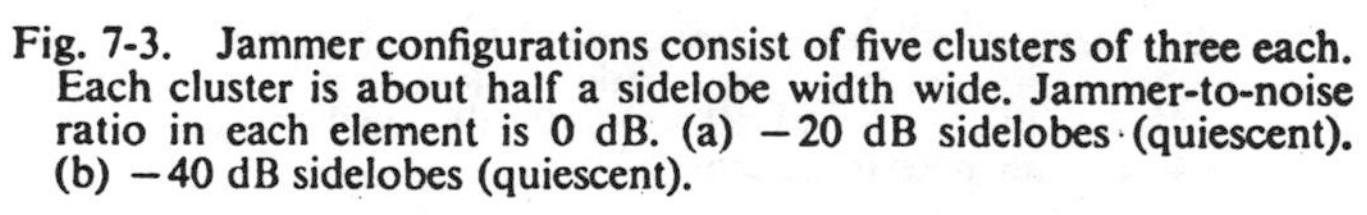

Fig. 7-3. Jammer configurations consist of five clusters of three each. Each cluster is about half a sidelobe width wide. Jammer-to-noise ratio in each element is 0 dB. (a) −20 dB sidelobes (quiescent). (b) −40 dB sidelobes (quiescent).

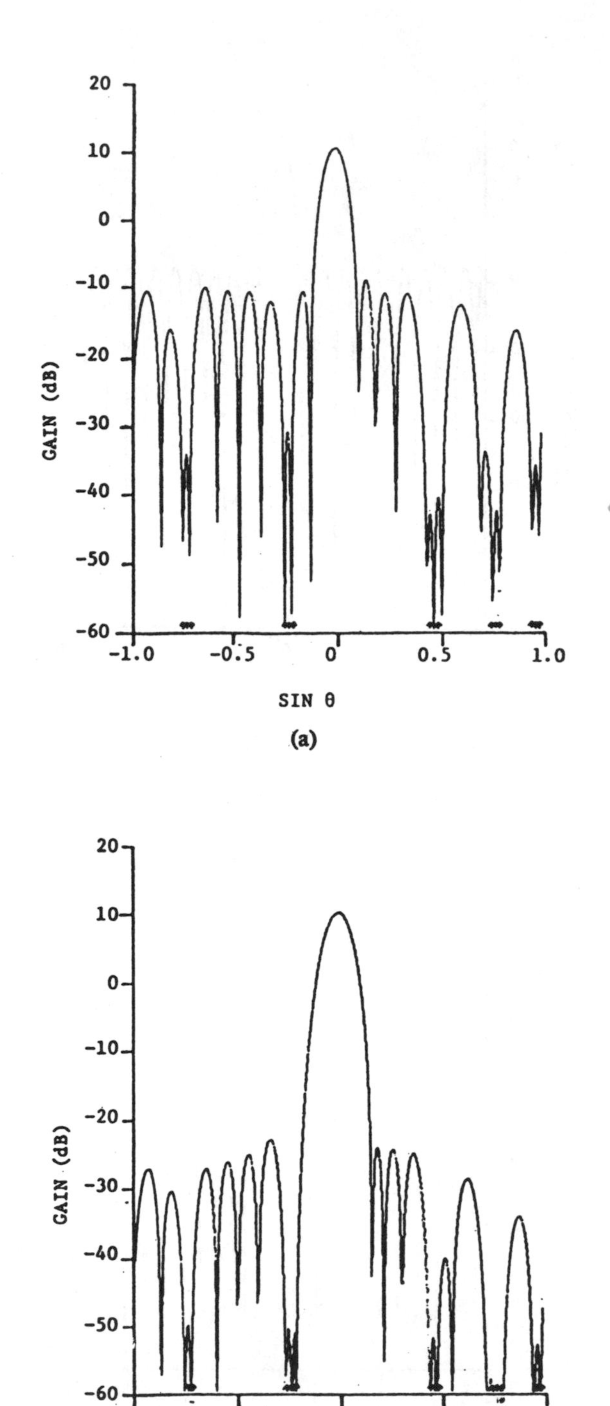

Fig. 7-4. Same as Fig. 7-3 with 15 dB jammer-to-noise ratio.

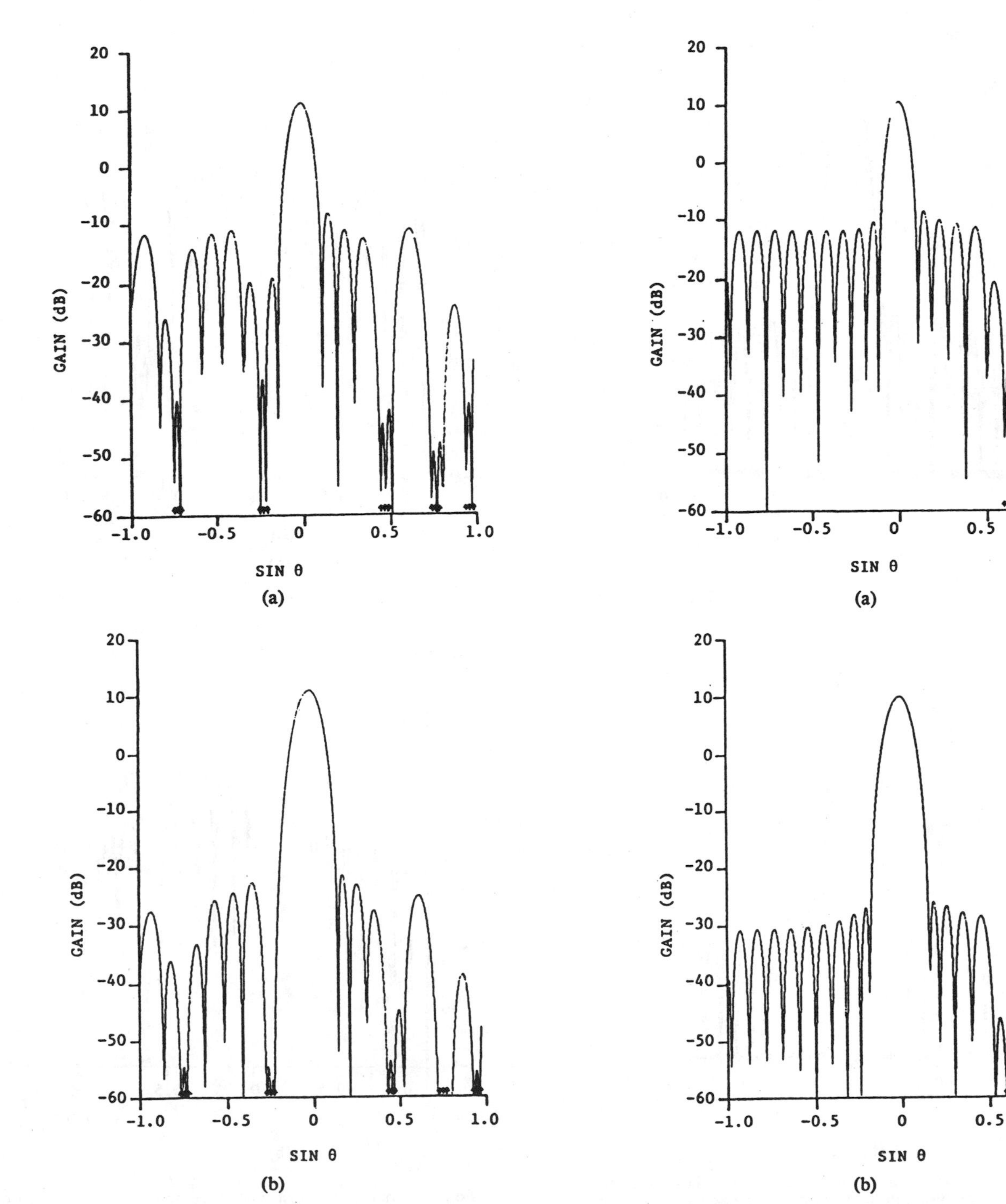

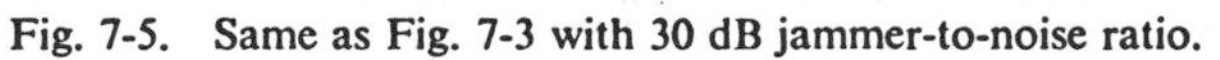

Fig. 7-5. Same as Fig. 7-3 with 30 dB jammer-to-noise ratio.

Fig. 7-6. Nine jammers are clustered and span almost two sidelobes.

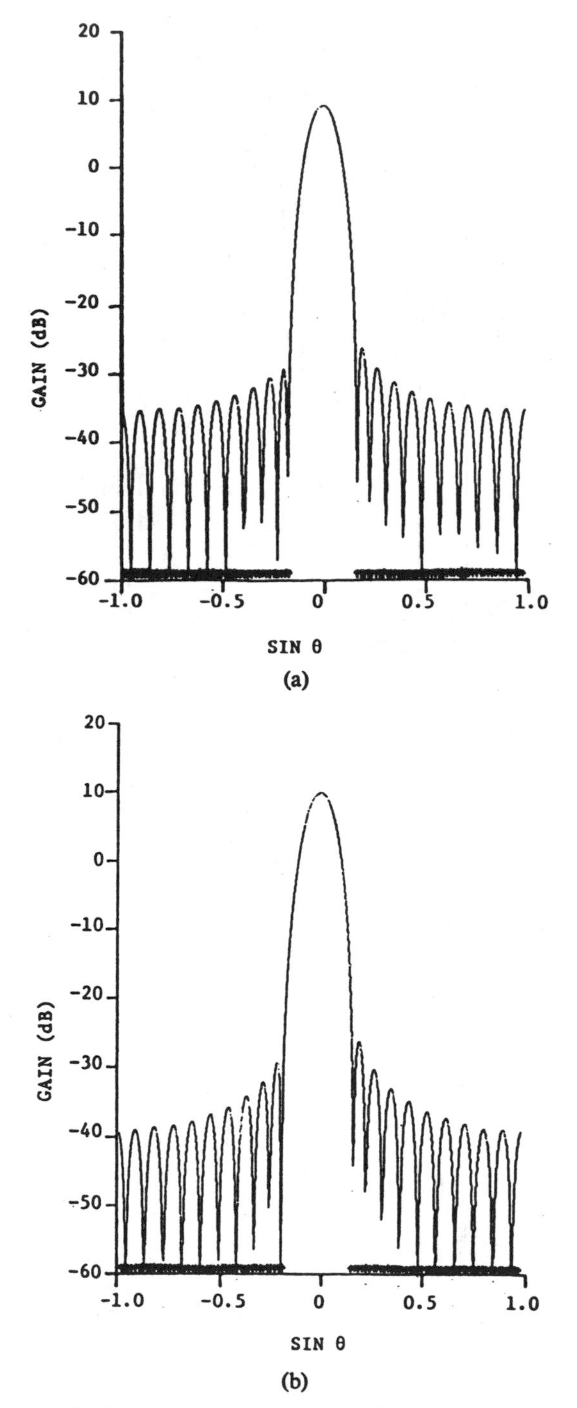

Fig. 7-7. 91 jammers span entire sidelobe region. Two resulting patterns are almost identical, indicating that under severe conditions choice of quiescent pattern has little effect. (a) −20 dB sidelobes (quiescent). (b) −40 dB sidelobes (quiescent).

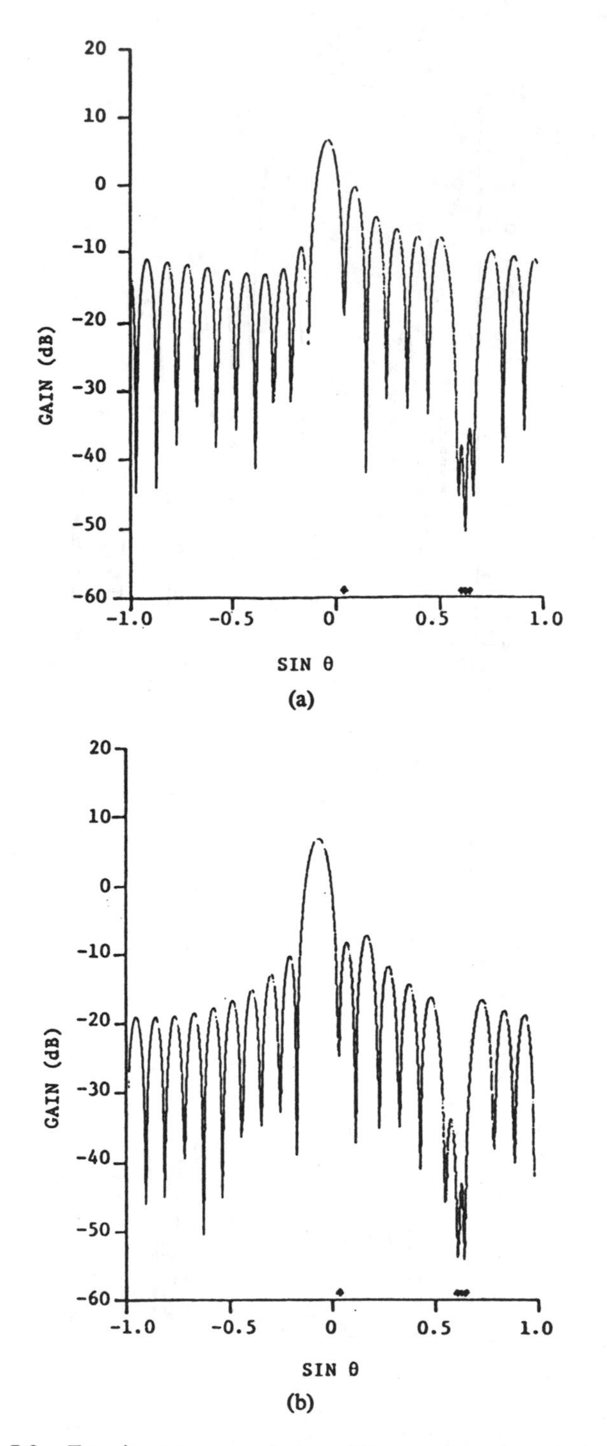

Fig. 7-8. Four jammers; one cluster of three and one in main beam. One in main beam apparently forces beam to move so as to put jammer in its first null. (a) −20 dB sidelobes (quiescent). (b) −40 dB sidelobes (quiescent).

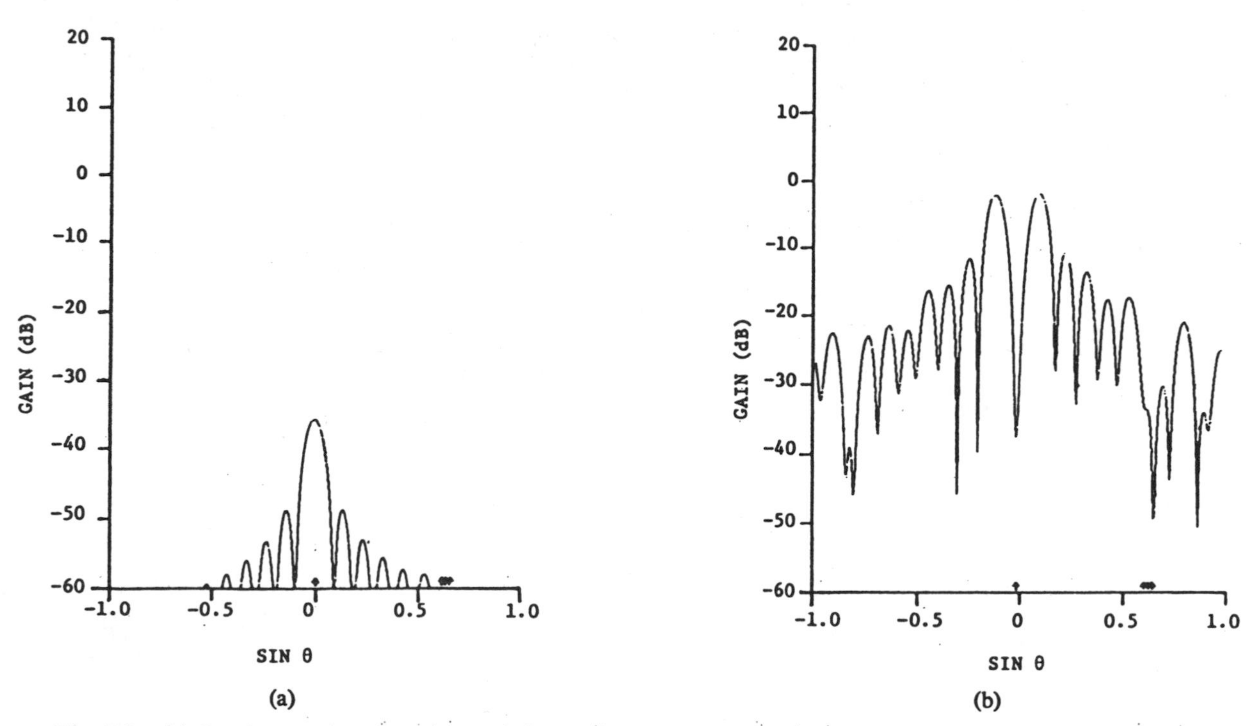

Fig. 7-9. (a) One jammer is at boresight and cluster of three is in sidelobe. Quiescent patter is (sin Kx)/(sin x) beam. Shape of main beam is unaltered by presence of jammer since cancellation beam has same shape. Main effect is drastic lowering of level of pattern. (b) −30 dB Dolph–Chebyshev quiescent pattern. Deep notch is produced in center of main beam. This occurs because main beam of quiescent pattern is broader than main beam of (sin Kx)/(sin x) "cancellation" beam. There is much less effect on pattern level than in first case.

Various vectors T were used to give different quiescent beams. The quiescent beams studied were a (sin Kx)/(sin x) beam and Dolph–Chebyshev beams with 20, 30, and 40 dB sidelobes. The matrix M was computed for various spatial configurations of jammers, and narrow bandwidth was assumed. Wide bandwidth can be simulated by using narrowband calculations and spreading the jammers spatially. Thus, many of the figures show configurations of clustered jammers. Each cluster may be regarded as a single wideband jammer. The equivalent bandwidth Δf of a cluster of jammers expressed as a fraction of the center frequency is

$$\frac{\Delta f}{f_c} = \frac{\Delta(\sin \theta)}{\sin \theta_c} \quad (7\text{-}2)$$

where

$\Delta(\sin \theta)$ spread of the cluster in $\sin \theta$ space,
$\sin \theta_c$ center of the cluster.

The results are shown as plots of the pattern of the array in the presence of jammers. The abscissa is $\sin \theta$ for $0 \leq \theta \leq \pi/2$ from $-1 \leq \sin \theta \leq 1$. The desired target direction corresponds to the center of the plots. Except for Fig. 7-9, only 20 and 40 dB Dolph–Chebyshev patterns are shown. For ease of comparison, two plots are given in each figure with the same jammer environment assumed in both cases.

Acknowledgment

This report has been reprinted because of the recurring interest in this study. Except for format changes in some figures, the text corresponds to that in previous editions. The antenna gain patterns in Section 7 were replotted on a Calcomp plotter (originally they were plotted using a line printer). The author extends his thanks to D. J. Chapman for handling these format changes. The author also wishes to thank J. W. Stauffer, presently of the General Electric Company, for useful discussions, suggestions, and help with the artwork of the original report. This report has also been identified as SPL TR 69-76 and SPL TR 66-1.

Part II
Adaptive Algorithms and Their Properties

THE key to successful adaptive signal processing is understanding the fundamental properties of adaptive algorithms. These properties are stability, speed of convergence, misadjustment errors, robustness, numerical complexity, and round-off error analysis of adaptive algorithms. This part presents a diverse collection of papers on these topics. To illustrate approximate evolution of these ideas, papers are presented in the chronological order. Much of the recent adaptive algorithmic research has focused on algorithms that are based on the lattice structures. Since there is a large body of recent literature on lattice algorithms, these papers comprise Part III.

The first paper in this chapter is Ungerboeck's "Theory on the Speed of Convergence in Adaptive Equalizers for Digital Communication," which presents an analysis of the convergence properties of adaptive transversal equalizers that minimize the mean square distortion. He develops the relationships between the speed of convergence and number of taps, the step size parameter, and the spectrum of the unequalized signal. Ungerboeck shows that the convergence speed of the expected mean square distortion depends to a large extent on the number of equalizer taps. He also develops a simple two-step procedure for determining a time-dependent step size which leads to the fastest convergence of the expected mean square distortion. Ungerboeck's paper is an extension of Gersho and Lucky's work discussed in Part I.

In the next paper, Godard shows how the Kalman-Bucy filtering theory can be applied to the problem of setting the tap gains of transversal equalizers to minimize mean square distortion. The Kalman-Bucy filter algorithm leads to faster convergence than other algorithms that have been applied to channel equalization. Godard shows that the speed of convergence is independent of the characteristics of the channel. A large spread of eigenvalues does not lead to slow convergence, as is the case with the steepest descent methods. Godard shows that his algorithm converges in less than $2N$ steps, where N is the number of tap weights. This result is in agreement with computer simulations. The disadvantage of the Kalman-Godard algorithm is computational complexity.

Although Godard used a Kalman filtering framework, his algorithm is essentially a recursive least squares (RLS) algorithm that can be implemented several ways. If one uses the matrix inversion lemma, as in Godard's paper, or LU (or LDU) decomposition as in Bierman's book [1], the computational complexity is proportional to N^2. However, in the "fast" RLS algorithms, the computational complexity is proportional to N. In the paper "Application of Fast Kalman Estimation to Adaptive Equalization," Falconer and Ljung show how "fast recursive estimation" techniques can be applied to the equalizer adjustment problem, resulting in the same fast convergence as the Kalman-Godard algorithm, but with the number of operations per iteration proportional only to N.

Carayannis, Manolakis, and Kalouptsidis, in their paper "A Fast Sequential Algorithm for Least Squares Filtering and Prediction," discuss the fast Kalman algorithm of Falconer and Ljung in parallel with their more efficient FAEST (*F*ast *a* *posteriori* *e*rror *s*equential *t*echnique) algorithm. The authors point out that in conventional recursive least squares algorithms, the inverse covariance matrix is updated in each recursion, and then a Kalman gain is computed by a vector matrix multiplication which requires an order of N operations. The fast Kalman algorithms exploit shift invariance of Toeplitz covariance matrices to reduce the computational complexity of updating the Kalman gain. The fast Kalman algorithm is mainly based on the *a priori* error formulation. The FAEST algorithm uses both *a priori* and *a posteriori* errors in its formulation. The FAEST algorithm achieves its computational efficiency by an alternative definition of the gain vector, and exploitation of the relationships between *a priori* and *a posteriori* errors. This paper is a good companion to Godard's classic paper because it clarifies the difference between adaptive algorithms and their efficient implementation using fast algorithms. Other papers of value to readers who are interested in fast algorithms are as follows: Cioffi and Kailath [2], Griffiths [3], Marple [4], Mendaugh and Griffiths [5], and the Ph.D. dissertation by Cioffi [6].

The next paper returns to the analysis of learning characteristics of the LMS (least mean square) adaptation algorithm of Widrow and Hoff. While less sophisticated than the previously discussed Godard's algorithm, the LMS algorithm is nevertheless simple, robust, and easily understandable. Because of these properties, the LMS algorithm has become popular in adaptive signal processing applications. The LMS algorithm is a stochastic version of a steepest descent algorithm. In the LMS algorithm, the changes in the adaptive filter weights are made proportional to the estimated negative gradient of the mean square error function. Mean square error is a quadratic function of the weights, and the stochastic gradient algorithm seeks the unique minimum of this quadratic error function. In the LMS algorithm, the gradient estimation process introduces noise into the weight vector, and the noise is proportional to the speed of adaptation and number of weights. The effect of this noise is expressed in terms of a dimensionless ratio of the average excess mean square error (MSE) to the minimum mean square error (optimum performance). This ratio is called the misadjustment. When the adaptive filter must track a nonstationary stochastic process, there is another contribution to the misadjustment due to the lag of the adaptive process in tracking the moving minimum point of the error surface. This additive error contribution is proportional to the number of weights, but inversely proportional to the speed of adaptation.

Widrow, *et al*, show that the sum of misadjustments can be minimized by choosing the speed of adaptation such that the two contributions are equal. This paper extends the results of Widrow's two other papers on the application of LMS algorithms, which are reprinted in Part I, to the analysis of nonstationary learning characteristics.

Rapid convergence of adaptive equalization algorithms is an important concern in data transmission applications. In the paper "Self-Orthogonalizing Adaptive Equalization Algorithms," Gitlin and Magee compare several self-orthogonalizing adaptive algorithms which accelerate the rate of convergence. The accelerated convergence of the estimated gradient algorithm is achieved by pre-multiplying the correction term in the algorithm by a matrix which is an estimate of the inverse of the channel correlation matrix. The essential idea in the self-orthogonalization algorithms is to find a sequence of matrices which pre-multiply the estimated gradient and converge to the inverse of the channel correlation matrix. Various self-orthogonalizing algorithms differ in the manner in which the estimate of the inverse of the channel covariance matrix is obtained. Self-orthogonalizing algorithms can converge an order of magnitude more rapidly than the simple gradient algorithm. The authors show that Godard's algorithm has the orthogonalizing property. We will return to the self-orthogonalizing concepts in the discussion of lattice algorithms.

In the following paper, Macchi and Eweda present a rigorous second-order convergence analysis of the constant step-size mean square error gradient algorithm. Their analysis considers the important case of highly correlated input sequences. Their results are useful for a wide class of practical applications.

In addition to the previously discussed topics of convergence rates, stability, and computational complexity, important characteristics of adaptive algorithms are propagation, decay or amplification of quantization, and round-off noise. The last three papers in this part deal with these topics. The paper "On the Design of Gradient Algorithms for Digitally Implemented Adaptive Filters" by Gitlin, Mazo, and Taylor show that digitally implemented adaptive gradient algorithms can have significantly different learning characteristics and excess mean square errors than the "infinite precision" (analog) algorithms. The difference between digital and analog algorithms is most clearly illustrated in the design of the time-varying step size of the stochastic approximation algorithm. The stochastic approximation theory states that, in the infinite precision case, the step size should decrease as l/n, where n is the number of iterations. In the infinite precision algorithm, this step-size sequence produces a mean square error that vanishes as n becomes large. This is not the case in digital algorithms. The authors show that in the digital case, the time-varying step-size sequence suggested by stochastic approximation cannot achieve as small a residual mean-square error as attained by the optimum constant step-size. It should be mentioned that the final step-size is also desirable because the stochastic gradient algorithm retains its capability to track a time-varying environment.

This paper also presents a good tutorial development of the design of adaptive gradient algorithms in the absence of quantization. It presents interesting results on the design of the optimum step-size sequence, but the main interest in this sequence of three papers is that it gives insights into the effect of digital implementation on the design and performance of the gradient algorithm.

The paper, "A Round-Off Error Analysis of the LMS Adaptive Algorithm" by Caraiscos and Liu, analyzes the steady state output error of the least mean square (LMS) due to the finite precision arithmetic in a digital processor. They investigate both fixed and floating point operations. The steady state output error consists of three terms caused by data quantization, round-off in the arithmetic operations, and filter coefficient quantization. The term of the output mean square error caused by the filter coefficient quantization is inversely proportional to the adaptation step size. Thus, the small step size for the purpose of reducing the excess mean square error may result in significant error due to filter coefficient quantization. The authors show that the excess mean square error is larger that the quantization error, as long as the step-size allows the algorithm to converge completely. The effect of filter quantization can be reduced if more bits are used for the filter coefficients than for the data.

The last paper in the series of papers on round-off error analysis is by S. Ljung and L. Ljung. In their paper "Error Propagation Properties of Recursive Least-Squares Adaptation Algorithms," they discuss the error propagation properties of the most common implementations of the recursive least squares method. The error propagation properties of adaptive algorithms are crucial to system design when these algorithms are used as essential components in adaptive control and signal processing systems. In these systems, algorithms must be resilient to abnormal data, disturbances, and round-off errors. The authors give a brief, but very readable, overview of the most common implementation of the recursive least squares method. This paper is worth reading if only for this overview, which makes it a nice supplement to the previously presented papers on recursive least squares (RLS) algorithms. The implementations of RLS algorithms that are analyzed by the authors are: the basic recursive least squares (RLS), conventional least squares (CLS), upper triangular diagonal factorization (UDLS), fast least squares (FLS), and fast lattice (FL) algorithms. The RLS, CLS, UDLS and FL algorithms are exponentially stable to round-off errors and other numerical disturbances. The base of the exponential decay is equivalent to the forgetting factor λ. Counterexamples show that FLS does not have this exponential decay property and may be unstable when $\lambda < l$. Without special fixes, it is therefore unsuitable for continuous adaptative algorithms.

In addition to papers reprinted here, there are five excellent textbooks that have very thorough treatments of adaptive algorithms and their properties [7–11]. Widrow and Stearn's book [7] presents a thorough treatment of the LMS algorithm and its applications. Honing and Messerschmitt [8] treat a broader class of adaptive algorithms. They include a very thorough discussion of lattice structures. Their book has a good discussion of adaptive equalization, including decision feedback equalization. S. Haykin [9] discusses and compares least mean square (LMS) and recursive least squares (RLS) algorithms. Goodwin and Sin's book [10] is a control system approach to the adaptive algorithm. S. T. Alexander, in his book *Adaptive Signal Processing*, covers both gradient-based and more modern least squares methods of adaptive signal processing [11]. Ljung and Söderström, in their book *Theory and Practice of Recursive Identification* [12], discuss application of various least squares techniques to system identification of more general models than AR-processes. There is currently a cross fertilization of ideas that has arisen in adaptive control and adaptive signal processing [13], [14]. In addition to the above mentioned books, there are a number of recent books on adaptive filters and array processing that are recommended: S. Haykin [15], R. T. Compton [16], and books edited by S. Haykin [17], and Cowan and Grant [18].

References

[1] Bierman, G. J., *Factorization Methods for Discrete-Sequential Estimation.* New York, NY: Academic Press, 1977.

[2] Cioffi, J. M., and Kailath, T., "Fast recursive-least-squares transversal filters for adaptive filtering," *IEEE Trans. Acous., Speech, and Signal Processing*, vol. ASSP-32, no. 2, pp. 304–337, Apr. 1984.

[3] Griffiths, L. J., "A continuously-adaptive filter implemented as a lattice structure," in *Proc. Int. Conf. Acoust., Speech, and Signal Processing*, 1977, pp. 683–686.

[4] Marple, S. L., "Efficient least-squares FIR system identification," *IEEE Trans. Acous., Speech, and Signal Processing*, vol. ASSP-29, no. 1, pp. 62–73, Feb. 1981.

[5] Medaugh, R. S., and Griffiths, L. J., "A comparison of two fast linear predictors," in *Proc. Int. Conf. Acoust., Speech, and Signal Processing,* 1981, pp. 293–296.

[6] Cioffi, J. M., "Fast transversal filters for communication applications," Ph.D. dissertation, Stanford University, 1984.

[7] Widrow, B., and Stearns, S. D., *Adaptive Signal Processing*. Englewood Cliffs, NJ: Prentice-Hall, 1985.

[8] Honig, M. L., and Messerschmitt, D., *Adaptive Filters, Structures, Algorithms and Applications*. Boston, MA: Kulwer Academic Publishers, 1984.

[9] Haykin, S., *Introduction to Adaptive Filters*. New York, NY: MacMillan, 1984.

[10] Goodwin, G. C., and Sin, K. S., *Adaptive Filtering Prediction and Control*. Englewood Cliffs, NJ: Prentice-Hall, 1984.

[11] Alexander, T. S., *Adaptive Signal Processing, Theory and Applications*. New York: Springer Verlag, 1986.

[12] Ljung, and Söderström, T., *Theory and Practice of Recursive Identification*. Cambridge, MA: MIT Press, 1983.

[13] *Adaptive Systems in Control and Signal Processing*, I. D. Landau, M. Tomizuka, and D. M. Auslander, Eds. Oxford, UK: Pergamon Press, 1983.

[14] *Adaptive Systems in Control and Signal Processing,* K. J. Åström, Ed. Oxford, UK: Pergamon Press, 1986.

[15] Haykin, S., *Adaptive Filter Theory*. Englewood Cliffs, NJ: Prentice-Hall, 1986.

[16] Compton, R. T., Jr., *Adaptive Antennas: Concepts and Applications.* Englewood Cliffs, NJ: Prentice-Hall, 1988.

[17] Haykin, S., Ed., *Array Signal Processing*. Englewood Cliffs, NJ: Prentice-Hall, 1984.

[18] Cowan, C. F. N., and Grant, P. M., Eds. *Adaptive Filters*. Englewood Cliffs, NJ: Prentice-Hall, 1985.

G. Ungerboeck

Theory on the Speed of Convergence in Adaptive Equalizers for Digital Communication*

Abstract: This paper presents an analysis of the convergence properties of adaptive transversal equalizers minimizing mean-square distortion. The intention is to reveal the influence on the speed of convergence exerted by the number of taps, the step-size parameter in the adjustment loops, and the spectrum of the unequalized signal. Attention is focused on the convergence of the expected mean-square distortion. Several approximations are made in the analysis, among them the approximation of higher-order statistics by second-order statistical parameters. Comparison with results obtained by computer simulation, however, shows that the theory developed renders a quite accurate picture of the convergence process.

Previous work in this field demonstrated the limits set to the speed of convergence by the extreme values of the power spectrum of the unequalized signal. It is shown here that, with regard to the mean-square distortion, the influence of the number of taps will usually dominate by far. The theory provides a simple criterion for convergence and answers the question of how to attain the fastest convergence.

Introduction

Synchronous data transmission over the existing telephone network at speeds of several kbps requires equalization. Most equalizers presently used in modem receivers are of the transversal filter type. Various methods for adjusting their tap gains automatically have been described in the literature [1-9]. Basically there are two kinds of automatic adjustment processes. The first involves sending a series of isolated test pulses prior to data transmission. The equalizer settings derived in this initial "training" period are kept constant during the subsequent period of data transmission. The second is known as *adaptive equalization.* Here the equalizer settings are directly derived from the received data signal. Adaptive equalizers seek continuously to minimize the deviation of their sampled output signal from a quantized reference signal that resembles the transmitted pulse amplitudes. During the initial training period an ideal reference signal can be made available to the equalizer by transmitting a known sequence of pseudorandom data over the channel and by generating internally in the receiver an identical sequence in proper synchronism. When actual data are transmitted, the residual distortion has usually decreased to a small value. The equalizer can then use the reconstructed output signal of the receiver as a reference signal. In this so-called *decision-directed mode* the effect of false decisions is usually negligible. Thus the adaptation mechanism continues to be effective during the entire period that data are transmitted.

In many practical applications the transmitted messages are short. The start-up time (during which the receiver locks on to the carrier, establishes bit synchronization, and performs automatic equalization) may constitute a substantial portion of the total holding time. Particularly in multiparty polling systems, the start-up time of the individual receivers can seriously affect the line utilization. For the early automatic equalizers, settling times in the order of seconds have been reported. In the meanwhile great improvements have been achieved.

In this paper we shall present a theory on the convergence process in adaptive equalizers which employ the mean-square (MS) algorithm [3-7]. The aim of the analysis is to reveal the influence of various system parameters on the convergence process and thus to establish the limits of the speed of convergence.

In a similar investigation Gersho [5] considered the expected tap-gain errors relative to their optimum settings. He showed that the optimum speed at which these expectations may converge to zero is largely determined by the maximum and minimum values of the power density spectrum of the unequalized signal. Gersho's analysis suggests that the number of taps apparently has little influence on the convergence process. Similar results, but not for exactly the same equalizer, have been reported by Chang [8] and Kobayashi [9]. Gersho also considered the expected variance of the tap-gain errors and presented a general proof of convergence. Here we shall extend Gersho's analysis. Con-

*This paper is based in part on the author's presentation at the IEEE International Conference on Communications, Philadelphia, Pennsylvania, June 1972.

Reprinted with permission from *IBM J. Res. Develop.*, Nov. 1972.

sidering the mean-square distortion, we show that the number of taps is generally much more important than the extreme values of the power density spectrum. A criterion for stability is derived and the question of how to attain fastest convergence is addressed.

We first review the fundamentals of this type of equalizer. Later on, a theory on the convergence of the expected mean-square distortion is developed. The theoretical results are then compared with results obtained by computer simulation.

Review of fundamentals

The equalizer considered in this paper is depicted in Fig. 1. It consists basically of a linear transversal filter that transforms the input signal $x(t)$ into the output signal $z(t)$. The equalized signal is sampled at regular intervals T. As the input signal we consider a PAM base-band signal

$$x(t) = \sum_{n=-\infty}^{+\infty} a_n h(t-nT) + w(t), \tag{1}$$

where the $\{a_n\}$ are quantized pulse amplitudes, $h(t)$ is the channel response, T the baud interval, and $w(t)$ the additive noise. Let $\mathbf{c}_n$ be the vector of tap gains of the transversal filter, and $\mathbf{x}_n$ the vector of tap output signals, both at the nth sampling instant. These vectors are N-dimensional, N being the number of taps. Throughout the paper a prime (′) denotes transposition. At the nth sampling instant the output signal of the equalizer reads $z_n = \mathbf{c}_n'\mathbf{x}_n$. It deviates from the originally transmitted pulse amplitude by $e_n = z_n - a_n$. In this study we adopt the familiar mean-square distortion criterion. For a particular $\mathbf{c}_n$ the mean-square distortion is defined as the average value of e_n^2 over all possible pulse amplitude and noise sequences [3–9]:

$$\langle e_n^2 \rangle = \langle (\mathbf{c}_n'\mathbf{x}_n - a_n)^2 \rangle = \mathbf{c}_n'\mathbf{R}\mathbf{c}_n - 2\mathbf{c}_n'\mathbf{b} + \langle a_n^2 \rangle. \tag{2}$$

Here $\mathbf{R}$ is a positive definite $N \times N$ matrix with elements

$$r_{ik} = \langle x_{ni} \cdot x_{nk} \rangle, \qquad 1 \le i,k \le N, \tag{3}$$

and $\mathbf{b}$ denotes an N-dimensional vector with elements

$$b_i = \langle a_n \cdot x_{ni} \rangle, \qquad 1 \le i \le N. \tag{4}$$

$\mathbf{R}$ and $\mathbf{b}$ do not depend on n since we assume that both the sequence $\{a_n\}$ and the noise $w(t)$ are stationary.

The mean-square distortion assumes its minimum value, $\langle e_{\text{opt}}^2 \rangle$, when $\mathbf{c}_n$ is chosen equal to

$$\mathbf{c}_{\text{opt}} = \mathbf{R}^{-1}\mathbf{b}. \tag{5}$$

Introducing a tap-gain error vector

$$\mathbf{p}_n = \mathbf{c}_n - \mathbf{c}_{\text{opt}} \tag{6}$$

we have

$$\langle e_n^2 \rangle = \mathbf{p}_n'\mathbf{R}\mathbf{p}_n + \langle e_{\text{opt}}^2 \rangle. \tag{7}$$

The MS algorithm [3–7], the convergence properties of which will be investigated in this paper, is intended to minimize the mean-square distortion. Note that this is equivalent to minimizing the positive definite quadratic term $\mathbf{p}_n'\mathbf{R}\mathbf{p}_n$ ("excess mean-square distortion"). New values are assigned to $\mathbf{p}_n$, and hence to $\mathbf{c}_n$, at each sampling instant by estimating the gradient $\partial\langle e_n^2 \rangle / \partial \mathbf{p}_n$ and modifying $\mathbf{p}_n$ accordingly. The MS algorithm results from taking $\partial e_n^2 / \partial \mathbf{p}_n = 2\, e_n \mathbf{x}_n$ as an unbiased estimate of $\partial\langle e_n^2 \rangle / \partial \mathbf{p}_n$. This leads, then, to the iterative formula

$$\mathbf{p}_{n+1} = \mathbf{p}_n - \alpha_{(n)} e_n \mathbf{x}_n. \tag{8}$$

Defining $e_{n\text{opt}} = \mathbf{c}_{\text{opt}}'\mathbf{x}_n - a_n$ we have

$$e_n = e_{n\text{opt}} + \mathbf{p}_n'\mathbf{x}_n. \tag{9}$$

Note that

$$\langle e_{n\text{opt}} \cdot \mathbf{x}_n \rangle = \mathbf{0}. \tag{10}$$

Fig. 1 shows the implementation of the algorithm.

The step-size parameter $\alpha_{(n)}$ may vary as a function of time. To begin with we shall consider a constant step-size parameter. For $\alpha > 0$ and sufficiently small, $\mathbf{p}_n$ converges in the mean towards $\mathbf{0}$ from arbitrary initial settings $\mathbf{p}_0$ because of the positive definite quadratic nature of the excess mean-square distortion. Noise and the finiteness of the number of taps make it impossible to attain zero mean-square distortion. Thus, in the steady state, finite output errors e_n cause $\mathbf{p}_n$ to fluctuate randomly about $\mathbf{0}$—with zero mean, but finite variance.

Our major concern in the analysis will be devoted to the speed of convergence of the expected mean-square distortion, denoted by $E(\langle e_n^2 \rangle)$. This quantity represents the ensemble mean value of $\langle e_n^2 \rangle$, subject to averaging $\langle e_n^2 \rangle$ over $\mathbf{p}_n$. $E(\langle e_n^2 \rangle)$ converges towards $\langle e_{\text{opt}}^2 \rangle$ as $\mathbf{p}_n$ converges towards $\mathbf{0}$, but because of the finite variance of $\mathbf{p}_n$ in the steady state, it will settle at a value greater than $\langle e_{\text{opt}}^2 \rangle$.

Analysis of the convergence process

We shall first introduce a transformation that considerably facilitates further analysis. The convergence properties of $E(\mathbf{p}_n)$ can then easily be examined. In the remainder of this section a theory on the convergence of $E(\langle e_n^2 \rangle)$ will be developed.

• *Coordinate transformation*

Since $\mathbf{R}$ is symmetric it can be represented in the form

$$\mathbf{R} = \mathbf{U}\,\text{Diag}\,(\boldsymbol{\rho})\,\mathbf{U}'. \tag{11}$$

($\boldsymbol{\rho}$ is the vector of the eigenvalues of $\mathbf{R}$; $\rho_i > 0$, $1 \le i \le N$, since $\mathbf{R}$ is positive definite; $\mathbf{U}$ is the unitary matrix whose

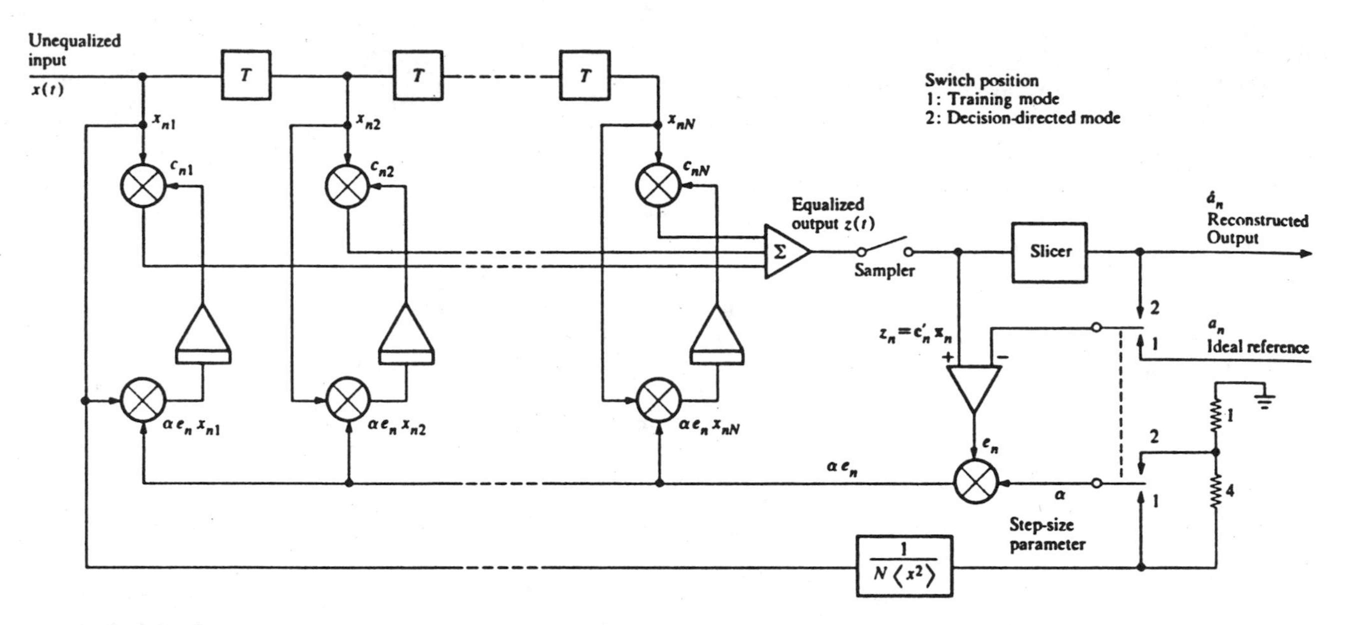

Figure 1 Adaptive transversal equalizer employing the MS algorithm with controlled step-size parameter.

ith column is the eigenvector $\mathbf{u}_i$ of $\mathbf{R}$, associated with ρ_i.) We now introduce

$$\mathbf{y}_n = \mathbf{U}'\mathbf{x}_n, \tag{12}$$

and

$$\mathbf{q}_n = \mathbf{U}'\mathbf{p}_n. \tag{13}$$

This transformation is equivalent to a rotation of the coordinate system. The elements of the modified tap-gain vector $\mathbf{y}_n$ are uncorrelated:

$$\begin{aligned}\langle y_{ni} \cdot y_{nk}\rangle &= 0, \qquad i \neq k\\ &= \rho_i, \qquad i = k.\end{aligned} \tag{14}$$

Multiplication of (8) by $\mathbf{U}'$ from the left yields

$$\mathbf{q}_{n+1} = \mathbf{q}_n - \alpha e_n \mathbf{y}_n. \tag{15}$$

Similarly, we obtain from (9) and (10),

$$e_n = e_{n\text{opt}} + \mathbf{q}_n'\mathbf{y}_n \tag{16}$$

and

$$\langle e_{n\text{opt}} \cdot \mathbf{y}_n\rangle = 0. \tag{17}$$

• *Convergence properties of* $E(\mathbf{p}_n)$ *and* $E(\mathbf{q}_n)$
From (15), (16), and (17) it follows that

$$E(\mathbf{q}_{n+1}) = E(\mathbf{q}_n) - \alpha\, E[(\mathbf{q}_n'\mathbf{y}_n)\, \mathbf{y}_n]. \tag{18}$$

In order to facilitate further mathematical treatment, $\mathbf{p}_n$ and $\mathbf{x}_n$ are assumed to be statistically independent of each other. The same applies then to $\mathbf{q}_n$ and $\mathbf{y}_n$ and thus $E(\mathbf{q}_n)$ can be extracted from the rightmost term in (18). Since $\mathbf{p}_n$ depends on $\mathbf{x}_{n-1}, \mathbf{x}_{n-2}, \cdots$, and $\mathbf{x}_n$ merely comprises the tap output signals after another baud interval ($x_{n(i+1)} = x_{(n-1)i}$, $1 \leq i < N$), the assumption is not strictly true. In view of small step-size parameters, however, Gersho [5] felt that the dependence between $\mathbf{p}_n$ and $\mathbf{x}_n$ is weak and can therefore be neglected. With this assumption, (18) assumes the form

$$E(\mathbf{q}_{n+1}) = \text{Diag}\,(1 - \alpha\rho)\, E(\mathbf{q}_n). \tag{19}$$

It should be mentioned here that in a recent paper [11] an attempt was made to include the dependency between $\mathbf{p}_n$ and $\mathbf{x}_n$ in the analysis.

Several authors [5,9] have shown that the eigenvalues of $\mathbf{R}$ are bounded by

$$\frac{1}{T}\,\text{Inf}\, P^*(\omega) < \rho_i < \frac{1}{T}\,\text{Sup}\, P^*(\omega), \quad 1 \leq i \leq N, \tag{20}$$

where $P^*(\omega)$ represents the periodic power density spectrum of the sampled unequalized signal: $P^*(\omega) = P^*[\omega + (2\pi/T)]$. The extreme eigenvalues approach the bounds as N goes to infinity.

Let $\rho_{\min}$ and $\rho_{\max}$ denote the smallest and the largest eigenvalues of $\mathbf{R}$. From (19) it follows that $E(\mathbf{q}_n)$ converges to 0 if

$$0 < \alpha < 2/\rho_{\max}. \tag{21}$$

Because of (13), the same applies to $E(\mathbf{p}_n)$. Gersho [5] has shown that for $\alpha = 2/(\rho_{\min} + \rho_{\max})$ fastest convergence takes place. The Euclidian norm of $E(\mathbf{q}_n)$, which is equal to the Euclidian norm of $E(\mathbf{p}_n)$, is then reduced at least by the factor $(\rho_{\max} - \rho_{\min})/(\rho_{\max} + \rho_{\min})$ in each iteration. Since $\rho_{\min}$ and $\rho_{\max}$ resemble in good approximation the extreme values of $P^*(\omega)$, a direct relationship between $P^*(\omega)$ and the optimum speed of convergence of $E(\mathbf{q}_n)$ results.

One might suspect that $\rho_{\min}$ and $\rho_{\max}$ determine to a similar extent also the optimum speed of convergence of the expected mean-square distortion, $E(\langle e_n^2\rangle)$. Later in this paper, however, we shall see that the convergence properties of $E(\langle e_n^2\rangle)$ depend, unlike $E(\mathbf{q}_n)$, also on the number of taps, N. In fact, for most practical cases N will be the dominating factor which, for convergence of $E(\langle e_n^2\rangle)$, imposes a condition much tighter than (21) on the values of α. There exist values of α for which $E(\mathbf{q}_n)$ converges, but $E(\langle e_n^2\rangle)$ diverges. This property suggests that, if both quantities converge, $E(\mathbf{q}_n)$ will generally converge much faster than $E(\langle e_n^2\rangle)$. In the following analysis we may therefore assume that $E(\mathbf{q}_n)$ becomes rapidly negligible during the equalization process if it were not already zero from the beginning, i.e.,

$$E(\mathbf{q}_n) \approx \mathbf{0}. \qquad (22)$$

• *Convergence properties of* $E(\langle e_n^2\rangle)$

Equation (13) enables us to decompose $\mathbf{p}_n'\mathbf{R}\mathbf{p}_n$ into N components. Actually we are interested in the expectation thereof:

$$E(\mathbf{p}_n'\mathbf{R}\mathbf{p}_n) = \sum_{i=1}^{N} \rho_i E(q_{ni}^2) = \rho' \mathbf{s}_n. \qquad (23)$$

Using (22) and the assumption that $\mathbf{q}_n$ and $\mathbf{y}_n$ were statistically independent of each other, and making some further approximations, we show in the Appendix that

$$\mathbf{s}_{n+1} \simeq \mathbf{A}\mathbf{s}_n + \alpha^2 \langle e_{\rm opt}^2\rangle \rho, \qquad (24)$$

where

$$\mathbf{A} = \begin{bmatrix} (1-\alpha\rho_1)^2 & \alpha^2\rho_1\rho_2 & \cdots & \alpha^2\rho_1\rho_N \\ \alpha^2\rho_2\rho_1 & (1-\alpha\rho_2)^2 & \cdots & \alpha^2\rho_2\rho_N \\ \vdots & & & \vdots \\ \alpha^2\rho_N\rho_1 & & & (1-\alpha\rho_N)^2 \end{bmatrix} \qquad (25)$$

The matrix $\mathbf{A}$ is symmetric and its elements are all positive. The matrix, however, is not necessarily positive definite. Similarily to (11), we introduce

$$\mathbf{A} = \mathbf{V}\ \mathrm{Diag}\ (\lambda)\ \mathbf{V}'. \qquad (26)$$

(λ is the vector of the eigenvalues of $\mathbf{A}$; $\mathbf{V}$ is the unitary matrix whose ith column is the eigenvector $\mathbf{v}_i$ of $\mathbf{A}$, associated with λ_i.) Let $\langle x^2\rangle$ denote the mean-square value of the tap output signals. Using the relation

$$N\langle x^2\rangle = \mathrm{trace}\ \mathbf{R} = \sum_{i=1}^{N} \rho_i, \qquad (27)$$

it can be shown that the solution of (24) reads

$$\mathbf{s}_n \simeq \sum_{i=1}^{N} \gamma_i \lambda_i^n \mathbf{v}_i + \frac{\alpha\langle e_{\rm opt}^2\rangle}{(2-\alpha N\langle x^2\rangle)} \mathbf{1}, \qquad (28)$$

where γ_i is determined by the initial conditions

$$\gamma_i = \mathbf{v}_i'\left(\mathbf{s}_0 - \frac{\alpha\langle e_{\rm opt}^2\rangle}{(2-\alpha N\langle x^2\rangle)}\mathbf{1}\right), \qquad 1 \le i \le N,$$

$$s_{0i} = E(q_{0i}^2) = E[(\mathbf{u}_i'\mathbf{p}_0)^2], \qquad 1 \le i \le N.$$

Substituting (28) into (23) and observing (7) we finally obtain

$$E(\langle e_n^2\rangle) \simeq \sum_{i=1}^{N} \delta_i \lambda_i^n + \frac{2\langle e_{\rm opt}^2\rangle}{(2-\alpha N\langle x^2\rangle)}, \qquad (29)$$

where

$$\delta_i = (\mathbf{v}_i'\rho)\cdot\gamma_i. \qquad (30)$$

In (29) the first term on the right-hand side describes the transient behavior of $E(\langle e_n^2\rangle)$, whereas the second term represents its steady-state value.

• *Transient behavior of* $E(\langle e_n^2\rangle)$

In investigating the properties of $\mathbf{A}$ we shall be able to make some observations concerning the transient behavior of $E(\langle e_n^2\rangle)$. Among them we present a new criterion for stability. We also indicate that a spread of the eigenvalues of $\mathbf{R}$ has not the strong influence on the speed of convergence of $E(\langle e_n^2\rangle)$ that might be expected from considering the convergence properties of $E(\mathbf{q}_n)$. Furthermore, for all eigenvalues of $\mathbf{R}$ being equal we show that only the largest eigenvalue of $\mathbf{A}$ determines the speed of convergence of $E(\langle e_n^2\rangle)$:

a) All eigenvalues of $\mathbf{A}$ are real numbers since $\mathbf{A}$ is symmetric. Hence, the transient of $E(\langle e_n^2\rangle)$ will exhibit no oscillations.
b) For $\alpha \to 0$ all eigenvalues of $\mathbf{A}$ approach unity.
c) The equalizer is stable and the expected mean-square distortion converges to a steady state if $|\lambda_i| < 1$, $1 \le i \le N$. This will be the case if α satisfies

$$0 < \alpha < 2/N\langle x^2\rangle = 2/\sum_{i=1}^{N}\rho_i. \qquad (31)$$

Proof: The N elements of the ith row of $\mathbf{A}$ add up to

$$\sum_{k=1}^{N} a_{ik} = 1 - \alpha\rho_i(2 - \alpha N\langle x^2\rangle).$$

If α satisfies (31), then each row sum of $\mathbf{A}$ is smaller than unity. A matrix which has this property and whose elements are all positive can have only eigenvalues with absolute value smaller than unity [10].

The criterion for stability thus found imposes a much narrower upper bound on α than (21). It clearly exhibits the significance of the number of taps while showing no dependence on the distribution of the eigenvalues of **R**. We recall that fastest convergence of $E(\mathbf{q}_n)$ takes place for $\alpha = 2/(\rho_{min} + \rho_{max})$. The new criterion indicates that with this step-size parameter $E(\langle e_n^2 \rangle)$ would diverge, provided $N > 2$. Since $E(\langle e_n^2 \rangle)$ is closely related to the probability of errors, (31) must be considered as a necessary condition. Intuitively this dependence on the number of taps could be expected, since for a given step-size parameter each additional tap increases through its tap-gain fluctuations the expected excess mean-square distortion, $E(\mathbf{p}_n' \mathbf{R} \mathbf{p}_n)$. Expansion of the number of taps without decreasing the step-size parameter must therefore lead to instability.

d) A small eigenvalue of **R** ($\rho_i \to 0$) leads to a slowly converging term in (29), ($\lambda_i \to 1$). But the slower the term converges relative to the other terms, the smaller the probability that this term contributes significantly to $E(\langle e_n^2 \rangle)$, ($\delta_i \to 0$).

Proof: For $\rho_i = 0$ the ith row of **A** reads $\{0, \cdots, 0, a_{ii} = 1, \cdots, 0\}$. Consequently, $\lambda_i = 1$ and $\mathbf{v}_i' = \{0, \cdots, 0, y_{ii} = 1, 0, \cdots, 0\}$. Since $\mathbf{v}_i' \boldsymbol{\rho} = 0$ it follows from (30) that $\delta_i = 0$.

Generally, it is true that a larger spread of the eigenvalues of **R** leads to slower convergence. But the fact that the slower-converging terms in (29) are usually given smaller weights acts to alleviate the effect. Thus a spread of the eigenvalues of **R** affects the convergence of $E(\langle e_n^2 \rangle)$ less than the convergence of $E(\mathbf{q}_n)$.

e) For all eigenvalues of **R** being equal, i.e., $\rho_i = \langle x^2 \rangle$, $1 \le i \le N$, the largest eigenvalue of **A** is given by

$$\lambda_{imax} = 1 - \alpha \langle x^2 \rangle (2 - \alpha N \langle x^2 \rangle) . \qquad (32)$$

The other eigenvalues of **A** have no influence on the transient behavior of $E(\langle e_n^2 \rangle)$ since $\delta_i = 0$, $i \ne i_{max}$.

Proof: It can easily be verified that λ_{imax} is an eigenvalue of **A** and that $\mathbf{v}_{imax}' = N^{-1/2} \{1, 1, \cdots, 1\}$ represents the associated eigenvector. It follows from the Perron-Frobenius theorem [10] on positive matrices that λ_{imax} is indeed the largest eigenvalue of the positive matrix **A**. The theorem says that the largest eigenvalue of a positive matrix is a positive real number and the associated eigenvector consists entirely of positive elements. Because the eigenvectors of **A** form a set of orthogonal vectors, only one eigenvector can have this property. Since $\mathbf{v}_{imax}$ consists entirely of positive elements, λ_{imax} must be the largest eigenvalue of **A**. Since $\mathbf{v}_{imax}$ is parallel to $\boldsymbol{\rho}$, the other eigenvectors of **A** are orthogonal to $\boldsymbol{\rho}$. Consequently, $\delta_i = 0$, $i \ne i_{max}$.

- *The steady state*

The rightmost term of (29) reveals a simple relationship between α and the steady-state value of $E(\langle e_n^2 \rangle)$. We again see that a steady state exists only if α satisfies (31). Equation (28) indicated that in the steady state all elements of $\mathbf{s}_n$ become equal. Taking into account $E(q_{ni} q_{nk}) \to 0$, $i \ne k$, as shown in the Appendix, we find that in the steady state the tap gains fluctuate with equal variance but in an uncorrelated fashion about their optimum settings.

- *Optimum speed of convergence for all eigenvalues of* **R** *being equal*

Data communication over telephone channels suffers generally more from phase distortion than from amplitude distortion. If the modulation scheme provides for a flat amplitude characteristic and the spectrum of the transmitted signal is not shaped by coding techniques, the eigenvalues of **R** will be clustered closely about $\langle x^2 \rangle$. Let us assume $\rho_i = \langle x^2 \rangle$, $1 \le i \le N$. It follows then from what has been stated in the discussion of (32), and from (29)

$$E(\langle e_{n+1}^2 \rangle) \simeq [1 - \alpha_n \langle x^2 \rangle (2 - \alpha_n N \langle x^2 \rangle)] \cdot E(\langle e_n^2 \rangle) + 2\alpha_n \langle x^2 \rangle \langle e_{opt}^2 \rangle . \qquad (33)$$

Equation (33) is written with a time-dependent step-size parameter. It can easily be verified that

$$\alpha_{nopt} = \frac{1}{N \langle x^2 \rangle} \cdot \frac{E(\langle e_n^2 \rangle) - \langle e_{opt}^2 \rangle}{E(\langle e_n^2 \rangle)} \qquad (34)$$

leads to fastest convergence.

Usually, $E(\langle e_n^2 \rangle) \gg \langle e_{opt}^2 \rangle$ at the beginning of the equalization process. Thus we have $\alpha_{nopt} \approx 1/N \langle x^2 \rangle$ and

$$E(\langle e_{n+1}^2 \rangle) \approx (1 - 1/N) \, E(\langle e_n^2 \rangle) . \qquad (35)$$

Approximately 2.3 N iterations are then required to reduce $E(\langle e_n^2 \rangle)$ by one order of magnitude.

Since $\langle e_{opt}^2 \rangle$ is generally unknown and estimation of $E(\langle e_n^2 \rangle)$ is time-consuming, the optimum step-size parameters given by (34) cannot be realized exactly. But the optimum trajectory of $E(\langle e_n^2 \rangle)$ can be closely approached if α is controlled in the following simple manner:

a) Measure $\langle x^2 \rangle$.

b) Use $\alpha = 1/N \langle x^2 \rangle$ during the entire training period. $E(\langle e_n^2 \rangle)$ converges towards $2\langle e_{opt}^2 \rangle$.

c) Reduce the step-size parameter to $\alpha = 1/5N \langle x^2 \rangle$ when the equalizer is switched into the decision-directed mode. $E(\langle e_n^2 \rangle)$ converges further towards $1.1 \langle e_{opt}^2 \rangle$ ($\langle e_{opt}^2 \rangle$ + 0.5 dB).

A step-wise reduction of the step-size parameter was already proposed by Lucky in his first paper on automatic equalization [1]. It is, however, still surprising to see how closely the optimum trajectory of $E(\langle e_n^2 \rangle)$ is approached by the simple two-step procedure suggested above. Figure 2 shows the comparison. The implementation of the procedure is indicated in Fig. 1. In practice modems are equipped with automatic gain control. If therewith $\langle x^2 \rangle$ is kept sufficiently constant, no further estimation of $\langle x^2 \rangle$ is required and the division by a variable $\langle x^2 \rangle$ in determining α is unnecessary.

The procedure proposed is also applicable when the eigenvalues of **R** are spread out over a rather wide range. This will be demonstrated in the following section by computer simulation.

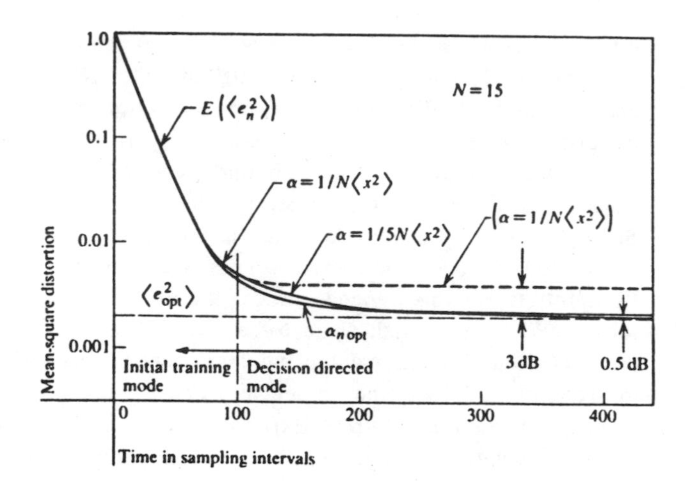

Figure 2 Speed of convergence with optimum sequence of step-size parameters ($\alpha_{n\,opt}$) and with the step-size parameters proposed ($\alpha = 1/N\langle x^2 \rangle$ and $\alpha = 1/5N\langle x^2 \rangle$).

Computer simulation

Various approximations had to be made in the theoretical analysis. We shall now check the validity of the theory by comparing the theoretical results with those obtained by computer simulation. The investigation was based on the following model.

A random sequence of polar binary-signals ($a_n = \pm 1$) is transmitted over a telephone channel at the speed of 3600 baud. Vestigal-sideband amplitude modulation is used with the carrier located at 2.7 kHz. The transmitter filter exhibits symmetrical cosine-roll-off characteristics with 6-dB points at 0.9 and 2.7 kHz. Three telephone channels with characteristics shown in Fig. 3 are considered. A signal-to-noise ratio of 30 dB caused by white Gaussian noise is assumed. The equalizer comprises $N = 15$ taps. Initially, the tap gains exhibit zero values. Thus we have $\mathbf{p}_0 = -\mathbf{c}_{opt}$ and $\langle e_0^2 \rangle = 1$. An ideal reference signal is assumed to be available in proper phase to the equalizer.

Figure 3 Telephone channel characteristics.

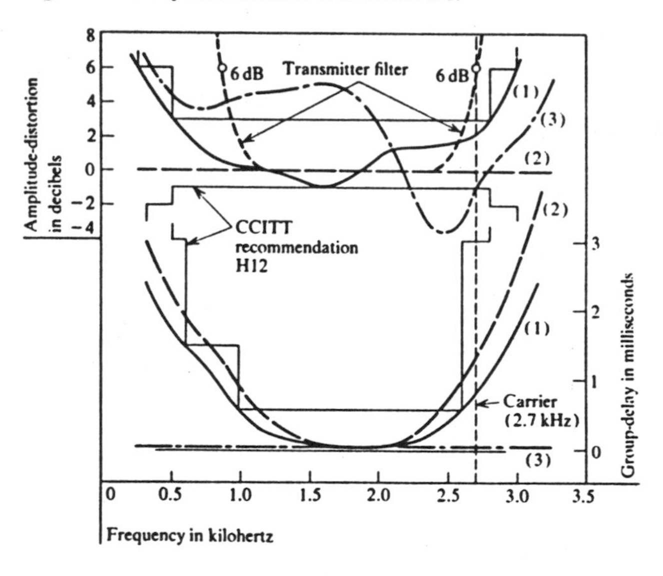

Two programs have been written. The first program calculates the sample values of the waveform $h(t)$ for the modulation scheme envisaged and a given telephone channel. The second program determines **R**, **b**, $\mathbf{e}_{opt}$, the eigenvalues of **R**, etc., and finds the theoretical values of $E(\langle e_n^2 \rangle)$ by iteratively applying (24). Furthermore, it generates a random data signal, adds noise to it, simulates the equalizer, and calculates $\langle e_n^2 \rangle$ at each sampling instant by evaluating (7).

At first we consider the results obtained for telephone channel-characteristic (1) (moderate amplitude and phase distortion). A step-size parameter $\alpha = 1/N\langle x^2 \rangle$ was chosen. The results of five program runs with different initializations of the random number source are presented in Fig. 4(a). Fairly good agreement of the theoretical and simulation results can be observed. On the average, however, the mean-square distortion obtained by simulation appears to converge slightly faster than is theoretically predicted. Looking for a reason, we found that this deviation can mainly be attributed to the assumption of statistical independence between $\mathbf{p}_n$ and $\mathbf{x}_n$ (equivalently $\mathbf{q}_n$ and $\mathbf{y}_n$). When additional baud intervals were introduced in the simulation between those sampling instants where tap-gain corrections are made, successive tap output signals were forced to be quasi-statistically independent of one another. In this way, without counting the additional baud intervals, a much better agreement between theory and simulation was obtained, as indicated in Fig. 4(b).

Further simulations with various step-size parameters were performed for the channel-characteristics (2) and (3) presented in Fig. 3. In order to obtain equal eigenvalues of **R** [$P^*(\omega)$ constant] with channel-characteristic (2) (phase distortion only), a transmitter filter with ideal bandpass filter characteristic had to be assumed, since otherwise aliasing would have converted phase

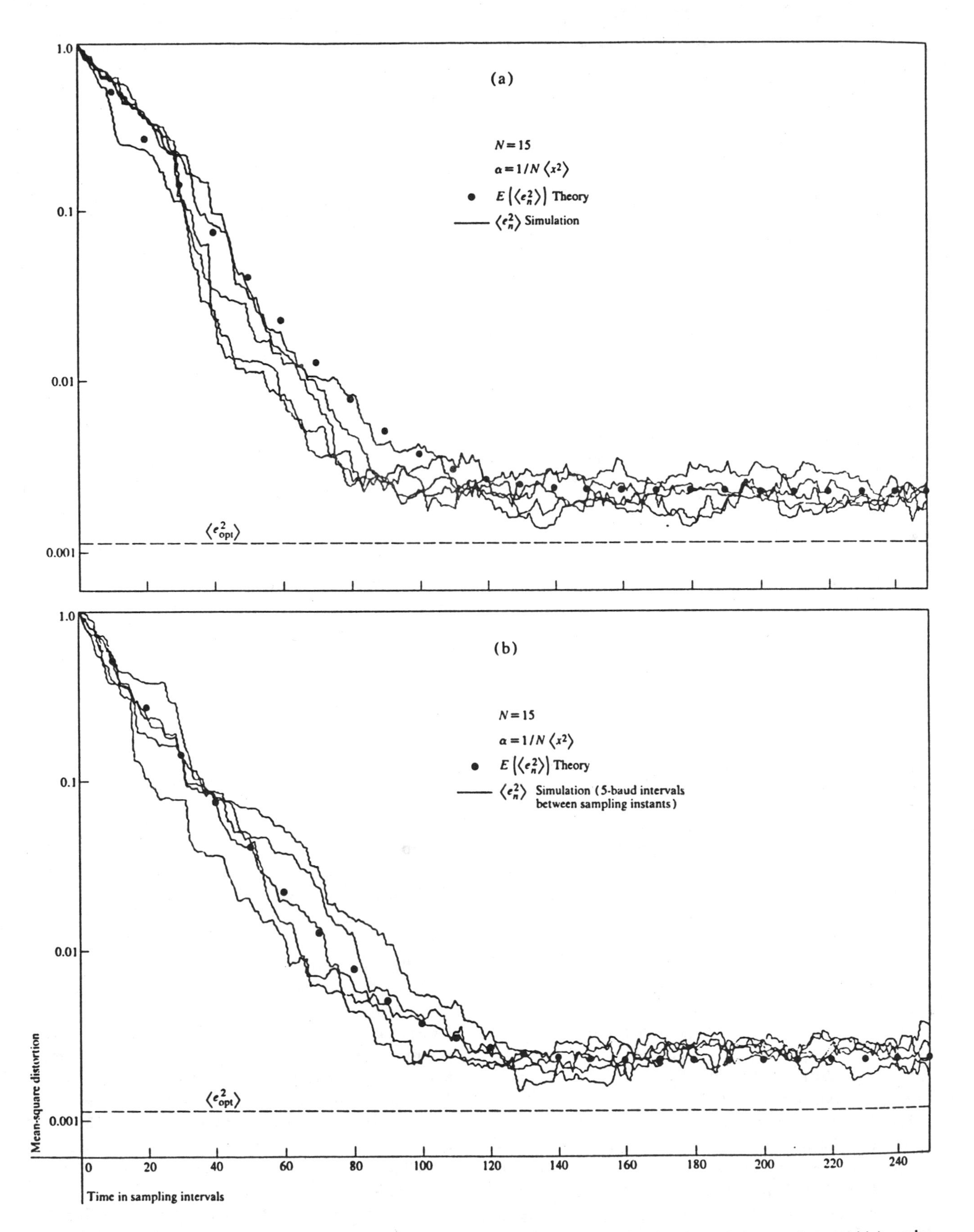

Figure 4 Theoretically predicted convergence and results obtained by computer simulation for channel-characteristic (1). (a) regular simulation; (b) additional baud intervals introduced between sampling instants.

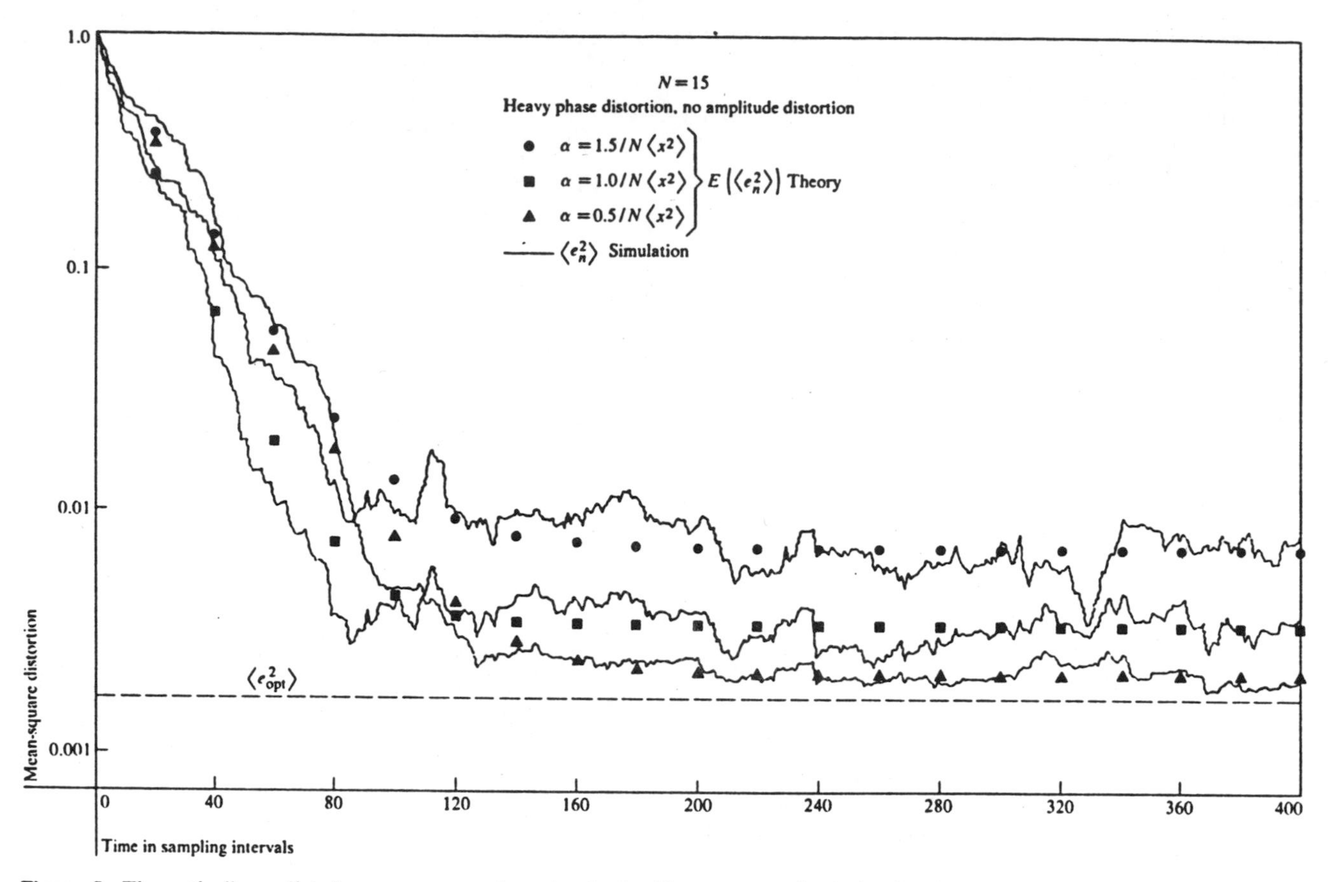

Figure 5 Theoretically predicted convergence and results obtained by computer simulation for channel-characteristic (2).

distortion into amplitude distortion. Figure 5 shows the results obtained for channel-characteristic (2). Corresponding results for channel-characteristic (3) (amplitude distortion only) are presented in Fig. 6. On the whole, the results confirm the validity of the theory, but slightly faster convergence than predicted is consistently obtained.

For channel-characteristic (3) the largest and the smallest eigenvalues of **R** differ approximately by the factor 10. This large eigenvalue spread, however, reduces the speed of convergence only by a factor of approximately 2, relative to the speed of convergence with channel-characteristic (2). Our theoretical finding that a spread of the eigenvalues of **R** affects the speed of convergence of $E(\langle e_n^2\rangle)$ less than the speed of convergence of $E(\mathbf{q}_n)$ is thus corroborated.

Figure 6 indicates that for a large spread of the eigenvalues of **R** instability occurs at a value of α smaller than $2/N\langle x^2\rangle$. In the example $\alpha = 1.5/N\langle x^2\rangle$ is close to the actual limit of stability. In this respect our theory fails for large-amplitude distortion. The discrepancy is again largely due to the assumption of statistical independence between $\mathbf{p}_n$ and $\mathbf{x}_n$.

The curves presented in Figs. 5 and 6 illustrate that in the initial phase fastest convergence is in both cases achieved by a step-size parameter close to $1/N\langle x^2\rangle$. The speed of convergence does not appear to be very sensitive to variations of α about this value. The procedure proposed at the end of the previous section for controlling α is therefore also applicable for channels that exhibit considerable amplitude distortion.

Summary and conclusion

A theory has been presented on the convergence of the expected mean-square distortion at the output of adaptive transversal equalizers that employ the well-known MS algorithm. Several approximations had to be made in the analysis, but simulation results show that quite an accurate picture of the convergence process can nevertheless be developed. The assumption of statistical independence between the tap output signals at successive sampling instants turned out to be the weakest of the approximations made.

Previous work in the field emphasized the influence of the relative difference between the largest and the smallest eigenvalue of **R** on the speed of convergence. In

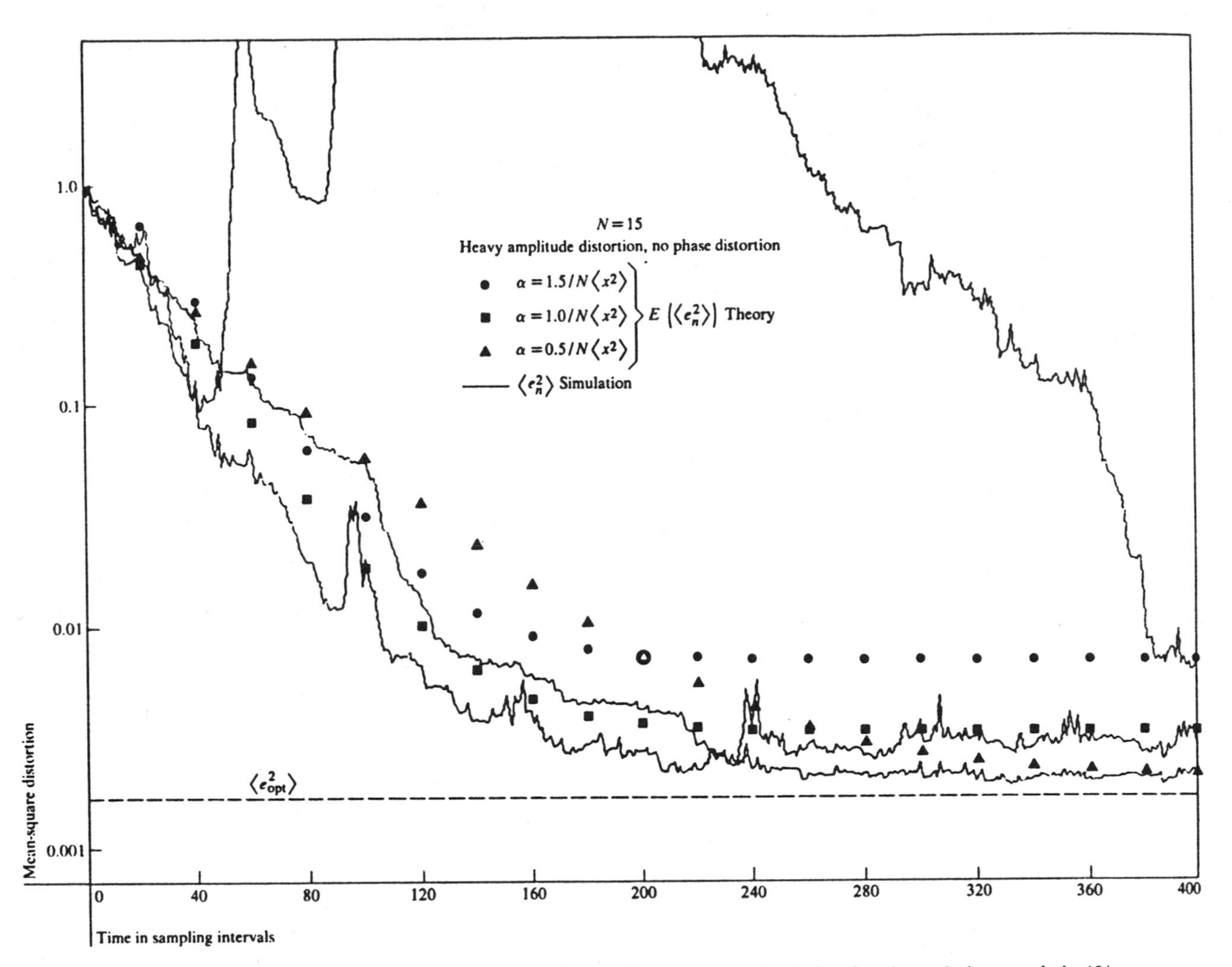

Figure 6 Theoretically predicted convergence and results obtained by computer simulation for channel-characteristic (3).

adaptive equalizers which employ the MS algorithm, these earlier results apply only to the expected tap-gain error vector, $E(\mathbf{p}_n)$. The convergence speed of the more important expected mean-square distortion, $E(\langle e_n^2\rangle)$, depends to a large extent on the number of taps: The larger N, the slower the speed of convergence. The eigenvalues of $\mathbf{R}$ have some influence on the speed of convergence of $E(\langle e_n^2\rangle)$, but this influence is not as distinct as in the case of $E(\mathbf{p}_n)$.

We have suggested a simple two-step procedure for controlling the step-size parameter in order to achieve fast convergence of the expected mean-square distortion. The true goal, however, should be to reduce as quickly as possible the expected probability of false decisions. This probability depends not only on the expectation of the mean-square distortion, but also to a certain extent on its variance. The analysis in this paper is limited in that it provides only the expectation of the mean-square distortion. It is, however, obvious that the variance of the mean-square distortion decreases monotonically with the step-size parameter. In this respect step-size parameters slightly smaller than we have proposed from the viewpoint of the expected mean-square distortion alone might be preferable.

We finally illustrate the theoretical results by a specific example. Assume an equalizer comprising 15 taps and a transmission speed of 3600 baud, as we did for the computer simulation. During the settling time the mean-square distortion should be reduced from 1 to 0.001, provided it does not level off at a larger value, i.e., $\langle e_{opt}^2\rangle > 0.001$. According to (35), with phase distortion only and the step-size parameter optimally adjusted, the equalizer settles in about 100 baud intervals, or 28 milliseconds. Moderate amplitude distortion will have no strong effect. With characteristic (1) of Fig. 3 the optimum settling time is still of the order of 30 milliseconds. For completeness, it should be noted that this does not include the additional time required for carrier acquisition, sampling clock adjustment, and synchronization with a known reference sequence.

Appendix

• *Derivation of Eq. (24)*

Analysis of the convergence properties of $s_{ni} = E(q_{ni}^2)$, $1 \leq i \leq N$, requires that we consider initially also the mixed quantities $E(q_{ni} \cdot q_{nk})$, $i \neq k$. Assume $\mathbf{y}_n$ and $\mathbf{q}_n$ were statistically independent of each other. From (15), observing (14), (16) and (17), it can then be shown that

$$E(q_{(n+1)i} \cdot q_{(n+1)k}) = E(q_{ni} \cdot q_{nk}) \cdot [1 - \alpha(\rho_i + \rho_k)]$$
$$+ \alpha^2 \sum_{l_1=1}^{N} \sum_{l_2=1}^{N} E(q_{nl_1} \cdot q_{nl_2}) \langle y_{ni} \cdot y_{nl_1} \cdot y_{nl_2} \cdot y_{nk} \rangle$$
$$+ 2\alpha^2 \sum_{l=1}^{N} E(q_{nl}) \langle e_{n\text{opt}} \cdot y_{ni} \cdot y_{nl} \cdot y_{nk} \rangle$$
$$+ \alpha^2 \langle e_{n\text{opt}}^2 \cdot y_{ni} \cdot y_{nk} \rangle, \quad 1 \leq i, k \leq N. \quad \text{(A1)}$$

According to (14) and (17) the quantities $y_{n1}, y_{n2}, \cdots, y_{nN}$, and $e_{n\text{opt}}$ are uncorrelated. Suppose they are almost statistically independent. Then

$$\langle e_{n\text{opt}} \cdot y_{ni} \cdot y_{nl} \cdot y_{nk} \rangle \approx 0.$$

Referring to (22),

$$E(q_{nl}) \approx 0, \qquad 1 \leq l \leq N.$$

Hence, the third line of (A1) consists of products of small quantities and can thus be neglected.

Similarly for $i \neq k$,

$$\langle y_{ni} \cdot y_{nl_1} \cdot y_{nl_2} \cdot y_{nk} \rangle \approx 0,$$
$$\langle e_{n\text{opt}}^2 \cdot y_{ni} \cdot y_{nk} \rangle \approx 0.$$

Consequently, for $i \neq k$ only the first line of (A1) is important. Convergence of $E(q_{ni} \cdot q_{nk})$ takes place if $0 < \alpha < 2/(\rho_i + \rho_k)$. We may therefore assume that these mixed terms become rapidly negligible during the equalization process if they were not already negligible from the beginning:

$$E(q_{ni} \cdot q_{nk}) \approx 0, \qquad i \neq k.$$

When this is applied to the second line of (A1), then for $l_1 \neq l_2$ we have products of two small quantities. Neglecting these products reduces the double sum to a simple summation. Considering now only the case $i = k$ we find

$$s_{(n+1)i} \simeq s_{ni}(1 - 2\alpha\rho_i) + \alpha^2 \sum_{l=1}^{N} s_{nl} \langle y_{nl}^2 \cdot y_{ni}^2 \rangle + \alpha^2 \langle e_{n\text{opt}}^2 \cdot y_{ni}^2 \rangle, \quad 1 \leq i \leq N. \quad \text{(A2)}$$

We now approximate the higher-order expectations by second-order statistical parameters

$$\langle y_{nl}^2 \cdot y_{ni}^2 \rangle \simeq \rho_l \rho_i, \qquad 1 \leq l, i \leq N, \quad \text{(A3)}$$
$$\langle e_{n\text{opt}}^2 \cdot y_{ni}^2 \rangle \simeq \langle e_{\text{opt}}^2 \rangle \rho_i, \qquad 1 \leq i \leq N. \quad \text{(A4)}$$

With these approximations (A2) reads

$$s_{(n+1)i} \simeq s_{ni}(1 - \alpha\rho_i)^2 + \alpha^2 \sum_{\substack{l=1 \\ l \neq i}}^{N} s_{nl} \rho_l \rho_i + \alpha^2 \langle e_{\text{opt}}^2 \rangle \rho_i, \qquad 1 \leq i \leq N. \quad \text{(A5)}$$

Writing (A5) in vector form, we finally obtain (24).

References

1. R. W. Lucky, "Automatic Equalization for Digital Communication," *Bell System Tech. J.* **44**, 547 (1965).
2. R. W. Lucky, "Techniques for Adaptive Equalization of Digital Communication Systems," *Bell System Tech. J.* **45**, 255 (1966).
3. R. W. Lucky and H. R. Rudin, "An Automatic Equalizer for General-Purpose Communication Channels," *Bell System Tech. J.* **46**, 2179 (1967).
4. R. W. Lucky, J. Salz and E. J. Weldon Jr., *Principles of Data Communication*, McGraw-Hill, Book Co., Inc., New York, 1968.
5. A. Gersho, "Adaptive Equalization of Highly Dispersive Channels for Data Transmission," *Bell System Tech. J.* **48**, 55 (1969).
6. D. Hirsch and W. J. Wolf, "A Simple Adaptive Equalizer for Efficient Data Transmission," *IEEE Trans. Commun. Technol.* **Com-18**, 5 (1970).
7. K. Möhrmann, "Einige Verfahren zur adaptiven Einstellung von Entzerrern für die schnelle Datenübertragung," *Nachrichtentechnische Zeitschrift* **24**, 18 (1971).
8. R. W. Chang, "A New Equalizer Structure for Fast Start-Up Digital Communication," *Bell System Tech. J.* **50**, (1971).
9. H. Kobayashi, "Application of Hestens-Stiefel Algorithm to Channel Equalization," *Conference Record, IEEE International Conference on Communications*, Montreal, 1971.
10. F. J. Gantmacher, *The Theory of Matrices*, Chelsea Publishing Company, New York, 1964, Vol. 2.
11. L. D. Davisson and J. Kim, "Convergence Properties of an Adaptive Equalization Algorithm for M-dependent Stationary Processes," *Conference Record, IEEE International Conference on Communications*, Philadelphia, 1972.

Received May 19, 1972

The author is located at the IBM Zurich Research Laboratory, 8803 Rüschlikon, Switzerland.

D. Godard

Channel Equalization Using a Kalman Filter for Fast Data Transmission

Abstract: This paper shows how a Kalman filter may be applied to the problem of setting the tap gains of transversal equalizers to minimize mean-square distortion. In the presence of noise and without prior knowledge about the channel, the filter algorithm leads to faster convergence than other methods, its speed of convergence depending only on the number of taps. Theoretical results are given and computer simulation is used to corroborate the theory and to compare the algorithm with the classical steepest descent method.

Introduction

Data transmission systems generally use voiceband communication channels. These are characterized by a relatively narrow bandwidth (300 to 3000 Hz), a high signal-to-noise ratio (about 20 to 30 dB), and amplitude and phase distortion slowly varying in time. High speed data transmission then requires equalization. Many presently used modem receivers are equipped with a matched filter to maximize the signal-to-noise ratio and an equalizer to minimize the inter-symbol interference due to distortion. Equalizers usually are of the transversal filter type (tapped delay line filter) with tap gains adjusted to minimize some error criterion. An automatic equalization process requires an initial training period during which the equalizer reduces the error. In "preset equalization," isolated pulses are transmitted prior to data transmission, and the derived tap-gain settings are kept constant during the data transmission itself. Periodically, a short training period may be entered to update the tap gains.

A second kind of equalization process is known as "adaptive equalization." Here the equalizer settings are derived from the received signal. During the training period, the equalizer continuously seeks to minimize the deviation of its sampled output signal from an ideal reference signal generated internally in proper synchronism in the receiver. When the residual distortion is small enough, actual data may be transmitted. The equalizer is then switched into the "decision-directed mode," using as reference a reconstructed signal obtained by thresholding the output signal of the equalizer. Adaptive equalization has many advantages over preset equalization, among them being the ability to adapt to changes in channel characteristics during the transmission.

Clearly, there is a delay in data transmission proportional to the length of the training period, and a decrease in this delay is desirable. Many adjustment algorithms [1-11] have been described in the literature, often emphasizing the speed of convergence. For the well-known mean-square algorithm Gersho [4] showed that the speed of convergence is largely determined by the maximum and minimum values of the power spectrum of the unequalized signal. Similar results for more sophisticated algorithms have been reported by Chang [5] and Kobayashi [6]. To achieve fast equalization, a new equalizer structure has been developed by Sha and Tang [11], although their theory can fail for large distortion. Devieux and Pickholtz [7] studied adaptive equalization with a second-order gradient algorithm, but that algorithm requires computation of the covariance matrix of the sampled received signal. More recently, Ungerboeck [8] showed that, in the speed of convergence of the mean-square algorithm, the influence of the number of taps usually dominates. He gave a new criterion for convergence and an optimal step size parameter in the adjustment loops.

In this paper, a new algorithm, based on Kalman filtering theory [12, 13] is proposed for obtaining fast convergence of the tap gains of transversal equalizers to their optimal settings. A Kalman filter had previously been applied to channel equalization by Lawrence and Kaufman [15] but in a quite different way, since in their study the equalizer is replaced by the filter.

It is shown here that the convergence of mean-square distortion is obtained, under noisy conditions, within a number of iterations determined only by the number of

Reprinted with permission from *IBM J. Res. Develop.*, vol. 18, May 1974.

taps, without prior information about the channel. First, fundamentals of Kalman's theory are reviewed. Then we show how to apply the Kalman filter to the equalizer and derive an expression for the speed of convergence. Finally, computer simulations are used to check the validity of the theory and to compare the proposed algorithm with the steepest descent method. Comparison is also made with other sophisticated algorithms.

Kalman-Bucy filtering theory

Over a decade ago, with publication of now classical papers, Kalman [12] and Kalman and Bucy [13] defined a recursive method and deeply transformed filtering and predicting theories.

Application of the Kalman filter supposes the studied system to be described by a set of linear difference equations, in the case of discrete systems that are of interest in this study. Let $\mathbf{x}_k$ be, at the kth sampling instant, the N-dimensional vector of the N state variables [14] of a system modeled as follows:

$$\mathbf{x}_k = \mathbf{\Phi}(k, k-1)\mathbf{x}_{k-1} + \mathbf{W}_k, \tag{1}$$

$$\mathbf{Z}_k = \mathbf{H}_k\mathbf{x}_k + \mathbf{V}_k, \tag{2}$$

where $\mathbf{\Phi}(k, k-1)$ is the $N \times N$ state transition matrix; $\mathbf{Z}_k$, the M-dimensional measurement vector; $\mathbf{H}_k$, the $M \times N$ measurement matrix; and $\mathbf{W}_k$ and $\mathbf{V}_k$, respectively, N- and M-dimensional vectors of white noise processes of zero mean. Later we use the covariance matrices $\mathbf{Q}_k$ and $\mathbf{R}_k$ of $\mathbf{W}_k$ and $\mathbf{V}_k$. It is assumed that the noise processes $\mathbf{W}$ and $\mathbf{V}$ are statistically independent; i.e., if E denotes expectation, then

$$\mathrm{E}[\mathbf{W}_k\mathbf{V}_j'] = 0, \quad k, j = 1, 2, \cdots.$$

Throughout the paper, a prime (′) denotes matrix transposition.

Assume we know at the $(k-1)$th sampling instant an estimate $\hat{\mathbf{x}}_{k-1}$ of the actual state vector $\mathbf{x}_{k-1}$ and the error covariance matrix

$$\mathbf{P}_{k-1} = \mathrm{E}[(\mathbf{x}_{k-1} - \hat{\mathbf{x}}_{k-1})(\mathbf{x}_{k-1} - \hat{\mathbf{x}}_{k-1})']. \tag{3}$$

It is possible to derive from (1) a predicted value of the state at the kth sampling instant,

$$\hat{\mathbf{x}}_{k,k-1} = \mathbf{\Phi}(k, k-1)\hat{\mathbf{x}}_{k-1}, \tag{4}$$

and a predicted error covariance matrix defined by

$$\mathbf{P}_{k,k-1} = \mathrm{E}[(\mathbf{x}_k - \hat{\mathbf{x}}_{k,k-1})(\mathbf{x}_k - \hat{\mathbf{x}}_{k,k-1})']. \tag{5}$$

It can be shown that

$$\mathbf{P}_{k,k-1} = \mathbf{\Phi}(k, k-1)\mathbf{P}_{k-1}\mathbf{\Phi}'(k, k-1) + \mathbf{Q}_k. \tag{6}$$

A predicted measurement $\hat{\mathbf{Z}}_k$ is derived from (2):

$$\hat{\mathbf{Z}}_k = \mathbf{H}_k\hat{\mathbf{x}}_{k,k-1}. \tag{7}$$

The deviation of the predicted measurement from the actual measurement $\mathbf{Z}_k$ is used to define a new estimate of the state vector as

$$\hat{\mathbf{x}}_k = \hat{\mathbf{x}}_{k,k-1} + \mathbf{K}_k(\mathbf{Z}_k - \hat{\mathbf{Z}}_k). \tag{8}$$

The $N \times M$ correction matrix $\mathbf{K}_k$ is computed in order to minimize the trace of the error covariance matrix $\mathbf{P}_k$. Then it can be shown that $\mathbf{K}_k$ and $\mathbf{P}_k$ are given by

$$\mathbf{K}_k = \mathbf{P}_{k,k-1}\mathbf{H}_k'(\mathbf{H}_k\mathbf{P}_{k-1}\mathbf{H}_k' + \mathbf{R}_k)^{-1} \tag{9}$$

and

$$\mathbf{P}_k = \mathbf{P}_{k,k-1} - \mathbf{K}_k\mathbf{H}_k\mathbf{P}_{k,k-1}. \tag{10}$$

Application of the Kalman filter requires an estimate $\hat{\mathbf{x}}(0)$ of the initial state vector and computation of the corresponding error covariance matrix $\mathbf{P}_0$. Usually $\hat{\mathbf{x}}(0)$ is chosen as the mean value of $\mathbf{x}(0)$. Then it can be shown that the subsequent estimates are unbiased.

The Kalman filter defined by recursive formulas (4) through (10) leads to minimization of the trace of the error covariance matrix. More generally, if $\mathbf{A}$ is a symmetric positive definite matrix, minimization of the product $(\mathbf{x}_k - \hat{\mathbf{x}}_k)'\mathbf{A}(\mathbf{x}_k - \hat{\mathbf{x}}_k)$ is obtained.

Application of a Kalman filter to equalizers

For simplicity we limit ourselves to a pulse-amplitude-modulated (PAM) baseband system. The equalizer is an N-tap delay line filter. The input signal of the equalizer is the PAM baseband signal

$$U(t) = \sum_{k=-\infty}^{\infty} a_k h(t - kT) + v(t), \tag{11}$$

where $\{a_k\}$ is the sequence of quantized pulses to be transmitted, $h(t)$ is the channel response, T is the intersymbol separation (or baud interval), and $v(t)$ is the additive noise. Both sequence $\{a_k\}$ and noise $v(t)$ are stationary. Let $\mathbf{u}_k$ be the vector of tap output signals and $\mathbf{c}_k$ be the vector of tap gains of the equalizer, both at the kth sampling instant. These two vectors are N-dimensional. The output signal of the equalizer is

$$s_k = \mathbf{u}_k'\mathbf{c}_k, \tag{12}$$

and we define the error signal as

$$e_k = a_k - s_k. \tag{13}$$

We choose as the error criterion the minimization of the expected mean-square distortion,

$$\mathscr{E}^2 = \mathrm{E}[e_k^2]. \tag{14}$$

With sequence $\{a_k\}$ and noise $v(t)$ being assumed stationary, $\mathscr{E}^2$ does not depend on k for a given tap-gain setting.

It can be shown [4] that the optimal tap gains $\mathbf{c}_{opt}$ are given by

$$\mathbf{c}_{opt} = \mathbf{A}^{-1}\mathbf{b}, \quad (15)$$

where A is the $N \times N$ symmetric positive definite matrix

$$\mathbf{A} = E[\mathbf{u}_k\mathbf{u}_k'], \quad (16)$$

and b is the N-dimensional vector

$$\mathbf{b} = E[a_k\mathbf{u}_k]. \quad (17)$$

When $\mathbf{c}_k$ is chosen equal to $\mathbf{c}_{opt}$, the mean-square distortion assumes its minimum value $\mathscr{E}^2_{opt}$. Even if no noise is present, $\mathscr{E}^2_{opt}$ is not zero because of the finiteness of the number of taps. Let $e_{k\ opt}$ be the error signal with the optimal adjustment of the tap gains. Then we have

$$a_k = \mathbf{u}_k'\mathbf{c}_{opt} + e_{k\ opt}, \quad (18)$$

with

$$\mathscr{E}^2_{opt} = E[e^2_{k\ opt}]. \quad (19)$$

During the equalization process, the expected mean-square distortion at the kth sampling instant is, using (18),

$$\mathscr{E}_k^2 = E[\{\mathbf{u}_k'(\mathbf{c}_{opt} - \mathbf{c}_k) + e_{k\ opt}\}^2]. \quad (20)$$

It is well known [3] that

$$E[e_{k\ opt}\mathbf{u}_k] = 0, \quad (21)$$

so the variables $\mathbf{u}_k'(\mathbf{c}_{opt} - \mathbf{c}_k)$ and $e_{k\ opt}$ are uncorrelated and (20) may be rewritten as

$$\mathscr{E}_k^2 = E[(\mathbf{c}_{opt} - \mathbf{c}_k)'\mathbf{u}_k\mathbf{u}_k'(\mathbf{c}_{opt} - \mathbf{c}_k)] + \mathscr{E}^2_{opt}. \quad (22)$$

As in [4] and [8] we assume that the dependence between $\mathbf{u}_k$ and $\mathbf{c}_k$ may be neglected. Denoting by $\mathbf{P}_k$ the covariance matrix,

$$\mathbf{P}_k = E[(\mathbf{c}_{opt} - \mathbf{c}_k)(\mathbf{c}_{opt} - \mathbf{c}_k)'], \quad (23)$$

and using (16), the expected mean-square distortion at the kth sampling instant is given by

$$\mathscr{E}_k^2 = \text{trace } \mathbf{P}_k\mathbf{A} + \mathscr{E}^2_{opt}. \quad (24)$$

Since A is a symmetric positive definite matrix, minimization of (24) is done by the algorithm by choosing the tap gains as state variables. To simplify the theory we assume the channel characteristics to be stationary over the training period, so that the optimum tap-gain values are constant during the settling time of the equalizer. This is not a severe assumption since, for normal transmission speeds, the equalizer settles in a few milliseconds.

The problem is now stated: One wants to identify the optimal equalizer characterized by the constant state variables $\mathbf{c}_{opt}$, knowing that the output signal $\mathbf{u}_k'\mathbf{c}_{opt}$ satisfies

$$a_k = \mathbf{u}_k'\mathbf{c}_{opt} + e_{k\ opt}. \quad (18)$$

Referring to the general state equations (1) and (2), one sees that the state transition matrix is here the identity matrix and that (18) is the measurement equation, $e_{k\ opt}$ appearing as the measurement noise. Clearly, sequence $\{a_k\}$ and noise $v(t)$ being of zero mean, $e_{k\ opt}$ is a random variable of zero mean. To apply the Kalman-filter approach, we assume that $e_{k\ opt}$ may be considered as a white noise process of zero mean and variance $\mathscr{E}^2_{opt}$. This seems to be a reasonable approximation because the optimal mean-square distortion is usually small, including intersymbol interference still present at the output of the optimal equalizers, so that the correlation between successive samples of the noise may be neglected. However, this assumption is not necessary for the convergence of the algorithm.

Let us assume we know at the $(k-1)$th sampling instant an estimate $\mathbf{c}_{k-1}$ of the state vector and the error covariance matrix $\mathbf{P}_{k-1}$. Obviously, since the state transition matrix is the identity matrix, the predicted state vector $\mathbf{c}_{k,k-1}$ is equal to $\mathbf{c}_{k-1}$, and the predicted matrix $\mathbf{P}_{k,k-1}$ is equal to $\mathbf{P}_{k-1}$. The predicted measurement is

$$\hat{s}_k = \mathbf{u}_k'\mathbf{c}_{k-1},$$

and the new estimate $\mathbf{c}_k$ is given by

$$\mathbf{c}_k = \mathbf{c}_{k-1} + \mathbf{K}_k(a_k - \hat{s}_k), \quad (25)$$

with

$$\mathbf{K}_k = \mathbf{P}_{k-1}\mathbf{u}_k(\mathbf{u}_k'\mathbf{P}_{k-1}\mathbf{u}_k + \mathscr{E}^2_{opt})^{-1} \quad (26)$$

and

$$\mathbf{P}_k = \mathbf{P}_{k-1} - \mathbf{K}_k\mathbf{u}_k'\mathbf{P}_{k-1}. \quad (27)$$

Computation of the Kalman gain $\mathbf{K}_k$, here reduced to an N-dimensional vector, involves only inversion of a scalar quantity, but requires prior knowledge of the optimal mean-square distortion. Clearly, $\mathscr{E}^2_{opt}$ cannot be known a priori. Usually, after equalization, the signal-to-noise ratio at the output of the equalizer is between 20 and 30 dB, so that to compute the Kalman gain one can use an estimated value $\mathscr{E}^2_{opt}$ of the optimal mean-square distortion included between 0.01 and 0.001. Later we show that the estimated value of $\mathscr{E}^2_{opt}$ has no influence on the successive estimates of the tap gains.

As initial estimate $\mathbf{c}_0$ of the tap gains, we choose $\mathbf{c}_0 = \mathbf{0}$. The optimal tap gains being assumed to be uniformly distributed between plus and minus 1.5, the initial covariance matrix $\mathbf{P}_0$ is the matrix with elements

$$(\mathbf{P}_0)_{ij} = 0.75\ \delta_{ij}, \quad i, j = 1, \cdots, N,$$

δ_{ij} being the Kronecker function. The off-diagonal terms of the matrix are zero, to reflect the statistical independence of the initial estimates.

• *Speed of convergence*

Equation (26) may be rewritten as

$$(\mathbf{I} - \mathbf{K}_k\mathbf{u}_k')\mathbf{P}_{k-1}\mathbf{u}_k = \mathbf{K}_k\mathscr{E}^2_{\text{opt}}, \tag{26'}$$

where $\mathbf{I}$ is the $N \times N$ identity matrix. Multiplying (27) on the right by $\mathbf{u}_k$ and using (26'), we obtain

$$\mathbf{P}_k\mathbf{u}_k = \mathbf{K}_k\mathscr{E}^2_{\text{opt}},$$

or

$$\mathbf{K}_k = \mathbf{P}_k\mathbf{u}_k/\mathscr{E}^2_{\text{opt}}. \tag{28}$$

Eliminating $\mathbf{K}_k$ from (27), we obtain

$$\mathbf{P}_k = \mathbf{P}_{k-1} - \mathbf{P}_k\mathbf{u}_k\mathbf{u}_k'\mathbf{P}_{k-1}/\mathscr{E}^2_{\text{opt}}. \tag{27'}$$

The $\mathbf{P}_k$ matrix, being a covariance matrix, is positive definite and has an inverse. Thus multiplication of (27') on the left by $\mathbf{P}_k^{-1}$ and on the right by $\mathbf{P}_{k-1}^{-1}$ leads to

$$\mathbf{P}_k^{-1} = \mathbf{P}_{k-1}^{-1} + \mathbf{u}_k\mathbf{u}_k'/\mathscr{E}^2_{\text{opt}}. \tag{29}$$

From (29) we derive

$$\mathbf{P}_k = \mathscr{E}^2_{\text{opt}}\left(\mathscr{E}^2_{\text{opt}}\mathbf{P}_0^{-1} + \sum_{i=1}^{k} \mathbf{u}_i\mathbf{u}_i'\right)^{-1}. \tag{30}$$

After a few iterations, $\mathscr{E}^2_{\text{opt}}$ usually being smaller than 0.01, the diagonal matrix $\mathscr{E}^2_{\text{opt}}\mathbf{P}_0^{-1}$ may be neglected in (30) so that $\mathbf{P}_k$ becomes

$$\mathbf{P}_k = \mathscr{E}^2_{\text{opt}}\left(\sum_{i=1}^{k} \mathbf{u}_i\mathbf{u}_i'\right)^{-1}. \tag{31}$$

and (28) becomes

$$\mathbf{K}_k = \left(\sum_{i=1}^{k} \mathbf{u}_i\mathbf{u}_i'\right)^{-1}\mathbf{u}_k, \tag{32}$$

showing that the dependence between the Kalman gains and $\mathscr{E}^2_{\text{opt}}$ may be neglected for usual values of $\mathscr{E}^2_{\text{opt}}$.

Clearly, the matrix $k^{-1}(\Sigma^k_{i=1}\ \mathbf{u}_i\mathbf{u}_i')$ converges to the $\mathbf{A}$ matrix when k goes to infinity. Nevertheless, one can estimate that after 30 or 40 iterations it is possible to write

$$\text{trace } \mathbf{P}_k\mathbf{A} \approx \mathscr{E}^2_{\text{opt}}Nk^{-1}. \tag{33}$$

Then, from (24), if the optimal mean-square distortion is known a priori, we derive

$$\mathscr{E}_k^{\,2} \approx \mathscr{E}^2_{\text{opt}}(1 + Nk^{-1}), \tag{34}$$

which means that convergence can theoretically be obtained within less than $2N$ steps.

• *Further analysis of the algorithm*

It has been seen that the algorithm can be applied to the identification of the optimal equalizer if the optimal error $e_{k\ \text{opt}}$ may be considered as white noise. We now show that this assumption is not necessary.

Denote by $\mathbf{S}_k$ the $N \times N$ matrix defined by

$$\mathbf{S}_k = \mathbf{P}_k/\mathscr{E}^2_{\text{opt}}, \quad k = 0, 1, 2, \cdots. \tag{35}$$

Then (26) and (27) become

$$\mathbf{K}_k = \mathbf{S}_{k-1}\mathbf{u}_k/(1 + \mathbf{u}_k'\mathbf{S}_{k-1}\mathbf{u}_k) \text{ and} \tag{36}$$

$$\mathbf{S}_k = \mathbf{S}_{k-1} - \mathbf{K}_k\mathbf{u}_k'\mathbf{S}_{k-1}. \tag{37}$$

It is easy to verify that equations (28) and (29) may be rewritten as

$$\mathbf{K}_k = \mathbf{S}_k\mathbf{u}_k \text{ and} \tag{38}$$

$$\mathbf{S}_k^{-1} = \mathbf{S}_{k-1}^{-1} + \mathbf{u}_k\mathbf{u}_k', \text{ or} \tag{39}$$

$$\mathbf{S}_k = \left(\mathbf{S}_0^{-1} + \sum_{i=1}^{k} \mathbf{u}_i\mathbf{u}_i'\right)^{-1}. \tag{40}$$

Substituting (38) into (25) we obtain

$$\mathbf{c}_k = \mathbf{c}_{k-1} + \mathbf{S}_k\mathbf{u}_k a_k - \mathbf{S}_k\mathbf{u}_k\mathbf{u}_k'\mathbf{c}_{k-1}. \tag{41}$$

Substituting $\mathbf{S}_k^{-1} - \mathbf{S}_{k-1}^{-1}$ from (39) for $\mathbf{u}_k\mathbf{u}_k'$ and multiplying both sides of (41) by $\mathbf{S}_k^{-1}$, we have

$$\mathbf{S}_k^{-1}\mathbf{c}_k = \mathbf{S}_{k-1}^{-1}\mathbf{c}_{k-1} + a_k\mathbf{u}_k. \tag{42}$$

With $\mathbf{c}_0 = \mathbf{0}$, (42) may be rewritten as

$$\mathbf{c}_k = \mathbf{S}_k \sum_{i=1}^{k} a_i\mathbf{u}_i$$

or, using (40),

$$\mathbf{c} = \left(\mathbf{S}_0^{-1} + \sum_{i=1}^{k} \mathbf{u}_i\mathbf{u}_i'\right)^{-1} \sum_{i=1}^{k} a_i\mathbf{u}_i. \tag{43}$$

For sufficiently large k, such as $k \geqslant N$, the diagonal matrix $\mathbf{S}_0^{-1}$ may be neglected in (43), so that the kth estimate of the tap gains is given by

$$\mathbf{c}_k = \left(\sum_{i=1}^{k} \mathbf{u}_i\mathbf{u}_i'\right)^{-1} \sum_{i=1}^{k} a_i\mathbf{u}_i \tag{44}$$

and is independent of the previous estimates if the initial estimate of $\mathbf{c}_0$ is $\mathbf{0}$.

From (44) it is obvious that the kth estimate $\mathbf{c}_k$ is the optimal one for the received sequence up to the kth sampling instant. For example, if the pseudorandom sequence used during the training mode is periodic, sending only one period of the sequence is sufficient. The algorithm must give good results even if the signal-to-noise ratio is small, since the algorithm builds up the inverse of the correlation matrix of the sampled received signal, which is corrupted by noise. The assumption that $e_{k\ \text{opt}}$ is a white noise process is necessary only to express the $\mathbf{P}_k$ matrix as the error covariance matrix of the tap gains.

• *Adaptation to slowly varying channels*

The optimal tap-gain values are time varying as a consequence of the amplitude and phase characteristics of real channels being not stationary. From (31) and (32) the P_k matrix elements and the Kalman gain K_k converge to zero when k goes to infinity, so the equalizer cannot adapt itself to changes in channel characteristics during the transmission. Nevertheless, one can easily derive adaptive techniques from previous theory:

1. One can assume that the optimal tap-gain values are randomly varying about a mean value. This leads to the state equation

$$c_{k\,opt} = c_{(k-1)\,opt} + \Delta c_k, \tag{45}$$

where Δc_k is considered as a white noise process. Then one has to calculate the correlation matrix

$$Q = E[\Delta c_k \Delta c_k']$$

and, at each step from (6), the predicted error covariance matrix

$$P_{k,k-1} = P_{k-1} + Q.$$

Although Eq. (45) does not describe the true situation, it could give good results in the case of rapidly varying channels.
2. One can freeze the P_k matrix after, say, $5N$ sampling intervals. The Kalman gain stays sufficiently large to ensure adaptation. (This procedure may be compared with the one used by Chang [5] when his prefixed weighting matrix is not perfectly suited to the A matrix.)
3. When the equalizer is switched into the decision-directed mode, the P_k matrix is restated and fixed to a diagonal matrix with elements $(P)_{ij} = \alpha\hat{\mathscr{E}}^2_{opt}\delta_{ij}$, where α is the step-size parameter usually used in the stochastic gradient method [8]. It is easy to verify, referring to (28) and (25), that in this case the equalization process becomes the same as in the steepest descent method, which gives good enough results when the equalizer has only to track slow changes in channel characteristics during data transmission.

This procedure would be attractive in a signal processor in which a large part of the computation power could be used during a brief portion of the start-up phase to achieve a fast reduction of mean-square distortion.

Computer simulation

During the equalization process, the expected mean-square distortion at the kth sampling instant is approximately given by

$$\mathscr{E}_k^2 = \mathscr{E}^2_{opt}(1 + Nk^{-1}), \tag{34}$$

showing that convergence must be obtained within less than $2N$ steps, and that the speed of convergence does not depend on the characteristics of the channel. Computer simulation has been used to check the validity of these assertions and to compare the speed of convergence of the proposed algorithm with that in the steepest descent method, where estimates of the tap gains are iteratively given by

$$c_{k+1} = c_k + \mu e_k u_k.$$

In addition, the influence of the estimated value $\hat{\mathscr{E}}^2_{opt}$ of the optimal mean-square distortion was investigated. As step-size parameter we chose the optimal one defined by Ungerboeck [8]:

$$\mu = 1/N\langle u^2\rangle,$$

where $\langle u^2\rangle$ denotes the energy of the unequalized signal.

The algorithms have been tested with three channels:

Channel 1 Moderate amplitude and phase distortion
Channel 2 Heavy amplitude distortion, no phase distortion
Channel 3 Heavy amplitude and phase distortion.

For Channels 2 and 3 a large spread of the eigenvalues of the A matrix occurs, leading to slow convergence with the steepest descent method.

Two programs have been written. The first one, for a given voiceband communications channel, determines A, b, c_{opt}, and $\mathscr{E}^2_{opt}$. The second program generates a random sequence of bipolar signals ($a_k = \pm 1$), simulates the channel, adds white Gaussian noise, and simulates the equalizer. A signal-to-noise ratio of 30 dB is assumed at the input of the equalizer. At each sampling instant the mean-square distortion $\langle e_k^2\rangle$ is computed from

$$\langle e_k^2\rangle = (c_k - c_{opt})A(c_k - c_{opt}) + \mathscr{E}^2_{opt}$$

for various estimates of the optimal mean-square distortion.

The results of the simulation are presented in Figs. 1 and 2. From them we draw the following conclusions:

1. The speed of convergence of $\langle e_k^2\rangle$ does not depend on the choice of $\hat{\mathscr{E}}^2_{opt}$, provided that the value chosen is reasonably small but not zero. A zero value would mean that no noise is present and that the equalizer is of infinite length. For a given channel, three computer runs with $\hat{\mathscr{E}}^2_{opt} = 0.1$, 0.001, and 0.0001 gave identical results for the same sequence $\{a_k\}$ and the same sequence of noise samples $v(kT)$.
2. The speed of convergence is independent of the characteristics of the channel. A large spread of the eigenvalues of the A matrix, as is the case for Channels 2 and 3, leads to a slow convergence with the steepest descent method but has no effect on the speed of convergence obtained through our algorithm.
3. Good agreement of the expected mean-square distortion theoretically predicted by (34) and $\langle e_k^2\rangle$ ob-

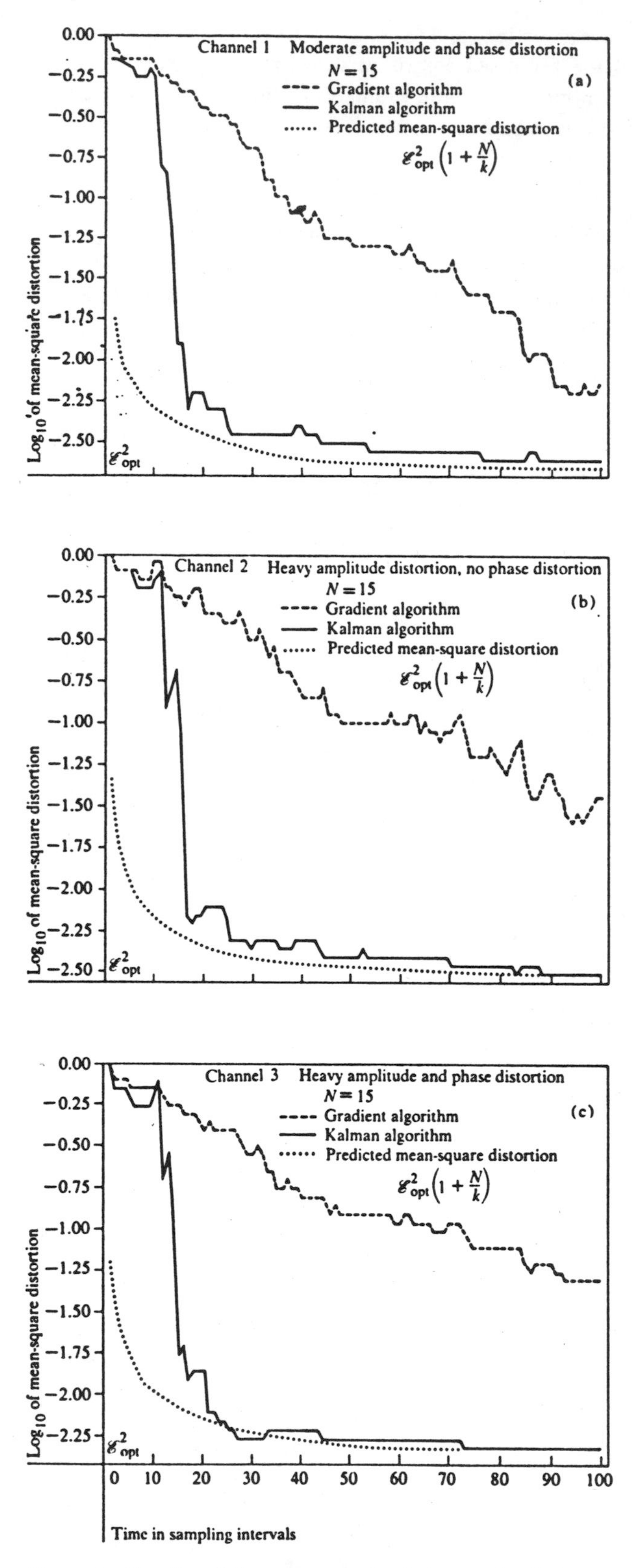

Figure 1 Results of the computer simulation of the Kalman-filter algorithm with three test channels: (a) Channel 1 – moderate amplitude and phase distortion; (b) Channel 2 – heavy amplitude distortion, no phase distortion; and (c) Channel 3 – heavy amplitude and phase distortion.

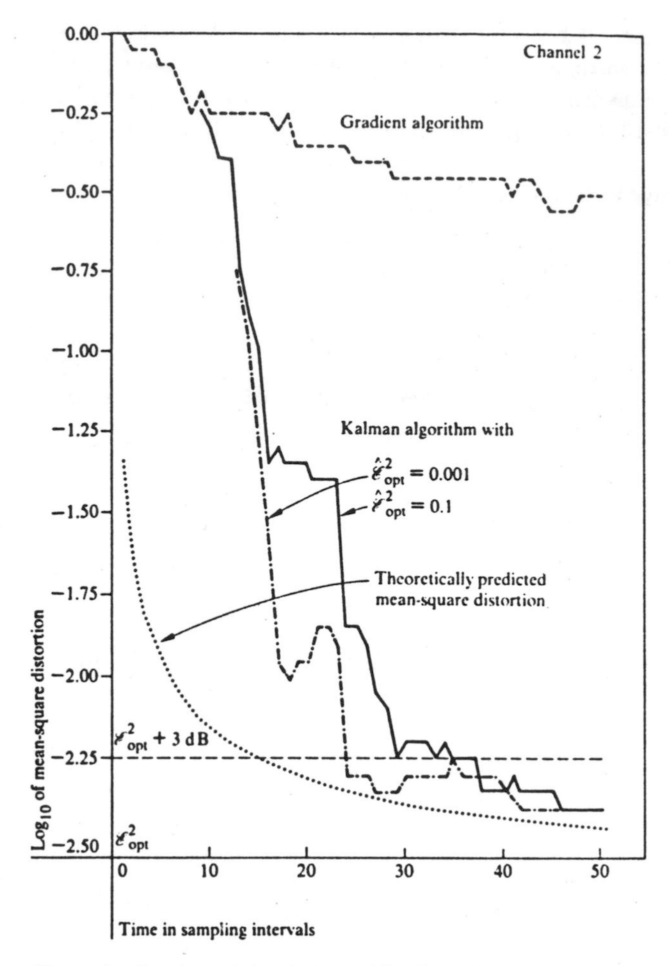

Figure 2 Results of simulation with Channel 2 to test the assumption of statistical independence of $\mathbf{c}_k$ and $\mathbf{u}_k$.

tained by simulation may be observed in spite of the various approximations that were made, among them being the statistical independence of $\mathbf{c}_k$ and $\mathbf{u}_k$.

To investigate the influence of this assumption, a computer run was made with five baud intervals introduced between sampling instants when tap-gain corrections are made. As a consequence the successive tap output signals are forced to be quasi-statistically independent of one another and the noise $e_{k\,\mathrm{opt}}$ at the output of the optimal equalizer is white noise. It can be seen in Fig. 2 that, without counting the additional baud intervals, the speed of convergence is unchanged.

All simulations showed that convergence towards $\mathscr{E}^2_{\mathrm{opt}}$ + 3 dB was obtained within less than $2N$ steps. The equalizer may be switched into the decision-directed mode after about $2N$ sampling intervals. With an N of 15 and a transmission speed of 2400 bauds, the settling time of the equalizer is about 12 ms.

It would be interesting to compare these results with those of Chang [5] and Sha and Tang [11]. Their structures, however, use equally spaced, isolated test pulses during the training period. For a given distortion smaller than one, the Sha and Tang equalizer is optimally settled when it has received, on the average, four or five isolated pulses. Thus the settling time is at least $4N$ or $5N$ sampling intervals. When the **A** matrix is perfectly known and under noise-free conditions, the Chang equalizer requires only one training pulse. But when the **A** matrix is not precisely known and with a high signal-to-noise ratio, the settling time is again about $4N$ or $5N$ sampling intervals.

Assume now that the equalizer using the Kalman filter algorithm receives one isolated test pulse and that no noise is present. If the diagonal elements of the $\mathbf{P}_0$ matrix, or of the $\mathbf{S}_0$ matrix, are chosen large enough, Eq. (44) holds and the equalizer is optimally settled when one test pulse has been received, without prior knowledge about the channel characteristics. The algorithm then leads to the optimal speed of convergence, but it requires a larger amount of computation than other methods.

We also made a computer run with the $\mathbf{P}_k$ matrix reduced to its nine main diagonals, its other elements being zero. This resulted for Channel 3 in only a small degradation of the speed of convergence, since convergence below $\mathscr{E}^2_{\text{opt}} + 3$ dB was obtained within about 60 sampling intervals, while halving the required amount of computation.

References

1. R. W. Lucky, "Automatic Equalization for Digital Communication," *Bell Syst. Tech. J.* **44**, 547 (1965).
2. R. W. Lucky and H. R. Rudin, "An Automatic Equalizer for General Purpose Communication Systems," *Bell Syst. Tech. J.* **45**, 225 (1966).
3. R. W. Lucky, J. Salz and E. J. Weldon, *Principles of Data Communication*, McGraw-Hill Book Co., Inc., New York, 1968.
4. A. Gersho, "Adaptive Equalization of Highly Dispersive Channels for Data Transmission," *Bell Syst. Tech. J.* **48**, 55 (1969).
5. R. W. Chang, "A New Equalizer Structure for Fast Start-Up Digital Communications," *Bell Syst. Tech. J.* **50**, 1969 (1971).
6. H. Kobayashi, "Application of Hestenes-Stiefel Algorithm To Channel Equalization," *Conference Record*, IEEE International Conference on Communications, Montreal, 1971.
7. C. Devieux and R. L. Pickholtz, "Adaptive Equalization With a Second Order Gradient Algorithm," *Proceedings of the Symposium on Computer Processing in Communications*, Polytechnic Press, Polytechnic Inst. of Brooklyn, New York, 1970.
8. G. Ungerboeck, "Theory on the Speed of Convergence in Adaptive Equalizers for Digital Communication," *IBM J. Res. Develop.* **16**, 546 (1972).
9. T. J. Schonfeld and M. Schwartz, "A Rapidly Converging First Order Training Algorithm for an Adaptive Equalizer," *IEEE Trans. Inf. Theory* **IT-17**, 431 (1971).
10. D. A. George, D. C. Coll, A. R. Kaye, and R. R. Bowen, "Channel Equalization for Data Transmission," *Proceedings of the 83rd Annual Meeting of the Engineering Institute of Canada*, Vancouver, B.C., September 1969, pp. 20-31.
11. R. T. Sha and D. T. Tang, "A New Class of Automatic Equalizers," *IBM J. Res. Develop.* **16**, 556 (1972).
12. R. E. Kalman, "A New Approach to Linear and Prediction Problems," *J. Basic Eng., Trans. ASME* **82**, 35 (1960).
13. R. E. Kalman and R. S. Bucy, "New Results in Linear Filtering and Prediction Theory," *J. Basic Eng., Trans. ASME* **95**, 107 (1961).
14. M. DeRusso, R. Ray and C. Close, *State Variables for Engineers*, John Wiley & Sons, Inc., New York, 1965.
15. R. E. Lawrence and H. Kaufman, "The Kalman Filter for the Equalization of a Digital Communications Channel," *IEEE Trans. Commun. Tech.* **COM-19**, 1137 (1971).

Received September 17, 1973

The author is located at the Centre d'Etudes et de Recherches, Compagnie IBM France, 06 La Gaude, Alpes Maritimes, France.

Application of Fast Kalman Estimation to Adaptive Equalization

DAVID D. FALCONER AND LENNART LJUNG

Abstract—Very rapid initial convergence of the equalizer tap coefficients is a requirement of many data communication systems which employ adaptive equalizers to minimize intersymbol interference. As shown in recent papers by Godard, and by Gitlin and Magee, a recursive least squares estimation algorithm, which is a special case of the Kalman estimation algorithm, is applicable to the estimation of the optimal (minimum MSE) set of tap coefficients. It was furthermore shown to yield much faster equalizer convergence than that achieved by the simple estimated gradient algorithm, especially for severely distorted channels. We show how certain "fast recursive estimation" techniques, originally introduced by Morf and Ljung, can be adapted to the equalizer adjustment problem, resulting in the same fast convergence as the conventional Kalman implementation, but with far fewer operations per iteration (proportional to the number of equalizer taps, rather than the square of the number of equalizer taps). These fast algorithms, applicable to both linear and decision feedback equalizers, exploit a certain shift-invariance property of successive equalizer contents. The rapid convergence properties of the "fast Kalman" adaptation algorithm are confirmed by simulation.

1. INTRODUCTION

Many modern data communication systems employ adaptive transversal equalizers whose tap coefficients can be adjusted to minimize intersymbol interference. Typically, an adaptive equalizer adjusts its tap coefficients during an initial "training" period in which a data sequence, known to the receiver, is transmitted. In many systems a requirement exists for as short a training period as possible. This has spurred considerable recent interest in equalizer adaptation algorithms that converge much faster than the simple and commonly-used stochastic approximation-type algorithms [1], [2], [3] even for severely distorted channels.

D. Godard, in a seminal paper, [4] regarded the equalizer adaptation problem as the minimum mean square estimation of the tap coefficient vector, given the sequences of equalizer inputs and of desired outputs. This formulation, together with an assumption on the statistics of the ideal equalizer's output error, and on independence between the equalizer's input and previous sets of tap coefficients, lead to a special case of the Kalman estimation algorithm for the optimum equalizer tap coefficients. This appears to be the fastest known equalizer adaptation algorithm. This assertion is supported by interpretation of the Kalman/Godard equalizer adaptation algorithm as an ideal self-orthogonalizing algorithm [5], [6] and also by telephone channel simulation results reported in [4] and [5]. A further significance of that work is its highlighting of a fruitful application of control theory to adaptation problems in communications and signal processing.

As pointed out in Section 3, this same equalizer adaptation algorithm arises, without invoking assumptions on statistics, from a classical least squares problem: after each new equalizer input and output, find the set of equalizer tap coefficients that minimizes the accumulation of the squared errors up to that time. The solution of this classical discrete-time Wiener-Hopf problem, implemented recursively, is a special case of the Kalman algorithm, with a constant state variable.

A disadvantage of the Kalman algorithm is its complexity. An N by N matrix must be adapted and stored once per iteration, where N, the number of equalizer tap coefficients, may typically be on the order of 50 or so. Thus on the order of N^2 operations must be performed per iteration, in contrast to the approximately N operations required by adaptation algorithms generally used in practice [1], [2], [3].

Recently, certain "fast algorithms" have been reported which reduce the number of operations per iteration in recursive least squares N-dimensional vector estimation algorithms

Paper approved by the Editor for Communication Theory of the IEEE Communications Society for publication after presentation at the International Symposium on Information Theory, Cornell University, Ithaca, NY, October 1977. Manuscript received July 2, 1977; revised May 10, 1978.

D. D. Falconer was on leave at the Department of Electrical Engineering, Linköping University, Linköping, Sweden. He is with Bell Laboratories, Holmdel, NJ 07733.

L. Ljung is with the Department of Electrical Engineering, Linköping University, Linköping, Sweden.

Reprinted from *IEEE Trans. Comm.*, vol. COM-26, no. 10, pp. 1439–1446, Oct. 1978.

to be proportional to N [7], [8]. This work has a conceptual relationship with Levinson's algorithm [9] for prediction problems. These fast recursive least squares algorithms exploit the "shifting property" of most sequential estimation problems. In equalization, this property expresses the fact that at each iteration the number of new samples entering and old samples leaving the equalizer is not N, but a much smaller integer p. For example, $p = 1$ for a conventional linear equalizer.

Reference [10] generalized the fast least squares estimation algorithm, widening the class of systems to which it is applicable. In this paper we show the applicability of the "fast Kalman" algorithm to various types of adaptive equalizers, and report the results of computer simulations. The algorithm is mathematically equivalent to the original Kalman/Godard equalizer adaptation algorithm (with some additional restrictions on starting conditions, given in Section 4). As well, a simple modification is shown, which enables the algorithm to follow channel time-variations.

2. FORMULATION OF EQUALIZER ADAPTATION

During the initial equalizer training period, a succession of data symbols $d(1), d(2), \cdots, d(n)$, known by the equalizer adjustment algorithm, is transmitted over a channel, resulting in a sequence of equalizer inputs $y(1), y(2), \cdots, y(n)$. At time n, the equalizer has the N latest inputs stored, denoted by a vector

$$x_N(n) \triangleq \begin{bmatrix} y(n-1) \\ y(n-2) \\ \cdot \\ \cdot \\ \cdot \\ y(n-N) \end{bmatrix} \tag{1a}$$

We shall indicate the dimensionality of vectors and matrices by subscripts. The absence of a subscript indicates dimension 1; e.g. $y(n)$ is a scalar, $x_N(n)$ is an N-dimensional vector, $A_{Np}(n)$ is a matrix with N rows and p columns. Matrix transpose will be represented by a superscript T.

The equalizer's tap coefficients at time n are represented by the vector $C_N(n-1)$, and its output is $C_N(n-1)^T x_N(n)$, which may differ from the *ideal* output $d(n)$ by an error $e(n)$

$$e(n) = d(n) - C_N(n-1)^T x_N(n). \tag{2}$$

This formulation of the equalizer's output and error encompasses several types of equalizers. The most obvious is the linear transversal equalizer with N taps, for which the $\{y(n)\}$ are channel outputs sampled at the data symbol rate. The linear equalizer structure is also relevant to receivers which employ Viterbi algorithm detection, in which case, $d(n)$ in (2) is replaced by a linear combination of recent data symbols. In a variation of the linear equalizer structure, in which the channel output is sampled at p times the symbol rate $1/T$, the vector $x_N(n)$ has the form

$$x_N(n) = \begin{bmatrix} y\left((n-1)T + \frac{p-1}{p}T\right) \\ y\left((n-1)T + \frac{p-2}{p}T\right) \\ \cdot \\ \cdot \\ \cdot \\ y((n-1)T) \\ y\left((n-2)T + \frac{p-1}{p}T\right) \\ \cdot \\ \cdot \\ \cdot \\ y((n-N)T) \end{bmatrix} \tag{1b}$$

(See reference [11].)

In the case of a decision feedback equalizer with N_1 forward and N_2 feedback taps $x_N(n)$ has the form

$$x_N(n) = \begin{bmatrix} y(n-1) \\ y(n-2) \\ \cdot \\ \cdot \\ \cdot \\ y(n-N_1) \\ \text{--------} \\ d(n-1) \\ d(n-2) \\ \cdot \\ \cdot \\ \cdot \\ d(n-N_2) \end{bmatrix} \tag{1c}$$

where $N = N_1 + N_2$, the $\{y(n)\}$ are the sampled channel outputs and the $\{d(n)\}$ are previously-decided data symbols.

A related application is one in which the $\{d(n-1), \cdots, d(n-N_2)\}$ appearing in (1c) are the N_2 latest data symbols emanating from a local, interfering transmitter during simultaneous two-way data communication, [12].

In these and in many other estimation and prediction applications, the input vectors $x_N(n)$ are such that $x_N(n+1)$ is obtained from $x_N(n)$ by shifting its components, introducing p new components and deleting the p oldest components. For example the $p = 2$ new elements $y(n), d(n)$ enter the decision feedback equalizer at time $n + 1$, while the elements $y(n - N_1)$ and $d(n - N_2)$ leave. *This shifting property is crucial to the "fast" Kalman algorithm.* It is not exploited by the conventionally-implemented Kalman algorithm.

To unify subsequent notation, we define a p-dimensional vector $\xi_p(n)$ specifying the p new elements and a vector $\rho_p(n)$ specifying the p deleted old elements at time $n + 1$; e.g. for the linear equalizer, $\xi_1(n) = y(n)$ and $\rho_1(n) = y(n - N)$. For the decision feedback equalizer,

$$\xi_2(n) = \begin{bmatrix} y(n) \\ d(n) \end{bmatrix} \quad \text{and} \quad \rho_2(n) = \begin{bmatrix} y(n-N_1) \\ d(n-N_2) \end{bmatrix}.$$

We also define an *extended* vector $\bar{x}_M(n)$, with $M = N + p$ dimensions, which contains the elements of $\xi_p(n)$ appended in proper order to the elements of $x_N(n)$. For example in the case of the decision feedback equalizer, with $p = 2$,

$$\bar{x}_M(n) = \begin{bmatrix} y(n) \\ \hdashline y(n-1) \\ \vdots \\ y(n-N_1) \\ \hdashline d(n) \\ \hdashline d(n-1) \\ \vdots \\ d(n-N_2) \end{bmatrix}. \tag{3}$$

Dotted lines denote partitioning of vectors and matrices.

In general there are obvious *permutation matrices* S_{MM} and Q_{MM} which rearrange the elements of the extended vector $\bar{x}_M(n)$ to display $\xi_p(n)$, $x_N(n)$, $\rho_p(n)$ and $x_N(n + 1)$ in simple partitioned form; i.e.

$$S_{MM}\bar{x}_M(n) = \begin{bmatrix} \xi_p(n) \\ \hdashline x_N(n) \end{bmatrix} \tag{4}$$

and

$$Q_{MM}\bar{x}_M(n) = \begin{bmatrix} x_N(n+1) \\ \hdashline \rho_p(n) \end{bmatrix}. \tag{5}$$

Each row and column of S_{MM} and Q_{MM} contains a single 1 and

$$S_{MM}^{-1} = S_{MM}^T; \qquad Q_{MM}^{-1} = Q_{MM}^T. \tag{6}$$

S_{MM} and Q_{MM} are identity matrices in the special case of the linear equalizer.

3. APPLICATION OF A RECURSIVE LEAST SQUARES ESTIMATION ALGORITHM TO EQUALIZER ADAPTATION

Casting the equalizer adaptation problem as a classical recursive least squares problem [13], [14] is a useful starting point. In particular we shall require the equalizer adaptation algorithm to generate that tap coefficient vector $C_N(n)$ at time n which minimizes the cumulative squared error

$$\sum_{k=1}^{n} [d(k) - C_N(n)^T x_N(k)]^2.$$

The minimizing vector is the solution of the Wiener-Hopf equation:

$$C_N(n) = R_{NN}(n)^{-1} \left[\sum_{k=1}^{n} d(k) x_N(k) \right] \tag{7}$$

where

$$R_{NN}(n) \equiv \sum_{k=1}^{n} x_N(n) x_N(n)^T + \delta I_{NN}, \tag{8}$$

and $\delta \geqslant 0$.

In practice, the parameter δ is fixed at a small positive constant to insure nonsingularity of the matrix $R_{NN}(n)$. Then it can be shown [4], [14] that this sequence $C_N(n)$ can be generated recursively as follows:

$$C_N(n) = C_N(n-1) + k_N(n) e(n), \tag{9}$$

where

$$k_N(n) \equiv R_{NN}(n)^{-1} x_N(n). \tag{10}$$

The inverse estimated covariance matrix $R_{NN}(n)^{-1}$ displayed in (10) greatly accelerates the equalizer's adaptation, independent of the channel's dispersion characteristics. [4], [5], [6]

The vector $k_N(n)$ can be generated by a recursive algorithm which yields the N by N matrix $R_{NN}(n)^{-1}$ without requiring explicit matrix inversion. The resulting algorithm, including (9), is the special case of the conventional Kalman algorithm for equalizer adaptation, reported by Godard. The parameter δ turns out to be an estimate of the final mean squared error; the algorithm's transient and steady state performance are not very sensitive to the choice of δ.

This algorithm's complexity, which is proportional to N^2, arises from the N by N matrix used to compute $k_N(n)$. The "fast Kalman" algorithm to be presented here is mathematically equivalent to it, but exploits the shifting property to compute the vectors $k_N(n)$ recursively, without needing to compute or store any N by N matrix.

In the Kalman algorithm (conventional or fast) for equalizer adaptation, all previous vectors $x_N(n)$ and previous errors $e(n)$ are accorded equal weight in determining the current tap coefficient estimates. This works well for a limited-duration initial start-up phase. However steady-state operation usually requires that vectors $\{x_N(k)\}$ in the distant past be "forgotten", in order to afford the possibility of following channel time variations and to avoid problems associated with digital round-off errors. A convenient performance criterion, meeting this steady state requirement, is minimization of an *exponentially-weighted* squared error at time n.

$$\sum_{k=1}^{n} \lambda^{n-k}[d(k) - C_N(n)^T x_N(k)]^2$$

where λ is some positive number close to, but less than 1. The inverse of $1 - \lambda$ is, roughly speaking, the memory of the algorithm. The minimizing vector $C_N(n)$ is now

$$C_N(n) = R_{NN}(n)^{-1} \left[\sum_{k=1}^{n} \lambda^{n-k} d(k) x_N(k) \right] \tag{11}$$

where

$$R_{NN}(n) \equiv \sum_{k=1}^{n} \lambda^{n-k} x_N(n) x_N(n)^T + \delta\lambda^n I_{NN}. \tag{12a}$$

Note that (12a) implies

$$R_{NN}(n) = \lambda R_{NN}(n-1) + x_N(n) x_N(n)^T. \tag{12b}$$

Then it is easy to show, as in references [4] and [14] that the recursive relations (9) and (10) still hold, with $R_{NN}(n)$ given by (12). The next section presents the fast algorithm for updating the vector $k_N(n)$ found in Equation (9).

4. THE "FAST" KALMAN ALGORITHM

The fast algorithm for computing the vector sequence $\{k_N(n)\}$ (and thus $\{C_N(n)\}$ in (9)) is specified below. A full derivation was given in reference [10] for $\lambda = 1$. The Appendix contains an abbreviated derivation, which emphasizes the use made of the shifting property of the input vectors and of certain optimal predictor equations and associated relationships.

In addition to quantities specified earlier, the following scalar, vector and matrix quantities are to be stored and updated during each iteration:

(1) N-by-p matrices $A_{Np}(n)$ and $D_{Np}(n)$ with initial values $A_{Np}(0) = D_{Np}(0) = 0_{Np}$.
(2) p-by-p matrix $E_{pp}(n)$ with initial value $E_{pp}(0) = \delta I_{pp}$.
(3) M-dimensional vector $\bar{k}_M(n)$ where $M = N + p$.
(4) p-dimensional vectors $\epsilon_p(n)$, $\epsilon_p(n)'$, $\eta_p(n)$, and $\mu_p(n)$.
(5) N-dimensional vector $m_N(n)$.

The initial value of the N-dimensional vector $k_N(n)$ is

$$k_N(1) = 0_N \tag{13}$$

and all $x(n) = 0$ for $n \leqslant 0$.

Then starting at $n = 1$, $k_N(n+1)$ is updated with the following algorithm:

$$\epsilon_p(n) = \xi_p(n) + A_{Np}(n-1)^T x_N(n) \tag{14}$$

$$A_{Np}(n) = A_{Np}(n-1) - k_N(n)\epsilon_p(n)^T \tag{15}$$

$$\epsilon_p(n)' = \xi_p(n) + A_{Np}(n)^T x_N(n) \tag{16}$$

$$E_{pp}(n) = \lambda E_{pp}(n-1) + \epsilon_p(n)'\epsilon_p(n)^T \tag{17}$$

$$\bar{k}_M(n) = S_{MM} \left[\begin{array}{c} E_{pp}(n)^{-1}\epsilon_p(n)' \\ \hdashline k_N(n) + A_{Np}(n)E_{pp}(n)^{-1}\epsilon_p(n)' \end{array} \right]. \tag{18}$$

(Multiplication by permutation matrices S_{MM} and Q_{MM} simply amounts to a permutation of vector components.) Dotted lines indicate partitioning.

Partition $Q_{MM}\bar{k}_M(n)$ as follows:

$$Q_{MM}\bar{k}_M(n) = \left(\begin{array}{c} m_N(n) \\ \hdashline \mu_p(n) \end{array} \right) \tag{19}$$

$$\eta_p(n) = \rho_p(n) + D_{Np}(n-1)^T x_N(n+1) \tag{20}$$

$$D_{Np}(n) = [D_{Np}(n-1) - m_N(n)\eta_p(n)^T][I_{pp} - \mu_p(n)\eta_p(n)^T]^{-1} \tag{21}$$

$$k_N(n+1) = m_N(n) - D_{Np}(n)\mu_p(n). \tag{22}$$

Updating of the equalizer tap coefficient vector $C_N(n+1)$ then takes place according to (2) and (9), using the vector $k_N(n+1)$. The existence of the inversion in (21) is proven in reference [10].

A count of the number of operations needed to calculate each equalizer output and update the set of tap coefficients according to this fast algorithm indicates $7Np + Np^2 + 3p^2 + 2N + \frac{4}{3}p^3 - p/3$ multiplications* and $7Np + Np^2 + p^2/2 + 4N + \frac{4}{3}p^3 + \frac{19}{6}p$ additions. This counts divisions as multiplications and assumes Gaussian elimination to invert p-by-p matrices. [15] The corresponding figures for the conventionally-implemented Kalman algorithm are $3N^2 + 3N$ multiplications and $2N^2 + 2N + 1$ additions. For the simple gradient algorithm, where $C_N(n)$ is updated according to

$$C_N(n) = C_N(n-1) + g e(n) x_N(n), \quad (g \text{ a constant}) \tag{23}$$

$2N$ multiplications and $2N$ additions are needed to compute the equalizer output and update the tap coefficients.

Thus for a linear equalizer ($p = 1$), the number of computations per symbol interval are:

Algorithm	Multiplications	Additions
simple gradient	$2N$	$2N$
fast Kalman	$10N + 4$	$12N + 5$
conventional Kalman	$3N^2 + 3N$	$2N^2 + 2N + 1$

For large N, the complexity of the equalizer with fast Kalman updating is thus about 5 times that of the equalizer with simple gradient updating. The conventionally-implemented Kalman algorithm however is about N times as complex as the simple gradient algorithm.

*Equations (2), (9) and (14) through (22).

5. SIMULATION RESULTS

Figure 1 shows a computer-simulated data communication system with a linear equalizer whose tap coefficients could be adapted according to either the fast Kalman algorithm or the simple gradient algorithm. The simulation program was written in BASIC, and run on a time-shared Data General "ECLIPSE" computer. Pseudo-random 4-level data symbols were generated and passed through the baseband equivalent sampled impulse response of a 6600 baud VSB-modulated data communication system with 3455 Hz carrier frequency, used over a fairly typical voiceband telephone channel. Thus the overall bit rate is 13200 bits/s. Additive Gaussian noise samples added to the filter output simulated a signal-to-noise ratio of 31 dB. This baseband equivalent impulse response and signal-to-noise ratio were identical to those used in reference [5] to test conventional Kalman equalizer adaptation. The telephone channel's frequency characteristic, plotted in reference [5], has deep nulls which cause slow convergence of the gradient algorithm, [2].

The linear equalizer had $N = 31$ taps. Its output $C_N(n-1)^T x_N(n)$ was subtracted from an ideal reference, the current transmitted data symbol $d(n)$, to form the error signal $e(n)$. The error sequence $\{e(n)\}$ and the equalizer input sequence $\{x(n)\}$ were used to adjust the equalizer tap coefficients adaptively, and the measured mean squared error during adaptation was displayed as a function of the number of iterations, as in Figure 2. Note $\langle e^2 \rangle > 1$ initially because data symbols are ± 1, ± 3, with equal probabilities.

The simple gradient adaptation algorithm was implemented as in equation (23), where the constant g, chosen for the best convergence rate, was 0.002. The channel impulse response was normalized to have unit energy.

The fast Kalman algorithm was implemented as specified in Section 4 with $\lambda = 1$. Because this algorithm is started with all vectors $k_N(0)$, $A_{Np}(0)$ etc. set to zero, and with $x(n) = 0$ for $n \leqslant 0$, it was found convenient to start up the updating of $k_N(n)$ before the leading edge of the first data pulse was received. However for both algorithms, best results were obtained by starting up the equalizer *tap coefficient* updating after the main part of the first data pulse was received. "Number of iterations" in Figure 2 is reckoned from the start of the equalizer tap coefficient updating.

The mean squared error was estimated at each iteration as an exponentially-weighted average of previous squared errors. The memory of the exponential weighting was on the order of 10 symbol intervals. For both equalizer adaptation algorithms, the initial value of all tap coefficients was zero. The estimated mean squared error versus number of iterations is plotted in Figure 2 for both algorithms. The convergence of the simple gradient algorithm was very slow because of the channel's frequency response nulls. [5] The striking improvement offered by the fast Kalman algorithm is apparent: convergence to a steady state mean value within about 80 symbol intervals. This convergence behavior is, as would be expected, similar to that of the conventionally-implemented Kalman/Godard algorithm reported by Gitlin and Magee, [5].

Interestingly, starting the equalizer's center tap coefficient at the value 1, did not lead to significant improvement in the Kalman algorithm's convergence time, but did improve that of the simple gradient algorithm. The results are shown in Figure 3. Again, the fast Kalman algorithm is clearly superior. Thus, prior knowledge of the location of the main equalizer tap coefficient is not necessary for this algorithm.

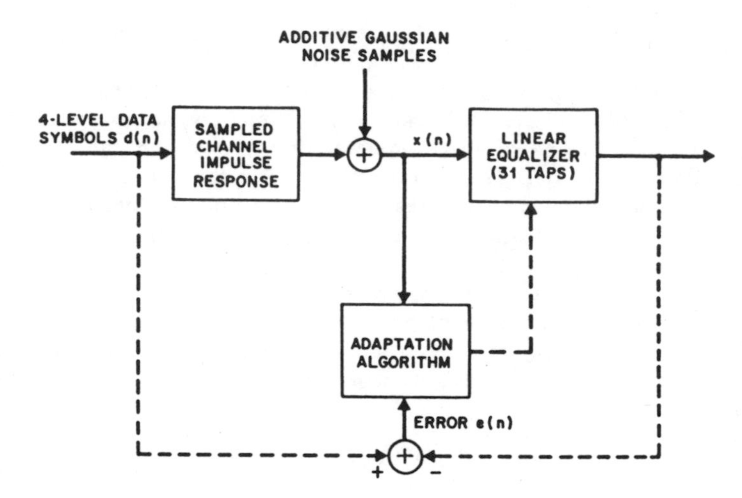

Fig. 1 Simulated Data Transmission System

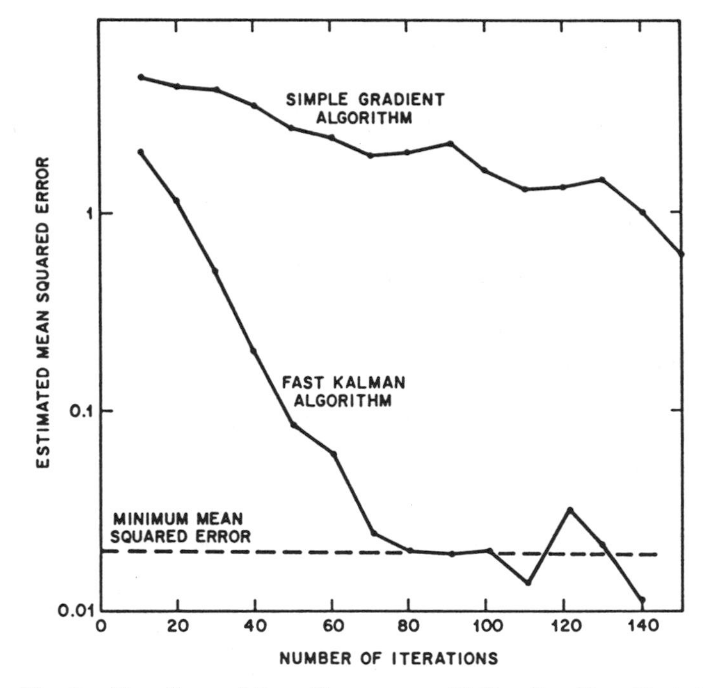

Fig. 2 Mean Squared Error Convergence, All Equalizer Taps Started at Zero

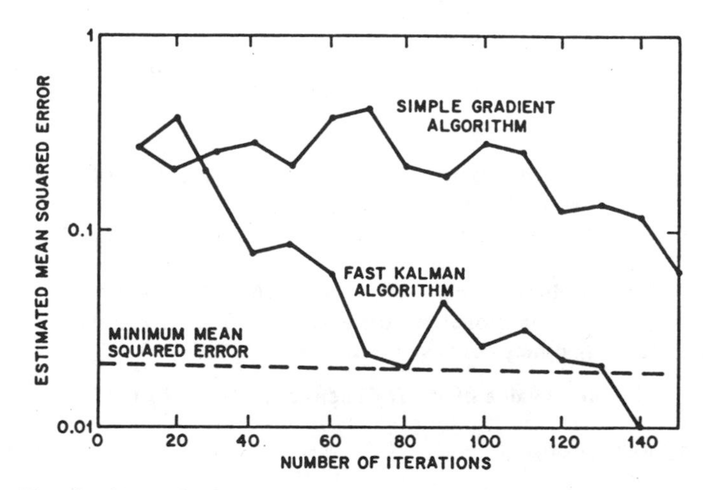

Fig. 3 Mean Squared Error Convergence, Center Equalizer Tap Started at 1

6. CONCLUSIONS

The fast Kalman algorithm for equalizer adaptation has a complexity proportional to N, just as in the case of the simple gradient algorithm, while offering the same fast convergence as the conventionally-implemented Kalman algorithm, whose complexity is proportional to N^2. The fast algorithm is in fact mathematically equivalent to the conventional Kalman algorithm; the only requirements are (1) the shifting property of the input vectors (satisfied in all known equalizer or echo canceller applications); and (2) that updating of the $k_N(n)$ vector start with all previous samples $x(n)$ assumed to be zero. The latter requirements guarantees a simple starting condition consistent with the induction argument used in the fast algorithm's derivation (see Appendix).

For a linear equalizer, the fast Kalman algorithm requires a number of operations per iteration only about 5 times that of the simple gradient algorithm. According to the count of numbers of operations tabulated in Section 4, a corresponding conventionally-implemented Kalman algorithm would require about 45 times as many operations as the simple gradient algorithm, if there are 30 equalizer taps and about 180 times as many if there are 60 equalizer taps.

APPENDIX

SUMMARY OF THE DERIVATION OF THE FAST ALGORITHM

A fast algorithm has previously been presented in reference [7] for the case where the input vectors $x_N(n)$ are partitioned into p *equal*-dimensioned parts (as in the $p = 2$ example of equation (1c) if N_1 and N_2 were equal). Reference [10] generalized the algorithm to apply where the p parts may have different dimensionalities. This is the case in general for decision feedback equalization. The following derivation illuminates the role of forward and backward prediction in the fast algorithm, and includes the exponentially-weighted error criterion ($\lambda \leqslant 1$).

To exploit the shifting property, we must relate the newest entering samples $\xi_p(n)$ and the oldest discarded samples $\rho_p(n)$ to the vectors $x_N(n)$ and $x_N(n+1)$, respectively. These relationships involve least squares forward and backward prediction. Thus we shall define a matrix $-A_{Np}(n-1)$ of forward predictor coefficients, and a forward prediction error $\epsilon_p(n)$ so that

$$\xi_p(n) = -A_{Np}(n-1)x_N(n) + \epsilon_p(n). \qquad \text{(A.1)}$$

The scalar product in (A.1) is the least squares *linear prediction* of $\xi_p(n)$ from the vector $x_N(n)$; i.e. the matrix $A_{Np}(n-1)$ is chosen to minimize the exponentially-weighted sum of squared prediction errors at time n

$$\sum_{k=1}^{n} \lambda^{n-k}[\xi_p(k) + A_{Np}(n-1)^T x_N(k)]^2. \qquad \text{(A.2)}$$

Then just as in (7) and (8), the least squares prediction coefficient matrix $A_{Np}(n)$ can be shown to satisfy the Wiener-Hopf relationship

$$R_{NN}(n)A_{Np}(n) = -B_{Np}(n), \qquad \text{(A.3)}$$

where

$$B_{Np}(n) \equiv \sum_{k=1}^{n} \lambda^{n-k} x_N(k)\xi_p(k)^T; \qquad \text{(A.4a)}$$

(A.4a) implies that

$$B_{Np}(n) = \lambda B_{Np}(n-1) + x_N(n)\xi_p(n)^T. \qquad \text{(A.4b)}$$

Analogous to (9) and (10), the solution to (A.3) is generated recursively by the following least squares recursive algorithm

$$A_{Np}(n) = A_{Np}(n-1) - k_N(n)\epsilon_p(n)^T, \qquad \text{(A.5)}$$

with

$$A_{Np}(0) = 0_{Np}.$$

Similarly, we shall define a matrix $-D_{Np}(n-1)$ of *backward* "prediction" coefficients and a backward "prediction" error $\eta_p(n)$ so that

$$\rho_p(n) = -D_{Np}(n-1)^T x_N(n+1) + \eta_p(n). \qquad \text{(A.6)}$$

The least squares linear backward predictor $D_{Np}(n-1)$ is chosen to minimize

$$\sum_{k=1}^{n} \lambda^{n-k}[\rho_p(k) + D_{Np}(n-1)^T x_N(k+1)]^2. \qquad \text{(A.7)}$$

Analogous to (A.3), (A.4), and (A.5) we have

$$R_{NN}(n+1)D_N(n) = -\tilde{B}_{Np}(n), \qquad \text{(A.8)}$$

where

$$\tilde{B}_{Np}(n) \equiv \sum_{k=1}^{n} \lambda^{n-k} x_N(k+1)\rho_p(k)^T, \qquad \text{(A.9a)}$$

implying

$$\tilde{B}_{Np}(n) = \lambda\tilde{B}_{Np}(n-1) + x_N(n+1)\rho_p(n)^T. \qquad \text{(A.9b)}$$

Also

$$D_{Np}(n) = D_{Np}(n-1) - k_N(n+1)\eta_p(n)^T \qquad \text{(A.10)}$$

with

$$D_{Np}(0) = 0_{Np}.$$

The initial vector $k_N(1)$ is

$$k_N(1) = 0_N. \qquad \text{(A.11)}$$

This is consistent with the assumption that all $x(n) = 0$ for $n \leqslant 0$. At the nth iteration ($n \geqslant 1$) we assume that $k_N(n)$ is available satisfying (10) and we seek a vector $k_N(n+1)$ satisfying

$$R_{NN}(n+1)k_N(n+1) = x_N(n+1). \qquad \text{(A.12)}$$

The extended vector $\bar{x}_M(n)$, with $M = N + p$, was defined by (3), (4), and (5). We also define an extended matrix $\bar{R}_{MM}(n)$ in an analogous way to (12b):

$$\bar{R}_{MM}(n) = \lambda \bar{R}_{MM}(n-1) + \bar{x}_M(n)\bar{x}_M(n)^T \tag{A.13}$$

with

$$\bar{R}_{MM}(0) = \delta I_{MM}.$$

Defined in this way, $\bar{R}_{MM}(n)$ contains the elements of both $R_{NN}(n)$ and of $R_{NN}(n+1)$. Using the permutation matrices S_{MM} and Q_{MM} to systematically display partitioning of $R_{MM}(n)$, we have

$$S_{MM}\bar{R}_{MM}(n)S_{MM}{}^T = \begin{bmatrix} \pi_{pp}(n) & B_{Np}(n)^T \\ B_{Np}(n) & R_{NN}(n) \end{bmatrix} \tag{A.14a}$$

and also

$$Q_{MM}\bar{R}_{MM}(n)Q_{MM}{}^T = \begin{bmatrix} R_{NN}(n+1) & \tilde{B}_{Np}(n) \\ \tilde{B}_{Np}(n)^T & \tilde{\pi}_{pp}(n) \end{bmatrix} \tag{A.14b}$$

$$\pi_{pp}(n) = \lambda \pi_{pp}(n-1) + \xi_p(n)\xi_p(n)^T \tag{A.15}$$

$$\tilde{\pi}_{pp}(n) = \lambda \tilde{\pi}_{pp}(n-1) + \rho_p(n)\rho_p(n)^T \tag{A.16}$$

with

$$\pi_{pp}(0) = \tilde{\pi}_{pp}(0) = \delta I_{pp}.$$

We shall first obtain an extended vector $\bar{R}_M(n)$ satisfying

$$\bar{R}_{MM}(n)\bar{k}_M(n) = \bar{x}_M(n). \tag{A.17}$$

Then the substitution of (A.14b) into (A.17) will lead to the desired expression for $k_N(n+1)$ that satisfies (A.12).

To find the extended vector $\bar{k}_M(n)$, we first note since $S_{MM}T = S_{MM}{}^{-1}$ that we can write (A.17) as

$$\bar{k}_M(n) = S_{MM}{}^T(S_{MM}\bar{R}_{MM}(n)S_{MM}{}^T)^{-1}S_{MM}\bar{x}_M(n), \tag{A.18}$$

assuming the inverse exists.

Then making use of the formula for the inverse of a partitioned matrix, equation (A.3) for $B_{Np}(n)$ and equation (4) for $S_{MM}\bar{x}_M(n)$, we get

$$\bar{k}_M(n) = S_{MM}{}^T \begin{bmatrix} E_{pp}(n)^{-1}\epsilon_p(n)' \\ k_N(n) + A_{Np}(n)E_{pp}(n)^{-1}\epsilon_p(n)' \end{bmatrix}, \tag{A.19}$$

where

$$E_{pp}(n) \equiv \pi_{pp}(n) + A_N(n)^T B_{Np}(n) \tag{A.20}$$

and

$$\epsilon_p(n)' \equiv \xi_p(n) + A_{Np}(n)^T x_N(n) \tag{A.21}$$

is the error between $\xi_p(n)$ and its *newly updated* predicted value $-A_{Np}(n)^T x_N(n)$.

The p by p matrix $E_{pp}(n)$ may be simplified, but first we remark that substitution of (A.3) into expression (A.2) defining the exponentially-weighted squared error results in an expression for the *minimum* exponentially weighted squared error which is identical to $E_{pp}(n)$. Thus $E_{pp}(n)$ is necessarily positive definite, and the inverse $E_{pp}(n)^{-1}$ exists. To simplify the computation of $E_{pp}(n)$, we note that it can be written recursively as

$$E_{pp}(n) = \lambda E_{pp}(n-1) + \xi_p(n)\xi_p(n)^T + A_{Np}(n)^T B_{Np}(n) - \lambda A_{Np}(n-1)^T B_{Np}(n-1). \tag{A.22}$$

Then application of (A.4b), (A.5), (A.3) and (A.21) yields

$$E_{pp}(n) = \lambda E_{pp}(n-1) + \epsilon_p(n)\epsilon_p(n)'^T. \tag{A.23}$$

We have thus found a vector $\bar{k}_M(n)$ which can be computed from the N-dimensional vectors $x_N(p)$ and $k_N(n)$, the N-by-p-dimensional matrices $A_{Np}(n)$ and $B_{Np}(n)$ which have been updated according to (A.5) and (A.4b) respectively, and the p-by-p-dimensional matrix $E_{pp}(n)$ updated according to (A.23). Multiplication by the M-by-M permutation matrix $S_M{}^T$ amounts to a simple permutation of vector components. The computational effort to obtain $\bar{k}_M(n)$ varies linearly with N. We shall presently see that the vector $k_N(n+1)$ can be obtained from $\bar{k}_M(n)$. Form

$$Q_{MM}\bar{k}_M(n)$$

(which again amounts to a simple permutation of vector components) and partition the result as follows:

$$\begin{bmatrix} m_N(n) \\ \mu_p(n) \end{bmatrix} \triangleq Q_{MM}\bar{k}_M(n). \tag{A.24}$$

We premultiply this by $Q_{MM}\bar{R}_{MM}(n)Q_{MM}{}^T$ of (A.14b), obtaining

$$\begin{aligned} Q_{MM}\bar{R}_{MM}(n)Q_{MM}{}^T Q_{MM}\bar{k}_M(n) &= Q_{MM}\bar{R}_{MM}(n)\bar{k}_M(n) \\ &= \begin{bmatrix} R_{NN}(n+1)m_N(n) + \tilde{B}_{Np}(n)\mu_p(n) \\ \tilde{B}_{Np}(n)^T m_N(n) + \tilde{\pi}_{pp}(n)\mu_p(n) \end{bmatrix}. \end{aligned} \tag{A.25}$$

Replacing $\bar{R}_{MM}(n)\bar{k}_M(n)$ by $\bar{x}_M(n)$, as in (A.17) and using the fact that

$$Q_{MM}\bar{x}_M(n) = \begin{bmatrix} x_N(n+1) \\ \rho_p(n) \end{bmatrix} \tag{A.26}$$

we have

$$R_{NN}(n+1)m_N(n) + \tilde{B}_{Np}(n)\mu_p(n) = x_N(n+1). \qquad (A.27)$$

But $\tilde{B}_{Np}(n)$ satisfies equation (A.8):

$$\tilde{B}_{Np}(n) = -R_{NN}(n+1)D_{Np}(n),$$

and so (A.27) can be written as

$$R_{NN}(n+1)[m_N(n) - D_{Np}(n)\mu_p(n)] = x_N(n+1), \quad (A.28)$$

which has the same form as (A.12) if

$$k_N(n+1) = m_N(n) - D_{Np}(n)\mu_p(n). \qquad (A.29)$$

Thus equation (A.29) updates the vector $k_N(n)$.

The updating of the prediction vector $D_{Np}(n)$, according to (A.10), requires the vector $k_N(n+1)$. This is avoided by solving (A.10) and (A.29) to obtain a recursion for $D_{Np}(n)$ not depending on $k_N(n+1)$. The result is

$$D_{Np}(n) = [D_{Np}(n-1) - m_N(n)\eta_p(n)^T][I_{pp} - \mu_p(n)\eta_p(n)^T]^{-1}. \qquad (A.30)$$

The invertibility of the p-by-p matrix in (A.30) is proven in reference [10].

ACKNOWLEDGMENT

We are grateful to R. D. Gitlin and F. R. Magee, Jr. for providing the sampled impulse response used in the simulations.

REFERENCES

[1] R. W. Lucky, "Automatic Equalization for Digital Communication", *Bell Sys. Tech. J.*, April, 1965, pp. 547-588.

[2] A. Gersho, "Adaptive Equalization of Highly Dispersive Channels for Data Transmission", *Bell Sys. Tech. J.*, January, 1969, pp. 55-70.

[3] B. Widrow and M. E. Hoff, Jr., "Adaptive Switching Circuits", *IRE Wescon Conv. Rec., pt. 4*, August, 1960, pp. 96-104.

[4] D. Godard, "Channel Equalization Using a Kalman Filter for Fast Data Transmission", *IBM J. Research and Development*, May, 1974, pp. 267-273.

[5] R. D. Gitlin and F. R. Magee, Jr., "Self-Orthogonalizing Adaptive Equalization Algorithms", *IEEE Trans. Communications*, July 1977, pp. 666-672.

[6] R. W. Chang, "A New Equalizer for Fast Start-Up Digital Communication", *Bell Sys. Tech. J.*, July-August, 1971, pp. 1969-2014.

[7] M. Morf, L. Ljung and T. Kailath, "Fast Algorithms for Recursive Identification", Proc. IEEE Conf. on Decision and Control, Clearwater Beach, Florida, December, 1976.

[8] M. Morf and L. Ljung, "Fast Algorithms for Recursive Identification", IEEE International Symposium on Information Theory, June, 1976, Ronneby, Sweden.

[9] N. Levinson, "The Wiener RMS (Root Mean Square) Error Criterion in Filter Design and Prediction", *J. Math. Phys.*, 1947, Vol. 25, pp. 261-278.

[10] L. Ljung, M. Morf, and D. D. Falconer, "Fast Calculation of Gain Matrices for Recursive Estimation Schemes", *International J. Control*, January, 1978.

[11] O. Macchi and L. Guidoux, "Un Nouvel Égaliseur: L' Égaliseur á Double Échantillonage", *Ann. Telecommun.*, Sept.-Oct. 1975, pp. 331-338.

[12] D. D. Falconer, K. H. Mueller and S. B. Weinstein, "Echo Cancellation Techniques for Full-Duplex Data Transmission on Two-Wire Lines", *Proc. Nat. Telecomm. Conf.*, Dallas, December 1976, pp. 877-883.

[13] T. Söderström, L. Ljung and I. Gustavsson, "A Comparative Study of Recursive Identification Methods", Report 7427, 1974, Dept. of Automatic Control, Lund Institute of Technology, Lund, Sweden.

[14] K. J. Aström and P. Eykhoff, "System Identification—A Survey", *Automatica*, Vol. 7, pp. 123-162, (1971).

[15] D. I. Steinberg, *Computational Matrix Algebra*, McGraw-Hill, 1974.

A Fast Sequential Algorithm for Least-Squares Filtering and Prediction

GEORGE CARAYANNIS, MEMBER, IEEE, DIMITRIS G. MANOLAKIS, STUDENT MEMBER, IEEE, AND NICHOLAS KALOUPTSIDIS, MEMBER, IEEE

***Abstract*—A new computationally efficient algorithm for sequential least-squares (LS) estimation is presented in this paper. This fast *a posteriori* error sequential technique (FAEST) requires $5p$ MADPR (multiplications and divisions per recursion) for AR modeling and $7p$ MADPR for LS FIR filtering, where p is the number of estimated parameters. In contrast the well-known fast Kalman algorithm requires $8p$ MADPR for AR modeling and $10p$ MADPR for FIR filtering. The increased computational speed of the introduced algorithm stems from an alternative definition of the so-called Kalman gain vector, which takes better advantage of the relationships between forward and backward linear prediction.**

I. Introduction

THE DESIGN of adaptive FIR filters with optimum learning, in the sense of minimizing the accumulated squared error between the output signal and a desired response, is of major importance in many areas of digital signal processing, estimation, and control.

Sequential LS algorithms are time recursive techniques which belong to this class of algorithms. Due to their extremely rapid convergence properties these schemes find many applications in speech processing, noise cancellation, design of fast start-up equalizers, event detection, spectral estimation, etc.

The conventional form of sequential LS algorithms requires $O(p^2)$ MADPR [1]-[3]. This complexity makes their real-time implementation a quite difficult task even with present day hardware. For multichannel signals, due to the enormous amount of computations, the problem is more crucial even in a parallel processing environment.

In contrast to the conventional implementation, the recently introduced [4], [5] fast Kalman algorithm requires $O(p)$ MAD per recursion. This high reduction of the computational complexity has raised a new growing interest in exact LS adaptive algorithms since they now require the same order of computations as the suboptimum gradient-type algorithms [6].

In conventional sequential LS algorithms the inverse covariance matrix is updated in each recursion and then a "Kalman type" gain is computed by a matrix by vector multiplication resulting in an $O(p^2)$ number of operations. The success of the fast Kalman algorithm in reducing the computational complexity to $O(p)$ operations is due to the fact that by exploiting some shift invariance properties a direct updating of the "Kalman type" gain is achieved, thus avoiding the matrix-by-vector multiplication. This fast gain updating requires $8p$ MADPR for the prewindowing signal case [4], [5] which implies a complexity of $8p$ ($10p$) MADPR for AR modeling (LS filtering). An algorithm for the more general unwindowed case has been developed in [7] and requires an additional amount of $5p$ MADPR.

In this paper a new fast sequential LS algorithm with increased computational speed is introduced. This fast *a posteriori* error sequential technique (FAEST) requires $5p$ MADPR for AR modeling and $7p$ MADPR for LS FIR filtering. The algorithm is developed for the prewindowed single-channel signal case. The more general unwindowed case for multichannel signals is treated in [10].

The complexity reduction in the FAEST algorithm is achieved through a different definition of the Kalman gain vector [3], [8], [9]. This definition is a natural consequence of the introduced "*a posteriori* error" formulation of the problem in contrast to the "*a priori* error" formulation in the fast Kalman algorithm. The time updating of the gain vector in FAEST requires only $5p$ MADPR.

Another interesting feature of the new algorithm is the balanced role, the forward and backward prediction play in its development.

The present paper is organized as follows. In Section II the principles of sequential least-squares filtering and prediction are introduced, thus preparing the ground for the subsequent development. Section III deals with the conventional sequential least-squares algorithm whereas Section IV introduces the fast Kalman algorithm. Next the FAEST algorithm is developed in Section V. The discussion of both schemes in a unified way reveals some interesting aspects of these algorithms and makes the comparison easier. Finally, the extension to the exponentially weighted case and some simulation results are given in Section VI.

II. Sequential Least-Squares: Principles and Definitions

The purpose of this section is twofold: first, to introduce the problems of least-squares FIR filtering, forward linear prediction (FLP), and backward linear prediction (BLP) for prewindowed signals; second, to obtain two alternative time updating procedures (i.e., sequential solutions) for the LS filter,

Manuscript received October 1, 1982; revised April 28, 1983.

G. Carayannis is with the Council of Europe, 67006 Strasbourg, France.

D. G. Manolakis is with the Department of Electronics, National Technical University of Athens, Athens, Greece.

N. Kalouptsidis is with the Department of Physics, Division of Electronics, University of Athens, Athens, Greece.

Reprinted from *IEEE Trans. Acoust., Speech, Signal Processing,* vol. ASSP-31, no. 6, pp. 1394-1402, Dec. 1983.

the FLP, and the BLP. Additionally, sequential relationships for the computation of the corresponding minimum total squared errors are derived. It is worth noting that the development of these recursions is independent to the existence or not of fast algorithms.

The time domain input-output relationship for an mth order FIR filter is

$$y_m(n) = -\sum_{j=1}^{m} c_j(n)x(n+1-j) = -\mathbf{c}_m^t(n)\mathbf{x}_m(n) \tag{1}$$

where

$$\mathbf{c}_m(n) = [c_1(n) \quad c_2(n) \quad \cdots \quad c_m(n)]^t \tag{2}$$

and

$$\mathbf{x}_m(n) = [x(n) \quad x(n-1) \quad \cdots \quad x(n-m+1)]^t \tag{3}$$

are the filter coefficient vector and the input signal vector, respectively.

Let $z(n)$ be a desired response signal and suppose that, at time instant n, we have the following finite record of data at our disposal:

$$\{x(j), z(j), j = M, M+1, \cdots, n\}.$$

In addition we assume that $x(j) = z(j) = 0$ for $j < M$ (prewindowing assumption).

Minimization of the total squared error

$$\sum_{j=M}^{n} \epsilon_m^2(j) \tag{4}$$

where

$$\epsilon_m(j) = z(j) + \mathbf{c}_m^t(n)\mathbf{x}_m(j) \tag{5}$$

leads to the following set of normal equations, which determines the LS filter $\mathbf{c}_m(n)$ at time n:

$$\mathbf{R}_m(n)\mathbf{c}_m(n) = -\mathbf{r}_m(n) \tag{6}$$

where

$$\mathbf{R}_m(n) = \sum_{j=M}^{n} \mathbf{x}_m(j)\mathbf{x}_m^t(j) = X_m^t(n)X_m(n) \tag{7}$$

$$\mathbf{r}_m(n) = \sum_{j=M}^{n} \mathbf{x}_m(j)z(j) = X_m^t(n)\zeta_M(n) \tag{8}$$

$$X_m(n) = \begin{bmatrix} x(M) & & \bigcirc \\ x(M+1) & x(M) & \\ \vdots & & \ddots \\ x(M+m-1) & \cdots & x(M) \\ \vdots & & \vdots \\ x(n) & x(n-1) & x(n-m+1) \end{bmatrix} \tag{9}$$

$$\zeta_M(n) = [z(M) \quad z(M+1) \quad \cdots \quad z(n)]^t. \tag{10}$$

The attained minimum value of the cost function is

$$E_m(n) = \mathbf{r}_m^t(n)\mathbf{c}_m(n) + \sum_{j=M}^{n} z^2(j). \tag{11}$$

Prediction is a special case of filtering. The problems of FLP and BLP are, respectively, defined by

$$e_m^f(n) = x(n) - \hat{x}(n) = x(n) + \mathbf{a}_m^t(n)\mathbf{x}_m(n-1) \tag{12}$$

$$\epsilon_m^b(n) = x(n-m) - \hat{x}(n-m) = x(n-m) + \mathbf{b}_m^t(n)\mathbf{x}_m(n). \tag{13}$$

It can be easily shown that the corresponding LS predictors are determined by the following systems of normal equations:

$$\mathbf{R}_m(n-1)\mathbf{a}_m(n) = -\mathbf{r}_m^f(n) \tag{14}$$

$$\mathbf{R}_m(n)\mathbf{b}_m(n) = -\mathbf{r}_m^b(n) \tag{15}$$

where

$$\mathbf{r}_m^f(n) = \sum_{j=M}^{n} \mathbf{x}_m(j-1)x(j) \tag{16}$$

$$\mathbf{r}_m^b(n) = \sum_{j=M}^{n} \mathbf{x}_m(j)x(j-m). \tag{17}$$

The minimum values of the total squared errors are

$$E_m^f(n) = \mathbf{a}_m^t(n)\mathbf{r}_m^f(n) + \sum_{j=M}^{n} x^2(j) \tag{18}$$

$$E_m^b(n) = \mathbf{b}_m^t(n)\mathbf{r}_m^b(n) + \sum_{j=M}^{n} x^2(j-m). \tag{19}$$

We stress once more that FLP, BLP, and their interrelationships play a central role in the development of fast algorithms for AR modeling and LS FIR filtering. Note that the problem of LS FLP is equivalent to that of fitting an mth-order AR model to the data set $\{x(j), j = M, M+1, \cdots, n\}$.

The key to the development of sequential algorithms is the existence of simple time updating formulas for the parameters of the normal equations (6), (14), and (15). Indeed using (7), (8), (16), and (17) it can be easily shown that

$$\mathbf{R}_m(n+1) = \mathbf{R}_m(n) + \mathbf{x}_m(n+1)\mathbf{x}_m^t(n+1) \tag{20}$$

$$\mathbf{r}_m(n+1) = \mathbf{r}_m(n) + \mathbf{x}_m(n+1)z(n+1) \tag{21}$$

$$\mathbf{r}_m^f(n+1) = \mathbf{r}_m^f(n) + \mathbf{x}_m(n)x(n+1) \tag{22}$$

$$\mathbf{r}_m^b(n+1) = \mathbf{r}_m^b(n) + \mathbf{x}_m(n+1)x(n-m+1). \tag{23}$$

Assume now that the filter $\mathbf{c}_m(n)$ is already available. The problem is to obtain $\mathbf{c}_m(n+1)$ in terms of $\mathbf{c}_m(n)$ and the new observations $x(n+1)$ and $z(n+1)$. The LS filter at time $n+1$ is determined by

$$\mathbf{R}_m(n+1)\mathbf{c}_m(n+1) = -\mathbf{r}_m(n+1). \tag{24}$$

Substitution of (20) and (21) into (6) gives

$$[\mathbf{R}_m(n+1) - \mathbf{x}_m(n+1)\mathbf{x}_m^t(n+1)]\mathbf{c}_m(n) = -[\mathbf{r}_m(n+1) - \mathbf{x}_m(n+1)z(n+1)]$$

or after some simple manipulations

$$\mathbf{R}_m(n+1)\mathbf{c}_m(n) - \mathbf{x}_m(n+1)e_m(n+1) = -\mathbf{r}_m(n+1) \tag{25}$$

where

$$e_m(n+1) = z(n+1) + c_m^t(n)x_m(n+1). \tag{26}$$

The quantity $e_m(n+1)$ will be called *a priori error* to distinguish it from the so-called *a posteriori error* defined by [see (5)]

$$\epsilon_m(n+1) = z(n+1) + c_m^t(n+1)x_m(n+1). \tag{27}$$

The reasons for this terminology will be justified later on.

We next define the so-called Kalman gain vector $w_m^*(n+1)$ by

$$R_m(n+1)w_m^*(n+1) = -x_m(n+1). \tag{28}$$

Substitution of (28) into (25) and comparison with (24) supplies the following time updating relation for the filter $c_m(n)$:

$$c_m(n+1) = c_m(n) + w_m^*(n+1)e_m(n+1). \tag{29}$$

This formula is widely used in sequential least-squares system identification in the context of the equation error formulation [11].

We next derive an alternative to (29) time updating procedure. Substitution of (20) and (21) into (24) and taking into account (27) gives

$$R_m(n)c_m(n+1) + x_m(n+1)\epsilon_m(n+1) = -r_m(n). \tag{29a}$$

If we define an "alternative Kalman gain" by

$$R_m(n)w_m(n+1) = -x_m(n+1) \tag{30}$$

and we combine (30) with (29a), we obtain

$$R_m(n)[c_m(n+1) - w_m(n+1)\epsilon_m(n+1)] = -r_m(n)$$

which compared to (6) supplies

$$c_m(n+1) = c_m(n) + w_m(n+1)\epsilon_m(n+1). \tag{31}$$

We note that $w_m(n+1)$ contains "in some way" less time domain information than $w_m^*(n+1)$. The errors $e_m(n+1)$ and $\epsilon_m(n+1)$ include the same information about the desired response but $\epsilon_m(n+1)$ is more "up-to-date" concerning the model evolution with time. Considering now the model updating it is quite natural to associate $w_m(n+1)$ with $\epsilon_m(n+1)$ and $w_m^*(n+1)$ with $e_m(n+1)$. Simple inspection of (27) and (31) indicates that (31) cannot be used to update the filter parameters since the unknown vector $c_m(n+1)$ appears also in (27). We will show in the next section how to overcome this difficulty.

Analogous results are easily obtained for the problems of FLP and BLP in a similar way.

For easy reference all these results are summarized in Table I.

The instantaneous estimation errors denoted by the letter e are called *a priori* since their computation at a given time requires the parameter vector already available, i.e., *before* the time updating. In contrast the so-called *a posteriori* errors, denoted by the Greek letter ϵ, supply the estimation error based on the parameter vector obtained *after* the time updating. Both types of errors and their relationships play an important role in the new fast algorithm introduced in this paper.

We derive next a sequential formula for the computation of the minimum total error $E_m(n)$ in terms of the *a priori* and *a posteriori* errors $e_m(n)$ and $\epsilon_m(n)$. Indeed, (11) implies that

$$E_m(n+1) = r_m^t(n+1)c_m(n+1) + \sum_{j=M}^{n+1} z^2(j) \tag{33}$$

which combined with (32a) [(32) is in Table I] gives

$$E_m(n+1) = \sum_{j=M}^{n} z^2(j) + z^2(n+1) + r_m^t(n+1)c_m(n) + r_m^t(n+1)w_m^*(n+1)e_m(n+1). \tag{34}$$

Using (21), (11), and (32h), we take

$$\begin{aligned}
&\sum_{j=M}^{n} z^2(j) + z^2(n+1) + r_m^t(n+1)c_m(n) \\
&\quad = \sum_{j=M}^{n} z^2(j) + r_m^t(n)c_m(n) \\
&\qquad + z(n+1)[z(n+1) + c_m^t(n)x_m(n+1)] \\
&\quad = E_m(n) + z(n+1)e_m(n+1).
\end{aligned}$$

Substitution of the last relation into (34) gives

$$\begin{aligned}
E_m(n+1) &= E_m(n) + [z(n+1) + r_m^t(n+1)w_m^*(n+1)] \\
&\qquad \cdot e_m(n+1) \\
&= E_m(n) + [z(n+1) + r_m^t(n+1)R_m^{-1}(n+1) \\
&\qquad \cdot R_m(n+1)w_m^*(n+1)]e_m(n+1) \\
&= E_m(n) + [z(n+1) + c_m^t(n+1)x_m(n+1)] \\
&\qquad \cdot e_m(n+1)
\end{aligned}$$

where we have used (24) and (32g). Taking into account (32i) we obtain

$$E_m(n+1) = E_m(n) + \epsilon_m(n+1)e_m(n+1) \tag{35}$$

which is the desired formula.

In a similar way we obtain for the FLP and the BLP cases

$$E_m^f(n+1) = E_m^f(n) + \epsilon_m^f(n+1)e_m^f(n+1) \tag{36}$$

$$E_m^b(n+1) = E_m^b(n) + \epsilon_m^b(n+1)e_m^b(n+1). \tag{37}$$

Concluding this section we note that the *a priori* formulation does not yet lead to an efficient recursive algorithm since it requires at each step the solution of a linear system for the computation of the Kalman gain vector. This implies a computational complexity of $O(m^3)$ MADPR.

III. The Conventional Time Recursive LS Algorithm

The algorithms developed so far are not yet fully recursive since they require at each recursion the solution of a linear system for the Kalman gain vector. This includes a matrix inversion and a matrix by vector multiplication with computational complexity of $O(m^3)$ and $O(m^2)$, respectively. The simplicity of recursive LS schemes is due to the fact that the matrix inversion at each step is replaced by a simple scalar division and thus complexity reduces to $O(m^2)$. This is achieved by exploiting the low rank updating property (20).

Indeed, using the matrix inversion lemma (or Sherman-Morrisson identity) [11], (20) gives

TABLE I

A Priori and *A Posteriori* Error Formulations for Sequential LS FIR Filtering, LS FLP and LS BLP

	Parameter Updating		Error Definition	
FIR Filtering	$c_m(n+1) = c_m(n) + w_m^*(n+1)\, e_m(n+1)$	(a)	$e_m(n+1) = z(n+1) + c_m^t(n)\, x_m(n+1)$	(h)
	$c_m(n+1) = c_m(n) + w_m(n+1)\, \epsilon_m(n+1)$	(b)	$\epsilon_m(n+1) = z(n+1) + c_m^t(n+1)\, x_m(n+1)$	(i)
FLP	$a_m(n+1) = a_m(n) + w_m^*(n)\, e_m^f(n+1)$	(c)	$e_m^f(n+1) = x(n+1) + a_m^t(n)\, x_m(n)$	(j)
	$a_m(n+1) = a_m(n) + w_m(n)\, \epsilon_m^f(n+1)$	(d)	$\epsilon_m^f(n+1) = x(n+1) + a_m^t(n+1)\, x_m(n)$	(k)
BLP	$b_m(n+1) = b_m(n) + w_m^*(n+1)\, e_m^b(n+1)$	(e)	$e_m^b(n+1) = x(n+1-m) + b_m^t(n)\, x_m(n+1)$	(l)
	$b_m(n+1) = b_m(n) + w_m(n+1)\, \epsilon_m^b(n+1)$	(f)	$\epsilon_m^b(n+1) = x(n+1-m) + b_m^t(n+1)\, x_m(n+1)$	(m)
	Gain Definition			
	$R_m(n+1)\, w_m^*(n+1) = -x_m(n+1)$	(g)	$R_m(n)\, w_m(n+1) = -x_m(n+1)$	(n)

(32)

$$R_m^{-1}(n+1) = R_m^{-1}(n) - \frac{R_m^{-1}(n)x_m(n+1)x_m^t(n+1)R_m^{-1}(n)}{1 + x_m^t(n+1)R_m^{-1}(n)x_m(n+1)}. \tag{38}$$

Defining the quantities

$$P_m(n) = R_m^{-1}(n) \tag{39}$$

$$\alpha_m(n+1) = 1 - x_m^t(n+1)\, w_m(n+1), \tag{40}$$

(38) becomes

$$P_m(n+1) = P_m(n) - w_m(n+1)\, w_m^t(n+1)/\alpha_m(n+1) \tag{41}$$

which constitutes a simple updating formula for the inverse covariance matrix.

Next a simple relationship between $w_m^*(n+1)$ and $w_m(n+1)$ is established. Indeed, (32g) gives

$$w_m^*(n+1) = -P_m(n+1)\, x_m(n+1)$$

and using (41) it follows that

$$w_m^*(n+1) = -[P_m(n) - w_m(n+1)\, w_m^t(n+1)\, \alpha_m^{-1}(n+1)] \cdot x_m(n+1).$$

Taking into consideration (32n) and (40), after some simple manipulations, the latter relation leads to

$$w_m^*(n+1) = w_m(n+1)/\alpha_m(n+1). \tag{42}$$

The gain vector $w_m^*(n)$ can be recursively updated in $O(m^2)$ operations using the following scheme:

$$w_m(n+1) = -P_m(n)x_m(n+1) \tag{43a}$$

$$\alpha_m(n+1) = 1 - w_m^t(n+1)\, x_m(n+1) \tag{43b}$$

$$w_m^*(n+1) = w_m(n+1)/\alpha_m(n+1) \tag{43c}$$

$$P_m(n+1) = P_m(n) - w_m^*(n+1)\, w_m^t(n+1). \tag{43d}$$

The initialization of (43) can be done either by evaluating P for an initial block of data or by simply setting $P_m = I/\sigma$, σ being a small positive constant and I the identity matrix.

It is worth noting that all recursive LS algorithms are based on the recursive updating of the positive definite inverse covariance matrix. This procedure is known to be sometimes ill-conditioned. A numerically more robust implementation can be obtained by updating the square-root of $P_m(n)$ which is taken from the LU or LDU decomposition of this matrix (square-root algorithms [12]).

We will now establish some useful relations between the *a priori* and *a posteriori* errors. Indeed comparison of (32a) with (32b), (32c) with (32d), and (32e) with (32f) in conjunction with (42) supplies

$$\epsilon_m(n+1) = \frac{e_m(n+1)}{\alpha_m(n+1)} \tag{44a}$$

$$\epsilon_m^f(n+1) = \frac{e_m^f(n+1)}{\alpha_m(n)} \tag{44b}$$

$$\epsilon_m^b(n+1) = \frac{e_m^b(n+1)}{\alpha_m(n+1)}. \tag{44c}$$

A remarkable consequence of (44) is that the *a posteriori* errors at time $n+1$ can be computed before the filter parameters producing them, i.e., before the computation of $c_m(n+1)$, $a_m(n+1)$, and $b_m(n+1)$!

Next we further explore the nature of the parameter $\alpha_m(n)$. Postmultiplying (41) by $R_m(n)$ we easily get

$$R_m^{-1}(n+1)R_m(n) = I + w_m(n+1)x_m^t(n+1)\,\alpha_m^{-1}(n+1).$$

The determinant of this relation is

$$[\det R_m(n+1)]^{-1} \det R_m(n) = \det [I + w_m(n+1)x_m^t(n+1)\,\alpha_m^{-1}(n+1)].$$

Using the identity

$$\det [I + BC] = \det [I + CB]$$

and (43b), we obtain

$$\alpha_m(n+1) = \frac{\det R_m(n+1)}{\det R_m(n)}. \tag{45}$$

Combining (32n) and (40) gives

$$\alpha_m(n+1) = 1 + x_m^t(n+1)R_m^{-1}(n)x_m(n+1). \tag{46}$$

Since $R_m(n)$ is positive definite, the same is true for its inverse; thus

$$0 \leqslant \frac{1}{\alpha_m(n+1)} \leqslant 1. \tag{47}$$

We establish now a simple relation between the parameter

$\alpha_m(n)$ and the so-called likelihood variable $\gamma_m(n)$ [13] defined by

$$\gamma_m(n+1) = x_m^t(n+1)R_m^{-1}(n+1)x_m(n+1) = -w_m^{*t}(n+1)x_m(n+1). \tag{48}$$

Let

$$\theta_m(n+1) = x_m^t(n+1)R_m^{-1}(n)x_m(n+1) = \alpha_m(n+1) - 1. \tag{49}$$

Pre- and postmultiplying (38) by $x_m^t(n+1)$ and $x_m(n+1)$ we obtain

$$\gamma_m(n+1) = \frac{\theta_m(n+1)}{1+\theta_m(n+1)}. \tag{50}$$

Since $R_m(n)$ is positive definite, $\theta_m(n+1) > 0$ and it follows that

$$0 \leqslant \gamma_m(n+1) \leqslant 1. \tag{51}$$

The last relation allows for the interpretation of $\gamma_m(n)$ as an angle variable [13], [14].

Substitution of (49) into (50) gives

$$\alpha_m(n+1) = \frac{1}{1-\gamma_m(n+1)}. \tag{52}$$

Thus $\alpha_m(n)$ is the optimal gain encountered in LS lattice filters [14]. On the other hand, (47) or (51) guarantees the invertibility of $R_m(n+1)$. In addition, (47) allows for the interpretation of $1/\alpha_m(n)$ as an angle variable.

Finally we note that (44a) and (47) easily imply the following intuitively expected remarks:

- the *a posteriori* errors are less in magnitude than the *a priori* errors;
- the variance of the *a posteriori* errors is less than that of the *a priori* errors.

IV. The Fast Kalman Algorithm

A closer look at the Kalman gain vector $w_m^*(n)$ defined at times n and $n+1$ by

$$R_m(n)\,w_m^*(n) = -x_m(n) \tag{53}$$

$$R_m(n+1)\,w_m^*(n+1) = -x_m(n+1) \tag{54}$$

indicates that successive gain vectors are obtained from the solution of linear systems whose right-hand side members are shifted versions of each other. Fast algorithms for such families of systems, in case the system matrix is time invariant with a Toeplitz or near-to-Toeplitz structure, have been developed in [15], [16].

Suppose now that $w_m^*(n)$ is known, the problem is to compute $w_m^*(n+1)$ using $w_m^*(n)$ or in other words to find a recursion for the Kalman gain vector.

To take advantage of the shifting nature of the right-hand side vector $x_m(n)$ we consider, as in [5], the following partitionings of $x_{m+1}(n)$:

$$x_{m+1}(j) = \begin{bmatrix} x_m(j) \\ x(j-m) \end{bmatrix} = \begin{bmatrix} x(j) \\ x_m(j-1) \end{bmatrix}. \tag{55}$$

Using (55), (7), (16), and (17) we obtain the following partitionings of $R_m(n+1)$ for prewindowed signals:

$$R_{m+1}(n) = \left[\begin{array}{c|c} r_{om}^f(n) & r_m^{ft}(n) \\ \hline r_m^f(n) & R_m(n-1) \end{array}\right] \tag{56}$$

$$R_{m+1}(n) = \left[\begin{array}{c|c} R_m(n) & r_m^b(n) \\ \hline r_m^{bt}(n) & r_{om}^b(n) \end{array}\right] \tag{57}$$

where

$$r_{om}^f(n) = \sum_{j=M}^{n} x^2(j) \tag{58}$$

$$r_{om}^b(n) = \sum_{j=M}^{n} x^2(j-m). \tag{59}$$

It is emphasized that the simple form taken by (56), (57) is exclusively due to the prewindowing condition. In the case of unwindowed signals these partitioning relations have a more complex form [7], [10].

Using the matrix inversion for partitioned matrices [8], [11], (56) and (57) give

$$R_{m+1}^{-1}(n+1) = \left[\begin{array}{c|c} 1/\alpha_m^f(n+1) & a_m^t(n+1)/\alpha_m^f(n+1) \\ \hline a_m(n+1)/\alpha_m^f(n+1) & R_m^{-1}(n) + a_m(n+1)a_m^t(n+1)/\alpha_m^f(n+1) \end{array}\right] \tag{60}$$

where

$$\alpha_m^f(n+1) = r_{om}^f(n+1) + a_m^t(n+1)r_m^f(n+1) = \frac{\det R_{m+1}(n+1)}{\det R_m(n)} \tag{61}$$

and

$$R_{m+1}^{-1}(n+1) = \left[\begin{array}{c|c} R_m^{-1}(n+1) + b_m(n+1)b_m^t(n+1)/\alpha_m^b(n+1) & b_m(n+1)/\alpha_m^b(n+1) \\ \hline b_m^t(n+1)/\alpha_m^b(n+1) & 1/\alpha_m^b(n+1) \end{array}\right] \tag{62}$$

where

$$\alpha_m^b(n+1) = r_{om}^b(n+1) + b_m^t(n+1)r_m^b(n+1) = \frac{\det R_{m+1}(n+1)}{\det R_m(n+1)}, \tag{63}$$

respectively.

To compute $w_m^*(n+1)$ we first determine the gain $w_{m+1}^*(n+1)$ defined by

$$w_{m+1}^*(n+1) = -R_{m+1}^{-1}(n+1)x_{m+1}(n+1). \tag{64}$$

Combining (64) with (62), (55), and (54) after some simple manipulations we obtain the following Levinson type recursion:

$$w^*_{m+1}(n+1) = \begin{bmatrix} w^*_m(n+1) \\ 0 \end{bmatrix} - \frac{\epsilon^b_m(n+1)}{\alpha^b_m(n+1)} \begin{bmatrix} b_m(n+1) \\ 1 \end{bmatrix}. \tag{65}$$

Similarly (64), (60), (55), and (53) give

$$w^*_{m+1}(n+1) = \begin{bmatrix} 0 \\ w^*_m(n) \end{bmatrix} - \frac{\epsilon^f_m(n+1)}{\alpha^f_m(n+1)} \begin{bmatrix} 1 \\ a_m(n+1) \end{bmatrix}. \tag{66}$$

Since $w^*_m(n)$ is known, we can compute $w^*_{m+1}(n+1)$ by (66) and then $w^*_m(n+1)$ by (65). Note that since (65) runs backwards the computation of $\epsilon^b_m(n+1)$, $\alpha^b_m(n+1)$ is not necessary. Indeed, if we consider the partitioning

$$w^*_{m+1}(n+1) = \begin{bmatrix} d^*_m(n+1) \\ \delta^*_m(n+1) \end{bmatrix}, \tag{67}$$

comparison with (65) gives

$$\frac{\epsilon^b_m(n+1)}{\alpha^b_m(n+1)} = -\delta^*_m(n+1). \tag{68}$$

Unfortunately we have not a closed recursion, because $b_m(n+1)$ is not yet available. To complete the recursion we exploit the coupling between $w^*_m(n+1)$ and $b_m(n+1)$ as it is expressed by the "system of equations" (65) and (32e). The solution of this system is easily found to be

$$w^*_m(n+1) = \frac{d^*_m(n+1) - \delta^*_m(n+1)\, b_m(n)}{1 + \delta^*_m(n+1)\, e^b_m(n+1)}. \tag{69}$$

Comparing (61), (63) with (18), (19) gives

$$\alpha^f_m(n) = E^f_m(n)$$

$$\alpha^b_m(n) = E^b_m(n). \tag{70}$$

Thus, these quantities can be computed sequentially by

$$\alpha^f_m(n+1) = \alpha^f_m(n) + \epsilon^f_m(n+1)\, e^f_m(n+1) \tag{71}$$

$$\alpha^b_m(n+1) = \alpha^b_m(n) + \epsilon^b_m(n+1)\, e^b_m(n+1). \tag{72}$$

The above algorithm is summarized in Table II which includes also an operations count. This scheme which is known as a fast Kalman algorithm was introduced in [4], [5] and needs $8m$ MAD per recursion for AR modeling and $10m$ MAD for LS FIR filtering.

V. The Fast *A Posteriori* Error Sequential Technique (FAEST)

A simple inspection of Table II reveals that the fast Kalman algorithm is mainly based on the *a priori* error formulation. In this section we will introduce a new algorithm based, to a great extent, on the *a posteriori* error formulation. In addition the new scheme makes a better exploitation of the relationships between *a priori* and *a posteriori* errors.

The key to the development of the fast sequential algorithm FAEST is the use of the gain vector $w_m(n)$ instead of $w^*_m(n)$. To establish a fast time updating procedure for $w_m(n)$ we obtain from (32n)

$$R_m(n-1)\, w_m(n) = -x_m(n) \tag{73}$$

$$R_m(n)\, w_m(n+1) = -x_m(n+1) \tag{74}$$

$$R_{m+1}(n)\, w_{m+1}(n+1) = -x_{m+1}(n+1). \tag{75}$$

Working as in Section IV, combination of (75), (74), and (62) with (55) yields the Levinson type recursion

$$w_{m+1}(n+1) = \begin{bmatrix} w_m(n+1) \\ 0 \end{bmatrix} - \frac{e^b_m(n+1)}{\alpha^b_m(n)} \begin{bmatrix} b_m(n) \\ 1 \end{bmatrix}. \tag{76}$$

In a similar way using (75), (73), (60), and (55), we obtain

$$w_{m+1}(n+1) = \begin{bmatrix} 0 \\ w_m(n) \end{bmatrix} - \frac{e^f_m(n+1)}{\alpha^f_m(n)} \begin{bmatrix} 1 \\ a_m(n) \end{bmatrix}. \tag{77}$$

Thus $w_m(n)$ can be updated through (76), (77) by the following scheme:

$$w_m(n) \xrightarrow{(77)} w_{m+1}(n+1) \xrightarrow{(76)} w_m(n+1).$$

Table I and (76), (77) indicate that the algorithm requires the computation of both *a priori* and *a posteriori* errors. To reduce the computations we use the definitions of *a priori* errors and then we compute the *a posteriori* ones by (44).

We next establish a recursive procedure for the computation of $\alpha_m(n+1)$ which offers an additional gain of m MADPR. This can be done in two steps as is indicated in the following. From (43b) we have

$$\alpha_{m+1}(n+1) = 1 - x^t_{m+1}(n+1)\, w_{m+1}(n+1). \tag{78}$$

Using (55) and (77), (78) gives

TABLE II

Computational Organization and Complexity of the Fast Kalman Algorithm, which Requires $8m + 2$ MAD per Recursion for AR Modeling and $10m + 2$ for FIR Filtering

The Fast Kalman Algorithm

Available at time n: $w^*_m(n), a_m(n), b_m(n), c_m(n), x_m(n)$
New Information: $x(n+1), z(n+1)$

Time Updating of the Gain Vector	MADPR
$e^f_m(n+1) = x(n+1) + a^t_m(n)\, x_m(n)$	m
$a_m(n+1) = a_m(n) + w^*_m(n)\, e^f_m(n+1)$	m
$\epsilon^f_m(n+1) = x(n+1) + a^t_m(n+1)\, x_m(n)$	m
$\alpha^f_m(n+1) = \alpha^f_m(n) + e^f_m(n+1)\, \epsilon^f_m(n+1)$	1
$w^*_{m+1}(n+1) = \begin{bmatrix} 0 \\ w^*_m(n) \end{bmatrix} - \frac{\epsilon^f_m(n+1)}{\alpha^f_m(n+1)} \begin{bmatrix} 1 \\ a_m(n+1) \end{bmatrix}$	$m + 1$
Partition: $w^*_{m+1}(n+1) = \begin{bmatrix} d^*_m(n+1) \\ \delta^*_m(n+1) \end{bmatrix}$	0
$e^b_m(n+1) = x(n+1-m) + b^t_m(n)\, x_m(n+1)$	m
$w^*_m(n+1) = \frac{d^*_m(n+1) - \delta^*_m(n+1)\, b_m(n)}{1 + \delta^*_m(n+1)\, e^b_m(n+1)}$	$2m$
$b_m(n+1) = b_m(n) + w^*_m(n+1)\, e^b_m(n+1)$	m
Total Number of Operations	$8m + 2$
Time Updating of the LS FIR Filter	**MADPR**
$e_m(n+1) = z(n+1) + c^t_m(n)\, x_m(n+1)$	m
$c_m(n+1) = c_m(n) + w^*_m(n+1)\, e_m(n+1)$	m

$$\alpha_{m+1}(n+1) = 1 - [x(n+1)x_m^t(n)]\begin{bmatrix} 0 \\ w_m(n) \end{bmatrix}$$

$$+ [x(n+1)x_m^t(n)]\begin{bmatrix} 1 \\ a_m(n) \end{bmatrix}\frac{e_m^f(n+1)}{\alpha_m^f(n)}$$

which due to (43b) and (32j) results in

$$\alpha_{m+1}(n+1) = \alpha_m(n) + \frac{e_m^f(n+1)}{\alpha_m^f(n)} e_m^f(n+1). \tag{79}$$

To obtain $\alpha_m(n+1)$ we recall from (78) that

$$\alpha_{m+1}(n+1) = 1 - x_{m+1}^t(n+1) w_{m+1}(n+1)$$

which using (55) and (76) becomes

$$\alpha_{m+1}(n+1) = 1 - [x_m^t(n+1)x(n+1-m)]\begin{bmatrix} w_m(n+1) \\ 0 \end{bmatrix}$$

$$+ [x_m^t(n+1)x(n+1-m)]\begin{bmatrix} b_m(n) \\ 1 \end{bmatrix}\frac{e_m^b(n+1)}{\alpha_m^b(n)}.$$

Taking into account (43b) and (32l) the last relation gives

$$\alpha_{m+1}(n+1) = \alpha_m(n+1) + \frac{e_m^b(n+1)}{\alpha_m^b(n)} e_m^b(n+1)$$

or

$$\alpha_m(n+1) = \alpha_{m+1}(n+1) - \frac{e_m^b(n+1)}{\alpha_m^b(n)} e_m^b(n+1). \tag{80}$$

Thus to time update $\alpha_m(n)$ we first compute $\alpha_{m+1}(n+1)$ by (79) and then $\alpha_m(n+1)$ by (80).

The exact sequence of computational procedures and the number of operations for the fast sequential algorithm are given in Table III.

Formula (80) can be further simplified, since from Table III we have

$$\delta_m(n+1) = -\frac{e_m^b(n+1)}{\alpha_m^b(n)}.$$

Thus (80) becomes

$$\alpha_m(n+1) = \alpha_{m+1}(n+1) + \delta_m(n+1) e_m^b(n+1). \tag{81}$$

From (45), (61), and (63) it follows that

$$\alpha_m(n+1) = \frac{\alpha_m^f(n+1)}{\alpha_m^b(n+1)}. \tag{82}$$

Combining (72) and (44c) gives

$$\alpha_m^b(n+1) = \alpha_m^b(n) + \frac{[e_m^b(n+1)]^2}{\alpha_m(n+1)}. \tag{83}$$

Substitution of (83) into (82) leads to

$$\alpha_m(n+1) = \frac{\alpha_m^f(n+1) - [e_m^b(n+1)]^2}{\alpha_m^b(n)}. \tag{84}$$

A simple inspection of (76) shows that there is no coupling anymore between the computation of the gain vector and the computation of the optimum backward predictor. This is due to the fact that $b_m(n)$ and *not* $b_m(n+1)$ appears in (76). Thus the new scheme involves only Levinson type recursions in contrast to the fast Kalman algorithm which includes more complex expressions [see (69)].

Finally we note that the FAEST algorithm needs $5m$ MADPR for AR modeling and $7m$ MADPR for LS FIR filtering.

TABLE III
COMPUTATIONAL ORGANIZATION AND COMPLEXITY OF THE FAEST ALGORITHM

The Fast *a posteriori* Error Sequential Technique
Available at Time n: $w_m(n), a_m(n), b_m(n), x_m(n), \alpha_m^f(n), \alpha_m^b(n), \alpha_m(n), c_m(n)$
New Information: $x(n+1), z(n+1)$

Time Updating of the Gain Vector	MADPR
$e_m^f(n+1) = x(n+1) + a_m^t(n)x_m(n)$	m
$\epsilon_m^f(n+1) = e_m^f(n+1)/\alpha_m(n)$	1
$a_m(n+1) = a_m(n) + w_m(n)\epsilon_m^f(n+1)$	m
$\alpha_m^f(n+1) = \alpha_m^f(n) + e_m^f(n+1)\epsilon_m^f(n+1)$	1
$w_{m+1}(n+1) = \begin{bmatrix} 0 \\ w_m(n) \end{bmatrix} - \frac{e_m^f(n+1)}{\alpha_m^f(n)}\begin{bmatrix} 1 \\ a_m(n) \end{bmatrix}$	$m+1$
Partition: $w_{m+1}(n+1) = \begin{bmatrix} d_m(n+1) \\ \delta_m(n+1) \end{bmatrix}$	0
$e_m^b(n+1) = -\delta_m(n+1)\alpha_m^b(n)$	1
$w_m(n+1) = d_m(n+1) - \delta_m(n+1)b_m(n)$	m
$\alpha_{m+1}(n+1) = \alpha_m(n) + \frac{e_m^f(n+1)}{\alpha_m^f(n)} \cdot e_m^f(n+1)$	1
$\alpha_m(n+1) = \alpha_{m+1}(n+1) + \delta_m(n+1) \cdot e_m^b(n+1)$	1
$\epsilon_m^b(n+1) = e_m^b(n+1)/\alpha_m(n+1)$	1
$\alpha_m^b(n+1) = \alpha_m^b(n) + e_m^b(n+1)\epsilon_m^b(n+1)$	1
$b_m(n+1) = b_m(n) + w_m(n+1)\epsilon_m^b(n+1)$	m
Total Number of Operations	$5m+8$
Time Updating of the LS FIR Filter	MADPR
$e_m(n+1) = z(n+1) + c_m^t(n)x_m(n+1)$	m
$\epsilon_m(n+1) = e_m(n+1)/\alpha_m(n+1)$	1
$c_m(n+1) = c_m(n) + w_m(n+1)\epsilon_m(n+1)$	m

VI. SOME EXTENSIONS AND COMPUTATIONAL CONSIDERATIONS

The algorithms described so far work well when the input signal and the desired response are generated from constant parameter models. In the case of time varying parameters the algorithms should be properly modified to track these variations. To allow for the tracking of time varying parameters we can use a fading memory window to give more weight to the effect of recent data, "forgetting" the data in the remote past. This can be achieved by minimizing the following exponentially weighted squared error at time n:

$$E_m(n) = \sum_{j=M}^{n} \lambda^{n-j}\epsilon_m^2(j) \tag{85}$$

where $0 < \lambda < 1$. Note that $\lambda = 1$ gives the cost function (4). Similar expressions are used for the problems of FLP and

BLP. The resulting algorithms for the fading memory case are obtained from those of the prewindowed case by the substitutions

$$\alpha_m^f(n) \to \lambda \alpha_m^f(n)$$
$$\alpha_m^b(n) \to \lambda \alpha_m^b(n). \tag{86}$$

The fast Kalman algorithm was generalized for the unwindowed signal case in [7] requiring $O(13m)$ MADPR for AR modeling and $O(15m)$ for LS filtering. A fast sequential algorithm for unwindowed multichannel signals was introduced in [10] and gives a complexity of $O(8p)$ and $O(10p)$, respectively. These algorithms are quite useful for applications like event detection and sequential spectral estimation.

The FAEST algorithm can be readily extended to multichannel filtering. In this case $R_m(n)$ is a block matrix and $\alpha_m^f(n)$, $\alpha_m^b(n)$ are matrices with scalar entries. All divisions by these quantities may then be substituted by matrix inversions. As an example, (77) in the multichannel case becomes

$$w_{m+1}(n+1) = \begin{bmatrix} 0 \\ w_m(n) \end{bmatrix} - \begin{bmatrix} 1 \\ a_m(n) \end{bmatrix} [\alpha_m^f(n)]^{-1} e_m^f(n+1).$$

We next note some remarks concerning the actual implementation of the FAEST algorithm.

Thus when we program the algorithm we prefer to use as an intermediate variable the "angle variable"

$$\psi_m(n+1) = \frac{1}{\alpha_m(n+1)} = 1 - \gamma_m(n+1). \tag{87}$$

Using (87) the algorithm requires only two divisions. On the other hand, from (46) and (47) it follows that

$$0 \leq \psi_{m+1}(n+1) \leq \psi_m(n+1) \leq 1. \tag{88}$$

From (44), (71), (72), and (86) we take

$$\alpha_m^f(n+1) = \lambda \alpha_m^f(n) + \frac{[e_m^f(n+1)]^2}{\alpha_m(n)} \tag{89}$$

$$\alpha_m^b(n+1) = \lambda \alpha_m^b(n) + \frac{[e_m^b(n+1)]^2}{\alpha_m(n+1)}. \tag{90}$$

Due to (88), $\alpha_m^f(n+1)$ and $\alpha_m^b(n+1)$ are positive as long as the initial conditions in (89), (90) are positive. The positive definiteness of $\alpha_m^f(n)$, $\alpha_m^b(n)$ guarantees the invertibility of the covariance matrix $R_m(n)$.

Violation of (88) and the positive definiteness of $\alpha_m^f(n)$, $\alpha_m^b(n)$ in a practical implementation provides information about the deterioration of the algorithm due to finite register effects, etc. In practice we set the initial values of $\alpha_m^f(n)$ and $\alpha_m^b(n)$ equal to a suitable small positive constant σ.

Finally, we give some simulation results from a system identification experiment conducted to study the behavior of the FAEST algorithm. The algorithm is realized by the Fortran subroutine FAEST provided in Fig. 1, which includes the time updating of the gain vector for the exponentially weighted criterion.

Subroutine FAEST has the following arguments.

```
C--------------------------------------------------------------
C       Fast A-posteriori Error Sequential Technique
C                       F A E S T
C       Fast gain updating : 5p MAD Per Recursion
C--------------------------------------------------------------
        SUBROUTINE FAEST(X, IP, W, A, B, ALPHA, XIN, SG, RL, JIM)
        DIMENSION X(IP), W(IP), A(IP), B(IP), WE(51)
        IF(JIM.EQ.1) GO TO 20
C       Initialization
        IP1=IP+1
        DO 10 J=1, IP
        X(J)=0.
        W(J)=0.
        A(J)=0.
        B(J)=0.
        WE(J)=0.
10      CONTINUE
        WE(IP1)=0.
        AF=SG
        AB=AF
        ALPHA=1.
        PSI=1.
        RETURN
20      CONTINUE
C       Time Recursion
        EF=XIN
        DO 30 J=1, IP
        EF=EF+A(J)*X(J)
30      CONTINUE
        HELP=EF/(RL*AF)
        WE(1)=-HELP
        DO 40 J=1, IP
        WE(J+1)=W(J)-HELP*A(J)
40      CONTINUE
        EFG=EF*PSI
        DO 50 J=1, IP
        A(J)=A(J)+W(J)*EFG
50      CONTINUE
        DO 60 J=1, IP
        W(J)=WE(J)-WE(IP1)*B(J)
60      CONTINUE
        EB=-RL*AB*WE(IP1)
        AF=RL*AF+EF*EFG
        ALPHA=ALPHA+WE(IP1)*EB+HELP*EF
        PSI=1./ALPHA
        EBG=EB*PSI
        AB=RL*AB+EB*EBG
        DO 70  J=1, IP
        B(J)=B(J)+W(J)*EBG
70      CONTINUE
        IPM1=IP-1
        DO 80  J=1, IPM1
        X(IP1-J)=X(IP-J)
80      CONTINUE
        X(1)=XIN
        RETURN
        END
```

Fig. 1. Fortran subroutine for fast gain updating using the FAEST algorithm.

Inputs:

$\mathrm{X} = x(n)$

$\mathrm{IP} = p$

$\mathrm{XIN} = x(n)$

$\mathrm{SG} = \sigma$

$\mathrm{RL} = \lambda.$

Outputs:

$\mathrm{W} = w_m(n)$

$\mathrm{A} = a_m(n)$

$\mathrm{B} = b_m(n)$

$\mathrm{ALPHA} = \alpha_m(n).$

For initialization we call once FAEST in the beginning with JIM = 0. Next we always set JIM = 1.

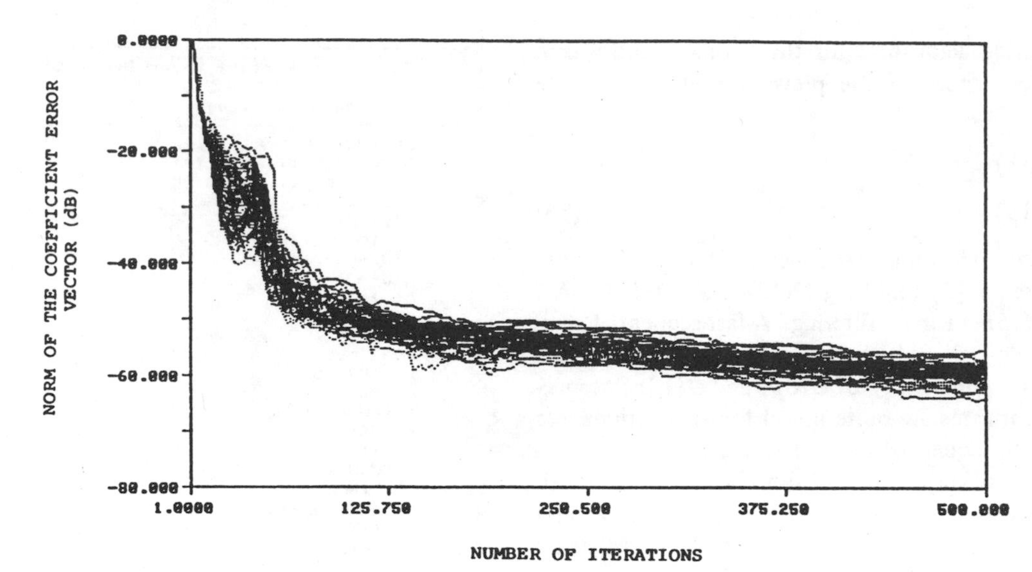

Fig. 2. An ensemble of 50 learning curves from a system identification experiment using the FAEST algorithm with $p = 40$, $S/N = 60$ dB, $\sigma = 0.001$, $\lambda = 1$.

Fig. 2 shows an ensemble of 50 learning curves from a system identification experiment. The plant was an FIR filter with order $\hat{p} = 40$. The FAEST algorithm was used for the adaptation of an FIR model with $p = 40$. The output signal-to-noise ratio was 60 dB, $\sigma = 0.001$, and $\lambda = 1$. As a criterion of performance, the norm of the coefficient error vector is used [17]. The numerical behavior of the FAEST algorithm is currently under investigation.

VII. Conclusions

In this paper a new very efficient sequential LS algorithm, named FAEST, has been developed. The updating of a Kalman type gain vector using FAEST requires $5p$ MADPR in contrast to the already known fast Kalman algorithm which requires $8p$ MADPR. The computational efficiency of the proposed algorithm is mainly due to an alternative definition of the gain vector which makes a better exploitation of the relationships between forward and backward linear prediction possible.

For simplicity the prewindowed and the exponentially weighted signal cases for single-channel signals have been treated. The more general multichannel and unwindowed case is examined in [10].

References

[1] D. Godard, "Channel equalization using a Kalman filter for fast data transmission," *IBM J. Res. Develop.*, pp. 267-273, May 1974.

[2] C. Gueguen and G. Carayannis, "Analyse de la parole par filtrage optimal de Kalman," *Automatisme*, vol. XVIII, pp. 99-105, Mar. 1973.

[3] G. Carayannis, "Different aspects of linear modeling for signal processing," in *Pattern Recognition and Signal Processing*, C. H. Chen, Ed. Groningen, The Netherlands: Sishoff and Noordhoff, 1978.

[4] L. Ljung, M. Morf, and D. D. Falconer, "Fast calculation of gain matrices for recursive estimation schemes," *Int. J. Contr.*, pp. 1-19, Jan. 1978.

[5] D. D. Falconer and L. Ljung, "Application of fast Kalman estimation to adaptive equalization," *IEEE Trans. Commun.*, vol. COM-26, pp. 1439-1446, Oct. 1978.

[6] B. Widrow *et al.*, "Adaptive noise cancelling: Principles and applications," *Proc. IEEE*, vol. 63, pp. 1702-1716, Dec. 1975.

[7] C. Halkias, G. Carayannis, J. Dologlou, and D. Emmanolopoulos, "A new generalized recursion for the fast computation of the Kalman gain to solve the covariance equations," in *Proc. ICASSP 82*, Paris, France, May 1982, pp. 1760-1763.

[8] G. Carayannis, N. Kalouptsidis, and D. Manolakis, "Fast recursive algorithms for a class of linear equations," *IEEE Trans. Acoust., Speech, Signal Processing*, vol. ASSP-20, pp. 227-239, Apr. 1982.

[9] G. Carayannis, D. Manolakis, and N. Kalouptsidis, "Fast Kalman type algorithms for sequential signal processing," in *Proc. ICASSP 83*, Boston, MA, Apr. 1983.

[10] N. Kalouptsidis, G. Carayannis, and D. Manolakis, "A fast covariance type algorithm for sequential least-squares filtering and prediction," in *Proc. IEEE Conf. Decision Contr.*, San Antonio, TX, 1983. Also *ASSP Trans. Automat. Contr.*, to be published.

[11] T. C. Hsia, *System Identification.* Lexington, MA: Lexington Books, 1978.

[12] G. J. Bierman, *Factorization Methods for Discrete Sequential Estimation.* New York: Academic, 1977.

[13] B. Friedlander, "Lattice filters for adaptive processing," *Proc. IEEE*, vol. 70, pp. 829-867, Aug. 1982.

[14] D. T. L. Lee, "Canonical ladder form realizations and fast estimation algorithms," Ph.D. dissertation, Dep. Elec. Eng., Stanford Univ., Stanford, CA, 1981.

[15] D. Manolakis, N. Kalouptsidis, and G. Carayannis, "Fast algorithms for discrete-time Wiener filters with optimum lag," *IEEE Trans. Acoust., Speech, Signal Processing*, vol. ASSP-31, Aug. 1983.

[16] N. Kalouptsidis, G. Carayannis, and D. Manolakis, "Fast design of FIR least-squares filters with optimum lag," *IEEE Trans. Acoust., Speech, Signal Processing*, to be published.

[17] L. Rabiner, R. Crochiere, and J. Allen, "FIR system modeling and identification in the presence of noise and with band-limited inputs," *IEEE Trans. Acoust., Speech, Signal Processing*, vol. ASSP-26, pp. 319-333, Aug. 1978.

Stationary and Nonstationary Learning Characteristics of the LMS Adaptive Filter

BERNARD WIDROW, FELLOW, IEEE, JOHN M. McCOOL, SENIOR MEMBER, IEEE,
MICHAEL G. LARIMORE, STUDENT MEMBER, IEEE, C. RICHARD JOHNSON, JR., STUDENT MEMBER, IEEE

Abstract—This paper describes the performance characteristics of the LMS adaptive filter, a digital filter composed of a tapped delay line and adjustable weights, whose impulse response is controlled by an adaptive algorithm. For stationary stochastic inputs, the mean-square error, the difference between the filter output and an externally supplied input called the "desired response," is a quadratic function of the weights, a paraboloid with a single fixed minimum point that can be sought by gradient techniques. The gradient estimation process is shown to introduce noise into the weight vector that is proportional to the speed of adaptation and number of weights. The effect of this noise is expressed in terms of a dimensionless quantity "misadjustment" that is a measure of the deviation from optimal Wiener performance. Analysis of a simple nonstationary case, in which the minimum point of the error surface is moving according to an assumed first-order Markov process, shows that an additional contribution to misadjustment arises from "lag" of the adaptive process in tracking the moving minimum point. This contribution, which is additive, is proportional to the number of weights but inversely proportional to the speed of adaptation. The sum of the misadjustments can be minimized by choosing the speed of adaptation to make equal the two contributions. It is further shown, in Appendix A, that for stationary inputs the LMS adaptive algorithm, based on the method of steepest descent, approaches the theoretical limit of efficiency in terms of misadjustment and speed of adaptation when the eigenvalues of the input correlation matrix are equal or close in value. When the eigenvalues are highly disparate ($\lambda_{max}/\lambda_{min} > 10$), an algorithm similar to LMS but based on Newton's method would approach this theoretical limit very closely.

I. Introduction

OUR PURPOSE IS to derive relationships between speed of adaptation and performance of adaptive systems. In general, faster adaptation leads to more noisy adaptive processes. When the input environment of an adaptive system is statistically stationary, best steady-state performance results from slow adaptation. However, when the input statistics are time variable, best performance is obtained by a compromise between fast adaptation (necessary to track variations in input statistics) and slow adaptation (necessary to contain the noise in the adaptive process). These issues will be studied both analytically and by computer simulation. The context of this study will be restricted to adaptive digital filters "driven" by the LMS adaptation algorithm of Widrow and Hoff [1], [2]. This algorithm and similar algorithms have been used for many years in a wide variety of practical applications [3]–[26].

We are attempting to formulate a "statistical theory of adaptation." This is a very difficult subject and the present work should be regarded as only a beginning. Stability and rate of convergence are analyzed first, then gradient noise and its effects upon performance are assessed. The concept of "misadjustment" is defined and used to establish design criteria for an adaptive predictor. Extension of the concept to the analysis of a useful but relatively simple form of nonstationary adaptation leads to criteria governing optimal choice of speed of adaptation.

The results reported here have been gradually developed in our laboratory during the past 15 years and are being extended and applied by ongoing research.

II. An Adaptive Filter

The filter considered here comprises a tapped delay line, variable weights (variable gains) whose input signals are the signals at the delay-line taps, a summer to add the weighted signals, and an adaptation process that automatically seeks an optimal impulse response by adjusting the weights. Fig. 1 illustrates the adaptive filter as used in modeling an unknown dynamic system.

In addition to the usual input signals, another input signal, the "desired response," must be supplied to the adaptive filter during the adaptation process. In Fig. 1, essentially the same input is applied to the adaptive filter as to the unknown system to be modeled. The output of this system provides the desired response for the adaptive filter. In other applications, considerable ingenuity may be required to obtain a suitable desired response for an adaptive process.

III. The Performance Surface

The analysis of the adaptive filter is developed by considering the "adaptive linear combiner" of Fig. 2, a subsystem of the adaptive filter of Fig. 1, comprising its most significant part.[1]

In Fig. 2, a set of input signals is weighted and summed to form an output signal. The inputs occur simultaneously and discretely in time. The jth input vector is

$$X_j = [x_{1j}, x_{2j}, \cdots, x_{lj}, \cdots, x_{nj}]^T.$$

The set of weights is designated by the vector $W^T = [w_1, w_2, \cdots, w_l, \cdots, w_n]$. The jth output signal is

$$y_j = \sum_{l=1}^{n} w_l x_{lj} = W^T X_j = X_j^T W. \tag{1}$$

The input signals and desired response are assumed to be stationary ergodic processes. Denoting the desired response as

Manuscript received September 29, 1975; revised March 9, 1976. This work was supported in part by the National Science Foundation under Grant ENGR 74-21752.

B. Widrow, M. Larimore, and C. R. Johnson are with the Information Systems Laboratory, Department of Electrical Engineering, Stanford University, Stanford, CA 94305.

J. M. McCool is with the Fleet Engineering Department, Naval Undersea Center, San Diego, CA 92132.

[1]This combinational system can be connected to the elements of a phased array antenna to make an adaptive antenna [5]–[9], or to a quantizer to form an adaptive threshold element ("Adaline" [1], [3] or TLU [2]) for use in adaptive logic and pattern-recognition systems. It can also be used as the adaptive portion of certain learning control systems [10], [11]; as a key portion of adaptive filters for channel equalization [12]–[16]; for adaptive noise cancelling [17], [18]; or for adaptive systems identification [19]–[26].

Reprinted from *Proc. IEEE*, vol. 64, no. 8, pp. 1151–1162, Aug. 1976.

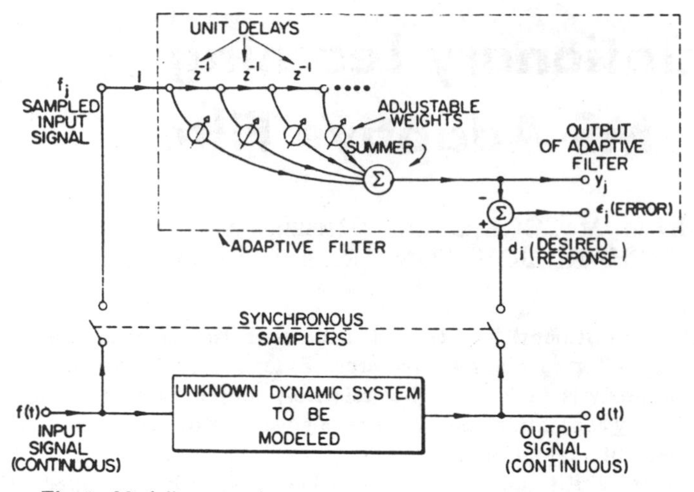

Fig. 1. Modeling an unknown system by a discrete adaptive filter.

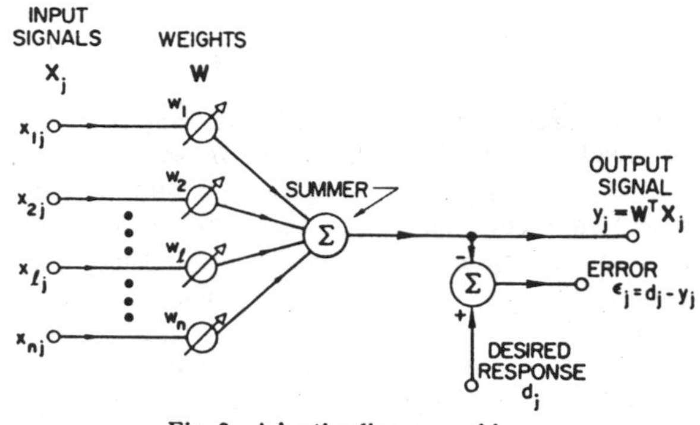

Fig. 2. Adaptive linear combiner.

d_j, the error at the jth time is

$$\epsilon_j = d_j - y_j = d_j - W^T X_j = d_j - X_j^T W. \tag{2}$$

The square of this error is

$$\epsilon_j^2 = d_j^2 - 2d_j X_j^T W + W^T X_j X_j^T W. \tag{3}$$

The mean-square error ξ, the expected value of ϵ_j^2, is

$$\begin{aligned}\xi \triangleq E[\epsilon_j^2] &= E[d_j^2] - 2E[d_j X_j^T]\, W + W^T E[X_j X_j^T] W \\ &= E[d_j^2] - 2P^T W + W^T R W\end{aligned} \tag{4}$$

where the cross correlation vector between the input signals and the desired response is defined as

$$E[d_j X_j] = E \begin{bmatrix} d_j x_{1j} \\ d_j x_{2j} \\ \cdot \\ \cdot \\ \cdot \\ d_j x_{nj} \end{bmatrix} \triangleq P \tag{5}$$

and where the symmetric and positive definite input correlation matrix R of the x-input signals is defined as

$$E[X_j X_j^T] = E \begin{bmatrix} x_{1j}x_{1j} & x_{1j}x_{2j} & \cdots \\ x_{2j}x_{1j} & x_{2j}x_{2j} & \cdots \\ \vdots & \vdots & \\ & \cdots & x_{nj}x_{nj} \end{bmatrix} \triangleq R. \tag{6}$$

It may be observed from (4) that the mean-square-error (mse) performance function is a quadratic function of the weights, a "bowl-shaped" surface; the adaptive process will be continuously adjusting the weights, seeking the bottom of the bowl. This may be accomplished by steepest descent methods [27], [28] discussed below.

In the nonstationary case, the adaptive process must track the bottom of the bowl, which may be moving. An analysis of a simple nonstationary case is presented in Section XI.

IV. The Gradient and the Wiener Solution

The method of steepest descent uses gradients of the performance surface in seeking its minimum. The gradient at any point on the performance surface may be obtained by differentiating the mse function, equation (4), with respect to the weight vector. The gradient vector is

$$\nabla = -2P + 2RW. \tag{7}$$

Set the gradient to zero to find the optimal weight vector W^*:

$$W^* = R^{-1} P \tag{8}$$

which is the Wiener–Hopf equation in matrix form.

The minimum mse is obtained from (8) and (4):

$$\xi_{\min} = E[d_j^2] - P^T W^*. \tag{9}$$

Substituting (9) into (4) yields a useful formula for mse:

$$\xi = \xi_{\min} + (W - W^*)^T R (W - W^*). \tag{10}$$

Define V as the difference between W and the Wiener solution W^*:

$$V \triangleq (W - W^*). \tag{11}$$

Therefore,

$$\xi = \xi_{\min} + V^T R V. \tag{12}$$

Differentiation of (12) yields another form for the gradient:

$$\nabla = 2RV. \tag{13}$$

The input correlation matrix, being symmetric and positive definite, may be represented as

$$R = Q \Lambda Q^{-1} = Q \Lambda Q^T \tag{14}$$

where Q is the orthonormal modal matrix of R and Λ is its diagonal matrix of eigenvalues:

$$\Lambda = \operatorname{diag}[\lambda_1, \lambda_2, \cdots, \lambda_p, \cdots, \lambda_n]. \tag{15}$$

Equation (12) may be reexpressed as

$$\xi = \xi_{\min} + V^T Q \Lambda Q^{-1} V. \tag{16}$$

Define a transformed version of V as

$$V' \triangleq Q^{-1} V \quad \text{and} \quad V = QV'. \tag{17}$$

Accordingly, equation (12) may be put in normal form as

$$\xi = \xi_{\min} + V'^T \Lambda V'. \tag{18}$$

The primed coordinates are therefore the principal axes of the quadratic surface. Transformation (17) may be applied to the weight vector itself,

$$W' = Q^{-1} W \quad \text{and} \quad W = QW'. \tag{19}$$

V. The Method of Steepest Descent

The method of steepest descent makes each change in the weight vector proportional to the negative of the gradient vector:

$$W_{j+1} = W_j + \mu(-\nabla_j). \tag{20}$$

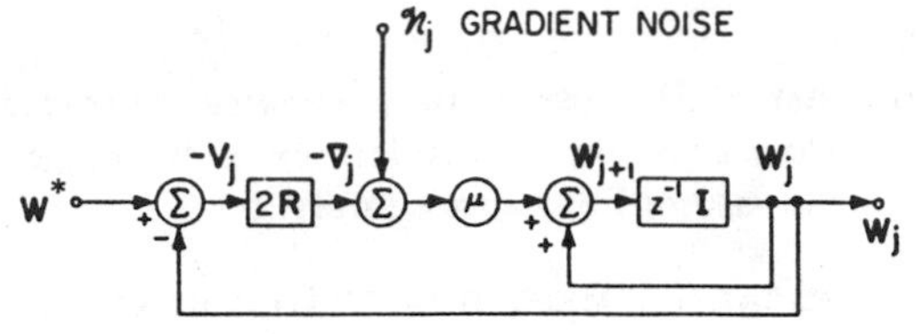

Fig. 3. Feedback model of steepest descent.

The scalar parameter μ is a convergence factor that controls stability and rate of adaptation. The gradient at the jth iteration is ∇_j. Using (13), (14), and (17), equation (20) becomes

$$V'_{j+1} - (I - 2\mu\Lambda) V'_j = 0. \tag{21}$$

This homogeneous vector difference equation is uncoupled. It has a simple geometric solution in the primed coordinates [5]:

$$V'_j = (I - 2\mu\Lambda)^j V_0 \tag{22}$$

where V'_0 is an initial condition:

$$V'_0 = W'_0 - W^{*\prime}. \tag{23}$$

For convergence, it is necessary that

$$1/\lambda_{\max} > \mu > 0 \tag{24}$$

where $\lambda_{\max}$ is the largest eigenvalue of R. From (22), we see that transients in the primed coordinates will be geometric; the geometric ratio of the pth coordinate is

$$r_p = (1 - 2\mu\lambda_p). \tag{25}$$

An exponential envelope can be fitted to a geometric sequence. If the basic unit of time is considered to be the iteration cycle, time constant τ_p can be determined as follows:

$$r_p = \exp\left(-\frac{1}{\tau_p}\right) = 1 - \frac{1}{\tau_p} + \frac{1}{2!\tau_p^2} - \cdots. \tag{26}$$

The case of general interest is slow adaptation; i.e., large τ_p. Therefore,

$$r_p = (1 - 2\mu\lambda_p) \simeq 1 - \frac{1}{\tau_p}$$

or

$$\tau_p \simeq \frac{1}{2\mu\lambda_p}. \tag{27}$$

Equation (27) gives the time constant of the pth mode.

Steepest descent can be regarded as a feedback process where the gradient plays the role of vector error signal. The process, if stable, tends to bring the gradient to zero.[2] Fig. 3 shows a feedback model for a stationary quadratic mse surface being searched by the method of steepest descent. The model is equivalent to the following set of relations.

$$W_j = W_{j+1}\,|_{\text{delayed one iteration}}$$

$$W_{j+1} = W_j + \mu(-\nabla_j)$$

$$\nabla_j = 2R(W_j - W^*) = 2RV. \tag{28}$$

This feedback model is used subsequently in a study of nonstationary adaptation. Notice an input not mentioned earlier, "gradient noise." Because gradients are estimated at each iteration cycle with finite amounts of input data, they will be imperfect or noisy.

[2]This has been called performance feedback [1], [29].

VI. The LMS Algorithm

The LMS algorithm is an implementation of steepest descent using measured or estimated gradients:

$$W_{j+1} = W_j + \mu(-\hat{\nabla}_j). \tag{29}$$

The estimate of the true gradient is $\hat{\nabla}$.

The gradient estimate used by LMS takes the gradient of the square of a single error sample. Thus

$$\hat{\nabla}_j = -2\epsilon_j X_j. \tag{30}$$

The LMS algorithm can be written as

$$W_{j+1} = W_j + 2\mu\epsilon_j X_j. \tag{31}$$

If we assume that X_j is uncorrelated over time (i.e., that $E[X_j X_{j+l}^T] = 0,\ \forall l \neq 0$), an assumption common in the field of stochastic approximation [30], [31], then the expected value of the gradient estimate equals the true gradient, and the weight-vector mean is convergent to the Wiener solution of (8), as shown in [4] and [5].

Condition (24) is necessary and sufficient for convergence of the LMS algorithm. However, in practice, the individual eigenvalues are rarely known so that (24) is not always easy to apply. Since tr R is the total input power to the weights, a generally known quantity, and since tr $R > \lambda_{\max}$ as R is positive definite, a sufficient condition for convergence is

$$1/\text{tr}\, R > \mu > 0. \tag{32}$$

VII. The Learning Curve and its Time Constants

During adaptation, the error ϵ_j is nonstationary as the weight vector adapts toward W^*. The mse can be defined only on the basis of ensemble averages. From (18), we obtain

$$\xi_j = \xi_{\min} + V_j'^T \Lambda V_j'. \tag{33}$$

Imagine an ensemble of adaptive processes, each having individual stationary ergodic inputs drawn from the same statistical population, with all initial weight vectors equal. The mse ξ_j is a function of iteration number j, obtained by averaging over the ensemble at iteration j.

Using (22), but assuming no noise in the weight vector, equation (33) becomes

$$\begin{aligned}\xi_j &= \xi_{\min} + V_0'^T \Lambda (I - 2\mu\Lambda)^{2j} V'_0 \\ &= \xi_{\min} + V_0^T (I - 2\mu R)^j R (I - 2\mu R)^j V_0.\end{aligned} \tag{34}$$

When the adaptive process is convergent, it is clear from (34) that

$$\lim_{j\to\infty} \xi_j = \xi_{\min}$$

and that the geometric decay in ξ_j going from ξ_0 to $\xi_{\min}$ will, for the pth mode, have a geometric ratio of r_p^2 and a time constant

$$\tau_{p_{\text{mse}}} \triangleq \frac{1}{2}\tau_p = \frac{1}{4\mu\lambda_p}. \tag{35}$$

The result obtained by plotting mse against number of iterations is called the "learning curve." Due to noise in the weight vector, actual practice will show ξ_j to be higher than indicated by (34).

VIII. Gradient and Weight-Vector Noise

Gradient noise will affect the adaptive process both during initial transients and in steady state. The latter condition is of particular interest here.

Assume that the weight vector is close to the Wiener solution. Assume, as before, that X_j and d_j are stationary and ergodic and that X_j is uncorrelated over time; i.e.,

$$E[X_j X_{j+k}] = 0, \qquad k \neq 0. \tag{36}$$

The LMS algorithm uses a gradient estimate

$$\hat{\nabla} = -2\epsilon_j X_j = \nabla_j + N_j \tag{37}$$

where ∇_j is the true gradient and N_j is a zero-mean gradient estimation noise vector. When $W_j = W^*$, the true gradient is zero, but the gradient would be estimated according to (30) and is equal to the gradient noise:

$$N_j = -2\epsilon_j X_j. \tag{38}$$

According to Wiener filter theory, when $W_j = W^*$, ϵ_j and X_j are uncorrelated. If they are assumed zero-mean Gaussian, ϵ_j and X_j are statistically independent. As such, the covariance of N_j is

$$\begin{aligned} \text{cov}\,[N_j] &= E[N_j N_j^T] = 4E[\epsilon_j^2 X_j X_j^T] \\ &= 4E[\epsilon_j^2]\,E[X_j X_j^T] \\ &= 4E[\epsilon_j^2]R. \end{aligned} \tag{39}$$

When $W_j = W^*$, $E[\epsilon_j^2] = \xi_{\min}$. Accordingly,

$$\text{cov}\,[N_j] = 4\xi_{\min} R. \tag{40}$$

As long as $W_j \simeq W^*$, we assume that the gradient noise covariance is given by (40) and that this noise is stationary and uncorrelated over time. The latter assumption is based on (36) and (38).

Projecting the gradient noise,

$$N_j' = Q^{-1} N_j \tag{41}$$

its covariance becomes

$$\begin{aligned} \text{cov}\,[N_j'] = E[N_j' N_j'^T] = E[Q^{-1} N_j N_j^T Q] &= Q^{-1}\,\text{cov}\,[N_j]Q \\ &= 4\xi_{\min} Q^{-1} R Q \\ &= 4\xi_{\min}\Lambda. \end{aligned} \tag{42}$$

Although the components of N_j are correlated with each other, those of N_j' are mutually uncorrelated and can, therefore, be handled more easily.

Gradient noise propagates and causes noise in the weight vector. Accounting for gradient noise, the LMS algorithm can be expressed as

$$W_{j+1}' = W_j' + \mu(-\hat{\nabla}_j') = W_j' + \mu(-\nabla_j' + N_j'). \tag{43}$$

This equation can be written in terms of V_j' as

$$V_{j+1}' = V_j' + \mu(-2\Lambda V_j' + N_j'). \tag{44}$$

Near the minimum point of the error surface in steady-state, the mean of V_j' is zero and the covariance of the weight-vector noise is [18, appendix D, section B]

$$\text{cov}\,[V_j'] = \mu\xi_{\min} I \tag{45}$$

where the components of the weight-vector noise are of equal variance and are mutually uncorrelated. It has been found, however, that (45) closely approximates measured weight-vector covariances under a considerably wider range of conditions than the assumptions above imply.

IX. Misadjustment Due to Gradient Noise

Random noise in the weight vector causes an excess mse. If the weight vector were noise free and converged such that $W_j = W^*$, then the mse would be $\xi_{\min}$. However, this does not occur in actual practice so that the weight vector is on the average "misadjusted" from its optimal setting.

An expression for mse in terms of V_j' is given by (33), from which we obtain an expression for excess mse:

$$(\text{excess mse}) = V_j'^T \Lambda V_j'. \tag{46}$$

The average excess mse is an important quantity:

$$E[V_j'^T \Lambda V_j'] = \sum_{p=1}^{n} \lambda_p E[(v_{pj}')^2] \tag{47}$$

where v_{pj}' is the pth component of V_j'. After adaptive transients die out, $E[V_j'] = 0$. Therefore, from (45) we have

$$E[(v_{pj}')^2] = \mu\xi_{\min}, \ \forall p. \tag{48}$$

Substitution into (47) yields the average excess mse,

$$E[V_j'^T \Lambda V_j'] = \mu\xi_{\min} \sum_{p=1}^{n} \lambda_p = \mu\xi_{\min}\,\text{tr}\,R. \tag{49}$$

We define the "misadjustment" due to gradient noise as the dimensionless ratio of the average excess mse to the minimum mse,

$$M \triangleq \frac{\text{average excess mse}}{\xi_{\min}}. \tag{50}$$

For the LMS algorithm, under the conditions assumed above,

$$M = \mu\,\text{tr}\,R. \tag{51}$$

This formula works well for small values of misadjustment, 25 percent or less, so that the assumption

$$W_j \simeq W^* \tag{52}$$

is satisfied. The misadjustment is a useful measure of the cost of adaptability. A value of $M = 10$ percent means that the adaptive system has a mse only 10 percent greater than $\xi_{\min}$.

It is useful to relate misadjustment to the speed of adaptation and the number of weights being adapted. Since $\text{tr}\,R$ equals the sum of the eigenvalues,

$$M = \mu \sum_{p=1}^{n} \lambda_p = \mu n \lambda_{\text{ave}} \tag{53}$$

where λ_{ave} is the average of the eigenvalues. From (35),

$$\lambda_p = \frac{1}{4\mu}\left(\frac{1}{\tau_{p_{\text{mse}}}}\right) \quad \text{or} \quad \lambda_{\text{ave}} = \frac{1}{4\mu}\left(\frac{1}{\tau_{p_{\text{mse}}}}\right)_{\text{ave}}. \tag{54}$$

Substituting into (53) yields

$$M = \frac{n}{4}\left(\frac{1}{\tau_{p_{\text{mse}}}}\right)_{\text{ave}}. \tag{55}$$

The special case where all eigenvalues are equal is an important one. The learning curve has only one time constant τ_{mse}, and the misadjustment is given by

$$M = \frac{n}{4\tau_{mse}}. \tag{56}$$

When the eigenvalues are sufficiently similar for the learning curve to be approximately fitted by a single exponential, its time constant may be applied to (56) to give an approximate value of M.

Since transients settle in about four time constants, equation (56) leads to an approximate "rule of thumb:" the misadjustment equals the number of weights divided by the settling time. A 10-percent misadjustment would be satisfactory for many engineering designs. Operation with 10-percent misadjustment can generally be achieved with an adaptive settling time equal to ten times the memory time span of the adaptive transversal filter.

X. A Design Example/Choosing Number of Filter Weights for an Adaptive Predictor

Fig. 4 is a block diagram of an adaptive predictor.[3] Its adaptive filter converts the delayed input $x_{j-\Delta}$ into x_j as best possible. If the adaptive-filter weights are copied into an auxiliary filter having a tapped delay-line structure identical to that of the adaptive filter and the input x_j is applied to this auxiliary filter, the resulting output will be a linear least squares estimate of $x_{j+\Delta}$ (limited by finite filter length and misadjustment).

A computer implementation of the adaptive predictor was made using a simulated input signal x_j obtained by bandpass filtering a white Gaussian signal and adding this to another independent white Gaussian signal. Prediction was one time sample in the future, i.e., $\Delta = 1$, using an adaptive filter with five weights, all initially set to zero.

Fig. 5 depicts three learning curves. For each adaptive step, the mse ξ_j corresponding to the current weight vector W_j was calculated from (10) using known values of R and ξ_{min}, giving the "individual learning curve." The smooth "ensemble average learning curve" is simply the average of 200 such individual curves and approximates the adaptive behavior in the mean. The third curve calculated from (34) shows how the process would evolve if perfect knowledge of the gradient were available at each step. It is a noiseless "steepest descent learning curve."

Of particular interest is the residual difference between the ensemble learning curve and the steepest descent learning curve after convergence. The latter, of course, converges to ξ_{min}. The difference is the excess mse due to gradient noise, in this case, giving a measured misadjustment of 3 percent. The theoretical misadjustment was $M = 2.5$ percent. The minor discrepancy is due mainly to the fact that the input samples are highly correlated in violation of the assumption that $E[X_j X_{j+k}^T] = 0, \forall k \neq 0$, used in the derivation of misadjustment formula (56).

The ensemble average learning curve has an effective measured time constant τ_{mse} of about 50 iterations since it falls to within 2 percent of its converged value at around iteration 200.

[3]This same predictor was described by Widrow in [5]; it has been used for data compression and speech encoding [32] and for "maximum entropy" spectral estimation [33].

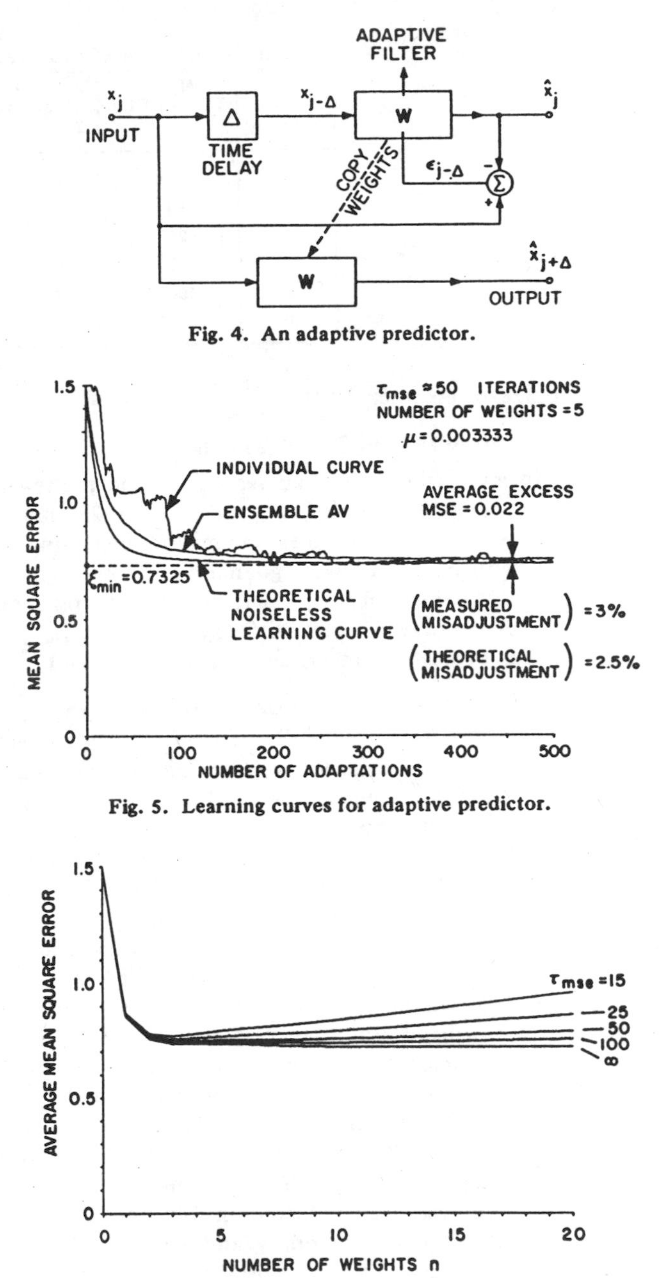

Fig. 4. An adaptive predictor.

Fig. 5. Learning curves for adaptive predictor.

Fig. 6. Performance versus number of weights and adaptive predictor time constant.

When all eigenvalues are equal, equation (35) becomes

$$\tau_{mse} = \frac{1}{4\mu\lambda} = \frac{n}{4\mu \operatorname{tr} R}. \tag{57}$$

Using (57) in the present case (although the eigenvalues range over a 10 to 1 ratio) yields $\tau_{mse} = 50$, which agrees with experiment. Equation (57) gives a formula for an "effective time constant," useful even when the eigenvalues are highly disparate.

The performance of the adaptive filter may improve with an increase in the number of weights. However, for a fixed rate of convergence, larger numbers of weights increase misadjustment. Fig. 6 shows these conflicting effects. The lowest curve for $\tau_{mse} = \infty$ represents idealized noise-free adaptation, providing the minimum mse $\xi_{min}(n)$ for each value of n. The other curves include average excess mse due to gradient noise. We define the "average mse" to be the sum of the minimum mse

TABLE I

COMPARISON OF THEORETICAL AND EXPERIMENTAL ADAPTIVE PREDICTOR PERFORMANCE

NUMBER OF WEIGHTS n	APPROX TIME CONSTANT τ_{mse}	AVERAGE MSE Theoretical/Experimental		MISADJUSTMENT Theoretical/Experimental	
5	100	.742	.751	1.3%	2.5%
5	50	.751	.754	2.5%	3.0%
5	25	.769	.781	5.0%	6.6%
5	15	.794	.824	8.3%	12.6%
10	100	.737	.745	2.5%	3.5%
10	50	.755	.764	5.0%	6.2%

and the average excess mse. Thus

$$(\text{average mse}) = [1 + M]\xi_{min}(n). \tag{58}$$

Using this formula, theoretical curves have been plotted in Fig. 6 for approximate values of τ_{mse} of 100, 50, 25, and 15 iterations. It is apparent from these curves that increasing the number of weights does not always guarantee improved system performance. Experimental points derived by computer simulation have compared very well with theoretical values predicted by (58). Typical results are summarized in Table I.

XI. RESPONSE OF THE LMS ADAPTIVE FILTER IN A NONSTATIONARY ENVIRONMENT

Filtering nonstationary signals is a major area of application for adaptive techniques, especially when the stochastic properties of the signals are unknown *a priori*. Although the utility of adaptive filters with nonstationary inputs has been demonstrated experimentally, very little of this work has been published, perhaps due to the inherently complex mathematics associated with such problems [34], [35]. The nonstationary situations to be studied here are highly simplified, but they retain the essence of the problem that is common to more complicated and realistic situations.

The example considered here involves modeling or identifying an unknown time-variable system by an adaptive LMS transversal filter. The unknown system is assumed to be a transversal filter of same length n whose weights (impulse response values) undergo independent stationary ergodic first-order Markov processes, as indicated in Fig. 7. The input signal x_j is assumed to be stationary and ergodic. Additive output noise, assumed to be stationary, of mean zero, and of variance ξ_{min}, prevents a perfect match between the unknown system and the adaptive system. The minimum mse is, therefore, ξ_{min}, achieved whenever the weights of the adaptive filter W_j match those of the unknown system. The latter are at every instant the optimal values for the corresponding weights of the adaptive filter and are designated W_j^*, the subscript indicating that the unknown "target" to be tracked is time variable.

According to the scheme of Fig. 7, minimizing mse causes the adaptive weight vector W_j to attempt to best match the unknown W_j^* on a continual basis. The R matrix, dependent only on the statistics of x_j, is constant even as W_j^* varies. The desired response of the adaptive filter d_j is nonstationary, being the output of a time-variable system. The minimum mse ξ_{min} is constant. Thus the mse function, a quadratic bowl, varies in position while its eigenvalues, eigenvectors, and ξ_{min} remain constant.

In order to study this form of nonstationary adaptation both analytically and by computer simulation, a model comprising an ensemble of nonstationary adaptive processes has been defined and constructed as illustrated in Fig. 8. The unknown

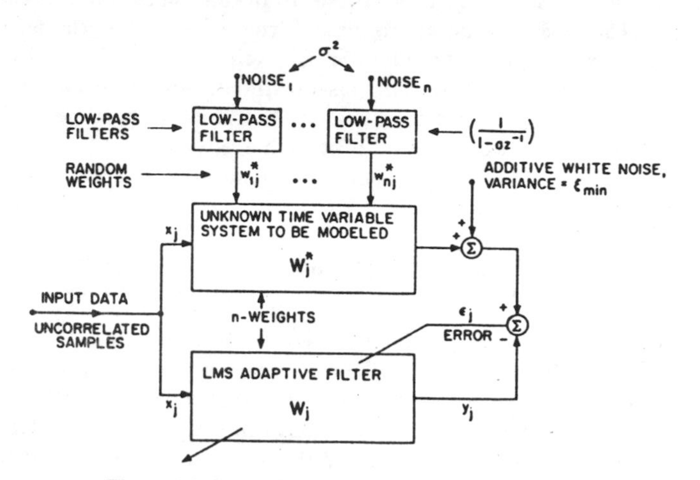

Fig. 7. Modeling an unknown time-variable system.

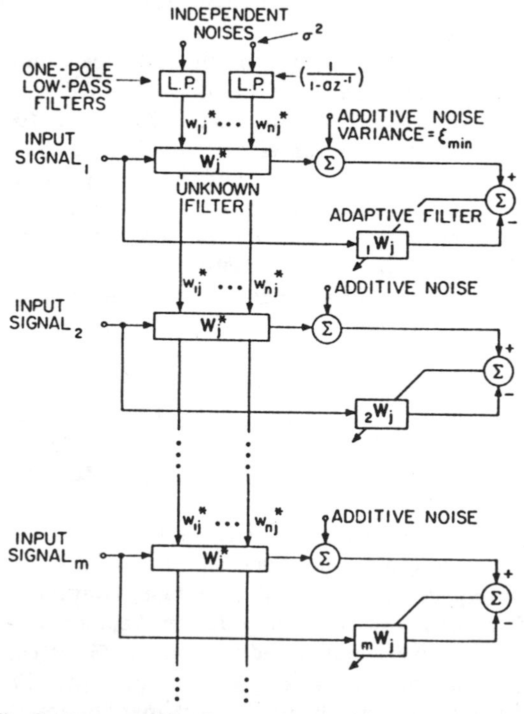

Fig. 8. An ensemble of nonstationary adaptive processes.

filters to be modeled are all identical and have the same time-variable weight vector W_j^* throughout the ensemble. Each ensemble member has its own independent input signal going to both the unknown system and the corresponding adaptive system. The effect of output noise in the unknown systems is obtained by the addition of independent noise of variance ξ_{min}. All of the adaptive filters are assumed to start with the same initial weight vector W_0; each develops its own weight vector over time in attempting to pursue the Markovian target W_j^*.

For a given adaptive filter, the weight-vector tracking error at the jth instant is $(W_j - W_j^*)$. This error is due to both the effects of gradient noise and weight-vector lag, and may be expressed as

$$(\text{Weight-vector error})_j = (W_j - W_j^*) \equiv \underbrace{(W_j - E[W_j])}_{\text{weight-vector noise}} + \underbrace{(E[W_j] - W_j^*)}_{\text{weight-vector lag}}. \quad (59)$$

The expectations are averages over the ensemble. The components of error are identified in (59). Any difference between the ensemble mean of the adaptive weight vectors and the target value W_j^* is due to lag in the adaptive process, while the deviation of the individual adaptive weight vectors about the ensemble mean is due to gradient noise.

Weight-vector error causes an excess mse. The ensemble average excess mse at the jth instant is

$$\begin{pmatrix}\text{average excess}\\ \text{mse}\end{pmatrix}_j = E[(W_j - W_j^*)^T R(W_j - W_j^*)]. \quad (60)$$

Using (59), this can be expanded as follows:

$$\begin{aligned}\begin{pmatrix}\text{average excess}\\ \text{mse}\end{pmatrix}_j &= E[(W_j - E[W_j])^T R(W_j - E[W_j])] \\ &\quad + E[(E[W_j] - W_j^*)^T R(E[W_j] - W_j^*)] \\ &\quad + 2E[(W_j - E[W_j])^T R(E[W_j] - W_j^*)]. \quad (61)\end{aligned}$$

Expanding the last term of (61) and simplifying since W_j^* is constant over the ensemble,

$$\begin{aligned}&2E[W_j^T RE[W_j] - W_j^T RW_j^* - E[W_j]^T RE[W_j] + E[W_j]^T RW_j^*] \\ &= 2[E[W_j]^T RE[W_j] - E[W_j]^T RE[W_j] \\ &\quad - E[W_j]^T RW_j^* + E[W_j]^T RW_j^*] \\ &= 0. \quad (62)\end{aligned}$$

Therefore, (61) becomes

$$\begin{aligned}\begin{pmatrix}\text{average excess}\\ \text{mse}\end{pmatrix}_j &= E[(W_j - E[W_j])^T R(W_j - E[W_j])] \\ &\quad + E[(E[W_j] - W_j^*)^T R(E[W_j] - W_j^*)]. \quad (63)\end{aligned}$$

The average excess mse is thus a sum of components due to both gradient noise and lag:

$$\begin{aligned}\begin{pmatrix}\text{average excess}\\ \text{mse due to lag}\end{pmatrix}_j &= E[(E[W_j] - W_j^*)^T R(E[W_j] - W_j^*)] \\ &= E[(E[W_j'] - W_j^{*\prime})^T \Lambda(E[W_j] - W_j^{*\prime})] \quad (64)\end{aligned}$$

$$\begin{aligned}\begin{pmatrix}\text{average excess}\\ \text{mse due to}\\ \text{gradient noise}\end{pmatrix}_j &= E[(W_j - E[W_j])^T R(W_j - E[W_j])] \\ &= E[(W_j' - E[W_j'])^T \Lambda(W_j' - E[W_j'])]. \quad (65)\end{aligned}$$

Fig. 9 is a feedback diagram adapted from Fig. 3, illustrating the two sources of weight-vector error. From the feedback diagram, it can be seen that the "output" W_j attempts to track the time variable "input" W_j^*. Tracking error $(W_j - W_j^*)$ is caused by the propagation of gradient noise and by the response of the adaptive process to the random variations of W_j^*. It will be shown that increasing the time constant of the adaptive process diminishes the propagation of gradient noise but simultaneously increases the lag error that results from the random changes in W_j^*.

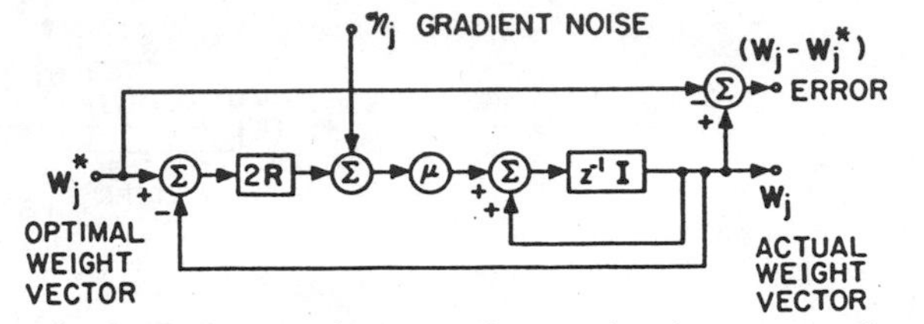

Fig. 9. Feedback diagram of steepest descent showing sources of weight tracking error.

The gradient-noise covariance for the stationary case (40) is a function of R. Since R is constant, equation (40) is a good representation of covariance for the type of nonstationarity under study. Furthermore, Fig. 9 shows that the propagation of gradient noise in the linear feedback system representing the adaptive process is not affected by variability of W_j^*. Therefore, equation (49) can be used to provide an evaluation of (65), the excess mse from gradient noise. The next step is an evaluation of (64), the excess mse due to lag. Statistical knowledge of $(E[W_j'] - W_j^{*\prime})$ will be required. In finding lag effects, we may eliminate gradient noise from consideration so that $E[W_j'] = W_j'$. Knowledge of $(W_j' - W_j^{*\prime})$ will be sufficient.

Without gradient noise, the method of steepest descent and the LMS algorithm are represented by (13) and (20). With variable W_j^*, they become

$$W_{j+1} - (I - 2\mu R) W_j = 2\mu R W_j^*. \quad (66)$$

Premultiplying both sides by Q^{-1} transforms (66) into the primed coordinates,

$$W_{j+1}' - (I - 2\mu\Lambda) W_j' = 2\mu\Lambda W_j^{*\prime}. \quad (67)$$

We have assumed for our present study that all components of W_j^* are stationary, ergodic, independent, and first-order Markov; they all have the same variances and the same autocorrelation functions. Since $W_j^{*\prime} = Q^{-1} W_j^*$ and Q^{-1} is orthonormal, all components of $W_j^{*\prime}$ are independent and have the same autocorrelation functions as the components of W_j^*. Therefore, equation (67), being in diagonal form and having a driving function whose components are independent, may be treated as an array of n independent first-order linear difference equations.

Let the z transform of W_j' be $\mathfrak{W}'(z)$. The z transform of (67) is then

$$z\mathfrak{W}'(z) - (I - 2\mu\Lambda)\mathfrak{W}'(z) = 2\mu\Lambda\mathfrak{W}^{*\prime}(z). \quad (68)$$

Solving (68) yields the transform of W_j':

$$\mathfrak{W}'(z) = 2\mu\Lambda(zI - I + 2\mu\Lambda)^{-1}\mathfrak{W}^{*\prime}(z). \quad (69)$$

The weight tracking error $(W_j' - W_j^{*\prime})$ is of direct interest. Its transform is obtained from (69) as

$$\mathfrak{W}'(z) - \mathfrak{W}^{*\prime}(z) = [2\mu\Lambda(zI - I + 2\mu\Lambda)^{-1} - I]\mathfrak{W}^{*\prime}(z). \quad (70)$$

The transfer function connecting $W_j^{*\prime}$ to the weight tracking error is thus

$$2\mu\Lambda(zI - I + 2\mu\Lambda)^{-1} - I. \quad (71)$$

Since (71) is diagonal, the scalar transfer function of its pth

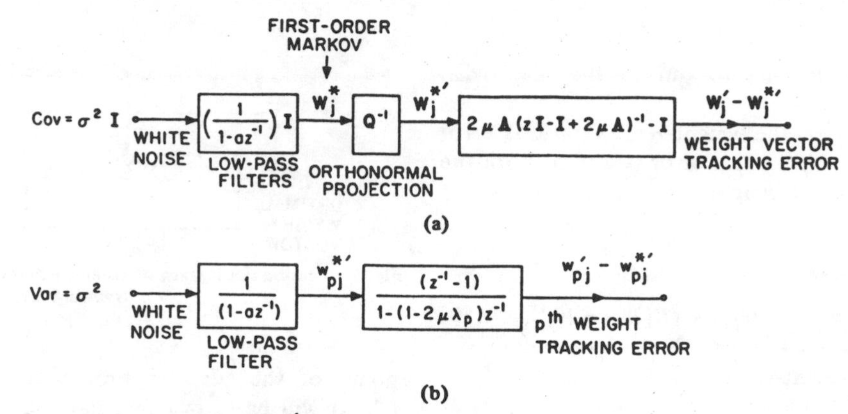

Fig. 10. Origin of W_j^* and its propagation into weight-lag error. (a) All channels. (b) pth channel.

diagonal element may be written as

$$2\mu\lambda_p(z - 1 + 2\mu\lambda_p)^{-1} - 1 = \frac{(z^{-1} - 1)}{1 - (1 - 2\mu\lambda_p)z^{-1}}. \quad (72)$$

This transfer function has a zero at $z = 1$ and a pole whose impulse response has a geometric ratio of $(1 - 2\mu\lambda_p) = r_p$.

Fig. 10(a) shows the origin of the vector W_j^* as a first-order Markov process and its propagation into the weight tracking error. W_j^* is assumed to originate from independent stationary ergodic white-noise excitation (of variance σ^2) to a bank of one-pole filters, all having transfer function $1/(1 - az^{-1})$. The pth channel of this process is shown in Fig. 10(b). Its scalar transfer function is

$$\frac{(z^{-1} - 1)}{(1 - az^{-1})(1 - (1 - 2\mu\lambda_p)z^{-1})} = \frac{(z^{-1} - 1)}{(1 - az^{-1})(1 - r_p z^{-1})}$$

$$= \frac{\left(\frac{1 - a}{a - r_p}\right)}{(1 - az^{-1})} + \frac{\left(\frac{r_p - 1}{a - r_p}\right)}{(1 - r_p z^{-1})}. \quad (73)$$

The sampled impulse response of this transfer function is obtained by inversion of (73) into the time domain. From it, the variance of the lag error of the pth component of the primed weight vector can be computed as the sum of the squares of the samples of the impulse response multiplied by σ^2. The sum of squares is given by

$$\text{sum squares} = \sum_{j=0}^{\infty}\left[\left(\frac{1 - a}{a - r_p}\right)a^j + \left(\frac{r_p - 1}{a - r_p}\right)r_p^j\right]^2$$

$$= \left(\frac{1}{a - r_p}\right)^2 \left[\left(\frac{1 - a}{1 + a}\right) + \left(\frac{1 - r_p}{1 + r_p}\right) + \frac{2(1 - a)(r_p - 1)}{(1 - ar_p)}\right]. \quad (74)$$

In cases of interest, τ_p is large so that $r_p \lesssim 1$. From (27),

$$\tau_p = \frac{1}{1 - r_p} = \frac{1}{2\mu\lambda_p}. \quad (75)$$

Furthermore, we assume that the time constant of nonstationarity τ_{W*} is also large, so that $a \lesssim 1$

$$\tau_{W*} = \frac{1}{1 - a}. \quad (76)$$

A common operating region would be where

$$\tau_{W*} >> \tau_p, \forall p. \quad (77)$$

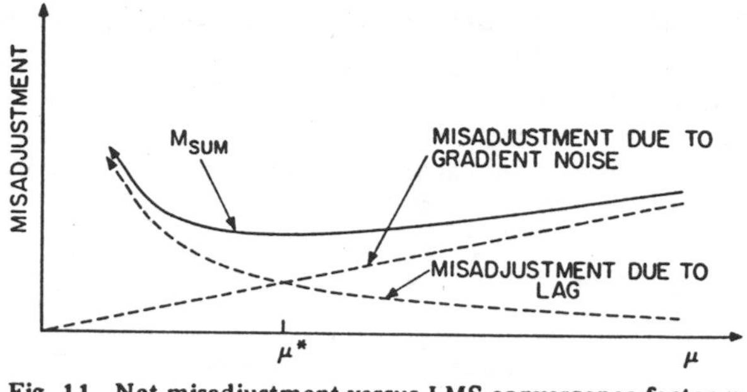

Fig. 11. Net misadjustment versus LMS convergence factor μ.

The value of μ is set so that the response times of the adaptive weights are short compared to the time constant of nonstationarity. Under these conditions, equation (74) reduces to

$$(\text{sum squares})_{\tau_{W*} >> \tau_p} = \frac{1}{2}\tau_p = \frac{1}{4\mu\lambda_p}. \quad (78)$$

Using this relation, the covariance of the lag error is obtained as

$$\text{cov}\,[W_j' - W_j^{*\prime}]\Big|_{\substack{N=0 \\ \tau_{W*} >> \tau_p}} = \frac{\sigma^2}{2}\begin{bmatrix} \tau_1 & & & & 0 \\ & \ddots & & & \\ & & \tau_p & & \\ & & & \ddots & \\ 0 & & & & \tau_n \end{bmatrix} = \frac{\sigma^2}{4\mu}\Lambda^{-1}. \quad (79)$$

Making use of (64),

$$(\text{average excess mse due to lag}) = \frac{\sigma^2}{2}\sum_{p=1}^{n}\tau_p\lambda_p = \frac{n\sigma^2}{4\mu}. \quad (80)$$

Because of the ergodic properties of W_j^*, this average is not time variable. The misadjustment due to lag is

$$(M_L)_{\tau_{W*} >> \tau_p} = \frac{\sigma^2}{2\xi_{\min}}\sum_{p=1}^{n}\tau_p\lambda_p = \left(\frac{n\sigma^2}{4\xi_{\min}}\right)\frac{1}{\mu}. \quad (81)$$

Under usual operating conditions, the misadjustment due to lag is inversely proportional to μ.

Set μ to a very small value so that the adaptive weight vector W_j does not track W_j^* but merely assumes the value of its time average. As $r_p \to 1$, equation (74) reduces to

$$(\text{sum squares})_{\mu \approx 0} = \tfrac{1}{2}\tau_{W*}. \quad (82)$$

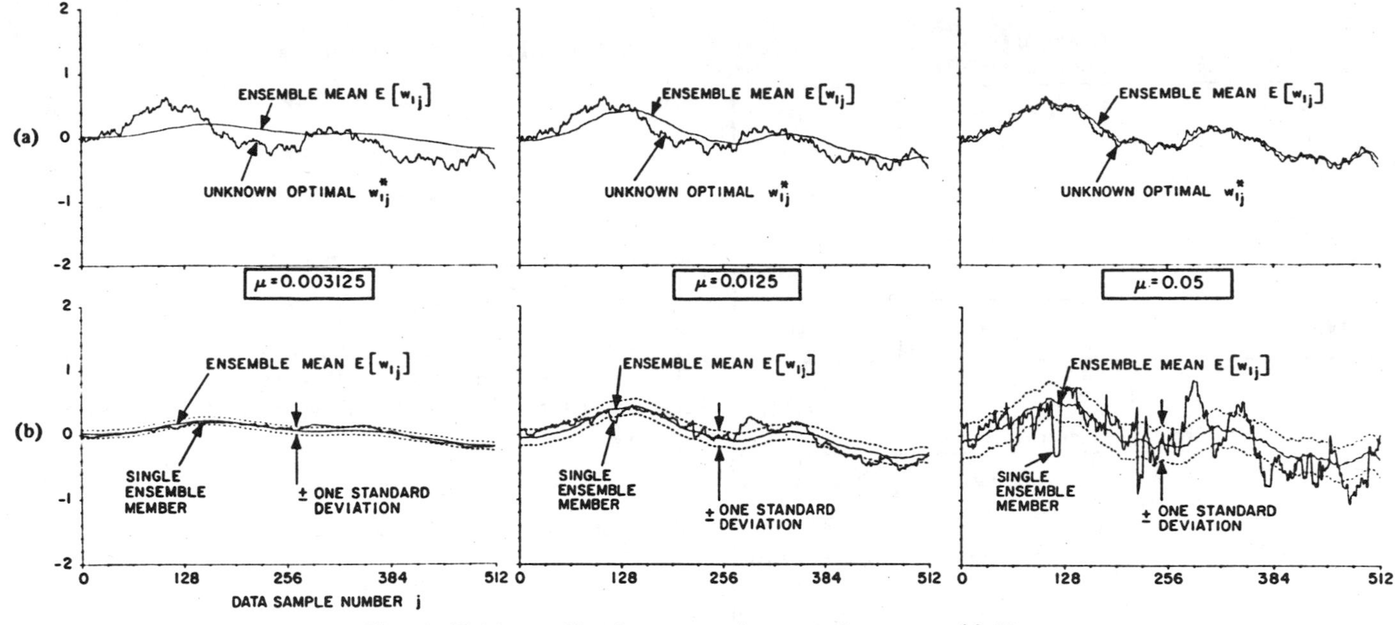

Fig. 12. Weight tracking in a nonstationary environment. (a) Plots demonstrate weight lag as a function of μ. (b) Plots demonstrate weight noise as a function of μ.

The misadjustment due to lag turns out to be

$$(NS) \triangleq (M_L)_{\mu \approx 0} = \frac{\sigma^2}{2\xi_{\min}} \tau_{W*} \operatorname{tr} R. \tag{83}$$

Since there is no tracking, the misadjustment for this case is a measure of the "nonstationarity," NS, of the randomly moving hyperparaboloidal bowl.

An interesting special case is that of all equal eigenvalues. Combining (81) with (83),

$$(M_L)_{\tau_{W*} >> \tau_p} = (NS)\left[\frac{\tau}{\tau_{W*}}\right] = (NS)\left[\frac{2\tau_{\text{mse}}}{\tau_{W*}}\right]. \tag{84}$$

This result has intuitive appeal. The misadjustment equals the product of nonstationarity and the ratio of the adaptive time constant to the time constant of nonstationarity.

From (63), the average excess mse is the sum of components due to gradient noise and lag. The total misadjustment is, therefore, the sum of two misadjustment components. Making use of (51) and (81),

$$(M_{\text{sum}})_{\tau_{W*} >> \tau_p} = (\mu) \operatorname{tr} R + \left(\frac{1}{\mu}\right) \frac{n\sigma^2}{4\xi_{\min}}. \tag{85}$$

Optimizing the choice of μ results in minimum M_{sum} when the two right-hand terms are equal. The speed of adaptation is optimized when the loss of performance due to gradient noise equals the loss in performance due to weight-vector lag.[4] The optimal μ is

$$\mu^*|_{\tau_{W*} >> \tau_p} = \left[\frac{n\sigma^2}{4\xi_{\min} (\operatorname{tr} R)}\right]^{1/2}. \tag{86}$$

A typical plot of M_{sum} versus μ is shown in Fig. 11, indicating the tradeoffs involved in adjusting μ for minimization of M_{sum}. In practice, μ^* might need to be approximated by trial and error, particularly when data are unavailable for application of (86).

The theory developed in this section has been tested extensively by computer simulation based on an ensemble of adaptive processes, as illustrated in Fig. 8. Every mathematical quantity discussed in this section has been measured. Typical experimental results are presented below.

Fig. 12 illustrates weight tracking and the associated errors. The adaptive filter had four weights. Responses are shown only for weight number one. The effects of weight lag are demonstrated by comparing the ensemble average of weight number one plotted over time against weight number one of W_j^*. Averages were taken over 128 ensemble members. The lag effect is highly evident in the first experiment with $\mu = 0.003125$. In the third experiment, with $\mu = 0.05$, the lag is quite small decreasing in proportion to μ.

The effects of gradient noise are demonstrated with the same experiment. The ensemble mean of weight number one is plotted as a function of time j. Theoretical one-standard-deviation lines for weight noise are shown about this mean. In addition, weight number one of W_j of a single ensemble member is plotted to indicate what occurred in an individual situation. It is clear that weight-noise power increases in proportion to μ.

In these experiments, the inputs were white and of unit power, so that $R = I$. The additive output noise power was $\xi_{\min} = 1$. Equation (85) has been used to obtain theoretical values of misadjustment and its components. Tables II, III, and IV summarize the results of three experiments, comparing theory and experiments for three values of μ, and fixing everything else. The input data were the same for all three experiments. Initial transients were allowed to die out before measurements were taken. Experimental values of misadjustment and its components were obtained by ensemble average measurements using (60), (64), and (65), normalizing with respect to $\xi_{\min}$, which in this case was 1. Theoretical and experimental results compared well, expect for lag misadjustment in the first experiment. In this case, where μ is very

[4]Another case has been analyzed by Widrow [29] where the fluctuation of W_j^* has a uniform low-pass power spectrum. In this case, the misadjustment due to lag is proportional to the square of μ; the speed of adaptation is optimized when the gradient-noise loss equals twice the loss due to lag. The misadjustment due to lag turns out to be quite sensitive to the spectral characteristics of the fluctuation of W_j^*.

TABLE II
FIRST EXPERIMENT, $\mu = 0.003125$

n = 4 weights τ_{mse} = 80 data samples τ_{W*} = 125 data samples (NS) = 24.9%	Misadjustment Due to Weight Lag	Misadjustment Due to Gradient Noise
Theoretical	32.0%	1.25%
Experimental	13.5%	1.5%

TABLE III
SECOND EXPERIMENT, $\mu = 0.0125$

n = 4 weights τ_{mse} = 20 data samples τ_{W*} = 125 data samples (NS) = 24.9%	Misadjustment Due to Weight Lag	Misadjustment Due to Gradient Noise
Theoretical	8.0%	5.0%
Experimental	5.6%	5.7%

TABLE IV
THIRD EXPERIMENT, $\mu = 0.05$

n = 4 weights τ_{mse} = 5 data samples τ_{W*} = 125 data samples (NS) = 24.9%	Misadjustment Due to Weight Lag	Misadjustment Due to Gradient Noise
Theoretical	2.0%	20.0%
Experimental	1.8%	28.3%

small, equation (78) is inaccurate since τ_{W*} is no longer much larger than τ_p.

Much more work needs to be done in the study of nonstationary adaptive behavior. We have presented a simplistic but meaningful beginning.

APPENDIX A
THE EFFICIENCY OF ADAPTIVE ALGORITHMS

We have analyzed the efficiency of the LMS algorithm from the point of view of misadjustment versus rate of adaptation. The question arises, could another algorithm be devised that would produce less misadjustment for the same rate of adaptation?

Suppose that an adaptive linear combiner is fed N independent input $n \times 1$ data vectors $X_1, X_2, \cdots, X_N$ drawn from a stationary ergodic process. Associated with these input vectors are their scalar desired responses $d_1, d_2, \cdots, d_N$, also drawn from a stationary ergodic process. Keeping the weights fixed, a set of N error equations can be written as

$$\epsilon_i = d_i - W^T X_i, \quad i = 1, 2, \cdots, N. \tag{A.1}$$

Let the objective be to find a weight vector that minimizes the sum of the squares of the error values based on a sample of N items of data.

Equation (A.1) can be written in matrix form as

$$\mathcal{E} = D - \mathfrak{X} W \tag{A.2}$$

where $\mathfrak{X}$ is an $N \times n$ rectangular matrix

$$\mathfrak{X} \triangleq [X_1 X_2 \cdots X_N]^T \tag{A.3}$$

and where $\mathcal{E}$ is an N element error vector

$$\mathcal{E} \triangleq [\epsilon_1 \epsilon_2 \cdots \epsilon_N]^T. \tag{A.4}$$

A unique solution of (A.1), bringing $\mathcal{E}$ to zero, exists only if $\mathfrak{X}$ is square and nonsingular. However, the case of greatest interest is that of $N >> n$. The sum of the squares of the errors is

$$\mathcal{E}^T \mathcal{E} = D^T D + W^T \mathfrak{X}^T \mathfrak{X} W - 2 D^T \mathfrak{X} W. \tag{A.5}$$

This sum multiplied by $1/N$ is an estimate $\hat{\xi}$ of the mse ξ. Thus

$$\hat{\xi} = \frac{1}{N} \mathcal{E}^T \mathcal{E} \quad \text{and} \quad \lim_{N \to \infty} \hat{\xi} = \xi. \tag{A.6}$$

Note that $\hat{\xi}$ is a quadratic function of the weights, the parameters of the quadratic form being related to properties of the N data samples. $(\mathfrak{X}^T \mathfrak{X})$ is square and positive semidefinite. $\hat{\xi}_{min}$ is the small-sample-size mse function, while ξ is the large-sample-size mse function. These functions are sketched in Fig. 13.

The function $\hat{\xi}$ is minimized by setting its gradient to zero:

$$\nabla \hat{\xi} = 2 \mathfrak{X}^T \mathfrak{X} W - 2 \mathfrak{X}^T D. \tag{A.7}$$

The "optimal" weight vector based only on the N data samples is

$$\hat{W}* \triangleq (\mathfrak{X}^T \mathfrak{X})^{-1} \mathfrak{X}^T D. \tag{A.8}$$

This formula gives the position of the minimum of the small-sample-size bowl. The corresponding formula for the large-sample-size bowl is the Wiener–Hopf equation (8).

We could calculate $\hat{W}*$ by a training process, a regression process, LMS, or some other optimization procedure. Taking the first block of N data samples, we obtain a small-sample-size function $\hat{\xi}_1$ whose minimum is at $\hat{W}_1^*$. This could be repeated with a second data sample, giving a function $\hat{\xi}_2$ whose minimum is at $\hat{W}_2^*$, etc. Typically, all the values of $\hat{W}*$ would differ from the true optimum $W*$ and would, thereby, be misadjusted.

To analyze the misadjustment, assume that N is large and that the typical small-size curve approximately matches the large-sample-size curve. Therefore,

$$\hat{\xi} \approx \xi \quad \text{and} \quad (\xi - \hat{\xi}) \triangleq d\xi. \tag{A.9}$$

The true large-sample-size function is

$$\xi = \xi_{min} + V'^T \Lambda V'.$$

The gradient of this function expressed in the primed coordinates is

$$\nabla' = 2 \Lambda V'.$$

A differential deviation in the gradient is

$$(d\nabla') = 2\Lambda (dV') + 2 (d\Lambda) V'. \tag{A.10}$$

This deviation could represent the difference in gradients between small- and large-sample-size curves.

Refer to Fig. 13. Let $W' = W*'$, then $V' = 0$. The gradient

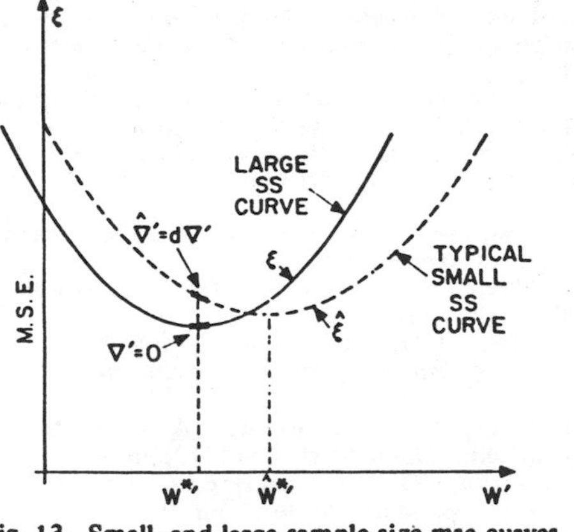

Fig. 13. Small- and large-sample-size mse curves.

of ξ is zero, while the gradient of $\hat{\xi}$ is $\hat{\nabla}' = d\hat{\nabla}$. Using (A.10),

$$(d\nabla') = 2\Lambda(dV'). \tag{A.11}$$

From (A.11), the deviation in gradient can be linked to the deviation in position of the small-sample-size curve minimum since $(dV') = (W^{*\prime} - \hat{W}^{*\prime})$. Taking averages of (A.11) over an ensemble of small-sample-size curves,

$$\text{cov}\,[d\nabla'] = 4\Lambda\,\text{cov}\,[dV']\Lambda. \tag{A.12}$$

Equation (42) indicates that the covariance of the gradient noise when $W' = W^{*\prime}$ is given by $4\xi_{\min}\Lambda$. If the gradient were estimated under the same conditions but using N independent error samples,

$$\text{cov}\,[d\nabla'] = \frac{4}{N}\xi_{\min}\Lambda. \tag{A.13}$$

Substituting this into (A.12) yields

$$\text{cov}\,[dV'] = \frac{1}{N}\xi_{\min}\Lambda^{-1}. \tag{A.14}$$

The average excess mse, an ensemble average, is

$$\begin{pmatrix}\text{average}\\ \text{excess}\\ \text{mse}\end{pmatrix} = E[(dV')^T\Lambda(dV')]. \tag{A.15}$$

Equation (A.14) shows cov $[dV']$ to be diagonal, so that

$$\begin{pmatrix}\text{average}\\ \text{excess}\\ \text{mse}\end{pmatrix} = \frac{n}{N}\xi_{\min}. \tag{A.16}$$

The misadjustment is, therefore,

$$M = \frac{n}{N} = \frac{(\text{number of weights})}{(\text{number of independent training samples})}. \tag{A.17}$$

This formula was first presented without detailed proof by Widrow and Hoff [1] in 1960. It has been used for many years in pattern recognition studies. For small values of M (less than 25 percent), it has proven to be very useful. A formula similar to (A.17), although based on somewhat different assumptions, was derived by Davisson [36] in 1970.

Although equation (A.17) has been derived for training with finite blocks of data, it can be used to assess the efficiency of steady-flow algorithms. Consider an adaptive transversal filter with stationary stochastic inputs, adapted by the LMS algorithm. For simplicity, let all eigenvalues of R be equal. As such

$$M = \frac{n}{4\tau_{\text{mse}}}. \tag{56}$$

The LMS algorithm exponentially weights its input data over time in determining current weight values. If an equivalent uniform averaging window is assumed equal to the adaptive settling time, approximately four time constants, the equivalent data sample taken at any instant by LMS is essentially $N_{\text{eq}} = 4\tau_{\text{mse}}$ samples. Accordingly for LMS,

$$M = \frac{n}{N_{\text{eq}}}. \tag{A.18}$$

A comparison of (A.18) and (A.17) shows that when eigenvalues are equal, LMS is about as efficient as a least squares algorithm can be.[5] However, with disparate eigenvalues, the misadjustment is primarily determined by the fastest modes while settling time is limited by the slowest modes. To sustain efficiency with disparate eigenvalues, algorithms similar to LMS have been devised based on Newton's method rather than on steepest descent [38], [39]. Such algorithms premultiply the gradient estimate each iteration cycle by an estimate of the inverse of R:

$$W_{j+1} = W_j + \mu\widehat{R^{-1}}\hat{\nabla}_j$$

or

$$W_{j+1} = W_j + 2\mu\widehat{R^{-1}}\epsilon_j X_j. \tag{A.19}$$

This process causes all adaptive modes to have essentially the same time constant. Algorithms based on this principle are potentially more efficient that LMS but are typically more difficult to implement.

ACKNOWLEDGMENT

The authors wish to acknowledge the helpful discussions and contributions of C. S. Williams and J. R. Treichler of Stanford University; Dr. O. L. Frost III of Argo Systems, Inc.; and Dr. M. E. Hoff, Jr. of the Intel Corporation. Special thanks are also due Diane Byron, who assisted in editing the paper.

REFERENCES

[1] B. Widrow and M. E. Hoff, "Adaptive switching circuits," in *1960 WESCON Conv. Rec.*, pt. 4, pp. 96-140.
[2] N. Nilsson, *Learning Machines*. New York: McGraw-Hill, 1965.
[3] J. Koford and G. Groner, "The use of an adaptive threshold element to design a linear optimal pattern classifier," *IEEE Trans. Inform. Theory*, vol. IT-12, pp. 42-50, Jan. 1966.
[4] B. Widrow, P. Mantey, L. Griffiths, and B. Goode, "Adaptive antenna systems," *Proc. IEEE*, vol. 55, pp. 2143-2159, Dec. 1967.
[5] B. Widrow, "Adaptive filters," in *Aspects of Network and System Theory*, R. Kalman and N. DeClaris, Eds. New York: Holt, Rinehart, and Winston, 1971, pp. 563-587.
[6] S. P. Applebaum, "Adaptive arrays," Special Projects Lab., Syracuse Univ. Res. Corp., Rep. SPL 769.
[7] L. J. Griffiths, "A simple adaptive algorithm for real-time processing in antenna arrays," *Proc. IEEE*, vol. 57, pp. 1696-1704, Oct. 1969.
[8] O. L. Frost III, "An algorithm for linearly constrained adaptive array processing," *Proc. IEEE*, vol. 60, pp. 926-935, Aug. 1972.
[9] W. F. Gabriel, "Adaptive arrays—An introduction," *Proc. IEEE*, vol. 64, pp. 239-272, Feb. 1976.

[5] Attempts have been made to devise algorithms more efficient than LMS by using variable μ [37]. Initial values of μ are chosen high for rapid convergence; final values of μ are chosen low for small misadjustment. This works as long as input statistics are stationary. This procedure and the methods of stochastic approximation on which it is based will not perform well in the nonstationary case.

[10] F. W. Smith, "Design of quasi-optimal minimum-time controllers," *IEEE Trans. Automat. Contr.*, vol. AC-11, pp. 71–77, Jan. 1966.
[11] B. Widrow, "Adaptive model control applied to real-time blood-pressure regulation," in *Pattern Recognition and Machine Learning*, Proc. Japan–U.S. Seminar on the Learning Process in Control Systems, K. S. Fu, Ed. New York: Plenum Press, 1971, pp. 310–324.
[12] R. Lucky, "Automatic equalization for digital communication," *Bell Syst. Tech. J.*, vol. 44, pp. 547–588, Apr. 1965.
[13] M. DiToro, "A new method of high-speed adaptive serial communication through any time-variable and dispersive transmission medium," in *Conf. Record*, 1965 IEEE Annual Communications Convention, pp. 763–767.
[14] R. Lucky and H. Rudin, "An automatic equalizer for general-purpose communication channels," *Bell Syst. Tech. J.*, vol. 46, pp. 2179–2208, Nov. 1967.
[15] R. Lucky *et al.*, *Principles of Data Communication*. New York: McGraw-Hill, 1968.
[16] A. Gersho, "Adaptive equalization of highly dispersive channels for data transmission," *Bell Syst. Tech. J.*, vol. 48, pp. 55–70, Jan. 1969.
[17] M. Soudhi, "An adaptive echo canceller," *Bell Syst. Tech. J.*, vol. 46, pp. 497–511, Mar. 1967.
[18] B. Widrow *et al.*, "Adaptive noise cancelling: Principles and applications," *Proc. IEEE*, vol. 63, pp. 1692–1716, Dec. 1975.
[19] P. E. Mantey, "Convergent automatic-synthesis procedures for sampled-data networks with feedback," Stanford Electronics Laboratories, Stanford, CA, TR no. 7663-1, Oct. 1964.
[20] P. M. Lion, "Rapid identification of linear and nonlinear systems," in *Proc. 1966 JACC*, Seattle, WA, pp. 605–615, Aug. 1966; also *AIAA Journal*, vol. 5, pp. 1835–1842, Oct. 1967.
[21] R. E. Ross and G. M. Lance, "An approximate steepest descent method for parameter identification," in *Proc. 1969 JACC*, Boulder, CO, pp. 483–487, Aug. 1969.
[22] R. Hastings-James and M. W. Sage, "Recursive generalized-least-squares procedure for online identification of process parameters," *Proc. IEE*, vol. 116, pp. 2057–2062, Dec. 1969.
[23] A. C. Soudack, K. L. Suryanarayanan, and S. G. Rao, "A unified approach to discrete-time systems identification," *Int. J. Control*, vol. 14, no. 6, pp. 1009–1029, Dec. 1971.
[24] W. Schaufelberger, "Der Entwurf adaptiver Systeme nach der direckten Methode von Ljapunov," *Nachrichtentechnik*, Nr. 5, pp. 151–157, 1972.
[25] J. M. Mendel, *Discrete Techniques of Parameter Estimation: The Equation Error Formulation*. New York: Marcel Dekker, Inc., 1973.
[26] S. J. Merhav and E. Gabay, "Convergence properties in linear parameter tracking systems," *Identification and System Parameter Estimation–Part 2*, Proc. 3rd IFAC Symp., P. Eykhoff, Ed. New York: American Elsevier Publishing Co., Inc., 1973, pp. 745–750.
[27] R. V. Southwell, *Relaxation Methods in Engineering Science*. New York: Oxford, 1940.
[28] D. J. Wilde, *Optimum Seeking Methods*. Englewood Cliffs, NJ: Prentice-Hall, 1964.
[29] B. Widrow, "Adaptive sampled-data systems," in *Proc. First Intern. Cong. Intern. Federation of Automatic Control*, Moscow, 1960.
[30] H. Robbins, and S. Monro, "A stochastic approximation method," *Ann. Math. Statist.*, vol. 22, pp. 400–407, 1951.
[31] A. Dvoretzky, "On stochastic approximation," in *Proc. Third Berkeley Symp. Math. Statist. and Probability*, J. Neyman, Ed. Berkeley, CA: University of California Press, 1956, pp. 39–55.
[32] J. Makhoul, "Linear prediction: A tutorial review," *Proc. IEEE*, vol. 63, pp. 561–580, Apr. 1975.
[33] L. J. Griffiths, "Rapid measurement of digital instantaneous frequency," *IEEE Trans. Acoust., Speech, Signal Processing*, vol. ASSP-23, pp. 207–222, Apr. 1975.
[34] Y. T. Chien, K. S. Fu, "Learning in non-stationary environment using dynamic stochastic approximation," in *Proc. 5th Allerton Conf. Circuit and Systems Theory*, pp. 337–345, 1967.
[35] T. P. Daniell and J. E. Brown III, "Adaptation in nonstationary applications," in *Proc. 1970 IEEE Symp. Adaptive Processes (9th)*, Austin, TX, paper no. XXIV-4, Dec. 1970.
[36] L. D. Davisson, "Steady-state error in adaptive mean-square minimization," *IEEE Trans. Inform. Theory*, vol. IT-16, pp. 382–385, July 1970.
[37] T. J. Schonfeld and M. Schwartz, "A rapidly converging first-order training algorithm for an adaptive equalizer," *IEEE Trans. Inform. Theory*, vol. IT-17, pp. 431–439, July 1971.
[38] K. H. Mueller, "A new, fast-converging mean-square algorithm for adaptive equalizers with partial-response signaling," *Bell Syst. Tech. J.*, vol. 54, pp. 143–153, Jan. 1975.
[39] L. J. Griffiths and P. E. Mantey, "Iterative least-squares algorithm for signal extraction," in *Proc. Second Hawaii Int. Conf. System Sciences*, Western Periodicals Co., pp. 767–770, 1969.

Self-Orthogonalizing Adaptive Equalization Algorithms

RICHARD D. GITLIN, SENIOR MEMBER, IEEE, AND FRANCIS R. MAGEE, JR., MEMBER, IEEE

***Abstract*—A comparison is made of several self-orthogonalizing adjustment algorithms for linear tapped delay line equalizers. These adaptive algorithms accelerate the rate of convergence of the equalizer tap weights to those which minimize the output mean-squared error of a data transmission system. Accelerated convergence of the estimated gradient algorithm is effected by premultiplying the correction term in the algorithm by a matrix which is an estimate of the inverse of the channel correlation matrix. The various algorithms differ in the manner in which this estimate is sequentially computed. Depending on the degree of complexity available, the equalizer convergence time may be reduced more than an order of magnitude from that required by the simple gradient algorithm.**

I. INTRODUCTION

ADAPTIVE gradient algorithms, for the rapid adjustment of the gains in a linear equalizer, are by now commonplace in the data transmission literature [1]-[11]. Nevertheless there is still great interest in this subject—since there exist channels whose distortion is not severe enough to preclude reliable data transmission, but yet such that currently used adaptation algorithms converge too slowly to be of practical value. The purpose of this paper is to compare several new adaptive adjustment algorithms which are capable of reducing the settling time of a Tapped Delay Line (TDL) equalizer. For the equalized mean-square error performance criterion it is known [9] that the rate of convergence (ROC) of the conventional estimated-gradient algorithm can be strongly influenced by the ratio of the maximum to minimum eigenvalues of the channel-correlation matrix. Convergence can be accelerated by orthogonalizing (i.e., making all the eigenvalues identical) the above matrix. When the channel is known *a priori* this works well, as Chang [11] has demonstrated for partial-response signaling. Until recently [12], [15], [16], the main impediment has been the inability to perform this orthogonalization adaptively (i.e., without knowing the channel and using an estimated rather than the true gradient)—short of using the received signal to explicitly measure the channel, and subsequently compute the optimum tap weights in a noniterative manner via an efficient matrix inversion algorithm [13].

While the mathematical literature [14] abounds in conjugate-gradient, or quasi-Newton, algorithms which promise convergence in N steps, where N is the number of variables being adjusted, the noisy environment in which these algorithms must operate has a profound effect on their performance. The inclusion of noise is no small matter, and Proakis [7] has demonstrated, via simulation, that conjugate-gradient algorithms are unstable even in the presence of an extremely small amount of noise. Stern [15] and Godard [16] have each recently proposed adaptively self-orthogonalizing gradient algorithms. The latter algorithm, through a novel application of Kalman filter theory, achieves extremely rapid convergence under some mild assumptions. In this study we present an alternative derivation of this latter algorithm which directly exhibits the orthogonalizing capability. In addition, we determine the initial rate of convergence of the Godard

Paper approved by the Editor for Data Communication Systems of the IEEE Communications Society for publication after presentation at the National Telecommunications Conference, Dallas, TX, December 1976. Manuscript received June 21, 1976; revised March 10, 1977.

The authors are with Bell Laboratories, Holmdel, NJ 07733.

Reprinted from *IEEE Trans. Comm.*, vol. COM-25, no. 7, pp. 666–672, July 1977.

algorithm and indicate the dramatic improvement it offers.[1] We also describe a new self-orthogonalizing adaptive-gradient algorithm which strikes a compromise between complexity and speed of convergence.

In addition to the situation where the channel is used at or near the Nyquist rate, another potential application of these new algorithms is the adaptive prefiltering of pulse-amplitude modulated (PAM) data signals in optimum (maximum likelihood) receivers [17]-[18]. Here the overall impulse response is adjusted to a shape that optimizes the performance of the maximum likelihood receiver. These shapes frequently have nulls in their spectrum [19] (at the band edges), and thus present convergence problems for the conventional estimated-gradient algorithm.

In Section II we present the communications model relevant for the equalizer adjustment problem and also review the conventional gradient algorithm. Sections III and IV develop, respectively, the algorithms proposed by Stern and Godard. A new tap adjustment procedure which accelerates the convergence of the gradient algorithm, while retaining a modest level of complexity, is presented in Section V. Simulations comparing the performance of these techniques are shown in Section VI.

II. SYSTEM MODEL AND THE ESTIMATED-GRADIENT TAP ADJUSTMENT ALGORITHM

In this section we review the application of the estimated-gradient algorithm to iteratively obtain the equalizer taps which minimize the Mean-Squared Error (MSE).

A. The Model

A simplified model of a baseband PAM data transmission system [1] is shown in Figure 1. The discrete-valued independent data sequence $\{a_m\}$ modulates, at the symbol rate $1/T$, the transmitting-filter/channel/receiving-filter cascade which is denoted by the pulse $h(t)$. At the receiver the filtered signal, which is corrupted by additive noise, $w(t)$, of variance σ^2, is sampled every T seconds. This sampled signal, $x(nT)$, is presented to the TDL equalizer shown in Figure 2 with the objective of having the equalizer output, $y(nT)$, be a good approximation to the transmitted symbol a_n. The equalizer output is quantized to the nearest data level to produce the decision $\hat{a}_n$. The equalizer input is given by[2]

$$x_n = \sum_m a_m h_{n-m} + w_n, \tag{1}$$

and the equalizer output by

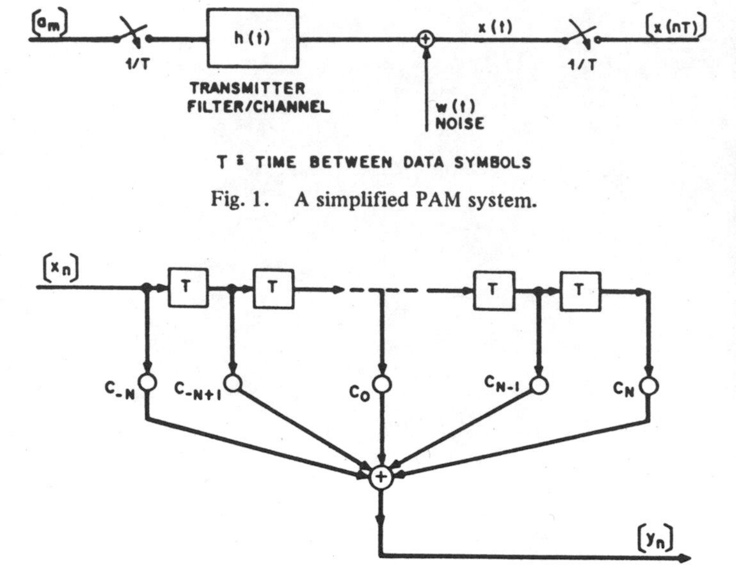

Fig. 1. A simplified PAM system.

Fig. 2. Tapped delay line equalizer.

$$y_n = \sum_{m=-N}^{N} c_m x_{n-m}, \tag{2}$$

where $\{c_m\}_{m=-N}^{N}$ are the $2N + 1$ equalizer tap weights. As has been mentioned, the performance measure is the equalized mean-squared error

$$\mathcal{E} = E((y_n - a_n)^2) = E(e_n{}^2) \tag{3}$$

where $E(\cdot)$ denotes expectation, e_n the equalizer error signal, and where it is noted that the stationarity of the data sequence and the noise implies that (3) is independent of time. Without any loss in generality we restrict the data symbols to the values ± 1, and by carrying out the indicated expectation, (3) can be written as

$$\mathcal{E}(c) = c'Ac - 2c'h + 1 \tag{4}$$

where the primes denote the vector transpose, A is the symmetric (Toeplitz) channel correlation matrix with elements[3]

$$A_{i-j} = \sum_{n=-\infty}^{\infty} h_{n-i} h_{n-j} + \sigma^2 \delta_{i-j}, \tag{5}$$

h is the vector of $2N + 1$ channel samples

$$h' = (h_N, \cdots, h_0, \cdots, h_{-N}), \tag{6}$$

and where c is the vector of tap weights

$$c' = (c_{-N}, \cdots, c_0, \cdots, c_N). \tag{7}$$

It is instructive to indicate what the receiver would do if the channel were known [3]. Differentiating (4) with respect to c, we see that the optimum tap vector is given by

[1] An "information-theoretic" lower bound on convergence time can be obtained as follows: Suppose that somehow the transmitter knew the equalizer tap weights to within the required digital precision; consequently, transmitting the total number of bits which convey this information, at the channel-capacity rate, should provide a lower bound on the equalizer settling time. It is indeed surprising that the Godard algorithm is capable of closely approaching this lower bound for a wide class of channels.

[2] Subscripted notation is employed to denote time samples, i.e., $x_n \equiv x(nT)$.

[3] The symbol δ_{i-j} is the Kronecker delta, i.e., $\delta_{i-j} = 1$ if $i = j$ and is zero otherwise.

$$c_{opt} = A^{-1}h, \tag{8}$$

while the residual mean-square error is

$$E(c_{opt}) = 1 - h'A^{-1}h. \tag{9}$$

Some results which are useful for the sequel are obtained when the number of taps becomes very large [20] (i.e., $N \to \infty$). In this case (8) becomes the convolution

$$\sum_{i=-\infty}^{\infty} A_{k-i}c_i = h_{-k} \qquad k = -\infty, \cdots, \infty \tag{10}$$

and taking the Fourier transform of the above gives

$$C_{opt}(\omega) = \frac{H^*(\omega)}{A(\omega)} = \frac{H^*(\omega)}{H(\omega)H^*(\omega) + N(\omega)} \tag{11}$$

where the Fourier transform of the weights is defined by

$$C(\omega) = T \sum_{n=-\infty}^{\infty} c_n e^{-j\omega nT}, \tag{12}$$

and $N(\omega)$ is the power spectrum of the additive noise.[4] Similarly, it is shown that the optimized MSE is

$$E(C_{opt}(\omega)) = \int_{-\pi/T}^{\pi/T} \frac{N(\omega)}{A(\omega) + N(\omega)} \frac{d\omega}{2\pi}. \tag{13}$$

The Fourier transform relationship, (11), between the various quantities, which is of course only valid for $N \to \infty$, will be very useful in suggesting and understanding several adaptive self-orthogonalizing algorithms. It should also be pointed out that when the input is periodic, similar relationships hold between the corresponding discrete Fourier Transforms (DFT). Thus all results and algorithms which are discussed for random data have an analog for a periodic training sequence [8]. In the next paragraph we will review how c_{opt} may be obtained in a recursive manner when the channel is unknown.

B. Estimated Gradient Algorithm

When a training sequence[5] is available at the receiver a widely used tap-adjustment algorithm can be obtained from (3) by interchanging the linear expectation and gradient operations, viz.

$$\nabla E = \nabla E((y_n - a_n)^2) = 2E((y_n - a_n) \cdot x_n), \tag{14}$$

where x_n is the vector of received samples stored in the equalizer delay line at time nT, and where the estimate of the gradient is given by

$$\hat{\nabla} E = (y_n - a_n)x_n. \tag{15}$$

Thus each component of the estimated gradient is obtained by correlating the instantaneous equalizer error signal, $y_n - a_n$, with the appropriate sample stored in the equalizer delay line. Using this estimate the estimated-gradient tap adjustment algorithm is [7], [9], [10], [12] (see Fig. 3)

$$c_{n+1} = c_n - \alpha_n e_n x_n, \tag{16}$$

where α_n is a positive quantity called the step-size, and where the mean-square error evolves according to

$$E(c_n) = E(c_{opt}) + \langle \epsilon_n' A \epsilon_n \rangle. \tag{17}$$

An exact analytical description of the convergence of $E[c_n]$ to $E[c_{opt}]$ is quite difficult; however, using several approximations it is shown in [9] that the evolution of the second term in (17) satisfies a difference equation whose initial behavior (i.e., for small n) is strongly dependent[6] on the ratio, ρ, of the minimum to maximum eigenvalue of the matrix A. Ungerboeck [10], using a slightly different approach than [9], indicates that the larger N (the number of taps) the slower the speed of convergence of the adaptive gradient algorithm. In fact, Ungerboeck's analysis suggests that at the beginning of the equalization process the optimum step size is on the order of $1/N \cdot Ex_n^2$; reference [9] suggests a step size on the order of $\lambda_{min}/\lambda_{min}^2$, where λ_{min} and λ_{max} are, respectively, the minimum and maximum eigenvalues of the channel correlation matrix, A. For a range of channels with average to mild amplitude distortion the eigenvalue ratio $\rho = \lambda_{min}/\lambda_{max}$ has been numerically computed and found to be approximately inversely proportional to N. Since λ_{max} is also on the order of Ex_n^2, the two approaches lead to pretty much the same step size, and suggest that the number of taps be kept to a minimum, consistent with the desired steady-state mean-squared error. All of the work done on optimizing the step size has not yielded any dramatic improvement in the rate of convergence (ROC), since the ROC will still be influenced by the eigenvalue spread; however, the concept of orthogonalizing the A matrix, i.e., premultiplying the gradient by A^{-1}, or more practically by an estimate of A^{-1}, potentially offers significant improvement.[7]

C. Self-Orthogonalizing Algorithms

The new ingredient in a self-orthogonalizing algorithm is to find a sequence of matrices, D_n, which premultiply the estimated gradient, and that rapidly converge to A^{-1}. The algorithms are thus of the form

$$c_{n+1} = c_n - \alpha_n D_n (y_n - a_n) x_n. \tag{18}$$

In the next sections we will present three such algorithms.

[4] The power spectrum $N(\omega)$ is the Fourier Transform of the noise-correlation function evaluated at the receiver sampling rate, and $H(\omega)$ is obtained from (12) with h_n replacing the c_n.

[5] By a training sequence we mean that the transmitted data symbols are known at the receiver.

[6] In [9], it is shown that for small n: $\langle \epsilon_{n+1}' A \epsilon_{n+1} \rangle < (1 - \rho^2) \epsilon_n A \epsilon_n$, i.e., the initial ROC is exponential, and if the step-size, α_n, is inversely proportional to n, then the mean-squared error ultimately decays like $1/n$.

[7] Note that this leaves both c_{opt} and $E(c_{opt})$ unchanged, and only affects the ROC.

III. STERN'S ALGORITHM

Recently Stern [15] has proposed a sequence of Toeplitz (but not symmetric) matrices, $D^{(n)}$, whose elements $D_{ij}{}^{(n)}$ are defined as the correlations

$$D_{ij}{}^{(n)} = D_{i-j}{}^{(n)} = \sum_{l=-N}^{N} c_l{}^{(n)} c_{l-(i-j)}{}^{(n)}, \tag{19}$$

where $c_{l-(i-j)} \equiv 0$ when $|l-(i-j)| > N$. The rationale for such a choice becomes clear by rewriting (11) as

$$A^{-1}(\omega) = \frac{|C(\omega)|^2}{H(\omega)C(\omega)} = \frac{|C(\omega)|^2}{G(\omega)}, \tag{20}$$

where $G(\omega)$ is the equalized system response $H(\omega)C(\omega)$. If it is assumed that equalization is perfect, i.e., $G(\omega) = 1$, and we let $D(\omega)$ be the corresponding value of $A^{-1}(\omega)$, viz.

$$D(\omega) = |C(\omega)|^2, \tag{21}$$

whose inverse transform gives D_{ij}. Thus the matrix which Stern has proposed is the inverse of the A matrix assuming 1) a large number of taps, 2) equalization is perfect, and 3) the taps have converged to their optimum values. As is evident, the resulting algorithm is highly nonlinear and presents enormous analytical difficulties. Simulation results concerning the rate of convergence of Stern's algorithm will be presented in a later section. It should be noted that in order to update each tap weight two matrix multiplications must be performed.

IV. GODARD'S ALGORITHM[8]

In this section we present a new derivation of the Godard algorithm which explicitly exhibits its orthogonalizing property. Also, we determine the approximate *initial* convergence rate of the algorithm.

A. An Alternative Derivation

The basic idea behind the accelerated convergence, or orthogonalization, algorithms considered here is to rewrite the estimated gradient algorithm as

$$c_{n+1} = c_n - \alpha_n D_n e_n x_n \tag{22}$$

where it is desired that D_n rapidly converge to A^{-1}. We begin by writing the well-known sample mean estimate [for the channel correlation matrix, $A \equiv E(xx')$]

$$\hat{A}_{n+1} = \frac{1}{n+1} \sum_{j=1}^{n+1} x_j x_j' \tag{23}$$

in the recursive form

$$\hat{A}_{n+1} = \frac{n}{n+1} \hat{A}_n + \frac{1}{n+1} x_{n+1} x_{n+1}'. \tag{24}$$

[8] The Godard tap-adjustment algorithm is readily extended to accommodate decision feedback equalization.

We will now develop an iterative equation for the inverse of A_n. Such an equation is provided by the matrix inversion lemma [23] (MIL) which has important application in estimation and control problems. Before applying the lemma we let $\tilde{A}_n = n\hat{A}_n$ so that (24) becomes

$$\tilde{A}_{n+1} = \tilde{A}_n + x_{n+1} x_{n+1}', \tag{25}$$

and now using the MIL we obtain the following recursion for the inverse of $\tilde{A}_n$:

$$\tilde{D}_{n+1} = \tilde{D}_n - \frac{\tilde{D}_n x_{n+1} x_{n+1}' \tilde{D}_n}{1 + x_{n+1}' \tilde{D} x_{n+1}} \tag{26}$$

where $\tilde{D}_n = [\tilde{A}_n]^{-1}$, and where A_0 has been taken to be the null matrix (Godard's algorithm [16] would be obtained[9] if A_0 were taken to be $\mathcal{E}_{opt}{}^2$ times the identity matrix and the step size α_n to be $[n(1 + x_{n+1}' \tilde{D}_n x_{n+1})]^{-1}$. Note that due to the absorption of the factor n in $\tilde{A}_n$, the matrix D_n, which is used in (22), is given by

$$D_n = \frac{1}{n} \tilde{D}_n. \tag{27}$$

B. Rate of Convergence

In this section we wish to determine, to first order, the initial rate of convergence associated with the adaptive Godard algorithm. Recall from (17) that the dynamics of the algorithm are governed by $\langle \epsilon_n' A \epsilon_n \rangle$. From (22) we have that

$$\begin{aligned} \epsilon_{n+1}' A \epsilon_{n+1} \equiv q_{n+1} &= (\epsilon_n' - \alpha_n e_n x_n') A (\epsilon_n - \alpha_n e_n D_n x_n) \\ &= \epsilon_n' A \epsilon_n - 2\alpha_n e_n x_n' D_n A \epsilon_n \\ &\quad + \alpha_n{}^2 e_n{}^2 x_n' D_n A D_n x_n. \end{aligned} \tag{28}$$

We will assume, as does Godard [16], that D_n converges to A^{-1} rather rapidly so that[10]

$$D_n A \approx \left(1 + \frac{\gamma}{\sqrt{n}}\right) I, \tag{29}$$

where the scalar γ measures the magnitude of the estimation error. Employing (29) and the techniques used in Appendix I of [9] we can bound (28) as

$$q_{n+1} \leqslant \left\{1 - 2\alpha_n \left(1 + \frac{\gamma}{\sqrt{n}}\right) + \alpha_n{}^2 \left(1 + \frac{\gamma}{\sqrt{n}}\right)^2\right\} q_n + \left(1 + \frac{\gamma}{\sqrt{n}}\right)^2 \sigma^2, \tag{30}$$

where σ^2 depends on the received signal-to-noise ratio and other system parameters (see [9]). Initially, it is reasonable to

[9] Thus, a slightly modified version of the Godard algorithm is considered.

[10] This assumption is based on the fact that sample-mean estimates, such as D_n, converge to the true parameter with a variance which is inversely proportional to time.

assume that the mean-squared error is much greater than the background noise, and thus

$$q_{n+1} \lesssim \left[1 - \alpha\left(1 + \frac{\gamma}{\sqrt{n}}\right)\right]^2 q_n = \left[(1-\alpha) - \frac{\alpha\gamma}{\sqrt{n}}\right]^2 q_n, \quad n \leqslant n_0. \quad (31)$$

In (31) we have let the step-size, α_n, be constant, and n_0 is a small integer. With $\alpha \ll 1$, but with $\alpha\gamma \approx 1$ the further approximation

$$q_{n+1} \lesssim \frac{\alpha^2\gamma^2}{n} q_n \qquad n \leqslant n_0 \quad (32)$$

will be accurate. Iterating (32) gives the *factorial* decay

$$q_n \leqslant \frac{(\alpha\gamma)^{2n}}{n!} q_0, \qquad n \leqslant n_0. \quad (33)$$

While many approximations have been made in deriving (33), the extraordinarily rapid factorial convergence will be verified in the computer simulations presented in the sequel. As n increases beyond n_0, i.e., when $(1-\alpha) > \alpha\gamma/\sqrt{n}$, the convergence becomes exponential and we have

$$q_{n+1} \leqslant [1-\alpha]^2 q_n \qquad n_0 \leqslant n \leqslant n_1 \quad (34)$$

or

$$q_{n+1} \leqslant (1-\alpha)^{2n} q_{n_0}. \quad (35)$$

If as q_n approaches its steady-state value and becomes comparable to σ^2, the step-size, α_n, were[11] to become proportional to $1/n$, then the convergence would be arithmetic [9], i.e., $q_n \sim 1/n$. Thus, under certain plausible assumptions the convergence of the mean-squared error to its minimum value has a factorial mode followed by an exponential node, and finally an arithmetic mode. As shall be evident from the simulation presented in the sequel, the arithmetic mode of convergence can be reached rather rapidly [16].

V. A NEW SELF-ORTHOGONALIZING ALGORITHM

The algorithm which we propose in this section makes use of the frequency-domain terminology introduced in Section III. These correspondences hold for infinitely long equalizers, and allow Toeplitz matrices as well as vectors to be identified with appropriate Fourier transforms. Here the essential idea will be to develop an additional error equation, based upon the current degree of distortion, to adjust the orthogonalizing matrix. Recall that when the taps are adjusted to their optimum value

$$A^{-1}(\omega) = \frac{C(\omega)}{H^*(\omega)}, \quad (11)$$

[11] This is exactly what happens in the Godard algorithm—since in [16], $\alpha_n = [n(1 + x_{n+1}'\tilde{D}x_{n+1})]^{-1}$.

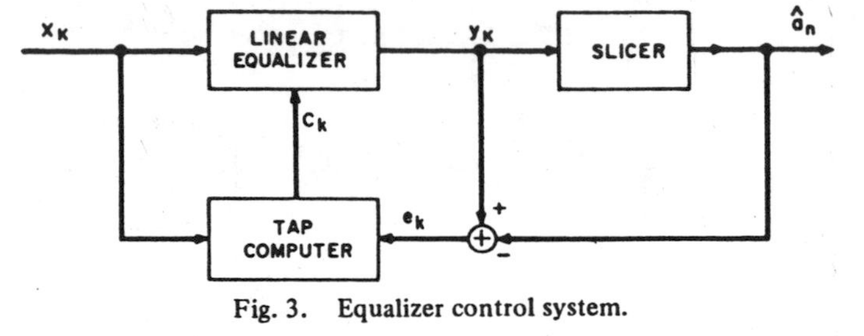

Fig. 3. Equalizer control system.

and by multiplying both sides of this equation by the spectrum of the transmitted signal, $S(\omega)$, we obtain

$$S(\omega)H(\omega)A^{-1}(\omega) = S(\omega)C^*(\omega). \quad (36)$$

Now $S(\omega)H(\omega)$ is simply the (noiseless) received signal, $X(\omega)$, while $S(\omega)C^*(\omega)$ represents the data bits convolved with the TDL equalizer, c^*, whose taps are reversed (i.e., c_n is replaced by c_{-n}) and where $A^{-1}(\omega)$ will be associated with the orthogonalizing matrix D. Thus the error signal

$$\tilde{E}(\omega) = X(\omega)D(\omega) - S(\omega)C^*(\omega) \quad (37)$$

is an additional measure of how close the equalizer taps are to their optimum values. Note that $A^{-1}(\omega)$ has been replaced by the transform $D(\omega)$, where the latter quantity is used to orthogonalize the channel correlation matrix. Interpreting (37) in vector notation gives the new error signal

$$\tilde{e}_n = x_n'd - a_n'c^*, \quad (38)$$

where $d = (d_{-N}, \cdots, d_0, \cdots, d_N)'$ is the vector corresponding to the spectrum $D(\omega)$. The orthogonalizing matrix, D, is now of necessity restricted to being the Toeplitz matrix whose diagonals are the components of d. Clearly there will be degradation associated with this procedure, since the inverse of Toeplitz matrix A is not necessarily Toeplitz.

Using the error signal (38), an additional mean-squared error is defined as

$$\tilde{E} = E((x_n'd - a_n'c^*)^2) = d'Ad - 2c'Hd + c'c, \quad (39)$$

where H is the Toeplitz matrix corresponding to the channel vector h. Taking the gradient of this expression with respect to d produces the algorithm

$$d_{n+1} = d_n - \beta_n \tilde{e}_n x_n \quad (40)$$

where β_n is a positive step-size. The recursions specified by (40) are to be combined with the corresponding portion of (18), where D_n is the Toeplitz matrix corresponding to d_n. An implementation of (40) is shown in Figure 4, where the new error signal is generated by first passing the data symbols through a TDL having the same weights as the actual equalizer but reversed in time sequence (i.e., c_{-n} replaces c_n, etc.), passing the received samples through another TDL having the current value, d_n, of the estimate of a row of the orthogonalizing matrix D, and finally forming the difference of these quantities.

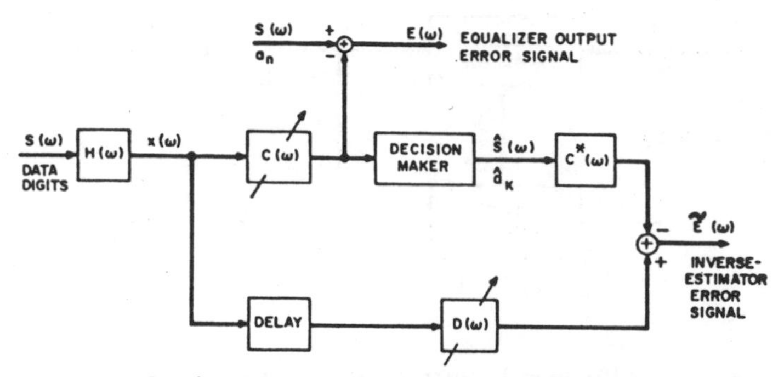

Fig. 4. Adaptive self-orthogonalizing algorithm.

The above algorithm is simpler to implement than the other algorithms considered, excepting the gradient, because it requires only the addition of another TDL equalizer to determine the orthogonalizing matrix, and the subsequent multiplication by this matrix during the computation of the c taps. Stern's proposal requires many multiplications for the formation of its orthogonalizing matrix in addition to the multiplication during the formation of the c tap estimate, while the Godard algorithm requires two matrix multiplications for each step in its operation.

VI. SIMULATION RESULTS

In order to compare the performance of the above algorithms in a noisy environment, they were simulated using the channel characteristics [24] shown in Figure 5. For the 31 tap equalizer used in the simulation the eigenvalue ratio of the baseband-equivalent channel-correlation (A) matrix is 18.6. The data rate considered in the simulation was 13,200 bits/s using four level Vestigial Sideband Modulation (VSB) with the carrier at 3455 Hz at a signal-to-noise ratio of 31 dB.

Figure 6 shows the simulation results for the various algorithms under discussion. For both the simple gradient and Stern's algorithm the step-size, α, was chosen to be 0.002. Experimentation has shown that this provides the quickest possible convergence, for this channel, while retaining a stable algorithm.[12] For the new algorithm (self-orthogonalizing algorithm) described in Section V the step-size was again chosen to be 0.002 in the gradient portion of the algorithm. The step-size β, which is used to adjust the estimate of the orthogonalizing matrix (D), was chosen as 0.001. These values were also chosen experimentally so as to optimize performance. In the case of the Godard algorithm several initial values were chosen as suggested in [16]. Such fine results were obtained that no further experimentation was necessary. Finally, the center tap c_0 of the c_0 vector was initially chosen to be unity with the rest of the vector equal to zero.

The conclusions to be drawn from Figure 6 are clear. The Godard algorithm has reached convergence within 60 iterations[13] (data symbols), while the fixed step size gradient algorithm requires about 900 relatively simple iterations. Stern's algorithm trims this convergence time to about 800 iterations but at a considerable increase in complexity. Finally, the self-orthogonalizing algorithm has converged within about 500 iterations, thus diminishing the convergence time considerably without adding a great amount of complexity. To further accelerate convergence in a practical system, the orthogonalizing matrix could be initially set to the inverse of the average channel correlation matrix.

[12] Note that this curve agrees rather well with the exponential decay suggested by (17).

[13] This is close to the lower bound discussed in the footnote in Section I.

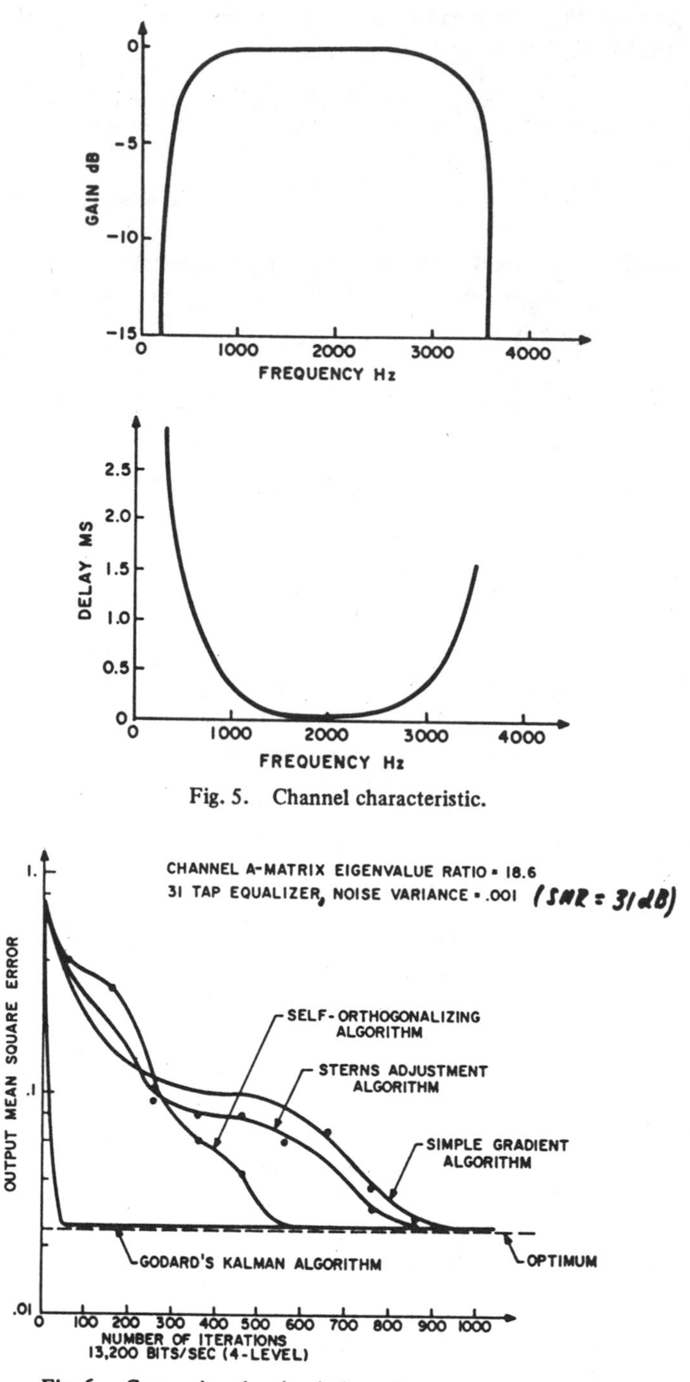

Fig. 5. Channel characteristic.

Fig. 6. Comparison by simulation of convergence properties.

VII. CONCLUSIONS

This paper has compared, through analysis and simulation, several recently proposed algorithms for rapid equalizer tap adjustment. In addition a new algorithm has been proposed which moderately improves the convergence time over the standard gradient algorithm, while not adding too much complexity to the adjustment scheme. It was shown that although Godard's algorithm provides the quickest convergence, typically by an order of magnitude over the gradient algorithm, it also requires the greatest amount of computation. The particular application at hand, and, most importantly, the current technology will determine whether or not a particular adaptive self-orthogonalizing algorithms will be implemented in practical systems.

ACKNOWLEDGMENT

The authors would like to thank D. D. Falconer for several helpful suggestions.

REFERENCES

[1] R. W. Lucky, J. Salz, and E. J. Weldon, Jr., *Principles of Data Communication*, New York: McGraw-Hill, 1968.

[2] B. Widrow and M. E. Hoff, Jr., "Adaptive Switching Circuits," *IRE Wescon Conv. Rec.*, pt. 4, pp. 96-104, August 1960.

[3] A. Gersho, "Adaptive Equalization of Highly Dispersive Channels for Data Transmission," *BSTJ*, 48, pp. 55-70, January 1969.

[4] T. J. Schonfeld and M. Schwartz, "Rapidly Converging First-Order Training Algorithm for an Adaptive Equalizer," *IEEE Trans. Info. Theory.*, Vol. 17, pp. 431-439, 1971.

[5] O. S. Kosovych, and R. L. Pickholtz, "Automatic Equalization Using a Successive Overrelaxation Iterative Technique," *IEEE Trans. on Information Theory*, Vol. IT-21, No. 1, Jan. 1975, pp. 51-58.

[6] H. Kobayashi, "Application of Hestenes-Stiefel Algorithm to Channel Equalization," *Proceedings Int. Conf. on Commun.*, 1971, pp. 21-25-21-30.

[7] J. G. Proakis, "Adaptive Filtering Techniques for Communication Through Time Dispersive Channels," Final Report NSF Grant GK-26329, Northeastern University, Boston, Mass., June 22, 1973.

[8] K. H. Mueller, D. A. Spaulding, "Cyclic Equalization–A New Rapidly-Converging Adaptive Equalization, Technique for Synchronous Data Communication," *B.S.T.J.*, Vol. 54, No. 2 pp. 369-406, Feb. 1975.

[9] R. D. Gitlin, J. E. Mazo, M. G. Taylor, "On the Design of Gradient Algorithms for Digitally Implemented Adjustment Filters," *IEEE Trans. on Circuit Theory*, Vol. CT-20, No. 2, pp. 125-136, March 1973.

[10] G. Ungerboeck, "Theory on the Speed of Convergence in Adaptive Equalizers for Digital Communication," *IBM J. RES. DEVELOP.*, November 1972, pp. 546-555.

[11] R. W. Chang, "A New Equalizer for Fast Start-Up Digital Communication," *BSTJ*, Vol. 50, No. 6, July-August, 1971, pp. 1969-2014.

[12] R. D. Gitlin and F. R. Magee, Jr., "Self-Orthogonalizing Algorithms for Accelerated Convergence of Adaptive Equalizer," Conference Record, *National Telecommunications Conference* (NTC '76), Dec. 1976, Dallas, Texas.

[13] P. Butler and A. Cantoni, "Noniterative Automatic Equalization," *IEEE Trans. on Communications*, Vol. COM-23, No. 6, pp. 621-634, June 1975.

[14] S. S. Oren, "Quasi-Newton Algorithms: Approaches and Motivations," *Proc. of 1973 IEEE CONF. on DECISION AND CONTROL*, pp. 422-427.

[15] T. E. Stern, "Some New Algorithms for Accelerating Convergence of Adaptive Equalizers," unpublished memorandum, 1974.

[16] D. Godard, "Channel Equalization Using a Kalman Filter for Fast Data Transmission," *IBM Journal of Research and Development*, May 1974, pp. 267-273.

[17] S. Qureshi and E. Newhall, "An Adaptive Receiver for Data Transmission Over Time Dispersive Channels," *IEEE Trans. Information Theory*, IT-19, July 1973.

[18] D. D. Falconer and F. R. Magee, Jr., "Adaptive Channel Memory Truncation for Maximum Likelihood Sequence Estimation," *BSTJ*, Vol. 52, No. 9, November 1973, pp. 1541-1562.

[19] G. J. Foschini, "Maximally Null Partial Response for a MLSE Digital Data System Involving High Loss Coaxial Cable," *IEEE Trans. on Information Theory*, Vol. IT-21, No. 6, November 1975.

[20] T. Berger and D. W. Tufts, "Optimum Pulse Amplitude Modulation, Parts I and II," *IEEE Trans. on Information Theory*, Vol. IT-13, April 1967.

[21] R. E. Kalman, "A New Approach to Linear Prediction Problems," *J. Basic Eng., Trans. ASME*, Vol. 82, 1960.

[22] U. Grenander and M. Rosenblatt, *Statistical Analysis of Stationary Time Series*, J. Wiley and Sons, 1967.

[23] A. P. Sage, *Optimum Systems Control*, Prentice-Hall, Englewood Cliffs, New Jersey, 1968.

[24] F. P. Duffy and T. W. Thatcher, Jr., "Analog Transmission Performance on the Switched Telecommunications Network," *BSTJ*, 50, No. 4, pp. 1311-1347.

Second-Order Convergence Analysis of Stochastic Adaptive Linear Filtering

ODILE MACCHI, MEMBER, IEEE, AND EWEDA EWEDA

Abstract—**The convergence of an adaptive filtering vector is studied, when it is governed by the mean-square-error gradient algorithm with constant step size. We consider the mean-square deviation between the optimal filter and the actual one during the steady state. This quantity is known to be essentially proportional to the step size of the algorithm. However, previous analyses were either heuristic, or based upon the assumption that successive observations were independent, which is far from being realistic. Actually, in most applications, two successive observation vectors share a large number of components and thus they are strongly correlated. In this work, we deal with the case of correlated observations and prove that the mean-square deviation is actually of the same order (or less) than the step size of the algorithm. This result is proved without any boundedness or barrier assumption for the algorithm, as it has been done previously in the literature to ensure the nondivergence. Our assumptions are reduced to the finite strong-memory assumption and the finite-moments assumption for the observation. They are satisfied in a very wide class of practical applications.**

I. Introduction

NOWADAYS, the adaptive filtering is an important matter, especially in the domain of detection and estimation. It consists of calculating iteratively the vector H_* that optimizes the linear estimation of a useful random signal, a, from an observation vector X having the same dimension N as H. This calculation is done, without knowing the statistics of a and X, on the basis of a stationary sequence of realizations (a_k, X_k) using the standard gradient algorithm

$$H_{k+1} = H_k + \mu X_k^*(a_k - X_k^T H_k) \tag{1}$$

of adaptive filtering, where the step size μ is a positive constant. In (1) X^* and X^T denote, respectively, the complex conjugate and the transpose of the vector X.

The algorithm (1) is intended to make H_k converge to the vector H_* that minimizes the quadratic mean deviation $f(H) = E(|a_k - X_k^T H|^2)$. Clearly, (1) represents the stochastic gradient algorithm for the function $f(H)$. The vector H_* is then the solution of the Wiener–Hopf equation

$$E(X_k^* X_k^T) H_* = E(a_k X_k^*). \tag{2}$$

The existence and uniqueness of H_* is ensured by the invertibility of the covariance matrix R

$$R \triangleq E(X_k^* X_k^T), \tag{3}$$

which we suppose implicitly in the following, together with the strict stationarity of the sequence (a_k, X_k).

The analysis presented in this work is concerned with the real version of the algorithm (1), i.e., we assume that a_k, X_k, and H_k are real valued. However, in a way similar to

Manuscript received April 13, 1981; revised November 30, 1981. Paper recommended by H. V. Poor, Chairman of the Estimation Committee.

O. Macchi is with C.N.R.S.—E.S.E., Plateau du Moulon, France.
E. Eweda is with the Department of Communication, Military Technical College, Cairo, Egypt.

Reprinted from *IEEE Trans. Automat. Contr.*, vol. AC-28, no. 1, pp. 76–85, Jan. 1983.

that in [1], it can be extended to cases in which a_k, X_k, and H_k are complex. An example of the latter case occurs in data transmission. It is the quadrature amplitude modulation (QAM) where the data signal a_k modulates the amplitude and phase of a carrier wave and where the received signal is demodulated with two quadrature carriers. In that case, both a_k and X_k, composed of N samples of the baseband channel output, are complex.

While the simulations and experimental works exhibit good convergence properties for the algorithm (1), there is not yet a satisfactory theoretical convergence proof, established under completely realistic hypothesis. Previous proofs, e.g., [1], were derived under the assumption of statistically independent pairs (a_k, X_k). In most applications, this assumption is far from being fulfilled because two successive vectors, X_k, share $(N-1)$ components. Indeed, X_k is made of N successive time samples of a signal, and X_{k+1} is simply a shifted version of X_k with an additional last component. The difficulty of the proof originates in this strong correlation. In a recent work, Ljung [2] has brought a very important contribution to the theory of the convergence under the realistic assumption of nonindependent observations. In his paper, he proves the almost sure convergence of an algorithm derived from (1) by replacing μ by a decreasing sequence μ_k tending to zero, when the observations are correlated. However, the proof uses a condition of nondivergence for the vector H_k, namely, that H_k will return indefinitely within a (random) compact set. One can guess that this condition will be ensured by putting a suitable barrier on the vector H_k. Unfortunately, the proof that the suitably bounded algorithm will converge to H_* is not yet established, even on the basis of Ljung's theorem.

Moreover, the almost sure convergence of the algorithm (1) with decreasing step size μ_k is not very valuable for applications because it cannot accommodate nonstationarities in the statistics of the pair (a_k, X_k). Indeed, H_k will essentially become constant after the convergence period, which precludes the tracking of time variations. Many authors have thus considered the algorithm (1) with a constant step size because it has the advantage of a limited memory, which enables it to track time fluctuations in H_*. It is now well known that the mean-square deviation (MSD) between the adaptive filter, H_k, and the optimum one, H_*, is then essentially proportional to the stepsize, μ. This result has already been stated by several authors, e.g., [3]–[5], and it is extensively used in the field of applications, for instance, in data transmissions, where it has a great importance. However, most previous proofs were heuristic except for the case of independent pairs (a_k, X_k) [1], [6]–[11], a requirement which is not fulfilled in many applications.

In the case of correlated pairs (a_k, X_k), several contributions to the problem have been published, and we now present some of them.

In [12], the convergence of (1) is analyzed with ergodic and uniformly asymptotically independent observations. In this paper it is proved that the asymptotic MSD can be made arbitrarily small by choosing a sufficiently small step size μ for the algorithm (1). However, that result is attained at the expense of a relatively strong assumption of uniformly bounded conditional moments of observations, given the past observations. Such an assumption is not satisfied for the important Gaussian example. Indeed, to satisfy that assumption, one is generally obliged to assume that observations are uniformly bounded. It should also be remarked upon that [12] does not offer a function of μ that bounds the asymptotic MSD, which is a point of great practical importance; for instance, for the evaluation of signal-to-noise ratios in communications [13].

In [14], a smoothed version of algorithm (1) is considered, namely,

$$H_{k+1} = H_k + \mu \left\{ \frac{1}{M} \sum_{n=kM}^{(k+1)M-1} \left(a_n - H_k^T X_n\right) X_n \right\}$$

where a block of pairs (a_n, X_n) of length M is used to estimate the deterministic gradient of $f(H)$.

The observations in that paper obey an M-dependent model similar to that used in this work. The analysis given in [14] is conditioned by a relatively large value of M.

Consequently, it cannot be applied to the algorithm (1) which corresponds to the case $M = 1$ and which is frequently used in applications because it has low computational complexity.

More recent contributions have brought better insight into the problem. For example, Farden [15] has analyzed the convergence problem with correlated observations under a realistic assumption of decreasing observation covariance.

In order to derive the bound, he assumes that useful signals, observations, and adaptive filters are uniformly bounded. To ensure the boundedness of H_k, a reflecting barrier is used to bring H_k back into a compact set every time it leaves that set. In the present paper, neither a_k, nor X_k, nor H_k are assumed bounded. Moreover, the asymptotic MSD obtained in our paper and given by (4) is more tight than that obtained in [15]. Indeed, the bound in (4) is, as usually expected, proportional to μ, while the bound in [15] is proportional to $\mu^{\nu/2}$, where ν is a given constant satisfying $0 < \nu < 1$.

We wish also to mention the study [16] of the asymptotic behavior of $|H_k - H_*|^2$. It is considered in a context wider than the sole algorithm (1).

It is shown that under the assumption of bounded observations (a_k, X_k), one has the weak convergence result that there exists a pair of real positive (μ_0, δ_ϵ) such that

$$P\{|H_k - H_*|^2 \geqslant \delta_\epsilon \mu\} \leqslant \epsilon; \quad \forall \mu \leqslant \mu_0; \ \forall k \geqslant N_\mu,$$

where the integer N_μ is a deterministic function of μ, essentially equal to $-\log \mu / \mu$. Moreover, in [16], it is shown that as $\mu \to 0$, the asymptotic deviation $|H_{N_\mu} - H_*|/\sqrt{\mu}$ converges weakly to a Gauss–Markov (diffusion) process. These results are of interest to approximate the tail of the asymptotic distribution of $|H_k - H_*|$ for infinitesimal step-size μ. However, they do not imply a bound for the limiting MSD which is one of the most simple quantity characteristics of convergence. Moreover, the eleven assumptions (A.2)–(A.12) added in [16] to the boundedness of observations are hard to understand. To find what kind of ergodic properties for the observations *do* ensure that these assumptions are fulfilled in simple practical applications such as the adaptive filter (1), is not a question dealt with in [16], and it has no obvious answer for us.

Finally, we present the important recent contribution brought by the exponential convergence analysis presented in [17] for the algorithm (1), and for two others which are useful generalizations of (1) encountered in the control field. Algorithm (1) is studied by considering the optimum estimation error $e_k^{\text{opt}} = a_k - X_k^T H_*$ as an additive noise, and distinguishing between homogeneous and nonhomogeneous cases according to whether or not this noise vanishes. It is proved that in the former case, algorithm (1) is exponentially convergent, provided the pairs (a_k, X_k) are uniformly bounded; i.e., there exists a fixed positive β such that $|H_k - H_*|$ goes to zero quicker than $(1+\beta)^{-k}$. Unfortunately, most applications of (1) are nonhomogeneous cases due to the fact that the relationship between useful signal a_k and observation vector X_k is not, in general, linear and/or involves an additive noise and/or has a greater degree of complexity than the dimension N of X_k. In this latter case, [17, eq. 23] provides a quick heuristic evaluation of the asymptotic MSD. Based on the idea that β will be roughly proportional to μ in the homogeneous case, it appears that the nonhomogeneous MSD will be proportional to μ.

In the present work, we give a strict mathematical proof of this intuitive result. We overcome the problem of the bound for the reference pairs (a_k, X_k), and of the barrier for the adaptive vector H_k, assuming successive correlated pairs (a_k, X_k) with finite memory and finite moments. These assumptions agree with the experimental and simulation results which have shown that neither the boundedness, nor the barrier, nor the independence are necessary for the convergence. Finally, the assumptions of finite memory and finite moments that we have used are justifiable in a very wide class of practical applications; two practical examples are given in Section III.

II. Theorem

Let N be the dimension of the vectors X_k, and let $|X_k|$ denote the modulus $(X_k^T X_k)^{1/2}$ of the vector X_k. If there exists a finite integer M such that

$$M > N/12, \qquad \text{(A.0)}$$

$$\{(a_i, X_i) : i \leqslant k\} \text{ and } \{(a_j, X_j) : j \geqslant k + M\} \qquad \text{(A.1)}$$

are statistically independent $\forall k$;

$$E\left(|X_k|^{2nM}\right) < \infty; \qquad \forall n \leqslant 12, \qquad \text{(A.2)}$$

$$E\left(|a_k|^{[4+4/(6M-1)]}\right) < \infty; \qquad \text{(A.3)}$$

then there exists a pair (μ_0, β) of real and positive numbers such that the vector H_k defined by the algorithm

$$H_{k+1} = H_k + \mu X_k \left(a_k - X_k^T H_k\right) \qquad (1)$$

satisfies

$$\limsup_{k \to \infty} E\left(|H_k - H_*|^2\right) \leqslant \beta\mu; \qquad \forall \mu \leqslant \mu_0. \qquad (4)$$

This theorem shows that the power of $|H_k - H_*|$ can be made arbitrarily small by choosing a sufficiently small step size. For practical applications, this type of convergence is sufficient. Indeed, (1) will be implemented on a digital computer with a finite accuracy which cannot be exceeded even with a very small step size. Now let us discuss the assumptions of this theorem.

Assumption (A.1) states that the sequence (a_k, X_k) has a finite strong-memory M. Assumption (A.0) states that this memory is more than $N/12$. As discussed in the next section, in most practical applications, M is more than N; thus, (A.0) is not a restrictive hypothesis. Now the finite strong memory is a special type of strong-mixing character. Thus, one could object to this theorem by saying that a more general type of strong-mixing property for the sequence (a_k, X_k) is not covered by it. For instance, if (a_k, X_k) is generated according to an ARMA model, it will not satisfy (A.1) exactly, but only approximately. This is true. But the previous theorem reflects, at present, the largest model for which we are able to derive the proof. We have underlined in the derivation the step that we cannot cross (up to now) with a general strong-mixing model for the sequence (a_k, X_k).

Assumptions (A.2), (A.3) state that the moments of $|a_k|$ and $|X_k|$ are finite up to certain orders. In many applications, such as those considered in Section III, all the moments of $|X_k|$ are bounded. Thus, Assumption (A.2) is more than fulfilled and (A.2) can be replaced by (A.2)′.

$$\text{All the moments of } |X_k| \text{ are finite.} \qquad \text{(A.2)}'$$

Notice that, in such a case, Assumption (A.3) can be enlarged to

$$\exists \epsilon > 0 \text{ such that } E\left(|a_k|^{4+\epsilon}\right) < \infty, \qquad \text{(A.3)}'$$

while Assumption (A.0) can be cancelled. These points are emphasized in Appendix I. However, in the cases where the probability density of $|X_k|$ is a rationally decreasing function, Assumption (A.2)′ is not satisfied, while (A.2) may hold.

We emphasize that, according to (A.1) and (A.2), the greater the memory, the higher the order of the $|X_k|$-moments that are required to be finite. Notice also that, according to (A.3), the order of bounded moments that are required for $|a_k|$ is greater than (but very near to) 4, and less than 5. In the particular case of independent successive vectors, X_k, it has been shown in [1] that the boundedness of fourth-order moments of $|X_k|$ and $|a_k|$ is sufficient for the result (4) of the theorem.

The reason that the orders of the moments that must be bounded are, respectively, $24M$ for $|X_k|$ and $4(1+1/(6M-1))$ for $|a_k|$, is a technical one. Moments of the order $24M$ for $|X_k|$ will appear in quantities such as (22), as well as in Lemma 2 of Appendix I. The order $4(1+1/(6M-1))$ for $|a_k|$ appears in Appendix II (Condition II.6).

Finally, notice that in the case where the components X_k^j, $j=1,2,\cdots,N$ of the vector X_k are not successive time samples of the same signal, the memory M' of (a_k, X_k) may be smaller than N, and even smaller than $N/12$. If, in such a case, we have $E(|X_k|^{2N+\epsilon})<\infty$ for a positive ϵ, then our theorem can be applied with $M=N/12$ instead of M' in (A.1) and (A.3). In practice, this case is not frequent (see the next section).

III. Practical Applications

In usual practical applications, the useful signal, a_k, is a time sequence $a_k = a(kT)$ that has to be estimated on the basis of another jointly stationary time sequence, $x(jT)$, which can be observed. The correlation width between $a(kT)$ and $x(jT)$ is finite, and the observation vector X_k is defined by the window of samples $x(jT)$ that have nonzero correlation with $a(kT)$: for some positive fixed integers K and L, one has

$$X_k^T = (x((k+K)T),\cdots,x(kT),\cdots,x((k-L)T)). \tag{5}$$

Thus, passing from $a(kT)$ to the following $a((k+1)T)$, the vector X_k is simply shifted by T, while an additional sample $x((k+1+K)T)$ is taken into consideration. Hence, X_k and X_{k+1} share all their components except the end ones, $x((k+K+1)T)$ and $x((k-L)T)$. Thus, the strong memory M cannot be less than N.

The linear estimation is a time-homogeneous filtering, having time response with length NT such that $N=K+L+1$. In that practical and important case, the sequence X_k has a memory M that cannot be less than N. Hence, the previous results such as [1], where independence is assumed, are not valid. That is why this theorem is an important contribution to the theory.

Now we give two practical examples, drawn from the context of data transmission, in which Assumptions (A) of the theorem are justifiable. In equalization, the first example, the signal a_k, is digital data, hence bounded; moreover, the sequence a_k is usually independent or with very short memory. The observation X_k results from filtering the sequence of data by a stable filter $\mathfrak{F}$ called "transmission channel" to which is added a Gaussian noise. Consequently, all the moments of $|X_k|$ are bounded. Finally, the memory M in Assumption (A.1) depends on the memory of $\mathfrak{F}$, which can be assumed finite without loss of generality in practical applications. The second example is the case of an echo canceller where a_k results from the digital data, X_k, by a stable filtering $\mathfrak{F}'$ which can, again, be assumed with finite memory without loss of generality in applications. These examples are further detailed in another paper intended for applications [13].

IV. Proof of the Theorem [18]

A. Basic Idea of the Proof

The following notations are used in this work:

$$v_k \triangleq H_k - H_*, \tag{6}$$

$$z_k \triangleq X_k(a_k - X_k^T H_*), \tag{7}$$

$$U_{k,t} \triangleq (I-\mu X_t X_t^T)(I-\mu X_{t-1}X_{t-1}^T)\cdots (I-\mu X_{k+1}X_{k+1}^T), \qquad t \geq k. \tag{8}$$

Finally, we denote by

$$\|U\| \triangleq \sup_{\{X:\,|X|=1\}} |UX|, \tag{9}$$

the norm of a matrix U.

The vector z_k represents the optimal increment in the algorithm. Indeed, z_k would be the increment in algorithm (1) if H_k were the optimum vector H_*. It follows from (2), (7) that z_k is zero mean:

$$E(z_k) = O, \tag{10}$$

which is used further in the proof. From (1) and (6), we can write

$$v_{k+1} = (I-\mu X_k X_k^T)v_k + \mu z_k. \tag{11}$$

To prove the theorem, we should prove that the quadratic mean of $|v_k|$ is bounded by a bound proportional to μ. For that purpose, we put (11) in the form

$$v_k = w_k + h_k, \tag{12}$$

where

$$w_k \triangleq U_{0,k} v_0, \tag{13}$$

$$h_k \triangleq \mu \sum_{j=1}^{k} U_{j,k} z_j, \tag{14}$$

and we consider the average quadratic norm of the components w_k and h_k. Both terms include matrices of the type $U_{j,k}$. If the algorithm (1) is convergent, v_k should forget its initial value v_0. Hence, according to (13), w_k and thus $U_{0,k}$ should tend to zero. At the same time, the series that appears in h_k, (14), should be convergent, e.g., thanks to an exponential decrease for $U_{j,k}$, when k tends to infinity. Taking the squared norm of (14) and averaging, we get

$$E(|h_k|^2) = \mu^2 \sum_{j,l=1}^{k} E(z_j^T U_{j,k}^T U_{l,k} z_l). \tag{15}$$

Now applying for $L=4$, the classical inequality

$$|E(y_1 y_2 \cdots y_L)| \leq [E(|y_1|^L)\cdot E(|y_2|^L)\cdots E(|y_L|^L)]^{1/L}, \tag{16}$$

which is a direct consequence of Hölder's inequality, we see that in order to bound (15), we must have a bound for the fourth-order moment $E(\|U_{j,k}\|^4)$. This discussion shows that a basic point in the theorem proof is to derive an exponentially decreasing bound for the latter moment. In fact, the first step of the proof consists of establishing the bound

$$E(\|U_{j,k}\|^4) \leqslant \begin{cases} C & \text{if } j \leqslant k, \\ C[\gamma(\mu)]^{k-j} & \text{if } j \leqslant k-2P, \text{ with } \gamma(\mu)<1, \end{cases} \tag{17}$$

where P is a fixed positive integer that will be determined. A straightforward consequence of (17) is the quadratic-mean convergence to zero of the first component, w_k.

The second (final) step is to prove that the power, given by (15), of the norm $|\boldsymbol{h}_k|$ of the second term $\boldsymbol{h}_k$, is essentially proportional to μ. In other words, the second step is the proof that $\mu^{-1} \cdot E(|\boldsymbol{h}_k|^2)$ is bounded.

The proof that the moments of $\|U_{j,k}\|$ decrease exponentially with $(k-j)$ relies upon two points. The first point is the factorization of that moment which is possible due to the independence (A.1).

This is the exact step where we use the finite strong-memory assumption. This is the point which prevents us from extending the proof to more general strong-mixing models. The second point is the choice of a sufficiently small μ such that the factors are strictly smaller than unity.

We want to emphasize that, besides the independence (A.1), we need the finite moments Assumption (A.2) to prove the boundedness of the moments of $\|U_{j,k}\|$. Assumption (A.3) is used to prove the boundedness of $E(|z_k|^4)$, the latter moment appearing after (15), at the same time as the moment $E(\|U_{j,k}\|^4)$.

B. Proof of the Theorem

The first step consists of establishing the bound (17). Let P be a fixed integer at least equal to M, and to be defined later (Appendix I, Lemma 3), and let us organize the sequence of indexes, k, within $[j,t]$ into three groups:

$$\begin{aligned} \Gamma_1 &= \{j+1,\cdots,j+P\}U\{j+2P+1,\cdots,j+3P\}U\cdots \\ \Gamma_2 &= \{j+P+1,\cdots,j+2P\}U\{j+3P+1,\cdots,j+4P\}U\cdots \\ \Gamma_3 &= \{j+2nP,\ j+2nP+1,\cdots,t\} \end{aligned} \tag{18}$$

where n is the integer part of $(t-j)/2P$. The groups Γ_1 and Γ_2 are interlaced, with an interval of P indexes. According to the multiplicative property,

$$U_{j,k} = U_{m,k} \circ U_{r,m} \circ U_{j,r}; \qquad j \leqslant r \leqslant m \leqslant k, \tag{19}$$

one gets

$$E(\|U_{j,t}\|^4) \leqslant E\left(\prod_{i=0}^{n-1} \|U_{j+2iP,\,j+(2i+1)P}\|^4 \cdot \|U_{j+(2i+1)P,\,j+(2i+2)P}\|^4 \|U_{j+2nP,\,t}\|^4 \right). \tag{20}$$

Using inequality (16) with $L=3$, one gets the bound

$$\begin{aligned} E(\|U_{j,t}\|^4) \leqslant & \left\{ E\left(\prod_{i=0}^{n-1} \|U_{j+2iP,\,j+(2i+1)P}\|^{12} \right) \right\}^{1/3} \\ & \cdot \left\{ E\left(\prod_{i=0}^{n-1} \|U_{j+(2i+1)P,\,j+(2i+2)P}\|^{12} \right) \right\}^{1/3} \\ & \cdot \left\{ E\left(\|U_{j+2nP,\,t}\|^{12} \right) \right\}^{1/3}. \end{aligned} \tag{21}$$

Thanks to the M-independence (A.1), and to the structure (18) of the subsets Γ_1 and Γ_2, each of the first two moments in (21) can be split into n factors of the type

$$E(\|U_{r,r+P}\|^{12}). \tag{22}$$

Each of the factors of $\|U_{r,r+P}\|$ is of order 2 with respect to any of the variables $|X_i|$. Thus, (22) is a moment of order less than or equal to 24, with respect to each $|X_i|$. In order to establish (17), it is thus required that the quantities (22) be uniformly bounded by a quantity smaller than one. This is the purpose of Lemma 1, given in Appendix I, which establishes the following.

Lemma 1: Under Assumptions (A.0), (A.2), and (A.1)′,

$$\forall k \geqslant 0, \text{ the sequences } \{X_i,\ i \leqslant k\} \text{ and } \{X_j, j \geqslant k+M\} \text{ are independent;} \tag{A.1$'$}$$

there exists a pair of positive numbers (δ, μ_0) and a positive integer $P \geqslant M$ such that

$$\forall r, \forall \mu \leqslant \mu_0 \qquad E(\|U_{r,r+P}\|^{12}) \leqslant \alpha(\mu), \tag{23}$$

where

$$\alpha(\mu) = 1-\mu\delta, \qquad \mu_0\delta<1. \tag{24}$$

The third moment in (21) also involves moments of the $|X_i|$'s of order $p \leqslant 24$. Thus, it is bounded for the same reason as (22). Hence, (17) follows, with

$$\gamma(\mu) = [\alpha(\mu)]^{1/3P}. \tag{25}$$

Applying the Schwarz inequality to w_k in (13), it follows from (17) that

$$\lim_{k\to\infty} E(|w_k|^2) = 0. \tag{26}$$

The second step of the theorem consists of finding a fixed positive number β' such that

$$\limsup_{k\to\infty} \mu^{-1} E(|\boldsymbol{h}_k|^2) \leqslant \beta'. \tag{27}$$

A detailed evaluation of that quantity is given in Appendix II. We get

$$\mu^{-1}E(|\boldsymbol{h}_k|^2) \leqslant \mu[D_1(\mu)+D_2(\mu)+D_3(\mu)+D_4(\mu)], \tag{28}$$

where

$$\begin{aligned} D_1(\mu) &= a_1[1-\gamma(\mu)^{q_1}]^{-1} + b_1, \\ D_2(\mu) &= \mu a_2[1-\gamma(\mu)^{q_2}]^{-2} \\ &\quad + \mu b_2[1-\gamma(\mu)^{q_2}]^{-1} + \mu c_2 \end{aligned}$$

$$D_3(\mu) = a_3[1-\gamma(\mu)^{q_3}]^{-1},$$
$$D_4(\mu) = a_4, \tag{29}$$

with a_1, a_2, a_3, a_4, b_1, b_2, q_1, q_2, q_3, and c_2 some positive constants given in Appendix II. With the help of (24), we show easily that

$$\lim_{\mu \to 0} \mu[1-\gamma(\mu)^q]^{-1} < \infty, \qquad \forall q > 0. \tag{30}$$

The last three relations (28)–(30) prove that there exists a positive number β' such that the bound (27) holds. Equations (26) and (27) complete the proof of the theorem.

Conclusion

The convergence of an adaptive filtering vector is proved when it is governed by the stochastic gradient algorithm with constant step size. Unlike the algorithm with decreasing step size, this algorithm is able to track time fluctuations in the optimum filtering vector, which is a valuable practical advantage. The convergence analysis presented in this work is done under assumptions nearer to practical applications than any presently existing analysis. We have considered the correlation, even strong, in the sequence of useful signal and the sequence of observations. Namely, we assume a finite memory model with finite moments for both sequences. This model suits a very wide class of practical applications. Under these assumptions, we have proved that the mean-square deviation between the optimal and actual filtering vectors tends to zero with the step size of the algorithm, and is essentially proportional to it.

Appendix I
Proof of Lemma 1

The proof of Lemma 1 involves two steps which we introduce as lemmas.

Lemma 2: Under Assumptions (A.1)′ and (A.2), for any integer $P > 0$, there exists a pair of positive numbers (μ_0, F) such that $\forall \mu \leq \mu_0$,

$$E(\|U_{r,r+P}\|^{12}) \leq \left|1 - 12\mu E\left(\lambda_{\min}\left(\sum_{i=r+1}^{r+P} X_i X_i^T\right)\right)\right| + F\mu^2. \tag{I.0}$$

Proof of Lemma 2: Consider the development of $U_{r,r+P}$ according to

$$\|U_{r,r+P}\| \leq \|I - \mu \sum_{i=r+1}^{r+P} X_i X_i^T\| + \mu^2\{\text{terms in } \mu^{p-2}|X_{i_1}|^2 \cdots |X_{i_p}|^2\}, \tag{I.1}$$

where $2 \leq p \leq P$ and where $i_1,\cdots,i_p$ are distinct integers.

By putting it to the power 12 and averaging, one gets

$$E(\|U_{r,r+P}\|^{12}) \leq E\left(\|I - \mu \sum_{i=r+1}^{r+P} X_i X_i^T\|^{12}\right) + \mu^2 E\left\{\text{terms in } \mu^{kp-2}|X_{i_1}|^{2k} \cdots |X_{i_p}|^{2k} \cdot \|I - \mu \sum_{i=r+1}^{r+P} X_i X_i^T\|^{12-k}\right\} \tag{I.2}$$

where $1 \leq k \leq 12$. Since

$$\|I - \mu X_i X_i^T\| \leq 1 + \mu |X_i|^2, \tag{I.3}$$

the bracketed average in the second term of (I.2) is bounded by a polynomial of the type

$$\sum \mu^m E\left(|X_{i_1}|^{2k} \cdots |X_{i_p}|^{2k} |X_i|^{2n}\right), \tag{I.4}$$

where $0 \leq n \leq 12 - k$. Hence, each coefficient of this polynomial is a moment of the type

$$E\left(|X_{i_1}|^{p_1} \cdots |X_{i_K}|^{p_K}\right); \qquad p_1,\cdots,p_K \leq 24, \tag{I.5}$$

where all the integers $i_1,\cdots,i_K$ are distinct and where the maximum exponent of any $|X_i|$ is 24. In (I.5), let us organize the indexes $i_1,\cdots,i_K$ into M groups $G_1,\cdots,G_M$ of interlaced indexes in such a way that inside each group G_j, the participating indexes, i, are separated by multiples of M (certain groups may be empty). Then, using inequality (16) for $L = M$, one gets

$$E\left[|X_{i_1}|^{p_1} \cdots |X_{i_K}|^{p_K}\right] \leq \left[\prod_{j=1}^{M} E\left(\prod_{i \in G_j} |X_i|^{p_i M}\right)\right]^{1/M}. \tag{I.6}$$

Hence (A.1)′ implies that

$$E\left[|X_{i_1}|^{p_1} \cdots |X_{i_K}|^{p_K}\right] \leq \prod_{j=1}^{M} \prod_{i \in G_j} \left[E(|X_i|)^{p_i M}\right]^{1/M}. \tag{I.7}$$

At this point, we use Assumption (A.2) to show that all the moments (I.5) are bounded:

$$E\left[|X_{i_1}|^{p_1} \cdots |X_{i_K}|^{p_K}\right] < \infty, \qquad \forall p_1,\cdots,p_K \leq 24. \tag{I.8}$$

Since μ is bounded, the polynomial (I.4) itself is bounded.

Hence, for some positive fixed constant F_1,

$$E(\|U_{r,r+P}\|^{12}) < E\left(\|I - \mu \sum_{i=r+1}^{r+P} X_i X_i^T\|^{12}\right) + F_1\mu^2. \tag{I.9}$$

Let $\lambda_{\min}(A)$ denote the minimum eigenvalue of a matrix A, and let

$$Y_r^P \triangleq \sum_{i=r+1}^{r+P} X_i X_i^T, \tag{I.10}$$

and let ω denote an arbitrary point in the space Ω of random events. Finally, let $P(d\omega)$ denote the probability measure on the space Ω; then we have

$$E(\|I-\mu Y_r^P\|^{12}) = \int_\Omega \|I-\mu Y_r^P\|^{12} P(d\omega)$$
$$\leqslant \int_{\{\omega:\, \mu\|Y_r^P\|<1\}} \left[1-\mu\lambda_{\min}(Y_r^P)\right]^{12} P(d\omega)$$
$$+ \int_{\{\omega:\, \mu\|Y_r^P\|\geqslant 1\}} (2\mu\|Y_r^P\|)^{12} P(d\omega). \tag{I.11}$$

Thus,

$$E(\|I-\mu Y_r^P\|^{12}) \leqslant E\left\{\left(1-\mu\lambda_{\min}(Y_r^P)\right)^{12}\right\}$$
$$+ (2\mu)^{12} E(\|Y_r^P\|^{12}). \tag{I.12}$$

Using the result (I.8) similarly for the moments of (I.10) as we have done for (I.3), one gets

$$E(\|Y_r^P\|^K) \leqslant F_2, \qquad \forall K \leqslant 12, \tag{I.13}$$

for some positive constant F_2. Now

$$E\left\{\left(1-\mu\lambda_{\min}(Y_r^P)\right)^{12}\right\}$$
$$\leqslant \left|1-12\mu E(\lambda_{\min}(Y_r^P))\right| + \binom{12}{2}\mu^2 E(\|Y_r^P\|^{12})$$
$$+ \cdots + \mu^{12} E(\|Y_r^P\|^{12}). \tag{I.14}$$

Therefore, from (I.12), (I.14), there exists $F_3 > 0$ such that $\forall r$; $\forall \mu \leqslant \mu_0$;

$$E(\|I-\mu Y_r^P\|^{12}) \leqslant \left|1-12\mu E\left(\lambda_{\min}(Y_r^P)\right)\right| + F_3\mu^2. \tag{I.15}$$

The combination of (I.9) and I.15) gives the results (I.0) of Lemma 2.

Comments: This lemma makes clear the relationship between the strong-memory M, and the order Q up to which the moment of $|X_i|$ must be finite. For the purpose of the technical evaluation, we have found that the order of bounded multivariate moments is 24. Along the lines (I.5)–(I.7), it appears that due to the strong memory M, this order is ensured if the univariate moments are bounded until the order $Q = 24M$. This is the explanation for Assumption (A.2).

Lemma 3: Under the assumptions (A.0), (A.1)′, (A.2), there exists an integer $P \leqslant NM$, and a real positive δ_0 such that $\forall r$,

$$E\left[\lambda_{\min}(Y_r^P)\right] \geqslant \delta_0. \tag{I.16}$$

Proof of Lemma 3: Since

$$\lambda_{\min}(Y_r^{P'}) \geqslant \lambda_{\min}(Y_r^P) \geqslant 0; \qquad \forall P' > P, \tag{I.17}$$

and due to stationarity, the LHS of (I.16) is solely a function of P, which is nondecreasing starting from zero. Thus, we are searching for an integer P at which this function departs from zero. To prove the lemma, one must show that

$$E\left[\lambda_{\min}(Y_r^{NM})\right] \geqslant \delta_1 > 0. \tag{I.18}$$

Now due to definition (I.10),

$$E\left[\lambda_{\min}(Y_r^{NM})\right] \geqslant E\left[\lambda_{\min}\left(\sum_{i=0}^{N-1} X_{r+1+iM} X_{r+1+iM}^T\right)\right]. \tag{I.19}$$

Consider the determinant of the matrix $(\sum_{i=0}^{N-1} X_{r+1+iM} X_{r+1+iM}^T)$ and denote by x_i^n the nth component of the vector X_i. Due to the multilinearity of the determinant with respect to each column, one gets

$$\det\left(\sum_{i=0}^{N-1} X_{r+1+iM} X_{r+1+iM}^T\right)$$
$$= \sum_{i_1=0}^{N-1} \cdots \sum_{i_N=0}^{N-1} x_{r+1+i_1M}^1 \cdots x_{r+1+i_NM}^N$$
$$\cdot d\left(X_{r+1+i_1M}, \cdots, X_{r+1+i_NM}\right), \tag{I.20}$$

where $d(U_1, \cdots, U_N)$ denotes the determinant of the matrix U with columns $U_1, \cdots, U_N$. If the indexes $i_1, \cdots, i_N$ are not all distinct, the latter determinant is zero. Consequently, one has

$$E\left\{\det\left[\sum_{i=0}^{N-1} X_{r+1+iM} X_{r+1+iM}^T\right]\right\}$$
$$= \sum_{(i_1,\cdots,i_N)\in\mathcal{P}} E\left\{d\left(x_{r+1+i_1M}^1 X_{r+1+i_1M}, \cdots,\right.\right.$$
$$\left.\left. x_{r+1+i_NM}^N X_{r+1+i_NM}\right)\right\}, \tag{I.21}$$

where $\mathcal{P}$ is the set of all permutations of $[0, 1, 2, \cdots, N-1]$. Using Assumption (A.1), we can show that each term on the RHS of (I.21) is $\det(R)$. Therefore,

$$E\left(\det\left\{\sum_{i=0}^{N-1} X_{r+1+iM} X_{r+1+iM}^T\right\}\right) = N!\det(R) > 0. \tag{I.22}$$

Denoting by $\lambda_1, \lambda_2, \cdots, \lambda_N$ the eigenvalues of $\sum_{i=0}^{N-1} X_{r+1+iM} X_{r+1+iM}^T$, arranged in increasing order, we get from (I.22)

$$E(\lambda_1\lambda_2\cdots\lambda_N) = N!\det(R) > 0. \tag{I.23}$$

Obviously, one has, $\forall \alpha > 0$,

$$0 < E(\lambda_1\lambda_2\cdots\lambda_N) \leqslant E(\lambda_1^\alpha \lambda_N^{N-\alpha}). \tag{I.24}$$

It follows from (I.24) and Hölder's inequality that

$$E(\lambda_1^\alpha \lambda_N^{N-\alpha}) \leqslant \left[E(\lambda_1)\right]^\alpha \cdot \left[E(\lambda_N^{(N-\alpha)/(1-\alpha)})\right]^{1-\alpha} \tag{I.25}$$

for any positive constant α. Suppose, in addition, that the following conditions are satisfied:

$$\frac{N-\alpha}{1-\alpha} \text{ is an integer } \leqslant 12M. \tag{I.26}$$

Then the obvious bound

$$\lambda_N \leqslant \sum_{i=0}^{N-1} |X_{r+1+iM}|^2, \tag{I.27}$$

together with Assumptions (A.1)′ and (A.2), will result in the inequality

$$E\left(\lambda_N^{(N-\alpha)/(1-\alpha)}\right) < \infty. \tag{I.28}$$

Combining the inequalities (I.24), (I.25), and (I.28), we obtain $E(\lambda_1) > 0$, i.e.,

$$E\left[\lambda_{\min}\left(\sum_{i=0}^{N-1} X_{r+1+iM} X_{r+1+iM}^T\right)\right] > 0. \tag{I.29}$$

Together with (I.19), this completes the lemma proof. Now the existence of a positive α satisfying (I.26) is ensured provided that

$$12M - N > 0. \tag{I.30}$$

For example, one can choose $\alpha = (12M - N)/(12M - 1)$.

The last condition (I.30) is fulfilled thanks to Assumption (A.0) of the theorem. Notice that in the slightly more restrictive version of the theorem, where Assumption (A.2) is replaced by (A.2)′, the condition (I.28) is fulfilled for any positive α for which $(N-\alpha)/(1-\alpha)$ is an integer. Hence, condition (I.30), i.e., Assumption (A.0) of the theorem, can be suppressed.

Proof of Lemma 1: (Lemma 1 is stated in Section IV-B). The result (23), (24) of Lemma 1 is a trivial consequence of the respective results (I.0) and (I.16) of Lemmas 2 and 3.

Appendix II
A Bound of $E(|h_k|^2)$

Let us define the following sets that are illustrated in Fig. 1.

$$S \triangleq \{(j,k): j=1,2,\cdots,t-1;\ k=2,3,\cdots,t;\ k>j\} \tag{II.1a}$$

$$S_2 \triangleq \{(j,k) \in S: k \geqslant j+M;\ j \leqslant t-2P\} \tag{II.1b}$$

$$S_3 \triangleq \{(j,k) \in S: k < j+M;\ j \leqslant t-2P\} \tag{II.1c}$$

$$S_4 \triangleq \{(j,k) \in S: j > t-2P\}. \tag{II.1d}$$

From (15), we have

$$E\left(|h_t|^2\right) = \mu^2 \sum_{j=1}^{t} E\left(|U_{j,t} z_j|^2\right) + 2\mu^2 \sum_{(j,k)\in S} E\left(z_j^T U_{j,t}^T U_{k,t} z_k\right). \tag{II.2}$$

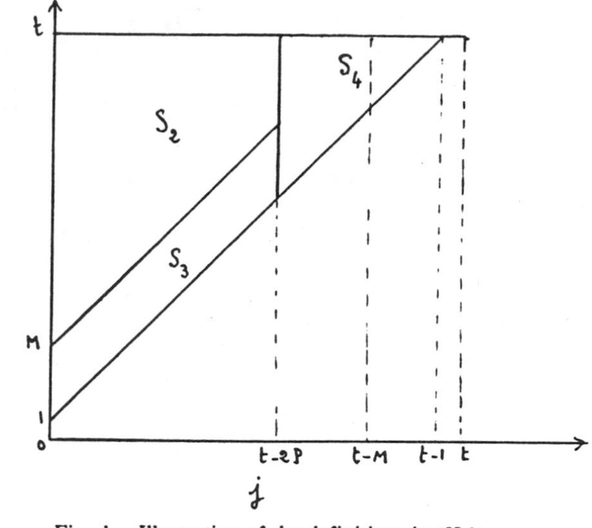

Fig. 1. Illustration of the definitions in (II.1).

Using the Schwarz inequality, we get

$$\sum_{j=1}^{t} E\left(|U_{j,t} z_j|^2\right) \leqslant \sum_{j=1}^{t} \left\{E\left(\|U_{j,t}\|^4\right) E\left(|z_j|^4\right)\right\}^{1/2}. \tag{II.3}$$

Developing z_j according to (7), we see that there exists a positive constant B such that

$$E\left(|z_j|^4\right) \leqslant B < \infty, \tag{II.4}$$

provided the quantities $E(|X_j|^8)$, $E(|X_j|^7 |a_j|), \cdots, E(|X_j|^4 |a_j|^4)$ are all finite. Thanks to (A.2), it is easily seen that a sufficient condition for the validity of (II.4) is

$$E\left(|X_j|^4 |a_j|^4\right) < \infty. \tag{II.5}$$

In turn, according to Hölder's inequality and to (A.2), (II.5) will hold provided

$$E\left(|a_j|^{24M/(6M-1)}\right) < \infty, \tag{II.6}$$

which is satisfied since it is equivalent to (A.3). Hence, the bound (II.4) is valid. Notice that the last condition (II.6) can be replaced by the less restrictive one

$$\exists \epsilon > 0 \text{ such that } E\left(|a_j|^{4+\epsilon}\right) < \infty \tag{II.7}$$

if Assumption (A.2) is replaced by Assumption (A.2)′. According to (II.3), (II.4), and (17), we have

$$\sum_{j=1}^{t} E\left(|U_{j,t} z_j|^2\right) \leqslant D_1(\mu) \tag{II.8}$$

where

$$D_1(\mu) \triangleq \left[2P + \left(1-\gamma(\mu)^{1/2}\right)^{-1}\right]\sqrt{BC}. \tag{II.9}$$

Now we split the second term of (II.2) into three parts according to the summing domains S_2, S_3, S_4:

$$\Sigma^2 \triangleq \sum_{(j,k)\in S_2} E\left(z_j^T U_{j,t}^T U_{k,t} z_k\right), \tag{II.10a}$$

$$\Sigma^3 \triangleq \sum_{(j,k)\in S_3} E\left(z_j^T U_{j,t}^T U_{k,t} z_k\right), \qquad \text{(II.10b)}$$

$$\Sigma^4 \triangleq \sum_{(j,k)\in S_4} E\left(z_j^T U_{j,t}^T U_{k,t} z_k\right). \qquad \text{(II.10c)}$$

Bounding Σ^2*:* Obviously, one has, for each term,

$$E\left(z_j^T U_{j,t}^T U_{k,t} z_k\right) = E\left(z_j^T U_{j,j+M-1}^T U_{j+M-1,t}^T U_{k,t} z_k\right). \qquad \text{(II.11)}$$

Let P_j be defined by

$$U_{j,j+M-1}^T \triangleq I + P_j. \qquad \text{(II.12)}$$

Because in S_2 one has $k \geqslant j+M$, then the independence (A.1), together with (II.11), implies that

$$E\left(z_j^T U_{j,t}^T U_{k,t} z_k\right) = E\left(z_j^T\right)\cdot E\left(U_{j+M-1,t}^T U_{k,t} z_k\right) + E\left(z_j^T P_j U_{j+M-1,t}^T U_{k,t} z_k\right) \qquad \text{(II.13)}$$

and according to the zero-mean property of z_j (10),

$$E\left(z_j^T U_{j,t}^T U_{k,t} z_k\right) = E\left(z_j^T P_j U_{j+M-1,t}^T U_{k,t} z_k\right). \qquad \text{(II.14)}$$

From the definition (II.12)

$$\|P_j\| \leqslant \Sigma \text{ terms in } \mu^p |X_{i_1}|^2 \cdots |X_{i_p}|^2, \quad \cdot 1 \leqslant p \leqslant M-1, \qquad \text{(II.15)}$$

where $i_1,\cdots,i_p$ are distinct integers. Putting (II.15) to the power 4 and using the result (I.8) in Appendix I, we see that there exists a finite positive constant C_1 such that

$$E\left(\|P_j\|^4\right) \leqslant C_1^4 \mu^4, \qquad \forall \mu \leqslant \mu_0. \qquad \text{(II.16)}$$

We can apply to (II.4) the inequality (16) with $L=4$. It becomes

$$\left|E\left(z_j^T U_{j,t}^T U_{k,t} z_k\right)\right| \leqslant \left\{E\left(|z_j|^4 |z_k|^4\right) E\left(\|P_j\|^4\right) \cdot E\left(\|U_{j+M-1,t}\|^4\right) E\left(\|U_{k,t}\|^4\right)\right\}^{1/4}. \qquad \text{(II.17)}$$

Since z_j and z_k are independent in S_2, it follows from (II.4), (II.16), and (II.17) that

$$\left|E\left(z_j^T U_{j,t}^T U_{k,t} z_k\right)\right| \leqslant \mu\sqrt{B}\, C_1 \left\{E\left(\|U_{j+M-1,t}\|^4\right) \cdot E\left(\|U_{k,t}\|^4\right)\right\}^{1/4}. \qquad \text{(II.18)}$$

A tedious, but straightforward, summation over S_2 of the terms (II.18), taking into account the bound (17) that has been derived in step 1), brings the result

$$|\Sigma^2| \leqslant \frac{1}{2} D_2(\mu) \qquad \text{(II.19)}$$

where

$$D_2(\mu) \triangleq \mu 2 C_1 \sqrt{BC}\left\{M^2 + \left(1-\gamma(\mu)^{1/4}\right)^{-2} + 2P\left(1-\gamma(\mu)^{1/4}\right)^{-1}\right\}. \qquad \text{(II.20)}$$

Bounding Σ^3*:* Applying (16) with $L=4$, we obtain

$$|\Sigma^3| \leqslant \sum_{(j,k)\in S_3} \left\{E\left(|z_j|^4\right) E\left(|z_k|^4\right) E\left(\|U_{j,t}\|^4\right) E\left(\|U_{k,t}\|^4\right)\right\}^{1/4} \qquad \text{(II.21)}$$

From (II.4) and step 1), (17), we deduce

$$|\Sigma^3| \leqslant \sum_{(j,k)\in S_3} \sqrt{BC}\,[\gamma(\mu)]^{(t-j)}. \qquad \text{(II.22)}$$

After the sum evaluation, it becomes

$$|\Sigma^3| \leqslant \frac{1}{2} D_3(\mu) \qquad \text{(II.23)}$$

where

$$D_3(\mu) \triangleq 2M\sqrt{BC}\left(1-\gamma(\mu)^{1/4}\right)^{-1}. \qquad \text{(II.24)}$$

Bounding Σ^4*:* Applying (16) with $L=4$, then using the bounds (17) and (II.4), it follows that

$$|\Sigma^4| \leqslant \sum_{(j,k)\in S_4} \sqrt{BC} = 2P^2\sqrt{BC} \triangleq \frac{1}{2} D_4(\mu). \qquad \text{(II.25)}$$

The three bounds (II.19), (II.23), and (II.25) show that

$$\left|\sum_{(j,k)\in S} E\left(z_j^T U_{j,t}^T U_{k,t} z_k\right)\right| \leqslant \frac{1}{2}\left(D_2(\mu) + D_3(\mu) + D_4(\mu)\right). \qquad \text{(II.26)}$$

Finally, according to (II.2), (II.8), and (II.26), we obtain

$$E\left(|h_t|^2\right) \leqslant \mu\eta(\mu), \qquad \text{(II.27)}$$

where

$$\eta(\mu) \triangleq \mu\{D_1(\mu) + D_2(\mu) + D_3(\mu) + D_4(\mu)\} \qquad \text{(II.28)}$$

which is the result (28), (29) announced in Section IV-B.

References

[1] O. Macchi, "Résolution adaptative de l'équation de Wiener Hopf," *Ann. Inst. Henri Poincaré*, vol. 14, pp. 357–379, 1978.

[2] L. Ljung, "Analysis of recursive stochastic algorithm," *IEEE Trans. Automat. Contr.*, vol. AC-22, pp. 551–575, Aug. 1977.

[3] G. Ungerboeck, "Theory of the speed of convergence in adaptive equalizers for digital communication," *IBM J. Res. Develop.*, pp. 546–555, 1972.

[4] B. Widrow, J. M. McCool, M. G. Larimore, and C. R. Johnson, "Stationary learning characteristics of the LMS adaptive filter," *Proc. IEEE*, vol. 64, no. 8, pp. 1151–1162, 1976.

[5] B. Widrow *et al.*, "Adaptive noise cancellation: Principles and applications," *Proc. IEEE*, vol. 63, pp. 1692–1716, 1975.

[6] E. Kiefer and J. Wolfowitz, "Stochastic estimation of the maximum of a regression function," *Ann. Math. Statist.*, vol. 23, no. 3, 1952.

[7] L. Schmetterer, "Stochastic approximation," in *Proc. 4th Berkeley Symp. Math. Statist. Probabil.*, pp. 587–609, 1961.

[8] D. Sakrison, "An adaptive receiver implementation for the Gaussian scatter channel," *IEEE Trans. Commun. Technol.*, vol. COM-17, no. 6, pp. 640–648, 1969.

[9] C. Macchi, "Itération stochastique et traitements numériques adaptatifs," thèse d'Etat, Paris, France, 1972.

[10] B. T. Polyak and Y. Z. Tsypkin, "Pseudo-gradient adaptation and training algorithms," *Automatika i Telemekhanika*, no. 3, pp. 45–68, 1973.

[11] J. E. Mazo, "On the independence theory of equalizer convergence," *Bell Syst. Tech. J.*, vol. 58, no. 5, pp. 963–993, 1979.
[12] T. P. Daniell, "Adaptive estimation with mutually correlated training sequences," *IEEE Trans. Syst. Sci. Cybern.*, vol. SSC-6, pp. 12–19, 1970.
[13] O. Macchi "Le filtrage adaptatif en télécommunications," *Ann. Télécomm.*, vol. 36, nos. 11–12, pp. 613–625, 1981.
[14] J. L. Kim and L. D. Davisson, "Adaptive linear estimation for stationary *M*-dependent processes," *IEEE Trans. Inform. Theory*, vol. IT-21, pp. 23–31, 1975.
[15] D. C. Farden, J. C. Goding, Jr., and K. Sayood, "On the desired behaviour of adaptive signal processing algorithms," in *Proc. 1979 IEEE Int. Conf. Acoust., Speech, Signal Processing*, 1979, pp. 941–944.
[16] H. J. Kushner and H. Huang, "Asymptotic properties of stochastic approximations with constant coefficients," *SIAM J. Contr. Optimiz.*, vol. 19, no. 1, pp. 87–105, 1981.
[17] R. R. Bitmead and B. D. O. Anderson, "Performance of adaptive estimation algorithms in dependent random environments," *IEEE Trans. Automat. Contr.*, vol. AC-25, no. 4, pp. 788–794, 1980.
[18] E. Eweda, "Egalisation adaptative d'un canal filtrant non stationnaire," thèse d'Ingénieur-Docteur, Orsay, France, 1980.
[19] D. D. Falconer, "Adaptive filter theory and applications," in *Proc. 4th Int. Conf. Analysis Optimiz. Syst.*, Versailles, France, 1980, pp. 163–188.
[20] I. D. Landau, "An extension of a stability theorem applicable to adaptive control," *IEEE Trans. Automat. Contr.*, vol. AC-25, pp. 814–817, 1980.
[21] V. Solo, "The convergence of AML," *IEEE Trans. Automat. Contr.*, vol. AC-24, pp. 958–963, 1979.

On the Design of Gradient Algorithms for Digitally Implemented Adaptive Filters

RICHARD D. GITLIN, J. E. MAZO, AND MICHAEL G. TAYLOR

Abstract—The effect of digital implementation on the gradient (steepest descent) algorithm commonly used in the mean-square adaptive equalization of pulse-amplitude modulated data signals is considered.

It is shown that digitally implemented adaptive gradient algorithms can exhibit effects which are significantly different from those encountered in analog (infinite precision) algorithms. This is illustrated by considering the often quoted result of stochastic approximation that to achieve the optimum rate of convergence in an adaptive algorithm the step size should be proportional to $1/n$, where n is the number of iterations. On closer examination one finds that this result applies only when n is large and is relevant only for analog algorithms. It is shown that as the number of iterations becomes large one should not continually decrease the step size in a digital gradient algorithm. This result is a manifestation of the quantization inherent in any digitally implemented system. A surprising result is that these effects produce a digital residual mean-square error that is minimized by making the step size as large as possible. Since the analog residual error is minimized by taking small step sizes, the optimum step-size sequence reflects a compromise between these competing goals.

The performance of a time-varying gain sequence suggested by stochastic approximation is contrasted with the performance of a constant step-size sequence. It is shown that in a digital environment the latter sequence is capable of attaining a smaller residual error.

I. Introduction and Summary

Because of the many attractive features of digital technology it is expected that most future signal processors, in particular the filters and equalizers found in communication systems, will be digitally implemented. The wide variation in channel characteristics often makes it necessary that communications processors be of the adaptive type. In order to clarify, at the outset, the framework of our discussion we introduce the baseband pulse-amplitude modulated (PAM) data-transmission system shown in Fig. 1. The discrete-valued data sequence $\{a_n\}$ is first impulse modulated at the symbol rate $1/T$, and then distorted by the linear channel[1] whose impulse response is denoted by $h(t)$. The received signal, which is also corrupted by additive noise, is sampled at the symbol rate and then processed by the tapped-delay-line equalizer, which adaptively compensates for the distortion introduced by the (unknown) channel. Adaptation is accomplished by adjusting the variable system (equalizer) parameters, using an appropriate algorithm, so as to continually decrease a suitable error measure. Our primary objective is to indicate the effects of digital implementation on the commonly used adaptive gradient algorithm. By digital implementation we have in mind that the adjustable parameters, as well as all internal signal levels, are quantized to within a least significant digit (LSD). The performance of the algorithm is measured in terms of the residual, or irreducible, error present when adaptation stops and by the rate of convergence to the final parameter settings. One might think, *a priori*, that the only consequence of digital implementation is that the variable parameters are truncated to within an LSD of the optimum settings; however, the digital effects are of a far more significant nature, and produce a residual error that is considerably larger than the error due solely to truncation.

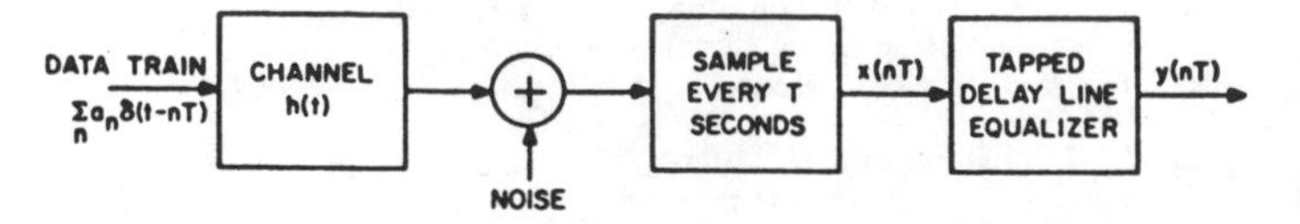

Fig. 1. A simplified PAM data system.

Since the adaptive gradient algorithm modifies the current parameter settings by adding a correction term, which is the product of a gradient estimate and a (possibly time-varying) step size, adaptation must stop when the correction term is smaller in magnitude than the LSD. Of course the algorithm might terminate at an earlier iteration, due to analog effects, and never display any digital peculiarities. A surprising result is that when adaptation is terminated by digital effects, the residual error can be further decreased by *increasing* the step size. Since the analog error is minimized by using a small step size, the best step size must reflect a compromise between these competing goals.

Further manifestations of digital implementation are observed when we consider the design of a time-varying step size. It is well known, from stochastic approximation theory, that the convergence properties of an analog (infinite precision) gradient algorithm can be improved by using a step size that decreases with time. Such a step-size sequence, when used in an analog environment, will produce an error that vanishes with

Manuscript received July 12, 1972.
R. D. Gitlin and M. G. Taylor are with Bell Telephone Laboratories, Inc., Holmdel, N. J. 07733.
J. E. Mazo is with the Mathematical Research Center, Bell Telephone Laboratories, Murray Hill, N. J.
[1] This filter, which actually represents the cascade of the transmitter shaping filter and the physical channel, is unknown at the receiver.

Reprinted from *IEEE Trans. Circuit Theory,* vol. CT-20, no. 2, pp. 125–136, Mar. 1973.

time. In order to contrast the two environments, we first present some new results on the design of the step-size sequence that minimizes the analog error at each iteration. It is found that when the initial error is substantially larger than the background noise variance the optimum step size is a constant while the error decreases exponentially. If the background noise is sufficiently small the constancy of the optimum step size persists until the error enters the quantization region; on the other hand, if the precision is fine enough, the error and step size will eventually become inversely proportional to the number of iterations until such time as adaptation is terminated by virtue of the error entering the quantization region. The performance of the variable step-size algorithm is contrasted with a fixed step-size algorithm, and somewhat surprisingly it turns out that the latter algorithm is capable of attaining a smaller residual error. An effective compromise would entail improving the rate of convergence by initially using the step size associated with the variable step-size algorithm and then "gear-shifting" to the step size that minimizes the residual error.[2]

Section II is partly tutorial and contains the development of the design of adaptive gradient algorithms in the absence of quantization; new results are presented concerning the design of the optimum step-size sequence. In Section III we consider the effect of digital implementation on the design and performance of the gradient algorithm.

II. Analog Design Considerations

In this section we consider the design of adaptive gradient algorithms in the absence of any digital or quantization effects. Thus the quantities used to update the algorithm, such as the received data and step size, as well as the operations of multiplication and addition, are assumed to be available or performed with infinite precision. It is well known from stochastic approximation theory [1] that the convergence rate is increased and the residual error is decreased when the step size is decreased with time. We will first present new results, of a more detailed nature than those previously reported, on the design of the step-size sequence that minimizes the error at each iteration. The insight gained into the behavior of such analog algorithms will make the effects of a digitally implemented version of such an algorithm on both the dynamic (rate of convergence) and steady-state performance (residual error) more readily appreciated.

For the sake of definiteness we consider the mean-square equalization of a PAM data signal. More specifically, we examine the adaptive gradient algorithm that uses a variable step size to rapidly attain the optimum structure. Considering this particular problem is not as restrictive as it might initially appear to be since the form of the algorithm is rather general, while the parameters that affect the algorithm's performance have physical significance. We now briefly describe the particular problem used to motivate our discussion.

A. The Equalization Problem

An essential component of any high-speed data-transmission system is an equalizer [2]. The equalizer adaptively compensates for the intersymbol interference introduced by the channel. For our discussion we only need to consider the sampled baseband received signal

$$x(nT) = \sum_{m} a_m h(nT - mT) + \nu(nT), \qquad n = 0, \pm 1, \pm 2, \cdots \tag{1a}$$

where a_m are the information symbols,[3] $h(\cdot)$ is the overall system impulse response, $\nu(\cdot)$ is the additive channel noise, and $1/T$ is the symbol as well as sampling rate. Rewriting the received samples as

$$x(nT) = a_n h_0 + \sum_{m \neq n} a_m h(nT - mT) + \nu(nT) \tag{1b}$$

clearly displays the second term as the intersymbol interference. The output of the familiar tapped-delay-line equalizer shown in Fig. 2 is

$$y(nT) = \sum_{m=-N}^{N} c_m x(nT - mT) = \mathbf{c}^T \mathbf{x}_n \tag{1c}$$

and is recognized as the convolution of the $2N+1$ tap weights $[c_i]$ and the received samples. The second equality makes use of the vector notation[4]

$$\mathbf{c} = \begin{pmatrix} c_N \\ c_{N-1} \\ \cdot \\ \cdot \\ c_0 \\ \cdot \\ c_{-N+1} \\ c_{-N} \end{pmatrix} \qquad \mathbf{x}_n = \begin{pmatrix} x_{n-N} \\ x_{n-N+1} \\ \cdot \\ \cdot \\ x_0 \\ \cdot \\ \cdot \\ x_{n+N-1} \\ x_{n+N} \end{pmatrix}$$

where the superscript T denotes transpose. The tap weights are iteratively adjusted so as to minimize the mean-square error at the equalizer output [3].

Since the desired output is a_n, the output error is given by

$$e_n = y_n - a_n \tag{2}$$

and the mean-square error is given by

$$\mathcal{E} = \langle [(y_n - a_n)^2] \rangle = \sum_{n \neq 0} g_n^2 + (g_0 - 1)^2 + \sigma^2 \tag{3}$$

[2] Certain system parameter values may dictate a "gear-shift" to a larger value.

[3] Binary symbols of value ± 1 will be assumed for convenience.

[4] We have used the subscripted notation $x_n \equiv x(nT)$. We freely apply this notation to all system variables, e.g., $a_n \equiv a(n)$.

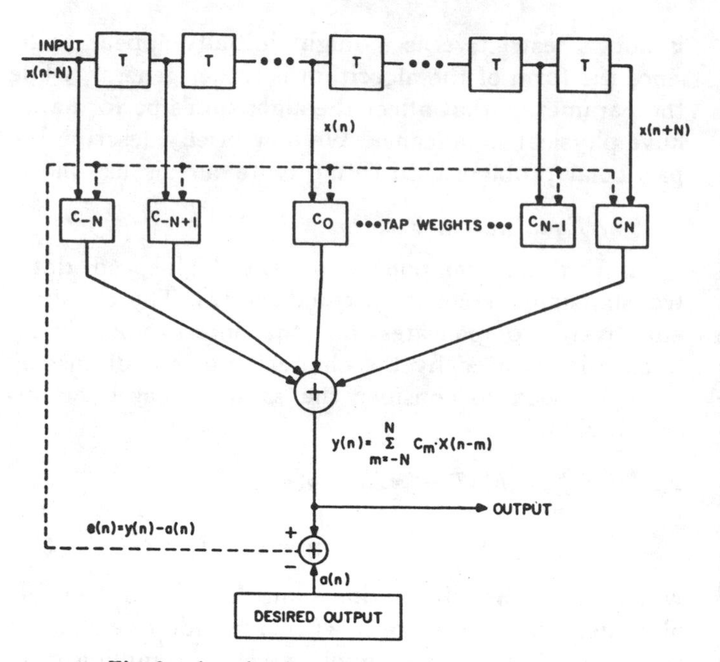

Fig. 2. An adaptive transversal digital equalizer.

where ⟨⟩ denotes the expectation operation, σ^2 the noise variance at the equalizer output, and g_n (which is the discrete convolution of h_n and c_n) represents the overall channel/equalizer impulse response. Perfect equalization would require that $g_0 = 1$ and $g_n = 0$ for $n \neq 0$.

To get a mathematical description of what the equalizer is trying to do, we write the mean-square error explicitly as a function of c, the tap-weight vector, and obtain

$$\mathcal{E}(\mathbf{c}) = \mathbf{c}^T A \mathbf{c} - 2\mathbf{c}^T \mathbf{v} + 1 \tag{4}$$

where $\mathbf{v}^T = (h_N, \cdots, h_0, \cdots, h_{-N})$ denotes the vector of the $2N+1$ center channel samples, and $A = \langle \mathbf{x}_n \mathbf{x}_n^T \rangle$ is recognized as the sum of the channel correlation matrix (whose ijth element is $\sum_m h_{m-i} h_{m-j}$) and the noise convariance matrix. Setting the gradient of the quadratic form given by (4) to zero gives the best tap setting $\mathbf{c}^*$ and the resulting residual error which are given, respectively, by the familiar expressions

$$A\mathbf{c}^* = \mathbf{v} \tag{5}$$

and

$$\mathcal{E}(c^*) = 1 - \mathbf{v}^T A^{-1} \mathbf{v}. \tag{6}$$

B. Deterministic Mode of Operation

In practice the optimum tap setting $\mathbf{c}^*$ is obtained in an iterative manner using some form of the gradient (steepest descent) algorithm, as opposed to solving (5).[5] To see how the system parameters affect the dynamic behavior of the gradient algorithm, we review the performance of this algorithm when the gradient of $\mathcal{E}$, with respect to the tap weights, is available.[6] The results obtained for this situation will provide a convenient reference for the more complicated case when the gradient is estimated directly from the received data. In the gradient, or steepest descent, algorithm [4], [5] the tap weights are adjusted according to the recursion

$$\mathbf{c}_{n+1} = \mathbf{c}_n - \Delta_n(A\mathbf{c}_n - \mathbf{v}), \qquad n = 0, 1, 2, \cdots \tag{7}$$

where Δ_n is a suitably chosen positive step size and $\mathbf{c}_n$ is the vector of tap weights at the nth iteration. Note that under the above assumptions the tap adjustments are done in a completely deterministic manner. Thus when the matrix A and the vector $\mathbf{v}$ are available the algorithm solves (5) in a recursive manner and there are no random fluctuations in the tap settings. The convergence properties of this algorithm are studied by introducing the tap-error vector

$$\boldsymbol{\varepsilon}_n = \mathbf{c}_n - \mathbf{c}^*. \tag{8}$$

Using (4)–(8) it is easy to see that the mean-square error when the taps are adjusted in accordance with (7) satisfies

$$\mathcal{E}(\mathbf{c}_n) \equiv \mathcal{E}_n = \mathcal{E}^* + \boldsymbol{\varepsilon}_n^T A \boldsymbol{\varepsilon}_n \tag{9}$$

where

$$\mathcal{E}^* \equiv \mathcal{E}(\mathbf{c}^*)$$

is the (minimum) mean-square error when the taps are at the optimum setting. From (9) we see that the dynamic behavior of (7) is determined by the quadratic term $\boldsymbol{\varepsilon}_n^T A \boldsymbol{\varepsilon}_n$, which can be shown to satisfy the recursion

$$\boldsymbol{\varepsilon}_{n+1}^T A \boldsymbol{\varepsilon}_{n+1} = \boldsymbol{\varepsilon}_n^T A \boldsymbol{\varepsilon}_n - 2\Delta_n \boldsymbol{\varepsilon}_n^T A^2 \boldsymbol{\varepsilon}_n + \Delta_n^2 \boldsymbol{\varepsilon}_n^T A^3 \boldsymbol{\varepsilon}_n. \tag{10}$$

To study this recursion we let α and β denote, respectively, the minimum and maximum eigenvalues of the symmetric positive-definite matrix A, and recall that

$$\alpha \mathbf{x}^T \mathbf{x} \leq \mathbf{x}^T A \mathbf{x} \leq \beta \mathbf{x}^T \mathbf{x}. \tag{11}$$

Applying (11) repeatedly, we obtain the upper bound

$$\boldsymbol{\varepsilon}_{n+1}^T A \boldsymbol{\varepsilon}_{n+1} \leq (1 - 2\alpha\Delta_n + \beta^2 \Delta_n^2) \boldsymbol{\varepsilon}_n^T A \boldsymbol{\varepsilon}_n \tag{12}$$

and we see that $\boldsymbol{\varepsilon}_{n+1}^T A \boldsymbol{\varepsilon}_{n+1}$ will approach zero as n becomes large (thus $\mathcal{E}_n$ will approach $\mathcal{E}^*$) provided

$$\gamma_n = 1 - 2\alpha\Delta_n + \beta^2\Delta_n^2 < 1. \tag{13}$$

This can be guaranteed by choosing $\Delta_n \leq (2\alpha/\beta^2)$.

The rate of convergence will be optimized in the sense that the step size Δ_n will be chosen to minimize γ_n at each n. Differentiating (13) we obtain the optimum step size, denoted by Δ^*, which is given by

$$\Delta^* = \frac{\alpha}{\beta^2} \cdot \tag{14}$$

[5] In actual operation of the equalizer the system impulse response is generally unknown so that A and $\mathbf{v}$ are not available.

[6] This would be the case when isolated test pulses are used to adapt the equalizer.

Notice that Δ^* is a constant, independent of n, and is half the maximum permissible step size. With $\Delta_n = \Delta^*$ we see that

$$\varepsilon_{n+1}^T A \varepsilon_{n+1} \leq \left[1 - \frac{\alpha^2}{\beta^2}\right] \varepsilon_n^T A \varepsilon_n. \tag{15}$$

Hence the convergence is exponentially bounded, since

$$\varepsilon_{n+1}^T A \varepsilon_{n+1} \leq \left[1 - \frac{\alpha^2}{\beta^2}\right]^{n+1} \varepsilon_0^T A \varepsilon_0 \tag{16}$$

and the rate of convergence is seen to depend on the ratio of the minimum-to-maximum eigenvalues of the A-matrix. Note that if all the eigenvalues of A are the same, then α equals β, and convergence is achieved in one step. This observation has been used by Chang [6] to design a rapidly converging equalizer structure.

In summary, if the actual gradient of the mean-square error is available the tap weights will converge (for any initial setting) to their optimum setting at an exponentially bounded rate, and the best step size is found to be a constant. The rate of convergence is seen to be a function of the "eigenvalue spread" of the channel-plus-noise correlation matrix.

C. Adaptive Mode of Operation

In actual operation the exact gradient of the mean-square error is not available and must be estimated from the received data. A well-known estimate is the gradient of the instantaneous squared error e_n^2, where e_n is given by (2). Using this estimate gives

$$\nabla e_n^2 = 2 e_n \mathbf{x}_n \tag{17}$$

where $\mathbf{x}_n$ is a vector whose entries are the $2N+1$ received samples $x_{n-N}, \cdots, x_0, \cdots, x_{n+N}$ that are "in" the equalizer at the time $t = nT$. The mathematical expectation of (17) is, in fact, the actual gradient. The adaptive or estimated gradient algorithm, suggested by Robbins and Monro [7] and explicitly described by Widrow [5], is then used to adjust the equalizer taps in accordance with

$$\begin{aligned} \mathbf{c}_{n+1} &= \mathbf{c}_n - \Delta_n \cdot e_n \cdot \mathbf{x}_n \\ &= \mathbf{c}_n - \Delta_n (\mathbf{c}_n^T \mathbf{x}_n - a_n) \mathbf{x}_n \end{aligned} \tag{18}$$

where we refer to Δ_n as the step size and to $\Delta_n \cdot e_n \cdot \mathbf{x}_n$ as the correction term. We note that due to the random nature of the correction term $\Delta_n \cdot e_n \cdot \mathbf{x}_n$ the tap vector is itself a random quantity, and it is desired to establish that the taps converge, in some probabilistic sense, to the optimum setting.

Before discussing the convergence problem it shall be necessary to make the following assumption. We assume that the sequence of vectors "in" the equalizer $\{\mathbf{x}_n\}$ is independent of $\mathbf{x}_m$ for $m \neq n$. The results obtained using this assumption agree well with observed laboratory performance of the algorithm, particularly in predicting the digital effects that are described in Section III.[7]

We note that $\mathbf{c}_{n+1}$, as given by (18), depends on $\mathbf{x}_0, \mathbf{x}_1, \cdots, \mathbf{x}_n$ and by the above assumption is then independent of $\mathbf{x}_{n+1}$. This is a very useful observation and one which will be used repeatedly in the sequel. Our first step in assessing the performance of the estimated gradient algorithm is to relate the mean-square error, at the nth iteration, to the tap error. The instantaneous squared error is given by

$$e_n^2 = (\mathbf{c}_n^T \mathbf{x}_n - a_n)^2 = (\varepsilon_n^T \mathbf{x}_n + \mathbf{x}_n^T \mathbf{c}^* - a_n)^2$$

and taking expectations gives the relation

$$\mathcal{E}_n = \langle \varepsilon_n^T A \varepsilon_n \rangle + \mathcal{E}(\mathbf{c}^*) \tag{19}$$

an expression closely related to (9) and one which reflects the random nature of the algorithm. We observe that the mean-square error is the sum of the irreducible error $\mathcal{E}^*$ given by (6) and the average power in a weighted tap-error vector. By considering the evolution of the term $\langle \varepsilon_n^T A \varepsilon_n \rangle$, we shall be able to describe the dynamic behavior of the estimated gradient algorithm. Subtracting the optimum tap setting from both sides of (18) gives

$$\varepsilon_{n+1} = \varepsilon_n - \Delta_n e_n \mathbf{x}_n \tag{20}$$

and we may easily establish that

$$\varepsilon_{n+1}^T A \varepsilon_{n+1} = \varepsilon_n^T A \varepsilon_n - 2\Delta_n e_n \mathbf{x}_n^T A \varepsilon_n + \Delta_n^2 e_n^2 \mathbf{x}_n^T A \mathbf{x}_n. \tag{21}$$

Applying the bounds developed in Appendix I we have the inequality

$$\langle \varepsilon_{n+1}^T A \varepsilon_{n+1} \rangle \equiv q_{n+1} \leq [1 - 2\alpha\Delta_n + \beta^2 \Delta_n^2] q_n + \Delta_n^2 \sigma^2 \tag{22}$$

where

$$\sigma^2 = 2\beta(2N+1)[\mu\rho + X_{\text{rms}}^2]. \tag{23}$$

The quantities μ and X_{rms} are defined by[8]

$$\mu = \langle (\mathbf{x}_n^T \mathbf{x}_n)^2 \rangle \tag{24a}$$

$$X_{\text{rms}} = \sqrt{\langle x_n^2 \rangle} \tag{24b}$$

and ρ is taken such that[9]

$$\mathbf{c}_n^T \mathbf{c}_n \leq (2N+1)\rho. \tag{24c}$$

It is interesting to note the similarity of (12) and (22). We see that the random nature of the algorithm

[7] This assumption permits us to easily take the average of both sides of (18). Doing this indicates that the evolution of the average tap vector $\langle c_n \rangle$ is governed by (7) with c_n replaced by $\langle c_n \rangle$. Since our interest is in studying the dynamics of the mean-square error we shall not say anything further about the average tap vector.

[8] Note that μ involves fourth-order statistics of the received signal.

[9] The ρ can be thought of as representing the dynamic range of the tap weight. See Appendix I for further discussion regarding the introduction of ρ.

is summarized by the σ^2 term.[10] The relation described by (22) is one which occurs frequently in the application of stochastic approximation techniques (it is [1, eq. (18)]), and is generally employed to develop conditions on the step-size sequence that are sufficient to quarantee that q_n will converge to zero with increasing n. Our emphasis in this investigation is slightly different since we wish to study the overall dynamic behavior of the sequence q_n in the presence of quantization. Consequently, we are very much interested in the transient behavior of q_n and Δ_n, i.e., how should we choose the step size initially and what is the resulting mean-square error? Our policy is to choose the step size so as to minimize q_n at each iteration.[11] Setting the derivative of the right-hand side (RHS) of (22), with respect to Δ_n, to zero gives the optimum step size

$$\Delta_n^* = \frac{\alpha q_n^*}{\sigma^2+\beta^2 q_n^*}, \qquad n = 0, 1, 2, \cdots \tag{25a}$$

or, solving for q_n^*,

$$q_n^* = \frac{\sigma^2\Delta_n^*}{\alpha - \beta^2\Delta_n^*} \cdot \tag{25b}$$

When Δ_n and q_n are taken to be related through (25) we say that the step size is chosen optimally. The algorithm, when initiated, has an arbitrary q_0. The initial tap vector is generally taken to be a reasonable *a priori* setting [e.g., $c_0^T = (0, 0, \cdots, 0, 1, 0, \cdots, 0, 0)$], and using (25a) the optimum Δ_0 could then be determined.

We of course would like a description of the sequences Δ_n and q_n as explicit functions of n. We conclude from (25a) that

$$\Delta_n^* \geq 0 \tag{26}$$

and in order that (25b) be consistent, i.e., that $q_n^* \geq 0$, it follows that

$$\Delta_n^* \leq \frac{\alpha}{\beta^2} \cdot \tag{27}$$

An iterative bound on the optimum step size is obtained by substituting (25a) in the RHS of (22) to give

$$q_{n+1}^* \leq q_n^* - \frac{(\alpha q_n^*)^2}{\sigma^2 + \beta^2 q_n^*} = (1 - \alpha\Delta_n^*)q_n^* \tag{28}$$

and combining (25) and (28) we have

$$\frac{\sigma^2\Delta_{n+1}^*}{\alpha - \beta^2\Delta_{n+1}^*} \leq (1 - \alpha\Delta_n^*)\frac{\sigma^2\Delta_n^*}{\alpha - \beta^2\Delta_n^*} \cdot \tag{29}$$

Transposing and solving for Δ_{n+1}^* gives

$$\Delta_{n+1}^* \leq \Delta_n^* \cdot \frac{1 - \alpha\Delta_n^*}{1 - \beta^2\Delta_n^{*2}} \tag{30}$$

which is valid in the range $0 \leq \Delta_n^* \leq \alpha/\beta^2$. Since in the above range the ratio appearing in (30) is less than unity (but greater than zero) we see that the optimum step size decreases monotonically with n. While an exact study of (30) appears to be difficult, it is easy to determine the behavior of Δ_n^* for small and large n. It is reasonable to assume that in most instances the initial tap error will be much larger than the background noise, i.e., $\beta^2 q_n^* \gg \sigma^2$, for small n. Under this assumption the optimum step size, from (25a), is

$$\Delta_n^* = \frac{\alpha}{\beta^2} \cdot \tag{31}$$

Since in obtaining (31) we have, in effect, neglected the "noise" it is not surprising that the optimum size is the same as that obtained in the absence of noise [see (14)]. Combining (28) and (31) we have

$$q_{n+1}^* \leq \left(1 - \frac{\alpha^2}{\beta^2}\right)q_n^* \tag{32}$$

thus the minimized mean-square error initially decays at an exponential rate while the optimum step size is a constant. This will be the mode of behavior until q_n is on the order of σ^2/β^2, at which time the "large initial error" assumption is no longer valid and the dynamic behavior is governed by (30). The above observation will be most important in interpreting the consequences of digital implementation on the behavior of the gradient algorithm.

When n is large we note that the monotonic decreasing nature of the Δ_n^* sequence implies that there is a number n_0 such that for $n > n_0$

$$\frac{1 - \alpha\Delta_n^*}{1 - \beta\Delta_n^*} \approx 1 \tag{33}$$

and (30) simplifies to

$$\Delta_{n+1}^* \leq \frac{\Delta_n^*}{1 + \beta\Delta_n^*}, \qquad n > n_0. \tag{34}$$

Iterating (34) yields

$$\Delta_{n+n_0}^* \leq \frac{\Delta_{n_0}}{1 + (n - n_0)\beta\Delta_{n_0}} \tag{35}$$

and using (25b), and the fact that $\beta^2\Delta_n \ll \alpha$, we have that q_n is proportional to the step size for large n. Equation (35) expresses the ubiguitous result of stochastic approximation that both the step size and mean-square error approach zero in a $1/n$ manner. The large initial step size insures rapid convergence to a region near the optimum setting while ultimately a small step size

[10] We sometimes casually refer to σ^2 as the variance of the background noise, but one should not regard this as implying that it is due solely to additive noise on the channel.

[11] More accurately, we minimize an upper bound on q_n at each iteration. We assume that this upper bound displays the essential relation between the system parameters.

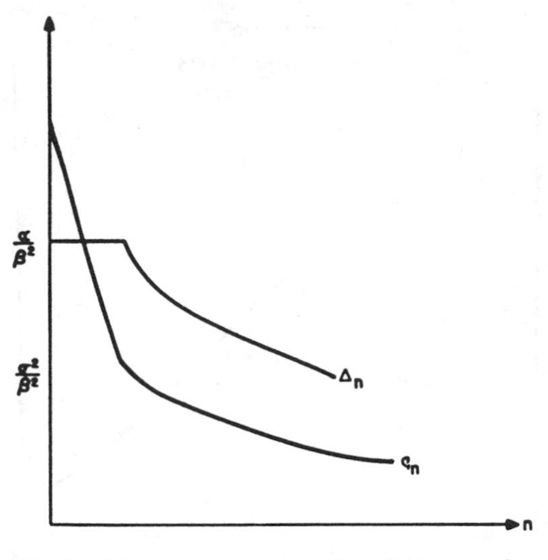

Fig. 3. Mean-square error and variable step size in the absence of quantization.

provides precise convergence by eliminating small fluctuations about the optimum point. Many practical algorithms employ a small but nonzero final step size to be able to track small drifts in the channel pulse response.

In summary we have shown that, for analog implementation, when the step size is chosen optimally, the step size is initially a constant and the mean-square error initially decreases at an exponential rate; the step size and mean-square error are ultimately (for large n) described by the familiar $1/n$ decay to zero. These relations are sketched in Fig. 3. In the next section we consider the effect of a digital implementation on the estimated gradient algorithm.

III. Digital Design Considerations

Having developed the necessary background in the design of analog adaptive gradient algorithms we are now ready to consider the main subject of this investigation—the design of digitally implemented adaptive gradient algorithms. Such algorithms are used when implementing adaptive systems whose adjustable parameters are quantized. We, of course, have in mind an adaptive equalizer whose tap weights are quantized. We shall see that the performance characteristics of digital algorithms are quite different from their analog counterparts. Our primary observation[12] is that the digitally implemented algorithm stops adapting whenever the correction term is smaller in magnitude than the LSD of the corresponding tap weight.[13] In general, quantization precludes the tap weights reaching the optimum (from analog considerations) setting, and as we shall see shortly, the digital effects significantly influence the design of the algorithm. Moreover, the quantization closely couples the steady-state and dynamic behavior of the digital estimated-gradient algorithm.

A. Digital Residual Error

In order to more easily contrast the analog and digital environments we first consider the case of constant step size. Suppose the (digital) algorithm is terminated due to quanitzation effects. Then we can estimate the value of the error when adaptation steps by setting the correction term (applied to the ith tap) to be less than or equal to the LSD. This gives the fundamental inequality

$$|\Delta \cdot e_{n_0} \cdot x_{n_0-i}| \leq \text{LSD} \tag{36}$$

which is valid when adaptation stops, where n_0 is the time at which the ith tap stops adapting. Suppose (36) is first satisfied for the ith tap, and as the particular input sample x_{n_0-i} propagates down the equalizer, the error and step size will further decrease in magnitude, thus insuring that this sample will turn off all the taps "down the line." Because of this observation we assume that all the taps stop adapting at the same time. To a first approximation, it is reasonable to replace $|x_{n-i}|$ by its rms value, X_{rms}, and this gives the following relation for the rms error[14] when adaptation stops:

$$|e_{n_0}| \leq \frac{\text{LSD}}{\Delta \cdot X_{\text{rms}}} \equiv e_d(\Delta). \tag{37}$$

We call $e_d(\Delta)$ the rms digital residual error (DRE). The above indicates that this error is inversely proportional to the step size. Therefore, it is clear that if adaptation is terminated due to digital effects one should try to make the step size as large as possible (while still guaranteeing convergence) in order to minimize the DRE. With a constant step size it is possible, however, that the error never enters the quantization region; consequently, in that case one could further reduce the error by decreasing the step size. We will have considerably more to say about the choice of step size in the sequel.

To clarify the practical significance of (37) let us look at a numerical example. Consider a 17-tap equalizer[15] with tap weights quantized to 12 bit ($\text{LSD} \cong 0.25 \times 10^{-3}$) with an input data stream having an rms value of unity, and with the step size fixed at 0.07. For this example, (37) estimates that the DRE will be about 0.35×10^{-2}. Let us compare this with the rms error that would be expected if the only source of error were the

[12] First made by Taylor [8].

[13] We also assume that before this situation occurs the algorithm is operating in the analog region where the taps are free to assume any value, and the correction term is computed with infinite precision. Of course, adaptation will only be terminated in this manner if the error is driven into the quantization region.

[14] It is reasonable to say that $|e_{n_0}|$ is approximately equal to $\sqrt{\mathcal{E}_{n_0}}$; we will use this approximation when discussing the dynamic behavior of the algorithm.

[15] With this length equalizer we can assume that the analog residual error $\mathcal{E}(c^*)$ is negligible. It will be convenient to make this assumption in the sequel.

quantization of the desired tap weights to 12 bit; that is, we assume that the ideal error-free equalizer has 17 taps, each of infinite precision. In Appendix II we show that the quantization error (QE) can be approximated by

$$\text{rms QE} \approx N^{1/2} \cdot \text{LSD} \cdot X_{\text{rms}}. \tag{38}$$

For the numerical example described above, this QE is about 10^{-3}. Thus the DRE due to the failure of the algorithm to find the best coefficient is roughly 3.5 times worse than the rms error would be if the best 12-bit coefficients had been found. As explained in Appendix III, the ratio of the DRE to the QE is proportional to $N^{1/2}$. Thus the residual error phenomenon will be even more pronounced in longer digital equalizers.

The ratio of the DRE to the QE clearly demonstrates the manifestations of digital implementation. The tap weights are, to a first approximation, trying to approach the quantized versions of the optimum settings; however, when the tap weights get close to the optimum setting the mean-square error and step size have decreased appreciably, and by the nature of the algorithm the taps try to approach the optimum setting by using very small correction terms. Once the correction term becomes smaller than the LSD, adaptation stops, and, as shown in Appendix III, the algorithm terminates while the taps are appreciably further away from the optimum setting than one LSD. Hence the quantization is enhanced, and produces the relatively large DRE.[16]

To see how the above observations are manifested in practice we show, in Fig. 4, the results of a computer experiment on a 17-tap digital adaptive equalizer. The experiment consisted of sending the same input stream into both the adaptive equalizer and the desired equalizer, a 17-tap equalizer with fixed optimum coefficients. The adaptive equalizer was adjusted by minimizing the mean-squared difference between their outputs. The final rms error[17] is plotted for various values of constant step size $\Delta(n) = k$. This error is normalized in the sense that the input data stream has an rms value of unity and the largest tap weight is one.

For very large values of k, the rms error due to the fluctuations of the tap weights dominates; hence choosing a smaller value for k improves the rms error. However, for all but very large values of k, the DRE phenomenon dominates so that making k smaller degrades the performance. In the example illustrated in Fig. 4, there is little to be gained from making $\Delta(n)$ a function of n, at least from the standpoint of minimizing the final rms error. The value of 0.07 is about as

[16] The preceding discussion suggests adaptive algorithms whcse correction term be in units of the LSD. Such an algorithm has been proposed by Lucky [9], and contrasted with the estimated gradient algorithm by Taylor [8].

[17] In this experiment the mean-squared error was approximated by averaging the instantaneous squared error over 4000 iterations.

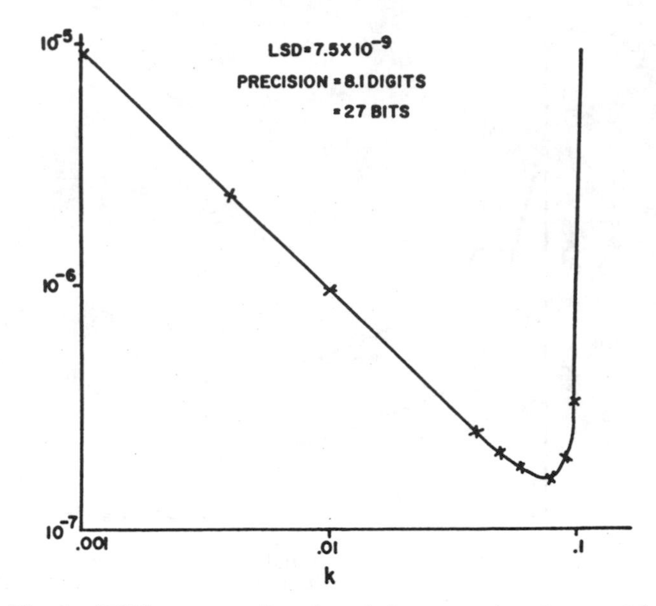

Fig. 4. RMS error as a function of the convergence constant k.

large as one would choose for initial rapid convergence and, as the results indicate, any change from this value of k will increase the residual error. One might think that once the minimum rms error had been achieved with one particular value of k, a subsequent decrease in k would have no effect since the corrections would be even less in magnitude than the LSD in the coefficients. This argument is not generally true, for although on the average the corrections are less than the LSD's, the fluctuations in the size of the instantaneous error may be sufficient to make some "corrections" large enough to perturb the coefficients. For example, it was observed that if the filter was adapted with $k = 0.07$ until the minimum rms error was attained and then k was decreased to 0.001, the rms error gradually increased up to the value of 0.95×10^{-5} as predicted by Fig. 3.

Since our primary objective is to obtain the minimum mean-squared error, the above experimental results suggest that the conventional stochastic approximation method, namely continually decreasing $\Delta(n)$ for each successive value of n, should *not* be used in a digital adaptive equalizer.

B. Choosing a Fixed Step Size

Applying the results obtained thus far we develop, in this section, a more complete understanding of the behavior of the constant step-size digital gradient algorithm. Suppose at $n = 0$ we begin with a large initial tap error. Then for any fixed step size $\Delta \leq (2\alpha/\beta^2)$, we have, by solving (22) with equality, that the error decreases exponentially to a value $q_\infty(\Delta)$ given by

$$q_\infty(\Delta) = \frac{\sigma^2 \Delta}{2\alpha - \beta^2 \Delta} \tag{39}$$

provided $q_\infty(\Delta)$ exceeds the corresponding digital residual mean-square error $e_d^2(\Delta)$. When $e_d^2(\Delta) > q_\infty(\Delta)$, adaptation will cease when the mean-square error $q_n(\Delta)$

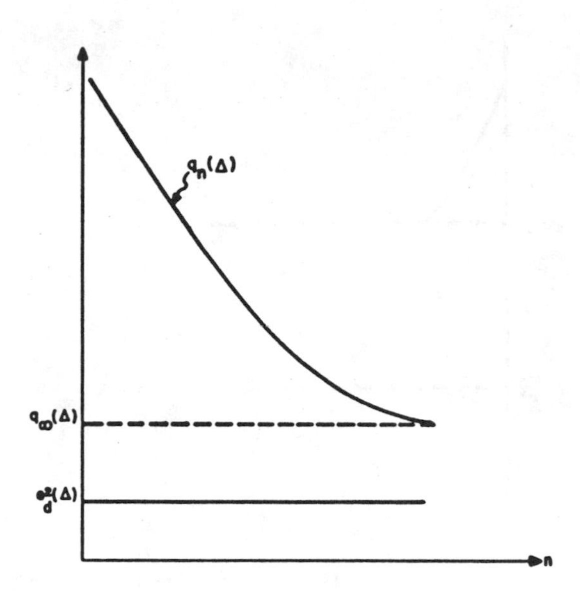

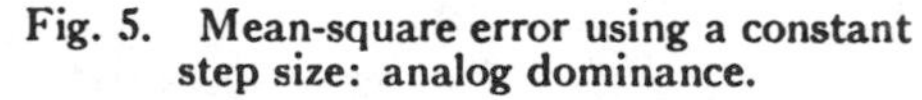

Fig. 5. Mean-square error using a constant step size: analog dominance.

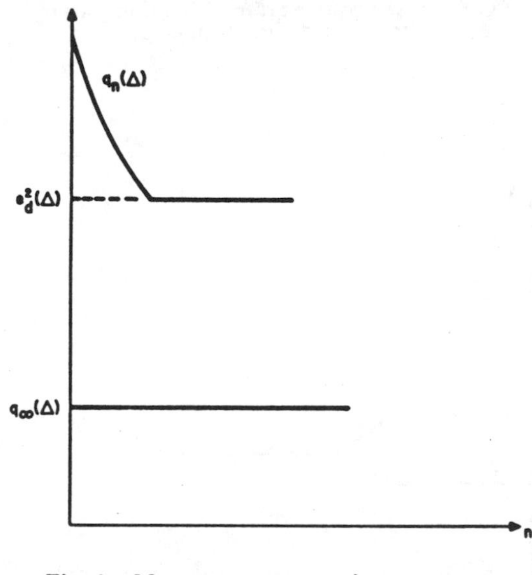

Fig. 6. Mean-square error using a constant step size: digital dominance.

decreases to the value $e_d^2(\Delta)$. Figs. 5 and 6 sketch the trajectory of $q_n(\Delta)$ under the respective conditions $q_\infty(\Delta) > e_d^2(\Delta)$ and $q_\infty(\Delta) < e_d^2(\Delta)$.

Two points might be mentioned about (39). The first is that $q_\infty(\Delta)$ increases or decreases as Δ increases or decreases. That is, for $\Delta < (2\alpha/\beta^2)$ we have

$$\frac{dq_\infty(\Delta)}{d\Delta} > 0. \tag{40}$$

A second observation is that when $\Delta < (\alpha/\beta^2)$, we have

$$q_\infty(\Delta) = \frac{\sigma^2}{\beta^2}\,\frac{\Delta}{\dfrac{2\alpha}{\beta^2} - \Delta} < \frac{\sigma^2}{\beta^2}\,\frac{\alpha/\beta^2}{\dfrac{2\alpha}{\beta^2} - \dfrac{\alpha}{\beta^2}} = \frac{\sigma^2}{\beta^2} \tag{41a}$$

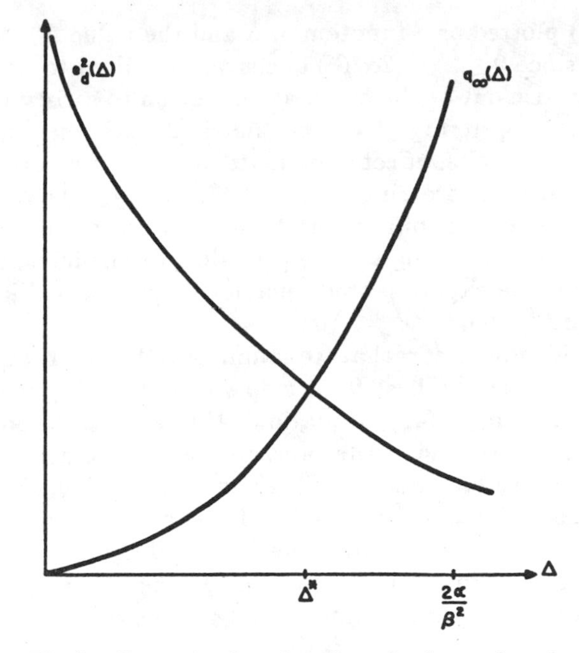

Fig. 7. Determination of the best fixed step size Δ*.

and when $(\alpha/\beta^2) \leq \Delta \leq (2\alpha/\beta^2)$, we have

$$q_\infty(\Delta) = \frac{\sigma^2}{\beta^2}\,\frac{\Delta}{\dfrac{2\alpha}{\beta^2} - \Delta} \geq \frac{\sigma^2}{\beta^2}\,\frac{\alpha/\beta^2}{\dfrac{2\alpha}{\beta^2} - \dfrac{\alpha}{\beta^2}} = \frac{\sigma^2}{\beta^2}\cdot \tag{41b}$$

Recall that σ^2/β^2 is roughly the value of q_n where, using stochastic approximation, the optimum step size would switch from constant to decreasing behavior.

If for the fixed step size Δ the magnitudes are such that[18]

$$\sqrt{q_\infty(\Delta)} > e_d(\Delta) \equiv \frac{\text{LSD}}{\Delta \cdot X_{\text{rms}}} \tag{42}$$

then, as shown in Fig. 5, the algorithm evolves to an error $q_\infty(\Delta)$ and digital effects are not seen. In fact, a further drop in rms error will be observed if the step size is decreased to a value Δ^* where

$$\sqrt{q_\infty(\Delta)} = e_d(\Delta)$$

namely the value of Δ that satisfies

$$\frac{\sigma^2\Delta^*}{2\alpha - \beta^2\Delta^*} = \left[\frac{\text{LSD}}{\Delta^* X_{\text{rms}}}\right]^2. \tag{43}$$

Since a decrease in step size is required, no new attention must be paid to convergence.

Suppose next that a constant step size is chosen so that $e_d > \sqrt{q_\infty}$; then, as shown in Fig. 6, digital effects stop adaptation at the level $e_d(\Delta)$. The final mean-square error may again be decreased by increasing the step size until (43) is satisfied. Fig. 7 shows $e_d^2(\Delta)$ and

[18] For simplicity, we are assuming an equalizer sufficiently long so that the irreducible error $\mathcal{E}(c^*)$ is negligible.

$q_\infty(\Delta)$ plotted as a function of Δ and the value Δ^*. Note that since $0 \leq \Delta^* \leq (2\alpha/\beta^2)$ one is not required to choose a step size out of the convergence region to satisfy (43). By looking at Fig. 7 we see that if $\Delta > \Delta^*$, then $q_\infty(\Delta) > e_d^2(\Delta)$, analog effects dominate, and the final error is reduced by decreasing Δ. If $\Delta < \Delta^*$, then $q_\infty(\Delta) < e_d^2(\Delta)$, digital effects dominate, and the final error can be reduced by increasing Δ. The preceding discussion should explain the experimental behavior displayed in Fig. 4, where $\Delta^* = 0.07$.

It is now clear that to minimize the final mean-square error the best fixed step size is $\Delta = \Delta^*$, given as the solution to (43). Equation (43) can also be interpreted as specifying the precision needed to achieve a certain residual error; i.e., given a desired $q_\infty(\Delta)$ we first solve for Δ^* using the left-hand side, and then we determine the LSD from the RHS. Of course the following question naturally arises: Is there anything to be gained by using a time-varying step size?

C. *Choosing a Time-Varying Step Size*

In an analog environment a time-varying step-size, such as that described in Section II-C, offers the possibility of rapid decay to the minimum attainable mean-square error. What we wish to do presently is to compare, in a digital environment, the residual mean-square error obtained by using the best constant step size Δ^* with the mean-square error resulting from using the optimum step-size sequence Δ_n.

The effect of quantization on the variable step-size algorithm will be to modify the trajectory of Fig. 3, reflecting the fact that adaptation ceases when $q_n(\Delta_n)$ falls below the quantization level $e_d^2(\Delta_n)$. The trajectory will have a different character depending upon whether or not σ^2/β^2 is larger or smaller than $e_d^2(\Delta = \alpha/\beta^2)$, where α/β^2 is the optimum initial step size suggested by stochastic approximation. Suppose $\sigma^2/\beta^2 < e_d^2(\alpha/\beta^2)$; then, as shown in Fig. 8, the mean-square error q_n will decay in an exponential manner until the quantization level $e_d^2(\alpha/\beta^2)$ is penetrated. During this time period the step size remains constant at α/β^2. Suppose we used a constant step-size algorithm with $\Delta = \alpha/\beta^2$. We note that since

$$q_\infty(\alpha/\beta^2) = \sigma^2/\beta^2$$

$q_\infty(\alpha/\beta^2)$ will be less than $e_d^2(\alpha/\beta^2)$ and the final error can be reduced further by increasing Δ (towards Δ^*). Thus $\Delta^* > \alpha/\beta^2$, which implies that

$$e_d^2(\Delta^*) \leq e_d^2(\alpha/\beta^2)$$

and hence the constant step-size algorithm is capable of attaining a smaller final value than the time-varying algorithm.[19] Ideally, if one knew the parameters of the system one might choose an initial step size of α/β^2, guaranteeing rapid initial decrease in mean-square error, and then "gear-shift" to the *larger* step size Δ.[19]

[19] We must still consider the case $\sigma^2/\beta^2 > e_d^2(\alpha/\beta^2)$.

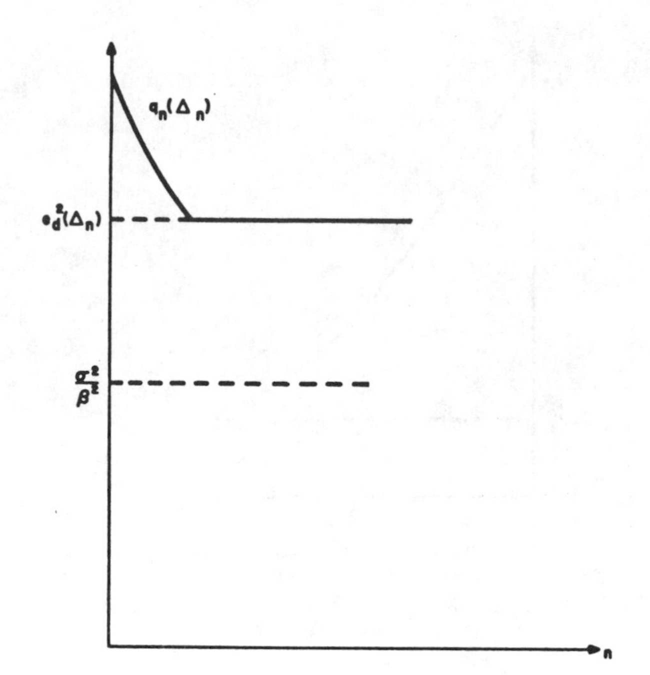

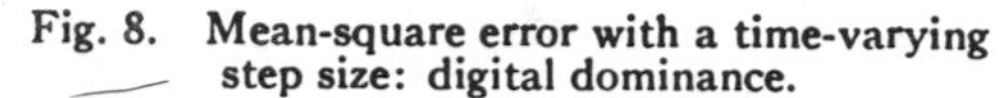

Fig. 8. Mean-square error with a time-varying step size: digital dominance.

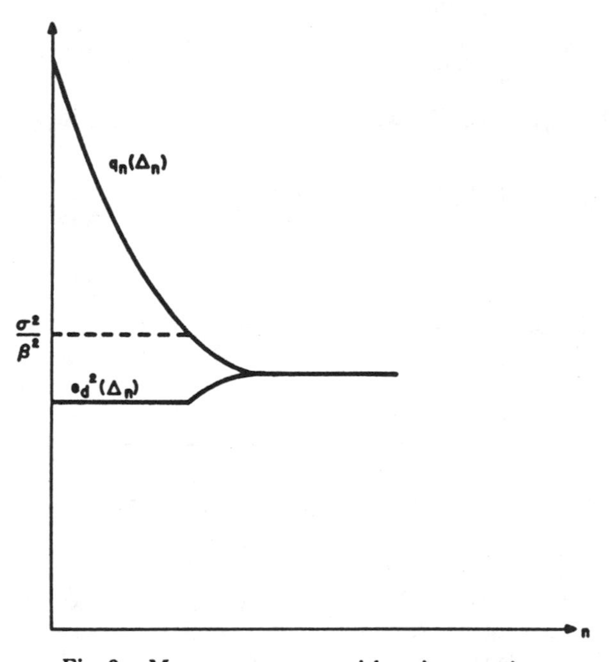

Fig. 9. Mean-square error with a time-varying step size: analog dominance.

Now consider the situation when $\sigma^2/\beta^2 > e_d^2(\alpha/\beta^2)$. The mean-square error q_n will decay in an exponential fashion until $q_n \approx \sigma^2/\beta^2$, at which time both q_n and Δ_n will decrease roughly as $1/n$. Note that as Δ_n decreases, the quantization level $e_d^2(\Delta_n)$ increases. As shown in Fig. 9, adaptation ceases when

$$q_n(\Delta_n) = e_d^2(\Delta_n) \tag{44a}$$

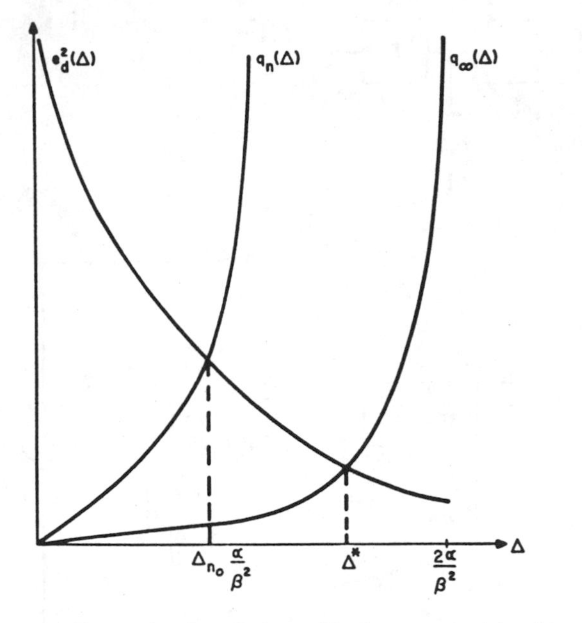

Fig. 10. Determination of the residual errors resulting from fixed and time-varying step-size sequences.

or, from (25b) and (42), when

$$\frac{\sigma^2 \Delta_{n_0}}{\alpha - \beta^2 \Delta_{n_0}} = \left[\frac{\text{LSD}}{X_{\text{rms}} \Delta_{n_0}}\right]^2 \tag{44b}$$

where Δ_{n_0} denotes the step size at stopping.[20] We again wish to compare Δ^* and Δ_{n_0}. This can be done by plotting the functions $q_n(\Delta)$, $q_\infty(\Delta)$, and $e_d^2(\Delta)$ versus Δ, and identifying the quantities Δ^* and Δ_{n_0}. Fig. 10 displays these functions and illustrates graphically the solution to (43) and (44b). Since $q_n(\Delta) > q_\infty(\Delta)$, it is clear that Δ_{n_0} is always less than Δ^*. Hence we again have

$$e_d^2(\Delta^*) \leq e_d^2(\Delta_{n_0})$$

i.e., the constant step-size algorithm can achieve a smaller final mean-square error. By noting that

$$q_\infty(\Delta = \alpha/\beta^2) = \frac{\sigma^2}{\beta^2} > e_d^2(\Delta = \alpha/\beta^2)$$

we see that the mean-square error is reduced by decreasing Δ, hence $\Delta^* < \alpha/\beta^2$. Thus the best step-size sequence would begin at α/β^2 and decrease to the value Δ^*, which is somewhat larger than that suggested by stochastic approximation.

IV. Conclusions

The effects of digital implementation on the performance and design of the adaptive gradient algorithm were shown to be significant. The residual error was shown to be several times that expected solely from quantizing the variable parameters to an LSD. Since this enhancement becomes more pronounced as the number of parameters is increased, it should be considered when designing large-scale adaptive digital systems. A surprising result is that the time-varying step-size sequence suggested by stochastic approximation cannot achieve as small a residual mean-square error as that attained by the best constant step size. Combining the desirability of an initial rapid decay and a minimum residual error suggests that as the algorithm converges the initial step size be "gear-shifted" to the value that minimizes the digital error.

[20] Recall that when $\sigma^2/\beta^2 < e_d^2(\alpha/\beta^2)$, $\Delta_{n_0} = \alpha/\beta^2$.

Appendix I

Derivation of (22)

The purpose of this appendix is to obtain (22) from (21). The second term on the RHS of (21) is lower bounded and averaged as follows:

$$\begin{aligned}\langle e_n \mathbf{x}_n{}^T A \boldsymbol{\varepsilon}_n \rangle &= \langle \mathbf{c}_n{}^T \mathbf{x}_n \mathbf{x}_n{}^T A \boldsymbol{\varepsilon}_n - a_n \mathbf{x}_n{}^T A \boldsymbol{\varepsilon}_n \rangle \\ &= \langle \mathbf{c}_n{}^T A^2 \boldsymbol{\varepsilon}_n \rangle - \langle v^T A \boldsymbol{\varepsilon}_n \rangle \\ &= \langle \mathbf{c}_n{}^T A^2 \boldsymbol{\varepsilon}_n \rangle - \langle \mathbf{c}^{*T} A^2 \boldsymbol{\varepsilon}_n \rangle \\ &= \langle \boldsymbol{\varepsilon}_n{}^T A^2 \boldsymbol{\varepsilon}_n \rangle \\ &> \alpha \langle \boldsymbol{\varepsilon}_n{}^T A \boldsymbol{\varepsilon}_n \rangle \end{aligned} \tag{45}$$

where we have used the positive definite property of A to define $A^{1/2}$ such that

$$\begin{aligned}\boldsymbol{\varepsilon}_n{}^T A^2 \boldsymbol{\varepsilon}_n = \boldsymbol{\varepsilon}_n{}^T A^{1/2} A A^{1/2} \boldsymbol{\varepsilon}_n &= (A^{1/2}\boldsymbol{\varepsilon}_n)^T A (A^{1/2}\boldsymbol{\varepsilon}_n) \\ &> \alpha (A^{1/2}\boldsymbol{\varepsilon}_n)^T (A^{1/2}\boldsymbol{\varepsilon}_n) = \alpha \boldsymbol{\varepsilon}_n{}^T A \boldsymbol{\varepsilon}_n. \end{aligned} \tag{46}$$

The third term on the RHS of (21) is treated as follows:

$$\begin{aligned}\langle e_n{}^2 \mathbf{x}_n{}^T A \mathbf{x}_n \rangle &= \langle (A^{1/2}\mathbf{x}_n e_n)^T (A^{1/2} \mathbf{x}_n e_n) \rangle \\ &= \langle z_n{}^T z_n \rangle \\ &= \sum_{i=-N}^{N} \langle [z_n{}^{(i)}]^2 \rangle \end{aligned} \tag{47}$$

where the components of $z_n = A^{1/2}\mathbf{x}_n e_n$ are denoted by $z_n{}^{(i)}$, $i = -N, -N+1, \cdots, N$.

Using the definition

$$\langle [z_n{}^{(i)}]^2 \rangle = \text{var } z_n{}^{(i)} + \langle z_n{}^{(i)} \rangle^2$$

we can write

$$\begin{aligned}\langle e_n{}^2 \mathbf{x}_n{}^T A \mathbf{x}_n \rangle &= \sum_{i=-N}^{N} \text{var } z_n{}^{(i)} + \sum_{i=-N}^{N} \langle z_n{}^{(i)} \rangle^2 \\ &= \sigma_n{}^2 + \langle z_n \rangle^T \langle z_n \rangle \end{aligned} \tag{48}$$

where $\sigma_n{}^2 = \sum_{i=-N}$ var $z_n{}^{(i)}$. We now want to obtain an upper bound on $\sigma_n{}^2$ that is independent of n, but we first note that

$$\begin{aligned}\langle z \rangle &= A^{1/2} \langle \mathbf{x}_n \mathbf{x}_n{}^T \mathbf{c}_n - a_n \mathbf{x}_n \rangle \\ &= A^{1/2}[A\mathbf{c}_n - A\mathbf{c}^*] = A^{3/2} \boldsymbol{\varepsilon}_n. \end{aligned} \tag{49}$$

Thus by applying (11) twice and using (49) we have

$$\langle z^T \rangle \langle z \rangle = \langle \boldsymbol{\varepsilon}_n{}^T A^{1/2} A^2 A^{1/2} \boldsymbol{\varepsilon}_n \rangle \leq \beta^2 \langle \boldsymbol{\varepsilon}_n{}^T A \boldsymbol{\varepsilon}_n \rangle. \tag{50}$$

Returning to the expression for σ_n^2 we first observe that

$$\sigma_n^2 \leq \left\langle \sum_{j=-N}^{N} [z_j^{(i)}]^2 \right\rangle = \langle \{c_n^T x_n^T A x_n x_n^T c_n - a_n c_n^T x_n x_n^T A x + a_n^2 x_n^T A x_n\} \rangle. \quad (51)$$

Considering the first item on the RHS of (51) gives

$$\langle c_n^T x_n x_n^T A x_n x_n^T c_n \rangle \leq \beta \langle c_n^T x_n x_n^T x_n x_n^T c_n \rangle = \beta \langle c_n^T X_n c_n \rangle \quad (52)$$

where the matrix X_n is defined by

$$X_n = x_n x_n^T x_n x_n^T. \quad (53)$$

Noting that the maximum eigenvalue of X_n is $[x_n^T x_n]^2$ and using the independence of x_n and c_n permits us to write

$$\langle c_n^T X_n c_n \rangle \leq \langle (x_n^T x_n)^2 \rangle \cdot \langle c_n^T c_n \rangle. \quad (54)$$

If we denote the fourth-order term by

$$\mu = \langle (x_n^T x_n)^2 \rangle \quad (55)$$

and require that[21]

$$c_n^T c_n \leq (2N+1)\rho \quad (56)$$

then the first term on the RHS of (51) is upper bounded by $N\rho\mu$. The last term is simply bounded by noting that

$$\langle a_n^2 x_n^T A x_n \rangle = \langle x_n^T A x_n \rangle \leq \beta \cdot \langle x_n^T x_n \rangle = \beta(2N+1) X_{rms}^2 \quad (57)$$

where the *mean-square input*

$$X_{rms}^2 \equiv \langle x_n^2 \rangle = \sum_{m=-\infty}^{\infty} h_m^2 + \langle \nu_n^2 \rangle \quad (58)$$

is the average power in the received samples. The second term in (51) is recognized as a cross term by letting

$$a = a_n \sqrt{x_n^T A x_n} \qquad b = c_n^T x_n \sqrt{x_n^T A x_n} \quad (59)$$

and can be absorbed by noting that

$$a_n c_n^T x_n x_n^T A x_n = ab \leq \frac{a^2 + b^2}{2}. \quad (60)$$

Combining the above inequalities we finally have that

$$\varepsilon_{n+1}^T A \varepsilon_{n+1} \leq [1 - 2\alpha\Delta_n + \beta^2 \Delta_n^2] \langle \varepsilon_n^T A \varepsilon_n \rangle + \sigma^2 \quad (61)$$

where

$$\sigma^2 = 2\beta(2N+1)[\mu\rho + X_{rms}^2].$$

[21] This would be true in any practical system; e.g., the dynamic range of the taps is taken to be limited to $\pm\sqrt{\rho}$. In making this assumption we implicitly assume that the optimum tap setting does, in fact, lie within the dynamic range of the taps. If this is indeed true, then the algorithm will converge to the proper tap settings even though at some intermediate range one or more taps might "saturate."

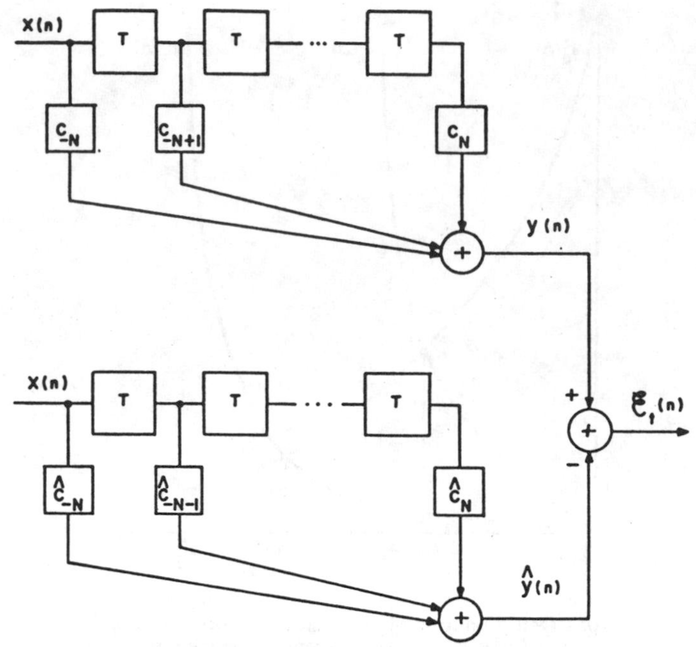

Fig. 11. A model for analyzing truncation error.

Appendix II

RMS Quantization Error

In this appendix we shall derive an expression for the rms QE. Let us start by considering the following two equalizers shown in Fig. 11. The first has tap weights $\{c_j\}$ that are assumed to be of infinite precision. The output of this equalizer is given by

$$y(i) = \sum_{j=-N}^{N} c_j \cdot x(i-j).$$

The tap weights in the second equalizer $\{\hat{c}_j\}$ are found by truncating the corresponding tap weights in the first filter. The output of the second equalizer equals

$$\hat{y}(i) = \sum_{j=-N}^{N} \hat{c}_j \cdot x(i-j).$$

The QE equals

$$\mathcal{E}_t(i) = y(i) - \hat{y}(i) = \sum_{j=-N}^{N} (c_j - \hat{c}_j) \cdot x(i-j)$$

and the mean-squared QE equals

$$\epsilon_t^2 = E[\mathcal{E}_t^2(i)] = E\left[\sum_{j=-N}^{N} \sum_{l=-N}^{N} (c_j - \hat{c}_j) \cdot (c_l - \hat{c}_l) \cdot x(i-j) \cdot x(i-l)\right].$$

Let us assume that the inputs $\{x_i\}$ are independent random variables with rms value equal to X_{rms}. Then

$$\epsilon_t^2 = \sum_{j=-N}^{N} (c_j - \hat{c}_j)^2 X_{rms}^2 \leq (2N+1) \cdot \text{LSD}^2 \cdot X_{rms}^2$$

where LSD is the value of the LSD in the truncated tap weights. The above expressions are exact for the case of statistically independent inputs and provide useful approximate results when the inputs are dependent.

Appendix III

Digital Residual Error versus Quantization Error

As explained in Section III, the rms DRE using the estimated gradient algorithm can be approximated by

$$e_d \approx \frac{\text{LSD}}{\Delta \cdot X_{\text{rms}}}. \tag{62}$$

In Appendix II, the rms QE was found to be upper bounded by

$$\epsilon_t^2 \leq (2N+1)^{1/2} \cdot \text{LSD} \cdot X_{\text{rms}}. \tag{63}$$

To relate these two errors we must express Δ as a function of N. This can be accomplished by means of the sufficient condition (14) for the convergence of the estimated gradient algorithm, namely

$$0 \leq \Delta \leq \frac{\alpha}{\beta^2}.$$

We now note that

$$\frac{\alpha}{\beta^2} = \frac{\alpha}{\beta}\frac{1}{\beta} \geq \frac{\alpha}{\beta} \cdot \frac{1}{\text{trace } A} = \frac{\alpha}{\beta} \frac{1}{(2N+1)X_{\text{rms}}^2} \tag{64}$$

is a lower bound on the maximum step size, where X_{rms} is defined by (58).

Combining the above we can lower bound the ratio of DRE to QE as follows:

$$\frac{e_d}{\epsilon_t} \geq \frac{\text{LSD} \cdot (2N+1) X_{\text{rms}}^2}{X_{\text{rms}} \frac{\alpha}{\beta} (2N+1)^{1/2} \cdot \text{LSD} \cdot X_{\text{rms}}}$$

$$= \frac{\beta}{\alpha}(2N+1)^{1/2} \tag{65}$$

a number that grows with the length of the equalizer.

References

[1] D. J. Sakrison, "Stochastic approximation: A recursive method for solving regression problems," in *Advances in Communication Theory*, vol. 2, A. V. Balakhrishnan, Ed. New York: Academic, 1966.

[2] R. W. Lucky, J. Salz, and E. J. Weldon, Jr., *Principles of Data Communications.* New York: McGraw-Hill, 1968.

[3] A. Gersho, "Adaptive equalization of highly dispersive channels for data transmission," *Bell Syst. Tech. J.*, vol. 48, no. 1, pp. 55–70, 1969.

[4] D. G. Luenberger, *Optimization by Vector Space Methods.* New York: Wiley, 1969.

[5] B. Widrow, *Adaptive Filters I: Fundamentals*, System Theory Lab., Stanford Electronics Labs., Stanford Univ., Stanford, Calif., TR 6764-6, Dec. 1966.

[6] R. W. Chang, "A new equalizer structure for fast start-up digital communication," *Bell Syst. Tech. J.*, vol. 50, no. 6, pp. 1969–2014, 1971.

[7] H. Robbins and S. Monro, "A stochastic approximation method," *Ann. Math. Statist.*, vol. 22, pp. 400–407, 1951.

[8] M. G. Taylor, "A comparison of algorithms for adapting digital filters," *Symp. Dig. 1970 Canadian Symp. Communications*, 1970.

[9] R. W. Lucky, "Techniques for adaptive equalization of digital communication systems," *Bell Syst. Tech. J.*, vol. 45, no. 2, 1966.

A Roundoff Error Analysis of the LMS Adaptive Algorithm

CHRISTOS CARAISCOS, MEMBER, IEEE, AND BEDE LIU, FELLOW, IEEE

Abstract—The steady state output error of the least mean square (LMS) adaptive algorithm due to the finite precision arithmetic of a digital processor is analyzed. It is found to consist of three terms: 1) the error due to the input data quantization, 2) the error due to the rounding of the arithmetic operations in calculating the filter's output, and 3) the error due to the deviation of the filter's coefficients from the values they take when infinite precision arithmetic is used. The last term is of particular interest because its mean squared value is inversely proportional to the adaptation step size μ. Both fixed and floating point arithmetics are examined and the expressions for the final mean square error are found to be similar. The relation between the quantization error and the error that occurs when adaptation possibly ceases due to quantization is also investigated.

I. Introduction

SINCE its introduction by Widrow [1], the least mean square (LMS) adaptive algorithm has found numerous applications. New lattice structure algorithms introduced recently [2], [3] have been shown to have certain advantages. For example, the rate of convergence is less sensitive to the spread of the eigenvalues of the input correlation matrix. Nevertheless, the LMS algorithm may still be appealing because of its simple structure which makes it easy to implement. There are also applications for which a lattice algorithm cannot be or has not been proposed yet, as for example, in adaptive array beamforming using only present samples of the array outputs. In the sequel, we are using bold lower case characters for vectors and bold upper case characters for matrices.

In its typical form, the LMS adaptive algorithm that we examine addresses the problem of estimating a sequence of zero mean scalars d_n from a sequence of zero mean, length N vectors $\mathbf{x}_n$. The vector $\mathbf{x}_n$ may be, for example, either the present samples of N data sequences, such as the outputs of an array of N elements, or the N most recent samples of a single sequence. The estimate of d_n is $\mathbf{w}_n^T\mathbf{x}_n$, where $\mathbf{w}_n$ is a length N weight vector which is updated at the nth time instance, according to the (LMS) algorithm

$$\mathbf{w}_{n+1} = \mathbf{w}_n + \mu \mathbf{x}_n(d_n - \mathbf{w}_n^T\mathbf{x}_n), \quad n = 0, 1, 2, \cdots. \tag{1}$$

The initial value $\mathbf{w}_0$ may be arbitrarily chosen, usually to be zero. Under the assumption that the sequence d_n and $\mathbf{x}_n$ are stationary and $\mathbf{x}_n$ independent in time (independence assumption), the expected value of $\mathbf{w}_n$ converges exponentially [1] to the optimal (Weiner) vector $\mathbf{w}^* = \mathbf{R}^{-1}\mathbf{p}$, where $R = E(\mathbf{x}_n\mathbf{x}_n^T)$ and $\mathbf{p} = E(\mathbf{x}_n d_n)$. Convergence is obtained if $0 < \mu < 2/\lambda_{max}$, where λ_{max} is the maximum eigenvalue of $\mathbf{R}$. Similar behavior of the algorithm has been observed when $\mathbf{x}_n$ is not an independent sequence. In the steady state, the weight vector $\mathbf{w}_n$ fluctuates around $\mathbf{w}^*$ causing an extra mean square error called *excess mean square error*, which is proportional to the adaptation step size μ [4]. In a digital implementation of the algorithm a quantization error is also present. It is the purpose of the present paper to analyze this quantization error and make some suggestions to combat it. Throughout the paper, we adopt the independence assumption on $\mathbf{x}_n$.

In Sections II and III, expressions are derived for the steady state mean square quantization error when fixed and floating point arithmetics are used. The methodology that we follow here is similar to the one presented in [5]. The relation of this error to a possibly early termination of the algorithm due to quantization effects [6] is examined in Section IV. Some simulation results and the conclusion are given in Sections V and VI.

II. Fixed Point Arithmetic

We assume that the input sequences have been properly scaled, so that their values lie between -1 and 1. Each data sample is represented by B_d bits plus sign, and each filter coefficient is represented by B_c bits plus sign. We assume no overflow occurs, hence, additions do not introduce any error, while each multiplication introduces an error after the product is quantized. We also assume that rounding is used in the quantization, thus the quantization error associated with a B_d-plus-sign bit number has variance $\sigma_d^2 = 2^{-2B_d}/12$, and the error associated with a B_c-plus-sign bit number has variance $\sigma_c^2 = 2^{-2B_c}/12$. Denoting the fixed point quantization of the quantity a by $fx[a]$, we can write $fx[a] = a + \eta$, where η is the quantization error.

We use unprimed and primed symbols to represent quantities of infinite and finite precision, respectively. A schematic representation of the LMS adaptive filter is shown in Fig. 1. The scaling factor a implemented by a shift to the right by one or more bits has been introduced to prevent overflow of the elements of the weight vector $\mathbf{w}_n'$ and the estimate $\hat{y}_n$ during the operation of the filter. It affects the magnitude of $\mathbf{w}_n'$ but not its rate of convergence. Usually, the required value of a is not expected to be very small.

For the input sequences we can write

$$\mathbf{x}_n' = \mathbf{x}_n + \boldsymbol{\alpha}_n, \quad y_n' = y_n + \beta_n, \quad n = 0, 1, 2, \cdots \tag{2}$$

Manuscript received December 20, 1982; revised September 29, 1983. This research was supported by the Air Force Office of Scientific Research, USAF, under Grant AFOSR-81-0186.

The authors are with the Department of Electrical Engineering and Computer Science, Princeton University, Princeton, NJ 08544.

Reprinted from *IEEE Trans. Acoust., Speech, Signal Processing,* vol. ASSP-32, no. 1, pp. 34–41 Feb. 1984.

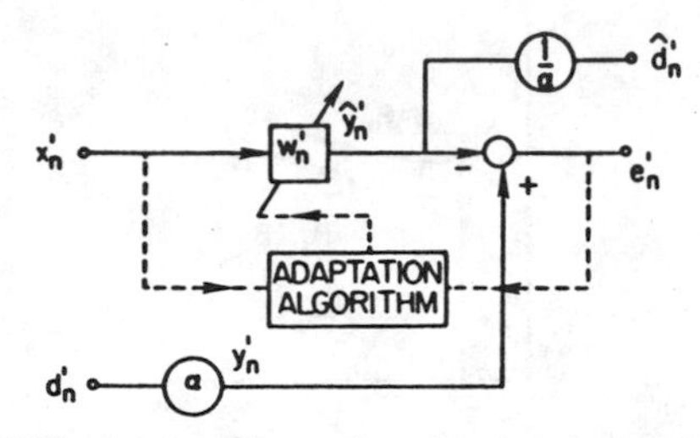

Fig. 1. The LMS adaptive filter when fixed point arithmetic is used. The multiplies by a and $1/a$ are absent for floating point arithmetic.

where α_n and β_n are quantization noises. Their elements are assumed to be white sequences independent of the signals and between each other. They have zero mean and variance σ_d^2 [7]. Because of the finite precision arithmetic used, the actual weight vector at time n can be written as

$$w_n' = w_n + \rho_n \tag{3}$$

where ρ_n is the error vector of length N caused by the quantization effects. In the sequel, we shall ignore all error terms higher than first order. The computed output $\hat{y}_n$ is given by

$$\hat{y}_n = fx(w_n'^T x_n') = w_n^T x_n + \rho_n^T x_n + w_n^T \alpha_n + \eta_n \tag{4}$$

where η_n is, approximately, a white sequence of quantization noise independent of the signals and the rest of the error sequences. Its mean value is zero and its variance is $c\sigma_d^2$ where the constant c depends on the way in which the inner product $w_n'^T x_n'$ is computed. If all N scalar products involved in $w_n'^T x_n'$ are computed without quantization, summed, and the final result is quantized in B_d bits, then the variance of η_n is equal to σ_d^2. If the scalar products are quantized individually and then summed, the variance of η_n is $N\sigma_d^2$.

The estimation error e_n' is

$$e_n' = y_n' - \hat{y}_n = y_n - w_n^T x_n - \rho_n^T x_n - w_n^T \alpha_n + \beta_n - \eta_n \tag{5}$$

and the total output error is equal to

$$d_n - \hat{d}_n' = \frac{1}{a} e_n - \frac{1}{a} (\rho_n^T x_n + w_n^T \alpha_n + \eta_n) \tag{6}$$

where $e_n = y_n - w_n^T x_n$. The term e_n/a is the estimation error of the infinite precision algorithm. The finite precision arithmetic is reflected in the second term $1/a\,(\rho_n^T x_n + w_n^T \alpha_n + \eta_n)$. Because α_n and η_n are quantization noise sequences independent of the data sequences and each other, the terms $\rho_n^T x_n$, $w_n^T \alpha_n$, and η_n are uncorrelated. For the same reason, e_n is uncorrelated to $w_n^T \alpha_n$ and η_n. The vector ρ_n depends on data up to time $n - 1$. Because of the independence assumption adopted throughout the paper, we have $E(e_n \rho_n^T x_n) = E(\rho_n^T)\,E(e_n x_n)$. By invoking the same assumption, one can easily see from (A6) in Appendix A that $E(\rho_n) = 0$. So, the terms e_n and $\rho_n^T x_n$ are uncorrelated. Therefore, e_n/a is uncorrelated to $1/a\,(\rho_n^T x_n + w_n^T \alpha_n + \eta_n)$. We note that multiplication by $1/a$ corresponds to a left shift, hence introduces no quantization error.

It is shown in Appendix A that the total output mean square error is given by

$$\xi = \xi_{\min} + \frac{\mu \xi_{\min} \operatorname{tr} R}{2 - \mu \operatorname{tr} R} + \frac{1}{a^2}\left[\frac{N\sigma_c^2 + \mu^2 [(1 + c + |w^*|^2) \operatorname{tr} R + \xi_{\min} N]\, \sigma_d^2}{2\mu - \mu^2 \operatorname{tr} R} + \left(\left|w^*\right|^2 + \frac{1}{2}\mu \xi_{\min} N\right)\sigma_d^2 + c\sigma_d^2\right] \tag{7}$$

where $\xi_{\min} = E(d_n^2) - p^T R^{-1} p$ is the mean square error of the optimal (Wiener) filter $w^* = R^{-1} p$. If the same word length is used for both data and coefficients ($B_d = B_c$) and μ is sufficiently small, (7) reduces to

$$\xi = \xi_{\min} + \frac{1}{2}\mu \xi_{\min} \operatorname{tr} R + \frac{N\sigma_c^2}{2a^2\mu} + \frac{1}{a^2}(|w^*|^2 + c)\,\sigma_d^2. \tag{8}$$

The significance of this result in the design is discussed in Section IV.

III. Floating Point Arithmetic

Each element of the input sequences x_n and d_n is expressed as $(sign)\, 2^a b$, where the exponent a is an integer and the mantissa b is either zero or a fraction between $\frac{1}{2}$ and 1. The floating point quantization of a number q can be expressed as $fl(q) = q(1 + \epsilon)$. The (relative) error ϵ is modeled as a random variable independent of q. Both addition and multiplication introduce errors when the results are quantized. We can write $fl(x + y) = (x + y)(1 + \gamma)$ and $fl(xy) = (xy)(1 + \delta)$ where γ and δ are zero mean random variables independent of x and y. We are primarily interested in the mean squared values of the quantization errors γ and δ. For rounding, it has been found [8], [9] that $E(\delta^2) \approx (0.18)\, 2^{-2B}$ where B is the number of bits used to represent the mantissas. The mean squared value of γ is somewhat smaller than this, but taking $E(\gamma^2) = E(\delta^2)$ will be sufficiently accurate for our purposes. We denote this value by σ_c^2 for a B_c-bit mantissa and by σ_d^2 for a B_d-bit mantissa.

There is no more need to scale the input sequence d_n as in the fixed point arithmetic case. Hence, the scaling (right shift) a and the left shift $1/a$ are absent from Fig. 1. It is assumed that B_d-bit mantissas are used for the data and B_c-bit mantissas for the coefficients and that enough bits have been allotted to the exponents to prevent overflow. For the actual input sequences, we have $x_n' = fl(x_n) = (I + A_n)\, x_n$ and $d_n' = fl(d_n) = (1 + \beta_n)\, d_n$. The elements of the diagonal noise matrix A_n and the scalar noise β_n are white sequences independent of the signals and each other with variance σ_d^2. The deviation of w_n' from the ideal w_n, due to quantization noise, is modeled as an additive vector sequence ρ_n, i.e., $w_n' = w_n + \rho_n$.

To compute the mean square value of the error we follow the same procedure as in Section II. However, it becomes more cumbersome because the additions introduce a quantization error also. From Fig. 1, we have

$$\hat{d}_n' = w_n'^T x_n + \eta_n = w_n^T x_n + \rho_n^T x_n + w_n^T A_n x_n + \eta_n. \tag{9}$$

The error sequence η_n depends on how the computation of the inner product $w_n'^T x_n'$ is performed. From (9), we get the total estimation error

$$d_n - \hat{d}_n' = e_n - (\rho_n^T x_n + w_n^T A_n x_n + \eta_n) \tag{10}$$

where $e_n = d_n - w_n^T x_n$. Because A_n and η_n are quantization noises independent of the data and each other one can find, in a way similar to the one followed in Section II, that e_n and $\rho_n^T x_n + w_n^T A_n x_n + \eta_n$ are uncorrelated. The same is true for $\rho_n^T x_n$, $w_n^T A_n x_n$ and η_n. The second term in (10) is the arithmetic error and has mean squared value

$$\xi_{ar} = E(\rho_n^T x_n)^2 + E(w_n^T A_n x_n)^2 + E(\eta_n^2). \tag{11}$$

The first term is the error due to the quantization noise of the weight vector w_n, the second reflects the input quantization noise, and the last one is the error associated with the computation of the inner product $w_n'^T x_n'$. Their mean squared values are calculated in Appendix B.

The total output quantization error consists, approximately, of one term inversely proportional to μ, one term independent of μ, and a third one proportional to μ. By neglecting the third term we arrive at the following expression for the total mean square error

$$\xi = \xi_{\min} + \frac{1}{2}\mu\xi_{\min} \operatorname{tr} R + \frac{|w^*|^2 \sigma_c^2}{2\mu} + \frac{1}{4}\xi_{\min} N\sigma_c^2 + \left[2\sum_{k=1}^{N} R_{kk}(w^{*k})^2 + C\right]\sigma_d^2 \tag{12}$$

where R_{kk} denotes the (k, k)th element of R and w^{*k} the kth element of w^*. The quantity C is defined in Appendix B.

Remark: For the LMS algorithm to be implemented in fixed point arithmetic, the input sequences have to be scaled so that their values lie inside the interval $(-1, 1)$. No such scaling is required when floating point arithmetic is used. However, the step size μ is not scale invariant; in other words, for the two implementations to have the same convergence rate, different values of μ have to be used. If μ is used in the floating point implementation, then μ/s^2 has to be used in the fixed point one, where s is the scaling factor which brings the data sequences inside the interval $(-1, 1)$. Since scaling affects also the auto- and cross-correlation of the input sequences, it is straightforward to see that, as expected, the convergence conditions for both implementations are the same. Similarly, the various simplifications made under the assumption that μ is sufficiently small are equally valid in both cases.

IV. Design Issues

There are some significant differences between the analog form (infinite precision) of the adaptation algorithm

$$w_{n+1} = w_n + \mu x_n e_n \tag{13}$$

and the corresponding digital (finite precision) algorithm

$$w'_{n+1} = w'_n + fx(\mu x'_n e'_n). \tag{14}$$

When some component of the product $\mu x'_n e'_n$ is less in magnitude than 2^{-B_c-1}, its quantized value is equal to zero and the corresponding filter coefficient remains unchanged. If μ is small enough, most of the time the coefficients are not updated, and the adaptation of the filter virtually stops. This phenomenon has been analyzed in [6], where it is shown that early termination of the adaptation due to quantization may result in a mean square error significantly larger than the error

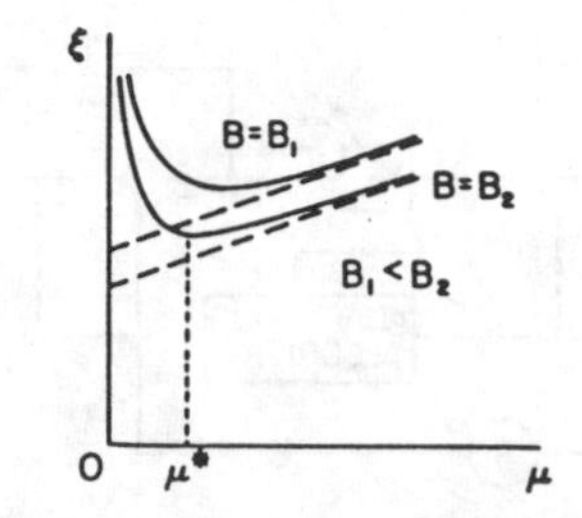

Fig. 2. The total output mean square error ξ as a function of the adaptation step size μ. B, $B1$, and $B2$ denote numbers of bits of both data and coefficients.

of the infinite precision algorithm. The smaller μ is, the larger this error becomes. Throughout the previous analysis we have assumed that μ is such that it allows the algorithm to converge completely.

For fixed point arithmetic, the total output mean squared error is given by (7) in which the term $N\sigma_c^2/(2\mu - \mu^2 \operatorname{tr} R)$ dominates the rest of the quantization errors if $B_c = B_d$. This dominance ceases if more bits are used for the coefficients than for the data. When $B_c = B_d$ and μ is sufficiently small, (7) can be approximated by (8).

Although a small value of μ reduces the excess mean square error $\frac{1}{2}\mu\xi_{\min} \operatorname{tr} R$, it may result in a large quantization error $(N\sigma_c^2/2a^2\mu)$. Fig. 2 shows two typical sketches of ξ versus μ according to (8). The error ξ is minimized for

$$\mu = \mu^* = \frac{\sigma_c}{a}\sqrt{\frac{N}{\xi_{\min} \operatorname{tr} R}} = \frac{2^{-B_c}}{2a}\sqrt{\frac{N}{3\xi_{\min} \operatorname{tr} R}}. \tag{15}$$

However, as we show next, this optimal value of μ is too small to allow the algorithm to converge completely.

The condition for the ith component of the vector w'_n not to be updated is

$$|\mu x_n'^i e_n'| < 2^{-B_c-1} \tag{16}$$

where $x_n'^i$ is the ith component of the vector x'_n. To get a condition for the overall algorithm to stop adapting, we replace $x_n'^i$ in (16) by $x_{\text{rms}} = [E(x_n^{i2})]^{1/2}$ and e'_n by $[E(e_n^2)]^{1/2}$. At this point, we are making the assumption, which is very often valid, that all components of the vector x_n have the same mean square value. Thus, the condition for the adaptation to stop becomes

$$\mu^2 x_{\text{rms}}^2 E(e_n^2) < \frac{2^{-2B_c}}{4}. \tag{17}$$

Using $x_{\text{rms}}^2 = \operatorname{tr} R/N$, $E(e_n^2) = a^2\xi_{\min}(1 + \frac{1}{2}\mu \operatorname{tr} R)$, and $\mu = \mu^*$, we get

$$\mu^2 x_{\text{rms}}^2 E(e_n^2) = \frac{1}{3}\left(1 + \frac{1}{2}\mu \operatorname{tr} R\right)\frac{2^{-2B_c}}{4}. \tag{18}$$

Since for convergence of the mean square error we must have $\mu < 2/\operatorname{tr} R$ [10], inequality (17) is clearly satisfied. So, if the value of μ used is equal to or less than the optimal value μ^*, the adaptation essentially stops before complete convergence of the algorithm is reached. In other words, (8) is a valid expression for the total output mean square error only for μ larger than some value $\mu_1 > \mu^*$. In this case, the excess mean square error is larger than the quantization error. As μ decreases from large values towards μ^* and beyond it, the role of quantization changes from simply adding a small term $(N\sigma_c^2/2a^2\mu)$ to the total output mean square error, to completely determining this error. In case μ does not allow the algorithm to converge completely, the steady state total output mean square error increases with decreasing μ [6] and the dependence of ξ on μ is still similar to the one shown in Fig. 2.

In the case of floating point arithmetic the situation is similar. The value of μ which minimizes the total error ξ in (12) is $\mu^* = \sigma_c |w^*|/\sqrt{\xi_{\min} \operatorname{tr} R} = 2^{-B_c}|w^*|\sqrt{0.18/\xi_{\min} \operatorname{tr} R}$. Now, the coefficient update equation is

$$w'_{n+1} = fl[w'_n + fl(\mu x'_n e'_n)] \tag{19}$$

and the condition for the ith component $w_n'^i$ of the vector w'_n not to be updated is

$$|\mu x_n'^i e'_n| < 2^{-B_c-1}|w_n'^i|. \tag{20}$$

By replacing $x_n'^i$ by x_{rms}, e'_n by $[E(e_n^2)]^{1/2}$, and $w_n'^i$ by w^{*i} we get the following overall condition.

$$\mu^2 x_{\text{rms}}^2 E(e_n^2) < \frac{2^{-2B_c}}{4}|w^{*i}|^2. \tag{21}$$

For $\mu = \mu^*$ we have

$$\mu^2 x_{\text{rms}}^2 E(e_n^2) = \frac{2^{-2B_c}(0.72)|w^*|^2}{4N}\left(1 + \frac{1}{2}\mu \operatorname{tr} R\right). \tag{22}$$

An upper bound for $(0.72|w^*|^2/N)(1 + \frac{1}{2}\mu \operatorname{tr} R)$ is $(1.44|w^*|^2/N)$. This is not smaller than the squares of all elements of w^*. Therefore, condition (21) is not necessarily satisfied for all elements w^{*i}. Nevertheless, the frequency of the instances when an element of the weight vector has not been updated is high enough, which leads to the conclusion that using a value of μ less than or equal to μ^* will make the algorithm essentially stop adapting before it converges completely.

V. Simulation Results

For both fixed point and floating point arithmetics, the component $E(\rho_n^T x_n)^2$ of the mean square error is of particular interest. It increases as μ is decreased. In this section, we present some simulation results to confirm this observation. Since this study is not oriented towards any specific application, simple cases of the LMS algorithm were used so that the simulation results may be checked against the corresponding theoretical results. It was more convenient to work with $P_n = \operatorname{cov}(\rho_n)$, instead of $E(\rho_n^T x_n)^2$. The behavior of P_n is directly reflected in $E(\rho_n^T x_n)^2$, since $E(\rho_n^T x_n)^2 = \operatorname{tr} P_n R)$.

For μ sufficiently small ($\mu \operatorname{tr} R \ll 2$), (A16) can be approximated by

$$\mu(RP_n + P_n R) = Q_n. \tag{23}$$

We shall work with input sequences such that $R = \sigma^2 I$, in which case the solution of the above equation is

$$P_n = \frac{1}{2\mu} R^{-1} Q_n. \tag{24}$$

A pseudorandom Gaussian sequence of zero mean and variance $\sigma^2 = \frac{1}{16}$ was generated, quantized, and used for the data sequences. When fixed point arithmetic was used, the few samples outside the interval (-1, 1.) were discarded. Estimates of P_n were derived as time averages in the steady state, using the difference between the weight vectors of the infinite precision arithmetic algorithm and the finite one. Double precision arithmetic was used as a quantization-error-free implementation of the algorithm.

In the first of our tests, the LMS algorithm was used to estimate the sequence $d_n = a_1 x_{1n} + a_2 x_{2n} + a_3 x_{3n}$ from the sequences x_{1n} and x_{2n}, where x_{1n}, x_{2n}, x_{3n} are independent subsequences of the generated pseudorandom sequence. Obviously, $w^* = [a_1\ a_2]^T$. The minimum mean square error $\xi_{\min}$ is equal to $a_3^2\sigma^2$. We used the values $a_1 = 0.25$, $a_2 = 0.25$, $a_3 = 0.0625$, and $\mu = 0.5$, 0.25, 0.125. In both single and double precision implementations, the same quantized sequences were used as input data. For the fixed point implementation we have from (A8)

$$b_n = -\mu x_n \eta_n + \sigma_n \tag{25}$$

and so

$$Q_n = \mu^2 c\sigma_d^2 R + \sigma_c^2 I = (2\mu^2\sigma_d^2\sigma^2 + \sigma_c^2)I. \tag{26}$$

Finally, from (24), we get

$$P_n = \left(\mu\sigma_d^2 + \frac{\sigma_c^2}{2\mu\sigma^2}\right)I. \tag{27}$$

For the values of μ which were used and for $B_c = B_d$, the second term dominates resulting in

$$P_n = \frac{\sigma_c^2}{2\mu\sigma^2} I. \tag{28}$$

The two terms in (27) become of comparable size if B_c is properly increased.

When floating point arithmetic is used, from (B11) we get

$$b_n = \Delta_n w_{n+1} + \mu[\Sigma_n x_n e_n + x_n(\kappa_n e_n + \vartheta_n e_n - \eta_n)] \tag{29}$$

which in the steady state and for small $\xi_{\min}$ can be approximated by

$$b_n = \Delta_n w^* - \mu x_n \eta_n. \tag{30}$$

Therefore,

$$Q_n = \sigma_c^2 \operatorname{diag}[a_1^2\ a_2^2] + \mu^2 E(\eta_n^2 x_n x_n^T). \tag{31}$$

As in the fixed point case, in the steady state we get

$$P_n = \frac{\sigma_c^2}{2\mu\sigma^2}\operatorname{diag}[a_1^2\ a_2^2] + \frac{\mu}{2\sigma^2}E(\eta_n^2 x_n x_n^T). \tag{32}$$

Using (B3), we find that the diagonal elements of $E(\eta_n^2 x_n x_n^T)$ are equal to $2\sigma^4(3a_1^2 + a_2^2)\sigma_d^2$ and $2\sigma^4(a_1^2 + 3a_2^2)\sigma_d^2$ and the off-diagonal ones are equal to $2\sigma^4 a_1 a_2 \sigma_d^2$. If $B_c = B_d$, (32) can be approximated by

$$P_n = \frac{\sigma_c^2}{2\mu\sigma^2}\operatorname{diag}[a_1^2 a_2^2]. \tag{33}$$

The algorithm ran for 15 000 iterations. We consider only the diagonal elements of P_n because these are the ones that determine the error term $E(\rho_n^T x_n)^2$, since R is diagonal. Table I shows the theoretical and the estimated values of a representative element of P_n, namely the first diagonal element $E(\rho_{1n}^2)$.

TABLE I
VALUES OF $E(\rho_{1n}^2)$ FOR THE FIRST TEST (2 ELEMENT ARRAY)
$a_1 = 0.25$, $a_2 = 0.25$, $a_3 = 0.0625$

	Fixed Point Arithmetic				Floating Point Arithmetic			
μ	B_d	B_c	Theor.	Simul.	B_d	B_c	Theor.	Simul.
0.5	12	12	0.79×10^{-7}	0.74×10^{-7}	12	12	0.10×10^{-7}	0.12×10^{-7}
0.25	12	12	0.16×10^{-6}	0.13×10^{-6}	12	12	0.21×10^{-7}	0.22×10^{-7}
0.125	12	12	0.32×10^{-6}	0.35×10^{-6}	12	12	0.43×10^{-7}	0.52×10^{-7}
0.5	12	15	0.37×10^{-8}	0.41×10^{-8}	12	16	0.13×10^{-9}	0.15×10^{-9}
0.25	12	15	0.37×10^{-8}	0.37×10^{-8}	12	17	0.63×10^{-10}	0.69×10^{-10}
0.125	12	16	0.19×10^{-8}	0.16×10^{-8}	12	18	0.31×10^{-10}	0.36×10^{-10}

TABLE II
VALUES OF $E(\rho_{1n}^2)$ FOR THE SECOND TEST (4 ELEMENT ARRAY)
$a_1 = 0.125$, $a_2 = 0.110$, $a_3 = 0.130$, $a_4 = 0.135$, $a_5 = 0.0625$, $B_c = B_d = 12$

	Fixed Point Arithmetic		Floating Point Arithmetic	
μ	Theor.	Simul.	Theor.	Simul.
0.5	0.79×10^{-7}	0.99×10^{-7}	0.27×10^{-8}	0.35×10^{-8}
0.25	0.16×10^{-6}	0.17×10^{-6}	0.54×10^{-8}	0.50×10^{-8}
0.125	0.32×10^{-6}	0.35×10^{-6}	0.11×10^{-7}	0.11×10^{-7}

TABLE III
VALUES OF $E(\rho_{1n}^2)$ FOR THE THIRD TEST (TRANSVERSAL FILTER)
$a_1 = 0.125$, $a_2 = 0.110$, $a_3 = 0.130$, $a_4 = 0.135$, $a_5 = 0.0625$, $B_d = B_c = 12$

	Fixed Point Arithm.	Floating Point Arithm.
μ	Simulation	Simulation
0.5	0.97×10^{-7}	0.35×10^{-8}
0.25	0.14×10^{-6}	0.68×10^{-8}
0.125	0.31×10^{-6}	0.88×10^{-8}

In the upper half of the table, (28) and (33) were used; it shows that the estimates vary roughly as $1/\mu$. In the lower half of the table, (27) and (32) were used and the difference $B_c - B_d$ was chosen so that the two terms in each one of the equations were of comparable size.

In the second test, the algorithm had to estimate the sequence $d_n = a_1 x_{1n} + \cdots + a_5 x_{5n}$ from $x_{1n}, \cdots, x_{4n}$, with $a_1 = 0.125$, $a_2 = 0.110$, $a_3 = 0.130$, $a_4 = 0.135$, and $a_5 = 0.0625$. Theoretical and estimated values of the representative $E(\rho_{1n}^2)$ are shown in Table II for $B_c = B_d = 12$.

In both cited examples, all assumptions made in the derivation of the theoretical results (Gaussian independent input vector sequence $\boldsymbol{x}_n$) are satisfied. A third test was performed concerning a transversal filter, in which the sequence $d_n = a_1 x_n + \cdots + a_4 x_{n-3} + a_5 s_n$ had to be estimated from the sequence x_n. Now x_n and s_n are independent Gaussian sequences and the values of the constants $a_1, \cdots, a_5$ and μ are the same as in the second test. The results of this test are shown in Table III and indicate that, again, the quantization error is larger for smaller values of μ.

VI. CONCLUSION

An algorithm of the roundoff error of the LMS adaptive algorithm in the steady state has been presented. It was found that the (accumulated) quantization noise of the filter's coefficients results in an output quantization error whose mean squared value is, approximately, inversely proportional to the adaptation step size μ. Taking μ to be very small, in order to reduce the excess mean square error, may result in a considerable quantization error. The excess mean square error is larger than the one due to quantization, as long as μ is given values larger than the one which makes the algorithm stop before convergence is obtained. The quantization error can be combated if more bits are used for the filter's coefficients than for the data.

APPENDIX A
DERIVATION OF (7)

The total output error is given by (6), where the two terms on the right-hand side are uncorrelated. As shown in [10], the mean squared value of the first term is $\xi_{\min} + \mu \xi_{\min} \operatorname{tr} \boldsymbol{R}/2 - \mu \operatorname{tr} \boldsymbol{R}$, where $\xi_{\min} = E(d_n^2) - \boldsymbol{p}^T \boldsymbol{R}^{-1} \boldsymbol{p}$ is the mean square error of the optimal (Wiener) filter.

To compute the mean squared value of the second term of (6), we note that the sequences $\boldsymbol{\alpha}_n$ and η_n have zero mean and are white and independent of the signals. Since the sequence $\boldsymbol{x}_n$ is assumed to be independent in time, this error term has zero mean for all n, which implies that the presence of quantization errors does not affect the learning characteristics of the algorithm. The mean squared value of the arithmetic error is equal to

$$\xi_{ar} = \frac{1}{a^2} \left[E(\boldsymbol{\rho}_n^T \boldsymbol{x}_n)^2 + E(\boldsymbol{w}_n^T \boldsymbol{\alpha}_n)^2 + E(\eta_n^2) \right]. \tag{A1}$$

The term $E(\eta_n^2)$ is due to the quantization performed in the computation of the inner product $\boldsymbol{w}_n'^T \boldsymbol{x}_n'$ and is always present, whether the filter is adaptive or not. Its value is equal to $c\sigma_d^2$ as referred in Section II. Hereafter, we assume that the step size μ has a value well below its upper bound $2/\lambda_{\max}$ dictated by the convergence in the mean requirements. In the steady state, using some results in [4], we get

$$E(\boldsymbol{w}_n^T \boldsymbol{\alpha}_n)^2 = E(\boldsymbol{w}_n^T \boldsymbol{w}_n)\, \sigma_d^2 = (|\boldsymbol{w}^*|^2 + \tfrac{1}{2} \mu \xi_{\min} N)\, \sigma_d^2. \tag{A2}$$

This term is the contribution of the quantization noise of the input sequences to the total mean square error at the output.

To derive an expression for $E(\boldsymbol{\rho}_n^T \boldsymbol{x}_n)^2$ in the steady state, we follow the approach of [10]. We begin with

$$E(\boldsymbol{\rho}_n^T \boldsymbol{x}_n)^2 = \operatorname{tr}\, [E(\boldsymbol{\rho}_n \boldsymbol{\rho}_n^T) \boldsymbol{R}]. \tag{A3}$$

Writing e_n for $y_n - \boldsymbol{w}_n^T \boldsymbol{x}_n$ and ζ_n for $\boldsymbol{\beta}_n - \boldsymbol{\eta}_n$ we get from (5),

$$e_n' = e_n - \rho_n^T x_n - w_n^T \alpha_n + \zeta_n \tag{A4}$$

where ζ_n has variance $(1+c)\sigma_d^2$. The digital counterpart of the coefficient update, (1) can now be written as

$$\begin{aligned} w_{n+1}' &= w_n' + fx(\mu x_n' e_n') = w_n' + \mu x_n' e_n' + \sigma_n \\ &= w_n' + \mu x_n e_n - \mu x_n x_n^T \rho_n - \mu x_n w_n^T \alpha_n \\ &\quad + \mu x_n \zeta_n + \mu \alpha_n e_n + \sigma_n. \end{aligned} \tag{A5}$$

By using (3) and (A5), we get

$$\rho_{n+1} = F_n \rho_n + b_n \qquad n = 0, 1, 2, \cdots \tag{A6}$$

where

$$F_n = I - \mu x_n x_n^T \tag{A7}$$

and

$$b_n = -\mu x_n w_n^T \alpha_n + \mu x_n \zeta_n + \mu \alpha_n e_n + \sigma_n. \tag{A8}$$

For simplicity, we assume that the algorithm starts with $w_0 = 0$ and, of course, $\rho_0 = 0$. The vector noise σ_n depends primarily on the way the term $\mu x_n' e_n'$ is computed. Often μ is a power of 2, which implies that the elements of the product $x_n' e_n'$ are shifted to the right and quantized. In this case, σ_n is a white quantization noise vector whose components have zero mean and variance σ_c^2. The same happens with μ has only a few significant bits, the multiplication $\mu e_n'$ is performed without quantization and then the product $x_n'(\mu e_n')$ is computed and quantized. If the product $\mu e_n'$ is first quantized and then multiplied by X_n', an extra error term $x_n' \epsilon_n$ has to be added to σ_n. The zero mean quantization error ϵ_n has a variance which depends on the number of bits to which $\mu e_n'$ is quantized. In the following, we shall be concerned only with σ_n.

By denoting $E(\rho_n \rho_n^T)$ by P_n and $E(b_n b_n^T)$ by Q_n, we get

$$P_{n+1} = E(F_n P_n F_n) + Q_n \tag{A9}$$

with the initial condition $P_0 = 0$.

From (A8), we have

$$\begin{aligned} Q_n &= \mu^2 E(w_n^T w_n)\, \sigma_d^2 R + \mu^2 (1+c)\, \sigma_d^2 R \\ &\quad + \mu^2 E(e_n^2)\, \sigma_d^2 I + \sigma_c^2 I. \end{aligned} \tag{A10}$$

In the steady state the following approximation is valid

$$Q_n = \mu^2 |w^*|^2 \sigma_d^2 R + \mu^2 (1+c) \sigma_d^2 R + \mu^2 \xi_{\min} \sigma_d^2 I + \sigma_c^2 I. \tag{A11}$$

For convenience, we shall drop the time subscripts. The ijth element of the matrix $E(FPF)$ is given by

$$\begin{aligned} E(FPF)_{ij} &= E \sum_k \sum_l f_{ik} p_{kl} f_{lj} \\ &= E \sum_k \sum_l (\delta_{ik} - \mu x_i x_k)\, p_{kl} (\delta_{lj} - \mu x_l x_j) \\ &= p_{ij} - \mu (RP)_{ij} - \mu (PR)_{ij} \\ &\quad + \mu^2 \sum_k \sum_l p_{kl} E(x_i x_k x_l x_j) \end{aligned} \tag{A12}$$

where δ_{ij} denotes the Kronecker delta. To make things manageable, we assume that the elements of the vectors x_n are jointly Gaussian. Thus,

$$E(x_i x_k x_l x_j) = R_{ij} R_{kl} + R_{ik} R_{lj} + R_{il} R_{kj} \tag{A13}$$

and

$$\begin{aligned} E(FPF)_{ij} &= (P)_{ij} - \mu (RP)_{ij} - \mu (PR)_{ij} + \mu^2 (R)_{ij} \operatorname{tr}(RP) \\ &\quad + 2\mu^2 (RPR)_{ij}. \end{aligned} \tag{A14}$$

The term RPR is potentially smaller than $R \operatorname{tr}(RP)$ and is neglected. Now, from (A9) we obtain the recursion

$$P_{n+1} = P_n - \mu(RP_n + P_n R) + \mu^2 R \operatorname{tr}(RP_n) + Q_n. \tag{A15}$$

It can be shown [10] that P_n is still a nonnegative definite matrix, although we have neglected the term $2\mu^2 RP_nR$. If, instead of being jointly Gaussian, the elements of the vectors x_n are the received baseband signal of a data transmission system where the transmitted symbols are independent binary random variables, then by keeping only the most significant terms we get the same equation (A15). By following the procedure described in [10], we find that the condition for P_n to converge is $0 < \mu < 2/\operatorname{tr} R$, which is the condition for convergence of the output mean square error of the infinite precision algorithm. Actually, a slightly improved upper bound for μ is $2/(\operatorname{tr} R + 2\lambda_{\min})$, where $\lambda_{\min}$ is the minimum eigenvalue of the autocorrelation matrix R [11].[1]

In the steady state, (A15) becomes

$$-\mu(RP_n + P_n R) + \mu^2 R \operatorname{tr}(RP_n) + Q_n = 0 \tag{A16}$$

which implies

$$\operatorname{tr}(RP_n) = \frac{\operatorname{tr} Q_n}{2\mu - \mu^2 \operatorname{tr} R}. \tag{A17}$$

From (A3), (A11), and (A17) we get

$$E(\rho_n^T x_n)^2 = \frac{N\sigma_c^2 + \mu^2 [(1 + c + |w^1|^2) \operatorname{tr} R + \xi_{\min} N]\, \sigma_d^2}{2\mu - \mu^2 \operatorname{tr} R}. \tag{A18}$$

Equations (A1), (A2), and (A18) lead to (7).

Appendix B
Derivation of (12)

The mean squared output error due to the input quantization noise, when floating point arithmetic is used, is given by the second term of (11)

$$\begin{aligned} E(w_n^T A_n x_n)^2 &= \sigma_d^2 E[(w_n^1 x_n^1)^2 + (w_n^2 x_n^2)^2 + \cdots + (w_n^N x_n^N)^2] \\ &= \sigma_d^2 \sum_{k=1}^{N} R_{kk} E(w_n^k)^2. \end{aligned} \tag{B1}$$

We know that $w_n = w^* + v_n$, where the deviation vector v_n, in the steady state and for sufficiently small μ, has covariance [4]

$$\operatorname{cov}(v_n) = \tfrac{1}{2} \mu \xi_{\min} I.$$

[1] Both of these conditions are approximate. An exact necessary and sufficient condition for convergence of the LMS algorithm under the independence assumption is given in [12].

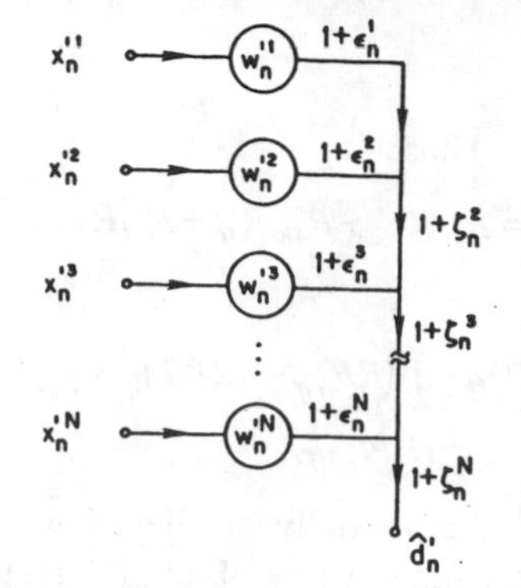

Fig. 3. A flow graph for the inner product computation $w_n'^T x_n'$ along with the associated quantization noises when floating point arithmetic is used.

Hence

$$E(w_n^T A_n x_n)^2 = \sigma_d^2 \left[\sum_{k=1}^{N} R_{kk} (w^{*k})^2 + \tfrac{1}{2} \mu \xi_{\min} \operatorname{tr} R \right]. \tag{B2}$$

where w^{*k} is the kth element of w^*.

To compute $E(\eta_n^2)$, we refer to the flow graph of Fig. 3 which shows the straightforward implementation of $w_n'^T x_n'$, along with the associated multiplication and addition quantization noise sequences. By neglecting all second order error terms, we find that

$$\eta_n = x_n^1 w_n^1 \epsilon_n^1 + \cdots + x_n^N w_n^N \epsilon_n^N + (x_n^1 w_n^1 + x_n^2 w_n^2) \zeta_n^2 + \cdots + (x_n^1 w_n^1 + \cdots + x_n^N w_n^N) \zeta_n^N. \tag{B3}$$

The error sequences ϵ_n^i and ζ_n^i are independent white sequences and so is η_n. The mean squared value of η_n depends on the statistics of x_n and w_n. Using the approximate expression (B3) for the output error due to the quantization in the filtering operation, we find its mean square value to be

$$\begin{aligned} E(\eta_n^2) &= \sigma_d^2 \sum_{k=1}^{N} R_{kk} E(w_n^k)^2 + \sigma_d^2 [E(x_n^1 w_n^1 + x_n^2 w_n^2)^2 \\ &\quad + \cdots + E(x_n^1 w_n^1 + \cdots + x_n^N w_n^N)^2] \\ &= \sigma_d^2 \left[\sum_{k=1}^{N} R_{kk} (w^{*k})^2 + C \right. \\ &\quad \left. + \tfrac{1}{2} \mu \xi_{\min} \left\{ \operatorname{tr} R + \sum_{k=1}^{N} (N-k+1) R_{kk} - R_{11} \right\} \right] \end{aligned} \tag{B4}$$

where the symbol C is used to denote the term $E(x_n^1 w^{*1} + x_n^2 w^{*2})^2 + \cdots + E(x_n^1 w^{*1} + \cdots + x_n^N w^{*N})^2$, which is a function of the autocorrelation matrix R and the optimal (Wiener) vector w^*.

Finally, for the error $E(\rho_n^T x_n)^2$ due to the quantization noise ρ_n of the weight vector w_n, we begin with $e_n' = fl(d_n' - \hat{d}_n')$ which, by using (9), becomes

$$\begin{aligned} e_n' &= fl(d_n' - \hat{d}_n') \\ &= (1 + \vartheta_n) [(1 + \beta_n) d_n - w_n^T x_n - \rho_n^T x_n \\ &\quad - w_n^T A_n x_n - \eta_n] \\ &= e_n + \vartheta_n e_n + \beta_n d_n - \rho_n^T x_n - w_n^T A_n x_n - \eta_n. \end{aligned} \tag{B5}$$

Also,

$$fl(\mu e_n') = (1 + \kappa_n) \mu e_n' \tag{B6}$$

and

$$\begin{aligned} fl[x_n' fl(\mu e_n')] &= (I + \Sigma_n) x_n' fl(\mu e_n') \\ &= \mu [x_n e_n + (\Sigma_n + A_n) x_n e_n \\ &\quad + x_n (\kappa_n e_n + \vartheta_n e_n + \beta_n d_n - \rho_n^T x_n \\ &\quad - w_n^T A_n x_n - \eta_n)]. \end{aligned} \tag{B7}$$

Very often the multiplication $\mu e_n'$ is free of quantization error, hence $\kappa_n = 0$. The weight vector update equation can be written as

$$\begin{aligned} w_{n+1}' &= fl[w_n' + fl\{x_n' fl(\mu e_n')\}] \\ &= (I + \Delta_n) [w_n' + fl\{x_n' fl(\mu e_n')\}] \end{aligned} \tag{B8}$$

$$\begin{aligned} &= w_n' + \Delta_n w_{n+1} + \mu [x_n e_n + (\Sigma_n + A_n) x_n e_n \\ &\quad + x_n (\kappa_n e_n + \vartheta_n e_n + \beta_n d_n - \rho_n^T x_n \\ &\quad - w_n^T A_n x_n - \eta_n)] \end{aligned} \tag{B9}$$

which may be put in the form

$$\rho_{n+1} = F_n \rho_n + b_n \qquad n = 0, 1, 2, \cdots \tag{B9}$$

where

$$F_n = I - \mu x_n x_n^T \tag{B10}$$

and

$$b_n = \Delta_n w_{n+1} + \mu [(\Sigma_n + A_n) x_n e_n + x_n (\kappa_n e_n + \vartheta_n e_n + \beta_n d_n - w_n^T A_n x_n - \eta_n)]. \tag{B11}$$

The initial values are $\rho_0 = 0$ and $b_0 = \mu[(\Sigma_0 + A_0) x_0 e_0 + x_0 (\kappa_0 e_0 + \beta_0 d_0)]$, and the algorithm starts with $w_0 = 0$. Equation (B9) is the same as its fixed point counterpart (A6), which implies that the condition for convergence of the mean square arithmetic error is, again, the condition for convergence of the output mean square error of the infinite precision algorithm $0 < \mu < 2/\operatorname{tr} R$.

The elements of the diagonal matrix sequences Δ_n, Σ_n, and A_n and the scalar sequences κ_n, ϑ_n, and β_n are modeled as zero mean white sequences independent of the signals and of each other. Each one of them has variance σ_c^2 or σ_d^2, depending on whether it corresponds to quantization of a coefficient or a data word. The elements of the vector sequence b_n are zero mean white sequences independent of the signals and of each other. Again, the presence of the quantization error does not affect the learning characteristics of the algorithm. An expression for the trace of $Q_n = E(b_n b_n^T)$ in the steady state is needed. From (B11), we get

$$\operatorname{tr} Q_n = (|w^*|^2 + \tfrac{1}{2} \mu \xi_{\min} N) \sigma_c^2 + \text{a term proportional to } \mu^2. \tag{B12}$$

As in Appendix A, we arrive at the approximation

$$E(\rho_n^T x_n)^2 = \frac{\operatorname{tr} Q_n}{2\mu - \mu^2 \operatorname{tr} R} \approx \left(\frac{|w^*|^2}{2\mu} + \frac{1}{4} \xi_{\min} N \right) \sigma_c^2. \tag{B13}$$

Equation (12) follows from (B2), (B4) and (B13).

REFERENCES

[1] B. Widrow, "Adaptive filters," in *Aspects of Network and System Theory*, R. Kalman and N. DeClaris, Eds. New York: Holt, Rinehart, and Winston, 1971, pp. 563–587.

[2] L. J. Griffiths, "A continuously adaptive filter implemented as a lattice structure," in *Proc. 1977 IEEE Int. Conf. Acoust., Speech, Signal Processing*, Hartford, CT, May 1977, pp. 683–686.

[3] D.T.L. Lee *et al.*, "Recursive least squares ladder estimation algorithms," *IEEE Trans. Acoust., Speech, Signal Processing*, vol. ASSP-29, pp. 627–641, June 1981.

[4] B. Widrow *et al.*, "Stationary and nonstationary learning characteristics of the LMS adaptive filter," *Proc. IEEE*, vol. 64, pp. 1151–1162, Aug. 1976.

[5] M. Andrews and R. Fitch, "Finite word length arithmetic computational error effects on the LMS adaptive weights," in *Proc. 1977 IEEE Int. Conf. Acoust., Speech, Signal Processing*, Hartford, CT, May 1977, pp. 628–631.

[6] R. D. Gitlin *et al.*, "On the design of gradient algorithms for digitally implemented adaptive filters," *IEEE Trans. Circuit Theory*, vol. CT-20, pp. 125–136, Mar. 1973.

[7] A. V. Oppenheim and R. W. Schafer, *Digital Signal Processing*. Englewood Cliffs, NJ: Prentice-Hall, 1975.

[8] T. Kaneko and B. Liu, "On local roundoff errors in floating-point arithmetic," *J. Ass. Comp. Mach.*, vol. 20, pp. 391-398, July 1973.

[9] A. B. Spirad and D. L. Snyder, "Quantization errors in floating-point arithmetic," *IEEE Trans. Acoust., Speech, Signal Processing*, vol. ASSP-26, pp. 456–463, Oct. 1978.

[10] J. E. Mazo, "On the independence theory of equalizer convergence," *Bell Syst. Tech. J.*, vol. 58, pp. 963-993, May–June, 1979.

[11] C. Caraiscos, "Implementation issues in digital signal processing," Ph.D. dissertation, Princeton Univ., Jan. 1984.

[12] A. Feuer and E. Weinstein, "Convergence analysis of the LMS filters with uncorrelated Gaussian data," *IEEE Trans. Acoust., Speech, Signal Processing*, submitted for publication.

Error Propagation Properties of Recursive Least-squares Adaptation Algorithms*

STEFAN LJUNG†‡ and LENNART LJUNG†

The propagation of a single numerical error has different stability properties for various recursive least-squares algorithms, some of which may not be suitable for continuous adaptation.

Key Words—System identification; recursive identification; least-squares algorithm; numerical properties.

Abstract—The numerical properties of implementations of the recursive least-squares identification algorithm are of great importance for their continuous use in various adaptive schemes. Here we investigate how an error that is introduced at an arbitrary point in the algorithm propagates. It is shown that conventional LS algorithms, including Bierman's UD-factorization algorithm are exponentially stable with respect to such errors, i.e. the effect of the error decays exponentially. The base of the decay is equal to the forgetting factor. The same is true for fast lattice algorithms. The fast least-squares algorithm, sometimes known as the 'fast Kalman algorithm' is however shown to be unstable with respect to such errors.

1. INTRODUCTION

RECURSIVE least-squares identification algorithms play a crucial role for many problems in adaptive control, adaptive signal processing and for general model building and monitoring problems. Therefore questions of how to implement these algorithms for best numerical properties have met considerable interest, and many different mechanizations have been suggested and used.

Behind the concept of 'numerical properties' hides several different aspects. We may include:

(a) The number of operations involved in making one iteration (or the required computing time to complete one time recursion).

(b) The required amount of memory locations to store the data and the program.

(c) The structure of the information flow in the algorithm, so as to allow for parallel processing/systolic arrays and/or VLSI chip-implementation.

(d) Programming investment required.

(e) Resilience to bad or abnormal data (such as poorly exciting input).

(f) Error propagation properties.

(g) Effect and amplification of round-off noise.

Which aspects are the most important ones in a given case depends on the application. A lot of attention has recently been given to (a) and (b). So-called fast algorithms have been developed that require an order of magnitude less computations and memory space. See, among many references, for example Lee *et al.* (1981), Ljung *et al.* (1978) and Griffiths (1977). Aspect (c) represents a challenging new problem and philosophy for algorithm design. For algorithms that are going to be in near continuous use, such as algorithms that are part of adaptive control or signal processing schemes, the aspects (e)–(g) are most important. Surprisingly little work has been done on these problems, considering their importance. Some references are Bierman (1977), Samson and Reddy (1983), Graupe *et al.* (1980), Ljung (1983), Cioffi and Kailath (1984) and Lin (1984).

The purpose of the present contribution is to study problem (f) more closely for the most common recursive least-squares implementations. In Section 2 we give a brief background about the different algorithms. A formalization of the error propagation and the round-off noise problems is given in Section 3. Then the different algorithm families are analyzed in Sections 4–6.

2. RECURSIVE LEAST-SQUARES ALGORITHMS

2.1. *The least-squares problem*

We here only give the formulation for estimation of the parameters of a scalar AR-process. This restriction is, however, totally immaterial for our

* Received 19 July 1983; revised 2 April 1984; revised 15 October 1984. The original version of this paper was presented at the 9th IFAC World Congress on A Bridge Between Control Science and Technology which was held in Budapest, Hungary during July 1984. The Published Proceedings of this IFAC Meeting may be ordered from Pergamon Press Limited, Headington Hill Hall, Oxford OX3 0BW, England. This paper was recommended for publication in revised form by associate editor G. C. Goodwin under the direction of editors B. D. O. Anderson and P. C. Parks.

† Division of Automatic Control, Department of Electrical Engineering, Linköping University, S-581 83 Linköping, Sweden.

‡ Currently with ASEA, Västerås, Sweden.

study and more general cases are entirely analogous. See, e.g. Ljung and Söderström (1983) for application of various least-squares schemes to more general models. This reference also gives all details of the algorithms to be quoted here.

Given an AR model

$$y(t) + a_1 y(t-1) + \cdots + a_n y(t-n) = e(t) \quad (1)$$

of a time series $\{y(t)\}$, we would like to estimate a_i from observations of $\{y(t)\}$. Introduce

$$\varphi(t) = (-y(t-1) \ldots -y(t-n))^{\mathrm{T}} \quad (2a)$$

$$\theta = (a_1 \ldots a_n)^{\mathrm{T}}. \quad (2b)$$

Then (1) can be written

$$y(t) = \theta^{\mathrm{T}}\varphi(t) + e(t) \quad (3)$$

and the LS estimate of θ is given by

$$\theta(N) = \arg\min \sum_{t=1}^{N} (y(t) - \theta^{\mathrm{T}}\varphi(t))^2 \lambda^{N-t} \quad (4)$$

based on data $\{y(t)\}_1^N$. Here λ is the so-called forgetting factor, that discounts older measurements, so as to make $\theta(N)$ representative for the current properties of the system. Simple calculations show that

$$\theta(N) = R^{-1}(N) f(N) \quad (5a)$$

$$R(N) = \sum_{t=1}^{N} \varphi(t)\varphi^{\mathrm{T}}(t)\lambda^{N-t} \quad (5b)$$

$$f(N) = \sum_{t=1}^{N} \varphi(t) y(t) \lambda^{N-t}. \quad (5c)$$

We shall generally assume that the forgetting factor is less than unity. Moreover, we shall restrict our attention to sequences $\{\varphi(t)\}$ such that with $\delta > 0$ we have

$$\delta I \leq R(t) \leq CI$$

for sufficiently large t. This assumption is a condition on persistence of excitation of the regressor sequence. One might add that several numerical problems may occur when $R(t)$ approaches a singular matrix (cf. problem (e) of the Introduction). These are usually dealt with by *regularization*, which means that the lower bound $\delta I \leq R(t)$ is enforced by some modification of the algorithm. In this treatment we do not consider such measures explicitly.

2.2. *The basic recursive least-squares (RLS) algorithm*

By simple manipulation (5) can be rearranged to

$$\theta(t) = \theta(t-1) + R^{-1}(t)\varphi(t)(y(t) - \theta^{\mathrm{T}}(t-1)\varphi(t)) \quad (6a)$$

$$R(t) = \lambda R(t-1) + \varphi(t)\varphi^{\mathrm{T}}(t). \quad (6b)$$

We call this algorithm the basic RLS algorithm.

2.3. *The conventional LS (CLS) algorithm*

By introducing

$$P(t) = R^{-1}(t)$$

and applying the matrix inversion lemma to (6b) we can rewrite (6a) as

$$\theta(t) = \theta(t-1) + L(t)(y(t) - \theta^{\mathrm{T}}(t-1)\varphi(t)) \quad (7a)$$

$$L(t) = \frac{P(t-1)\varphi(t)}{\lambda + \varphi^{\mathrm{T}}(t)P(t-1)\varphi(t)} \quad (7b)$$

$$P(t) = \frac{1}{\lambda}\left[P(t-1) - \frac{P(t-1)\varphi(t)\varphi^{\mathrm{T}}(t)P(t-1)}{\lambda + \varphi^{\mathrm{T}}(t)P(t-1)\varphi(t)}\right] \quad (7c)$$

This is perhaps the most common form of the RLS algorithm, and we shall call it 'conventional least-squares', CLS.

2.4. *UD-factorization algorithms (UDLS)*

A numerically favourable version of CLS is obtained by factoring the matrix $P(t)$ into an upper triangular matrix $U(t)$ with '1's along the diagonal and a diagonal matrix $D(t)$

$$P(t) = U(t)D(t)U^{\mathrm{T}}(t). \quad (8)$$

Then the factors U and D are updated instead of P in (7c). This gives a somewhat lengthy set of expressions which we do not give here (see Bierman, 1977). Conceptually we have

$$(U(t), D(t)) = f(U(t-1), D(t-1), \varphi(t)). \quad (9)$$

The UD-factorization algorithm is an example of a square-root-type algorithm, in the sense that square-root factors of P are used instead of P itself. That will make the matrices better conditioned. Another square-root algorithm has been given by Potter (1963).

2.5. *The fast least-squares (FLS) algorithm*

In the algorithms discussed so far, the sequence of vectors $\{\varphi(t)\}$ could be arbitrary. The special structure (2a) has not been utilized. Taking such a shift structure into account, a faster algorithm can be developed. It is given by

$$\theta(t) = \theta(t-1) + L(t)(y(t) - \theta^{T}(t-1)\varphi(t))$$
$$e_0(t) = -y(t) + A^{T}(t-1)\varphi(t)$$
$$A(t) = A(t-1) - L(t)e_0(t)$$
$$e(t) = -y(t) + A^{T}(t)\varphi(t)$$
$$R_e(t) = \lambda R_e(t-1) + e(t)e_0(t)$$
$$R_e(t)w(t) = e(t) \tag{10}$$
$$\begin{pmatrix} m(t) \\ \mu(t) \end{pmatrix} = \begin{pmatrix} w(t) \\ L(t) + A(t)w(t) \end{pmatrix} \quad (\dim m = n, \dim \mu = 1)$$
$$\eta_0(t) = -y(t-n) + D^{T}(t-1)\varphi(t+1)$$
$$D(t) = [D(t-1) - m(t)\eta_0(t)][1 - \mu(t)\eta_0(t)]^{-1}$$
$$L(t+1) = m(t) - D(t)\mu(t).$$

See Ljung *et al.* (1978) for details.

Variants of this algorithm have been given by Carayannis *et al.* (1983), Cioffi and Kailath (1984) and Proakis (1983).

2.6. *Fast ladder and lattice algorithms* (*FL*)

Ladder (or lattice) implementations of linear filters have long been used in signal theory for less coefficient sensitivity in high order filters. The recursive identification of the coefficients (known as 'the reflection coefficients') of these filters was studied by several authors. See Lee *et al.* (1981) and Griffiths (1977). A ladder representation will not give us the AR-parameters θ, but the equivalent reflection coefficients. As long as we are primarily interested in the predictions

$$y^{p}(t) = \theta^{T}(t-1)\varphi(t) \tag{11}$$

this internal parametrization is of course immaterial.

The ladder algorithm has the following structure:

For $k = 0$ to $n-1$

$$K_k(t-1) = F_k(t-1)/R_k^{r}(t-2)$$
$$K_k^{*}(t-1) = F_k(t-1)/R_k^{e}(t-1)$$
$$y_0^{p}(t) = 0$$
$$y_{k+1}^{p}(t) = y_k^{p}(t) + K_k(t-1)r_k(t-1)$$
$$y^{p}(t) = y_n^{p}(t)$$
$$R_k^{r}(t-1) = \lambda R_k^{r}(t-2) + (1 - \beta_k(t))r_k(t-1)^2$$
$$\beta_0(t) = 0 \tag{12}$$
$$\beta_{k+1}(t) = \beta_k(t) + (1 - \beta_k(t))^2 r_k^2(t-1)/R_k^{r}(t-1)$$
$$r_0(t) = y(t)$$
$$r_{k+1}(t) = r_k(t-1) - K_k^{*}(t-1)e_k(t)$$
$$e_k(t) = y(t) - y_k^{p}(t)$$
$$R_k^{e}(t) = \lambda R_k^{e}(t-1) + (1 - \beta_k(t))e_k(t)^2$$
$$F_k(t) = \lambda F_k(t-1) + (1 - \beta_k(t))r_k(t-1)e_k(t).$$

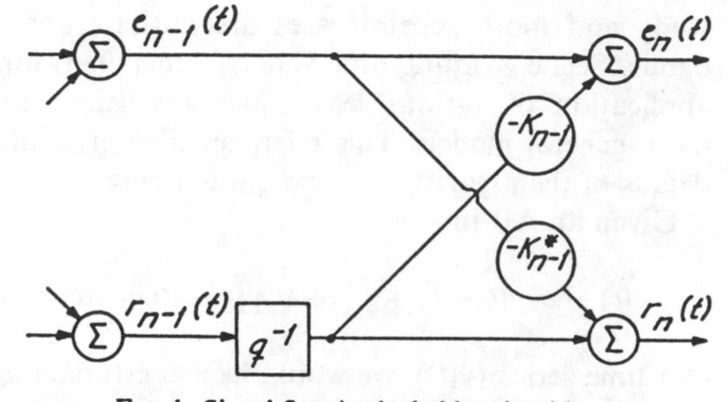

FIG. 1. Signal flow in the ladder algorithm.

The algorithm can be depicted as in Fig. 1. Notice that the signal flow is such that higher steps ('rungs') do not at all affect the previous ones.

There are several variants of this ladder/lattice form. See e.g. Cioffi and Kailath (1984) for a comparative study.

3. ERROR PROPAGATION AND NUMERICAL PROPERTIES

All the algorithms we discussed, can be viewed as discrete-time, non-linear dynamical systems. The input will then be

$$z(t) = \begin{pmatrix} y(t) \\ \varphi(t) \end{pmatrix} \tag{13}$$

and the output can be taken as

$$\eta(t) = \theta(t) \tag{14}$$

or

$$\eta(t) = L(t) \tag{15}$$

or

$$\eta(t) = y(t) - \theta^{T}(t-1)\varphi(t). \tag{16}$$

Conceptually, the algorithms can be written as

$$\xi(t) = f(\xi(t-1), z(t)) \tag{17a}$$
$$\eta(t) = h(\xi(t), z(t)). \tag{17b}$$

Here $\xi(t)$ is the state vector. For the basic RLS algorithm (6) we take

$$\xi_{RLS}(t) = \begin{pmatrix} \theta(t) \\ \mathrm{col}\, R(t) \end{pmatrix}$$
$$\xi_{RLS} \in \Omega_1 \subset R^{l} \quad l = n + (n+1)n/2 \tag{18}$$

where $\Omega_1 = \{\xi \mid \delta I \le R \le CI\}$.

('col A' means the vector that is obtained by stacking the columns of A on top of each other.

Since R is symmetric, we include in ξ only its 'upper part'). For CLS (7) we take

$$\xi_{CLS}(t) = \begin{pmatrix} \theta(t) \\ \text{col}\, P(t) \end{pmatrix}$$

$$\xi_{CLS} \in \Omega_2 \subset R^l \qquad (19)$$

$$\Omega_2 = \{\xi \mid (1/C)I \le P \le (1/\delta)I\}.$$

For the UDLS algorithm (8) we have

$$\xi_{UD}(t) = \begin{pmatrix} \theta(t) \\ \text{col}\, U(t) \\ \text{col}\, D(t) \end{pmatrix}$$

$$\xi_{UD} \in \Omega_3 \subset R^l \qquad (20)$$

$$\Omega_3 = \{\xi \mid (1/C)I \le UDU^T \le (1/\delta)I\}$$

while for FLS we would use

$$\xi_{FLS}(t) = \begin{pmatrix} A(t) \\ L(t) \\ D(t) \\ R_e(t) \end{pmatrix}$$

$$\xi_{FLS} \in \Omega_4 \subset R^m; \quad m = 3n + 1 \qquad (21)$$

$$\Omega_4 = \{\xi \mid \delta \le R_e \le C\}.$$

Finally, for the FL algorithm we have

$$\xi_{FL}^T(t) = (R_0^e(t) \ldots R_{n-1}^e(t) \quad R_0^r(t-1) \ldots R_{n-1}^r(t-1) \quad F_0(t-1) \ldots F_{n-1}(t-1) \quad r_0(t-1) \ldots r_{n-1}(t-1))$$

$$\xi_{FL} \in \Omega_5 \subset R^p \qquad p = 4n \qquad (22)$$

$$\Omega_5 = \{\xi \mid \delta \le R_i^e \le C; \delta \le R_i^r \le C\}.$$

The different algorithms that we have listed thus correspond to different choices of state vector and associated functions f and h. Since the algorithms are all algebraically equivalent when $\varphi(t)$ is subject to (2a) it follows that the input–output relationship given by (17) is the same for each choice of representation. The latter statement requires some qualification. The algorithms (6)–(9) are all valid for arbitrary sequences of vectors $\{\varphi(t)\}$. Moreover, the corresponding state space representation (17) with state vectors (18)–(20) all relate to each other by a change of basis in the state space. We thus have, e.g.

$$\text{for} \quad \xi_{UD}(t) \in \Omega_3 \quad \text{and} \quad \xi_{RLS}(t) \in \Omega_1 \qquad (23)$$

$$\xi_{RLS}(t) = g\{\xi_{UD}(t)\}; \quad \xi_{UD}(t) = g^{-1}\{\xi_{RLS}(t)\}$$

and similarly for CLS.

The mutual invertability of these mappings follows since θ is a common part to all state vectors and since the association

$$P(t) = R^{-1}(t) = U(t)D(t)U^T(t) \qquad (24)$$

is unique (with the given specifications of U and D).

We have no such change-of-basis relationships for the FL and FLS algorithms, since these algorithms utilize the particular shift structure in (2a) and are algebraically equivalent to RLS, CLS and UDLS only for such inputs.

Now, while the system (17) represents the ideal algorithms, any numerical implementation will generate inaccuracies due to round-off errors and representation errors in the computer. This means that instead of the exact values $\xi(t)$ and $\eta(t)$ we will obtain values $\hat{\xi}(t)$ and $\hat{\eta}(t)$. Let $\delta\xi(t)$ denote the error generated in one step of the algorithm:

$$\delta\xi(t) = \hat{\xi}(t) - f(\hat{\xi}(t-1), \hat{z}(t)).$$

Here $\hat{z}(t) = z(t) + \delta z(t)$ is the value by which the time variable $z(t)$ is represented in the computer. The quantity $\delta\hat{\xi}(t)$ arises from round-off errors in the calculations corresponding to the function f. The size of $\delta\xi(t)$ (or the relative size $|\delta\xi(t)|/|\hat{\xi}(t)|$) depends on how these calculations are organized.

The actual, numerically implemented, algorithm will thus be described by the system

$$\hat{\xi}(t) = f(\hat{\xi}(t-1), z(t) + \delta z(t)) + \delta\xi(t) \quad (25a)$$

$$\hat{\eta}(t) = h(\hat{\xi}(t), z(t) + \delta z(t)) + \delta\eta(t), \qquad (25b)$$

subject to disturbances δz, $\delta\xi$ and $\delta\eta$.

The ultimate goal is of course to study the discrepancy

$$\Delta\eta(t) = \hat{\eta}(t) - \eta(t), \qquad (26)$$

analyze its properties and give advice for the choice of state representation and organization of the calculations so that $|\Delta\eta|$ becomes as small as possible for a given number representation. That problem contains the following subproblems:

(i) Study the relative size of the momentary error $\delta\xi(t)$. This depends on the choice of state vector and on the organization of the calculations corresponding to the function f.

(ii) Study the propagation of the errors $\delta z(t)$ and $\delta\xi(t)$ to future time instants.

(iii) Study the cumulative effects of the errors δz and $\delta\xi$, i.e. how $\Delta\eta$ builds up from the momentary errors.

The last problem (iii), depends both on the error propagation properties (ii) and on the relationships between the terms in the momentary error sequence $\{\delta\xi\}$. It has been studied for recursive least-squares algorithms with a mixture of analysis, heuristics and simulations in, e.g. Samson and Reddy (1983), Graupe *et al.* (1980) and Ljung (1983).

In the present paper we shall focus on the problem (ii). We then consider the propagation of a single error in the state vector occurring at time t_0. If this error is denoted by $\delta\xi(t_0)$ we thus have the system

$$\hat{\xi}(t) = f(\hat{\xi}(t-1), z(t)) \qquad \hat{\xi}(t_0) = \xi(t_0) + \delta\xi(t_0) \tag{27a}$$

$$\hat{\eta}(t) = h(\hat{\xi}(t), z(t)). \tag{27b}$$

To study the effect of $\delta\xi(t_0)$ on $\hat{\eta}(t)$ it is sufficient to consider the effect on $\hat{\xi}(t)$, since h in (27b) is continuous in $\hat{\xi}$.

We are thus faced with a classical Lyapunov stability problem for the non-linear difference equation (27a): How does the state react on perturbations in the initial conditions?

Briefly, the solution $\xi(t)$ (corresponding to $\delta\xi(t_0) = 0$) is said to be *stable* if for all $\varepsilon \geq 0$ there exists a δ, such that $|\delta\xi(t_0)| < \delta$ implies that $|\hat{\xi}(t) - \xi(t)| < \varepsilon$ for all $t \leq t_0$. The solution is said to be *asymptotically stable* if it is stable and there exists a $\delta > 0$ such that $|\delta\xi(t_0)| < \delta$ implies that $|\hat{\xi}(t) - \xi(t)| \to 0$ as $t \to \infty$. We say that the solution is *exponentially stable*, if it is asymptotically stable and there exists C and $0 < \lambda < 1$, such that

$$|\hat{\xi}(t) - \xi(t)| \leq C\lambda^{t-t_0}, \quad t \geq t_0 \quad \text{if} \quad |\delta\xi(t_0)| < \delta. \tag{28}$$

Notice that these properties relate to the system (27) as such and do not depend on the numerical representation. In particular, they are invariant under a change of basis, such as (23). This can be seen as follows. Suppose that $\hat{\xi}$ and ξ are the perturbed and unperturbed states is one state representation while $\hat{\zeta}$ and ζ correspond to another basis in the state space, such that $\zeta = g(\xi)$ with g continuously differentiable with bounded derivative. Then

$$\hat{\zeta}(t) - \zeta(t) = g(\hat{\xi}(t)) - g(\xi(t)) = g'(\rho(t))[\hat{\xi}(t) - \xi(t)].$$

Since g' is bounded, ζ will obey the same stability conditions as ξ. Notice, though, that this is valid only as long as the disturbance does not force the solution outside the region where the change of basis is well-defined. This means, in view of (23), that the algorithms RLS, CLS and UDLS all will have the same error propagation properties, as long as the errors keep the states within Ω_i; $i = 1, 2, 3$.

We shall in this paper encounter systems of the following kind:

$$\xi(t) = F_1(\rho(t))\xi(t-1) + G_1(\rho(t))z(t) \tag{29a}$$

$$\rho(t) = F_2(t)\rho(t-1) + G_2(t)z(t). \tag{29b}$$

Then the following result is useful.

Lemma. Consider the system (29) where $F_1(\rho)$ and $G_1(\rho)$ are continuously differentiable with respect to ρ, with bounded derivatives in a neighbourhood of a nominal trajectory $\rho(t)$. Suppose that (29b) is exponentially stable and that (29a) is exponentially stable for $\rho(t)$ being the nominal solution to (29b). Then the solution $(\rho(t), \xi(t))$ to (29) is exponentially stable with respect to perturbations in the initial state $(\rho(t_0), \xi(t_0))$.

Proof: We have, with our standard notation

$$\xi(t+1) = F_1(\rho(t) + \Delta\rho(t))\xi(t) + G(\rho(t) + \Delta\rho(t))z(t)$$

and

$$\Delta\xi(t+1) = [F_1(\rho(t)) + F_1'(\tilde{\rho}(t))\Delta\rho(t)] \times \Delta\xi(t) + G'(\tilde{\rho}(t))\Delta\rho(t)z(t)$$

where $\tilde{\rho}(t)$ is a value in a $\Delta\rho$-neighbourhood of $\rho(t)$. Since (29b) is exponentially stable we have that $\Delta\rho(t) \to 0$. The result now follows from Theorem 32.1 in Brockett (1970). □

Before studying the error propagation properties for the different algorithms of Section 2, let us comment on what such results imply. First, if a system is found to be unstable, so that the effect of a single disturbance tends to infinity, then the corresponding algorithm cannot be used in continuous operation. A smaller error will always arise sooner or later, and eventually it will be amplified to large values. There is nothing in terms of numerical tricks and sophisticated calculations that can be done to avoid this. All numerical 'rescues' must thus involve actual changes in the underlying algorithm. Second, it is well known that (uniform) exponential stability of a system will imply bounded-input–bounded-output stability. This should thus mean that if the error propagation is found to be exponentially stable, then also in problem (iii) above bounded output errors would be guaranteed, at least as long as the trajectories stay within the areas Ω_i in (18)–(22).

4. ERROR PROPAGATION FOR RLS, CLS AND UDLS

We shall now study the algorithms (6), (7) and (9). The parameter update step (7a):

$$\theta(t) = \theta(t-1) + L(t)(y(t) - \varphi^{\mathrm{T}}(t)\theta(t-1)) \tag{30}$$

is common to all the algorithms. They differ in the way the gain $L(t)$ is computed.

$$\begin{aligned} L(t) &= R^{-1}(t)\varphi(t) = P(t)\varphi(t) \\ &= \frac{P(t-1)\varphi(t)}{\lambda + \varphi^{\mathrm{T}}(t)P(t-1)\varphi(t)} \\ &= U(t)D(t)U^{\mathrm{T}}(t)\varphi(t). \end{aligned} \tag{31}$$

Let us therefore first consider the part (30) for which we consider $\theta(t)$ as the state variable and output and $L(t)$, $y(t)$ and $\varphi(t)$ as the inputs. We then have

$$\theta(t) = (I - L(t)\varphi^{\mathrm{T}}(t))\theta(t-1) + L(t)y(t). \tag{32}$$

If an error $\delta\theta(t_0)$ is introduced at time t_0 we find from (32) that

$$\Delta\theta(t) = \prod_{k=t_0}^{t} (I - L(k)\varphi^{\mathrm{T}}(k))\delta\theta(t_0). \tag{33}$$

The error propagation properties are thus dependent on the matrix

$$\phi(t, t_0) = \prod_{k=t_0}^{t} (I - L(k)\varphi^{\mathrm{T}}(k)). \tag{34}$$

Now, suppose that the algorithm (32) is fed with the correct $\{L(t)\}$-sequence:

$$L(t) = R^{-1}(t)\varphi(t)$$

with $R(t)$ given by (5b). It is then easy to verify that

$$\phi(t, t_0) = \lambda^{t-t_0}R^{-1}(t)R(t_0). \tag{35}$$

As long as $R^{-1}(t)$ remains uniformly bounded and $\lambda < 1$ we thus conclude that the effect of a single error in (30) decays exponentially. In case $\lambda = 1$ we have $R^{-1}(t)R(t_0) \leq I$. Then we have stability. If, in addition, $R^{-1}(t)$ tends to zero we have asymptotic, but not exponential stability (see also Anderson and Johnson (1982) for a related stability result).

We now turn to the question of the stability properties of the different ways of computing $L(t)$ (RLS, CLS and UDLS). First we note the important fact that these algorithms all correspond to different choices of state vectors. They are thus mutually subject to (23). In view of what was said in Section 3 they will then also have the same error propagation properties. It is consequently sufficient to analyze one of them and we choose the simplest one, (6):

$$\begin{aligned} R(t) &= \lambda R(t-1) + \varphi(t)\varphi^{\mathrm{T}}(t) \\ L(t) &= R^{-1}(t)\varphi(t). \end{aligned} \tag{36}$$

The state vector here is $\operatorname{col} R(t)$, and we see immediately that (36) is exponentially stable. The effect of a single error $\delta R(t_0)$ decays like

$$\Delta R(t) = \lambda^{t-t_0}\delta R(t_0) \tag{37}$$

and

$$\begin{aligned} \Delta L(t) &= [(R(t) + \Delta R(t))^{-1} - R^{-1}(t)]\varphi(t) \\ &\approx R^{-1}(t)\Delta R(t)R^{-1}(t)\varphi(t) \\ &= \lambda^{t-t_0}R^{-1}(t)\delta R(t_0)L(t) \end{aligned} \tag{38}$$

which decays exponentially, provided $R^{-1}(t)$ remains bounded.

Any of the discussed algorithms can now be considered as a concatenation of two steps as depicted in Fig. 2. We thus have an algorithm of the type (29). According to the Lemma it now follows that the error propagation in the resulting full algorithm (RLS, CLS or UDLS) is exponentially stable when $\lambda < 1$ and $\{R^{-1}(t)\}$ is bounded.

5. ERROR PROPAGATION IN THE FAST LATTICE ALGORITHMS

An important feature with the fast lattice algorithm (12) is, as indicated in Fig. 1, that the signal flow in a certain level does not affect previous ones. Consider the kth level ('rung'):

The input is $\{r_{k-1}(t-1)\}$, $\{e_{k-1}(t)\}$, $\{\beta_k(t)\}$, $\{K^*_{k-1}(t)\}$ and $\{K_{k-1}(t)\}$. The state vector is

$$\xi_k(t) = \begin{pmatrix} R^e_k(t) \\ R^r_k(t-1) \\ F_k(t) \end{pmatrix}.$$

The output is $\{r_k(t)\}$, $\{e_k(t)\}$, $\{\beta_{k+1}(t)\}$, $\{K^*_k(t)\}$ and $\{K_k(t)\}$. We have

$$\begin{aligned} r_k(t) &= r_{k-1}(t-1) - K^*_{k-1}(t-1)e_{k-1}(t) \\ e_k(t) &= e_{k-1}(t) - K_{k-1}(t-1)r_{k-1}(t-1). \end{aligned} \tag{39}$$

The sequences $\{r_k(t)\}$ and $\{e_k(t)\}$ are thus formed by direct transformation of the input variables. The state variables are updated as

$$\begin{aligned} \xi_k(t) &= \lambda\xi_k(t-1) + (1 - \beta_k(t)) \\ &\times \begin{pmatrix} e^2_k(t) \\ r^2_k(t-1) \\ e_k(t)r_k(t-1) \end{pmatrix}. \end{aligned} \tag{40}$$

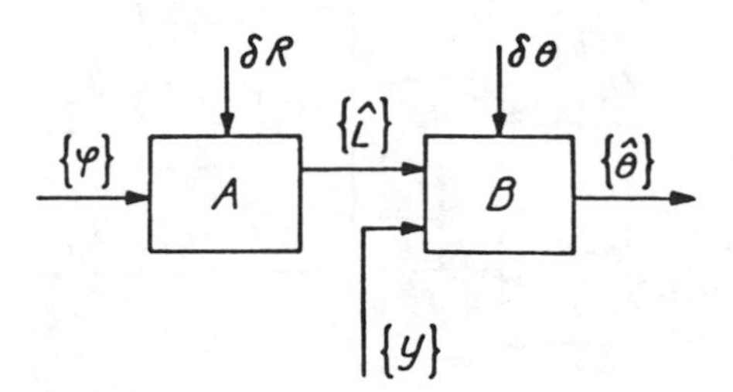

FIG. 2. Signal flow in (30), (36).

The output is then formed as

$$K_k(t) = F_k(t)/R_k^r(t-1)$$
$$K_k^*(t) = F_k(t)/R_k^e(t) \tag{41}$$

$$\beta_{k+1}(t) = \beta_k(t) + (1-\beta_k(t))^2 r_k^2(t-1)/R_k^r(t-1). \tag{42}$$

The signal flow of (39)–(42) can be depicted as in Fig. 3. We notice first that any disturbance in the state variables will decay exponentially as λ^{t-t_0}, according to (40). This also implies that $\Delta K_k(t)$, $\Delta K_k^*(t)$ and $\Delta\beta_{k+1}(t)$ will decay exponentially as λ^{t-t_0}, provided $1/R_k^r(t)$ and $1/R_k^e(t)$ remain bounded. The link k is thus exponentially stable with respect to momentary errors.

Let us now consider what happens when the actual input variables $\hat{r}_{k-1}$, $\hat{e}_{k-1}$, $\hat{K}_{k-1}$, $\hat{K}_{k-1}^*$ and $\hat{\beta}_k$ are subject to errors Δr_{k-1}, etc. We have, e.g. from (39) that

$$\begin{aligned}\Delta r_k(t) = \Delta r_{k-1}(t-1) - \Delta K_{k-1}^*(t-1)e_{k-1}(t) \\ + K_{k-1}^*(t-1)\Delta e_{k-1}(t) + \\ \Delta K_{k-1}^*(t-1)\Delta e_{k-1}(t).\end{aligned}$$

We see that if the input errors Δr_{k-1}, Δe_{k-1} and ΔK_{k-1}^* decay exponentially, so will Δr_k, provided K_{k-1}^* and e_{k-1} remain bounded. The case for Δe_k is of course analogous. Similarly, an exponentially decaying input disturbance on (40) will produce an exponentially decaying error $\Delta\xi_k(t)$. We thus conclude the following about link k:

> If the link is subject to a momentary error of some sort, then this produces an exponentially decaying (base λ) error on the output variables. Similarly, an exponentially decaying (base λ) error in the input variables will produce a similarly decaying error in the output variables of the link. (43)

For the whole lattice we know that there is no feedback from higher order links to lower ones. Therefore (43) implies that a momentary disturbance inflicted anywhere in the lattice will produce exponentially decaying errors. The error propagation in the FL algorithm is thus exponentially stable.

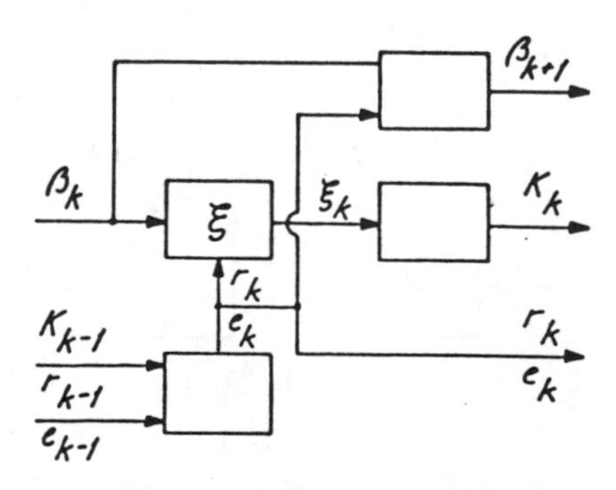

FIG. 3. Signal flow in one link.

We may note that what we have studied is the error propagation form time instant to time instant. We have not studied the error propagation from rung to rung at a fixed time instant [$(k-1)$ to k in (39)]. For high order filters, this could be an important issue, as pointed out by Cioffi (1984).

6. ERROR PROPAGATION IN THE FLS ALGORITHM

While the calculations for the algorithms so far have been straightforward enough, the FLS algorithm is considerably more difficult to analyze. Numerical experience with this algorithm has also been mixed: when λ is less than unity, unexplained 'explosions' in the algorithm have been noticed to occur under some circumstances, Ljung (1983), Soong (1981), Lin (1984), Mueller (1981) and Cioffi and Kailath (1984).

Since a general analysis is difficult we shall here only consider a counterexample where the nonlinear difference equation (10) indeed is unstable, and the linearized difference equation is exponentially unstable.

One might discuss what insight is gained by such a counterexample, which, admittedly, is somewhat contrived. The important message is that FLS does not have the same nice error propagation properties as the other algorithms considered here. The problems, mentioned above, are therefore probably not caused by numerically unsound implementations of FLS, and there is nothing in terms of numerical tricks and improved organization of calculations that can be done to remove the bad error propagation properties. That can only be done by changing the algorithm itself. See the end of this section for such a device.

Let us apply (10) to a simple case with $n = 1$, viz. to calculate the scalar

$$L(t) = \left[\sum_{k=1}^{t} u^2(k-1)\cdot\lambda^{t-k}\right]^{-1} u(t-1). \tag{44}$$

Here $u(t-1) = \varphi(t)$ is a scalar. The algorithm (10) gives

$$A(t) = A(t-1)(1 - L(t)u(t-1)) - L(t)u(t) \tag{45a}$$

$$\begin{aligned}R_e(t) = \lambda R_e(t-1) + (u(t) + A(t-1)u(t-1)) \\ \times (u(t) + A(t)u(t-1))\end{aligned} \tag{45b}$$

$$m(t) = R_e^{-1}(t)(u(t) + A(t)u(t-1)) \tag{45c}$$

$$\mu(t) = L(t) + A(t)m(t) \tag{45d}$$

$$\begin{aligned}D(t) = [D(t-1) - m(t)(u(t-1) + D(t-1)u(t))] \\ \times [1 - \mu(t)(u(t-1) + D(t-1)u(t))]^{-1}\end{aligned} \tag{45e}$$

$$L(t+1) = m(t) - D(t)u(t). \tag{45f}$$

Introduce the state vector

$$\xi(t) = \begin{pmatrix} A(t) \\ D(t) \\ L(t+1) \\ R_e(t) \end{pmatrix}. \quad (46)$$

Then (45) can be rewritten as

$$\xi(t) = f(\xi(t-1), u(t), u(t-1)). \quad (47)$$

The system matrix of the linearized difference equation is

$$F(t) = \frac{\partial}{\partial \xi} f(\xi, u(t), u(t-1))|_{\xi=\xi(t-1)} \quad (48)$$

It is still difficult to determine $\{F(t)\}$ and analyze its stability properties for general sequences $\{u(t)\}$. We therefore choose, after some thought, the particular sequence

$$u(t) = \begin{cases} 1 & \text{if } t = 4\cdot k \quad k \text{ integer} \\ 0 & \text{if } t = 4\cdot k - 1 \quad k \text{ integer} \\ -1 & \text{if } t = 4\cdot k - 2 \quad k \text{ integer} \\ 0 & \text{if } t = 4\cdot k - 3 \quad k \text{ integer} \end{cases}. \quad (49)$$

This gives the nominal trajectory

$$\xi(4k) = \begin{pmatrix} 0 \\ 0 \\ 0 \\ 1/(1-\lambda^2) \end{pmatrix}, \quad \xi(4k-1) = \begin{pmatrix} 0 \\ 0 \\ -(1-\lambda^2) \\ \lambda/(1-\lambda^2) \end{pmatrix}$$

$$\xi(4k-2) = \begin{pmatrix} 0 \\ 0 \\ 0 \\ 1/(1-\lambda^2) \end{pmatrix}$$

$$\xi(4k-3) = \begin{pmatrix} 0 \\ 0 \\ (1-\lambda^2) \\ \lambda/(1-\lambda^2) \end{pmatrix}$$

for large k. The details of the calculation of (48) for the case (49) are given in the Appendix, where it is shown that we always have

$$F(t)F(t-1) = \begin{pmatrix} 1-(1-\lambda^2)+\dfrac{(1-\lambda^2)}{\lambda} & \dfrac{-(1-\lambda^2)}{\lambda^2} & 0 & 0 \\ -\lambda(1-\lambda^2) & 1 & 0 & 0 \\ 0 & 0 & 0 & -\lambda(1-\lambda^2)^2 \\ 0 & 0 & 0 & \lambda^2 \end{pmatrix} \quad (50)$$

when t is large (i.e. we neglect a transient of the form λ^t). The matrix $F(t)$ is time-varying, but the product $F(t)F(t-1)$ is time-invariant. The stability properties of $\{F(t)\}$ can thus be investigated in terms of the eigenvalues of $F(t)F(t-1)$. The matrix (50) has one eigenvalue at the origin and one λ^2. The sum of the two others is

$$2 + (1-\lambda^2)\left(\frac{1}{\lambda} - 1\right)$$

which is larger than 2 if $\lambda < 1$. Hence at least one eigenvalue is strictly outside the unit circle, and the linearized equation (47), (48) is exponentially unstable.

This means that round-off errors will be amplified arbitrarily much and eventually cause degradation of the algorithm. Since the unstable eigenvalue (in this case) is $\approx 1 + (1-\lambda)\sqrt{5}$, one might suspect that this effect is more pronounced the smaller λ is. This is confirmed by simulations for general cases; Ljung (1983). From that reference we quote Table 1. The FLS algorithm was applied to the identification of the following pole-zero model.

$$y(t) + \sum_{k=1}^{5}\left(\frac{6-k}{6}\right)y(t-k) + \frac{1}{6}y(t-6)$$

$$= u(t-1) + \sum_{k=1}^{5} d_k u(t-k-1).$$

Here the numerical values of d_k were chosen at random between 0 and 3. The 12 coefficients of a model

$$y(t) + \sum_{k=1}^{6} a_k y(t-k) = \sum_{k=1}^{6} b_k u(t-k)$$

were estimated. A binary white noise input was used. We might add that the algorithm behaves well up to

TABLE 1. NUMERICAL DEGRADATION OF THE FLS ALGORITHM

λ	0.99	0.98	0.97	0.96	0.95
Degradation at time step	8700	3600	1300	800	400

a few steps before the explosion time, given in Table 1. For $\lambda = 1$ tests with a million steps have been run without problems.

On Lin's rescue device

Lin (1984) has studied the FLS algorithm experimentally. He found that the rapid degradation of the behaviour is immediately preceded by the event that the variable $[1 - \mu(t)\eta_0(t)]$ (second last formula in (10)) takes a negative value. This variable should theoretically be positive, and the wrong sign must be a result of accumulated numerical errors. Lin (1984) thus suggests that the algorithm is reinitialized when $[1 - \mu(t)\eta_0(t)]$ is negative. This is 'Lin's rescue device', and it has been used also by Cioffi and Kailath (1984). The penalty with the rescue is that the accumulated information in the memory variables $A(t)$, $L(t)$, $D(t)$ and $R_e(t)$ is sacrificed. Since the estimates are based on a criterion (4) with forgetting factor λ the 'memory period' of the estimates is about $1/(1 - \lambda)$ samples (see Ljung and Söderström (1983), Chap. 5). This means that for $\lambda = 0.98$ the 50 estimates following a rescue have worse statistical properties than necessary, but thereafter the difference becomes insignificant. Comparing with Table 1 we see that the rescue is necessary much more seldom, so the overall degradation of the statistical properties of the estimates may be negligible.

7. CONCLUSIONS

We have discussed the error propagation properties of the most common implementations of the recursive least squares method. We have described how such a property is fundamental to the numerical behaviour in presence of round-off errors and other numerical disturbances.

The conventional methods, RLS, CLS, UDLS as well as the fast lattice algorithms were shown to be exponentially stable with respect to such errors. The base of the exponential decay equals the forgetting factor.

By means of counterexamples we showed that the FLS method does not have this property and may be unstable when $\lambda < 1$. It is therefore not suited for continuous adaptation algorithms, unless special rescue devices are included.

REFERENCES

Anderson, B. D. O. and C. R. Johnson Jr (1982). Exponential convergence of adaptive identification and control algorithms. *Automatica*, **18**, 1.

Bierman, G. J. (1977). *Factorization Methods for Discrete Estimation*. Academic Press, New York.

Brockett, R. W. (1970). *Finite-dimensional Linear Systems*. John Wiley, New York.

Carayannis, G., D. Manolakis and N. Kalouptsidis (1983). A fast sequential algorithm for least squares filtering and prediction. *IEEE Trans.*, **ASSP-31**.

Cioffi, J. (1984). Private communication.

Cioffi, J. and T. Kailath (1984). Fast recursive, least-squares transversal filters for adaptive filtering. *IEEE Trans.*, **ASSP-32**, 304.

Graupe, D., V. K. Jain and J. Salahi (1980). A comparative analysis of various least-squares identification algorithms. *Automatica*, **16**, 663.

Griffiths, L. J. (1977). A continuously adaptive filter implemented as a lattice structure, *Proc. 1977 IEEE International Conf. on Acoustics, Speech and Signal Processing*, Hartford, Conn., p. 683.

Lee, D. T., M. Morf and B. Friedlander (1981). Recursive least squares ladders estimation algorithms. *IEEE Trans.*, **ASSP-29**, 627.

Lin, D. W. (1984). On digital implementation of the fast Kalman algorithm. *IEEE Trans.*, **ASSP-32**, 998.

Ljung, L., M. Morf and D. Falconer (1978). Fast calculation of gain matrices for recursive estimation schemes. *Int. J. Control*, **27**, 1.

Ljung, L. and T. Söderström (1983). *Theory and Practice of Recursive Identification*. MIT Press, Cambridge, Mass.

Ljung, S. (1983). Fast algorithms for integral equations and least squares identification problems, Linköping Studies in Science and Technology, Dissertation No 93, Linköping University, Sweden.

Mueller, M. S. (1981). Least-squares algorithms for adaptive equalizers. *Bell. Syst. Tech. J.*, **60**, 1905.

Potter, J. E. (1963). New statistical formulas. Memo 40, Instrumentation Laboratory, Massachusetts Institute of Technology.

Proakis, J. G. (1983). *Digital Communications*. McGraw Hill, New York.

Samson, C. and V. U. Reddy (1983). Fixed point error analysis of the normalized ladder algorithm. *IEEE Trans.* **ASSP-31**, 1177.

Soong, F. (1981). Private communication.

APPENDIX: DERIVATION OF THE LINEARIZED DIFFERENCE EQUATION, CORRESPONDING TO FLS

Let us introduce the convention that

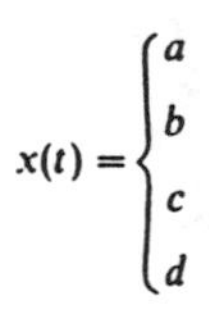

$$x(t) = \begin{cases} a \\ b \\ c \\ d \end{cases}$$

is short for

'$x(4k) = a$; $x(4k - 1) = b$; $x(4k - 2) = c$; $x(4k - 3) = d$; where k is an integer'.

We thus have

$$u(t) = \begin{cases} 1 \\ 0 \\ -1 \\ 0 \end{cases}$$

The interpretation of $A(t)$ and $D(t)$ is that they are the forward and backward predictors, respectively for the $\{u(t)\}$-sequence. We have then from (10) (think also of $A(t)$ as the LS estimate of a model $u(t) = Au(t - 1)$).

$$A(t) = \left[\sum_{k=1}^{t} \lambda^{t-k} u^2(k-1)\right]^{-1} \sum_{k=1}^{t} \lambda^{t-k} u(k-1) u(k).$$

The nominal trajectory consequently is

$$\xi_1(t) = A(t) = \begin{cases} 0 \\ 0 \\ 0 \\ 0 \end{cases}; \quad \xi_2(t) = D(t) = \begin{cases} 0 \\ 0 \\ 0 \\ 0 \end{cases}$$

$$\xi_3(t) = L(t) = \left\{\begin{matrix} 0 \\ -\gamma \\ 0 \\ \gamma \end{matrix}\right. \qquad \xi_4(t) = R_e(t) = \left\{\begin{matrix} 1/\gamma \\ \lambda/\gamma \\ 1/\gamma \\ \lambda/\gamma \end{matrix}\right. \tag{A1}$$

$$m(t) = \left\{\begin{matrix} \gamma \\ 0 \\ -\gamma \\ 0 \end{matrix}\right. \qquad \mu(t) = \left\{\begin{matrix} 0 \\ -\gamma \\ 0 \\ \gamma \end{matrix}\right. \qquad \gamma = (1 - \lambda^2)$$

where we neglect the transient influence from the initial condition, e.g. we set

$$\sum_{k=1}^{t} \lambda^{t-k} u^2(k-1) = \frac{1}{1-\lambda^2} \text{ for } t \text{ odd.}$$

The equations can be written

$$\xi_1(t) = \xi_1(t-1)(1 - \xi_3(t-1)u(t-1)) - \xi_3(t-1)u(t)$$

$$m(t) = \frac{1}{\xi_4(t)}\{u(t) + [\xi_1(t-1)(1 - \xi_3(t-1)u(t-1)) - \xi_3(t-1)u(t)]u(t-1)\}$$

$$u(t) = \xi_3(t-1) + [\xi_1(t-1)(1 - \xi_3(t-1)u(t-1)) - \xi_3(t-1)u(t)]m(t)$$

$$\xi_2(t) = [\xi_2(t-1) - m(t)(u(t-1) + \xi_2(t-1)u(t))] \times [1 - \mu(t)(u(t-1) + \xi_2(t-1)u(t))]^{-1}$$

$$\xi_3(t) = m(t) - \xi_2(t)\mu(t)$$

$$\xi_4(t) = \lambda\xi_4(t-1) + (u(t) + \xi_1(t-1)u(t-1)) \{u(t) + [\xi_1(t-1)(1 - \xi_3(t-1)u(t-1)) - \xi_3(t-1)u(t)]\}$$

It is now a straightforward, but lengthy exercise to calculate the respective derivatives and evaluate them in (A1). It gives

$$\frac{\partial\xi_1(t)}{\partial\xi_1(t-1)} = 1 - \xi_3(t-1)u(t-1) = \left\{\begin{matrix} 1 \\ 1-\gamma \\ 1 \\ 1-\gamma \end{matrix}\right.$$

$$\frac{\partial\xi_1(t)}{\partial\xi_2(t-1)} = 0$$

$$\frac{\partial\xi_1(t)}{\partial\xi_3(t-1)} = -u(t) = \left\{\begin{matrix} -1 \\ 0 \\ 1 \\ 0 \end{matrix}\right.$$

$$\frac{\partial\xi_1(t)}{\partial\xi_4(t-1)} = 0$$

$$\frac{\partial m(t)}{\partial\xi_1(t-1)} = \xi_4^{-1}(t)(1 - \xi_3(t-1)u(t-1))u(t-1)$$

$$= \left\{\begin{matrix} 0 \\ (-1+\gamma)\gamma/\lambda \\ 0 \\ (1-\gamma)\gamma/\lambda \end{matrix}\right.$$

$$\frac{\partial m(t)}{\partial\xi_2(t-1)} = 0; \quad \frac{\partial m(t)}{\partial\xi_3(t-1)} = 0;$$

$$\frac{\partial m(t)}{\partial\xi_4(t-1)} = \frac{1}{\xi_4(t)} m(t) = \left\{\begin{matrix} -\gamma^2 \\ 0 \\ -\gamma^2 \\ 0 \end{matrix}\right.$$

$$\frac{\partial\mu(t)}{\partial\xi_1(t-1)} = (1 - \xi_3(t-1)u(t-1))m(t) = \left\{\begin{matrix} \gamma \\ 0 \\ -\gamma \\ 0 \end{matrix}\right.$$

$$\frac{\partial\mu(t)}{\partial\xi_2(t-1)} = 0$$

$$\frac{\partial\mu(t)}{\partial\xi_3(t-1)} = 1 + (-u(t)m(t)) = \left\{\begin{matrix} 1-\gamma \\ 1 \\ 1-\gamma \\ 1 \end{matrix}\right.$$

$$\frac{\partial\mu(t)}{\partial\xi_4(t-1)} = 0$$

$$\frac{\partial\xi_2(t)}{\partial\xi_1(t-1)} = \frac{\partial m(t)}{\partial\xi_1(t-1)} u(t-1)/(1 - \mu(t)u(t-1))$$

$$= \left\{\begin{matrix} 0 \\ -\gamma/\lambda \\ 0 \\ -\gamma/\lambda \end{matrix}\right.$$

$$\frac{\partial\xi_2(t)}{\partial\xi_2(t-1)} = (1 - m(t)u(t))/(1 - \mu(t)u(t-1))$$

$$= \left\{\begin{matrix} 1-\gamma \\ 1/(1-\gamma) \\ 1-\gamma \\ 1/(1-\gamma) \end{matrix}\right.$$

$$\frac{\partial\xi_2(t)}{\partial\xi_3(t-1)} = 0$$

$$\frac{\partial\xi_2(t)}{\partial\xi_4(t-1)} = \frac{\partial m(t)}{\partial\xi_4(t-1)} u(t-1)/(1 - \mu(t)u(t-1)) = 0$$

$$\frac{\partial\xi_3(t)}{\partial\xi_1(t-1)} = \frac{\partial m(t)}{\partial\xi_1(t-1)} - \mu(t)\frac{\partial\xi_2(t)}{\partial\xi_1(t-1)}$$

$$= \left\{\begin{matrix} 0 \\ -\gamma/\lambda \\ 0 \\ \gamma/\lambda \end{matrix}\right.$$

$$\frac{\partial\xi_3(t)}{\partial\xi_2(t-1)} = \mu(t)\frac{\partial\xi_2(t)}{\partial\xi_2(t-1)} = \left\{\begin{matrix} 0 \\ \gamma/(1-\gamma) \\ 0 \\ -\gamma/(1-\gamma) \end{matrix}\right.$$

$$\frac{\partial\xi_3(t)}{\partial\xi_3(t-1)} = 0; \quad \frac{\partial\xi_3(t)}{\partial\xi_4(t-1)} = \left\{\begin{matrix} -\gamma^2 \\ 0 \\ -\gamma^2 \\ 0 \end{matrix}\right.$$

$$\frac{\partial\xi_4(t)}{\partial\xi_k(t-1)} = 0 \quad k = 1, 2, 3$$

$$\frac{\partial\xi_4(t)}{\partial\xi_4(t-1)} = \lambda.$$

The linearized matrix is consequently

$$
F(t) = \left[\begin{array}{c|c|c|c}
1 & & -1 & \\
1-\gamma & 0 & 0 & 0 \\
1 & & 1 & \\
1-\gamma & & 0 & \\
\hline
0 & 1-\gamma & & \\
-\gamma/\lambda & 1/(1-\gamma) & 0 & 0 \\
0 & 1-\gamma & & \\
-\gamma/\lambda & 1/(1-\gamma) & & \\
\hline
0 & 0 & & -\gamma^2 \\
-\gamma/\lambda & \gamma/(1-\gamma) & 0 & 0 \\
0 & 0 & & -\gamma^2 \\
\gamma/\lambda & -\gamma/(1-\gamma) & & 0 \\
\hline
0 & 0 & 0 & \lambda
\end{array}\right]
$$

We thus have

$$
F(4k)F(4k-1) = \begin{bmatrix} 1 & 0 & -1 & 0 \\ 0 & 1-\gamma & 0 & 0 \\ 0 & 0 & 0 & -\gamma^2 \\ 0 & 0 & 0 & \lambda \end{bmatrix}
\times \begin{bmatrix} 1-\gamma & 0 & 0 & 0 \\ -\gamma/\lambda & 1/(1-\gamma) & 0 & 0 \\ -\gamma/\lambda & \gamma/(1-\gamma) & 0 & 0 \\ 0 & 0 & 0 & \lambda \end{bmatrix}
= \begin{bmatrix} 1-\gamma+\gamma/\lambda & -\gamma/(1-\gamma) & 0 & 0 \\ -\gamma(1-\gamma)/\lambda & 1 & 0 & 0 \\ 0 & 0 & 0 & -\gamma^2\lambda \\ 0 & 0 & 0 & \lambda^2 \end{bmatrix}
$$

and similarly for

$$F(4k-2)F(4k-3).$$

Consequently we have proved (50).

Part III
Lattice Filters and Their Properties

A review of articles in the previous part shows the need for adaptive algorithms which have self-orthogonalization properties. The adaptive algorithms that are represented by the lattice filter structures have this desired self-orthogonalization property. Lattice filters arise naturally when one uses the Gram-Schmidt orthogonalization procedure to replace a data sequence by an orthonormal set of variables (backward prediction errors). As a result of this stage-by-stage orthogonalization, lattice filters have the advantages of modularity, speed of adaptation, easy testing for minimum phase condition, and good numerical conditioning.

In the linear predictor application, the optimal lattice filter coefficients at each stage, or so-called PARCOR coefficient, are independent of the final predictor order. The order of the predictor can be increased by simply adding lattice filter stages. The linear predictors, which are based on the transversal filter structures, do not have this nice modularity. In the case of the transversal filter, when the predictor order is increased, all the filter coefficients must be recomputed. When the required predictor order is not known in advance, the lattice filter has clear advantages over transversal filters. The fact that self-orthogonalization speeds up adaptation has been demonstrated by many computer simulations reported in several papers presented in this part.

The lattice filter is a minimum phase filter (all its zeros are on or inside unit circle), if and only if the lattice filter coefficients k_m satisfy the inequality $|k_m| < 1$. For transversal filters, there is no such simple test for the minimum phase condition [1]. The minimum phase condition for an all-zero filter is important when an all-zero filter is used in a feedback loop to generate an all-pole filter. If the all-zero filter is not of minimum phase, the all-pole filter is unstable.

The disadvantages of the lattice filter algorithms are their numerical complexity and the mathematical sophistication that is required for thorough understanding of derivations of various lattice filter algorithms. There are several non-trivial derivations of lattice filter algorithms, and there is a perplexing variety of lattice adaptive filter algorithms. Papers have been selected from a large body of literature to guide the reader through this bewildering forest.

The first paper in this part is "Stable and Efficient Lattice Methods for Linear Prediction" by J. Makhoul. This paper presents a class of lattice methods for linear prediction that guarantees the stability of the all-pole filter, even with finite word length computation, independently of the stationarity and duration of the signal. Makhoul gives several methods for the determination of lattice filter coefficients, and he shows that the original methods of Itakura [2] and Burg [3] are special cases of the class of methods he considers.

The next paper is a major review paper entitled "Lattice Filters for Adaptive Processing" by B. Friedlander. This paper is the most complete overview of the state of the art of lattice filters as applied to adaptive signal processing. The lattice filter structures are developed for the linear prediction of time series. Friedlander shows that many current lattice methods are actually approximations to the stationary least squares solution. He uses the projection operator method to derive time and order update algorithms for normalized and unnormalized least squares lattice filters. The presentation in this paper allows the reader to understand and compare several least squares lattice algorithms and gradient lattice and transversal filter algorithms. The projection operator approach is one of the possible approaches for the derivation of lattice filter update algorithms. Mastery of this approach is well rewarded. Systematic development of least squares algorithms, as presented in this paper, will guide the reader through the thicket of lattice filter algorithms. The reader is urged to devote serious study to this paper. Friedlander also shows the connections between the lattice filter structures' fast Cholesky factorization, the Gram-Schmidt orthogonalization, the efficient solution of linear equations, and the various canonical forms that arise in the system theory. This paper contains a very complete bibliography on lattice filters, their theory, applications, and conceptual development. This bibliography is a good source of material for further reading and research on adaptive lattice filters and related topics.

The paper by M. J. Shensa uses what is known as the geometrical approach to derive recursive least squares lattice algorithms. This paper uses an inner product formalism to solve the least squares problem. This approach provides an alternate approach to the derivation of least squares algorithms which leads to some new insights. The expressions for lattice gain coefficients in terms of inner products provide important insights into the algorithm's behavior in regard to its numerical stability, convergence, and implementation.

In the paper "Adaptive Tracking of Multiple Sinusoids Whose Power Levels are Widely Separated," Hodgkiss and Presley present an interesting comparison of the gradient transversal (LMS), the gradient lattice (GL), and the least squares lattice (LSL) algorithms when used to track multiple sinusoids whose power levels are widely separated. Computer simulations showed that all three algorithms nicely tracked high-power sinusoids. However, the adaptive transversal filter (LMS) was unable to cope with the large eigenvalue spread—it could not track the weaker sinusoid. Two lattice algorithms were able to track both the weaker and the stronger sinusoid.

The last paper by Sohie and Sibul is a theoretical study of stochastic convergence properties (in a mean square sense) of the adaptive gradient lattice filter. A stochastic fixed point theorem is used to derive conditions for stepsize in the adaptive algorithm, and to derive analytical expressions for misadjustment and convergence rate. It was shown that in the gradient lattice, the total output misadjustment varies exponentially with the filter order, while in the LMS algorithm, the output misadjustment varies linearly with the number of taps.

This reprint book presents only a small sampling from the large body of literature on adaptive lattice structures. As stated before, Friedlander's paper contains a very large list of references. For additional reading on lattice filters and their applications, the reader is referred to two textbooks [1], [4], the Joint Special Issue on Adaptive Signal Processing [5], and papers by Shichor [6], Carayannis *et al* [7], and McWhirter and Shepherd [8]. These three papers give an alternative discussion of lattice structures that may be helpful to the reader. Shichor concludes that the lattice algorithm gives results identical to the fast Kalman algorithm when both algorithms have the same number of coefficients. The number of multiplications for the two algorithms is about the same, but the lattice requires more divisions for normalization by the residual error energy at each stage. In a limited-precision implementation, the lattice algorithm is expected to give better performance. The invited paper by G. Carayannis *et al* gives a unified development of fast order-recursive schemes and sequential methods for transversal and lattice implementations. In this paper, the derivation of all algorithms using a unified approach reveals the relationships among the various variables and between fast algorithms for transversal and lattice structures. McWhirter and Shepherd also discuss the relation between the "fast Kalman" algorithm and the exact least squares lattice algorithm. This paper presents a simplified derivation of the exact least squares algorithms. An interesting paper that is worth pointing out is "A Generalized Orthogonalization Technique with Applications to Time Series Analysis and Signal Processing," by G. Cybenko [9]. In this paper, the author uses a new orthogonalization technique for solving linear least-squares problems. Under specific assumptions his algorithm simplifies significantly and results in the lattice algorithm for solving linear prediction problems. From this point of view it is seen that the lattice algorithm is really an efficient way of solving specially structured least-squares problems by orthogonalization, as opposed to solving the normal equations by fast Toeplitz algorithms. This paper is significant in the sense that it establishes a clear connection between lattice filters and orthogonal decompositions. Orthogonal decompositions have wide applications in applied mathematics, theoretical physics, and system theory. Orthogonal expansions are building blocks for optimum estimators and array processors. Thus, understanding of the exact least squares algorithm from the point of view of orthogonal decomposition gives considerable insight into fundamental signal processing structures.

The application of adaptive filtering techniques to spectrum analysis is not represented in this book since two good collections of tutorial and review papers have been published: the *IEEE Proceedings* Special Issue on Spectral Estimation [10] and an IEEE PRESS reprint book on spectrum analysis [11]. In the context of the application of lattice methods for spectral estimation, B. Friedlander's paper in this special issue is called to the reader's attention [12]. He shows that lattice forms provide convenient parameterization of rational spectra of stationary processes, and that various well-known spectral estimation techniques such as the minimum entropy method (MEM) and maximum likelihood method (MLM), can be efficiently computed from lattice parameters. Original work on maximum entropy spectral analysis was done by Burg [3]. The rationale and status of maximum-entropy methods is reviewed by Jaynes in the same special issue of IEEE *Proceedings* [13]. This issue contains several other interesting articles related to adaptive signal processing.

To illustrate various interrelations between diverse adaptive signal processing topics, it should be mentioned that there is a one-to-one correspondence between super-resolution adaptive beamforming for angle estimation and high resolution spectrum estimation techniques. This relationship is discussed in a group of four papers on spatial spectrum estimation that have been published in the Special Issue on Adaptive Processing Antenna Systems [14]. In the same issue K. Gerlach discusses application of orthogonalization networks to multiple input and output adaptive array processors [15]. His algorithm is a type of adaptive lattice filter. This again illustrates how the basic orthogonalization concept arises in many adaptive signal processing applications.

Adaptive signal processing concepts have matured to the degree that they find applications in many practical systems. In practical application, numerical stability and error propagation properties of the adaptive algorithms are crucial for success. For this reason it is important to understand the numerical analysis aspects of matrix computation and least squares problems. These issues are discussed in a recent book by Golub and Van Loan [16], and in older books by Stewart [17], and Lawson and Hanson [18]. Robust and efficient implementation of adaptive signal processing algorithms remains an important research topic.

Another topic that we have not covered in this book is VLSI implementation of adaptive filters. This is an evolving field of research, and in sufficient time should have a reprint book of its own.

References

[1] Honig, M. L., and D. G. Messerschmitt, *Adaptive Filters, Structures Algorithms, and Applications.* Boston, MA: Kulwer Academic Publishers, 1984.

[2] Stakura, F., and S. Saito, "Digital filtering techniques for speech analysis and synthesis," in *Proc. 7th Int. Cong. Acoustics,* 1971, paper 25-C-1.

[3] Burg, J., "Maximum entropy spectral analysis," Ph.D. dissertation, Stanford University, Stanford, CA, May 1975.

[4] Haykin, S., *Introduction to Adaptive Filters.* New York, NY: MacMillan, 1984, ch. 6.

[5] Joint Special Issue on Adaptive Signal Processing, *IEEE Trans. Circuit Syst.,* vol. CAS-28, no. 6, and *IEEE Trans. Acoust., Speech, Signal Processing,* vol. ASSP-29, no. 3, June 1981.

[6] Shichor, E., "Fast recursive estimation using the lattice structure," *Bell Syst. Tech. J.,* vol. 61, no. 1, pp. 97–115, Jan. 1982.

[7] Carayannis, G., D. Manolakis, and N. Kalouptsidis, "A unified view of parametric processing algorithms for prewindowed signals," *Signal Processing,* vol. 11, no. 2, pp. 335–368, Oct. 1986.

[8] McWhirter, J. G. and T. J. Shepherd, "Least squares lattice algorithm for adaptive channel equalisation—A simplified derivation," *Proc. IEEE,* vol. 130, part F, pp. 532–542, Oct. 1983.

[9] Cybenko, G., "A general orthogonalization technique with applications to time series analysis and signal processing," *Mathematics of Computation,* vol. 40, no. 161, pp. 323–336, Jan. 1983.

[10] Special Issue on Spectral Estimation, *Proc. IEEE,* vol. 70, no. 9, Sept. 1982.

[11] Childers, D. G., Ed., *Modern Spectrum Analysis.* New York, NY: IEEE PRESS, 1978.
[12] Friedlander, B., "Lattice methods for spectral estimation," *Proc. IEEE,* vol. 70, no. 9, pp. 990–1017, Sept. 1982.
[13] Jaynes, E. T., "On the rationale of maximum-entropy methods," *Proc. IEEE,* vol. 70, no. 9, pp. 939–952, Sept. 1982.
[14] Spatial Spectrum Estimation (four papers), *IEEE Trans. Antennas Propagat.* (Special issue on adaptive processing antenna systems), vol. AP-34, no. 3, Mar. 1986.
[15] Gerlach, K., "Fast orthogonalization networks," *IEEE Trans. Antennas Propagat.* (Special issue on adaptive processing antenna systems), vol. AP-34, no. 3, pp. 458–462, Mar. 1986.
[16] Golub, G. H. and C. F. Van Loan, *Matrix Computations.* Baltimore, MD: Johns Hopkins University Press, 1983.
[17] Stewart, G. W., *Introduction to Matrix Computations.* New York, NY: Academic Press, 1973.
[18] Lawson, C. L. and R. L. Hanson, *Solving Least Squares Problems.* Englewood Cliffs, NJ: Prentice-Hall, 1974.

Stable and Efficient Lattice Methods for Linear Prediction

JOHN MAKHOUL, MEMBER, IEEE

Abstract—A class of stable and efficient recursive lattice methods for linear prediction is presented. These methods guarantee the stability of the all-pole filter, with or without windowing of the signal, with finite wordlength computations, and at a computational cost comparable to the traditional autocorrelation and covariance methods. In addition, for data-compression purposes, quantization of the reflection coefficients can be accomplished within the recursion, if desired.

I. Introduction

THE autocorrelation method of linear prediction [1] guarantees the stability of the all-pole filter, but has the disadvantage that windowing of the signal causes a reduction in spectral resolution. In practice, even the stability is not always guaranteed with finite wordlength (FWL) computations [2]. On the other hand, the covariance method [1], [3] does not guarantee the stability of the filter, even with floating-point computation, but has the advantage that there is no windowing of the signal. One solution to these problems was given by Itakura [4] in his lattice formulation. In this method, filter stability is guaranteed with no windowing and with much smaller sensitivity to FWL computations. Unfortunately, this is accomplished with about a fourfold increase in computation over the other two methods. A similar method was independently proposed by Burg [5], [6].

This paper presents a class of lattice methods that guarantees the stability of the all-pole filter, independently of the stationarity properties and the duration of the signal. It is shown that the methods of Itakura and Burg are special cases of this class of methods. Furthermore, a procedure is given that reduces the number of computations to values comparable to those in the autocorrelation and covariance methods. In this procedure, the "forward" and "backward" residuals are not computed; the reflection coefficients are computed directly from the covariance of the input signal.

Section II presents the class of lattice methods for computing the reflection coefficients, along with conditions for ensuring stability. Section III describes a procedure, termed the *covariance-lattice* method, for performing the necessary computations efficiently. Computational issues are then discussed in Section IV, followed in Section V by a step-by-step procedure for one of the promising lattice methods for linear predictive analysis.

Manuscript received May 5, 1976; revised September 9, 1976, and May 13, 1977. This work was supported by the Information Processing Techniques Branch of the Advanced Research Projects Agency under Contracts MDA903-75-C-0180 and N00014-75-C-0533.

The author is with Bolt Beranek and Newman Inc., Cambridge, MA 02138.

II. Lattice Formulations

In linear prediction, the signal spectrum is modeled by an all-pole spectrum with a transfer function given by

$$H(z) = \frac{G}{A(z)} \tag{1}$$

where

$$A(z) = \sum_{k=0}^{p} a_k z^{-k}, \qquad a_0 = 1 \tag{2}$$

is known as the inverse filter, G is a gain factor, a_k are the predictor coefficients, and p is the number of poles or predictor coefficients in the model. If $H(z)$ is stable (minimum phase), $A(z)$ can be implemented as a lattice filter [4], as shown in Fig. 1. The reflection (or partial correlation) coefficients K_m in the lattice are uniquely related to the predictor coefficients. Given K_m, $1 \leq m \leq p$, the set $\{a_k\}$ is computed by the recursive relation

$$\begin{aligned} a_m^{(m)} &= K_m \\ a_j^{(m)} &= a_j^{(m-1)} + K_m a_{m-j}^{(m-1)}, \qquad 1 \leq j \leq m-1, \end{aligned} \tag{3}$$

where the equations in (3) are computed recursively for $m = 1, 2, \cdots, p$. After each recursion, the coefficients $a_j^{(m)}$, $1 \leq j \leq m$, are the desired coefficients for the mth-order predictor. The final solution is given by $a_j = a_j^{(p)}$, $1 \leq j \leq p$. For a stable $H(z)$, one must have

$$|K_m| < 1, \qquad 1 \leq m \leq p. \tag{4}$$

In the lattice formulation, the reflection coefficients can be computed by minimizing some norm of the forward residual $f_m(n)$ or the backward residual $b_m(n)$, or a combination of the two. From Fig. 1, the following relations hold:

$$f_0(n) = b_0(n) = s(n) \tag{5a}$$

$$f_{m+1}(n) = f_m(n) + K_{m+1} b_m(n-1) \tag{5b}$$

$$b_{m+1}(n) = K_{m+1} f_m(n) + b_m(n-1) \tag{5c}$$

where $s(n)$ is the input signal and $e(n) = f_p(n)$ is the output residual. In z-transform notation: $E(z) = A(z) S(z)$.

We shall give several methods for the determination of the reflection coefficients. These methods depend on different ways of correlating the forward and backward residuals. Below, we shall make use of the following definitions:

$$F_m(n) = E[f_m^2(n)] \tag{6a}$$

$$B_m(n) = E[b_m^2(n)] \tag{6b}$$

Reprinted from *IEEE Trans Acoust., Speech, Signal Processing,* vol. ASSP-25, no. 5, pp. 423–428, Oct. 1977.

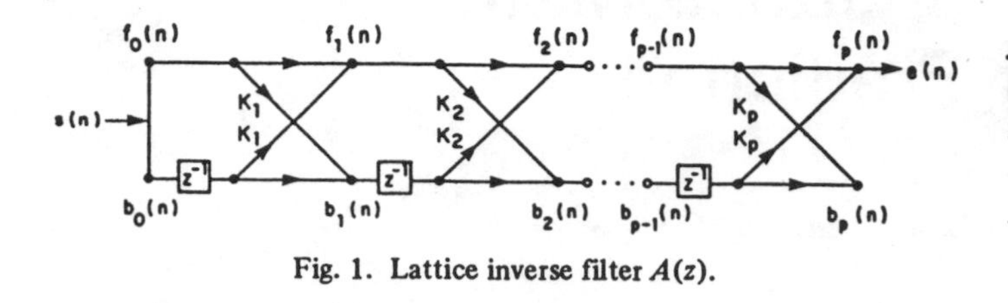

Fig. 1. Lattice inverse filter $A(z)$.

$$C_m(n) = E[f_m(n)b_m(n-1)], \tag{6c}$$

where $E(\cdot)$ denotes the expected value. The left-hand side of each of the equations in (6) is a function of n because we are making the general assumption that the signals are nonstationary. (Subscripts, etc., will be dropped sometimes for convenience.)

A. *Forward Method*

In this method, the reflection coefficient at stage $m+1$ is obtained as a result of the minimization of an error norm given by the variance (or mean square) of the forward residual

$$F_{m+1}(n) = E[f_{m+1}^2(n)]. \tag{7}$$

By substituting (5b) in (7) and differentiating with respect to K_{m+1}, one obtains

$$K_{m+1}^f = -\frac{E[f_m(n)b_m(n-1)]}{E[b_m^2(n-1)]} = -\frac{C_m(n)}{B_m(n-1)}. \tag{8}$$

This method of computing the filter parameters is similar to the autocorrelation and covariance methods in that the mean-squared *forward* residual is minimized.

B. *Backward Method*

In this case, the minimization is performed on the variance of the backward residual at stage $m+1$. From (5c) and (6b), the minimization of $B_{m+1}(n)$ leads to

$$K_{m+1}^b = -\frac{E[f_m(n)b_m(n-1)]}{E[f_m^2(n)]} = -\frac{C_m(n)}{F_m(n)}. \tag{9}$$

Note that, since $F_m(n)$ and $B_m(n-1)$ are both nonnegative and the numerators in (8) and (9) are identical, K^f and K^b always have the same sign S

$$S = \text{sign } K^f = \text{sign } K^b. \tag{10}$$

C. *Geometric-Mean Method (Itakura)*

The main problem in the previous two techniques is that the computed reflection coefficients are not always guaranteed to be less than 1 in magnitude; i.e., the stability of $H(z)$ is not guaranteed. One solution to this problem was offered by Itakura [4] where the reflection coefficients are computed from

$$K_{m+1}^I = -\frac{E[f_m(n)b_m(n-1)]}{\sqrt{E[f_m^2(n)]\,E[b_m^2(n-1)]}} = -\frac{C_m(n)}{\sqrt{F_m(n)B_m(n-1)}}. \tag{11}$$

K_{m+1}^I is the negative of the statistical correlation between $f_m(n)$ and $b_m(n-1)$; hence, property (4) follows. To the author's knowledge, (11) cannot be derived directly by minimizing some error criterion. However, from (8), (9), and (11), one can easily show that K^I is the geometric mean of K^f and K^b

$$K^I = S\sqrt{K^f K^b} \tag{12}$$

where S is given by (10), and we have omitted the subscript $m+1$. From the properties of the geometric mean, it follows that

$$\min\,[|K^f|, |K^b|] \leq |K^I| \leq \max\,[|K^f|, |K^b|].$$

Now, since $|K^I| < 1$, it follows that *if the magnitude of either K^f or K^b is greater than 1, the magnitude of the other is necessarily less than 1.* This important property can be summarized by the following.

$$\text{If } |K^f| > 1, \text{ then } |K^b| < 1,$$

or

$$\text{if } |K^b| > 1, \text{ then } |K^f| < 1. \tag{13}$$

Property (13) immediately brings to mind another possible definition for the reflection coefficient that guarantees stability.

D. *Minimum Method*

$$K^M = S \min\,[|K^f|, |K^b|]. \tag{14}$$

This says that at each stage, compute K^f and K^b and choose as the reflection coefficient the one with the smaller magnitude. Property (13) guarantees that K^M satisfies (4).

E. *General Method*

Between K^M and K^I there are an infinity of values that can be chosen as valid reflection coefficients (i.e., $|K| < 1$). These can be conveniently defined by taking the generalized rth mean of K^f and K^b

$$K^r = S\,[\tfrac{1}{2}(|K^f|^r + |K^b|^r)]^{1/r}. \tag{15}$$

As $r \to 0$, $K^r \to K^I$, the geometric mean. For $r > 0$, K^r cannot be guaranteed to satisfy (4). Therefore, for K^r to be a reflection coefficient, we must have $r \leq 0$. In particular

$$K^0 = K^I, \quad K^{-\infty} = K^M. \tag{16}$$

If the signal is stationary, one can show that $K^f = K^b$, and that

$$K^r = K^f = K^b, \quad \text{all } r \text{ (Stationary Case)}. \tag{17}$$

F. *Harmonic-Mean Method (Burg)*

There is one value of r for which K^r has some interesting properties, and that is $r = -1$. K^{-1}, then, would be the harmonic mean of K^f and K^b

$$K_{m+1}^B = K^{-1} = \frac{2K^f K^b}{K^f + K^b} = -\frac{2C_m(n)}{F_m(n) + B_m(n-1)}. \tag{18}$$

One can show that

$$|K^M| \leq |K^B| \leq |K^I|. \tag{19}$$

TABLE I
COMPUTATIONAL COST FOR TRADITIONAL LINEAR-PREDICTION METHODS AS COMPARED TO THE NEW AUTOCORRELATION-LATTICE AND COVARIANCE-LATTICE METHODS

	AUTOCORRELATION METHOD	COVARIANCE METHOD	REGULAR LATTICE (WITH RESIDUALS)
TRADITIONAL METHODS	$pN + p^2$	$pN + \frac{1}{6}p^3 + \frac{3}{2}p^2$	5 pN
NEW LATTICE METHODS	$pN + \frac{1}{6}p^3 + \frac{3}{2}p^2$	$pN + \frac{1}{2}p^3 + 2p^2$	5 pN

Note: Terms of order p have been neglected.

In fact, Itakura used K^B as an approximation to K^I in (11) to avoid computing the square root.

One important property of K^B that is not shared by K^I and K^M, is that K^B results directly from the minimization of an error criterion. The error is defined as the sum of the variances of the forward and backward residuals

$$E_{m+1}(n) = F_{m+1}(n) + B_{m+1}(n). \tag{20}$$

Using (5) and (6), one can show that the minimization of (20) indeed leads to (18). One can also show that the forward and backward minimum errors at stage $m + 1$ are related to those at stage m by the following:

$$F_{m+1}(n) = [1 - (K_{m+1}^B)^2]\, F_m(n) \tag{21a}$$

$$B_{m+1}(n) = [1 - (K_{m+1}^B)^2]\, B_m(n - 1). \tag{21b}$$

This formulation is originally due to Burg [5], [6].

G. Discussion

Note that, in general, lattice methods do not minimize any global error criterion, such as the variance of the final forward residual, etc. Any minimization that might take place is done stage by stage. If the signal $s(n)$ is truly stationary, the stage-by-stage minimization gives the same result as global minimization. In fact, for a stationary signal, all the lattice methods previously described, as well as the autocorrelation and covariance methods, give the same result. However, in general, the signal cannot be assumed to be stationary and the different lattice methods will give different results, which are still different from the covariance-method result. The lattice methods will indeed give suboptimal solutions; solutions that tend to an optimal solution as the signal becomes more stationary. Which lattice method to choose in a particular situation, then, is not clear cut. We tend to prefer the use of K^B in (18) because it minimizes a reasonable and well-defined error criterion.

III. THE COVARIANCE-LATTICE METHOD

If linear predictive analysis is to be performed on a regular computer, the number of computations for the lattice methods given far exceeds that of the autocorrelation and covariance methods (see the first row of Table I). This is unfortunate since, otherwise, lattice methods generally have superior properties when compared to the autocorrelation and covariance methods (see Table II). Below, we derive a new method, called the *covariance-lattice* method, which has all the advantages of a regular lattice, but with an efficiency comparable to the two nonlattice methods.

TABLE II
COMPARISON BETWEEN DIFFERENT PROPERTIES OF VARIOUS LINEAR-PREDICTION METHODS

PROPERTY	LINEAR PREDICTION METHOD: AUTOCORRELATION	COVARIANCE	REGULAR LATTICE	COVARIANCE LATTICE
WINDOWING	NECESSARY	NONE	NOT NECESSARY	NONE
STABILITY	THEORETICALLY GUARANTEED	NOT GUARANTEED	CAN BE GUARANTEED	
STABILITY WITH FINITE WORDLENGTH COMPUTATIONS	NOT GUARANTEED		CAN BE GUARANTEED	
COMPUTATIONAL EFFICIENCY	EFFICIENT		EXPENSIVE	EFFICIENT
LEAST-SQUARES OPTIMALITY	OPTIMAL		POSSIBLY ONLY SUBOPTIMAL	
QUANTIZATION OF REFLECTION COEFFICIENTS WITHIN RECURSION	NOT POSSIBLE		POSSIBLE	
NUMBER OF SAMPLES FOR ANALYSIS	N	CAN BE REDUCED TO ~0.7N FOR THE SAME RESOLUTION		

From the recursive relations in (3) and (5), one can show that

$$f_m(n) = \sum_{k=0}^{m} a_k^{(m)} s(n - k) \tag{22a}$$

$$b_m(n) = \sum_{k=0}^{m} a_k^{(m)} s(n - m + k). \tag{22b}$$

Squaring (22a) and taking the expected value, there results

$$F_m(n) = \sum_{k=0}^{m} \sum_{i=0}^{m} a_k^{(m)} a_i^{(m)} \phi(k, i) \tag{23}$$

where

$$\phi(k, i) = E\,[s\,(n - k)\, s\,(n - i)] \tag{24}$$

is the nonstationary autocorrelation (or covariance) of the signal $s(n)$. ($\phi(k, i)$ in (24) is technically a function of n, which has been dropped for convenience.) In a similar fashion one can show from (22b), with n replaced by $n - 1$, that

$$B_m(n - 1) = \sum_{k=0}^{m} \sum_{i=0}^{m} a_k^{(m)} a_i^{(m)} \phi(m + 1 - k, m + 1 - i) \tag{25}$$

$$C_m(n) = \sum_{k=0}^{m} \sum_{i=0}^{m} a_k^{(m)} a_i^{(m)} \phi(k, m + 1 - i). \tag{26}$$

Given the covariance of the signal, the reflection coefficient at stage $m + 1$ can be computed from (23), (25), and (26) by substituting them in the desired formula for K_{m+1}. The name "covariance-lattice" stems from the fact that this is basically a lattice method that is computed from the covariance of the signal; it can be viewed as a way of stabilizing the covariance method. One salient feature is that the forward and backward residuals are never actually computed in this method. But this is not different from the nonlattice methods.

In the harmonic-mean method (18), $F_m(n)$ need not be computed from (23); one can use (21a) instead, with m replaced by $m - 1$. However, one must use (25) to compute $B_m(n - 1)$;

(21b) cannot be used because $B_{m-1}(n-2)$ would be needed and it is not readily available.

A. Stationary Case

For a stationary signal, the covariance reduces to the autocorrelation

$$\phi(k,i) = R(i-k) = R(k-i) \qquad \text{(Stationary)}. \quad (27)$$

From (23)-(27), it is clear that

$$F_m = B_m = \sum_{k=0}^{m}\sum_{i=0}^{m} a_k^{(m)} a_i^{(m)} R(i-k) \quad (28)$$

and

$$C_m = \sum_{k=0}^{m}\sum_{i=0}^{m} a_k^{(m)} a_i^{(m)} R(m+1-i-k). \quad (29)$$

Making use of the normal equations [1]

$$\sum_{i=0}^{m} a_i^{(m)} R(i-k) = 0, \qquad 1 \leq k \leq m \quad (30)$$

and of (21), one can show that the stationary reflection coefficient is given by

$$K_{m+1} = -\frac{C_m}{F_m} = -\frac{\sum_{k=0}^{m} a_k^{(m)} R(m+1-k)}{(1-K_m^2)F_{m-1}} \quad (31)$$

with $F_0 = R_0$. Equation (31) is exactly the equation used in the autocorrelation method.

B. Quantization of Reflection Coefficients

One of the features of lattice methods is that the quantization of the reflection coefficients can be accomplished within the recursion, i.e., K_m can be quantized before K_{m+1} is computed. In this manner, it is hoped that some of the effects of quantization can be compensated for.

In applying the covariance-lattice procedure to the harmonic-mean method, one must be careful to use (23) and *not* (21a) to compute $F_m(n)$. The reason is that (21a) is based on the optimality of K^B, which would no longer be true after quantization.

Similar reasoning can be applied to the autocorrelation method. Those who have tried to quantize K_m inside the recursion have no doubt been met with serious difficulties. The reason is that (31) assumes the optimality of the predictor coefficients at stage m, which no longer would be true if K_m were quantized. The solution is to use (28) and (29), which make no assumptions of optimality. Thus we have what we shall call the *autocorrelation-lattice* method, where there is only one definition of K_{m+1}

$$K_{m+1} = -\frac{C_m}{F_m} = -\frac{\sum_{k=0}^{m}\sum_{i=0}^{m} a_k^{(m)} a_i^{(m)} R(m+1-i-k)}{\sum_{k=0}^{m}\sum_{i=0}^{m} a_k^{(m)} a_i^{(m)} R(i-k)}. \quad (32)$$

IV. Computational Issues

A. Simplifications

Equations (23), (25), and (26) can be rewritten to reduce the number of computations by about one half. The results for $C_m(n)$ and $F_m(n) + B_m(n-1)$ can be shown to be as follows:

$$\begin{aligned} C_m(n) = {} & \phi(0,m+1) + \sum_{k=1}^{m} a_k^{(m)} [\phi(0,m+1-k) \\ & + \phi(k,m+1)] + \sum_{k=1}^{m} [a_k^{(m)}]^2 \phi(k,m+1-k) \\ & + \sum_{k=1}^{m-1}\sum_{i=k+1}^{m} a_k^{(m)} a_i^{(m)} [\phi(k,m+1-i) \\ & + \phi(i,m+1-k)] \end{aligned} \quad (33)$$

$$\begin{aligned} F_m(n) + B_m(n-1) = {} & \phi(0,0) + \phi(m+1,m+1) \\ & + 2\sum_{k=1}^{m} a_k^{(m)} [\phi(0,k) + \phi(m+1,m+1-k)] \\ & + \sum_{k=1}^{m} [a_k^{(m)}]^2 \cdot [\phi(k,k) + \phi(m+1-k,m+1-k)] \\ & + 2\sum_{k=1}^{m-1}\sum_{i=k+1}^{m} a_k^{(m)} a_i^{(m)} \\ & \cdot [\phi(k,i) + \phi(m+1-k,m+1-i)]. \end{aligned} \quad (34)$$

The third term in (33) can be computed more efficiently as follows:

$$\begin{aligned} & \sum_{k=1}^{m} [a_k^{(m)}]^2 \phi(k,m+1-k) \\ & = \sum_{k=1}^{m/2} \{[a_k^{(m)}]^2 + [a_{m+1-k}^{(m)}]^2\} \phi(k,m+1-k) \\ & \underbrace{+ [a_{(m+1)/2}^{(m)}]^2 \phi\left(\frac{m+1}{2}, \frac{m+1}{2}\right)}_{\text{only if } m \text{ odd}}. \end{aligned} \quad (35)$$

A similar simplification can be used in (34).

For the stationary case, (28) can be rewritten as

$$F_m = \sum_{k=-m}^{m} b_k R(k) = b_0 + 2\sum_{k=1}^{m} b_k R(k) \quad (36)$$

where

$$b_k = \sum_{l=0}^{m-|k|} a_l^{(m)} a_{l+|k|}^{(m)} \quad (37)$$

is the autocorrelation of the impulse response of $A(z)$. By setting $l = i + k$ in (29), one can show that C_m is reduced to

$$C_m = \sum_{l=0}^{2m} c_l R(m+1-l) \quad (38)$$

where

$$c_l = \sum_{k=0}^{m} a_k^{(m)} a_{l-k}^{(m)}, \quad 0 \leq l \leq 2m \tag{39}$$

is the convolution of the impulse response of $A(z)$ with itself. Equation (39) assumes that $a_k^{(m)} = 0$ for $k < 0$ and $k > m$. Equation (38) can be rewritten as

$$C_m = R(m+1) + 2a_1^{(m)} R(m) + c_{m+1} R(0) + \sum_{k=1}^{m-1} (c_{m+1-k} + c_{m+1+k}) R(k). \tag{40}$$

Equation (39) can also be rewritten to reduce the computations further.

B. Covariance Computation

The covariance $\phi(k, i)$ of the signal is defined in (24) as a nonstationary autocorrelation, which, strictly speaking, should be estimated by averaging over an ensemble of the random process. In practice, however, it is often the case that such averaging is neither feasible nor desirable. For example, in most speech applications, one is interested in analyzing the time-varying properties of a particular utterance and not the whole ensemble of speech that a speaker might utter. In the case where a single time history of a random process is available for analysis, it is common to describe that single time record as nonstationary if its short-term sample properties (such as mean and autocorrelation) vary significantly with time [8]. For this situation, we give below two methods for computing the covariance of a signal that is known, say, for $0 \leq n \leq N-1$.

Method 1:

$$\phi(k, i) = \sum_{n=p}^{N-1} s(n-k) s(n-i), \quad 0 \leq k, i \leq p \tag{41}$$

where p is the order of the predictor, and the customary division by the number of terms in the summation (in this case $N - p$) has been omitted since it does not affect the solution for the reflection coefficients. If we assume that (41) estimates the covariance at time $t = 0$, then the covariance at any other time t can be estimated by setting the lower and upper limits of the summation in (41) to $p + t$ and $N - 1 + t$, respectively. Note that (41) makes no assumptions about the signal outside the given range and, hence, is especially useful for short durations [6] and nonstationary signals. On the other hand, if the signal is assumed to be zero outside the given range (i.e., the signal is windowed), then the signal is effectively forced to be stationary, with an associated autocorrelation given by

$$R(i) = \sum_{n=0}^{N-1-|i|} s_n s_{n+|i|}, \quad 0 \leq i \leq p. \tag{42}$$

Method 2: The second method makes maximum use of the data in the range $0 \leq n \leq N-1$. This is accomplished by recomputing the covariance for each new lattice stage as follows:

$$\phi_m(k, i) = \sum_{n=m}^{N-1} s(n-k) s(n-i), \quad 0 \leq k, i \leq m \tag{43}$$

where $\phi_m(k, i)$ is the covariance used in computing K_m. The computations in (43) can be simplified considerably by noting that

$$\phi_{m+1}(k, i) = \phi_m(k, i) - s(m-k) s(m-i), \quad 0 \leq k, i \leq m. \tag{44}$$

Therefore, the covariance coefficients for stage $m + 1$ can be computed from those for stage m using (44) in the range $0 \leq k, i \leq m$. For $k = m + 1$ or $i = m + 1$, (43) needs to be used.

It can be shown that when Method 2 for computing the covariance is used in conjunction with the harmonic-mean computation in (18), the results for the reflection coefficients are identical to Burg's method as described in [6]. However, our results here are obtained at a much lower computational cost.

For the case where $N \gg p$, Methods 1 and 2 should give similar results. However, if N is not much greater than p, then it would seem reasonable to utilize the given data maximally by using Method 2.

There are other possible methods for computing the covariance or the autocorrelation of the signal. Irrespective of which method one chooses, it is important to make sure that the resulting covariance or autocorrelation function is positive definite. Otherwise, filter stability cannot be guaranteed.

C. Computational Cost

Table I shows a comparison of the number of computations for the different methods, where terms of order p have been neglected. The computations for the autocorrelation-lattice and covariance-lattice methods are on the order of $pN + O(p^3)$, as compared to $5pN$ for the regular lattice methods where the residuals are computed. For $N \gg p$, the new lattice methods typically offer a 3-4-fold saving over the regular lattice methods.

When compared to nonlattice methods, the increase in computation for the covariance-lattice method is not significant if N is large compared to p, which is usually the case (compare the first and second rows in Table I). Furthermore, in the covariance-lattice method, the number of signal samples can be reduced to about half that used in the autocorrelation method. This not only reduces the number of computations but also improves spectral resolution by reducing the amount of averaging.

D. FWL Computations

One point of comparison between the different methods is the stability of the all-pole filter when FWL computations are used. The main comparison here is between the autocorrelation method and the lattice methods (the covariance method cannot guarantee stability, in general, even with floating-point computations). Under FWL conditions, we expect filter stability to be ensured more with the lattice methods than with the autocorrelation method. If, at some stage of the recursion, K_m turns out to be greater than one because of FWL computations, it can be artificially set to a value less than one to ensure stability. Such a scheme would work well with the lattice methods, but not with the autocorrelation method because in the latter, global optimality of each K is assumed at every stage. Lack of optimality leads to error propagation, which in turn makes later stages more susceptible to instability. The problem does not exist to the same magnitude in the lattice methods since consecutive stages are "decoupled," with no assumptions of global optimality being made. This phenomenon is the same as that discussed in Section III-B, which allows the quantization of the reflection coefficient inside the recursion of the lattice methods.

V. Procedure

Below is the complete algorithm for what we believe currently to be one of the more promising methods for linear predictive analysis. It comprises the harmonic-mean definition (18) for the reflection coefficients, and the covariance-lattice method.

a) Compute the covariances $\phi(k, i)$ for $k, i = 0, 1, \cdots, p$.
b) $m \leftarrow 0$.
c) Compute $C_m(n)$ and $F_m(n) + B_m(n-1)$ from (33) and (34), or from (23), (25), and (26).
d) Compute K_{m+1} from (18).
e) Quantize K_{m+1}, if desired (perhaps using log area ratios [7] or some other technique).
f) Using (3), compute the predictor coefficients $\{a_k^{(m+1)}\}$ from $\{a_k^{(m)}\}$ and K_{m+1}. Use the quantized value, if K_{m+1} was quantized in d).
g) $m \leftarrow m + 1$.
h) If $m < p$, go to c); otherwise exit.

We have used this procedure to analyze speech signals, with the signal covariance estimated by (41). In general, the results were somewhere between those using the autocorrelation and covariance methods. In particular, the pole bandwidths were usually less than those from the autocorrelation method, but greater than those from the covariance method. In all cases where the covariance method gave unstable results, the covariance-lattice method gave stable results.

While the performance of all linear prediction methods tends to deteriorate (in terms of spectral accuracy) as the number of signal samples N is sharply reduced, we believe that the procedure given above should continue to give better resolution than the autocorrelation method, and should continue to guarantee stability, unlike the covariance method, which tends to become unstable for short durations.

VI. Conclusions

This paper presented a class of lattice methods for linear prediction that guarantees the stability of the all-pole filter, with or without windowing of the signal, and with FWL computations. Also, for data-compression purposes, quantization of the reflection coefficients can be accomplished within the recursion, if desired, without affecting the stability of the filter. It was shown that the methods of Itakura and Burg are special cases of this class of lattice methods.

A procedure was derived to make these lattice methods more efficient computationally, with a cost comparable to the traditional autocorrelation and covariance methods. The procedure, named the *covariance-lattice* method, computes the reflection coefficients recursively in terms of the covariance of the signal and the filter parameters at each stage. When used with speech signals, this method gave results somewhere in between the autocorrelation and covariance methods.

Acknowledgment

The author wishes to thank R. Viswanathan for implementing the covariance-lattice method and for his discussions and comments on this paper.

References

[1] J. Makhoul, "Linear prediction: A tutorial review," *Proc. IEEE*, vol. 63, pp. 561-580, Apr. 1975.

[2] J. Markel and A. H. Gray, Jr., "Fixed-point truncation arithmetic implementation of a linear prediction autocorrelation vocoder." *IEEE Trans. Acoust., Speech, Signal Processing*, vol. ASSP-22, pp. 273-281, Apr. 1974.

[3] B. S. Atal and S. L. Hanauer, "Speech analysis and synthesis by linear prediction of the speech wave," *J. Acoust. Soc. Amer.*, vol. 50, pp. 637-655, Aug. 1971.

[4] F. Itakura and S. Saito, "Digital filtering techniques for speech analysis and synthesis," presented at the 7th Int. Cong. Acoustics, Budapest, 1971, Paper 25-C-1.

[5] J. Burg, "Maximum entropy spectral analysis," Ph. D. dissertation, Stanford Univ., Stanford, CA, May 1975.

[6] D. E. Smylie, G. K. C. Clarke, and T. J. Ulrych, "Analysis of Irregularities in the earth's rotation," in *Methods in Computational Physics*, vol. 13. New York: Academic, 1973, pp. 391-430.

[7] R. Viswanathan and J. Makhoul, "Quantization properties of transmission parameters in linear predictive systems," *IEEE Trans. Acoust., Speech, Signal Processing*, vol. ASSP-23, pp. 309-321, June 1975.

[8] J. Bendat and A. Piersol, *Random Data: Analysis and Measurement Procedures.* New York: Wiley-Interscience, 1971, pp. 13-14.

Lattice Filters for Adaptive Processing

BENJAMIN FRIEDLANDER, SENIOR MEMBER, IEEE

Abstract—This paper presents a tutorial review of lattice structures and their use for adaptive prediction of time series. Lattice filters associated with stationary covariance sequences and their properties are discussed. The least squares prediction problem is defined for the given data case, and it is shown that many of the currently used lattice methods are actually approximations to the stationary least squares solution. The recently developed class of adaptive least squares lattice algorithms are described in detail, both in their unnormalized and normalized forms. The performance of the adaptive least squares lattice algorithm is compared to that of some gradient adaptive methods. Lattice forms for ARMA processes, for joint process estimation, and for the sliding-window covariance case are presented. The use of lattice structures for efficient factorization of covariance matrices and solution of Toeplitz sets of equations is briefly discussed.

I. Introduction

A. Linear Prediction

THE AREAS OF adaptive signal processing and adaptive control have grown at a fast rate during the last decade. The demand for high-performance systems combined with the availability of ever-increasing computational power, motivated the search for more sophisticated processing algorithms capable of operating in uncertain, time-varying environments. A significant part of the work in these areas is based on modeling the process of interest as an output of a finite-order discrete-time linear system driven by white noise. Such processes were called autoregressive (AR) or autoregressive moving average (ARMA).

Many adaptive signal-processing techniques are based on the solution of the following prototype linear prediction problem: Let y_T be a (scalar) discrete-time stationary zero-mean process. We are interested in predicting the current value of this process from past measurements. A linear predictor of order N will have the form

$$\hat{y}_{T|T-1} = -\sum_{i=1}^{N} A_{N,i} y_{T-i} \tag{1}$$

where $\hat{y}_{T|T-1}$ is the predicted value of y_T based on data up to time $T-1$ and $\{A_{N,i}, i = 1, \cdots, N\}$ are the predictor coefficients. The difference between the actual value of the process and its predicted value will be called the prediction error (of order N)

$$\epsilon_{N,T} = y_T - \hat{y}_{T|T-1} = y_T + \sum_{i=1}^{N} A_{N,i} y_{T-i}. \tag{2}$$

The predictor can be written in z-transform notation as

$$A_N(z) = 1 + A_{N,1} z^{-1} + \cdots + A_{N,N} z^{-N} \tag{3a}$$

and (2) can be written more compactly as

$$\epsilon_{N,T} = A_N(z) y_T \tag{3b}$$

using the convention that $z^{-1} y_T = y_{T-1}$. An equivalent form of this equation is $\epsilon_N(z) = A_N(z) y(z)$, where $\epsilon_N(z)$ and $y(z)$ are the z-transforms of $\epsilon_{N,T}$ and y_T. The predictor coefficients need to be adjusted to make the prediction errors, in some well-defined sense, as small as possible. The optimal least squares predictor minimizes the mean-square error $E\{\epsilon_{N,T}^2\}$. The coefficients $A_{N,i}$ of this optimal predictor are uniquely determined by the second-order statistics of the process y_T, i.e., by the autocorrelation coefficients $\{R_i, i = 0, \cdots, N\}$ where $R_i = E\{y_T y_{T-i}\}$. These coefficients can be obtained by solving a set of linear equations called the Yule–Walker equations [164]. Levinson [156], Wiggins [163], Whittle [159], Durbin [152], Robinson [162], and others have provided efficient computational procedures for solving these equations.

The linear predictor has found many applications. In the Linear Predictive Coding (LPC) method of speech analysis/synthesis, it is used to reconstruct a speech waveform. In seismic data processing the prediction filter is used for deconvolution. Other applications in which the linear predictor plays a key role include: high-resolution spectral estimation [165] adaptive line enhancement [166], adaptive noise canceling [167], and adaptive array processing [168]. See [53] for a survey on linear prediction.

The linear predictor (1) can be considered as a finite-impulse response (FIR) filter acting on the data. As is known from the theory of digital signal processing such filters can be realized in many different ways: direct realization, cascade form, and parallel form [27]. In most cases the predictor is assumed to be implemented in direct (or tapped-delay-line) form as depicted in Fig. 1. In some cases, we find predictors implemented in a particular cascade form: the so-called lattice or ladder structure depicted in Fig. 2.

B. Lattice Prediction Filters

The transfer function of the lattice filter in Fig. 2 is determined by the values of the parameters K_i which are called reflection coefficients or PARtial CORrelation (PARCOR) coefficients, for reasons to be explained later. The values of these reflection coefficients are uniquely determined by the transfer function $A_N(z)$, or equivalently, by the autocorrelation sequence $\{R_i\}$. The transformation between the set of reflection coefficients $\{K_i\}$ and the parameters $\{A_{N,i}\}$ is highly nonlinear. Algorithms for computing reflection coefficients were developed by Schur [169], Levinson [156], and others [14], [15], [55]. Both the direct realization and the lattice form can be used to implement a given prediction filter on a digital computer or in special-purpose hardware. While the two implementations are mathematically equivalent, there are many practical differences which lead us to favor the lattice structure. Some of these differences are discussed next.

Two important issues in digital filtering are roundoff noise

Manuscript received October 13, 1981; revised June 21, 1982. The submission of this paper was encouraged after the review of an advance proposal. This work was supported by the Office of Naval Research under Contract N00014-81-C-0300.

The author is with Systems Control Technology, Inc., Palo Alto, CA 94304.

Reprinted from *Proc. IEEE,* vol. 70, no. 8, pp. 829–867, August 1982.

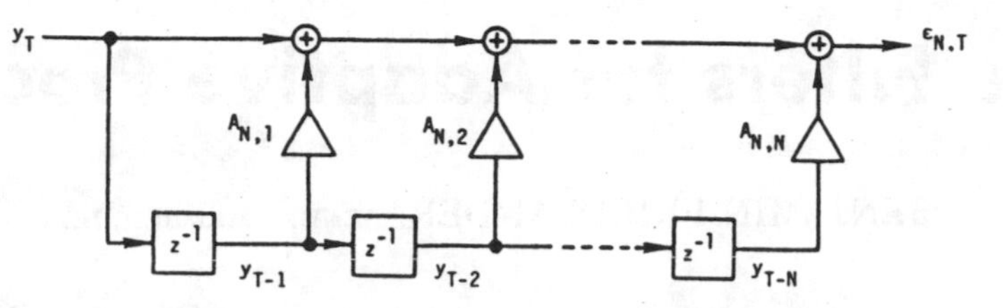

Fig. 1. Direct realization of a FIR least squares prediction error filter.

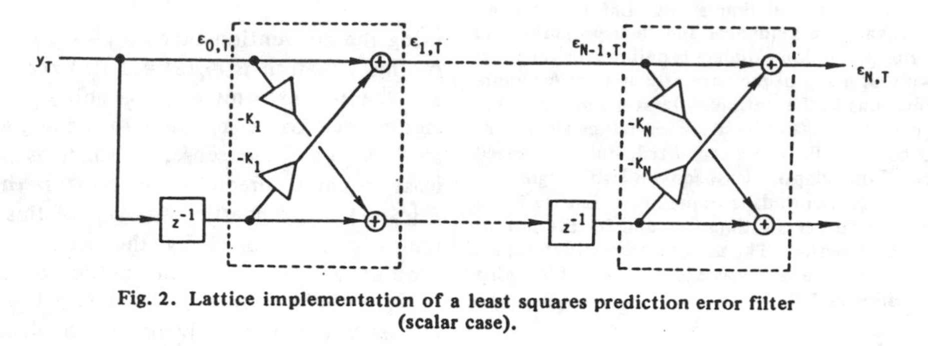

Fig. 2. Lattice implementation of a least squares prediction error filter (scalar case).

and the sensitivity of filter performance to perturbations in its parameter values. Roundoff noise arises inevitably in digital computations due to the need to represent numerical values by a finite number of digits. A commonly held opinion, based on experience with direct realizations, is that roundoff noise in fixed-point digital filters increases in severity as the width of the passband is made very small. Several studies have shown the existence of filter structures whose output noise is essentially independent of bandwidth and center frequency [15], [16], [19], [160], [161]. Lattice filters of the type discussed in this paper share this very useful property.

The parameters of any digital filter (e.g., $A_{p,i}$ or K_i) will be stored and used in quantized form. Typically, a coarser quantization (i.e., shorter word length) leads to simpler and cheaper hardware. This raises the question of how sensitive is the performance of the filter to parameter inaccuracies caused by quantization. In several studies it was shown that lattice filters can achieve a given performance criterion (such as the maximum ripple in the passband) using parameters specified by shorter word lengths than a corresponding direct realization [3], [5], [13]. This means that under similar hardware constraints, the performance of a lattice filter will generally be better than that of a direct realization.

In the discussion above we considered lattice implementations of digital filters in general. In the specific context of linear prediction, lattice structures have some particularly useful properties. One such property is that the Nth order least squares lattice prediction filter contains in it prediction filters of all lower orders. More precisely, the first p sections of the Nth order lattice predictor (cf. Fig. 2) form the pth order prediction filter. The signals $\{\epsilon_{1,T}, \cdots, \epsilon_{N,T}\}$ propagating in this filter are, therefore, least squares prediction errors. These prediction errors can be shown to be mutually orthogonal, i.e., $E\{\epsilon_{i,T}\epsilon_{j,T}\} = 0$ for $i \neq j$. The lattice filter can be viewed as performing Gram–Schmidt orthogonalization on the incoming data. Each lattice section performs one step in this Gram–Schmidt procedure as will be shown later. The usefulness of this orthogonality property manifests itself in many ways which will become clear only after a more detailed discussion of the lattice predictor. At this point, we want only to emphasize that the orthogonality property discussed above is not shared by direct realizations of the prediction filter: The first p stages of the filter in Fig. 1 do not form a least squares prediction filter, and the signals propagating in it are not orthogonal. To specify prediction, filters of all orders up to N will require knowledge of $N(N+1)/2$ predictor parameters ($A_{p,i}$ for $1 \leqslant i \leqslant p \leqslant N$), whereas only N reflection coefficients are needed (K_i, $i = 1, \cdots, N$) when a lattice filter is used. This seems to indicate that lattice structures provide a natural (minimal) parametrization of linear predictors.

Another area where lattice filters are of special interest is in adaptive prediction. Because of the importance of this topic we will briefly digress from the discussion of lattice filters.

C. Adaptive Prediction

As was mentioned before, the least squares predictor is determined by the second-order statistics of the observed process y_T. In most practical applications these statistics are not known *a priori* and have to be estimated from data. If the process is known to be stationary and if a large amount of data is available, it is possible to estimate the autocorrelation sequence and compute the least squares predictor once and for all. In practice, signals are often nonstationary, having time-varying statistics. In such situations, the process of estimating second-order statistics and computing predictor coefficients needs to be repeated often, to track the changing statistics. This problem of predicting a time series without prior knowledge of its statistics is called adaptive prediction.

Adaptive processing techniques can be roughly divided into two types: block processing and recursive methods. In the first, incoming data are divided into blocks, and each block is processed as a whole to estimate predictor parameters (or other information). In the second, predictor parameters are updated as each new data point becomes available. Recursive techniques are better suited for real-time applications since they utilize new data as soon as they become available, whereas in block processing the estimated parameters are updated only once per block period. Block methods play an important role in speech processing, spectral estimation, and other applications. However, in this paper we will be concerned mainly with recursive techniques.

Recursive algorithms for computing the least squares predictor coefficients $\{A_{N,i}\}$ have been known for quite some time. The so-called Recursive Least Squares (RLS) algorithm is widely used for system identification [158]. This algorithm is closely related to the Kalman filter which was developed in the early 1960's [170]. These algorithms are conceptually quite simple but require a fair amount of computations. Each update of the parameters of an Nth order predictor using the RLS algorithm requires in the order of N^2 operations (multiplies and adds). Various suboptimal techniques were developed to reduce the amount of computation. Perhaps the best known of these is a gradient-search technique for solving the underlying least squares optimization problem, called the Widrow–Hoff LMS (for Least Mean Squares) algorithm [167], [171], [172]. This algorithm requires only in the order of N operations per time step. Algorithms based on gradient-search procedures were also developed for updating the reflection coefficients of lattice prediction filters [46]–[49], [58], [59].

The simplicity and ease of implementation of gradient-search adaptive algorithms led to their widespread use in the past two decades. While gradient techniques have provided successful solutions for many practical adaptive signal-processing problems, their performance is inferior to that of true least squares techniques. An important manifestation of this suboptimal behavior is the relatively slow convergence rate of gradient-search algorithms (by "convergence rate" we mean the time it takes the algorithm to respond to sudden changes in the process statistics). We will discuss this issue in more detail later.

In the past ten years there have been some developments in the area of recursive least squares estimation that have important consequences for adaptive processing. A class of so-called "fast" least squares algorithms was introduced by Morf and his colleagues [82], [118], [121], [127], [128]. These algorithms are capable of recursively updating the least squares Nth order predictor coefficients $A_{p,i}$ using in the order of N operations per time step. In other words, these algorithms are capable of performing *least squares adaptive prediction at a computational price comparable with that of the suboptimal gradient techniques.* These algorithms have convergence properties superior to those of gradient-search techniques, since they are true least squares techniques. More recently, it was shown that the reflection coefficients of the least squares lattice predictor can also be updated with only order of N operations per time step [108]–[110], [122], [130]. In fact, the computational requirements for the adaptive lattice predictor are somewhat smaller than those for the corresponding direct realization. We will refer to this lattice predictor as the least squares lattice (LSL) form. In this paper we will deal almost exclusively with lattice predictors, motivated by their many attractive features. Readers interested in the area of "fast" algorithms for the direct realization of the linear predictor are referred to [82]–[84], [99], [100], [106], [107], [118], [121], [123], [125]–[128].

D. Paper Outline

The development of recursive algorithms for adaptive least squares lattice predictors is fairly recent. Results are scattered in various papers, reports, and theses. This paper represents a first attempt to present these results in a coherent framework and to put them into perspective with respect to earlier work on lattice filters, especially that related to speech processing.

We start in Section II by developing the Levinson algorithm for efficiently computing predictor parameters from the covariance sequence of a stationary process. Two lattice prediction filters are presented: one unnormalized, and one in which the signals propagating in the filter are normalized to have unit variance. Next we introduce a number of efficient computational schemes based on these lattice structures: Cholesky decomposition of a covariance matrix and its inverse (Cholesky decomposition of a matrix is its factorization into a product of a lower triangular times an upper triangular matrix, or vice versa), transforming reflection coefficients to direct realization parameters, and recovering the covariance sequence from knowledge of the related lattice predictor.

In Section III we consider the lattice predictor for the given data case, i.e., when the covariance sequence is not known *a priori*. The least squares prediction problem is precisely defined. A number of block processing techniques for computing reflection coefficients that have appeared in the literature are described. It is shown that these techniques can be interpreted as various approximations derived from the least squares lattice predictor for the given covariance case.

The basic least squares lattice form is developed in Section IV. The update equations for the reflection coefficients of this adaptive lattice predictor are derived. Some elementary notions from linear algebra and projections on linear vector spaces are required for this development. Two types of lattice structures are involved: one for computing lattice parameters from the observed data, and one for implementing the least squares lattice predictor. Several versions of the least squares lattice form are presented: unnormalized, normalized, and variance normalized.

In Section V we present some of the gradient-based adaptive filters. The performance of these algorithms is compared to that of the least squares lattice in terms of convergence rate, parameter tracking capability, and computational complexity.

Section VI summarizes some useful extensions of the basic least squares lattice filter. A lattice predictor for ARMA processes, a lattice predictor for one process from measurements of a related process, and the covariance lattice form are presented. This section may be skipped without loss of continuity.

The least squares lattice filter can be used for various applications outside the area of adaptive processing. Some of these applications are discussed in Section VII. The reader interested primarily in adaptive processing may want to skip this section as well as Sections II-D–II-F. Readers interested primarily in applications of lattice forms may want to go directly to Section V. The various algorithms discussed in the paper were summarized in tables for ease of reference and coding.

E. Notation

Throughout the paper we will use matrix notation. The following conventions will be used:

1) The transpose of matrix M is denoted M'.

2) Vectors are defined in row form, e.g., $v = [v_0, v_1, \cdots, v_T]$. Column vectors will be transposed row vectors, e.g., v'.

3) The square root $M^{1/2}$ of a square matrix M is any matrix (of the same dimensions as M) with the property $M^{1/2}(M^{1/2})' = M$. In this paper we will be especially interested in the case where $M^{1/2}$ is a lower triangular or an upper triangular matrix. When not otherwise specified, any matrix square root can be used for $M^{1/2}$. The transpose of a matrix square root $(M^{1/2})'$ will be denoted more compactly as $M^{T/2}$.

4) The inverse of a superscripted matrix $(M^s)^{-1}$ will be denoted more compactly as M^{-s}. The same convention will be used for the matrix square root, e.g., $(M^{1/2})^{-1} \triangleq M^{-1/2}$, $(M^{T/2})^{-1} \triangleq M^{-T/2}$.

The paper treats multichannel prediction problems. Much of

the notational burden can be ignored when the scalar case is considered. Finally, a list of the most commonly used symbols in this paper is presented.

F. List of Symbols

"—"	A variance normalized variable.
"~"	A normalized variable.
ϵ	Forward prediction error.
r	Backward prediction error.
R^ϵ	Forward prediction error variance.
R^r	Backward prediction error variance.
A	Forward prediction filter.
B	Backward prediction filter.
Δ	Cross correlation of forward and backward prediction errors.
K^ϵ, K^r	Unnormalized reflection coefficients.
K	Normalized reflection coefficients.
R	Autocorrelation coefficients of the data y.
$\mathbf{R}$	Covariance matrix of the data y.
γ	Likelihood variable.
ρ	Cross-correlation coefficient.
P	Projection operator.
θ	Transfer matrix of a lattice section.
λ	Exponential weighting factor.
$U_{\bar{A}}, U_\alpha$	Upper triangular factor: $\mathbf{R} = U_\alpha U'_\alpha$, $\mathbf{R}^{-1} = U'_{\bar{A}} U_{\bar{A}}$.
$L_{\bar{B}}, L_\beta$	Lower triangular factor: $\mathbf{R} = L_\beta L'_\beta$, $\mathbf{R}^{-1} = L'_{\bar{B}} L_{\bar{B}}$.
I	Identity matrix = 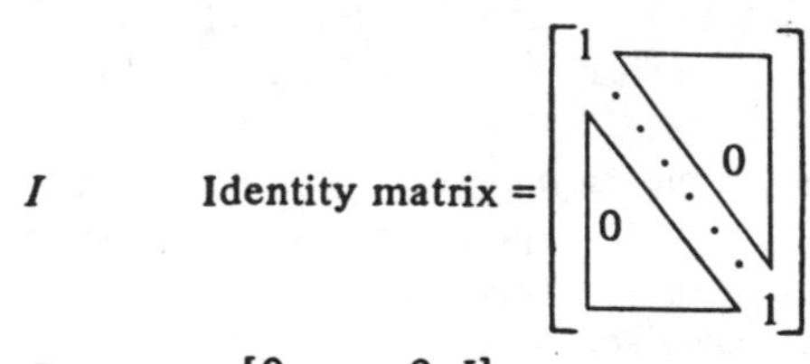
π	$= [0, \cdots, 0, I]$.
σ	$= [I, 0, \cdots, 0]$.
y_T	Data.
$y_{m:n}$	$= [y_m, y_{m+1}, \cdots, y_n]$ Data vector.
$y^p_{m:n}$	$= [0, \cdots, 0, \ y_m, \cdots, y_{n-p}]$ Shifted data vector.
Y	Data matrix.

II. The Stationary Case

To see how lattice structures arise in least squares estimation we start with the simple problem of predicting a multichannel stationary time series given its second-order statistics, i.e., the covariance sequence (the mean is assumed to be zero). It is worthwhile to consider in some detail the "given covariance" case since it provides the basis for many of the currently used lattice techniques and provides valuable insights into the more complex "given data" case.

A. The Levinson Algorithm

Consider the following estimation problem: Given measurements of a multichannel stationary time series $\{y_T\}$ (where y_T are m-vectors), find the least squares one-step ahead predictor, defined by (1). Let

$$R_{i-j} \triangleq E\{y_i y'_j\}, \qquad \text{correlation sequence.} \qquad (4)$$

The coefficients of the least squares predictor can be computed by using the following orthogonality property of the least squares prediction error [154]:

$$E\{\epsilon_{p,T} y'_{T-j}\} = 0, \qquad \text{for } j = 1, \cdots, p \qquad (5)$$

where

$$\epsilon_{p,T} = y_T + \sum_{i=1}^{p} A_{p,i} y_{T-i}, \qquad A_{p,i} \text{ are } m \times m \text{ matrices.} \qquad (6)$$

Post-multiplying the equation above by y'_{T-j} and taking expected values gives

$$R_j + \sum_{i=1}^{p} A_{p,i} R_{j-i} = 0, \qquad \text{for } i, j = 1, \cdots, p. \qquad (7a)$$

The mean-square error is given by

$$R^\epsilon_p = E\{\epsilon_{p,T} \epsilon'_{p,T}\} = R_0 + \sum_{i=1}^{p} A_{p,i} R_{-i}. \qquad (7b)$$

Equations (6), (7) can be rearranged in a single (block)-Toeplitz[1] matrix equation

$$[I, A_{p,1}, \cdots, A_{p,p}] \underbrace{\begin{bmatrix} R_0 & R_1 & \cdots & R_p \\ R_{-1} & R_0 & \ddots & \vdots \\ \vdots & \ddots & \ddots & R_1 \\ R_{-p} & \cdots & R_{-1} & R_0 \end{bmatrix}}_{\mathbf{R}_p} = [R^\epsilon_p, 0, \cdots, 0]. \qquad (8)$$

The above is called the multichannel Yule–Walker equation. The matrix is symmetric since $R_{-i} = R'_i$, as can be verified from the definition of the correlation sequence (4). As we shall see, efficient computation of the predictor coefficients $\{A_{p,i}\}$ involves some auxiliary quantities such as the backward predictor

$$\hat{y}_{T-p-1|T-1} = -\sum_{i=1}^{p} B_{p,p+1-i} y_{T-i}. \qquad (9)$$

Following the same procedure as before we can show that the backward predictor coefficients can be obtained as the solution of the following matrix equation:

$$[B_{p,p}, \cdots, B_{p,1}, \ I] \mathbf{R}_p = [0, \cdots, 0, \ R^r_p] \qquad (10)$$

where

$$R^r_p = E\{r_{p,T-1} r'_{p,T-1}\}, \qquad \text{mean-square error of the backward prediction} \qquad (11a)$$

and

$$r_{p,T-1} = y_{T-p-1} - \hat{y}_{T-p-1|T-1}, \qquad \text{backwards prediction error.} \qquad (11b)$$

An efficient solution of the matrix equation (8) was provided by Levinson [156] and later by Durbin [152] for scalar time series. These results were extended by Whittle [159] to the multivariate case. See [153] for a more complete perspective on Levinson's work.

The derivation of these recursions starts by considering a trial solution and then correcting for the error made in our initial attempt. Assuming that we know the pth order predic-

[1] A Toeplitz matrix has the property that all the entries along diagonals are the same, i.e., $R = [R_{i-j}]$.

tor we try the following $(p+1)$th order solution

$$\begin{bmatrix} I, & A_{p,1}, \cdots, A_{p,p}, & 0 \\ 0, & B_{p,p}, \cdots, B_{p,1}, & I \end{bmatrix} R_{p+1}$$

$$= \begin{bmatrix} R_p^\epsilon, & 0, \cdots, 0, & \Delta_{p+1}^\epsilon \\ \Delta_{p+1}^{r'}, & 0, \cdots, 0, & R_p^r \end{bmatrix} \tag{12a}$$

where

$$\Delta_{p+1}^\epsilon = R_{p+1} + \sum_{i=1}^{p} A_{p,i} R_{p-i+1} \tag{12b}$$

$$\Delta_{p+1}^{r'} = \sum_{i=1}^{p} B_{p,i} R_{-p-1+i} + R_{-p-1} \tag{12c}$$

are, in general, some nonzero terms. The trial solution thus fails to give the correct predictors. However, a simple linear combination of these solutions will force zeroes into the positions occupied by Δ_{p+1}^ϵ and Δ_{p+1}^{r}, and give the desired predictor coefficients

$$\begin{bmatrix} I & -\Delta_{p+1}^\epsilon R_p^{-r} \\ -\Delta_{p+1}^{r'} R_p^{-\epsilon} & I \end{bmatrix} \begin{bmatrix} I, & A_{p,1}, \cdots, A_{p,p}, & 0 \\ 0, & B_{p,p}, \cdots, B_{p,1}, & I \end{bmatrix} R_{p+1}$$

$$= \begin{bmatrix} I, & A_{p+1,1}, \cdots, A_{p+1,p+1} \\ B_{p+1,p+1}, & B_{p+1,p}, \cdots, I \end{bmatrix} R_{p+1}$$

$$= \begin{bmatrix} R_{p+1}^\epsilon, & 0, \cdots, 0, & 0 \\ 0, & 0, \cdots, 0, & R_{p+1}^r \end{bmatrix} \tag{13}$$

where

$$R_{p+1}^\epsilon = R_p^\epsilon - \Delta_{p+1}^\epsilon R_p^{-r} \Delta_{p+1}^{r'} \tag{14a}$$

$$R_{p+1}^r = R_p^r - \Delta_{p+1}^{r'} R_p^{-\epsilon} \Delta_{p+1}^\epsilon. \tag{14b}$$

The following useful identity:

$$\Delta_{p+1}^\epsilon = \Delta_{p+1}^r \triangleq \Delta_{p+1} \tag{15}$$

follows from

$$[I, \ A_{p,1}, \cdots, A_{p,p}, \ 0] \, R_{p+1} \, [0, \ B_{p,p}, \cdots, B_{p,1}, \ I]' = \Delta_{p+1}^\epsilon = \Delta_{p+1}^r. \tag{16}$$

The quantity Δ_{p+1} can be interpreted as the cross correlation of the forward and backward prediction errors at one unit of delay. To see this, rewrite (16) as

$$\Delta_{p+1} = E\{[I, \ A_{p,1}, \cdots, A_{p,p}, \ 0] [y_T', \cdots, y_{T-p-1}']' \cdot [y_T', \cdots, y_{T-p-1}'] [0, \ B_{p,p}, \cdots, B_{p,1}, \ I]'\}$$
$$= E\{\epsilon_{p,T} r_{p,T-1}'\}. \tag{17}$$

Eliminating the (nonsingular) covariance matrix R_{p+1} from (13) and making use of the identity above, leads to the multichannel Levinson recursions

$$\begin{bmatrix} I & -K_{p+1}^r \\ -K_{p+1}^{\epsilon'} & I \end{bmatrix} \begin{bmatrix} I, & A_{p,1}, \cdots, A_{p,p}, & 0 \\ 0, & B_{p,p}, \cdots, B_{p,1}, & I \end{bmatrix} = \begin{bmatrix} I, & A_{p+1,1}, \cdots, A_{p+1,p+1} \\ B_{p+1,p+1}, & B_{p+1,p}, \cdots, I \end{bmatrix}. \tag{18}$$

The quantities $K_{p+1}^r \triangleq \Delta_{p+1} R_p^{-r}$, $K_{p+1}^\epsilon = R_p^{-\epsilon} \Delta_{p+1}$ are the so-called reflection coefficients. Looking at the first and last block columns in (18) we note that $K_{p+1}^r = -A_{p+1,p+1}$ and $K_{p+1}^\epsilon = -B_{p+1,p+1}'$. Post-multiplying (18) by $[I, Iz^{-1}, \cdots, Iz^{-p-1}]'$ leads to the more familiar form of the Levinson algorithm in z-transform notation

$$\begin{bmatrix} A_{p+1}(z) \\ B_{p+1}(z) \end{bmatrix} = \underbrace{\begin{bmatrix} I & -K_{p+1}^r \\ -K_{p+1}^{\epsilon'} & I \end{bmatrix}}_{\theta_{p+1}} \begin{bmatrix} A_p(z) \\ z^{-1} B_p(z) \end{bmatrix}$$

$$= \theta_{p+1} \begin{bmatrix} I & 0 \\ 0 & z^{-1} \end{bmatrix} \cdots \theta_1 \begin{bmatrix} I & 0 \\ 0 & z^{-1} \end{bmatrix} \begin{bmatrix} A_0(z) \\ B_0(z) \end{bmatrix}, \quad A_0(z) = B_0(z) = I \tag{19}$$

where

$$A_p(z) \triangleq I + A_{p,1} z^{-1} + \cdots + A_{p,p} z^{-p} \tag{20a}$$

$$B_p(z) \triangleq B_{p,p} + B_{p,p-1} z^{-1} + \cdots + I z^{-p}. \tag{20b}$$

We note that in the scalar case the backward predictor is simply the reversed order forward predictor, i.e., $B_p(z) = z^{-p} A_p(z^{-1})$ or $B_{p,i} = A_{p,i}$, with $A_{p,0} = B_{p,0} \triangleq I$. In this case, $R_p^\epsilon = R_p^r$ and $K_p^\epsilon = K_p^r = K_p$. Much of the literature on these recursions and the associated lattice filters treats only the scalar case.

B. The Basic Lattice Filter

The Levinson recursions derived above induce directly a lattice structure as depicted in Fig. 3. This digital filter has the property that the transfer function from its input to its upper (lower) output is the forward (backward) predictor $A_p(z)$ $(B_p(z))$. Note that increasing the predictor order is achieved by adding one more lattice section, without changing any of the previous sections. This nesting property leads to the conclusion that given a lattice implementation of a predictor of some order, say N, we have in fact an implementation of all predictors up to that order $\{A_p(z), p = 1, \cdots, N\}$. This useful property is not shared by tapped-delay-line implementations of the least squares predictor, since the set of coefficients $\{A_{p,i}\}$ is generally different from the set of coefficients $\{A_{p+1,i}\}$. The predictor coefficients can be easily recovered from the reflection coefficients by looking at the impulse response of the lattice filter. In other words, if we feed the sequence $\{I, 0, \cdots\}$ to the input of the filter in Fig. 3, the outputs of section p will be $\{I, A_{p,1}, \cdots, A_{p,p}, 0, \cdots\}$ and $\{B_{p,p}, \cdots, B_{p,1}, I, 0, \cdots\}$. To make this scheme work, we need to "turn on" the sections of this lattice filter one-by-one. In other words, at $t = 0$ (when the impulse is applied) all reflection coefficients are set to zero. At $t = 1$ we set the reflection coefficients of the first section to K_1^ϵ, K_1^r, and so on. This procedure follows from careful examination of (19) and Fig. 3.

The set of prediction coefficients computed by the Levinson recursions can be arranged in the following upper triangular (block) matrix:

$$U_A \triangleq \begin{bmatrix} I & A_{N,1} & \cdots\cdots & A_{N,N} \\ & \ddots & & \vdots \\ & 0 & I & A_{1,1} \\ & & & I \end{bmatrix}. \tag{21}$$

Post-multiplying this matrix by R_N and using the defining property of the predictor (8) leads to the conclusion that $U_A R_N$ is a lower triangular matrix, with $\{R_N^\epsilon, \cdots, R_0^\epsilon\}$ as its main diagonal elements. From this it is straightforward to show that

$$R_N^{-1} = U_A' R^{-\epsilon} U_A \tag{22a}$$

where

$$R^\epsilon = \operatorname{diag}\{R_N^\epsilon, \cdots, R_0^\epsilon\}. \tag{22b}$$

In other words, the set of forward predictor coefficients provides a lower X diagonal X upper decomposition of the inverse of the covariance matrix. A similar reasoning leads to the conclusion that the backward predictors provide an upper X diagonal X lower decomposition of the same matrix

$$R_N^{-1} = L_B' R^{-r} L_B \tag{23a}$$

where

$$R^r = \operatorname{diag}\{R_0^r, \cdots, R_N^r\} \tag{23b}$$

and

$$L_B = \begin{bmatrix} I & & 0 & \\ B_{1,1} & I & & \\ \vdots & & \ddots & \\ B_{N,N} & \cdots & \cdots & I \end{bmatrix}. \tag{23c}$$

The set of predictors $\{A_p(z)\}$ defines a "whitening" filter for the stationary process $\{y_T\}$. Feeding this process into the causal filter U_A gives a set of white forward prediction errors at the output

$$E\{U_A[y_T', \cdots, y_{T-N-1}']'[y_T', \cdots, y_{T-N-1}']U_A'\} = U_A R_N U_A' = R^\epsilon. \tag{24}$$

Similarly, feeding data into the anticausal filter L_B will generate a set of white backward prediction errors.

In other words, if we apply the sequence $\{y_0, y_1, \cdots\}$ to the input of the lattice filter in Fig. 3, the quantities propagating in the filter will be the forward and backward prediction errors. To see this more clearly, post-multiply (18) by $[y_T', \cdots, y_{T-p-1}']'$ to get

$$\begin{aligned} \epsilon_{p+1,T} &= \epsilon_{p,T} - K_{p+1}^r r_{p,T-1} \\ r_{p+1,T} &= r_{p,T-1} - K_{p+1}^{\epsilon'} \epsilon_{p,T}. \end{aligned} \tag{25}$$

It is important to note that the true whitening filter for a stationary process is a time-varying filter, as indicated by the non-Toeplitz structure of U_A. In lattice form this time variation is of a particularly simple type: turning on the lattice sections one by one. If y_T is a true autoregressive process of order N, say, the whitening filter will become time invariant after $(N+1)$ time steps.

C. The Normalized Lattice Filter

An important consideration in the implementation of practical digital filters is the scaling or normalization of the quantities propagating in the filter. Fixed point arithmetic is usually used to perform the computations in such filters. To minimize roundoff errors it is desirable to have the range of values of the filter coefficients and the propagating quantities be as large as possible without creating an overflow (i.e., one wants to use the full "dynamic range" of the multipliers). This can be achieved by proper scaling of all the variables. Consider, for example, the lattice filter described above. The forward and backward prediction errors $\epsilon_{p,T}, r_{p,T-1}$ are zero-mean random variables with different variances R_p^ϵ, R_p^r. These variances may vary over a wide range of values. A natural idea would be to scale all the variables to have equal variance, say, unity. A normalized lattice filter based on this idea was developed by Gray and Markel [16]. This filter was reported to have roundoff noise properties superior to those of unnormalized lattice filters [19]. Another striking feature of the normalized lattice filter will become apparent in Section IV where it will be shown that proper normalization simplifies the adaptive lattice algorithm.

To derive the normalized lattice prediction filter we consider the consequences of normalizing the variance of the forward and backward prediction errors to unity. Let us define the normalized variables

$$\bar{\epsilon}_{p,T} = R_p^{-\epsilon/2} \epsilon_{p,T} \qquad \bar{r}_{p,T} = R_p^{-r/2} r_{p,T} \tag{26}$$

the associated normalized whitening filters

$$\bar{A}_p(z) = R_p^{-\epsilon/2} A_p(z) \qquad \bar{B}_p(z) = R_p^{-r/2} B_p(z) \tag{27}$$

and a normalized reflection coefficient

$$K_{p+1} = R_p^{-\epsilon/2} \Delta_{p+1} R_p^{-r/2'}. \tag{28}$$

Inserting the normalized filters into (19) gives

$$\begin{bmatrix} \bar{A}_{p+1}(z) \\ \bar{B}_{p+1}(z) \end{bmatrix} = \begin{bmatrix} R_{p+1}^{-\epsilon/2} R_p^{\epsilon/2} & 0 \\ 0 & R_{p+1}^{-r/2} R_p^{r/2} \end{bmatrix} \begin{bmatrix} I & -K_{p+1} \\ -K_{p+1}' & I \end{bmatrix} \cdot \begin{bmatrix} \bar{A}_p(z) \\ z^{-1}\bar{B}_p(z) \end{bmatrix}. \tag{29}$$

Next note that (14) can be factored as follows:

$$\begin{aligned} R_p^{-\epsilon/2} R_{p+1}^{\epsilon/2} &= (I - K_{p+1}K_{p+1}')^{1/2} \\ R_p^{-r/2} R_{p+1}^{r/2} &= (I - K_{p+1}'K_{p+1})^{1/2}. \end{aligned} \tag{30}$$

Combining these equations leads to the normalized lattice structure depicted in Fig. 4,

$$\begin{bmatrix} \bar{A}_{p+1}(z) \\ \bar{B}_{p+1}(z) \end{bmatrix} = \underbrace{\begin{bmatrix} (I - K_{p+1}K_{p+1}')^{-1/2} & 0 \\ 0 & (I - K_{p+1}'K_{p+1})^{-1/2} \end{bmatrix} \begin{bmatrix} I & -K_{p+1} \\ -K_{p+1}' & I \end{bmatrix}}_{\bar{\theta}(K_{p+1})} \begin{bmatrix} \bar{A}_p(z) \\ z^{-1}\bar{B}_p(z) \end{bmatrix}. \tag{31}$$

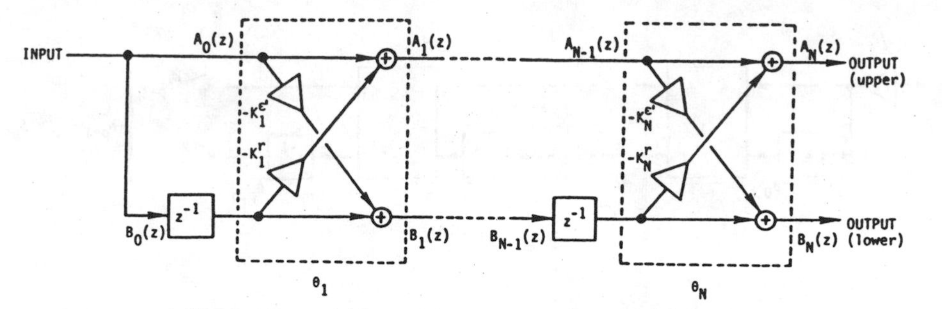

Fig. 3. The multichannel (unnormalized) prediction filter.

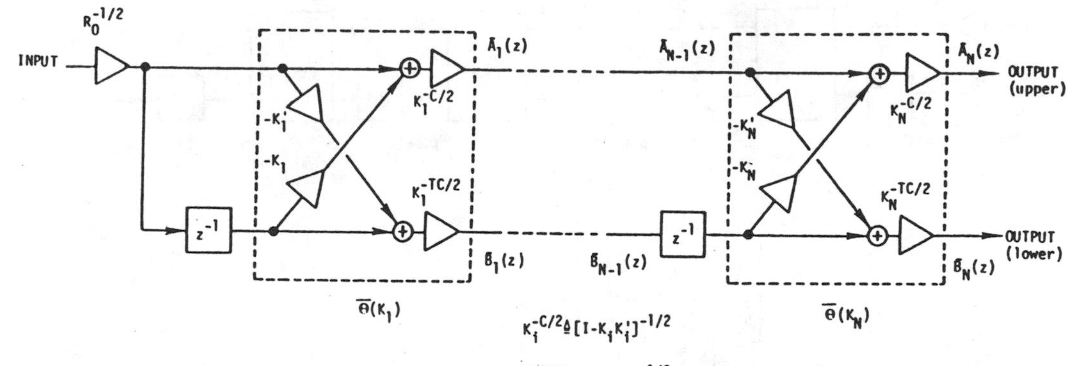

Fig. 4. The normalized lattice prediction filter.

A similar recursion will hold for the normalized prediction errors $\bar{\epsilon}_{p,T} = \bar{A}_p(z)\, y_T$, $\bar{r}_{p,T} = \bar{B}_p(z)\, y_T$. Note that the normalized recursions involve only one set of reflection coefficients $\{K_p\}$ compared to two sets in the unnormalized multichannel version $\{K_p^{\epsilon}, K_p^{r}\}$.

Note that (26)–(28) and (30) make it possible to go back and forth between normalized and unnormalized quantities. In the following section we will find it convenient to work mostly with the normalized version.

D. Fast Cholesky Factorization of Stationary Covariance Matrices

Lattice structures can be used for purposes other than least squares prediction. In this section we derive an efficient algorithm for factoring a Toeplitz covariance matrix, using the lattice filter presented earlier. The factorization of a matrix into a product of a lower triangular and upper triangular matrices is often called Cholesky factorization [155]. Standard techniques require in the order of N^3 operations (and $\sim N^2$ storage) to factor an $N \times N$ matrix. When the matrix has a Toeplitz structure it can be factored in only $\sim N^2$ operations (and $\sim N$ storage). Cholesky factorization is useful in many applications such as factoring the spectrum of a moving average process or solving least squares estimation problems. Before deriving the so-called fast Cholesky algorithm note that the Levinson algorithm provides a factorization of the *inverse* of the covariance matrix R_N. To see this, let us arrange the normalized predictor coefficients in the following matrices:

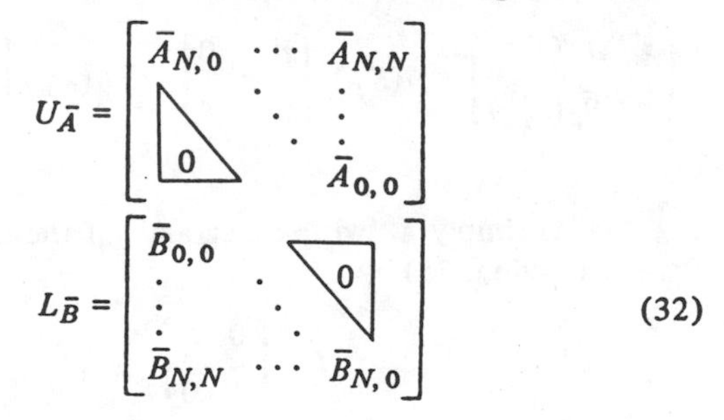

(32)

Comparison to (22) and (23) shows that these matrices provide a lower $\times$ upper and upper $\times$ lower decomposition of R^{-1}

$$R_N^{-1} = U_{\bar{A}}' U_{\bar{A}} = L_{\bar{B}}' L_{\bar{B}}. \tag{33}$$

The impulse response matrices $U_{\bar{A}}$, $L_{\bar{B}}$ can be efficiently inverted (because of their triangular form) to compute the factors U_α, L_β of the covariance matrix

$$R_N = U_{\bar{A}}^{-1} U_{\bar{A}}^{-T} = U_\alpha U_\alpha' \tag{34a}$$

$$R_N = L_{\bar{B}}^{-1} L_{\bar{B}}^{-T} = L_\beta L_\beta'. \tag{34b}$$

It is also possible to compute directly the factors of R_N from the covariance sequence $\{R_i\}$. The reflection coefficients $\{K_p\}$ are obtained as part of these computations.

To see this we examine more carefully the entries of $L_{\bar{B}} R_N = L_\beta'$

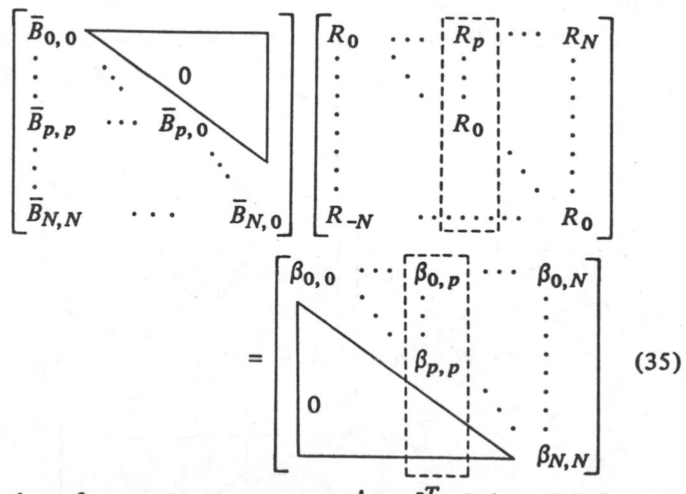

(35)

where $\beta_{i,j}$ are the elements of $L_\beta' = L_{\bar{B}}^{-T}$. It is straightforward to check that $\beta_{p,p}' = R_p^{r/2}$. Note that the pth column of L_β' is obtained by passing the sequence $\{R_0, R_1, \cdots, R_p\}$ through the sequence of filters $\bar{B}_0(z), \bar{B}_1(z), \cdots, \bar{B}_p(z)$. In other words, if we feed the covariance sequence through the lattice

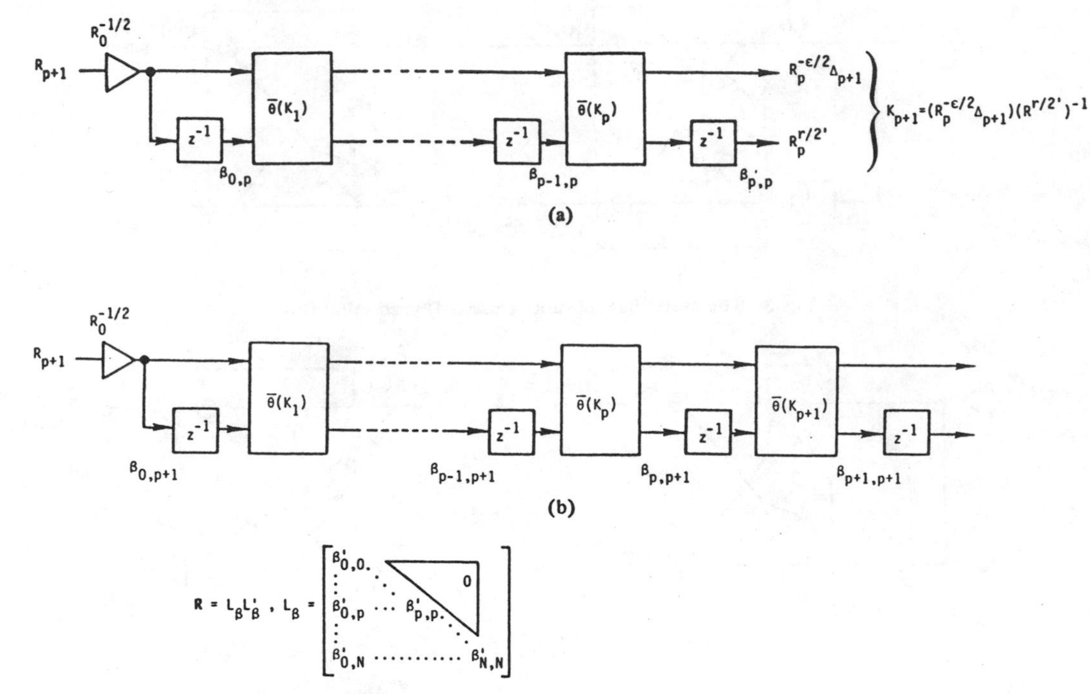

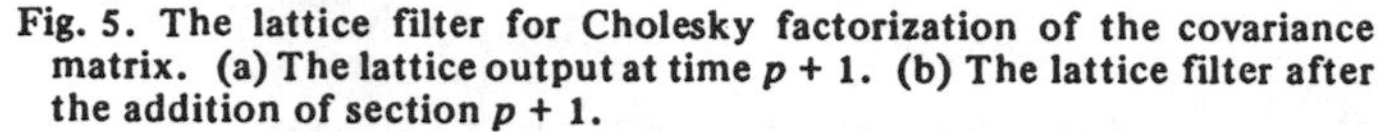

Fig. 5. The lattice filter for Cholesky factorization of the covariance matrix. (a) The lattice output at time $p + 1$. (b) The lattice filter after the addition of section $p + 1$.

filter depicted in Fig. 4, the quantities propagating on the lower line at time p, will be the entries of the pth row of L_β. Note also that the lower output of the pth section at time p is $R_p^{r/2'}$. Examining the normalized version of (12b) indicates that

$$R_p^{-\epsilon/2}\Delta_{p+1} = \bar{A}_{p,0}R_{p+1} + \bar{A}_{p,1}R_p + \cdots + \bar{A}_{0,0}R_1 \quad (36)$$

which means that the upper output of the pth section at time $p + 1$ will be $R_p^{-\epsilon/2}\Delta_{p+1}$. Therefore, the reflection coefficient K_{p+1} of the $(p + 1)$th section is the "ratio" of the upper output and delayed lower output of section p. This fact leads to a recursive procedure for computing reflection coefficients from the covariance sequence, as depicted in Fig. 5. The procedure lends itself easily to computer implementation. We emphasize the fact that this single filter provides a factorization of the covariance matrix and its inverse (obtained as the impulse response of the filter) and computes the reflection coefficients.

Similar reasoning leads to a procedure for computing the upper X lower factorization of R_N. To derive this procedure we first recall that $U_A R_N = U'_\alpha$, i.e.,

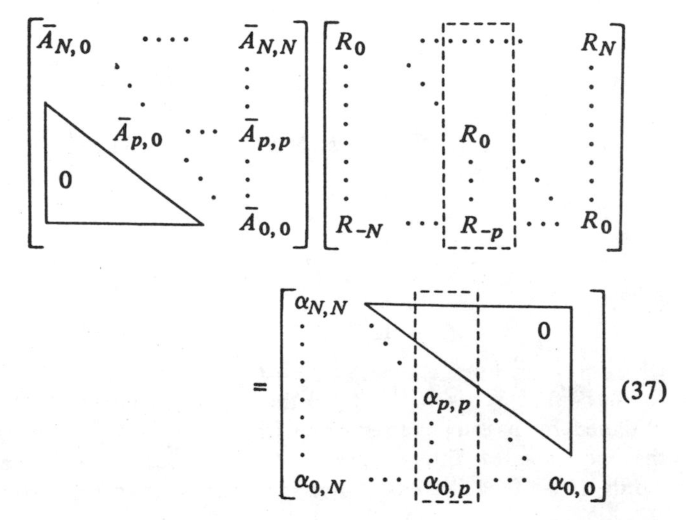

(37)

where $\alpha_{i,j}$ are the entries of $U'_\alpha = U_{\bar{A}}^{-T}$. It is straightforward to check that $\alpha'_{p,p} = R_p^{\epsilon/2}$ and that

$$R_p^{-r/2}\Delta'_{p+1} = \bar{B}_{p,p}R_{-1} + \bar{B}_{p,p-1}R_{-2} + \cdots + \bar{B}_{p,0}R_{-p-1}. \quad (38)$$

Note that the columns of U'_α can be obtained by filtering the sequence $\{R_0, R_{-1}, \cdots, R_{-p}, \cdots\}$ through the *reversed order* predictors $\bar{A}_0(z^{-1}), z^{-1}\bar{A}_1(z^{-1}), \cdots, z^{-N}\bar{A}_N(z^{-1})$ where

$$z^{-p}\bar{A}_p(z^{-1}) = \bar{A}_{p,p} + \bar{A}_{p,p-1}z^{-1} + \cdots + \bar{A}_{p,0}z^{-p}. \quad (39)$$

Similarly, the quantity $R_p^{-r/2}\Delta'_{p+1}$ is obtained by passing the same sequence through the reversed-order backward predictor $z^{-p}\bar{B}_p(z)$. We therefore need a lattice associated with these reversed-order predictors. Fig. 6 depicts two different representations of the desired lattice filter. To prove this, rewrite (31) with z replaced by z^{-1}

$$\begin{bmatrix} \bar{A}_p(z^{-1}) \\ \bar{B}_p(z^{-1}) \end{bmatrix} = \bar{\theta}(K_p)\begin{bmatrix} I & 0 \\ 0 & z \end{bmatrix}\cdots\bar{\theta}(K_1)\begin{bmatrix} I & 0 \\ 0 & z \end{bmatrix}\begin{bmatrix} I \\ I \end{bmatrix}. \quad (40)$$

Pre-multiply this equation by

$$\begin{bmatrix} z^{-p} & 0 \\ 0 & z^{-p} \end{bmatrix}$$

to give

$$\begin{bmatrix} z^{-p}\bar{A}_p(z^{-1}) \\ z^{-p}\bar{B}_p(z^{-1}) \end{bmatrix} = \bar{\theta}(K_p)\begin{bmatrix} z^{-1} & 0 \\ 0 & I \end{bmatrix}\cdots\bar{\theta}(K_1)\begin{bmatrix} z^{-1} & 0 \\ 0 & I \end{bmatrix}\begin{bmatrix} I \\ I \end{bmatrix}. \quad (41)$$

Fig. 5(a) is simply a "wiring diagram" of the equation above. Pre-multiplying (35) by

$$\tilde{I} = \begin{bmatrix} 0 & I \\ I & 0 \end{bmatrix}$$

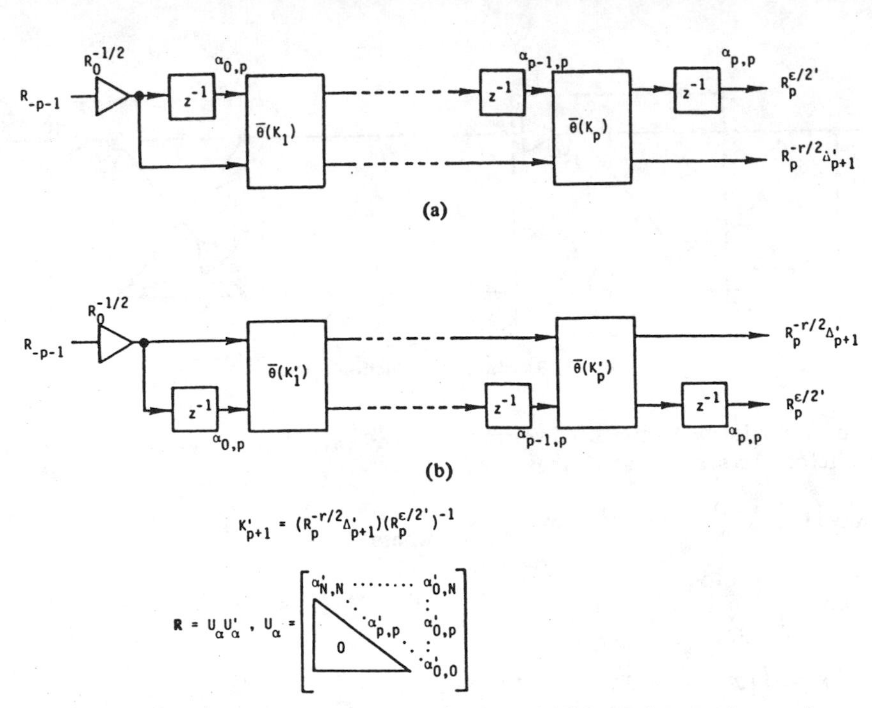

Fig. 6. The lattice filter for Cholesky factorization of the covariance matrix (dual form). (a) Delays on upper line. (b) Delays on lower line.

gives

$$\begin{bmatrix} z^{-p}\bar{B}_p(z^{-1}) \\ z^{-p}\bar{A}_p(z^{-1}) \end{bmatrix} = \bar{\theta}(K'_p)\begin{bmatrix} I & 0 \\ 0 & z^{-1} \end{bmatrix} \cdots \bar{\theta}(K'_1)\begin{bmatrix} I & 0 \\ 0 & z^{-1} \end{bmatrix}\begin{bmatrix} I \\ I \end{bmatrix} \tag{42}$$

where we used the fact that

$$\tilde{I}\bar{\theta}(K)\tilde{I} = \bar{\theta}(K') \tag{43}$$

which can be easily verified. In summary, the "dual" lattice structure depicted in Fig. 6 provides a procedure for computing the same set of reflection coefficients we discussed earlier, and the quantities propagating in it (on the line with the delays) give the upper X lower factorization of R_N.

The efficient procedure for covariance matrix factorization presented here has its origins in the work by Morf [121] on the fast Cholesky algorithm. The Cholesky factorization performed by the lattice structure computes successively the *rows* of the matrix factors. Algorithm for computing the *columns* of the Cholesky factor are presented in [67], [121].

Note that doing an upper X lower factorization of R is equivalent to doing a lower X upper factorization of a matrix $\tilde{R}$ whose elements are the elements of R in reversed order, i.e.,

$$\tilde{R} = \begin{bmatrix} R_0 & \cdots & R_{-1} & \cdots & R_{-p} \\ \vdots & \ddots & & \ddots & \vdots \\ R_1 & & R_0 & & R_{-1} \\ \vdots & \ddots & & \ddots & \vdots \\ R_p & \cdots & R_1 & \cdots & R_0 \end{bmatrix} \tag{44}$$

(compare to (8)). From this it follows that feeding the sequence $\{R_0, R_{-1}, R_{-2}, \cdots\}$ into the lattice of Fig. 5 will give an upper X lower factorization of R. Similarly, feeding the sequence $\{R_0, R_1, R_2, \cdots\}$ into the lattice of Fig. 6 will give a lower X upper factorization of R. We note, however, that the resulting reflection coefficients will generally be different from those obtained by the procedure described earlier (except in the scalar case).

E. The Inverse Prediction Filter

The lattice predictor in Fig. 4 is a time-varying filter with impulse response matrices $U_{\bar{A}}$ (upper output) and $L_{\bar{B}}$ (lower output). The time-varying nature of the filter is due to the act of turning on the reflection coefficients one-by-one. By rewiring the prediction filter, we obtain the inverse filter depicted in Fig. 7, whose impulse response matrix is $U_{\bar{A}}^{-1} = U_\alpha$. This filter can be used for synthesizing stationary AR processes, by feeding a stationary white noise process to the input. The time-varying inverse filter will produce an output sequence that is truely stationary. The common practice of using a fixed filter ($1/A_N(z)$) produces transients due to nonstationary initial conditions. It is then necessary to "run" the filter for a while before its output becomes stationary. The time-varying lattice inverse filter will generate a stationary sequence from the very first sample.

The inverse filter can also be used to efficiently compute the covariance sequence $\{R_1, R_2, \cdots\}$ given the reflection coefficients $\{K_1, \cdots, K_N\}$ of the least squares predictor. (We assume without loss of generality that $R_0 = I$.) The procedure consists of initializing the states (i.e., delay elements) of the inverse filter to $\{I, 0, \cdots, 0\}$ and then letting it operate with zero input. The output of the filter can be shown to be the covariance sequence $\{R_1, R_2, \cdots\}$. This fact has been known for some time (for the scalar case) and found applications in seismic deconvolution [66]. A proof for the multichannel case is presented in [185].

F. Relationship Between Variables of Lattice Filters and Direct Realizations

Finally, we note that the following identity relates directly reflection coefficients and predictor coefficients:

$$[K_1^r, \cdots, K_p^r] = [A_{p,1}, \cdots, A_{p,p}]\, L_{\bar{B}}^{-1} \tag{45}$$

(see Appendix I for a proof). This provides an interpretation

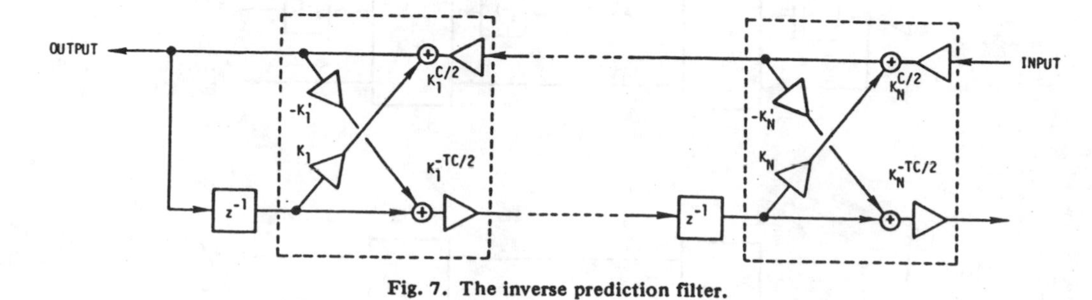

Fig. 7. The inverse prediction filter.

of the lattice filter as a Gram–Schmidt orthogonalized version of the tapped delay line predictor. To see this, recall that

$$y_T = -[A_{N,1}, \cdots, A_{N,N}][y'_{T-1}, \cdots, y'_{T-N}]' + \epsilon_T.$$

From Fig. 3 we can see that

$$\begin{aligned} y_T &= -[K_1^r, \cdots, K_N^r][r'_{1,T-1}, \cdots, r'_{N,T-1}]' + \epsilon_T \\ &= -[A_{N,1}, \cdots, A_{N,N}] L_B^{-1} L_B [y'_{T-1}, \cdots, y'_{T-N}]' + \epsilon_T. \end{aligned} \tag{46}$$

In other words, the lattice filter arises naturally when we try to replace the data $\{y_{T-1}, \cdots, y_{T-N}\}$ by an orthonormal set of variables obtained by applying a Gram–Schmidt procedure

$$\begin{aligned} r_{1,T-1} &= y_{T-1} \\ r_{2,T-1} &= y_{T-2} - B_{1,1} y_{T-1} \\ r_{3,T-1} &= y_{T-3} - B_{2,1} y_{T-1} - B_{2,2} y_{T-2} \end{aligned} \tag{47}$$

etc. This type of procedure is known to have superior numerical properties in solving least squares problems [150], [155].

III. Linear Prediction in the Given Data Case

The problem of finding the least squares predictor was solved in the previous section under the assumption that the second-order statistics of the process were known. Here we want to consider the more practical case where no prior knowledge is available about the process. All the information we have is contained in a finite number of data points $\{y_0, y_1, \cdots, y_T\}$. In this case it is still possible to formulate a well defined least squares prediction problem. In this section, nonrecursive block processing techniques for linear prediction will be considered.

A. The Least Squares Predictor

Let us consider a pth order forward predictor, and recall that the prediction error is given by

$$\epsilon_{p,t} = y_t + \sum_{i=1}^{p} A_{p,i} y_{t-i}. \tag{48}$$

We define a least squares predictor by choosing a set of coefficients $\{A_{p,i}\}$ which minimizes the sum of squared prediction errors

$$\mathrm{tr}\left\{\sum_{t=0}^{T+p} \epsilon_{p,t}\epsilon'_{p,t}\right\}$$

over the available data. Next we write (48) (for $t = 0, \cdots, T+p$) in matrix form

$$[\epsilon_{p,0}, \cdots, \epsilon_{p,T+p-1} 0] = y + [A_{p,1}, \cdots, A_{p,p}] Y_{p,T} \tag{49a}$$

where

$$y = [y_0, \cdots, y_T, \ 0, \cdots, 0], \quad \text{an } m \times (T+p+1) \text{ matrix} \tag{49b}$$

$$Y_{p,T} = \begin{bmatrix} y_0 & \cdots & y_{p-1} & \cdots & y_{T-1} & \cdot & 0 \\ & \ddots & \vdots & & \vdots & \ddots & \\ 0 & & y_0 & \cdots & y_{T-p} & \cdots & y_{T-1} \end{bmatrix},$$

$$\text{an } mp \times (T+p+1) \text{ data matrix.} \tag{49c}$$

Note that in writing this equation, we had to make some choice regarding the data outside the observation interval $[0, T]$. Here we made the choice that the data prior to time $t = 0$ are zero and that the data after $t = T$ are zero. This particular choice of the data matrix $Y_{p,T}$ is called the autocorrelation form [53], [61]. In this section we consider only this form since it is the one most closely related to the stationary case discussed earlier.

In later sections we will consider other choices of the data matrix such as: i) the pre-windowed form which assumes that data prior to $t = 0$ are zero but makes no assumptions about the data after $t = T$; ii) the covariance method which makes no assumptions about the data outside $[0, T]$; and iii) the post-windowed form [128]. The matrices for all four methods are defined below.

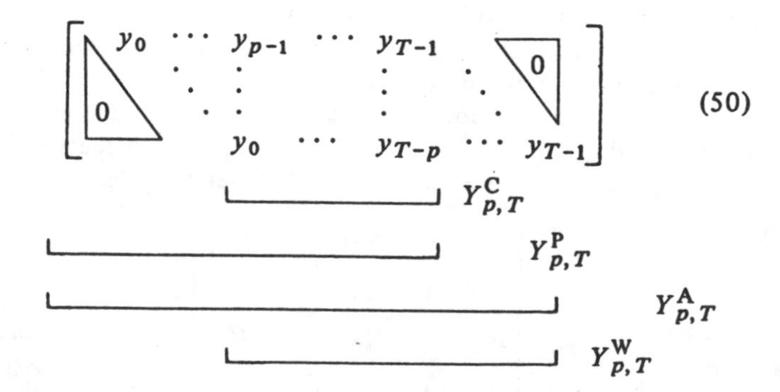

where the superscripts represent the type of windowing (Autocorrelation, Pre-windowed, post-Windowed, Covariance). We will drop the superscripts whenever it is clear from the context which case is being discussed.

It is well known [154], [155] that the solution of the overdetermined set of (49) which minimizes the L_2 norm of its left-hand side (i.e., trace $\{\Sigma \epsilon_{p,t}\epsilon'_{p,t}\}$) is obtained as the *unique* solution of

$$[A_{p,1}, \cdots, A_{p,p}](Y_{p,T} Y'_{p,T}) = -y Y'_{p,T}. \quad (51)$$

The matrix $(Y_{p,T} Y'_{p,T})$ has the interpretation of a sample covariance matrix and $y Y'_{p,T}$ is a set of estimated correlation coefficients. In fact, these equations have the same form as the Yule-Walker equations (8), which were derived from a stochastic point of view. The difference is that the true covariance matrix is replaced by a sample covariance matrix

$$Y^{A} Y^{A'} = \begin{bmatrix} \hat{R}_0 & & \hat{R}_{p-1} \\ & \ddots & \\ \hat{R}_{-p+1} & & \hat{R}_0 \end{bmatrix}, \quad \text{sample covariance matrix for autocorrelation case} \quad (52a)$$

$$\hat{R}_i = \sum_{t=i}^{T} y_{t-i} y_t. \quad (52b)$$

Note that the sample covariance matrix $Y^{A} Y^{A'}$ has a Toeplitz structure. This will *not* be the case for the covariance, pre-windowed, and post-windowed forms. The sample covariance matrices $Y^{C} Y^{C'}$, $Y^{P} Y^{P'}$, and $Y^{W} Y^{W'}$ are no longer Toeplitz. but they do have a very special structure whose significance will be explored later. We note here only that the least squares predictor obtained as the solution of (51) is guaranteed to be stable for the autocorrelation and post-windowed cases but not for the pre-windowed and covariance methods.

A set of equations similar to (49) defines a backward predictor. Let $r_{p,t-1}$ denote a backward prediction error

$$r_{p,t-1} = y_{t-p-1} + \sum_{i=1}^{p} B_{p,p+1-i} y_{t-i} \quad (53)$$

which can be written in matrix form

$$[0, \; r_{p,0}, \ldots, r_{p,T+p-1}] = [0, \cdots, 0, \; y_0, \cdots, y_{T-1}] + [B_{p,p}, \cdots, B_{p,1}] \; Y_{p,T}. \quad (54)$$

We consider here, again, the autocorrelation case. The least squares predictor is then given by the solution to the following set of linear equations:

$$[B_{p,p}, \cdots, B_{p,1}](Y_{p,T} Y'_{p,T}) = -[0, \cdots, 0, \; y_0, \cdots, y_{T-1}] \; Y'_{p,T}. \quad (55)$$

The discussion above provides a precise answer to least squares prediction in the given data case. In principle, all we need to do is to solve (51) or (55) and use the predictor coefficients to compute a set of prediction errors (48), (53). It would be useful to be able to associate a simple lattice structure with such a solution. The possibility of doing that is explored next.

B. Lattice Implementation of the Least Squares Predictor

The derivation presented in Section II associated a multichannel lattice structure with the recursive solution of the Yule-Walker equation. This solution followed directly from the Toeplitz structure of the covariance matrix. Since the autocorrelation method leads to a Toeplitz sample covariance matrix, it is clear that all the results of Section II apply to this case as well. In particular, the same lattice structures will provide an implementation of the least squares predictors $A_p(z)$ and $B_p(z)$.

If the autocorrelation method were the only one to be considered, there would be little motivation to explore alternative lattice structures. As it happens, both the covariance and pre-windowed methods are often used in spectral estimation, speech processing, and other applications. Furthermore, we will see later that the adaptive (recursive) least squares lattice can be derived for the pre-windowed and covariance forms but *not* for the autocorrelation form.

The Levinson recursion, which provided the basis for the lattice structure presented earlier, does not hold for the non-Toeplitz sample covariance matrices. The question, therefore, arises whether a lattice filter exists in these cases. The precise (and positive) answer to this question will be given in Section IV. Here we note only that a recursive solution of the Yule-Walker equation is possible in the non-Toeplitz case, using a generalization of the Levinson algorithm [99], [100]. This algorithm is somewhat more complex than the one presented in Section II, and the associated lattice prediction filter is correspondingly more complex.

A large number of lattice techniques have appeared in the literature. Most of these techniques are derived from the stationary (Toeplitz) case. The complexities related to the non-Toeplitz case are handled by *forcing* the lattice filter to have the same structure as in the stationary case and *assuming* that the methods for computing reflection coefficients from Toeplitz covariance matrices still hold. This leads to lattice prediction filters that are approximations to the true least squares predictor. As the number of data points increases, all of these methods converge to the least squares solution. In many applications, the approximate solutions are very close to the optimum.

C. Computing Reflection Coefficients

This section reviews a number of (nonrecursive) techniques that have been proposed to compute reflection coefficients from data. We consider here the unnormalized scalar case which appears to be the one most widely used. To present these various methods in a common framework we first discuss several methods for computing reflection coefficients for the stationary case. It is then shown that many of the proposed techniques can be obtained by making some natural approximations to the stationary methods.

1) Prediction Error Methods: The reflection coefficients can be defined as a cross correlation of forward and backward prediction errors. Recall that

$$K^{\epsilon}_{p+1} = R_p^{-\epsilon} \Delta_{p+1} = [E\{\epsilon_{p,t} \epsilon'_{p,t}\}]^{-1} \; E\{\epsilon_{p,t} r'_{p,t-1}\} \quad (56a)$$

$$K^{r}_{p+1} = \Delta_{p+1} R_p^{-r} = E\{\epsilon_{p,t} r'_{p,t-1}\} \; [E\{r_{p,t-1} r'_{p,t-1}\}]^{-1}. \quad (56b)$$

Replacing expected values by time averages gives a practical way of computing these coefficients

$$K^{\epsilon}_{p+1} = \left[\sum_t \epsilon_{p,t} \epsilon'_{p,t}\right]^{-1} \left[\sum_t \epsilon_{p,t} r'_{p,t-1}\right] \quad (57a)$$

$$K^{r}_{p+1} = \left[\sum_t \epsilon_{p,t} r'_{p,t-1}\right] \left[\sum_t r_{p,t-1} r'_{p,t-1}\right]^{-1}. \quad (57b)$$

We omitted the limits on the summations since these will depend on the particular method used ($t = 0$ to $t = T$ for the pre-windowed case, $t = p$ to $t = T$ for the covariance case, etc.). This procedure will require multiple passes on the data: once for every reflection coefficient. Data are always available at the output of the pth lattice section to compute the reflection coefficient for the next section. The procedure starts by computing the first reflection coefficient directly from the data

$$K_1^\epsilon = \left[\sum_t y_t y_t'\right]^{-1} \left[\sum_t y_t y_{t-1}'\right] \tag{58a}$$

$$K_1^r = \left[\sum_t y_t y_{t-1}'\right] \left[\sum_t y_t y_t'\right]^{-1}. \tag{58b}$$

Next the data are passed through the first-order lattice filter to obtain $\{\epsilon_{1,t}, r_{1,t}\}$. The second reflection coefficient can now be computed from these prediction errors using (57), and so on. The two reflection coefficients K_p^ϵ and K_p^r should be the same in the stationary scalar case (as in the lattice filter depicted in Fig. 2). However, when computed from finite data sets they will generally be different. In other words, even if

$$E\{\epsilon_{p,t}\epsilon_{p,t}'\} = E\{r_{p,t-1}r_{p,t-1}'\}$$

we will generally have

$$\sum_t \epsilon_{p,t}\epsilon_{p,t}' \neq \sum_t r_{p,t-1}r_{p,t-1}'.$$

The approximation made in going from (56) to (57) introduces two different reflection coefficients where there should be one. To get back a single reflection coefficient, some choice has to be made.

2) Evaluating the Reflection Coefficient: Makhoul [55] used the name "forward method" for the choice

$$K_{p+1} = K_{p+1}^r = \sum r_p \epsilon_p / \sum r_p r_p$$

and "backward method" for

$$K_{p+1} = K_{p+1}^\epsilon = \sum r_p \epsilon_p / \sum \epsilon_p \epsilon_p.$$

The forward method is related to the minimization of the mean square of the forward residuals while the backward method is related to minimizing the mean square of the backward residuals. Itakura and Saito [52] used the geometric mean method for computing the reflection coefficients, which they call PARCOR coefficients

$$K_{p+1} = \sum r_p \epsilon_p \Big/ \sqrt{\sum \epsilon_p^2 \sum r_p^2} = S \cdot \sqrt{K_{p+1}^\epsilon K_{p+1}^r} \tag{59a}$$

where

$$S = \text{sign}\, K_{p+1}^\epsilon = \text{sign}\, K_{p+1}^r. \tag{59b}$$

The PARCOR coefficients are guaranteed to be less than 1 in magnitude and this ensures the stability of the lattice filter (i.e., the inverse prediction filter used for synthesis). From the properties of the geometric mean it follows that

$$\min\{|K^\epsilon|, |K^r|\} \leqslant \sqrt{K^\epsilon K^r} \leqslant 1. \tag{60}$$

This property leads to another choice for the reflection coefficients that guarantees stability; namely, the minimum method

$$K_{p+1} = S \min\{|K_{p+1}^\epsilon|, |K_{p+1}^r|\}. \tag{61}$$

A general method of defining reflection coefficients is by taking a generalized nth mean of K^ϵ and K^r

$$K_{p+1} = S[1/2(|K^\epsilon|^n + |K^r|^n)]^{1/n}. \tag{62}$$

The choice of $n \to 0$ leads to the geometric mean while $n \to -\infty$ will give the minimum method. The choice $n = -1$ leads to the harmonic-mean method which has been used by Burg [34] in the maximum-entropy method of spectral estimation

$$K_{p+1} = \frac{2K^\epsilon K^r}{K^\epsilon + K^r} = \frac{2\sum \epsilon_p r_p}{\sum \epsilon_p^2 + \sum r_p^2}. \tag{63}$$

It can be shown that the harmonic-mean method is related to the minimization of $E\{\epsilon_p^2\} + E\{r_p^2\}$, see [55].

3) Sample Covariance Matrix Methods: Evaluating Δ_{p+1}, R_p^ϵ, and R_p^r from the prediction errors requires passing the data several times through a lattice filter of increasing order. More efficient methods for estimating these quantities involve a single pass through the data to generate a sample covariance matrix, followed by computations in which only correlation coefficients are used. The first such method [61] follows from (12), by replacing R_{p+1} with the sample covariance matrix $\hat{R}_{p+1} = Y_{p+1,T}Y_{p+1,T}'$ obtained by the autocorrelation, pre-windowed, or covariance methods. We denote the entries of the sample covariance matrix by $\hat{R}_{i,j}$, i.e.,

$$\hat{R}_{p+1} = \begin{bmatrix} \hat{R}_{p+1,p+1} & \cdots & \hat{R}_{p+1,0} \\ \vdots & & \vdots \\ \hat{R}_{0,p+1} & \cdots & \hat{R}_{0,0} \end{bmatrix}. \tag{64}$$

In the autocorrelation case $\hat{R}_{i,j} = \hat{R}_{i-j}$ and the matrix $\hat{R}_{p+1}$ is Toeplitz. Inserting (64) in (12) will give

$$R_p^\epsilon = \sum_{i=0}^{p} A_{p,i}\hat{R}_{p-i+1,p+1}, \qquad A_{p,0} \triangleq I \tag{65a}$$

$$R_p^r = \sum_{i=0}^{p} B_{p,i}\hat{R}_{i,0}, \qquad B_{p,0} \triangleq I \tag{65b}$$

$$\Delta_{p+1}' = \sum_{i=0}^{p} B_{p,i}\hat{R}_{i,p+1} \quad \text{or} \quad \Delta_{p+1} = \sum_{i=0}^{p} A_{p,i}\hat{R}_{p-i+1,0}. \tag{65c}$$

Note that this method involves only the first and last columns of the covariance matrix. A more "symmetric" solution, that uses all the entries of the covariance matrix, follows from the fact that in the stationary case

$$\begin{bmatrix} I, & A_{p,1}, \cdots, A_{p,p}, & 0 \\ 0, & B_{p,p}, \cdots, B_{p,1}, & I \end{bmatrix} R_{p+1} \begin{bmatrix} I, & A_{p,1}, \cdots, A_{p,p}, & 0 \\ 0, & B_{p,p}, \cdots, B_{p,1}, & I \end{bmatrix}' = \begin{bmatrix} R_p^\epsilon & \Delta_{p+1} \\ \Delta_{p+1}' & R_p^r \end{bmatrix}. \tag{66}$$

Replacing R_{p+1} by $\hat{R}_{p+1}$ gives

$$R_p^\epsilon = \sum_{i=0}^{p}\sum_{j=0}^{p} A_{p,i}\hat{R}_{p-i+1,p-j+1}A_{p,j}', \qquad A_{p,0} \triangleq I \tag{67a}$$

$$R_p^r = \sum_{i=0}^{p}\sum_{j=0}^{p} B_{p,i}\hat{R}_{i,j}B_{p,j}', \qquad B_{p,0} \triangleq I \tag{67b}$$

$$\Delta_{p+1}' = \sum_{i=0}^{p}\sum_{j=0}^{p} B_{p,i}\hat{R}_{i,p-j+1}A_{p,j}'$$

or

$$\Delta_{p+1} = \sum_{i=0}^{p}\sum_{j=0}^{p} A_{p,i}\hat{R}_{p-i+1,j}B_{p,j}'. \tag{67c}$$

The idea of using all the entries of the covariance matrix [55] is intuitively appealing. If one thinks of $\hat{R}_{i,j}$ as noisy versions of the true covariance R_{i-j}, a better estimate will be obtained by averaging more terms.

The computation of the reflection coefficients proceeds recursively as follows: given that we know K_p, use the Levinson recursions (18) to update the predictors $A_{p-1}(z) \to A_p(z)$, $B_{p-1}(z) \to B_p(z)$. The new predictor coefficients are then used in (65) or (67) to compute the quantities needed to get the next reflection coefficient K_{p+1}, and so on. The process is started by setting $A_{0,0} = B_{0,0} = I$. Note that using the Levinson recursions to update the predictors is not really correct, unless the sample covariance matrix is Toeplitz. Using it in the non-Toeplitz case will yield filters $A_p(z) B_p(z)$ that are approximations to the true least squares predictor. In the scalar autocorrelation case (65) and (67) will give $R_p^\epsilon = R_p^r$ and thus $K_p^\epsilon = K_p^r = K_p$. In the non-Toeplitz case, the two reflection coefficients will be different, requiring the use of one of the methods presented earlier, to obtain a single reflection coefficient.

A fairly large number of combinations of these various steps is possible, leading to many ways for computing reflection coefficients: i) the sample covariance matrix can be computed by the autocorrelation, pre-windowed, or covariance methods; ii) either (65) or (67) can be used to compute R^ϵ, R^r, Δ; iii) any of the methods described by (59)-(63) can be used to compute the reflection coefficient K. Most methods appearing in the literature can be interpreted as particular choices of these steps. For example, either the covariance or the autocorrelation methods combined with (67) and the harmonic mean method (63) define the covariance or the autocorrelation-lattice methods, respectively, presented by Makhoul [55]. Many of these methods, while suboptimum, perform quite well in speech applications.

We have seen that a somewhat bewildering number of ways is available to compute reflection coefficients from data. This is due to the fact that the different algorithms provide different approximations to the least squares predictor. Only in the scalar autocorrelation case do some of them compute the true least squares solution. As mentioned before, all of these methods converge to the same least squares predictor when the amount of data becomes sufficiently large. The techniques described above are all of the block-processing type: the prediction error methods require multiple passes through the data, the sample covariance matrix methods require computation of the sample correlation coefficients based on the entire data. These techniques have been used to develop a number of (recursive) gradient adaptive lattice algorithms. We defer discussion of these gradient techniques to Section V and present in Section IV the adaptive least squares lattice (LSL). In the process of deriving the LSL we will gain some insight into lattice structures associated with non-Toeplitz covariance matrices.

Finally we present a few comments on the history of lattice filters.

D. Lattice Filters for Signal Modeling

Much of the early work on lattice filters for modeling and prediction of signals was done in the context of speech processing. The lattice structure provides a piecewise-constant acoustic tube model of the vocal tract as observed by Kelly and Lochbaum (see [61, ch. 4] and Wiakita [181], [182]). Many alternative two-port network models were later proposed to model the vocal tract. Taking a statistical point of view, Itakura and Saito [51], [52] derived a new digital filter structure for time-domain speech analysis, which is called PARCOR (for partial correlation) lattice form. This filter structure matched nicely the physical models of the vocal tract developed by Kelly and Lochbaum and Atal [33]. The connection between the PARCOR lattice filter and the vocal tract model was established by Wakita. See [61] for a more complete discussion on lattice forms in speech processing.

Motivated by the successful use of lattice filters in speech analysis/synthesis, several researchers studied the properties of these filters and derived new algorithms for computing reflection coefficients from data. Markel and Gray derived several lattice structures [15], [60] and presented scaling conventions [20] and roundoff error characteristics [19] for finite-word-length implementations. Their results indicate that the two-multiplier lattice has worse roundoff characteristics than the one-multiplier lattice. They also presented the normalized lattice [16] mentioned earlier. Mullis and Roberts [160], [161] presented an analysis of optimal (lattice) structure for roundoff error effects.

An improved procedure for computing reflection coefficients from data was presented by Makhoul [54], [55] under the name of the covariance-lattice method. Properties of lattice digital filters were studied in [56] where one-, two-, three-, and four-multiplier lattice structures are considered. Applications to adaptive linear prediction adaptive Wiener filters and a fast start-up equalizer can be also found in [56].

In the works of Markel and Gray and others, extensive use was made of the theory of orthogonal polynomials and its close relationship to least squares estimation of stationary processes [61], [107], [153]. Here we will use a different approach based on the properties of projection operators on linear spaces.

Another area in which lattice structures have been implicitly used is high-resolution spectral estimation. The maximum-entropy method proposed by Burg [34], [151] involves computation of partial correlation coefficients and the forward and backward prediction errors typically associated with a lattice structure. The role of lattice forms in autoregressive spectral estimation is discussed in [57], [93], [96], [137].

The literature on lattice filters for estimation and prediction treats many different topics in addition to the ones mentioned above. Some examples: computing reflection coefficients from hard limited prediction errors [42], [73], by Cholesky decomposition of the covariance matrix and by other techniques [32], [38]-[41], [43], [44], [62], [67]. The statistics of the estimated reflection coefficients are discussed in [197].

IV. The Adaptive Least Squares Lattice

The history of the least squares lattice form for the nonstationary case, can be traced back to the work of Morf [121], [126] on efficient solutions for least squares predictors, for covariance matrices which are not Toeplitz but have a special structure (in effect they are sums of products of lower and upper triangular Toeplitz matrices; see Section VII). This work led to the generalization of the Levinson algorithm to the pre-windowed case, the covariance case, and more general matrix situations [83], [84], [97], [99], [100], [105], [148]. The generalized Levinson recursions induce a lattice structure as described in [109], [128]. As we noted in the stationary case, the Levinson recursions can be normalized, leading to a variance normalized lattice form. Carrying out a similar procedure on the generalized Levinson recursions leads to the variance normalized version of the LSL which was apparently first derived in [135], [137], [148]. Using a different type of normalization, Lee [110] derived a lattice form in which all quantities are in magnitude less than one. The original derivations involving the predictors $A_p(z)$, $B_p(z)$ were rather cumbersome. Since then, a number of different approaches were developed for deriving the lattice recursions directly including: a projection operator framework [140], the Hilbert array method [134], and a geometric approach [109], [146]; see also [79], [80]. In this paper, we chose to use the projection framework, which involves only some elementary notions from linear algebra.

A key feature of the generalized Levinson algorithm and the LSL is the existence of *efficient time-update* formulas for the predictor coefficients. The coefficients of the predictor are updated recursively as each data point becomes available, leading to a class of adaptive least squares algorithms. It is interesting to note that efficient time update formulas of this type do not seem to exist for the Toeplitz (autocorrelation) case. The simplest form of the adaptive LSL is obtained for the pre-windowed case. In this section we present a detailed discussion of the pre-windowed LSL, leaving the more complex covariance case to Section VI.

A. The Unnormalized Pre-Windowed Lattice Form

The lattice form provides a recursive procedure for computing prediction errors. Following (49), we define the forward prediction errors by

$$[\epsilon_{p,0}(T), \cdots, \epsilon_{p,T}(T)] = y_{0:T} + [A_{p,1}, \cdots, A_{p,p}]\ Y_{p,T} \tag{68a}$$

where

$$y_{0:T} = [y_0, \cdots, y_T], \qquad \text{an } m \times (T+1) \text{ matrix} \tag{68b}$$

$$Y_{p,T} = \begin{bmatrix} y_0 & \cdots & y_{p-1} & \cdots & y_{T-1} \\ & \ddots & \vdots & & \vdots \\ & & \ddots & & \vdots \\ 0 & & y_0 & \cdots & y_{T-p} \end{bmatrix}, \qquad \text{an } mp \times (T+1) \text{ data matrix.} \tag{68c}$$

The notation $\epsilon_{p,t}(T)$ means the pth order prediction error for y_t, using a least squares predictor based on all the data up to and including time T. Inserting the least squares predictor coefficients defined by (51), we can rewrite (68a) as

$$[\epsilon_{p,0}(T), \cdots, \epsilon_{p,T}(T)] = y_{0:T}[I - Y'_{p,T}(Y_{p,T}Y'_{p,T})^{-1}\ Y_{p,T}]. \tag{69}$$

Note that the matrix

$$P_{Y_{p,T}} \triangleq Y'_{p,T}(Y_{p,T}Y'_{p,T})^{-1}\ Y_{p,T} \tag{70}$$

is a projection operator, i.e., $P_{Y_{p,T}}P_{Y_{p,T}} = P_{Y_{p,T}}$. The last prediction error $\epsilon_{p,T} \triangleq \epsilon_{p,T}(T)$ can now be written as[2]

$$\epsilon_{p,T} = y_{0:T}(I - P_{Y_{p,T}})\ \pi' \tag{71a}$$

where

[2] To simplify the notation we will usually delete the index of ϵ and r.

TABLE I
DEFINITION OF VARIABLES FOR THE UNNORMALIZED LATTICE
$S = Y_{p,T}$

V	W	$V(I-P_S)W'$	Comment
$y_{0:T}$	π	$\epsilon_{p,T}$	Forward prediction error
$y^{p+1}_{0:T}$	π	$r_{p,T-1}$	Backward prediction error
$y_{0:T}$	$y^{p+1}_{0:T}$	$\Delta_{p+1,T}$	Cross correlation coefficient
$y_{0:T}$	$y_{0:T}$	$R^{\epsilon}_{p,T}$	Forward prediction error covariance
$y^{p+1}_{0:T}$	$y^{p+1}_{0:T}$	$R^{r}_{p,T-1}$	Backward prediction error covariance
π	π	$1-\gamma_{p-1,T-1}$	Likelihood variable

$$\pi = [0, \cdots, 0, I]. \tag{71b}$$

A similar result holds for the backward prediction errors

$$\begin{aligned}[0, r_{p,0}(T-1), \cdots, r_{p,T-1}(T-1)] &= y^{p+1}_{0:T} + [B_{p,p}, \cdots, B_{p,1}]\ Y_{p,T} \\ &= y^{p+1}_{0:T}[I - Y'_{p,T}(Y_{p,T}Y'_{p,T})^{-1}\ Y_{p,T}]\end{aligned} \tag{72a}$$

where

$$y^{p+1}_{0:T} = [0, \cdots, 0, \ y_0, \cdots, y_{T-p-1}], \qquad \text{shifted version of } y_{0:T}. \tag{72b}$$

The last backward prediction error $r_{p,T-1} \triangleq r_{p,T-1}(T-1)$ is then given by

$$r_{p,T-1} = y^{p+1}_{0:T}(I - P_{Y_{p,T}})\ \pi'. \tag{73}$$

Note that both $\epsilon_{p,T}$ and $r_{p,T-1}$ are given by expressions of the type $V(I - P_S)W'$ where V, W are some vectors (or matrices) and P_S is a projection operator, projecting on the space spanned by the rows of the matrix S. Recursions for expressions of this type can be obtained by considering what happens when we change the projection space. Let $\{S + X\}$ denote the space spanned by the rows of S and the rows of X. In Appendix II we prove the following update formula for arbitrary matrices V, W, S, X (of compatible dimensions)

$$\begin{aligned}V(I - P_{\{S+X\}})\ W' &= V(I - P_S)\ W' \\ &\quad - V(I - P_S)\ X'\ [X(I - P_S)\ X']^{-1}\ X(I - P_S)\ W'.\end{aligned} \tag{74}$$

Using this formula with proper choices for V, W, S, and X will generate the desired lattice recursions. In order to get a complete set of recursions we will find it necessary to define several variables in addition to $\epsilon_{p,T}$ and $r_{p,T-1}$. All of the relevant variables are summarized in Table I.

Various choices of X will give different types of updates. The following are the most useful:

i) Order Update: Note that

$$Y_{p+1,T} = \begin{bmatrix} Y_{p,T} \\ y^{p+1}_{0:T} \end{bmatrix}. \tag{75}$$

Therefore, choosing $X = y^{p+1}_{0:T}$ means that we are increasing the predictor order by one. Similar observations lead to other types of updates.

ii) Time and Order Update:

$$Y_{p+1,T+1} = \begin{bmatrix} y_{0:T} \\ Y_{p,T} \end{bmatrix}. \tag{76}$$

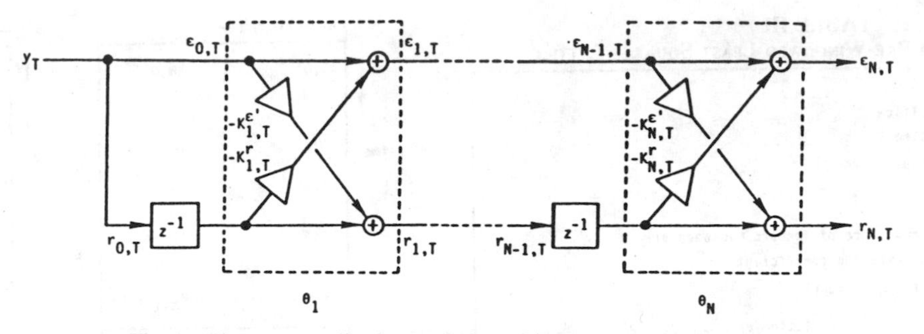

Fig. 8. The unnormalized pre-windowed least squares lattice filter.

TABLE II
UPDATE FORMULAS FOR THE UNNORMALIZED LATTICE
$S = Y_{p,T}$

X	V	W	$V(I-P_{(S+X)})W'$	$V(I-P_S)W'$	$V(I-P_S)X'$	$X(I-P_S)X'$	$X(I-P_S)W'$
$y^{p+1}_{0:T}$	$y_{0:T}$	π	$\epsilon_{p+1,T}$	$\epsilon_{p,T}$	$\Delta_{p+1,T}$	$R^r_{p,T-1}$	$r_{p,T-1}$
$y_{0:T}$	$y^{p+1}_{0:T}$	π	$r_{p+1,T}$	$r_{p,T-1}$	$\Delta'_{p+1,T}$	$R^\epsilon_{p,T}$	$\epsilon_{p,T}$
π	$y_{0:T}$	$y^{p+1}_{0:T}$	$\Delta_{p+1,T-1}$	$\Delta_{p+1,T}$	$\epsilon_{p,T}$	$1-\gamma_{p-1,T-1}$	$r'_{p,T-1}$
π	$y_{0:T}$	$y_{0:T}$	$R^\epsilon_{p,T-1}$	$R^\epsilon_{p,T}$	$\epsilon_{p,T}$	$1-\gamma_{p-1,T-1}$	$\epsilon'_{p,T}$
π	$y^{p+1}_{0:T}$	$y^{p+1}_{0:T}$	$R^r_{p,T-2}$	$R^r_{p,T-1}$	$r_{p,T-1}$	$1-\gamma_{p-1,T-1}$	$r^{\smile}_{p,T-1}$
$y_{0:T}$	π	π	$1-\gamma_{p,T}$	$1-\gamma_{p-1,T-1}$	$\epsilon'_{p,T}$	$R^\epsilon_{p,T}$	$\epsilon_{p,T}$
$y^{p+1}_{0:T}$	π	π	$1-\gamma_{p,T-1}$	$1-\gamma_{p-1,T-1}$	$r'_{p,T-1}$	$R^r_{p,T-1}$	$r_{p,T-1}$
$y^{p+1}_{0:T}$	$y_{0:T}$	$y_{0:T}$	$R^\epsilon_{p+1,T}$	$R^\epsilon_{p,T}$	$\Delta_{p+1,T}$	$R^r_{p,T-1}$	$\Delta'_{p+1,T}$
$y_{0:T}$	$y^{p+1}_{0:T}$	$y^{p+1}_{0:T}$	$R^r_{p+1,T}$	$R^r_{p,T-1}$	$\Delta'_{p+1,T}$	$R^\epsilon_{p,T}$	$\Delta_{p+1,T}$

iii) Time Update: In Appendix II, it is shown that choosing $X = \pi$ leads to the identity

$$I - P\{Y_{p,T+\pi}\} = \begin{bmatrix} I - P_{Y_{p,T-1}} & 0 \\ & \vdots \\ 0 \cdots\cdots & 0 \end{bmatrix}. \tag{77}$$

Using (75)–(77) and the definitions of Table I, we can now fill in the entries of Table II, which contain the various quantities involved in the update formula. This table is nothing more than a collection of definitions of variables. However, when used in conjunction with the update formula (74), each line in the table translates into an update equation. For example, the first two lines give

$$\epsilon_{p+1,T} = \epsilon_{p,T} - \Delta_{p+1,T} R^{-r}_{p,T-1} r_{p,T-1} \tag{78a}$$

$$r_{p+1,T} = r_{p,T-1} - \Delta'_{p+1,T} R^{-\epsilon}_{p,T} \epsilon_{p,T}. \tag{78b}$$

This equation defines a lattice filter for computation of the prediction errors, as depicted in Fig. 8. By analogy to the stationary case, we define $K^r_{p+1,T} = \Delta_{p+1,T} R^{-r}_{p,T}$ and $K^\epsilon_{p+1,T} = R^{-\epsilon}_{p,T} \Delta_{p+1,T}$. This lattice looks very much like the one developed for the stationary case. There are, however, some subtle differences: the reflection coefficients $K^\epsilon_{p,T}$ and $K^r_{p,T}$ will be different even in the scalar case. More importantly, the lattice of Fig. 8 has *time-varying* parameters, whereas the filter in Fig. 3 has a set of fixed, time-invariant coefficients. We found a simple lattice structure for computing exact prediction errors, but we seem to need a much larger number of parameters to specify it, namely $\{K_{1,t}, \cdots, K_{N,t}\}$ $t = 0, \cdots, T$. Fortunately, this apparent complexity is not as bad as it first seems since these reflection coefficients are computed recursively at the same time the prediction errors are being computed. No more than N reflection coefficients need to be retained at one time.

Exponential Weighting: So far we discussed update equations for a predictor which minimizes a least squares criterion of the type

$$\text{tr}\left\{\sum_{t=0}^{T} \epsilon_{p,t}\epsilon'_{p,t}\right\}.$$

In other words, the predictor is based on the entire data record in the interval $[0, T]$. In many applications, the statistics of the observed process y_T are slowly time-varying. It is therefore necessary to build some forgetting factor into the algorithm so that it could track these time variations. There are many ways of doing this, and one of them is based on replacing the least squares criterion with the exponentially *weighted least squares* criterion

$$\text{tr}\left\{\sum_{t=0}^{T} \lambda^{T-t} \epsilon_{p,t}\epsilon'_{p,t}\right\}.$$

The effect of this weighting is equivalent to an exponential windowing of the data, where "old" data points are given smaller weights than more recent data points.

The effect of exponential weighting on the update equations

TABLE III

THE UNNORMALIZED PRE-WINDOWED LEAST SQUARES LATTICE

Input parameters:

N = maximum order of lattice

y_T = data sequence at time T

λ = exponential weighting factor

Variables:

$R^{\epsilon}_{p,T}$, $R^{r}_{p,T-1}$ = sample covariance of forward/backward errors

$\Delta_{p,T}$ = sample partial correlation coefficient

$\gamma^{c}_{p,T} \triangleq 1 - \gamma_{p,T}$ = likelihood variable

$\epsilon_{p,T}$, $r_{p,T-1}$ = forward/backward prediction errors

$K^{\epsilon}_{p,T}$, $K^{r}_{p,T}$ = forward/backward reflection coefficients

These computations will be performed once for every time step (T=0,...,TMAX).

Initialize:

$$\epsilon_{0,T} = r_{0,T} = y_T$$

$$R^{\epsilon}_{0,T} = R^{r}_{0,T} = \lambda R^{\epsilon}_{0,T-1} + y_T y'_T$$

$$\gamma^{c}_{-1,T} = 1$$

Do for p=0 to min (N,T) - 1

$$\Delta_{p+1,T} = \lambda \Delta_{p+1,T-1} + \epsilon_{p,T} r'_{p,T-1} / \gamma^{c}_{p-1,T-1} \quad *$$

$$\gamma^{c}_{p,T} = \gamma^{c}_{p-1,T} - r'_{p,T} R^{-r}_{p,T} r_{p,T} \quad *$$

$$K^{r}_{p+1,T} = \Delta_{p+1,T} R^{-r}_{p,T-1} \quad *$$

$$\epsilon_{p+1,T} = \epsilon_{p,T} - K^{r}_{p+1,T} r_{p,T-1}$$

$$R^{\epsilon}_{p+1,T} = R^{\epsilon}_{p,T} - K^{r}_{p+1,T} \Delta'_{p+1,T}$$

$$K^{\epsilon}_{p+1,T} = R^{-\epsilon}_{p,T} \Delta_{p+1,T} \quad *$$

$$r_{p+1,T} = r_{p,T-1} - K^{\epsilon'}_{p+1,T} \epsilon_{p,T}$$

$$R^{r}_{p+1,T} = R^{r}_{p,T-1} - \Delta'_{p+1,T} K^{\epsilon}_{p+1,T}$$

Note: Only the variables Δ, R^{ϵ}, R^{r}, γ^{c}, r need to be stored from one time step to the other. They are all set initially to zero. The quantities R^{r}, γ^{c}, r need to be stored twice to avoid "overwriting".

In the scalar case when the divisor $x = \gamma^{c}$, R^{r}, R^{ϵ} is very small, set $1/x = 0$ in the equations marked with *. In the multichannel case see [140].

turns out to be very simple: the time update needs to be modified by multiplying the right-hand side of (77) by λ [115]. This, in turn, will introduce the factor λ in the time updates for R^{ϵ}, R^{r}, and Δ in Table II. For example, the third line in this table will read

$$\Delta_{p+1,T} = \lambda \Delta_{p+1,T-1} + \epsilon_{p,T} r'_{p,T-1} / (1 - \gamma_{p-1,T-1}). \quad (79)$$

Reading off the lines in Table II we can now get update formulas for all the variables needed to compute recursively the prediction errors and the lattice parameters. These recursions can be organized in many different ways to provide a complete adaptive algorithm. One particular implementation of the algorithm which we have been using is summarized in Table III. We refer the interested reader to [109], [115], [122], [128], [130] for more details.

The lattice filter described above provides a recursive solution to the least squares prediction problem. As mentioned before, the same problem can be solved by a block processing method, e.g., using (69) for $p = 0, \cdots, N$. To see more clearly how these two types of solutions are related, we consider the prediction errors generated in two different ways: i) computing optimal predictors of orders $p = 1, \cdots, N$ based on data $\{y_t\}$ $t = 0, \cdots, T$ and using them to filter the data; and ii) using the recursive filter on the same data set. The results are summarized in Fig. 9. Note that the sequences of prediction errors are different for the two methods except at the last step T.

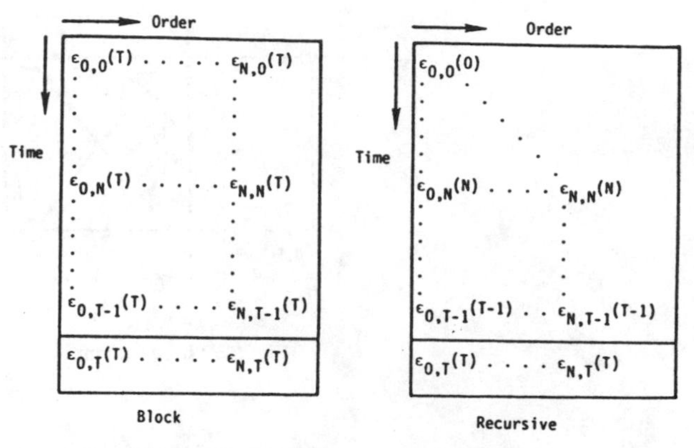

Fig. 9. Block versus recursive prediction.

The Likelihood Variable: The variable $\gamma_{N,T}$ which appears in the algorithm has an important interpretation as an approximate log-likelihood variable related to the process y_T. Consider the case where y_T is a scalar Gaussian random variable. The joint probability distribution for the last $(N+1)$ observations is given by

$$p\{y_{T-N}, \cdots, y_T\} = |2\pi R_N|^{-1/2} \cdot \exp\{-1/2 y_{T-N:T} R_N^{-1} y'_{T-N:T}\}. \quad (80)$$

The log-likelihood function L associated with (80) is given by

$$L = \ln |R_N| + y_{T-N:T} R_N^{-1} y'_{T-N:T}. \quad (81)$$

It can be shown that the second term in (81) is precisely $\gamma_{N,T}$ and that

$$\ln |R_N| = \sum_{p=0}^{N} R_p^{\epsilon} = \ln |R_0| + \sum_{p=1}^{N} \ln (1 - |K_{p,T}|^2). \quad (82)$$

Thus the exact log-likelihood function is easily computable from the lattice parameters. The variable $\gamma_{N,T}$ can be used as a good detection statistic for the "unexpectedness" of the recent data points $\{y_{T-N}, \cdots, y_T\}$. As long as the data come from a Gaussian distribution with second-order statistics R_N the variable $\gamma_{N,T}$ will be small. If the recent data come from a different distribution $\gamma_{N,T}$ will tend to be near unity [128]. This will cause the gain factor $1/(1 - \gamma_{N,T})$ appearing in the time-update formulas to be large, causing the lattice parameters Δ, R^{ϵ}, R^{r} to change quickly. In other words, $1/(1 - \gamma_{N,T})$ acts as an adaptive gain which facilitates fast tracking of changes in the statistics of the observed data. This variable was recently used to develop a sensitive pitch detector for speech analysis [111].

B. The Unnormalized Prediction Filter

The LSL derived in the previous section is a time-varying filter operating on the input data. This filter computes a set of prediction errors $\{\epsilon_{p,t}, r_{p,t}\}$. In some applications (spectral estimation, speech processing, etc.) it is desired to estimate the parameters of the least squares predictor, i.e., the solution of (51), which we denote $A_{p,T}(z)$. The subscript T denotes that this predictor was computed based on the data

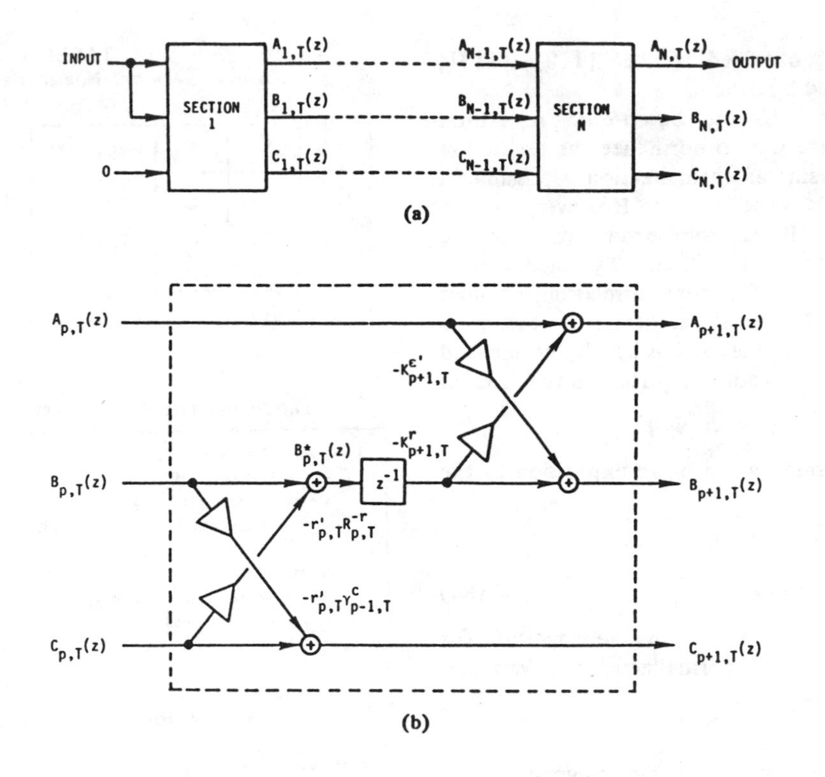

Fig. 10. The normalized pre-windowed least squares lattice predictor. (a) The overall filter structure. (b) A single section.

$\{y_t, t = 0, \cdots, T\}$. The question then is to specify the (time-invariant) lattice filter whose transfer function is $A_{p,T}(z)$.

A natural *approximation* to the least squares predictor is obtained by taking the LSL with its reflection coefficients *fixed* at their values at time T (i.e., set the coefficients of the filter in Fig. 3 to be $K_p^\epsilon = K_{p,T}^\epsilon, K_p^r = K_{p,T}^r$). It is important to note that the transfer function of this filter is not $A_{p,T}(z)$, although it is generally very close to it. A detailed discussion of this point is beyond the scope of this paper. We present here only a heuristic argument. Let

$$\epsilon_{p,T} = A_{p,T}(z)\, y_T \triangleq y_T + \sum_{i=1}^{p} A_{p,i}(T)\, y_{T-i}$$

$$r_{p,T} = B_{p,T}(z)\, y_T \triangleq y_{T-p-1} + \sum_{i=1}^{p} B_{p,i}(T)\, y_{T-i} \quad (83)$$

and insert ϵ and r into the lattice recursions (78). Assume for the moment that the predictor coefficients are time invariant, i.e., $A_{p,i}(T) = A_{p,i}$. Then it is easy to show (by comparing coefficients of y_{T-i}) that $A_{p,i}$, $B_{p,i}$ obey the same recursions as the prediction errors, leading to the conclusion that the corresponding lattice filter (Fig. 8) has the transfer function $A_{p,T}(z)$. (Note that this is the inverse of the argument used in Section II to show that the Levinson recursions induce a lattice structure.) Indeed, as the number of data points T increases, the predictor will approximately converge and $A_{p,i}(T) \simeq A_{p,i}$ will be a good assumption. However, for short data records the predictor coefficient can change significantly with time. The argument presented above thus fails, indicating that "freezing" the reflection coefficients in the LSL is not the right answer.

Using the projection framework presented earlier, it is possible to derive several lattice implementations of the least squares predictor $A_{p,T}(z)$. One such filter is depicted in Fig. 10. This lattice filter uses the parameters $\{K_{p,T}^\epsilon, K_{p,T}^r, r_{p,T}, \gamma_{p,T}\}$ provided by the LSL at time T. This filter can be translated directly into an algorithm for computing the predictor coefficients, summarized in Table IV. The proof of these results follows from the derivation presented in [140] for the normalized case. See also Section IV-D.

TABLE IV
COMPUTING PREDICTOR COEFFICIENTS FROM THE LATTICE PARAMETERS

Initialize: $B^*_{p,-1} = 0$ for $p = 0, \ldots, N-1$

$C_{0,0} = 0$

For $i = 0, \ldots, N$

$A_{0,i} = B_{0,i} = 1 \quad i = 0$

$A_{0,i} = B_{0,i} = 0 \quad i > 0$

Do for $p = 0, \ldots, N-1$

$B^*_{p,i} = B_{p,i} - r_p C_{p,i} / \gamma^c_{p-1}$

$C_{p+1,i} = C_{p,i} - r'_p R^{-r}_p B_{p,i}$

$A_{p+1,i} = A_{p,i} - K^r_{p+1} B^*_{p,i-1}$

$B_{p+1,i} = B^*_{p,i-1} - K^{\epsilon'}_{p+1} A_{p,i}$

In summary: the computation of the least squares predictor consists of two parts: i) the LSL in Table III computes prediction errors and a set of lattice parameters; and ii) these parameters can be sampled at any point in time and used to compute the predictor coefficients via Table IV.

C. The Normalized Lattice Form

The normalized lattice filter which was derived for the stationary case motivated the search for a similar structure in the given data case. A normalized lattice form with a strikingly simple structure was first obtained by Lee [110], [113], [115] and since then treated by others.

The Normalized Operator Update Equations: A natural choice in the stationary case was to normalize the prediction errors to unit variance. A similar normalization is possible in the given data case, as we will see shortly. However, a different normalization suggests itself from examination of the variables propagating in the lattice filter. As noted earlier, these variables have the form of a cross correlation, or inner product of two vectors. In statistical literature, the cross correlation of two random variables v, w is usually normalized by their standard deviations to produce a correlation coefficient

$$\rho = E\{u^2\}^{-1/2}\, E\{vw\}\, E\{w^2\}^{-1/2}.$$

By analogy, the proper normalization of an expression of the type $V(I - P_s)\, W'$ will be

$$\rho_S(V, W) = [V(I - P_S)\, V']^{-1/2}\, [V(I - P_S)\, W'] \cdot [W(I - P_S)\, W']^{-T/2}. \tag{84}$$

Note that this normalized cross correlation is in magnitude (or norm) always less than one. The normalized lattice variables are

$$\tilde{\epsilon}_{p,T} = \rho_{Y_{p,T}}(y_{0:T}, \pi) \quad \text{normalized forward prediction error}$$

$$\tilde{r}_{p,T} = \rho_{Y_{p,T}}(y_{0:T}^{p+1}, \pi) \quad \text{normalized backward prediction error}$$

$$K_{p+1,T} = \rho_{Y_{p,T}}(y_{0:T}, y_{0:T}^{p+1}) \quad \text{reflection coefficient.} \tag{85}$$

The explicit relationship between normalized and unnormalized variables is presented in Appendix III. To derive a set of lattice recursions, we first need to find an update formula for $\rho_{\{S+X\}}(V, W)$. As we say in the previous section, all the necessary recursions can be obtained by making the proper substitutions in that formula. In Appendix II we prove the following: for V, W, X, S being vectors or matrices in a given vector space

$$\rho_{\{S+X\}}(V, W) = [I - \rho_S(V, X)\, \rho_S(X, V)]^{-1/2} \cdot [\rho_S(V, W) - \rho_S(V, X)\, \rho_S(X, W)] \cdot [I - \rho_S(W, X)\, \rho_S(X, W)]^{-T/2}. \tag{86}$$

Equations of this type appear in the statistical literature for updating correlation coefficients. To avoid cumbersome expressions of this type we will find it convenient to define

$$F(a, b, c) = [I - cc']^{-1/2}\, [a - cb]\, [I - b'b]^{-T/2} \tag{87a}$$

and its "inverse"

$$F^{-1}(a, b, c) = [I - cc']^{1/2}\, a[I - b'b]^{T/2} + cb. \tag{87b}$$

The lattice recursions can now be obtained by making proper substitutions into the update formula (86), as depicted in Table V. The resulting algorithm is summarized in Table VI.

The normalized lattice form involves only three variables compared to six variables in the unnormalized version. The algorithm consists almost entirely of repeated calls of the function $F(\cdot, \cdot, \cdot)$ and its inverse, as can be seen from Table VI. Using subroutines to compute F and F^{-1} leads to very convenient coding of the algorithm.

TABLE V
DERIVATION OF THE NORMALIZED LATTICE FORM
$S = Y_{p,T}$

X	V	W	$\rho_{\{S+X\}}(V,W)$	$\rho_S(V,W)$	$\rho_S(X,W)$	$\rho_S(V,X)$
$y_{0:T}^{p+1}$	$y_{0:T}$	π	$\tilde{\epsilon}_{p+1,T}$	$\tilde{\epsilon}_{p,T}$	$\tilde{r}_{p,T-1}$	$K_{p+1,T}$
$y_{0:T}$	$y_{0:T}^{p+1}$	π	$\tilde{r}_{p+1,T}$	$\tilde{r}_{p,T-1}$	$\tilde{\epsilon}_{p,T}$	$K'_{p+1,T}$
π	$y_{0:T}$	$y_{0:T}^{p+1}$	$K_{p+1,T-1}$	$K_{p+1,T}$	$\tilde{r}'_{p,T-1}$	$\tilde{\epsilon}_{p,T}$

TABLE VI
THE NORMALIZED PRE-WINDOWED LATTICE ALGORITHM

Input parameters

N = maximum order of lattice

λ = exponential weighting factor

y_T = data sequence

Variables

S_T = estimated covariance of y_T

$K_{p,T}$ = reflection coefficients

$\tilde{\epsilon}_{p,T}$, $\tilde{r}_{p,T-1}$ = normalized forward and backward prediction errors

Initialization

K, $\tilde{r}$, S are set to zero

Main Loop

$S_T = \lambda S_{T-1} + y_T y'_T$

$\tilde{\epsilon}_{0,T} = \tilde{r}_{0,T} = S_T^{-1/2}\, y_T$

For $p = 0, \ldots, \min(N,T) - 1$

$K_{p+1,T} = F^{-1}(K_{p+1,T-1}, \tilde{r}'_{p,T-1}, \tilde{\epsilon}_{p,T})$

$\tilde{\epsilon}_{p+1,T} = F(\tilde{\epsilon}_{p,T}, \tilde{r}_{p,T-1}, K_{p+1,T})$ *

$\tilde{r}_{p+1,T} = F(\tilde{r}_{p,T-1}, \tilde{\epsilon}_{p,T}, K'_{p+1,T})$ *

$F(a,b,c) = [I - cc']^{-1/2}\, [a - cb][I - b'b]^{-T/2}$

$F^{-1}(a,b,c) = [I - cc']^{1/2}\, a[I - b'b]^{T/2} + cb$

The function $F(\cdot)$ involves division. In the scalar case, when the divisor x is small, set $1/x = 1$ in the equations marked by *. In the multichannel case, see [140].

D. The Variance Normalized Lattice

The normalized lattice form presented above is a time-varying nonlinear filter operating on the data sequence. The nonlinearity arises from the data-dependent normalization. To see this, we rewrite explicitly the normalized update formulas for the prediction errors

$$\tilde{\epsilon}_{p+1,T} = [I - K_{p+1,T}K'_{p+1,T}]^{-1/2}\, [\tilde{\epsilon}_{p,T} - K_{p+1,T}\tilde{r}_{p,T-1}] \cdot [I - \tilde{r}'_{p,T-1}\tilde{r}_{p,T-1}]^{-T/2} \tag{88a}$$

$$\tilde{r}_{p+1,T} = [I - K'_{p+1,T}K_{p+1,T}]^{-1/2}\, [\tilde{r}_{p,T-1} - K'_{p+1,T}\tilde{\epsilon}_{p,T}] \cdot [I - \tilde{\epsilon}'_{p,T}\tilde{\epsilon}_{p,T}]^{-T/2}. \tag{88b}$$

Note the normalization by the signals $\tilde{r}_{p,T-1}$, $\tilde{\epsilon}_{p,T}$. When the lattice recursions are used for data analysis/modeling, this nonlinear normalization is very useful. It guarantees that all the lattice variables are in magnitude less than one. This normalization works properly only if the reflection coefficients K are

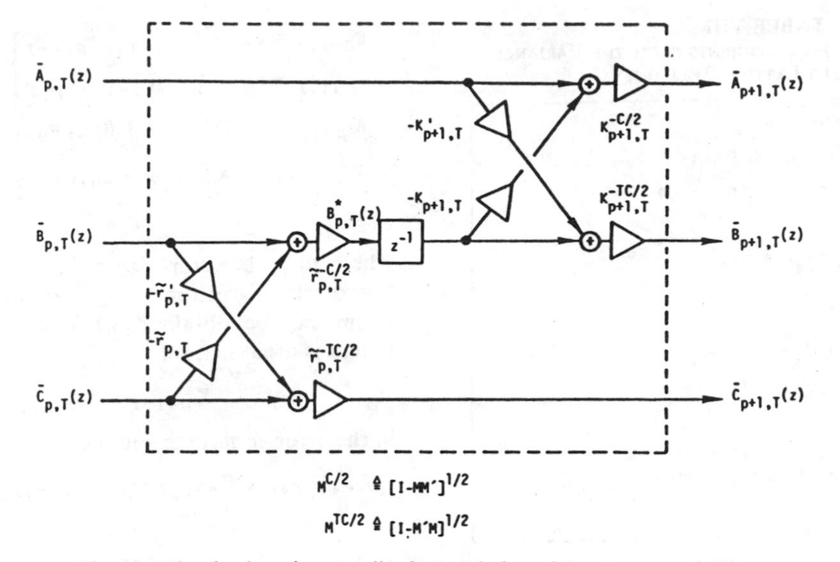

Fig. 11. The (variance) normalized pre-windowed least squares lattice predictor.

updated simultaneously with the prediction errors $\tilde{\epsilon}$, $\tilde{r}$. In some applications, such as synthesis or linear prediction, it is desired to use a filter with a *fixed* set of reflection coefficients. Attempting to use (88) in this case can cause problems: $(I - \tilde{\epsilon}'\tilde{\epsilon})$ or $(I - \tilde{r}'\tilde{r})$ may become negative, in which case the update equations become invalid (since they involve square roots of these quantities). In these applications we need to consider the variance normalized lattice filter. Note that the data normalization is due to the term $[W(I - P_S)\, W']^{-1/2}$ in the definition of the normalized correlation coefficient $\rho_S(V, W)$. Accordingly, we define a variance normalized correlation coefficient $\bar{\rho}_S(V, W)$ by

$$\bar{\rho}_S(V, W) = [I - V(I - P_S)\, V']^{-1/2}\, V(I - P_S)\, W'. \tag{89}$$

In Appendix II we prove the following update formula for this quantity:

$$\bar{\rho}\{S+X\}(V, W) = [I - \rho_S(V, X)\, \rho_S(X, V)]^{-1/2} \cdot [\bar{\rho}_S(V, W) - \rho_S(V, X)\, \bar{\rho}(X, W)]. \tag{90}$$

To avoid repeating this expression we define

$$G(a, b, c) = [I - cc']^{-1/2}\, [a - cb]. \tag{91}$$

The following variance normalized lattice variables are defined

$$\begin{aligned} \bar{\epsilon}_{p,T} &= \bar{\rho}_{Y_{p,T}}(y_{0:T}, \pi) && \text{forward prediction error} \\ \bar{r}_{p,T-1} &= \bar{\rho}_{Y_{p,T}}(y^{p+1}_{0:T}, \pi) && \text{backward prediction error} \\ \bar{\gamma}^c_{p-1,T-1} &= \bar{\rho}_{Y_{p,T}}(\pi, \pi) && \text{auxiliary variable.} \end{aligned} \tag{92}$$

Table VII summarizes various lattice recursions obtained by proper substitutions in the update formula (90). The first two lines of Table VII give the following recursions for the prediction errors:

$$\bar{\epsilon}_{p+1,T} = G(\bar{\epsilon}_{p,T}, \bar{r}_{p,T-1}, K_{p+1,T}) \tag{93a}$$

$$\bar{r}_{p+1,T} = G(\bar{r}_{p,T-1}, \bar{\epsilon}_{p,T}, K'_{p+1,T}). \tag{93b}$$

These equations define the lattice filter depicted in Fig. 4. The other lines in Table VII lead to two additional sets of recursions which as discussed can be used to derive the variance normalized least squares prediction filter. Lines 5 and 6 translate into

$$\bar{r}_{p,T-1} = G(\bar{r}_{p,T}, \bar{\gamma}^c_{p-1,T}, \tilde{r}_{p,T}) \tag{94a}$$

$$\bar{\gamma}^c_{p,T} = G(\bar{\gamma}^c_{p-1,T}, \bar{r}_{p,T}, \tilde{r}'_{p,T}). \tag{94b}$$

In analogy to (83) note that

$$\bar{\epsilon}_{p,T} = \bar{A}_{p,T}(z)\, y_T \tag{95a}$$

$$\bar{r}_{p,T} = \bar{B}_{p,T}(z)\, y_T. \tag{95b}$$

We define the transfer function $\bar{C}_{p,T}(z)$ by

$$\bar{\gamma}^c_{p-1,T} = \bar{C}_{p,T}(z)\, y_T. \tag{95c}$$

By inserting $\bar{\epsilon}$, $\bar{r}$, and $\bar{\gamma}$ from (95) into (93), (94), it is straightforward to show that the predictor coefficients obey the same recursions as these variables. From this we conclude that the variance normalized lattice predictor has the structure depicted in Fig. 10(a) with each section as in Fig. 11. An algorithm for computing the predictor coefficients is summarized in Table VIII.

The lattice predictor presented here was derived from a set of order update equations (the inputs and outputs of each lattice section have the same time indices). Another useful

TABLE VII
DERIVATION OF THE VARIANCE NORMALIZED LATTICE
$S = Y_{p,T}$

X	V	W	$\bar{\rho}_{\{S+X\}}(V,W)$	$\bar{\rho}_S(V,W)$	$\bar{\rho}_S(X,W)$	$\rho_S(V,X)$
$y^{p+1}_{0:T}$	$y_{0:T}$	π	$\bar{\epsilon}_{p+1,T}$	$\bar{\epsilon}_{p,T}$	$\bar{r}_{p,T-1}$	$K_{p+1,T}$
$y_{0:T}$	$y^{p+1}_{0:T}$	π	$\bar{r}_{p+1,T}$	$\bar{r}_{p,T-1}$	$\bar{\epsilon}_{p,T}$	$K'_{p+1,T}$
π	$y_{0:T}$	π	$\bar{\epsilon}_{p,T-1}$	$\bar{\epsilon}_{p,T}$	$\bar{\gamma}^c_{p-1,T-1}$	$\tilde{\epsilon}_{p,T}$
$y_{0:T}$	π	π	$\bar{\gamma}^c_{p,T}$	$\bar{\gamma}^c_{p-1,T-1}$	$\bar{\epsilon}_{p,T}$	$\tilde{\epsilon}'_{p,T}$
π	$y^{p+1}_{0:T}$	π	$\bar{r}_{p,T-2}$	$\bar{r}_{p,T-1}$	$\bar{\gamma}^c_{p-1,T-1}$	$\tilde{r}_{p,T-1}$
$y^{p+1}_{0:T}$	π	π	$\bar{\gamma}^c_{p,T-1}$	$\bar{\gamma}^c_{p-1,T-1}$	$\bar{r}_{p,T-1}$	$\tilde{r}'_{p,T-1}$

TABLE VIII
COMPUTING PREDICTOR COEFFICIENTS FROM THE VARIANCE NORMALIZED LATTICE PARAMETERS

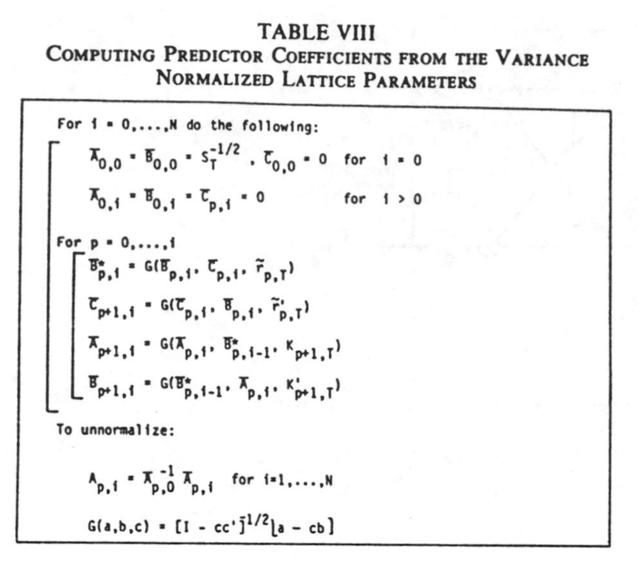

For $i = 0,\ldots,N$ do the following:

$\bar{A}_{0,0} = \bar{B}_{0,0} = S_T^{-1/2}$, $\bar{C}_{0,0} = 0$ for $i = 0$

$\bar{A}_{0,i} = \bar{B}_{0,i} = \bar{C}_{p,i} = 0$ for $i > 0$

For $p = 0,\ldots,i$

$\bar{B}^*_{p,i} = G(\bar{B}_{p,i}, \bar{C}_{p,i}, \tilde{r}_{p,T})$

$\bar{C}_{p+1,i} = G(\bar{C}_{p,i}, \bar{B}_{p,i}, \tilde{r}'_{p,T})$

$\bar{A}_{p+1,i} = G(\bar{A}_{p,i}, \bar{B}^*_{p,i-1}, K_{p+1,T})$

$\bar{B}_{p+1,i} = G(\bar{B}^*_{p,i-1}, \bar{A}_{p,i}, K'_{p+1,T})$

To unnormalize:

$A_{p,i} = \bar{A}^{-1}_{p,0}\bar{A}_{p,i}$ for $i=1,\ldots,N$

$G(a,b,c) = [I - cc']^{-1/2}[a - cb]$

TABLE IX
THE GRADIENT LEAST MEAN SQUARES (LMS) ADAPTIVE ALGORITHM

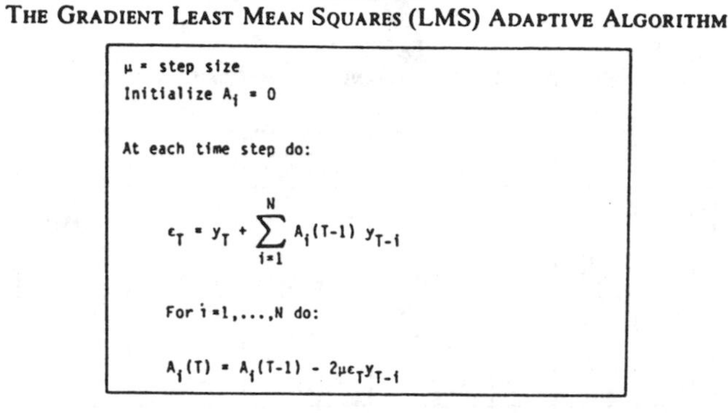

μ = step size

Initialize $A_i = 0$

At each time step do:

$\epsilon_T = y_T + \sum_{i=1}^{N} A_i(T-1)\, y_{T-i}$

For $i=1,\ldots,N$ do:

$A_i(T) = A_i(T-1) - 2\mu\epsilon_T y_{T-i}$

lattice predictor related to time and order update equations will be discussed in Section VII.

V. COMPARISON OF GRADIENT AND LEAST SQUARES ADAPTIVE ALGORITHMS

A number of time-recursive versions of the lattice algorithms presented in Section III were developed for adaptive processing applications using gradient descent techniques. In Section V-A we present a brief description of some of these algorithms, as well as a non-lattice adaptive predictor. A comparison of gradient and least squares adaptive algorithms is presented in Section V-B. Only the scalar case is considered in these sections.

A. Gradient Adaptive Prediction Filters

Adaptive prediction filters have been derived for a variety of signal-processing applications. Perhaps the most widely used technique is the (non-lattice) gradient-search least mean squares (LMS) algorithm developed by Widrow *et al.* [167], [171], [172]. This algorithm provides a simple update formula for the predictor coefficients $A_{p,i}$, as shown in Table IX. The properties of the LMS algorithm have been studied extensively and we will summarize some of them later.

Gradient adaptive lattice algorithms have also been derived. These algorithms were developed prior to the least squares lattice forms [35], [46]-[48], [58], [63], [183]. The following gradient adaptive lattice (GAL) algorithm was proposed by Griffiths [46], [48]:

$$\left.\begin{aligned} \epsilon_{p+1,T} &= \epsilon_{p,T} - K^r_{p+1,T} r_{p,T-1} \\ r_{p+1,T} &= r_{p,T-1} - K^\epsilon_{p+1,T}\epsilon_{p,T} \end{aligned}\right\} \text{usual error update} \tag{96a}$$

$$\left.\begin{aligned} K^r_{p+1,T+1} &= K^r_{p+1,T} + \mu_{p+1}\,\epsilon_{p+1,T} r_{p,T-1} \\ K^\epsilon_{p+1,T+1} &= K^\epsilon_{p+1,T} + \mu_{p+1}\,\epsilon_{p,T} r_{p+1,T} \end{aligned}\right\} \text{approximate gain update} \tag{96b}$$

where μ_{p+1} is a step size controlling the convergence properties of the algorithm. An even simpler version of this algorithm can be obtained when only a single reflection coefficient is used, i.e., set

$$K^\epsilon_{p+1,T} = K^r_{p+1,T} = K_{p+1,T}$$

in the error equations and let

$$K_{p+1,T+1} = K_{p+1,T} + \mu_{p+1}[\epsilon_{p+1,T} r_{p,T-1} + \epsilon_{p,T} r_{p+1,T}]. \tag{97}$$

In later papers [47], [49] the step size μ_{p+1} was chosen to be time varying and inversely proportional to the prediction error power (see also [183]). The one and two reflection coefficient GAL algorithms with power normalization are summarized in Tables X and XI.

Recently it was shown that this power normalization of the gradient lattice algorithm (with $\beta = 1$) gives a set of equations that has almost the same form as the unnormalized least squares lattice algorithm [120]. The difference is that the gradient-based algorithm does not contain the likelihood variable $\gamma_{p,T}$. The absence of this gain factor degrades somewhat the initial convergence of the algorithm, as we will see in the next section. Note also that the predictor coefficients are computed by the stationary lattice filter which is an approximation of the least squares predictor as was discussed in Section IV-B.

A very similar type of adaptive form was suggested by Makhoul [58] with the following gain update formula:

$$\begin{aligned} \Delta_{p+1,T} &= \lambda\Delta_{p+1,T-1} + 2\epsilon_{p,T} r_{p,T-1} \\ R^2_{p,T} &= \lambda R^2_{p,T-1} + \epsilon^2_{p,T} + r^2_{p,T-1} \\ K_{p+1,T} &= \Delta_{p+1,T}/R^2_{p,T}. \end{aligned} \tag{98}$$

Simple manipulation of these equations shows that they reduce to the power-normalized algorithm presented by Griffiths (Table XI) for a particular choice of the step size ($\beta = 1$). Recursive techniques for updating reflection coefficients were also developed by Parker [64], [65].

B. A Performance Comparison

To illustrate the behavior of the adaptive algorithms described above, we performed a number of simple simulation experiments. In the first set of experiments, stationary data were generated by passing a white noise sequence through a second-order all-pole filter $1/A(z)$ with the following parameters:

Case 1: $A(z) = 1 - 1.6z^{-1} + 0.95z^{-2}$
Case 2: $A(z) = 1 - 1.8z^{-1} + 0.95z^{-2}$.

The output data were normalized to have unit variance.

Initial Convergence: The ratio of the eigenvalues ($\lambda_{MAX}/\lambda_{MIN}$, where λ_{MAX} is the largest eigenvalue and λ_{MIN} is the smallest) of the data covariance matrix was

Case 1: $\lambda_{MAX}/\lambda_{MIN} = 10$

TABLE X
A GRADIENT ADAPTIVE LATTICE ALGORITHM WITH ONE REFLECTION COEFFICIENT PER SECTION (GAL1)

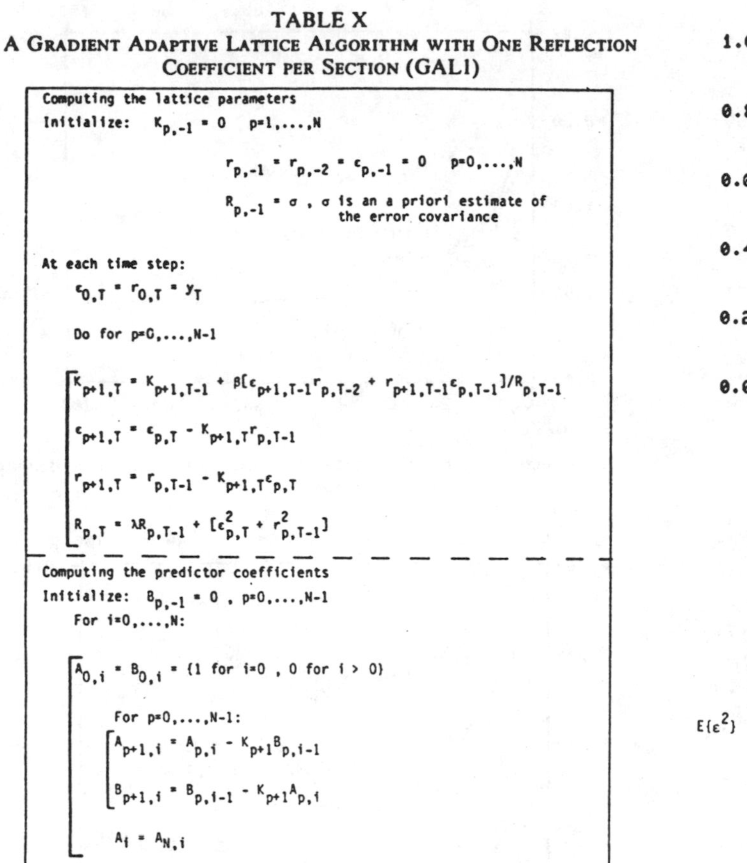

Computing the lattice parameters

Initialize: $K_{p,-1} = 0 \quad p=1,\ldots,N$

$r_{p,-1} = r_{p,-2} = \epsilon_{p,-1} = 0 \quad p=0,\ldots,N$

$R_{p,-1} = \sigma$, σ is an a priori estimate of the error covariance

At each time step:

$\epsilon_{0,T} = r_{0,T} = y_T$

Do for $p=0,\ldots,N-1$

$K_{p+1,T} = K_{p+1,T-1} + \beta[\epsilon_{p+1,T-1}r_{p,T-2} + r_{p+1,T-1}\epsilon_{p,T-1}]/R_{p,T-1}$

$\epsilon_{p+1,T} = \epsilon_{p,T} - K_{p+1,T}r_{p,T-1}$

$r_{p+1,T} = r_{p,T-1} - K_{p+1,T}\epsilon_{p,T}$

$R_{p,T} = \lambda R_{p,T-1} + [\epsilon^2_{p,T} + r^2_{p,T-1}]$

Computing the predictor coefficients

Initialize: $B_{p,-1} = 0$, $p=0,\ldots,N-1$

For $i=0,\ldots,N$:

$A_{0,i} = B_{0,i} = (1 \text{ for } i=0,\ 0 \text{ for } i>0)$

For $p=0,\ldots,N-1$:

$A_{p+1,i} = A_{p,i} - K_{p+1}B_{p,i-1}$

$B_{p+1,i} = B_{p,i-1} - K_{p+1}A_{p,i}$

$A_i = A_{N,i}$

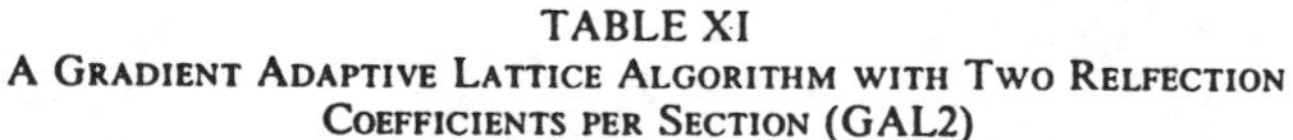

TABLE XI
A GRADIENT ADAPTIVE LATTICE ALGORITHM WITH TWO RELFECTION COEFFICIENTS PER SECTION (GAL2)

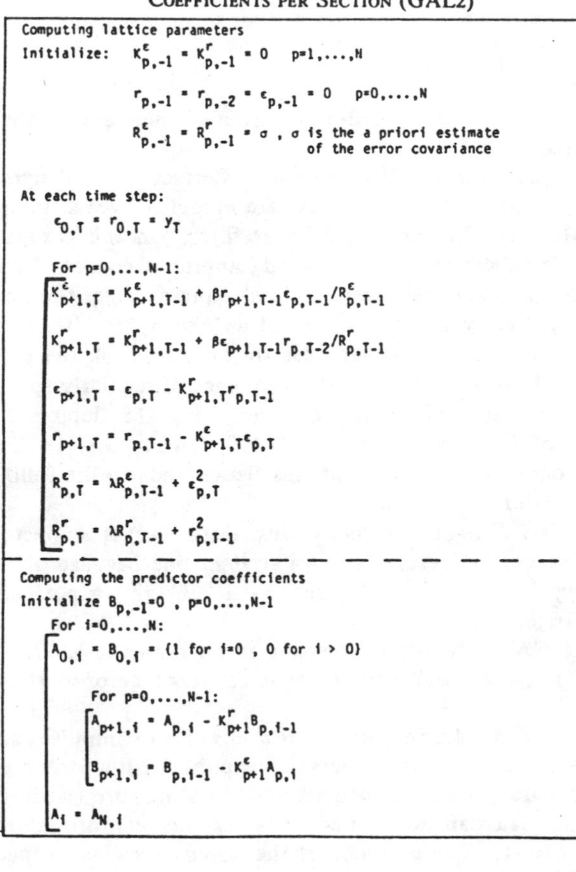

Computing lattice parameters

Initialize: $K^{\epsilon}_{p,-1} = K^{r}_{p,-1} = 0 \quad p=1,\ldots,N$

$r_{p,-1} = r_{p,-2} = \epsilon_{p,-1} = 0 \quad p=0,\ldots,N$

$R^{\epsilon}_{p,-1} = R^{r}_{p,-1} = \sigma$, σ is the a priori estimate of the error covariance

At each time step:

$\epsilon_{0,T} = r_{0,T} = y_T$

For $p=0,\ldots,N-1$:

$K^{\epsilon}_{p+1,T} = K^{\epsilon}_{p+1,T-1} + \beta r_{p+1,T-1}\epsilon_{p,T-1}/R^{\epsilon}_{p,T-1}$

$K^{r}_{p+1,T} = K^{r}_{p+1,T-1} + \beta\epsilon_{p+1,T-1}r_{p,T-2}/R^{r}_{p,T-1}$

$\epsilon_{p+1,T} = \epsilon_{p,T} - K^{r}_{p+1,T}r_{p,T-1}$

$r_{p+1,T} = r_{p,T-1} - K^{\epsilon}_{p+1,T}\epsilon_{p,T}$

$R^{\epsilon}_{p,T} = \lambda R^{\epsilon}_{p,T-1} + \epsilon^2_{p,T}$

$R^{r}_{p,T} = \lambda R^{r}_{p,T-1} + r^2_{p,T-1}$

Computing the predictor coefficients

Initialize $B_{p,-1}=0$, $p=0,\ldots,N-1$

For $i=0,\ldots,N$:

$A_{0,i} = B_{0,i} = (1 \text{ for } i=0,\ 0 \text{ for } i>0)$

For $p=0,\ldots,N-1$:

$A_{p+1,i} = A_{p,i} - K^{r}_{p+1}B_{p,i-1}$

$B_{p+1,i} = B_{p,i-1} - K^{\epsilon}_{p+1}A_{p,i}$

$A_i = A_{N,i}$

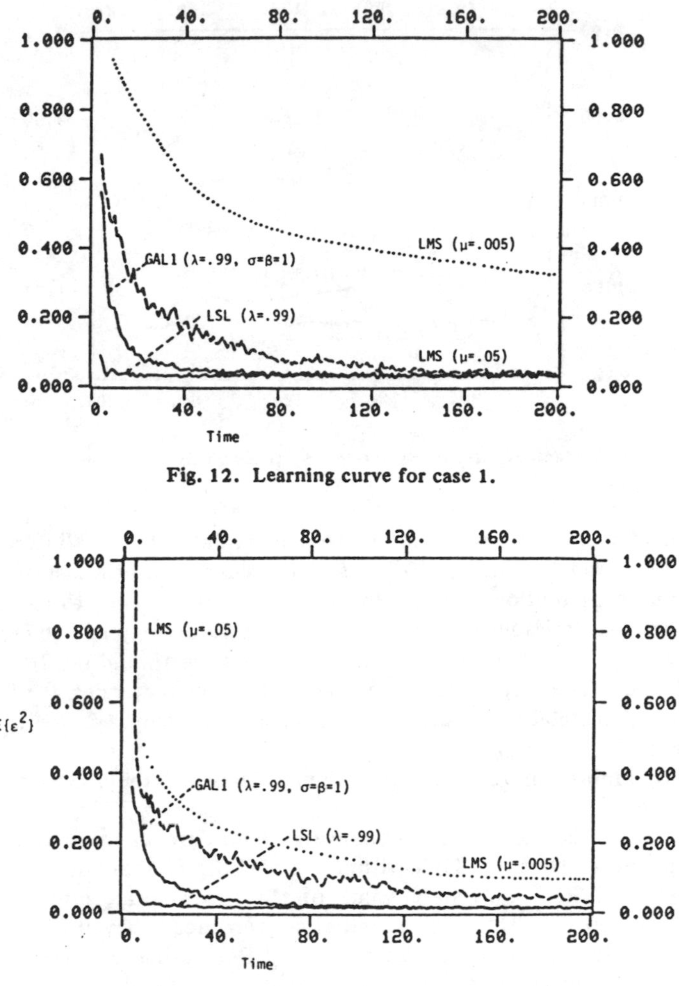

Fig. 12. Learning curve for case 1.

Fig. 13. Learning curve for case 2.

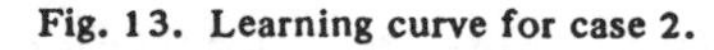

Case 2: $\lambda_{MAX}/\lambda_{MIN} = 77.$

This ratio has a strong effect on the convergence properties of the LMS, as discussed in [171]. The step size μ of the LMS algorithm has to be smaller than $1/\lambda_{MAX}$ to ensure stability of the algorithm. On the other hand, the mean-squared prediction error $E\{\epsilon_T^2\}$ of the LMS converges exponentially with the dominant (i.e., largest) time constant τ being given by

$$\tau = 1/4\mu\lambda_{MIN}. \tag{99}$$

When μ is chosen to ensure stability we have

$$\tau \geqslant \lambda_{MAX}/4\lambda_{MIN}. \tag{100}$$

Thus the larger the eigenvalue ratio, the slower will be the convergence of the LMS.

Figs. 12 and 13 depict the behavior of the mean-square error $E\{\epsilon_T^2\}$ versus time (called "learning curves") for the two cases. Each curve was obtained by averaging 200 independent computer runs. The curves in each figure correspond to the LMS algorithm (Table IX), the GAL1 algorithm (Table X), and the LSL algorithm (Table III). In our tests, GAL1 and GAL2 gave very similar results, and we have, therefore, presented only one of them. The lattice algorithms were run with a forgetting factor of $\lambda = 0.99$. (GAL had $\sigma = \beta = 1$ in all the cases presented in this section.) The LMS algorithm was run with two different step sizes: $\mu = 0.005$ and $\mu = 0.05$. The

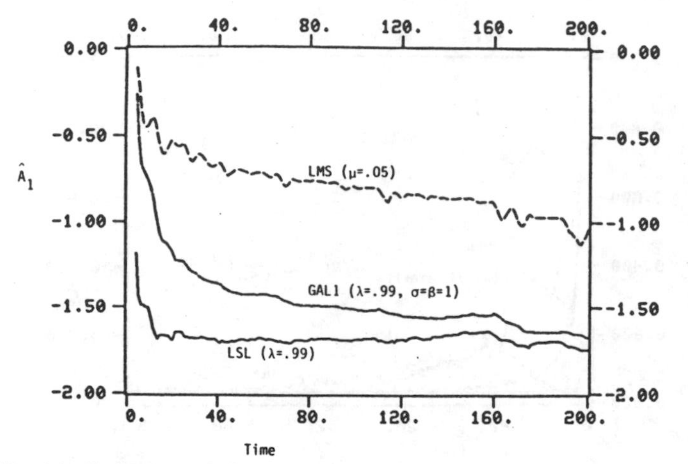

Fig. 14. Trajectory of the parameter A_1 (case 2) for Guassian driving noise.

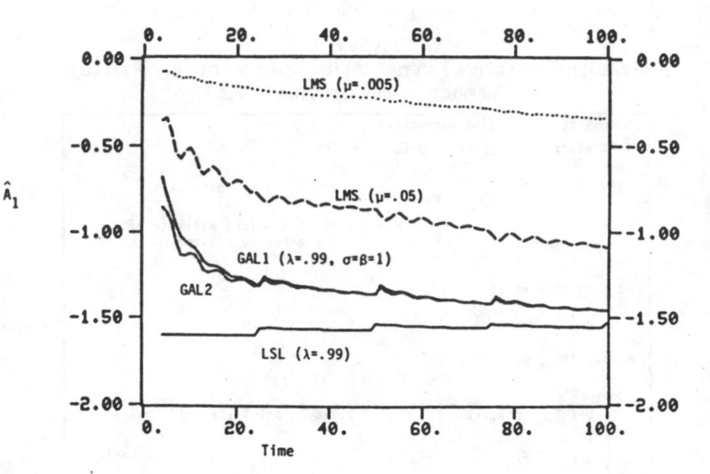

Fig. 15. Trajectory of the parameter A_1 (case 1) for impulsive driving noise.

smaller step size was chosen from formulas given in Widrow [171] and Griffiths [120] to give the same asymptotic mean-square prediction error as the lattice algorithm (i.e., to have the same misadjustment noise, in the terminology of [167], [171]). The theoretical results state that we should set $2\mu = 1 - \lambda$. The larger step size was chosen by increasing μ till initial instability occurred and then dividing it by a factor of two.

Examination of these two figures leads to the following observations:

i) The lattice algorithms converge considerably faster than the non-lattice LMS algorithm. The LSL is the fastest; it reaches the steady-state value of the mean-square error in about 10 time steps. (The initial convergence behavior of the GAL depends on the choice of σ. Decreasing σ tends to increase the convergence rate.)

ii) The convergence of the LMS algorithm is clearly effected by the eigenvalue ratio, while the lattice algorithms are quite insensitive to it.

It should be noted that the slower convergence of the GAL (compared to the LSL) is largely due to the particular initialization procedure used in Tables X and XI. The GAL algorithm can be easily modified to resemble more closely the LSL algorithm. Griffiths and Medaugh have developed order update formulas for the prediction error variance ($R_{p,T}$ in Table X and $R^{e}_{p,T}$, $R^{r}_{p,T}$ in Table XI). Replacing the time update of the prediction error variance by an order update makes it possible to apply the LSL initialization procedure (turning on the reflection coefficients one-by-one) to the modified GAL algorithm. This modification, suggested by Griffiths, is expected to make the initial convergence behavior of the GAL very close to that of the LSL. This conjecture awaits experimental verification at this time.

The mean-square prediction error is commonly used as a measure of the performance of adaptive filters. In our opinion, a more sensitive performance measure is given by the trajectories of the predictor parameters. Fig. 14 depicts the time history of the parameter estimate A_1 as computed by the different algorithms in a single run. Note the significant difference in the convergence rates of the LMS versus the lattice algorithms. In longer runs it was possible to see quite clearly that the LMS parameter trajectories for $\mu = 0.05$ were "noisier" than the lattice parameter trajectories. (This is due to the fact that $\mu = 0.05$ corresponds to a higher misadjustment noise than $\lambda = 0.99$.) See [50], [102], [128],

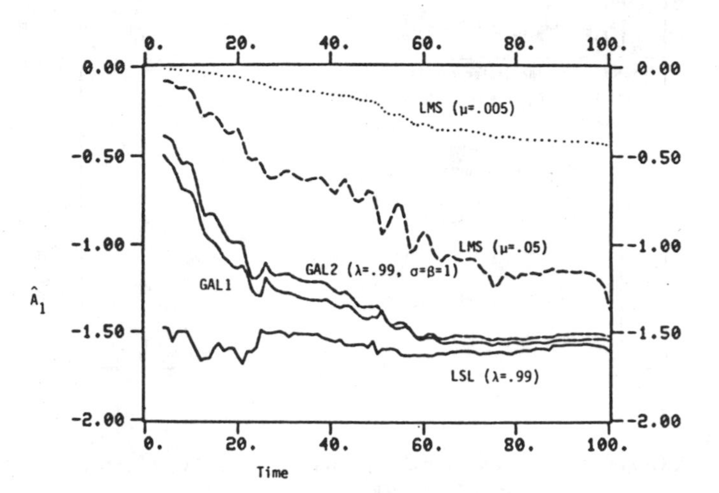

Fig. 16. Trajectory of the parameter A_1 (case 1) for driving noise which is a mixture of Gaussian and impulsive noise.

[130] for discussions regarding the convergence rates of these algorithms.

Effects of Impulsive Driving Noise: Certain types of signals (such as speech and seismic data) are modeled by a sequence of impulses passing through a linear filter. In this case, the output is a superposition of delayed impulse responses. Given a "clean" impulse response, a least squares algorithm can estimate perfectly the parameters of an Nth order filter from only N data points. To demonstrate this, we took the filter of case 1 above and fed it with a sequence of regularly spaced (at 25-point intervals) unit impulses. Fig. 15 depicts the response of the four algorithms (LMS, GAL1, GAL2, LSL) to these data. Examination of this figure leads to the following observations:

i) The LSL algorithm converges instantly, as expected. The parameter trajectory is not a straight line because of the forgetting factor $\lambda < 1$. Setting $\lambda = 1$ will give a perfectly straight trajectory.

ii) The GAL algorithms converge faster than the LMS, but they do not share the instantaneous convergence property of the LSL.

Sometimes the driving process is a mixture of impulses and Gaussian noise. Fig. 16 depicts the response of the form algorithm to data generated by passing such a mixture (with impulses and Gaussian noise of equal average power) through the filter of case 1. The presence of the impulses seems to speed

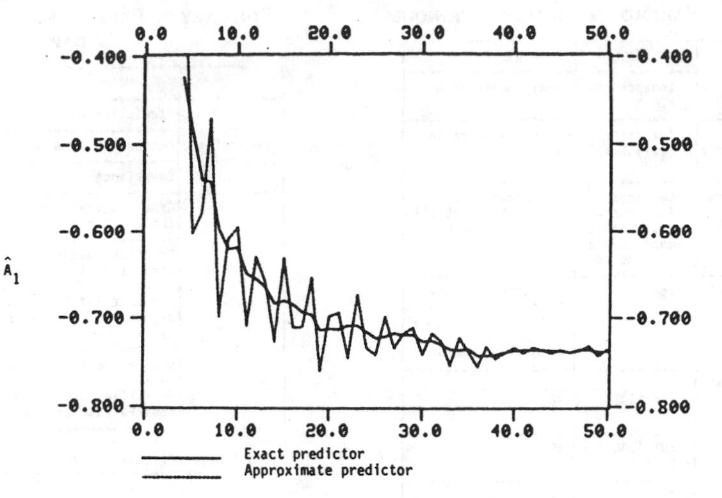

Fig. 17. Trajectory of the parameter A_1 for the exact and approximate lattice predictor.

up somewhat the initial convergence of the LSL while having little or no effect on the gradient algorithms. Figs. 15 and 16 also depict the similarity of GAL1 and GAL2.

Exact Versus Approximate Lattice Predictors: In Section IV-B we discussed the difference between the exact lattice predictors (Table IV) and the approximate predictor obtained by "freezing" the reflection coefficients of the LSL. Fig. 17 depicts the parameter (A_1) trajectories computed by these two methods for data generated by passing white noise through the filter $1/A(z)$ with

$$A(z) = 1 - 0.8z^{-1} + 0.95z^{-2}.$$

The LSL algorithm was run with $\lambda = 0.99$. Note that the difference between the two trajectories becomes very small after only a few data points. Thus in many practical applications it is possible to use the simpler lattice predictor with minimal loss in performance. In problems involving higher order predictors, the difference between the two lattice predictors may be more noticeable. We prefer to always use the exact lattice predictor since the increase in computational requirements is minor.

Parameter Tracking: The capability of these adaptive algorithms to track time-varying parameters was tested by the following experiment: data were generated by passing white noise through the filter $1/A(z)$ where

$$A(z) = 1 - 1.4z^{-1} + 0.975z^{-2}.$$

After 200 time steps, the filter parameters abruptly change to

$$A(z) = 1 - 1.9z^{-1} + 0.975z^{-2}.$$

Fig. 18 depicts the parameter trajectories for the algorithms LMS, GAL1, and LSL. The LMS tracks reasonably well but with a considerable bias (caused probably by the fact that the algorithm has not fully converged). Setting μ to a smaller value reduces the bias but makes the tracking very slow. Fig. 19 depicts the tracking behavior for data generated by the same filter with an impulsive driving noise. Note the superior performance of the lattice algorithms in comparison to the LMS. For a more detailed comparison of these algorithms in the tracking of noisy narrow-band signals see the work by Hodgkiss [101], [173], [194]. The tracking rate of these algorithms is determined by μ (for LMS) and λ (for GAL, LSL). The setting of μ is constrained by the requirement that

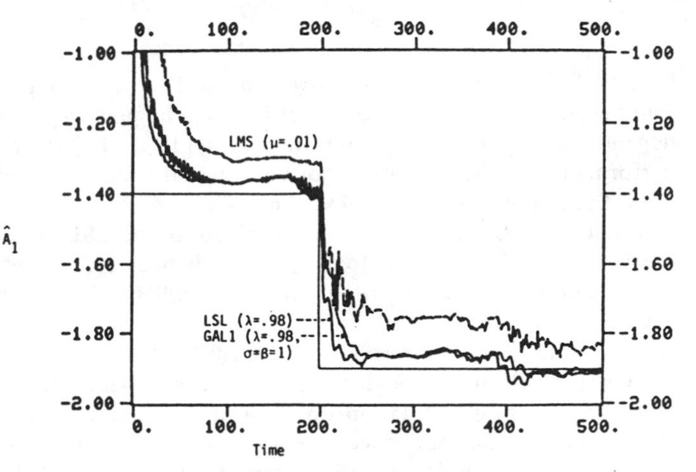

Fig. 18. Trajectory of A_1 in the presence of a parameter jump (Gaussian input).

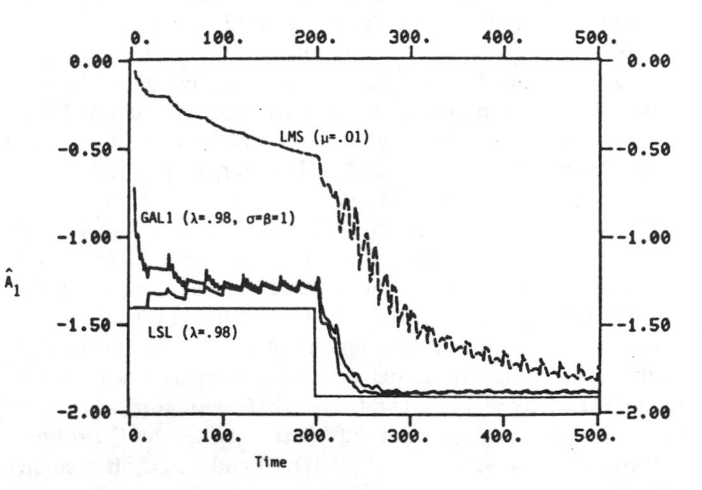

Fig. 19. Trajectory of A_1 in the presence of a parameter jump (impulsive input).

the algorithm stay stable, whereas λ can be set to achieve a desired window length (which equals $-1/\ln \lambda$).

Finally, we note the similarity in the tracking behavior of the LSL and GAL algorithms. After the initial convergence phase is over, the two algorithms are practically the same [120] (unless γ becomes very close to unity). The tracking rate of both algorithms is determined mainly by the effective

TABLE XII
COMPUTATIONAL REQUIRMENTS OF VARIOUS PREDICTION TECHNIQUES

Property		LMS	GAL	LSL
Stability		Depends on μ [171] (99)	Independent of λ	Independent of λ
Initial convergence and parameter tracking [173]		Sensitive to eigenvalue ratio [171] (100)	Insensitive to eigenvalue ratio [101],[143],[144]	
		Relatively slow convergence	Converges faster than LMS. Initial convergence depends on σ	Converges faster than LMS. Initially faster than GAL.
Effects of impulsive driving noise		Relatively small effect		Can speed up convergence. Instantaneous convergence for noiseless impulse response
Misadjustment noise		$M = \mu$ trace $(\underline{R})$ [171]	$M = \frac{1-\lambda}{1+\lambda} N$ [120]	
Computational Requirements		2NT	6NT (GAL1) 10NT (GAL2)	10NT
Simultaneous estimation of predictors of all orders		No	Yes	Yes

length of the exponential window (≈50 in Figs. 18 and 19). The question of which lattice algorithm is to be preferred will depend on the particular application. In [101], [173] the performance of the LSL was reported to be superior to that of the GAL, in the context of tracking noisy sinusoids. In [120], [184] the GAL was reported to outperform the LSL when the step-size β was chosen to be greater than unity. If fast initial convergence is important, the LSL will be the better choice.

Computational Requirements: It is difficult to compare the computational complexity of algorithms without considering a specific implementation on a specific computer. However, a rough comparison can be achieved by counting the number of arithmetic operations needed to carry out the computations. In Table XII we summarize the operation count for several of the algorithms mentioned in this paper. The count is made for an Nth order predictor and for T data points. In the block-processing algorithms we ignored terms that are independent of T. In the adaptive algorithms we considered only the computations involved in updating lattice parameters. To recover the predictor parameters $\{A_{p,i}\}$ extra computations will be required. However, this is often either unnecessary or is performed only infrequently.

The comparison based on operation count should be considered with some caution, since there are other issues involved. For example, the normalized lattice filter has the potential of fixed-point implementation due to the fact that all internal variables are in magnitude less than one. The normalized lattice recursions can actually be expressed as a sequence of circular and hyperbolic rotations. Efficient computation of these rotations can be achieved by the use of the COordinate Rotation DIgital Computer (CORDIC) and other bit-recursive algorithms; see, e.g., [74]–[78], [192]. The possibility of performing the basic lattice recursions using special-purpose hardware, e.g., a VLSI chip, makes the operation count less important than, say, the architecture of the computations.

A selective summary of the properties of the algorithms discussed in this section is given in Table XIII. The term misadjustment noise M is defined in [167], [171] as the relative increase in the mean-square prediction error $E\{\epsilon^2\}$ due to the presence of a forgetting factor, compared to the least squares prediction error $E\{\epsilon_{LS}^2\}$: $M = E\{\epsilon^2\}/E\{\epsilon_{LS}^2\} - 1$.

TABLE XIII
SUMMARY OF PROPERTIES OF GRADIENT AND LEAST SQUARES ADAPTIVE ALGORITHMS

Method		Number of Operations*
Block Processing [55]	Prediction error	5NT
	Autocorrelation	NT
	Covariance	NT
Adaptive	Gradient least mean squares (LMS), (Table 9)	2NT
	Gradient lattice, one coefficient (GAL1), (Table 10)	6NT
	Gradient lattice, two coefficients (GAL2), (Table 11)	10NT
	Unnormalized pre-windowed least squares lattice (Table 3)	10NT
	Normalized pre-windowed least squares lattice (Table 6)	12NT 3NT square roots

* Multiplies/divides and adds

One of the difficulties in comparing the LSL algorithm to gradient algorithms is that while there is essentially only one least squares algorithm, there are numerous approximate least squares gradient algorithms. It was not our intention to present here a comprehensive comparison of gradient techniques. We, therefore, limited our attention to only three algorithms (LMS, GAL1, GAL2) which seem to be representative of lattice and non-lattice adaptive gradient techniques. Other versions of these algorithms will perform somewhat differently. Thus the properties summarized in Table XIII should be interpreted in their proper context.

The adaptive lattice algorithms discussed in this section have been used in a variety of applications, including: speech processing [59], [92], [111], [128], [130], [136], [140], channel equalization [45], [68], [69], [88], [143], [144], [147], line enhancement [91], [142], noise cancelation [71], spectral estimation [57], [93], [101], [112], [173], [193], adaptive control [89], and adaptive filtering in general [70], [72], [133], [138], [145].

VI. A SELECTION OF ADAPTIVE LEAST SQUARES LATTICE ALGORITHMS

In the earlier sections, we presented the simplest of the adaptive LSL filters: the pre-windowed form. The projection approach (or alternative methods) can be used to derive a number of variations of this basic algorithm, obtained by: i) using the covariance instead of the pre-windowed method for computing the least squares predictor (50), (51); ii) developing a predictor for estimating one process from past measurements of a related process (rather than for estimating a process from its own past values); iii) using an Infinite Impulse Response (IIR) predictor instead of a Finite Impulse Response (FIR) predictor; iv) forcing the predictor to have a specific structure (e.g., $A_{N,i} = A_{N,N-i}$ which leads to linear-phase characteristics [119]). Both unnormalized and normalized versions of each algorithm can be derived. Other variations are also possible; see, e.g., [94]. The number of possible combinations of these features is quite large, and it is beyond the scope of this paper to explore them all. The three LSL filters which are described in this section are representative of this family of adaptive algorithms.

A. The Covariance Lattice Form

Adaptive prediction of data with time-varying statistics requires that the prediction algorithm be able to "forget" old data. The exponentially weighted lattice form achieves this

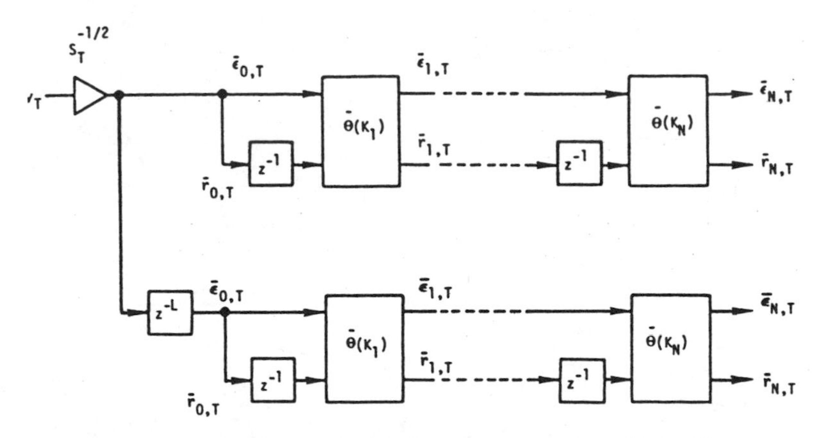

Fig. 20. The variance normalized sliding-window covariance lattice filter.

goal by giving an ever decreasing weight to past data. This type of forgetting mechanism works well when the underlying model is changing slowly and smoothly. Consider, however, a situation where the statistics of the data change abruptly. In this case, we would like to discard as soon as possible the data collected prior to the change.

Exponential windowing will allow the effects of the old data to linger on since it has, in effect, infinite memory. An alternative approach which seems better in such a situation is to base our estimates on a fixed number of data points. In other words, consider a fixed-length window sliding over the data. Only data within this window will be used to compute the predictor. This is a finite-memory strategy in which old data are completely discarded.

The forward prediction errors associated with the sliding-window problem are defined by

$$[\epsilon_{p,T-L}, \cdots, \epsilon_{p,T}] = y_{T-L:T} + [A_{p,1}, \cdots, A_{p,p}]\, Y^L_{p,T} \qquad \text{forward errors} \quad (101a)$$

where

$$y_{T-L:T} = [y_{T-L}, \cdots, y_T],$$

$$m \times (L+1) \text{ matrix} \quad (101b)$$

$$Y^L_{p,T} = \begin{bmatrix} y_{T-L-1} & \cdots & y_{T-1} \\ \vdots & & \vdots \\ y_{T-L-p} & \cdots & y_{T-p} \end{bmatrix},$$

$$mp \times (L+1) \text{ data matrix} \quad (101c)$$

and L is the window length. This definition is similar to (50) and $Y^L_{p,T}$ has the same form as the covariance data matrix $Y^C_{p,T}$. In fact, choosing a *time-varying* window length $L = T - p$ makes the two matrices identical. We call the LSL corresponding to this choice the *growing memory covariance* form [140]. In this section, we consider only the case of a *fixed* window length which is called the *sliding-window covariance* form.

The backward prediction errors are similarly defined

$$[r_{p,T-L-1}, \cdots, r_{p,T-1}] = [y_{T-L-p-1}, \cdots, y_{T-p-1}] + [B_{p,p}, \cdots, B_{p,1}]\, Y^L_{p,T} \qquad \text{backward errors.} \quad (102)$$

The least squares predictor minimizes the sum of squared errors over the window, i.e.,

$$\operatorname{tr}\left\{ \sum_{t=T-L}^{T} \epsilon_{p,t}\epsilon'_{p,T} \right\}.$$

TABLE XIV
THE NORMALIZED SLIDING-WINDOW COVARIANCE LATTICE

Initialization: Set S, $\tilde{r}$, $\breve{r}$, K, $\bar{K}$ to zero

For $T=0,1,\ldots,TMAX$:

$$S_T = \sum_{t=T-L}^{T} y_t y'_t = S_{T-1} + y_T y'_T - y_{T-L-1} y'_{T-L-1}$$

$$\tilde{\epsilon}_{0,T} = \tilde{r}_{0,T} = S_T^{-1/2} y_T$$

$$\breve{\epsilon}_{0,T} = \breve{r}_{0,T} = S_T^{-1/2} y_{T-L}$$

For $p=0,\ldots,\min\{N,T\}-1$:

$$K_{p+1,T} = F^{-1}(\bar{K}_{p+1,T-1}, \tilde{r}'_{p,T-1}, \tilde{\epsilon}_{p,T})$$

$$\bar{K}_{p+1,T} = F\,(K_{p+1,T-1}, \breve{r}'_{p,T-1}, \breve{\epsilon}_{p,T})$$

$$\tilde{\epsilon}_{p+1,T} = F(\tilde{\epsilon}_{p,T}, \tilde{r}_{p,T-1}, K_{p+1,T}) \quad *$$

$$\tilde{r}_{p+1,T} = F(\tilde{r}_{p,T-1}, \tilde{\epsilon}_{p,T}, K'_{p+1,T}) \quad *$$

$$\breve{\epsilon}_{p+1,T} = F(\breve{\epsilon}_{p,T}, \breve{r}_{p,T-1}, K_{p+1,T}) \quad *$$

$$\breve{r}_{p+1,T} = F(\breve{r}_{p,T-1}, \breve{\epsilon}_{p,T}, K'_{p+1,T}) \quad *$$

$$F(a,b,c) = [I - cc']^{-1/2}[a - cb][I - b'b]^{-T/2}$$

$$F^{-1}(a,b,c) = [I - cc']^{1/2} a [I - b'b]^{T/2} + cb$$

* The function $F(\cdot)$ involves division. Whenever the divisor x is very small, set $1/x=1$.

As in the pre-windowed case, the prediction errors are computed through a projection operator. In this case

$$P^L_{p,T} \triangleq Y^{L'}_{p,T}(Y^L_{p,T} Y^{L'}_{p,T})^{-1} Y^L_{p,T} \quad (103)$$

and

$$\epsilon_{p,T} = y_{T-L:T}(I - P^L_{p,T})\,\pi'$$

$$r_{p,T-1} = y_{T-L-p-1:T-p-1}(I - P^L_{p,T})\,\pi'. \quad (104)$$

Using proper substitutions in the projection update formulas given in Section IV leads to a set of lattice recursions. We shall omit the details which can be found in [140], and present only the final results.

The sliding-window covariance lattice form is presented in Fig. 20, for the variance-normalized case. It consists of two identical lattice filters of the pre-windowed type, with a common set of reflection coefficients. The variables in the lower lattice filter are needed for updating the reflection coefficients. The normalized algorithm is summarized in Table XIV. Its

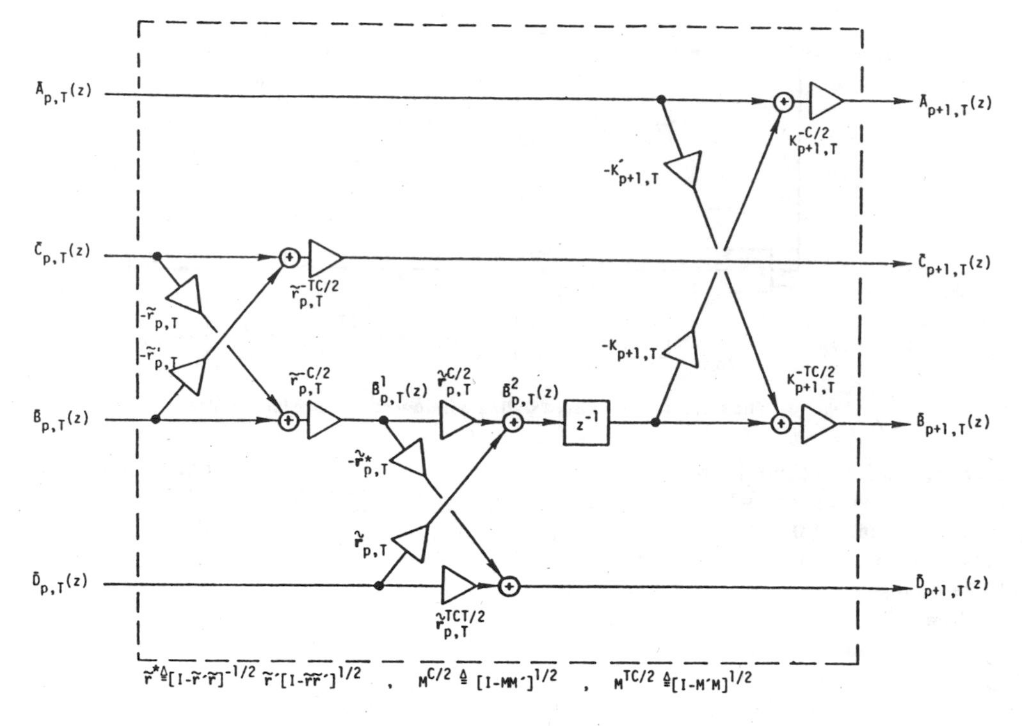

Fig. 21. The variance normalized sliding-window covariance lattice prediction filter.

unnormalized version is discussed in [128]. The covariance lattice can be derived from the pre-windowed form, as was shown in [141].

The sliding-window lattice form has twice the complexity of the prewindowed form. In addition, it requires storage of the last L data points, whereas the pre-windowed lattice needs only the current data point y_T. To alleviate the storage requirements of this type of filter, Makhoul proposed to use a pre-windowed lattice in which the single-pole filter associated with exponential weighting is replaced by a two- or three-pole filter [59].

The parameters computed by the covariance lattice can be sampled at any time, say T, and used to form the least squares prediction filter depicted in Fig. 21. Note that this filter resembles the (pure order-update) pre-windowed prediction filter in Fig. 21 except that it has one more line (associated with $\bar{D}_p(z)$). The algorithm for computing the predictor coefficients is summarized in Table XV. See [96], [140] for a more detailed discussion.

TABLE XV
THE NORMALIZED SLIDING-WINDOW COVARIANCE LATTICE PREDICTION FILTER

For $i=0,\ldots,N$ do:

$\bar{A}_{0,i} = \bar{B}_{0,i} = S_T^{-1/2}$, $\bar{C}_{0,i} = \bar{D}_{0,i} = 0$ $i=0$

$\bar{A}_{0,i} = \bar{B}_{0,i} = \bar{C}_{0,i} = \bar{D}_{0,i} = 0$ $i > 0$

For $p=0,\ldots,N-1$:

$\bar{B}^1_{p,i} = G(\bar{B}_{p,i}, \bar{C}_{p,i}, \tilde{r}_{p,T})$

$\bar{C}_{p+1,i} = G(\bar{C}_{p,i}, \bar{B}_{p,i}, \tilde{r}'_{p,T})$

$\bar{B}^2_{p,i} = G^{-1}(\bar{B}^1_{p,i}, \bar{D}_{p,i}, \hat{r}_{p,T})$

$\bar{D}_{p+1,i} = \tilde{G}^{-1}(\bar{D}_{p,i}, \bar{B}^1_{p,i}, -\hat{r}'_{p,T})$

$\bar{A}_{p+1,i} = G(\bar{A}_{p,i}, \bar{B}^2_{p,i-1}, K_{p+1,T})$

$\bar{B}_{p+1,i} = G(\bar{B}^2_{p,i-1}, \bar{A}_{p,i}, K'_{p+1,T})$

To unnnormalize: $A_p(z) = \bar{A}^{-1}_{p,0}\,\bar{A}_p(z)$

$G(a,b,c) = [I - cc']^{-1/2}\,[a - cb]$

$G^{-1}(a,b,c) = [I - cc']^{1/2}\,a + cb$

$\tilde{G}^{-1}(a,b,c) = [I - cc']^{T/2}a + [I - cc']^{-1/2}\,c[I - c'c]^{1/2}\,b$

In the scalar case: $\tilde{G}^{-1}(a,b,c) = G^{-1}(a,b,c)$

B. Lattice for Joint Estimation

So far we considered the prediction of a process y_T from its past values. A more general prediction problem which often arises is the prediction of one process x_T from measurements of a related process, e.g.,

$$\hat{x}_{T|T-1} = -\sum_{i=1}^{p} W_{p,i}\,y_{T-i}. \tag{105}$$

The prediction error is defined by

$$\epsilon^x_{p,T} = x_T - \hat{x}_{T|T-1} = x_T + \sum_{i=1}^{p} W_{p,i}\,y_{T-i} \tag{106}$$

which can be written in matrix form as

$$[\epsilon^x_{p,0}, \cdots, \epsilon^x_{p,T}] = x_{0:T} + [W_{p,1}, \cdots, W_{p,p}]\,Y_{p,T} \tag{107a}$$

$$x_{0:T} = [x_0, \cdots, x_T] \tag{107b}$$

and $Y_{p,T}$ is the pre-windowed data matrix, as in (50).

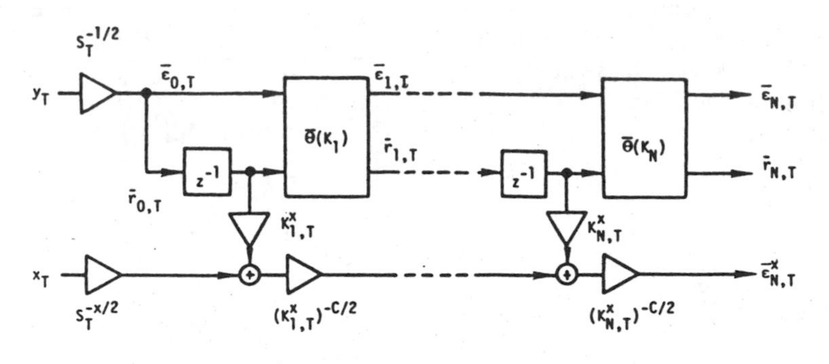

Fig. 22. The variance normalized pre-windowed lattice for joint process prediction.

The least squares prediction error $\epsilon^x_{p,T}$ is given by

$$\epsilon^x_{p,T} = x_{0:T}(I - P_{Y_{p,T}})\,\pi', \qquad \text{as in (71a)} \tag{108}$$

or in normalized form

$$\tilde{\epsilon}^x_{p,T} \triangleq \rho_{Y_{p,T}}(x_{0:T}, \pi), \qquad \text{as in (85).} \tag{109}$$

Applying the projection update formulas developed in Sections IV-A and IV-C leads directly to the unnormalized and normalized recursions [115], [139]. The additional equations for the normalized form are summarized below.

$$V = x_{0:T} \qquad W = \pi \qquad S = Y_{p,T} \qquad X = y^{p+1}_{0:T}$$

gives

$$\tilde{\epsilon}^x_{p+1,T} = F(\tilde{\epsilon}^x_{p,T}, \tilde{r}_{p,T-1}, K^x_{p+1,T}) \tag{110}$$

where

$$K^x_{p+1,T} \triangleq \rho_{Y_{p,T}}(x_{0:T}, y^{p+1}_{0:T}) \tag{111}$$

and

$$V = x_{0:T} \qquad W = y^{p+1}_{0:T} \qquad S = Y_{p,T} \qquad X = \pi$$

gives

$$K^x_{p+1,T} = F^{-1}(K^x_{p+1,T-1}, \tilde{r}'_{p,T-1}, \tilde{\epsilon}^x_{p,T}). \tag{112}$$

Adding (110) and (112) to the lattice algorithm for the prediction of the process y_T in Table VI gives a complete set of recursions for the joint estimation problem. The resulting lattice structure is depicted in Fig. 22 (for the variance normalized case) and is summarized in Table XVI (for the normalized case).

In some situations we want to estimate x_T not only from past values but also from the current values of y_T

$$\hat{x}_{T|T} = -\sum_{i=0}^{p-1} W_{p,i}\, y_{T-i}. \tag{113}$$

It is straightforward to check that the same algorithm will work, provided that the order in which quantities are being updated is changed. Specifically K^x and $\tilde{\epsilon}^x$ will be updated after K, $\tilde{\epsilon}$, and $\tilde{r}$.

A lattice prediction filter can be derived for the joint process estimation problem along the lines discussed in Section IV-D. The filter is based on the pre-windowed lattice predictor with one more line which involves the parameters K^x, $\tilde{\epsilon}^x$. See [96] for details.

TABLE XVI
THE JOINT PROCESS NORMALIZED PRE-WINDOW LATTICE ALGORITHM

Initialize: S, S^x, $\tilde{r}$, K, K^x to zero

Main loop:

$S_T = \lambda S_{T-1} + y_T y'_T$

$S^x_T = \lambda S^x_{T-1} + x_T x'_T$

$\tilde{\epsilon}_{0,T} = \tilde{r}_{0,T} = S_T^{-1/2}\, y_T$

$\tilde{\epsilon}^x_{0,T} = (S^x_T)^{-1/2}\, x_T$

For $p = 0, \ldots, \min(N,T) - 1$ do:

$K_{p+1,T} = F^{-1}(K_{p+1,T-1}, \tilde{r}'_{p,T-1}, \tilde{\epsilon}_{p,T})$

$K^x_{p+1,T} = F^{-1}(K^x_{p+1,T-1}, \tilde{r}'_{p,T-1}, \tilde{\epsilon}^x_{p,T})$

$\tilde{\epsilon}_{p+1,T} = F(\tilde{\epsilon}_{p,T}, \tilde{r}_{p,T-1}, K_{p+1,T})$ *

$\tilde{\epsilon}^x_{p+1,T} = F(\tilde{\epsilon}^x_{p,T}, \tilde{r}_{p,T-1}, K^x_{p+1,T})$ *

$\tilde{r}_{p+1,T} = F(\tilde{r}_{p,T-1}, \tilde{\epsilon}_{p,T}, K'_{p+1,T})$ *

$F(a,b,c) = [I - cc']^{-1/2}\,[a - cb][I - b'b]^{-T/2}$

$F^{-1}(a,b,c) = [I - cc']^{1/2}\, a[I - b'b]^{T/2} + cb$

* When dividing by a small number x set 1/x=1.

C. An Infinite Impulse Response Lattice

The lattice filters discussed so far were all FIR filters. Such filters have the proper structure for modeling and prediction of autoregressive (AR) processes. Such processes can be generated by passing white noise through an all-pole filter. In this section, we consider a more general IIR lattice structure which is related to the prediction of autoregressive moving average (ARMA) processes. These processes can be generated by passing white noise through a pole-zero filter. In other words, a process with a finite-dimensional rational spectrum is represented as an ARMA process [149], [174]. The class of AR processes is only a special case of the more general ARMA processes. In spite of this, AR models are more widely used in time-series analysis, spectral estimation, and signal-processing applications. The main reasons for this are: i) many ARMA processes can be closely approximated by an AR process of sufficiently high order; and ii) the problem of fitting an ARMA model to an observed time series is much more difficult than the AR modeling problem, as we will see next.

Consider an IIR prediction filter of the form

$$\epsilon_T = \frac{M_p(z)}{N_q(z)} y_T \tag{114a}$$

or

$$\epsilon_T = y_T + \sum_{i=1}^{p} M_{p,i} y_{T-i} - \sum_{i=1}^{q} N_{q,i} \epsilon_{T-i} \tag{114b}$$

where ϵ_T is the least squares prediction error sequence. The feedback structure of this filter makes the problem of computing the predictor coefficients $\{M_{p,i}, N_{q,i}\}$ quite difficult. In the FIR case (e.g., (48)) the predicted value $\hat{y}_{T|T-1}$ of the data y_T is a *linear* function of the unknown predictor parameters. Estimating the predictor coefficients involves the relatively easy solution of a set of linear equations, such as the Yule–Walker equations. The estimation of the IIR predictor parameters, however, involves a nonlinear problem as can be seen from (114). If the sequence of least squares prediction errors were known *a priori*, the problem would still be linear. In fact, the IIR predictor could be computed using any of the techniques presented in earlier sections by embedding the IIR problem in a larger FIR problem. To see this, rewrite (114) as

$$\underbrace{\begin{bmatrix} \epsilon_T \\ \epsilon_T \end{bmatrix}}_{\epsilon^z_{p,T}} = \underbrace{\begin{bmatrix} y_T \\ \epsilon_T \end{bmatrix}}_{z_T} + \sum_{i=1}^{p} \underbrace{\begin{bmatrix} M_{p,i} & -N_{p,i} \\ 0 & 0 \end{bmatrix}}_{A_{p,i}} \underbrace{\begin{bmatrix} y_{T-i} \\ \epsilon_{T-i} \end{bmatrix}}_{z_{T-i}}. \tag{115}$$

We have chosen here the case where $p = q$ to simplify the presentation. The more general case of $p \neq q$ is treated in [94]. Note that the predictor for the process $z_T = [y'_T, \epsilon'_T]'$ is an FIR predictor whose coefficients $A_{p,i}$ contain the parameters of the IIR predictor $M_{p,i}, N_{p,i}$. The difficulty is, of course, that the prediction errors ϵ_T are not available and their computation (via (114) requires the unknown predictor parameters. Consider, however, the following "boot-strapping" procedure. Let us say that we have somehow obtained an estimate $\hat{M}_{p,i}$, $\hat{N}_{p,i}$ of the predictor parameters. We can then compute an estimate $\hat{\epsilon}_T$ of the prediction error sequence, using (114)

$$\hat{\epsilon}_T = y_T + \sum_{i=1}^{p} \hat{M}_{p,i} y_{T-i} - \sum_{i=1}^{p} \hat{N}_{p,i} \hat{\epsilon}_{T-i}. \tag{116}$$

These prediction errors will then be used to update the predictor coefficients, and the procedure will be repeated. As the parameter estimates become better, $\hat{\epsilon}_T$ will become closer to ϵ_T, causing an improvement in future parameter estimates.

This procedure can be implemented in a time-recursive manner using for example the normalized pre-windowed adaptive LSL. At each time point, we use the lattice filter to compute the prediction error $\hat{\epsilon}_T$. We then use this prediction error, as if it were the true one (ϵ_T), to update the lattice parameters. Table XVII summarizes this algorithm. To recover the predictor coefficients we use the prediction filter described in Section IV-D and read off $N_{p,i}$, $M_{p,i}$ from the proper entries of $A_{p,i}$.

In the discussion above, various details related to the fine structure of the IIR lattice were omitted. The special structure of the predictor coefficients (115) imposes a certain structure on the lattice filter. Also, different choices of the matrix

TABLE XVII
THE IIR PRE-WINDOWED LATTICE FORM

Initialize S, $\tilde{r}^z$, K to zero

$\tilde{\epsilon}^z_{0,T} = \tilde{r}^z_{0,T} = S^{-1/2}_{T-1} [y'_T, 0]'$

For $p = 0, \ldots, N-1$ do:

$$\left[\begin{aligned} \tilde{\epsilon}^z_{p+1,T} &= F(\tilde{\epsilon}^z_{p,T}, \tilde{r}^z_{p,T-1}, K_{p+1,T-1}) \\ \tilde{r}^z_{p+1,T} &= F(\tilde{r}^z_{p,T-1}, \tilde{\epsilon}^z_{p,T}, K'_{p+1,T-1}) \end{aligned}\right.$$

compute prediction error

Set $\hat{\epsilon}_T = [I \ 0]\, \tilde{\epsilon}^z_{N,T}$

- - - - - - - - - - - - - - - - - - - -

$S_T = \lambda S_{T-1} + [y'_T, \hat{\epsilon}'_T]' [y'_T, \hat{\epsilon}'_T]$

$\tilde{\epsilon}^z_{0,T} = \tilde{r}^z_{0,T} = S^{-1/2}_T [y'_T, \hat{\epsilon}'_T]'$

for $p = 0, \ldots, N-1$ do:

$$\left[\begin{aligned} K_{p+1,T} &= F^{-1}(K_{p+1,T-1}, \tilde{r}^{z\prime}_{p,T-1}, \tilde{\epsilon}^z_{p,T}) \\ \tilde{\epsilon}^z_{p+1,T} &= F(\tilde{\epsilon}^z_{p,T}, \tilde{r}^z_{p,T-1}, K_{p+1,T}) \\ \tilde{r}^z_{p+1,T} &= F(\tilde{r}^z_{p,T-1}, \tilde{\epsilon}^z_{p,T}, K'_{p+1,T}) \end{aligned}\right.$$

update lattice variables

$F(a,b,c) = [I - cc']^{-1/2} [a - cb][I - b'b]^{-T/2}$

$F^{-1}(a,b,c) = [I - cc']^{1/2} a[I - b'b]^{T/2} + cb$

square roots that appear in the lattice updates lead to different lattice structures. See [108], [114], [132] for more details.

The derivation of the IIR lattice form presented above was based on a heuristic boot-strapping idea. A key question is, of course, whether this procedure would converge. A similar algorithm for non-lattice IIR predictors was developed in the system identification literature under the name of the Extended Least Squares (ELS) algorithm [157], [158]. The following conditions were shown to be sufficient for the convergence of this algorithm:

i) The order of the IIR predictor is at least as large as the order of the observed ARMA process.

ii) The numerator polynomial related to the ARMA process is positive real. (In other words, if $N(z)/M(z)$ is the true ARMA model of the process y_T, then we require that $\mathrm{Re}\,\{N(e^{j\omega})\} > 0$ for $0 \leq \omega \leq \pi$.)

In some recent work it was shown that the lattice algorithm described above has the same asymptotic properties as the ELS algorithm [87]. We have also observed the similarity in the behavior of the ELS and the IIR lattice in simulations. The positive-real condition is not very restrictive and is often met in practice. An IIR lattice algorithm which does not require this condition was developed in [98].

The IIR lattice form is useful in many applications involving adaptive filtering such as: line enhancement [91], adaptive noise canceling, and parametric spectral estimation [93], [96]. Relatively little work has been done on IIR lattice forms prior to the development of the LSL [31], [63]–[65], [188]. The class of IIR lattice forms is still in an active research and development stage.

The ELS algorithm is only one of many techniques developed for estimating the parameters of IIR predictors. A different approach known in the system identification literature is the Instrumental Variable (IV) Method [158] which provides unbiased estimates of the AR parameters $\{M(z)\}$ of an ARMA process. The MA parameters $\{N(z)\}$ need to be estimated separately. A recursive lattice implementation of the IV method is presented in [94]. See also [81]. The derivation of the algorithm involves a nonsymmetric projection operator and a more complicated version of the projection update formulas of Section IV.

The nonrecursive IV method is closely related to the modified Yule–Walker equations for estimating the AR parameters of ARMA processes from the sample covariance matrix [149], [174]. See [175] for a relevant discussion. A lattice algorithm for solving these equations is discussed in Section VII-C, and its application to high-resolution spectral estimation can be found in [96], [175]. Finally, we note that many other variations of the LSL can be developed: see [186], [187], [189], [191] for some recent examples.

VII. Other Applications of Lattice Forms

The least squares lattice structure is useful for applications other than linear prediction. In Section II-D it is shown how it can be used for efficient factorization of positive definite Toeplitz matrices and their inverses. In Sections VII-A and VII-B we briefly describe a generalized lattice structure which can be used to factor a certain class of matrices which do not have a Toeplitz structure. In Section VII-C it is shown how to use lattice algorithms to efficiently solve sets of linear equations.

Lattice forms also provide a convenient canonical form for treating various problems in system theory. In Section VII-D we present a brief overview of selected topics. Finally, in Section VII-E, some references are given to the applications of lattice structures in digital filtering.

A. The Generalized Levinson Algorithm

α-Stationary Matrices: In Section III we considered several ways of forming the sample covariance matrix. The autocorrelation method gives a Toeplitz covariance matrix, while the other methods involve matrices of a different structure. A closer examination of the pre-windowed and covariance forms leads to the following observations:

Denote by T the Toeplitz sample covariance matrix obtained by the autocorrelation method, i.e.,

$$T = Y^{A}_{p,T} Y^{A'}_{p,T},$$

$$Y^{A}_{p,T} = \begin{bmatrix} y_0 & \cdots & y_{p-1} & \cdots & y_{T-1} & & \\ & \ddots & \vdots & & \vdots & \ddots & 0 \\ 0 & & \vdots & & \vdots & & \ddots \\ & & y_0 & \cdots & y_{T-p} & \cdots & y_{T-1} \end{bmatrix}. \quad (117)$$

Thus the pre-windowed sample covariance matrix will have the form

$$Y^{P}_{p,T} Y^{P'}_{p,T} = T + U_1 U_1', \quad \text{see (50)} \quad (118a)$$

where

$$U_1 = \begin{bmatrix} y_{T-p-1} & \cdots & y_{T-1} \\ & \ddots & \vdots \\ 0 & & y_{T-p-1} \end{bmatrix},$$

strictly upper triangular Toeplitz matrix (118b)

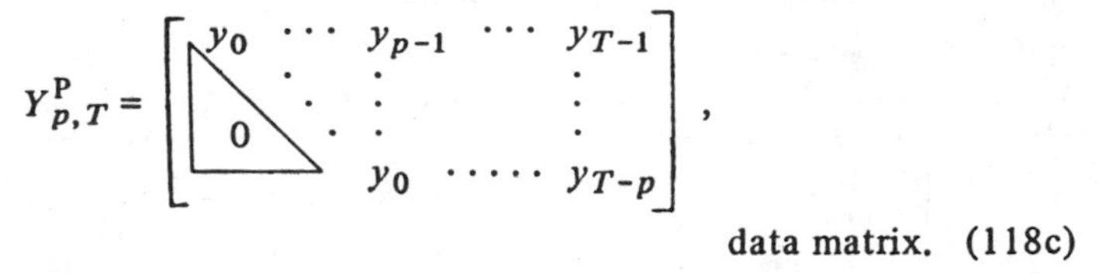

$$Y^{P}_{p,T} = \begin{bmatrix} y_0 & \cdots & y_{p-1} & \cdots & y_{T-1} \\ & \ddots & \vdots & & \vdots \\ 0 & & y_0 & \cdots & y_{T-p} \end{bmatrix},$$

data matrix. (118c)

The covariance form will be

$$Y^{C}_{p,T} Y^{C'}_{p,T} = T + U_1 U_1' - U_2 U_2', \quad \text{see (50)} \quad (119a)$$

where

$$U_2 = \begin{bmatrix} y_0 & \cdots & y_{p-2} \\ & \ddots & \vdots \\ 0 & & y_0 \end{bmatrix},$$

strictly upper triangular Toeplitz matrix (119b)

$$Y^{C}_{p,T} = \begin{bmatrix} y_{p-1} & \cdots & y_{T-1} \\ \vdots & & \vdots \\ y_0 & \cdots & y_{T-p} \end{bmatrix}, \quad \text{data matrix.} \quad (119c)$$

The special structure of these matrices can be generalized to include sums of products of upper times lower triangular Toeplitz matrices

$$R = T + \sum_{i=1}^{\alpha} \sigma_i U_i U_i', \qquad \sigma_i = \pm 1. \quad (120)$$

We call a matrix R which can be represented in this form an α-stationary matrix [100], [103], [104]. These matrices are also called "low shift rank" matrices [121] because of the following property: Let R_{ij} denote the entries of the matrix R

$$R = \begin{bmatrix} R_{N,N} & \cdots & R_{N,0} \\ \vdots & & \vdots \\ R_{0,N} & \cdots & R_{0,0} \end{bmatrix}. \quad (121)$$

Define a "shifted difference matrix" by

$$\delta R = \begin{bmatrix} R_{N,N} & \cdots & R_{N,1} \\ \vdots & & \vdots \\ R_{1,N} & \cdots & R_{1,1} \end{bmatrix} - \begin{bmatrix} R_{N-1,N-1} & \cdots & R_{N-1,0} \\ \vdots & & \vdots \\ R_{0,N-1} & \cdots & R_{0,0} \end{bmatrix}. \quad (122)$$

Let u^i_j be the entries of U_i

$$U_i = \begin{bmatrix} u^i_1 & \cdots & u^i_N \\ & \ddots & \vdots \\ 0 & & u^i_1 \end{bmatrix} \quad (123)$$

and define a so-called "generator matrix" g which contains all the last columns of U_i

$$g \triangleq \begin{bmatrix} g_N' \\ \vdots \\ g_1' \end{bmatrix} = \begin{bmatrix} u^1_N & \cdots & u^{\alpha}_N \\ \vdots & & \vdots \\ u^1_1 & \cdots & u^{\alpha}_1 \end{bmatrix},$$

an $mN \times \alpha$ matrix.

Then it can be shown [100] that

$$\delta R = g \operatorname{diag}\{\sigma_i\} g'$$

and therefore rank $\{\delta R\} = \alpha$ (assuming that g is full rank).

In [99], [100], [103], [104], [118], [125], [127], it was shown that for such matrices it is possible to derive generalized Levinson-type recursions and efficient matrix inversion formulas. (The continuous-time equivalent of these results is treated in [106].) Next we present a special case of the generalized Levinson: the pre-windowed form.

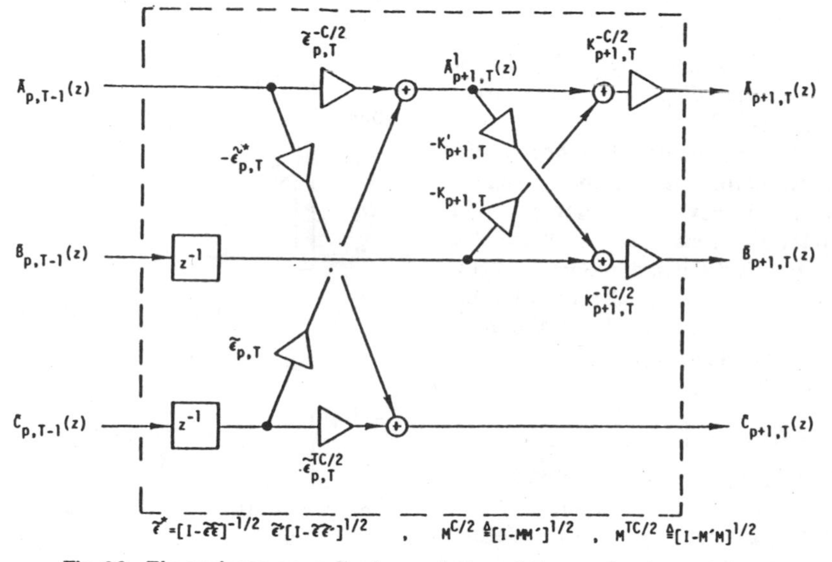

Fig. 23. The variance normalized pre-windowed time-and-order update lattice prediction filter.

The Pre-Windowed Levinson Algorithm: In Section IV-D we derived the pure-order update normalized lattice prediction filter. In a similar way it is possible to derive a time-and-order update filter which can be considered as a generalization of the Levinson algorithm. Reading off lines 1–4 in Table VII we obtain the following update equations for the prediction errors:

$$\bar{\epsilon}_{p+1,T} = G(\bar{\epsilon}_{p,T}, \bar{r}_{p,T-1}, K_{p+1,T}), \quad \text{see (93)} \tag{124a}$$

$$\bar{r}_{p+1,T} = G(\bar{r}_{p,T-1}, \bar{\epsilon}_{p,T}, K'_{p+1,T}), \quad \text{see (93)} \tag{124b}$$

$$\bar{\epsilon}_{p,T-1} = G(\bar{\epsilon}_{p,T}, \bar{\gamma}^c_{p-1,T-1}, \tilde{\epsilon}_{p,T}) \tag{124c}$$

$$\bar{\gamma}^c_{p,T} = G(\bar{\gamma}^c_{p-1,T-1}, \bar{\epsilon}_{p,T}, \tilde{\epsilon}'_{p,T}). \tag{124d}$$

By eliminating $\bar{\epsilon}_{p,T}$ from the equations above, we get

$$\bar{\gamma}^c_{p,T} = [I - \tilde{\epsilon}_{p,T}\tilde{\epsilon}'_{p,T}]^{T/2}\bar{\gamma}^c_{p-1,T-1} \underbrace{- [I - \tilde{\epsilon}'_{p,T}\tilde{\epsilon}_{p,T}]^{-1/2}\tilde{\epsilon}'_{p,T}[I - \tilde{\epsilon}_{p,T}\tilde{\epsilon}'_{p,T}]^{1/2}}_{\tilde{\epsilon}^*_{p,T}}\bar{\epsilon}_{p,T-1}. \tag{125}$$

The predictors $\bar{A}_p(z)$, $\bar{B}_p(z)$, $\bar{C}_p(z)$ (cf. (95)) can be shown to obey the same recursions as the signals $\bar{\epsilon}$, $\bar{r}$, $\bar{\gamma}^c$. These recursions induce the lattice structure depicted in Fig. 23. The corresponding algorithm for computing the predictor coefficients is summarized in Table XVIII. Such recursions were originally derived by normalizing the generalized Levinson algorithm [135], [137], [148] which is obtained by the methods of [121], [128]. Since then, derivations were presented using an embedding technique [82], [83], [96], a projection approach which is very closely related to the one used here [140] and other techniques [85], [86], [116], [117], [176], [195], [196], [198], [199].

To see more clearly the resemblence to the Levinson algorithm we note that the recursions for the predictors $\bar{A}$, $\bar{B}$, $\bar{C}$ can be rewritten (after some straightforward algebraic manipulations) as

TABLE XVIII
COMPUTING PREDICTOR COEFFICIENTS FROM THE NORMALIZED PRE-WINDOWED TIME-AND-ORDER UPDATE LATTICE FILTER

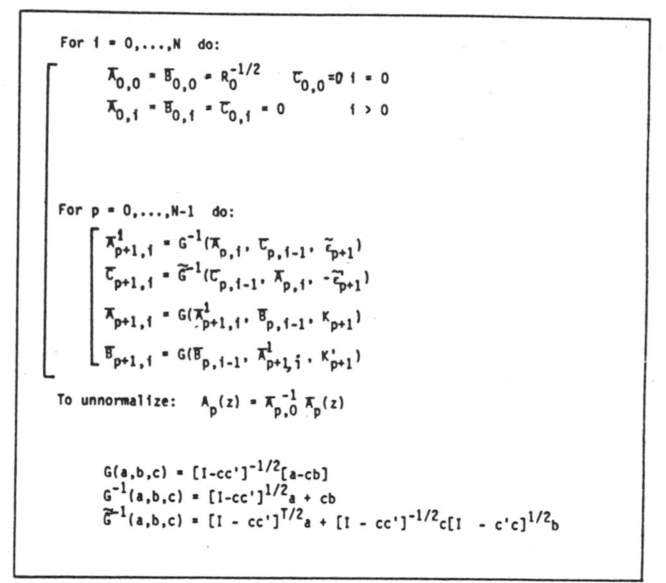

For $i = 0,\ldots,N$ do:

$\bar{A}_{0,0} = \bar{B}_{0,0} = R_0^{-1/2} \quad \bar{C}_{0,0} = 0 \quad i = 0$

$\bar{A}_{0,i} = \bar{B}_{0,i} = \bar{C}_{0,i} = 0 \quad i > 0$

For $p = 0,\ldots,N-1$ do:

$\bar{A}^1_{p+1,i} = G^{-1}(\bar{A}_{p,i}, \bar{C}_{p,i-1}, \tilde{\epsilon}_{p+1})$

$\bar{C}_{p+1,i} = \tilde{G}^{-1}(\bar{C}_{p,i-1}, \bar{A}_{p,i}, -\tilde{\epsilon}_{p+1})$

$\bar{A}_{p+1,i} = G(\bar{A}^1_{p+1,i}, \bar{B}_{p,i-1}, K_{p+1})$

$\bar{B}_{p+1,i} = G(\bar{B}_{p,i-1}, \bar{A}^1_{p+1,i}, K'_{p+1})$

To unnormalize: $A_p(z) = \bar{A}^{-1}_{p,0}\bar{A}_p(z)$

$G(a,b,c) = [I - cc']^{-1/2}[a - cb]$

$G^{-1}(a,b,c) = [I - cc']^{1/2}a + cb$

$\tilde{G}^{-1}(a,b,c) = [I - cc']^{T/2}a + [I - cc']^{-1/2}c[I - c'c]^{1/2}b$

where

$$K_{p+1,T} = [I - \tilde{\epsilon}_{p,T}\tilde{\epsilon}'_{p,T}]^{-1/2}[K_{p+1,T}, \tilde{\epsilon}_{p,T}],$$

an $m \times (m+1)$ generalized reflection coefficient (126b)

$$\bar{B}_{p+1,T}(z) = [\bar{B}'_{p+1,T}(z)\ \bar{C}'_{p+1,T}(z)]',$$

a generalized backward predictor. (126c)

Comparison to (31) shows that the form of this equation is identical to the (stationary) Levinson algorithm. The only differences are the dimensions of $\boldsymbol{K}$ and $\bar{\boldsymbol{B}}$.

The Levinson Algorithm for the α-Stationary Case: Using a similar approach it is possible to derive Levinson-type recursions for the covariance case and for the more general α-stationary case. We state here without proof some of the basic results.

$$\begin{bmatrix} \bar{A}_{p+1,T}(z) \\ \bar{\boldsymbol{B}}_{p+1,T}(z) \end{bmatrix} = \begin{bmatrix} (I - \boldsymbol{K}_{p+1,T}\boldsymbol{K}'_{p+1,T})^{-1/2} & 0 \\ 0 & (I - \boldsymbol{K}'_{p+1,T}\boldsymbol{K}_{p+1,T})^{-1/2} \end{bmatrix} \begin{bmatrix} I & -\boldsymbol{K}_{p+1,T} \\ -\boldsymbol{K}'_{p+1,T} & I \end{bmatrix} \begin{bmatrix} \bar{A}_{p,T-1}(z) \\ z^{-1}\bar{\boldsymbol{B}}_{p,T-1}(z) \end{bmatrix} \tag{126a}$$

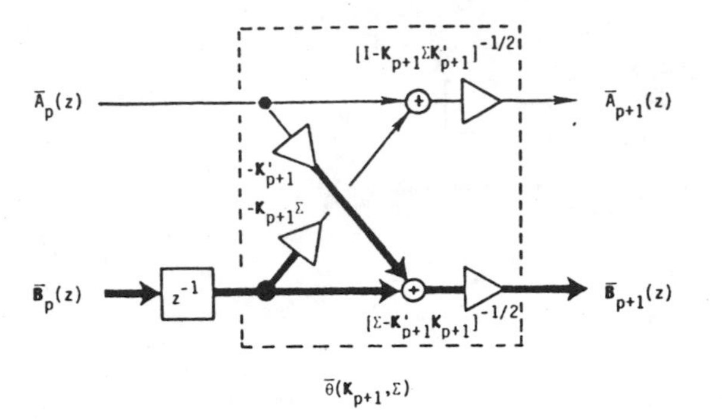

Fig. 24. The generalized (normalized) lattice predictor for the α-stationary case.

For a more comprehensive treatment of the generalized Levinson algorithm see [82]-[84], [96], [100], [176].

The least squares predictor $\bar{A}_p(z)$ associated with an α-stationary covariance matrix (cf. (120)) can be computed by the following algorithm:

$$\begin{bmatrix} \bar{A}_{p+1}(z) \\ \bar{B}_{p+1}(z) \end{bmatrix} = \underbrace{\begin{bmatrix} (I - K_{p+1}\Sigma K'_{p+1})^{-1/2} & 0 \\ 0 & (\Sigma - K'_{p+1}K_{p+1})^{-1/2} \end{bmatrix} \begin{bmatrix} I & -K_{p+1}\Sigma \\ -K'_{p+1} & I \end{bmatrix}}_{\bar{\theta}(K_{p+1},\Sigma)} \begin{bmatrix} \bar{A}_p(z) \\ z^{-1}\bar{B}_p(z) \end{bmatrix} \tag{127}$$

$$\bar{A}_0(z) = I \qquad \bar{B}_0(z) = [I \ \ 0 \cdots 0]'$$

where $\bar{A}_p(z)$ is $m \times m$, $\bar{B}'_p(z)$ and K_{p+1} are $m \times (m + \alpha)$, and Σ is a "signature matrix": $\Sigma = \text{diag}\{I, -\sigma_1, \cdots, -\sigma_\alpha\}$. The generalized Levinson is associated with the lattice algorithm depicted in Fig. 24. Note that the pre-windowed predictor presented earlier is obtained as a special case of this algorithm for $\alpha = 1$ and $\Sigma = [I, -1]$. The time and order update covariance predictor can be shown to be a special case with $\alpha = 2$ and $\Sigma = \text{diag}\{I, -1, 1\}$.

The reflection coefficients which define this generalized lattice filter can be computed in several ways:

i) From the data. For example: the adaptive pre-windowed LSL (Table VI) can be used to compute a set of parameters $K_{p+1,T}$, $\tilde{e}_{p,T}$ which define the generalized reflection coefficients (for $\alpha = 1$) via (126b). Similarly, the adaptive covariance LSL (Table XIV) computes $K_{p+1,T}$, $\tilde{e}_{p,T}$, $\tilde{\tilde{e}}_{p,T}$ which define a generalized reflection coefficient for $\alpha = 2$ as shown in [96]. Adaptive LSL forms for the α-stationary case were developed in [109], [129]. The stochastic interpretation of these results can be found in [85], [86].

ii) From the covariance matrix. In Section II we presented the fast Cholesky algorithm which, among other things, provided an efficient method for computing reflection coefficients from the covariance sequence. An extension of this method to the α-stationary case is presented in the next section.

The generalized Levinson algorithm has various applications including: inversion of α-stationary covariance matrices [100], factorization of the inverse of a covariance matrix (as in (32), (34a)), and recursive computation of the least squares predictor $\bar{A}_p(z)$. The generalized Levinson is a computationally efficient algorithm: it requires only $(\alpha + 1)$ times as many computations as the usual (stationary) Levinson algorithm. In other words, given an $N \times N$ α-stationary covariance matrix (scalar case), the generalized Levinson will require $\sim(\alpha + 1)N^2$ operations (compared to $\sim N^3$ operations using standard techniques to do the same computation).

The form of the general Levinson algorithm discussed above is nonminimal: for certain covariance matrices (such as $R = LL'$, where L is a triangular Toeplitz matrix) it is possible to implement the predictor using a somewhat simpler structure than the one depicted in Fig. 24. The question of the minimality of the generalized Levinson algorithm was discussed in [83] and [176]. Note that the lattice predictor in Fig. 24 has $\alpha + 1$ delays per section. An alternative predictor structure with only one delay per section was presented in [177]. This structure is closely related to the dual lattice form discussed in Section II-D, which has the delay on the upper, rather than lower, line.

B. The Generalized Fast Cholseky Algorithm

In Section II we have seen that the lattice structure related to the (stationary) Levinson algorithm can be used to perform Cholesky factorization of the covariance matrix and to compute the reflection coefficients. The same turns out to be true for the α-stationary case. Consider the "dual" lattice depicted in Fig. 25. This filter has the same sections as the normalized lattice predictor in Fig. 24, but the delay element is on the upper rather than the lower line. Feeding the covariance sequence $\{R_{0,1}, \cdots, R_{0,N}\}$ and the "generator sequence" $\{g_1, \cdots, g_N\}$ into this lattice filter provides a recursive procedure for computing the reflection coefficients, as depicted in Fig. 25. The signals propagating on the upper line of the filter are precisely the rows of the Cholesky factor. Comparison of Figs. 6 and 25 shows the similarity to the stationary case. Table XIX summarizes the generalized Cholesky algorithm related to this lattice filter. Derivations of the generalized Cholesky algorithm can be found in [82], [177], [176]. For a derivation that is closest to the one presented in Section II, see [96]. Reference [96] is a complementary paper treating the given covariance case, rather than the given data case which is the main theme of this paper. Setting $\alpha = 0$ in the generalized Cholesky will specialize the algorithm in Table XIX to the stationary case. In this case $\delta_{0,m} = \alpha_{0,m}$ since $g_m = 0$, and $\bar{\theta}(K_p, \Sigma)$ is replaced by $\bar{\theta}(K_p)$. We note that the pre-windowed case (with $g'_m = y_{T-N+m}$) and the covariance case (with $g'_m = [y_{T-N+m}, y_{m-1}]$) can both be obtained as special cases of these generalized recursions; see [96] for details. We note also that feeding the reversed order sequences $\{R_{0,N}, \cdots, R_{0,1}\}$ and $\{g_N, \cdots, g_1\}$ into the same lattice structure will give the lower times upper triangular factors of the covariance matrix.

It is worthwhile to note the simplicity of the generalized Cholesky algorithm in Table XIX. This algorithm is easy to code and is computationally efficient. Factorization of an $N \times N$ α-stationary covariance matrix (scalar case) will require proportional to $(\alpha + 1)N^2$ operations (compared to $\sim N^3$ operations using standard techniques [155]). When the covariance matrix has a banded structure (i.e., $R_{i,j} = 0$ for $|i - j| > n$), the Cholesky algorithm can be further simplified leading to an interesting feedback lattice structure [95], [178]. The covariance matrix of an MA process has such a structure.

The lattice structure associated with the generalized Cholesky leads to pipelined computational architectures which are con-

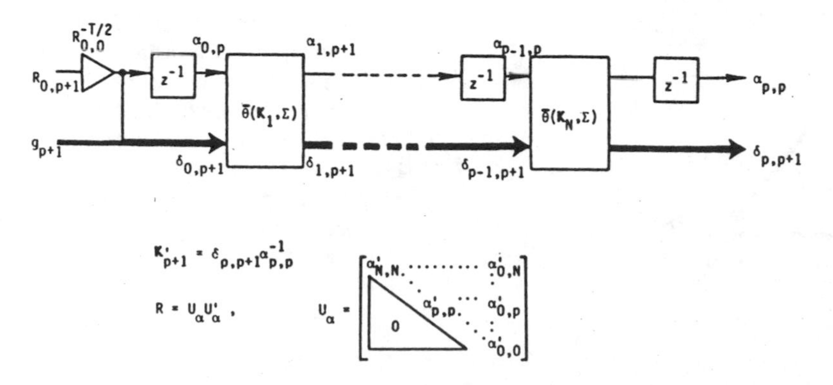

Fig. 25. The lattice structure for the generalized fast Cholesky algorithm.

TABLE XIX

THE GENERALIZED FAST CHOLESKY ALGORITHM

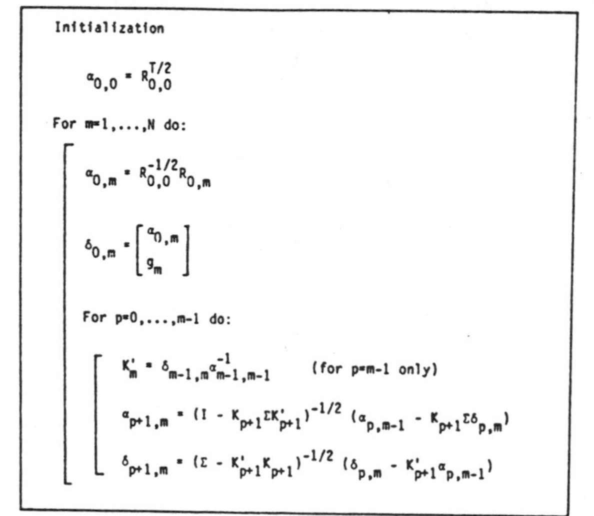

Initialization

$\alpha_{0,0} = R_{0,0}^{T/2}$

For m=1,...,N do:

$\alpha_{0,m} = R_{0,0}^{-1/2} R_{0,m}$

$\delta_{0,m} = \begin{bmatrix} \alpha_{0,m} \\ g_m \end{bmatrix}$

For p=0,...,m-1 do:

$K'_m = \delta_{m-1,m}\alpha^{-1}_{m-1,m-1}$ (for p=m-1 only)

$\alpha_{p+1,m} = (I - K_{p+1}\Sigma K'_{p+1})^{-1/2} (\alpha_{p,m-1} - K_{p+1}\Sigma\delta_{p,m})$

$\delta_{p+1,m} = (\Sigma - K'_{p+1}K_{p+1})^{-1/2} (\delta_{p,m} - K'_{p+1}\alpha_{p,m-1})$

venient for parallel processing [82], [124] and for VLSI implementations.

As mentioned before, the generalized fast Cholesky provides an alternative way of computing reflection coefficients, different from the LSL algorithms presented in Section IV. The recursive LSL is well suited for adaptive applications, while the Cholesky algorithm is better suited for off-line applications since it requires the computation of the sample covariance matrix. The two methods are, of course, mathematically equivalent.

C. Efficient Solution of Linear Equation

Least squares prediction problems are intimately related to solutions of linear equations such as the Yule-Walker equations (8) or their non-Toeplitz counterparts. The lattice algorithms provide efficient solution methods for such equations. Their use is not restricted to prediction problems; they can be applied whenever the structure of the equations is the same as in the prediction problem. Consider, for example, the following overdetermined Toeplitz set of equations:

$$[M_{p,1}, \cdots, M_{p,p}] \begin{bmatrix} R_p & R_{p+1} & \cdots & R_{k-1} \\ \vdots & \vdots & & \vdots \\ R_1 & R_2 & \cdots\cdots & R_{k-p} \end{bmatrix} = -[R_{p+1}, \cdots, R_k]. \quad (128)$$

As was mentioned in Section VI-C, these so-called modified Yule-Walker equations arise in the problem of estimating the AR parameters of an ARMA process. A comparison of (128) and the predictor equation for the covariance case (101) reveals that the two are identical (except for interchanging $M_{p,i}$ with $A_{p,i}$ and R_i with y_{i-1}). Therefore, the covariance lattice form can be used to efficiently solve the modified Yule-Walker equations which are commonly used in spectral estimation problems. Note that the right-hand side of (128) is a shifted version of the upper row of the matrix on the left. To solve a Toeplitz set of equations with an arbitrary right-hand side we will need the joint process covariance lattice [139].

Similar sets of equations arise in various problems such as: fitting a finite-order rational transfer function to a given impulse response sequence [180] and finding a low-order approximation to a high-order model. The lattice algorithms are particularly useful for handling the complexities of the multivariable case [179]. Other problems such as testing for co-primeness of matrix polynomials and transforming left matrix fraction description (MFD) of multivariable systems to right MFD can also be treated [179], [180].

D. Interconnections with System Theory

Lattice forms provide a useful canonical form for treating various problems in system theory. Consider, for example, the following state-space representation of the unnormalized lattice form [131] depicted in Fig. 3:

$$x_{t+1} = F_L x_t + G_L y_t$$
$$\epsilon_t = H_L x_t + y_t \quad (129)$$

where

$$F_L = \begin{bmatrix} I & & & & 0 \\ K_2^{\epsilon'} K_1^r & I & & & \\ \vdots & & \ddots & & \\ \vdots & & & I & \\ K_{N-1}^{\epsilon'} K_1^r & \cdots & & K_{N-1}^{\epsilon'} K_{N-2}^r & I \end{bmatrix}$$

$$G_L = [I - K_1^{\epsilon'} \ \cdots \ -K_{N-1}^{\epsilon'}]^T$$

$$H_L = [-K_1^r \ \cdots \ -K_N^r].$$

The state vector x_t contains the backward prediction errors $\{r_{p,t}\}$. Since the prediction errors are uncorrelated, the state covariance matrix $\Pi_t = \epsilon\{x_t x_t'\}$ will be diagonal. It is well known that the state covariance of any state-space realization $\{F, G\}$ can be found as the solution of the following Lyaponov matrix equation:

$$\Pi - F\Pi F' = GG'. \quad (130)$$

The need for solving such equations arises in problems such as stochastic realization and optimal control and estimation. The diagonal structure of Π for the lattice canonical form leads to easy solutions of this equation. Note furthermore that for the square-root normalized lattice form the state covariance is the identity matrix, $\Pi = I$. Lyapunov theory provides the basis for studying stability and related issues such as the Routh-Hurwitz test and the Schur-Cohn matrix. Various classical results on the stability of linear time-invariant systems can be proven very easily by using the lattice canonical form [109], [179].

Another interesting state-space realization is related to the covariance matrix of an α-stationary process [82], [86]. Consider the following discrete Lyapunov equation:

$$R - ZRZ' = G\Sigma G' \tag{131}$$

where Z is the shift operator

$$Z = \begin{bmatrix} 0 & & & \\ 1 & 0 & & \\ & \ddots & \ddots & \\ & & 1 & 0 \end{bmatrix} \tag{132}$$

and $\{G, \Sigma\}$ are the generators of the covariance matrix. (This definition is slightly different than the one introduced in Section VII-A.) It is always possible to choose matrices H and J to define an extended matrix Q

$$Q = \begin{bmatrix} F & G \\ H & J \end{bmatrix} \tag{133}$$

which has the property that

$$Q \begin{bmatrix} R & 0 \\ 0 & \Sigma \end{bmatrix} Q' = \begin{bmatrix} R & 0 \\ 0 & \Sigma \end{bmatrix}. \tag{134}$$

We can associate with Q a transfer function of the form

$$T(z) = J + H(zI - F)^{-1}G \tag{135}$$

where $T(z)$ is Σ para-unitary on the unit circle, that is to say

$$T(z)\,\Sigma T'(z) = \Sigma \tag{136}$$

and Σ-contractive outside the unit circle, meaning that

$$T(z)\,\Sigma T'(z) \leqslant \Sigma, \qquad \text{for } |z| \geqslant 1. \tag{137}$$

Transfer functions with these properties have been studied extensively in the area of network synthesis and circuit theory. It is known that $T(z)$ can be factored as a product of elementary Σ-contractive transfer functions. This cascade factorization forms a lattice realization of $T(z)$ and establishes a connection between the theory of lattice forms and the study of Σ-lossless systems. Delosme *et al.* [86] have shown that the generalized Levinson and fast Cholesky algorithms can be interpreted as two different ways of factoring a given transfer matrix $T(z)$; see also [6], [7], [36], [37]. These results can be generalized to the case where Z is an arbitrary lower triangular Toeplitz matrix, rather than the shift operator.

E. Digital Lattice Filters

Lattice structures appear in many contexts such as the scattering of waves in a stratified medium, electric transmission lines, and network synthesis [1]. The low-sensitivity properties of analog LC two-ports motivated researchers to explore similar structures in digital filtering. This led to the development of the so-called wave digital filter first introduced by Fettweis [8], [9], which has since received much attention in the circuit theory literature [3]. The properties of these filters were studied by Chrochiere [3], Fettweis [12], [13], Nouta [26], and others [2], [14]-[16], [18], [30]. Their work verified the exceptional insensitivity of digital wave filters to perturbations in the multiplier coefficients and their low quantization noise [4], [5], [19], [20], [28]. Concepts from analog network theory, such as power, passivity, and losslessness were successfully utilized in these studies [10]. Various modifications were introduced in the basic filter structure to reduce the number of multipliers and delays [11], [29] and to replace first-order lattice sections by second-order sections [21]. Other types of digital lattice networks were introduced by Hwang [17] and Mitra *et al.* [22]-[25].

VIII. Conclusions

In this survey we described several well-established and widely used results on lattice forms as well as some topics which are still under active investigation. One point we tried to make is that the adaptive LSL has a number of practical advantages compared to gradient adaptive techniques. The LSL represents a turning point in adaptive processing: it provides the superior performance of least squares techniques at a cost comparable to that of the widely used gradient techniques. We believe that the LSL and related algorithms may gradually replace gradient methods as the standard adaptive processing technique. Practical applications of the LSL have been fairly limited so far. This is in part due to the fact that many of the results presented here were relatively inaccessible to the signal-processing community. We hope that this tutorial paper will contribute to a more widespread dissemination of these algorithms and will encourage the testing and application of the least squares lattice.

In addition to their applications to adaptive processing, lattice structures play a key role in a variety of problems in least squares estimation and linear system theory. Because of the limited scope of this paper we presented only a very cursory treatment of these issues (in Section VII). We also limited the discussion of the non-lattice equivalents of all the "fast" adaptive algorithms presented in this paper. These topics deserve a more thorough treatment than the one presented here.

Finally, a word of caution regarding the performance of lattice forms. It should always be remembered that the lattice algorithms presented here solve a *linear least squares* estimation problem. Therefore, their performance can be no better than that predicted by the theory of such estimators. Furthermore, given sufficient computational power and accuracy, other (non-lattice) least squares estimators will perform as well. The advantages of lattice forms become apparent in the presence of practical constraints such as finite-word-length computations, limited storage and computing power, and limited hardware complexity.

Appendix I
Relating the AR Parameters and the Reflection Coefficients

Writing down the Levinson recursions for $p = 0, \cdots, N$ in matrix form gives

$$\begin{bmatrix} I & & & \\ B_{1,1} & I & & 0 \\ \vdots & & \ddots & \\ B_{N,N} & \cdots & B_{N,1} & I \end{bmatrix} = \underbrace{\begin{bmatrix} I & & & & \\ 0 & I & & & 0 \\ 0 & B_{1,1} & I & & \\ \vdots & & & \ddots & \\ 0 & B_{N-1,N-1} & \cdots & B_{N-1,1} & I \end{bmatrix}}_{L_B} - \begin{bmatrix} 0 & & & \\ & K_1^{\epsilon'} & & 0 \\ & & \ddots & \\ 0 & & & K_N^{\epsilon'} \end{bmatrix} \begin{bmatrix} I & & & \\ I & A_{1,1} & & 0 \\ \vdots & \vdots & \ddots & \\ I & A_{N,N}, & \cdots, & A_{N,1} \end{bmatrix}. \quad \text{(A1)}$$

Post-multiplying by $[R'(0), \cdots, R'(-N)]'$ we get

$$\begin{bmatrix} R(0) \\ 0 \\ \vdots \\ 0 \end{bmatrix} = \begin{bmatrix} R(0) & \\ L_B & \begin{bmatrix} R(1) \\ \vdots \\ R(-N) \end{bmatrix} \end{bmatrix} - \begin{bmatrix} 0 \\ K_1^{\epsilon'} R_0^{\epsilon} \\ \vdots \\ K_N^{\epsilon'} R_{N-1}^{\epsilon} \end{bmatrix} \quad \text{(A2)}$$

which can be rewritten as

$$[\Delta_1, \cdots, \Delta_N] = [R(1), \cdots, R(N)]\, L_B'. \quad \text{(A3)}$$

We used here the fact that $R'(-p) = R(p)$. Next recall that

$$[A_{N,1}, \cdots, A_{N,N}]\, R_{N-1} = [R(1), \cdots, R(N)]. \quad \text{(A4)}$$

From the last two equations we conclude that

$$\begin{aligned} [\Delta_1, \cdots, \Delta_N] &= [A_{N,1}, \cdots, A_{N,N}]\, [L_B^{-1} R^{-r} L_B^{-T}]\, L_B' \\ &= [A_{N,1}, \cdots, A_{N,N}]\, L_B^{-1} R^{-r} \end{aligned} \quad \text{(A5)}$$

or

$$[K_1^r, \cdots, K_N^r] = [A_{N,1}, \cdots, A_{N,N}]\, L_B^{-1}. \quad \text{(A6)}$$

Equation (A4) above is often used for computing reflection coefficients in Linear Predictive Coding systems. The lower triangular matrix L_B^{-1} is obtained by performing an LDU decomposition of the sample covariance matrix.

Appendix II
Update Formulas for Projection Operators

A. Unnormalized

To derive the update formula for $P\{S+X\}$ we first note that the space spanned by $S + X$ is the same as the space spanned by $S + X(I - P_S)$, where $X(I - P_S)$ is orthogonal to S. Projection on orthogonal subspace can be easily decomposed

$$\begin{aligned} P\{S+X\} &= P_S + P\{X(I-P_S)\} \\ &= P_S + (I - P_S)\, X'[X(I - P_S)\, X']^{-1} X(I - P_S) \end{aligned} \quad \text{(A7)}$$

or

$$\begin{aligned} (I - P\{S+X\}) &= (I - P_S) \\ &\quad - (I - P_S)\, X'[X(I - P_S)\, X']^{-1} X(I - P_S). \end{aligned} \quad \text{(A8)}$$

The update formula presented in Section IV-A is obtained by pre- and post-multiplying by V and W', respectively.

B. Normalized

To obtain the normalized update formula note that

$$\begin{aligned} V(I - P\{S+X\})\, W' &= [V(I - P_S)\, V']^{1/2} \\ &\cdot [\rho_S(V, W) - \rho_S(V, X)\, \rho_S(X, V)] \\ &\cdot [W(I - P_S)\, W']^{T/2} \end{aligned} \quad \text{(A9)}$$

where

$$\begin{aligned} \rho_S(V, W) &\triangleq [V(I - P_S)\, V']^{-1/2}\, V(I - P_S)\, W' \\ &\cdot [W(I - P_S)\, W']^{-T/2}. \end{aligned} \quad \text{(A10)}$$

Replacing W by V in (A9) gives

$$\begin{aligned} [V(I - P\{S+X\})\, V']^{1/2} &= [V(I - P_S)\, V']^{1/2} \\ &\cdot [I - \rho_S(V, X)\, \rho_S(X, V)]^{1/2}. \end{aligned} \quad \text{(A11)}$$

Replacing V by W in (A9) gives

$$\begin{aligned} [W(I - P\{S+X\})\, W']^{1/2} &= [W(I - P_S)\, W']^{1/2} \\ &\cdot [I - \rho_S(W, X)\, \rho_S(X, W)]^{1/2}. \end{aligned} \quad \text{(A12)}$$

Combining (A9), (A11), and (A12) gives the update formula presented in Section IV-D.

C. Variance Normalized

Let us define the "variance normalized" correlation coefficient

$$\bar{\rho}_S(V, W) = [V(I - P_S)\, V']^{-1/2}\, [V(I - P_S)\, W'] \quad \text{(A13)}$$

which is related to the "doubly normalized" coefficient by

$$\rho_S(V, W) = \bar{\rho}_S(V, W)[W(I - P_S)\, W']^{-T/2}. \quad \text{(A14)}$$

Using (A12), (A14) we get

$$\begin{aligned} \bar{\rho}_{S+X}(V, W) &= \rho_{S+X}(V, W)[I - \rho_S(W, X)\, \rho_S(X, W)]^{T/2} \\ &\cdot [W(I - P_S)\, W']^{T/2} \\ &= [I - \rho_S(V, X)\, \rho_S(X, V)]^{-1/2} \\ &\cdot [\rho_S(V, W) - \rho_S(V, X) \\ &\cdot \rho_S(X, W)]\, [W(I - P_S)\, W']^{T/2} \end{aligned} \quad \text{(A15)}$$

and, therefore,

$$\begin{aligned} \bar{\rho}_{S+X}(V, W) &= [I - \rho_S(V, X)\, \rho_S(X, V)]^{-1/2} \\ &\cdot [\bar{\rho}_S(V, W) - \rho_S(V, X)\, \bar{\rho}(X, W)] \end{aligned} \quad \text{(A16)}$$

which proves the update formula presented in Section IV-C.

D. Time Update

$$\bar{Y}_{p,T-1}\left\{\left\{\begin{bmatrix} y_0 & \cdots & y_{T-2} & 0 \\ & \ddots & \vdots & \vdots \\ 0 & y_0 & \cdots \; y_{T-p-1} & 0 \end{bmatrix}\right.\right.$$

$$\pi\{[0 \quad \cdots\cdots \quad 0 \quad I]$$

$$\left.\left.\begin{bmatrix} y_0 & \cdots & y_{T-1} \\ & \ddots & \vdots \\ 0 & y_0 & \cdots \; y_{T-p} \end{bmatrix}\right\}Y_{p,T}\right.$$

$$[0 \quad \cdots\cdots \quad 0 \quad I]\}\;\pi.$$

The space spanned by $Y_{p,T}+\pi$ and $\bar{Y}_{p,T}+\pi$ can be seen to be the same. Therefore,

$$P\{Y_{p,T}+\pi\} = P\{\bar{Y}_{p,T-1}+\pi\} = P_{\bar{Y}_{p,T-1}} + P_\pi$$

$$= \begin{bmatrix} P_{Y_{p,T-1}} & 0 \\ & \vdots \\ 0 \;\cdots & 0 \end{bmatrix} + \pi(\pi'\pi)^{-1}\pi$$

$$= \begin{bmatrix} P_{Y_{p,T-1}} & 0 \\ & \vdots \\ 0 \;\cdots & I \end{bmatrix}$$

$$(I - P\{Y_{p,T}+\pi\}) = \begin{bmatrix} I - P_{Y_{p,T-1}} & 0 \\ & \vdots \\ 0 \;\cdots & 0 \end{bmatrix}.$$

Appendix III
Normalized Versus Unnormalized Variables

While we often prefer to use the normalized lattice form, it is sometimes useful to compute unnormalized prediction errors for comparison with other methods. The following identities follow from the update formulas for R^ϵ, R^r, and γ in Section IV-A

$$[I - K_{p+1,T}K'_{p+1,T}]^{1/2} = R^{-\epsilon/2}_{p,T} R^{\epsilon/2}_{p+1,T}$$

$$[I - K'_{p+1,T}K_{p+1,T}]^{1/2} = R^{-r/2}_{p,T-1} R^{r/2}_{p+1,T}$$

$$(1 - \gamma_{p,T-1}) = (1 - \gamma_{p-1,T-1}) \cdot (1 - \tilde{r}'_{p,T-1}\tilde{r}_{p,T-1}). \quad \text{(A17)}$$

Using these identities, the normalizing coefficients can be evaluated

$$R^{\epsilon/2}_{p,T} = R^{1/2}_T \prod_{i=1}^{p} (I - K_{i,T}K'_{i,T})^{1/2}$$

$$R^{r/2}_{p,T} = R^{1/2}_T \prod_{i=1}^{p} (I - K'_{i,T-p+i}K_{i,T-p+i})^{1/2}$$

$$(1 - \gamma_{p-1,T-1}) = \prod_{i=0}^{p-1} (1 - \tilde{r}'_{i,T-1}, \tilde{r}_{i,T-1}). \quad \text{(A18)}$$

The normalized lattice variables are finally given by

$$K^{\epsilon'}_{p+1,T} = R^{r/2}_{p,T-1} K'_{p+1,T} R^{-\epsilon/2}_{p,T}$$

$$K^{r}_{p+1,T} = R^{\epsilon/2}_{p,T} K_{p+1,T} R^{-r/2}_{p,T-1} \quad \text{(A19)}$$

$$\epsilon_{p,T} = (1 - \gamma_{p-1,T-1})^{1/2} R^{\epsilon/2}_{p,T} \tilde{\epsilon}_{p,T}$$

$$r_{p,T-1} = (1 - \gamma_{p-1,T-1})^{1/2} R^{r/2}_{p,T-1} \tilde{r}_{p,T-1}. \quad \text{(A20)}$$

Note also that the variance normalized variables are given by

$$\bar{\epsilon}_{p,T} = (1 - \gamma_{p-1,T-1})^{1/2} \tilde{\epsilon}_{p,T}$$

$$\bar{r}_{p,T-1} = (1 - \gamma_{p-1,T-1})^{1/2} \tilde{r}_{p,T-1}. \quad \text{(A21)}$$

We see here that the normalized lattice variables completely specify all the quantities in the unnormalized lattice as well.

Acknowledgment

The author wishes to thank Prof. M. Morf for providing his unique insights on the topics discussed in this paper. Helpful discussions with B. Porat are gratefully acknowledged. Finally, the author wishes to thank the anonymous reviewers who made numerous suggestions which were helpful in rewriting and improving the first draft.

References

Digital Lattice Filters

[1] B. Belevitch, *Classical Network Theory.* San Francisco, CA: Holden-Day, 1969.

[2] L. T. Bruton, "Low-sensitivity digital ladder filters," *IEEE Trans. Circuits Syst.*, vol. CAS-22, no. 3, pp. 168-176, Mar. 1975.

[3] L. T. Bruton and D. A. Vaughan-Pope, "Synthesis of digital ladder filters from LC filters," *IEEE Trans. Circuits Syst.*, vol. CAS-23, no. 6, pp. 395-402, June 1976.

[4] P. L. Chu and D. G. Messerschmitt, "Zero sensitivity of the digital lattice filter," in *Proc. IEEE Int. Conf Acoust., Speech Signal Process.*, pp. 89-93, 1980.

[5] R. E. Crochiere, "Digital ladder structures and coefficient sensitivity," *IEEE Trans. Audio Electroacoust.*, vol. AU-20, no. 4, p. 240-246, Oct. 1972.

[6] E. Deprettere and P. Dewilde, "Orthogonal cascade realization of real multiport digital filter," Tech Rep., Dept. Elec. Eng., Delft University, Delft, The Netherlands, 1980.

[7] P. Dewilde, A. C. Vieira, and T. Kailath, "On a generalized Szegö–Levinson realization algorithm for optimal linear predictors based on a network synthesis approach," *IEEE Trans. Circuits Syst.*, vol. CAS-25, no. 9, pp. 663-675, Sept. 1978.

[8] A. Fettweis, "Digital filter structures related to classical filter networks," *Arch. Elek. Ubertragung*, vol. 25, pp. 78-89, 1971.

[9] ——, "Some principles of designing digital filters imitating classical filter structures," *IEEE Trans. Circuit Theory*, vol. CT-18, pp. 314-316, Mar. 1971.

[10] ——, "Pseudopassivity, sensitivity, and stability of wave digital filters," *IEEE Trans. Circuit Theory*, vol. CT-19, no. 6, pp. 668-673, Nov. 1972.

[11] ——, "Wave digital filters with reduced number of delays," *Int. J. Circuit Theory Appl.*, vol. 2, no. 4, pp. 319-320, 1974.

[12] ——, "On sensitivity and roundoff noise in wave digital filters," *IEEE Trans. Acoust., Speech Signal Process.*, vol. ASSP-22, no. 5, pp. 383-384, Oct. 1974.

[13] A. Fettweis, H. Levin, and A. Sedlmeyer, "Wave digital lattice filters," *Int. J. Circuit Theory Appl.*, vol. 2, pp. 203-211, 1974.

[14] A. H. Gray, Jr., "Passive cascade lattice digital filters," *IEEE Trans. Circuits Syst.*, vol. CAS-27, no. 5, pp. 337-344, May 1980.

[15] A. H. Gray, Jr., and J. D. Markel, "Digital lattice and ladder filter synthesis," *IEEE Trans. Audio Electroacoust.*, vol. Au-21, no. 6, pp. 491-500, 1973.

[16] ——, "A normalized digital filter structure," *IEEE Trans. Acoust., Speech Signal Process.*, vol. ASSP-23, no. 3, pp. 268-277, 1975.

[17] S. Y. Hwang, "Realization of canonical digital networks," *IEEE Trans. Acoust., Speech Signal Process.*, vol. ASSP-22, no. 1, pp. 27-39, Feb. 1974.

[18] W. H. Ku and S. M. Ng, "Floating-point coefficient sensitivity

and roundoff noise of recursive digital filters realized in ladder structures," *IEEE Trans. Circuits Syst.*, vol. CAS-22, no. 12, pp. 927–936, Dec. 1975.

[19] J. D. Markel and A. H. Gray, Jr., "Roundoff noise characteristics of a class of orthogonal polynomial structures," *IEEE Trans. Acoust., Speech Signal Process.*, vol. ASSP-23, pp. 473–486, 1975.

[20] ——, "Fixed-point implementation algorithms for a class of orthogonal polynomial filter structures," *IEEE Trans. Acoust., Speech Signal Process.*, vol. ASSP-23, 486–494, 1975.

[21] D. G. Messerschmitt, "A class of generalized lattice filters," *IEEE Trans. Acoust., Speech Signal Process.*, vol. ASSP-28, no. 2, pp. 198–204, Apr. 1980.

[22] S. K. Mitra and R. J. Sherwood, "Canonical realizations of digital filters using the continued fraction expansion," *IEEE Trans. Audio Electroacoust.*, vol. AU-20, no. 3, pp. 185–194, Aug. 1972.

[23] ——, "Digital ladder networks," *IEEE Trans. Audio Electroacoust.*, vol. AU-21, pp. 30–36, 1973.

[24] S. K. Mitra, P. S. Kamat, and D. C. Huey, "Cascaded lattice realization of digital filters," *Circuit Theory Appl.*, vol. 3, pp. 3–11, 1977.

[25] K. Mondal and S. K. Mitra, "p-normalized digital two-pairs," *IEEE Trans. Acoust., Speech Signal Process.*, vol. ASSP-26, no. 4, pp. 374–376, Aug. 1978.

[26] R. Nouta. "Studies in wave digital filter theory and design," Ph.D. dissertation, Delft University, Delft, The Netherlands, 1979.

[27] A. V. Oppenheim and R. W. Schafer, *Digital Signal Processing*. Englewood Cliffs, NJ: Prentice-Hall, 1975.

[28] K. Renner and S. C. Gupta, "On the design of wave digital filters with low sensitivity properties," *IEEE Trans. Circuit Theory*, vol. CT-20, no. 5, pp. 555–567, Sept. 1973.

[29] A. Sedlmeyer and A. Fettweis, "Digital filters with true ladder configuration," *Int. J. Circuit Theory Appl.*, vol. 1, no. 1, pp. 5–10, Mar. 1973.

[30] W. Wegener, "On the design of wave digital lattice filters with short coefficient word length and optimal dynamic range," *IEEE Trans. Circuits, Syst.*, vol. CAS-25, no. 12, p. 1091, Dec. 1978.

Lattice Forms for Estimation and Prediction

[31] M. A. Alam, "Orthonormal lattice filter: A multi-stage multichannel estimation technique," *Geophys.*, vol. 43, no. 7, pp. 1368–1383, Dec. 1978.

[32] M. A. Alam and A. P. Sage, "Sequential estimation and identification of reflection coefficients by minimax entropy inverse filtering," *Comput. Elec. Eng.*, vol. 2, pp. 315–338, 1975.

[33] B. S. Atal, "On determining partial correlation coefficients by the covariance method of linear prediction," presented at 94th Meeting, Acoust. Soc. Amer., 1977.

[34] J. P. Burg, "Maximum entropy spectral analysis," Ph.D. dissertation, Stanford University, Stanford, CA, 1975.

[35] T. E. Carter, "Study of an adaptive lattice structure for linear prediction analysis of speech," in *Proc. IEEE ICASSP* (Tulsa, OK, 1978), pp. 27–30.

[36] E. Deprettere, "Orthogonal digital cascade filters and recursive construction algorithms for stationary and nonstationary fitting of orthogonal models," Ph.D. dissertation, Delft University of Technology, Delft, The Netherlands, 1980.

[37] E. Deprettere and P. Dewilde, "Generalized orthogonal filters for stochastic prediction and modeling," in *Digital Signal Processing*, V. Cappellini and A. G. Constantinides, Eds. New York: Academic Press, 1980, pp. 35–48.

[38] B. W. Dickinson, "An approach to stationary autoregressive estimation," in *Proc. Conf. Information Sciences and Systems* (Johns Hopkins University, Baltimore, MD, 1977), pp. 507–511.

[39] ——, "Autoregressive estimation using residual energy ratios," *IEEE Trans. Informat. Theory*, vol. IT-24, no. 4, pp. 503–506, July 1978.

[40] ——, "Estimation of partial correlation matrices using Cholesky decomposition," *IEEE Trans. Automat. Contr.*, vol. AC-24, no. 2, pp. 302–305, Apr. 1979.

[41] B. W. Dickinson and J. M. Turner, "Reflection coefficient estimation using Cholesky decomposition," *IEEE Trans. Acoust., Speech Signal Process.*, vol. ASSP-27, no. 2, pp. 146–149, Apr. 1979.

[42] ——, "Reflection coefficient estimates based on a Markov chain model," in *Proc. IEEE Int. Conf. Acoust., Speech Signal Process.* (Washington, DC, Apr. 1979), pp. 727–730.

[43] D. Gibson, "On reflection coefficients and the Cholesky decomposition," *IEEE Trans. Acoust., Speech Signal Process.*, vol. ASSP-25, no. 1, pp. 93–96, 1977.

[44] C. J. Gibson and S. Haykin, "A comparison of algorithms for the calculation of adaptive lattice filter," in *Proc. IEEE Int. Conf. on Acoust., Speech Signal Process.*, 1979.

[45] R. D. Gitlin and F. R. Magee, "Self-orthogonalizing adaptive equalization algorithms," *IEEE Trans. Commun.*, vol. COM-26, no. 7, pp. 666–672, July 1977.

[46] L. J. Griffiths, "A continuously adaptive filter implemented as a lattice structure," in *Proc. IEEE Conf. Acoust., Speech Signal Process.*, pp. 683–686, May 1977.

[47] ——, "Adaptive structures for multiple-input noise cancelling applications," in *Proc. IEEE Int. Conf. Acoust., Speech Signal Process.* (Washington, DC, Apr. 1979), pp. 925–928.

[48] ——, "An adaptive lattice structure for noise cancelling applications," in *IEEE Int. Conf. Acoust., Speech Signal Process.* (Tulsa, OK, Apr. 1978), pp. 87–90.

[49] L. J. Griffiths and R. S. Medaugh, "Convergence properties of an adaptive noise cancelling lattice structure," in *IEEE Conf. Decision and Control* (San Diego, CA, Jan. 1979), pp. 1357–1361.

[50] M. Honig and D. G. Messerschmitt, "Convergence properties of an adaptive digital lattice filter," in *Proc. IEEE Int. Conf. Acoust., Speech Signal Process.*, 1979.

[51] F. Itakura and S. Saito, "A statistical method for estimation of speech spectral density and formant frequencies," *Electron. Commun.*, vol. 53-A, pp. 36–43, 1970.

[52] ——, "Digital filtering techniques for speech analysis and synthesis," paper 25-C-1, in *Proc. 7th Int. Conf. Acoust.* (Budapest, Hungary, 1971), pp. 261–264.

[53] J. Makhoul, "Linear prediction: A tutorial review," *Proc. IEEE*, vol. 63, pp. 501–580, 1975.

[54] ——, "New lattice methods for linear prediction," in *IEEE Int. Conf. Acoust., Speech Signal Process.* (Philadelphia, PA, Apr. 1976), pp. 462–465.

[55] ——, "Stable and efficient lattice methods for linear prediction," *IEEE Trans. Acoust., Speech Signal Process.*, vol. ASSP-25, pp. 423–428, Oct. 1977.

[56] ——, "A class of all-zero lattice digital filters: Properties and applications," *IEEE Trans. Acoust., Speech Signal Process.*, vol. ASSP-26, no. 4, pp. 304–314, Aug. 1978.

[57] ——, "Lattice methods in spectral estimation," in *Proc. RADC Spectrum Estimation Workshop*, pp. 159–174, May 1978.

[58] J. Makhoul and R. Viswanathan, "Adaptive lattice methods for linear prediction," in *Proc. IEEE Int. Conf. Acoust., Speech Signal Process.* (Tulsa, OK, Apr. 1978), pp. 83–86.

[59] J. I. Makhoul and L. K. Cosell, "Adaptive lattice analysis of speech," *IEEE Trans. Acoust., Speech Signal Process.*, vol. ASSP-29, no. 3, pp. 654–658, June 1981.

[60] J. D. Markel and A. H. Gray, Jr., "On autocorrelation equations as applied to speech analysis," *IEEE Trans. Audio Electroacoust.*, vol. AU-21, no. 2, pp. 69–79, Apr. 1973.

[61] ——, *Linear Prediction of Speech*. New York: 1976.

[62] O. Barndorff-Nielsen and G. Schou, "On the parametrization of autoregressive models by partial autocorrelation," *J. Multivariate Analysis*, vol. 3, pp. 408–419, 1973.

[63] D. Parikh, N. Ahmed, and S. D. Stearns, "An adaptive lattice algorithm for recursive filters," *IEEE Trans. Acoust., Speech Signal Process.*, vol. ASSP-28, no. 1, pp. 110–111, Feb. 1980.

[64] F. A. Perry and S. R. Parker, "Recursive solutions for zero-pole modeling," in *Proc. 13th Asilomar Conf. Circuits Systems and Computers* (Pacific Grove, Monterey, CA, Nov. 1979), pp. 509–512.

[65] ——, "Adaptive solution of multichannel lattice models for linear and non-linear systems," in *Proc. IEEE Int. Symp. Circuits Syst.*, pp. 744–747, 1980.

[66] E. A. Robinson and S. Treitel, "Maximum entropy and the relationship of the partial autocorrelation to the reflection coefficients of a layered system," *IEEE Trans. Acoust., Speech Signal Process.*, vol. ASSP-28, no. 2, pp. 224–235, Apr. 1980.

[67] J. Le Roux and C. Gueguen, "A fixed point computation of partial correlation coefficients," *IEEE Trans. Acoust., Speech Signal Process.*, vol. ASSP-25, pp. 257–259, 1977.

[68] E. H. Satorius and S. T. Alexander, "Channel equalization using adaptive lattice algorithms," *IEEE Trans. Commun.*, vol. COM-27, no. 6, pp. 899–905, June 1979.

[69] E. Satorius and J. Pack, "On the application of lattice algorithms to data equalization," in *Proc. 13th Asilomar Conf. Circuits Systems and Computers* (Pacific Grove, CA, Nov. 1979), pp. 363–366.

[70] E. H. Satorius and M. J. Shensa, "On the application of recursive least squares methods to adaptive processing," presented at Int. Workshop on Applications of adaptive Control, Yale University, New Haven, CT, Aug. 1979.

[71] E. H. Satorius, J. D. Smith, and P. M. Reeves, "Adaptive noise cancelling of a sinusoidal interference using a lattice structure," in *Proc. IEEE Int. Conf. Acoust., Speech Signal Process.* (Washington, DC, Apr. 1979), pp. 929–932.

[72] J. M. Turner, "The lattice structure and its use in estimation and filtering," in *Proc. Conf. Information Sciences and Systems* (Princeton Univ., Princeton, NJ, Mar. 1980).

[73] J. Turner, B. Dickinson, and D. Lai, "Characteristics of reflection

coefficient estimates based on a Markov chain model," in *Proc. Int. Conf. Acoust., Speech Signal Process.* (Denver, CO, 1980), pp. 131-134.

Exact Least Squares Techniques

[74] H. M. Ahmed and M. Morf, "VLSI array architecture for matrix factorization," in *Proc. Workshop on Fast Algorithms for Linear Systems* (Aussois, France, 1981).

[75] ——, "Synthesis and control of signal processing architectures based on rotations," preprint, 1981.

[76] H. M. Ahmed, P. H. Ang, and M. Morf, "A VLSI speech analysis chip set utilizing coordinate rotation arithmetic," in *Proc. Conf. Circuits Systems* (Chicago, IL, 1981).

[77] H. M. Ahmed, J. M. Delosme, and M. Morf, "Highly concurrent computing structures for matrix arithmetic and signal processing," *IEEE Comput. Mag.*, Jan. 1982.

[78] H. M. Ahmed, M. Morf, D. T. Lee, and P. H. Ang, "A VLSI speech analysis chip set based on square root normalized ladder forms," in *Proc. Int. Conf. Acoust., Speech, Signal Process.*, Mar. 1981.

[79] A. Benveniste and C. Chaure, "Une methode rapide pour arriver a des algorithms d'estimation rapides pour des fonctions de transfert AR et ARMA synthetisees en treillis," *IRISA Int. Pub.*, no. 116, Rennes, France, May 1979.

[80] ——, "AR and ARMA identification algorithms of Levinson type: An innovations approach," *IEEE Trans. Automat. Cont.*, vol. AC-26, no. 6, pp. 1243-1260, Dec. 1981.

[81] J. A. Cadzow and R. L. Moses, "An adaptive ARMA spectral estimator, Parts 1 and 2," in *Proc. 1st ASSP Workshop on Spectral Estimation* (McMaster Univ., Hamilton, Ont., Canada, Aug. 1982).

[82] J. M. Delosme, "Algorithms and implementations for linear least-squares estimation," Ph.D. dissertation, Stanford University, Stanford, CA, 1982.

[83] J. M. Delsome and M. Morf, in "Mixed and minimal representations for Toeplitz and related systems," in *Proc. 14th Asilomar Conf. Circuits, Systems and Computers* (Pacific Grove, CA, Nov. 1980), pp. 19-24.

[84] ——, "A tree classification of algorithms for Toeplitz and related equations including generalized Levinson and doubling type algorithms," in *Proc. 19th IEEE Conf. Decision and Control*, (Dec. 1980), pp. 42-66.

[85] ——, "A unified stochastic description of efficient algorithms for second order processes," in *Proc. Workshop on Fast Algorithms for Linear Systems* (Aussois, France, 1981).

[86] J. M. Delosme, Y. Genin, M. Morf, and P. Van Dooren, "Σ contractive embeddings and interpretation of some algorithms for recursive estimation," in *Proc. 14th Asilomar Conf. Circuits Systems and Computers* (Pacific Grove, CA, Nov. 1980), pp. 25-28.

[87] B. Egardt and M. Morf, "Asymptotic analysis of a ladder algorithm for ARMA models," preprint, July 1980.

[88] D. D. Falconer and L. Ljung, "Application of fast Kalman estimation to adaptive equalization," *IEEE Trans. Commun.*, vol. COM-26, no. 10, pp. 1439-1446, Oct. 1976.

[89] B. Friedlander, "Recursive lattice forms for adaptive control," in *Proc. Joint Automatic Control Conf.* (San Francisco, CA, paper WP2-E, Aug. 1980).

[90] ——, "Recursive lattice forms for spectral estimation and adaptive control," in *Proc. 19th IEEE Conf. Decision and Control* (Dec. 1980), pp. 466-471.

[91] ——, "A pole-zero lattice form for adaptive line enhancement," in *Proc. 14th Asilomar Conf. Circuits, Systems, and Computers* (Pacific Grove, CA, Nov. 1980), pp. 380--384.

[92] ——, "A modified lattice algorithm for deconvolving filtered impulsive processes," in *Proc. Int. Conf. Acoust., Speech Signal Process.* (Atlanta, GA, Mar. 1981), pp. 865-868.

[93] ——, "Recursive lattice forms for spectral estimation," in *Proc. ASSP Workshop on Spectral Estimation* (McMaster Univ., Hamilton, Ont., Canada, Aug. 1981). Also to be published in *IEEE Trans. Acoust., Speech, Signal Processing.*

[94] ——, "Lattice implementations of some recursive parameter estimation algorithms," in *6th IFAC Symp. on Identification and System Parameter Estimation* (June 1982), pp. 481-486.

[95] ——, "A lattice algorithm for factoring the spectrum of a moving average process," in *Proc. Conf. Information Sciences and Systems* (Princeton Univ., Princeton, NJ, Mar. 1982).

[96] ——, "Lattice methods for spectral estimation," *Proc. IEEE*, vol. 70, no. 9, Sept. 1982, to appear.

[97] B. Friedlander and M. Morf, "Efficient inversion formulas for sums of products of Toeplitz and Hankel matrices," in *Proc. 18th Annual Allerton Conf. Communication, Control and Computing*, Oct. 1980.

[98] B. Friedlander, L. Ljung, and M. Morf, "Lattice implementation of the recursive maximum likelihood algorithm," in *Proc. 20th IEEE Conf. Decision and Control* (Dec. 1981), pp. 1083-1084.

[99] B. Friedlander, T. Kailath, M. Morf, and L. Ljung, "Extended Levinson and Chandrasekhar equations for general discrete-time linear estimation problems," *IEEE Trans. Automat. Cont.*, vol. AC-23, no. 4, pp. 653-659, Aug. 1978.

[100] B. Friedlander, M. Morf, T. Kailath, and L. Ljung, "New inversion formulas for matrices classified in terms of their distance from Toeplitz Matrices," *Linear Algebra and Its Applications*, vol. 27, pp. 31-60, 1979.

[101] W. S. Hodgkiss and D. Alexandrou, "Poser normalization sensitivity of adaptive frequency tracking architectures," in *Proc. 1st ASSP Workshop on Spectral Estimation* (McMaster Univ., Hamilton, Ont., Canada, Aug. 1981).

[102] M. L. Honig and D. G. Messerschmitt, "Convergence models for adaptive gradient and least-squares algorithms," in *Proc. Int. Conf. Acoust., Speech Signal Process.* (Atlanta, GA, Mar. 1981), pp. 267-270.

[103] T. Kailath, S-Y Kung, and M. Morf, "Displacement ranks of matrices and linear equations," *J. Math. Analy. Appl.*, vol. 68, no. 2, pp. 395-407, Apr. 1979.

[104] ——, "Displacement ranks of a matrix," *Bull. Amer. Math. Soc.*, vol. 1, no. 5, pp. 769-773, Sept. 1979.

[105] T. Kailath, B. Levy, L. Ljung, and M. Morf, "The factorization and representation of operators in the algebra generated by Toeplitz operators," *SIAM J. Appl. Math.*, vol. 37, no. 3, pp. 467-484, Dec. 1979.

[106] T. Kailath, L. Ljung, and M. Morf, "Generalized Krein-Levinson equations for efficient calculation of Fredholm resolvents of nondisplacement kernels," in *Topics in Functional Analysis* (essays in honor of M. G. Krein, Advances in Mathematics Supplementary Studies, vol. 3). New York: Academic 1978.

[107] T. Kailath, A. Vieira, and M. Morf, "Inverses of Toeplitz operators, innovations, and orthogonal polynomials," *SIAM Rev.*, vol. 20, no. 1, pp. 1006-1019, 1978.

[108] D.T.L. Lee, B. Friedlander, and M. Morf, "Recursive ladder algorithms for ARMA modeling," to appear in *IEEE Trans. Automat. Contr.*, Aug. 1982.

[109] D. Lee, "Canonical ladder form realizations and fast estimation algorithms," Ph.D. dissertation, Dept. Elec. Eng., Stanford Univ., Stanford, CA, Aug. 1980.

[110] D. Lee and M. Morf, "Recursive square-root estimation algorithms," in *Proc. IEEE Int. Conf. Acoust., Speech Signal Process.* (Denver, CO, Apr. 1980), pp. 1005-1017.

[111] D.T.L. Lee and M. Morf, "A novel innovations based time-domain pitch detector," in *Proc. IEEE Int. Conf. Acoust., Speech Signal Process.* (Denver, CO, Apr. 1980), pp. 40-44.

[112] ——, "Distortion measures via ladder forms," presented at the IEEE Information Theory Symp., Santa Monica, CA, 1981.

[113] ——, "Recursive square-root ladder estimation algorithms," in *IEEE Int. Conf. Acoust., Speech Signal Process.* (Denver, CO, Apr. 1980), pp. 1005-1017.

[114] D. Lee, B. Friedlander, and M. Morf, "Recursive ladder algorithms for ARMA modeling," in *Proc. 19th Conf. Decision and Control* (Dec. 1980), pp. 1225-1231.

[115] D. Lee, M. Morf, and B. Friedlander, "Recursive square-root ladder estimation algorithms," *IEEE Trans. Acoust., Speech Signal Process.*, vol. ASSP-29, no. 3, pp. 627-641, June 1981.

[116] H. Lev-Ari and T. Kailath, "Schur and Levinson algorithms for nonstationary processes," in *Proc. IEEE Int. Conf. Acoust., Speech Signal Process.* (Atlanta, GA, Mar. 1981), pp. 860-864.

[117] ——, "Ladder form filters for nonstationary processes," in *Proc. 19th IEEE Conf. Decision and Control* (Dec. 1980), pp. 960-961.

[118] L. Ljung, M. Morf, and D. Falconer, "Fast calculations of gain matrices for recursive estimation schemes," *Int. J. Cont.*, vol. 27, no. 1, pp. 1-19, 1978.

[119] L. Marple, "Linear phase lattice filter," preprint, 1981.

[120] R. S. Medaugh and L. J. Griffiths, "A comparison of two fast linear predictors," in *Proc. Int. Conf. Acoust., Speech Signal Process.* (Atlanta, GA, Mar. 1981), pp. 293-296.

[121] M. Morf, "Fast algorithms for multivariable systems," Ph.D. dissertation, Dept. Elec. Eng., Stanford Univ., Stanford, CA, 1974.

[122] ——, "Ladder forms in estimation and system identification," in *Proc. 11th Asilomar Conf. Circuits Systems and Computers* (Monterey, CA, Nov. 1977), pp. 424-429.

[123] ——, "Doubling algorithms for Toeplitz and related equations," in *Proc. IEEE Int. Conf. Acoust., Speech Signal Process.*, pp. 954-959, Apr. 1980.

[124] M. Morf and J. M. Delosme, "Matrix decompositions and inversions via elementary signature-orthogonal transformations," presented at the Int. Symp. Mini- and Micro-Computers in Control and Measurements, San Francisco, CA, May 1981.

[125] M. Morf, B. Dickinson, T. Kailath, and A. Vieira, "Efficient solutions of covariance equations for linear prediction," *IEEE Trans. Acoust., Speech Signal Process.*, vol. ASSP-25, no. 5, pp. 429-435, 1977.

[126] M. Morf, T. Kailath, and B. W. Dickinson, "General speech models and linear estimation," in *Speech Recognition*, R. Reddy, Ed. New York: Academic Press, 1975.
[127] M. Morf, T. Kailath, and L. Ljung, "Fast algorithms for recursive identification," in *Proc. IEEE Conf. Decision and Control* (Clearwater Beach, FL, Dec. 1976), pp. 916-921.
[128] M. Morf and D.T.L. Lee, "Fast algorithms for speech modeling," Tech. Rep. M303-1, Information Systems Lab., Stanford Univ., Stanford, CA, Dec. 1978.
[129] M. Morf and D. Lee, "Recursive spectral estimation of α-stationary processes," in *Proc. Rome Air Development Center Spectrum Estimation Workshop* (Griffiss Air Force Base, NY, May 1978), pp. 97-108.
[130] —, "Recursive least squares ladder forms for fast parameter tracking," in *Proc. IEEE Conf. Decision and Control* (San Diego, CA, Jan. 1979), pp. 1362-1367.
[131] M. Morf and D.T.L. Lee, "State-space structures of ladder canonical forms," in *Proc. 19th IEEE Conf. Decision and Control* (Albuquerque, NM, Dec. 1980), pp. 1221-1224.
[132] M. Morf, D. T. Lee, J. R. Nickolls, and A. Vieira, "A classification of algorithms for ARMA models and ladder realizations," in *Proc. IEEE Conf. Acoust., Speech Signal Process.* (Hartford, CT, Apr. 1977), pp. 13-19.
[133] M. Morf, D. T. Lee, and A. Vieira, "Ladder forms for estimation and detection," abstracts of papers, in *IEEE Int. Symp. Information Theory* (Ithaca, NY, Oct. 1977), pp. 111-112.
[134] M. Morf, C. H. Muravchik, and D. T. Lee, "Hilbert space array methods for finite rank process estimation and ladder realizations for adaptive signal processing," in *Proc. IEEE Int. Conf. Acoust., Speech Signal Process.* (Atlanta, GA, Mar. 1981), pp. 856-859.
[135] M. Morf, A. Vieira, and T. Kailath, "Covariance characterization by partial autocorrelation matrices," *Annals Stat.*, vol. 6, pp. 643-645, May 1978.
[136] M. Morf, A. Vieira, and D. Lee, "Ladder forms for identification and speech processing," in *Proc. IEEE Conf. Decision and Control* (New Orleans, LA, Dec. 1977), pp. 1074-1078.
[137] M. Morf, A. Vieira, D. Lee, and T. Kailath, "Recursive multichannel maximum entropy spectral estimation," *IEEE Trans. Geosci. Electron.*, vol. GE-16, no. 2, pp. 85-94, Apr. 1978.
[138] J. D. Pack and E. H. Satorius, "Least squares, adaptive lattice algorithms," NOSC Tech. Rep. TR 423, Apr. 1979.
[139] B. Porat and T. Kailath, "Normalized lattice algorithms for least-squares FIR system identification," submitted for publication.
[140] B. Porat, B. Friedlander, and M. Morf, "Square-root covariance ladder algorithms," in *Proc. Conf. Acoust., Speech Signal Process.* (Atlanta, GA, Mar. 1981), pp. 877-880.
[141] B. Porat, M. Morf, and D. Morgan, "On the relationship among several square-root normalized ladder algorithms," in *Proc. Conf. Information Sciences and Systems* (Johns Hopkins Univ., Baltimore, MD, Mar. 1981).
[142] V. U. Reddy, B. Egardt, and T. Kailath, "Optimized lattice form adaptive line enhancer for a sinusoidal signal in broadband noise," *IEEE Trans. Acoust., Speech Signal Process.*, to appear.
[143] E. H. Satorius and J. D. Pack, "A least-squares adaptive lattice equalizer algorithm," NOSC Tech. Rep. 575, Sept. 2, 1980.
[144] —, "Application of least-squares lattice algorithms to adaptive equalization," *IEEE Trans. Commun.*, vol. COM-29, pp. 136-142, Feb. 1981.
[145] E. H. Satorius and M. J. Shensa, "Recursive lattice filters—A brief overview," in *Proc. 19th IEEE Conf. Decision and Control*, pp. 955-959, Dec. 1980.
[146] M. J. Shensa, "Recursive least-squares lattice algorithms: A geometrical approach," *IEEE Trans. Automat. Contr.*, vol. AC-26, pp. 695-702, June 1981.
[147] M. J. Shensa, "A least-squares lattice decision feedback equalizer," in *Proc. IEEE Int. Commun. Conf.* (Seattle, WA, Aug. 1980), pp. 57.61-57.65.
[148] A.C.G. Vieira, "Matrix orthogonal polynomials, with applications to autoregressive modeling and ladder forms," Ph.D. dissertation, Stanford Univ., Stanford, CA, Aug. 1977.

General

[149] G.E.P. Box and G. M. Jenkins, *Time Series Analysis: Forecasting and Control*. San Francisco, CA: Holden-Day, 1970.
[150] G. J. Biermen, *Factorization Methods for Discrete Sequential Estimation*. New York: Academic Press, 1977.
[151] *Modern Spectrum Estimation*, D. G. Childers, Ed. New York: IEEE Press, 1978.
[152] J. Durbin, "The fitting of time series models," *Rev. L'Institut Intl. de Statisque*, vol. 28, pp. 233-243, 1960.
[153] T. Kailath, "A view of three decades of linear filtering theory," *IEEE Trans. Inform. Theory*, vol. IT-20, no. 2, pp. 145-181, Mar. 1974.
[154] —, *Lectures on Linear Least-Squares Estimation* (CISM Courses and Lectures, no. 140). New York: Springer, 1976.
[155] C. L. Lawson and R. J. Hanson, *Solving Least Squares Problems*. Englewood Cliffs, NJ: Prentice-Hall, 1974.
[156] N. Levinson, "The Wiener RMS (root-mean-square) error criterion in filter design and prediction," *J. Math. Phys.*, vol. 25, pp. 261-278, 1947.
[157] L. Ljung, "Analysis of recursive stochastic algorithms," *IEEE Trans. Automat. Contr.*, vol. AC-22, pp. 551-575, 1977.
[158] G. C. Goodwin and R. L. Payne, *Dynamic System Identification, Experimental Design and Data Analysis*. New York: Academic Press, 1977.
[159] P. Whittle, "On the fitting of multivariable autoregressions and the approximate canonical factorization of a spectral density matrix," *Biometrika*, vol. 50, pp. 129-134, 1963.
[160] C. T. Mullis and R. A. Roberts, "Roundoff noise in digital filters: Frequency transformations and invariants," *IEEE Trans. Acoust., Speech Signal Process.*, vol. ASSP-24, no. 6, pp. 538-550, Dec. 1976.
[161] —, "Synthesis of minimum roundoff noise fixed-point digital filters," *IEEE Trans. Circuits Syst.*, vol. CAS-23, pp. 501-512, Sept. 1976.
[162] E. A. Robinson, *Multichannel Time-Series Analysis with Digital Computer Programs*. San Francisco, CA: Holden-Day, 1967.
[163] R. A. Wiggins and E. A. Robinson, "Recursive solution to the multichannel filtering problem," *J. Geophys. Res.*, vol. 70, pp. 1885-1891, Apr. 1966.
[164] U. G. Yule, "On a method for investigating periodicities in disturbed series with special reference to Wolfer's sunspot numbers," *Philos. Trans. Roy. Soc. London, Ser. A*, vol. 226, pp. 267-298, 1927.
[165] S. M. Kay and S. L. Marple, Jr., "Spectrum analysis—A modern perspective," *Proc. IEEE*, vol. 69, no. 11, pp. 1380-1419, Nov. 1981.
[166] J. Treichler, "The spectral line enhancer: The concept, and implementation, and an application," Ph.D. dissertation, Stanford Univ., Stanford, CA, June 1977.
[167] B. Widrow, J. R. Glover, Jr. *et al.*, "Adaptive noise cancelling: Principles and applications," *Proc. IEEE*, vol. 63, pp. 1692-1716, Dec. 1975.
[168] R. A. Monzingo and T. W. Miller, *Introduction to Adaptive Arrays*. New York: Wiley, 1980.
[169] A. Viera and T. Kailath, "On another approach to the Schur-Cohn criterion," *IEEE Trans. Circuits Syst.*, vol. CAS-24, pp. 218-220, Apr. 1977.
[170] R. E. Kalman, "A new approach to linear filtering and prediction problems," *J. Basic Eng.*, vol. 82, pp. 342-345, Mar. 1960.
[171] B. Widrow *et al.*, "Stationary and nonstationary learning characteristics of the LMS adaptive filter," *Proc. IEEE*, vol. 64, no. 8, pp. 1151-1162, Aug. 1976.
[172] B. Widrow and J. M. McCool, "A comparison of adaptive algorithms based on the methods of steepest descent and random search," *IEEE Trans. Antennas Propagat.*, vol. AP-24, no. 5, pp. 615-637, Sept. 1976.
[173] W. S. Hodgkiss, Jr., and J. A. Presley, Jr., "Adaptive tracking of multiple sinusoids whose power levels are widely separated," *IEEE Trans. Acoust., Speech Signal Process.*, vol. ASSP-29, no. 3, pp. 710-721, June 1981.
[174] T. W. Anderson, *The Statistical Analysis of Time Series*. New York: Wiley, 1971.
[175] B. Friedlander, "Instrumental variable methods for ARMA spectral estimation," in *Proc. IEEE Int. Conf. Acoust, Speech Signal Process.* (Paris, France, May 1982), pp. 248-251.
[176] H. Lev-Ari and T. Kailath, "Parametrization and modeling of nonstationary processes," *Proc. IEEE*, to appear.
[177] J-M. Delosme and M. Morf, "Constant-gain filters for finite shift-rank processes," in *Proc. IEEE Int. Conf. Acoust., Speech Signal Process.* (Paris, France, May 1982), pp. 1732-1735.
[178] M. Morf, C. H. Muravchik, P. H. Ang, and J-M. Delosme, "Fast Cholesky algorithms and adaptive feedback filters," in *Proc. IEEE Int. Conf. Acoust., Speech Signal Process.* (Paris, France, May 1982), pp. 1727-1731.
[179] B. Porat, "Contributions to the theory and applications of lattice forms," Ph.D. dissertation, Stanford Univ., Stanford, CA, 1982.
[180] T. Kailath, *Linear Systems*. Englewood Cliffs, NJ: Prentice-Hall, 1980.
[181] H. Wakita, "Direct estimation of the vocal tract shape by inverse filtering of acoustic speech waveform," *IEEE Trans. Audio Electroacoust.*, vol. AU-21, pp. 417-427, Oct. 1973.
[182] H. Wakita and A. H. Gray, Jr., "Numerical determination of the lip impedance and vocal tract area functions," *IEEE Trans. Acoust., Speech Signal Process.*, vol. ASSP-23, pp. 574-558, 1975.
[183] M. D. Srinath and M. M. Viswanathan, "Sequential algorithm for identification of parameters of an autoregressive process," *IEEE Trans. Automat. Contr.*, vol. AC-20, no. 4, pp. 542-546, Aug. 1975.
[184] R. S. Medaugh, "A comparison of two fast linear predictors,"

Ph.D. dissertation, Dep. Elec. Eng., Univ. Colorado, Boulder, CO, 1981.

[185] B. Friedlander, "Efficient computation of the covariance sequence of an autoregressive process," *IEEE Trans. Automat. Contr.*, to appear in Dec. 1982.

[186] J. D. Klein and B. W. Dickinson, "A normalized ladder form of the residual energy ratio algorithms for PARCOR estimation via projections," submitted for publication.

[187] W. S. Hodgkiss and J. A. Presley, "The complex adaptive least-squares lattice," *IEEE Trans. Acoust., Speech Signal Process.*, vol. ASSP-30, no. 2, pp. 330–333, Apr. 1982.

[188] I. L. Ayala, "On a new adaptive lattice algorithm for recursive filters," *IEEE Trans. Acoust., Speech Signal Process.*, vol. ASSP-30, no. 2, pp. 316–319, Apr. 1982.

[189] H. Sakai, "Circular lattice filtering using Pagano's method," *IEEE Trans. Acoust., Speech Signal Process.*, vol. ASSP-30, no. 2, pp. 279–287, Apr. 1982.

[190] J. M. Turner, "Application of recursive exact least-squares ladder estimation algorithm for speech recognition," in *Proc. IEEE Int. Conf. Acoust., Speech Signal Process.* (Paris, France, May 1982), pp. 543–545.

[191] J. M. Turner, "Fast approximate whitening ladder filters," in *Proc. IEEE Int. Conf. Acoust., Speech Signal Process.* (Paris, France, May 1982), pp. 655–658.

[192] D.T.L. Lee and M. Morf, "Generalized CORDIC for digital signal processing," in *Proc. IEEE Int. Conf. Acoust., Speech Signal Process.* (Paris, France, May 1982).

[193] B. Friedlander, "An efficient algorithm for ARMA spectral estimation," in *Proc. Conf. Spectral Analysis and Its Use in Underwater Acoustics* (London, England, Apr. 1982). Also to appear in *Proc. Inst. Elec. Eng.*

[194] W. S. Hodgkiss and D. Alexandrou, "Power normalization sensitivity of adaptive lattice structures," submitted for publication.

[195] H. Lev-Ari and T. Kailath, "On generalized Schur and Levinson-Szego algorithms for quasi-stationary processes," in *Proc. 20th IEEE Conf. Decision and Control* (San Diego, CA, Dec. 1981), pp. 1077–1080.

[196] H. Lev-Ari, "Parametrization and modeling of nonstationary processes," Ph.D. dissertation, Stanford Univ., Stanford, CA, 1982.

[197] S. Kay and J. Makhoul, "On the statistics of the estimated reflection coefficients of an autoregressive process," submitted for publication.

[198] J-M. Delosme and M. Morf, "Fast algorithms for finite shift-rank processes: A geometric approach," to appear in CNRS editions.

[199] T. Kailath, "Time-variant and time-invariant lattice filters for nonstationary processes," submitted for publication.

Recursive Least Squares Lattice Algorithms—A Geometrical Approach

M. J. SHENSA

Abstract—**Several time-recursive least squares algorithms have been developed in recent years. In this paper a geometrical formalism is defined which utilizes a nested family of metric spaces indexed by the data time interval. This approach leads to a simplified derivation of the so-called recursive least squares lattice algorithms (recursive in time and order). In particular, it is found that the resulting structure provides a single framework which encompasses an entire family of fairly complex algorithms, as well as providing geometrical insight into their behavior.**

I. INTRODUCTION

Least squares criteria for the estimation of linear systems have been a recurrent theme in signal processing for years; nevertheless, new algorithms continue to be introduced. To a large extent, this may be attributed to a search for recursive procedures, more efficient computational techniques, and improved numerical properties. Traditional on-line methods [1] involve extensive matrix manipulations which results in $0(p^2)$ operations or more per update where p is the order of the model. In contrast, the last decade has seen the appearance of the so-called lattice algorithms [2]-[14] which reduce the computational complexity to $0(p)$. Many of these algorithms possess the additional advantage of being recursive in order as well as time.

In 1973 Markel and Gray [3] presented a geometrical formulation of the lattice approach to regression analysis (also used by Itakura and Saito [2]). Their goal was the solution of a stochastic problem (least mean squares or so-called autocorrelation or normal equations), and consequently the relations derived were functions of expected values of variables. In particular, the results rely on the true autocorrelation matrix. More recently, Morf *et al.* [7]-[11] have developed time-recursive algorithms for the solution of the least squares problem, i.e., the minimization of a finite time average of errors, rather than the expectation of the error. The least squares problem also involves a set of normal equations, but in this case the relevant matrix is a finite-data approximation to the autocorrelation matrix.

In this paper we utilize an inner-product formalism similar to that of [3] to solve the least squares problem. The structure so obtained is of considerable interest, inasmuch as it forms the basis for a whole class of recursive least squares lattice algorithms [7]-[14]. This approach also enjoys several other advantages. First, it provides a simplified derivation of the above-mentioned algorithms. (It is especially noteworthy that the inversion lemma for a partitioned matrix is not needed.) Second, the geometrical nature of the variables defined makes their role in the algorithms readily apparent, and leads some physical intuition as well. Finally, many of the relations proved in [3] for expected values are shown to hold true for the actual variables. These relations, such as the expression of the lattice gain coefficients in terms of inner products, can provide important insights into the algorithms' behavior, i.e., regarding numerical stability, convergence (effect of $T \to \infty$), implementation, etc. [3], [5], [6], [12].

The unfamiliar reader is forewarned, however, that, despite the highly structured nature of the problem, the actual algorithm is quite complex, involving a large number of variables. It is further complicated by the interweaving of recursions in both time and order. The situation is in many ways similar to the DFT or linear programming, where the concept is simple but the implementation (FFT or simplex method) is fairly involved. It is sincerely hoped that these aspects will not overly tax the reader's patience.

Manuscript received August 22, 1979; revised June 5, 1980. Paper recommended by A. Z. Manitius, Past Chairman of the Optimal Systems Committee. This work was supported by the Naval Ocean Systems Center under Contract N00123-78-C-1005.

The author is with the Naval Ocean Systems Center, San Diego, CA 92152.

II. DEFINITIONS

We begin with some vector notation. Let $x(t)$ be a discrete time series, and define $\bar{x}$ to be the vector whose tth component is given by

$$[\bar{x}]_t = x(t) \qquad t = 0, 1, \cdots. \tag{1}$$

Although this vector lies in an infinite dimensional space ($\bar{x} \in \mathbf{R}^\infty$), all our operations shall be on finite dimensional subspaces, thus alleviating the need for sophisticated mathematical techniques. However, if so desired, the reader may assume an upper bound T_0 such that $0 \leqslant t \leqslant T_0$, i.e., $\bar{x} \in \mathbf{R}^{T_0}$. By shifting $\bar{x}$, we obtain a family of vectors $\bar{x}^i$

$$[\bar{x}^i]_t = \begin{cases} x(t-i) & t \geqslant i \\ 0 & 0 \leqslant t < i. \end{cases} \tag{2}$$

This family generates a nested set of subspaces S^p defined by

$$S^p = \{\bar{x}^0, \bar{x}^1, \cdots, \bar{x}^p\} \qquad p = 0, \cdots, p_0 \tag{3}$$

where the brackets indicate the linear space spanned by the vectors it encloses. It is assumed that the $\bar{x}^i$'s are linearly independent for $0 \leqslant i \leqslant p_0$; thus, S^p is $p+1$ dimensional for $p \leqslant p_0$. Note also that $S^p \subset S^r$ for $p \leqslant r$.

We next introduce a family of pseudometrics on $\mathbf{R}^\infty$ defined as follows.

$$\begin{aligned} \langle \bar{x}, \bar{y} \rangle_T &= \sum_{t=0}^{T} w^{T-t} [\bar{x}]_t [\bar{y}]_t \\ \| \bar{x} \|_T^2 &= \langle \bar{x}, \bar{x} \rangle_T \end{aligned} \tag{4}$$

where $0 < w \leqslant 1$ is constant factor which "exponentially windows" the data [11], [12]. Although $\langle \rangle_T$ is singular on R^∞, it is a true metric on S^p for sufficiently large T. More precisely, a necessary and sufficient condition for $\langle \rangle_T$ to be a metric on S^p for all $p \leqslant p_0$ is that the truncated vectors $\bar{x}^{i'}$ defined by

$$[\bar{x}^{i'}]_t = \begin{cases} [\bar{x}^i]_t & t \leqslant T \\ 0 & t > T \end{cases} \tag{5}$$

be linearly independent for $0 \leqslant i \leqslant p_0$ (see Appendix). We shall always assume this to be the case. Note that a necessary condition is $T > p_0$.

It is also convenient to define the linear shift operator z^i by

$$[z^i \bar{x}]_t = \begin{cases} [x]_{t+i} & t+i \geqslant 0 \\ 0 & t+i < 0 \end{cases} \tag{6}$$

where i may be any integer. It follows that

$$z^{-i} \bar{x} = \bar{x}^i. \tag{7}$$

Note that $zz^{-1} = I$, but $z^{-1}z \neq I$ where I is the identity. A short calculation utilizing definitions (6) and (4) yields the relation

$$\langle z^{-1}\bar{x}, \bar{y} \rangle_T = \langle \bar{x}, z\bar{y} \rangle_{T-1}. \tag{8}$$

Equation (8) will prove important, inasmuch as it relates time shift properties to order (i.e., to the index i of the family $\bar{x}^i$). It differs from the stochastic case of Markel and Gray [3], inasmuch as z and z^{-1} are not quite adjoint operators (T is replaced by $T-1$ in (8)]. However, the shift properties are sufficiently powerful to allow $\langle \rangle$ to replace

Reprinted from *IEEE Trans. Automat. Contr.*, vol. AC-26, no. 3, pp. 695-702, June 1981.

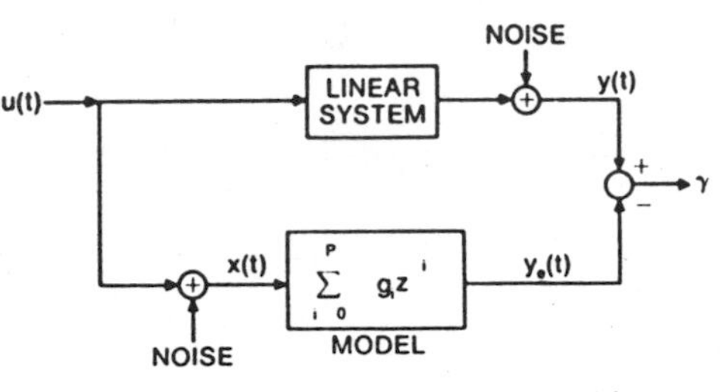

Fig. 1. Block diagram of least squares model.

expectation in the formalism. Physically, the shift operator as reflected in (8) has the effect of prewindowing the data ($x(t) = 0$ for $t < 0$).

The least squares problem whose solution we are ultimately seeking is the following (Fig. 1). Given sequences $x(t)$ and $y(t)$, find g_i^p such that

$$\left\| \bar{y} - \sum_{i=0}^{p} g_i^p \bar{x}^i \right\|_T^2 = \min_{q_i} \left\| \bar{y} - \sum_{i=0}^{p} q_i \bar{x}^i \right\|_T^2. \tag{9}$$

Actually, this is a family of problems where the order p may range over $0 \leq p \leq p_0$ and the time T over $p_0 \leq T \leq T_0$. Note that the weights w appearing in the definition of the metric (4), may be thought of as simply error weights, or as a sliding exponential window ($x(t) \rightarrow (\sqrt{w})^{T-t} x(t)$) of the data [12].

III. LATTICE STRUCTURE

We shall repeatedly have need of the phrase "orthogonal projection with respect to the metric $\langle\,\rangle_T$," and hence shall use the abbreviation "proj_T." Let $\sum_{i=0}^{p-1} b_i^p(T)\bar{x}^i$ be the proj_T of $\bar{x}^p$ onto S^{p-1} (where S^{-1} = empty set). Then we define $\bar{\beta}^p(T)$ by

$$\bar{\beta}^p(T) = \bar{x}^p - \sum_{i=0}^{p-1} b_i^p(T)\bar{x}^i. \tag{10}$$

These vectors have several important properties.

We first note that $\bar{\beta}^p(T)$ is the proj_T of $\bar{x}^p$ onto the orthogonal complement of S^{p-1} in S^p, i.e., $\langle \bar{\beta}^p(T), \bar{x}^p \rangle_T = \langle (S^{p-1})^\perp, \bar{x}^p \rangle_T$. Also, we may write S^p as the direct sum $S^p = S^{p-1} \oplus \bar{\beta}^p$ so that $\bar{\beta}^i$ for $0 \leq i \leq p$ form an orthogonal basis of S^p. In physical terms we may consider $\bar{\beta}^p(T)$ as the error residual of a least squares backward predictor (estimator of $\bar{x}^p$ by $\bar{x}^i$'s for $i < p$), since

$$\left\| \bar{x}^p - \sum_{i=0}^{p-1} b_i^p(T)\bar{x}^i \right\|^2 = \min_{q_i} \left\| \bar{x}^p - \sum_{i=0}^{p-1} q_i \bar{x}^i \right\|^2. \tag{11}$$

For future convenience, we define the magnitude-squared of $\bar{\beta}^p(T)$ by

$$\epsilon_\beta^p(T) = \| \bar{\beta}^p(T) \|_T^2. \tag{12}$$

It follows from the definition of $\bar{\beta}^p(T)$ that it is uniquely determined by the three conditions

$$\bar{\beta}^p(T) \in S^p \tag{13a}$$

$$\bar{\beta}^p(T) - \bar{x}^p \in S^{p-1} \tag{13b}$$

$$\langle \bar{\beta}^p(T), \bar{x}^i \rangle_T = 0 \qquad 0 \leq i \leq p-1. \tag{13c}$$

Equations (13) merely state that the set of $\bar{\beta}^i(T)$ are obtained by applying the Gram-Schmidt orthogonalization procedure to $\bar{x}^i$, $0 \leq i \leq p$ with respect to the metric $\langle\,\rangle_T$.

As in [3], we introduce a set of auxiliary vectors $\bar{\alpha}^p(T)$ to aid in the order recursion. Let Q^p be the subspace

$$Q^p = \{\bar{x}^1, \cdots, \bar{x}^p\} \tag{14}$$

and let $\sum_{i=1}^{p} a_i^p(T)\bar{x}^i$ be the proj_T of $\bar{x}^0$ onto Q^p. Then we define $\bar{\alpha}^p(T)$ by

$$\bar{\alpha}^p(T) = \bar{x}^0 - \sum_{i=1}^{p} a_i^p \bar{x}^i \tag{15}$$

with magnitude-squared

$$\epsilon_\alpha^p(T) = \| \bar{\alpha}^p(T) \|_T^2. \tag{16}$$

These $\bar{\alpha}^p(T)$ represent forward predictor residuals and are uniquely determined by the conditions

$$\bar{\alpha}^p(T) \in S^p \tag{17a}$$

$$\bar{\alpha}^p(T) - \bar{x}^0 \in Q^p \tag{17b}$$

$$\langle \bar{\alpha}^p(T), \bar{x}^i \rangle_T = 0 \text{ for } 1 \leq i \leq p. \tag{17c}$$

Finally, we generalize the output of the least squares model (9) to

$$\bar{y}_e^p(T) = \sum_{i=0}^{p} g_i^p(T)\bar{x}^i \tag{18}$$

which is simply the proj_T of $\bar{y}$ onto S^p. The subscript e is introduced to reflect the fact that $\bar{y}_e^p$ is the pth order estimate of the system output $\bar{y}$. Note that the tth component of this vector equation is the output of the model with input at time t, but uses the least squares model as determined by the data at time T (i.e., $[\bar{y}_e^p(T)]_t = \sum_{i=0}^{p} g_i^p(T)x(t-i)$). The error residual corresponding to (18) is

$$\bar{\gamma}^p(T) = \bar{y} - \bar{y}_e^p(T). \tag{19}$$

In brief, our approach shall be as follows. We wish to project $\bar{y}$ onto S^p, the space spanned by the input $\bar{x}^i, i = 0, \cdots, p$ (c.f. (9)). In order to aid in this process, we find an orthogonal basis for that space namely, $\bar{\beta}^i, i = 0, \cdots, p$. This basis may be extended to $i = p + 1$ using a Gram-Schmidt procedure which is faciliated by a set of auxiliary vectors $\bar{\alpha}^i$. The result is the so-called lattice decomposition of the input (see Fig. 2) and is recursive in p. The projection of $\bar{y}$ onto the $\bar{\beta}^i$ is realized via the vector $\bar{\gamma}^p$. The vectors $\bar{\alpha}$, $\bar{\beta}$, and $\bar{\gamma}$ all represent orthogonal complements of projections onto similar spaces and hence, as we shall see, have similar update properties. Recursive relations also will be developed which update the model from time $T-1$ to time T.

Some applications require the coefficients $g_i^p(T)$, a situation which may be dealt with by the isomorphism of Section V. More frequently, it is the filtered version of the input $[y_e^p(T)]_T = \sum_{i=0}^{p} g_i\ (T)x(T-i)$ that is desired. For example, in the case of equalization, the proper output variable is $[\bar{y}_e^p(T-1)]_T$. The $T-1$, which implies use of the filter model derived at time $T-1$, is necessary, since for this application $y(T)$ is not known and must be estimated by thresholding $[\bar{y}_e^p(T-1)]_T = \sum_{i=0}^{p} g_i^p(T-1)x(T-i)$. Similarly, in noise canceling (with primary input y and reference x), the output variable is $[\bar{\gamma}^p(T-1)]_T$. Here, $T-1$ is used in order to avoid cancellation of a portion of the signal. The particular algorithm derived in Section IV will use these variables.

We now state two theorems which contain the basic recursions for the order and time updates. Their derivations are treated in the next section:

Theorem I: The vectors defined above satisfy the following order update recursions:

$$\bar{\alpha}^{p+1}(T) = \bar{\alpha}^p(T) + K_\alpha^p(T)z^{-1}\bar{\beta}^p(T-1) \tag{20a}$$

$$\bar{\beta}^{p+1}(T) = z^{-1}\bar{\beta}^p(T-1) + K_\beta^p(T)\bar{\alpha}^p(T) \tag{20b}$$

$$\bar{\gamma}^{p+1}(T) = \bar{\gamma}^p(T) + K_\gamma^{p+1}(T)\bar{\beta}^{p+1}(T) \tag{20c}$$

$$\bar{\xi}^{p+1}(T) = \bar{\xi}^p(T) + K_\xi^{p+1}(T)\bar{\beta}^{p+1}(T) \tag{20d}$$

$$\epsilon_\alpha^{p+1}(T) = \epsilon_\alpha^p(T) + K_\alpha^p(T)k^p(T) \tag{21a}$$

$$\epsilon_\beta^{p+1}(T) = \epsilon_\beta^p(T) + K_\beta^p(T)k^p(T) \tag{21b}$$

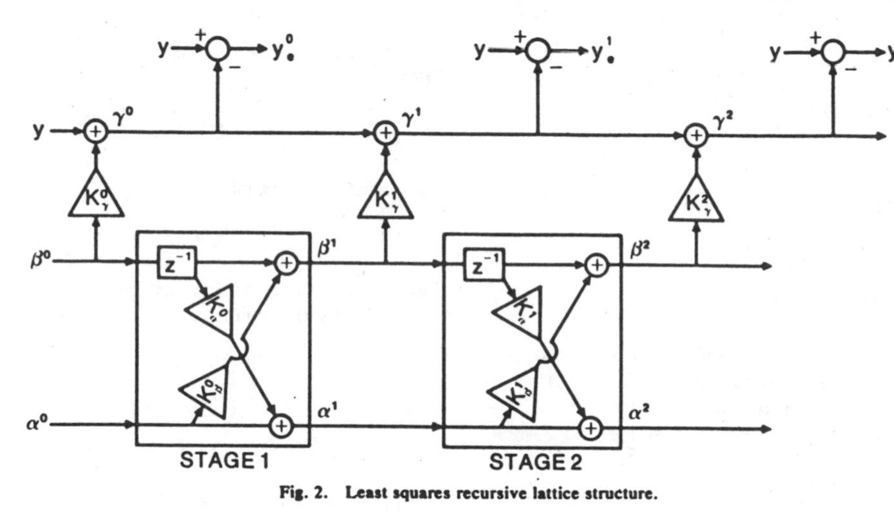

Fig. 2. Least squares recursive lattice structure.

where

$$K_\alpha^p(T) = -k^p(T)/\epsilon_\beta^p(T-1);$$
$$K_\beta^p(T) = -k^p(T)/\epsilon_\alpha^p(T);$$
$$K_\gamma^p(T) = -k_\gamma^p(T)/\epsilon_\beta^p(T)$$

and

$$K_\xi^p(T) = [\bar\beta^p(T)]_T/\epsilon_\beta^p(T);$$

with

$$k^p(T) \triangleq \langle \bar\alpha^p(T), z^{-1}\bar\beta^p(T-1)\rangle_T$$

$$k_\gamma^p(T) \triangleq \langle \bar\gamma^{p-1}(T), z^{-1}\bar\beta^{p-1}(T-1)\rangle_T.$$

$\bar\xi^p(T)$, the projection of the unit T-axis onto S^p, is an auxiliary variable fundamental to the time updates [see (39)-(40)]. Its magnitude is simply $[\bar\xi^p(T)]_T$, so that an order update for $\|\bar\xi^p(T)\|^2$ may be obtained by taking the Tth component of (20d).

Theorem II: The following time update recursions hold:

$$\bar\alpha^p(T) = \bar\alpha^p(T-1) - [\bar\alpha^p(T-1)]_T z^{-1}\bar\xi^{p-1}(T-1) \quad (22a)$$

$$\bar\beta^p(T) = \bar\beta^p(T-1) - [\bar\beta^p(T-1)]_T\bar\xi^{p-1}(T) \quad (22b)$$

$$\bar\gamma^p(T) = \bar\gamma^p(T-1) - [\bar\gamma^p(T-1)]_T\bar\xi^p(T) \quad (22c)$$

$$\epsilon_\alpha^p(T) = w\epsilon_\alpha^p(T-1) + [\bar\alpha^p(T-1)]_T[\bar\alpha^p(T)]_T \quad (23a)$$

$$\epsilon_\beta^p(T) = w\epsilon_\beta^p(T-1) + [\bar\beta^p(T-1)]_T[\bar\beta^p(T)]_T \quad (23b)$$

$$k^p(T) = wk^p(T-1) + [\bar\alpha^p(T-1)]_T[\bar\beta^p(T-1)]_{T-1} \quad (24)$$

$$k_\gamma^p(T) = wk_\gamma^p(T-1) + [\bar\gamma^{p-1}(T-1)]_T[\bar\beta^p(T)]_T \quad (25)$$

$$\bar\xi^p(T) = z^{-1}\bar\xi^p(T-1) - \frac{[\beta^p(T-1)]_{T-1}}{\epsilon_\beta^p(T-1)}\bar\beta^p(T-1) + \frac{[\bar\alpha^p(T)]_T}{\epsilon_\alpha^p(T)}\bar\alpha^p(T). \quad (26)$$

Those variables in Theorems I and II which have not appeared previously are auxiliary variables which aid in the recursions. Generally, they have geometrical interpretations (such as k^p/ϵ^p, which approximate the PARCOR coefficients [3], [5], [6], and $\bar\xi^p(T)$, which projects vectors onto the space of new observations), some of which are discussed briefly in the next section. Further comments may be found in [12].

Most least squares recursive lattice algorithms may be derived by considering the Tth component of (20)-(26). However, the occasional appearance of $p+1$ on the right-hand side of some of the relations in Theorem I and the appearance of T on the right-hand sides in Theorem II imply that any algorithm derived from these recursions must take careful account of the order in which the updates are performed. The particular algorithm derived in this paper updates only k^p, k_γ^p, ϵ_α^0, and ϵ_β^0 with respect to time and obtains the other variables at time T through an order update (their values for $p = 0$ are known functions of the input). Such an algorithm does not require the variables $\bar\xi^p$. We note that other schemes are possible. For example, if p is fixed and only the time updates are used, the isomorphism of Section V yields the "fast Kalman" algorithm [8].

IV. DERIVATIONS

In this section, we verify the relations found in Theorems I and II. From these recursions, we then derive a least squares lattice algorithm following the procedure outlined above.

A. Order Updates for α and β

We first verify relations (20a) and (20b), where $K_\alpha^p(T)$ and $K_\beta^p(T)$ are constants which we shall subsequently determine. To prove (20a), we note that $z^{-1}\bar\beta^p(T-1) \in Q^{p+1} \subset S^{p+1}$, and from (8)

$$\begin{aligned}\langle z^{-1}\bar\beta^p(T-1), \bar x^i\rangle_T &= \langle \bar\beta^p(T-1), z\bar x^i\rangle_{T-1} \qquad \text{for } i > 0\\ &= \langle \bar\beta^p(T-1), \bar x^{i-1}\rangle_{T-1}\\ &= 0 \qquad \text{for } 1 \le i \le p.\end{aligned} \quad (27)$$

Thus, for any constant K, $\bar\phi \triangleq \bar\alpha^p(T) + Kz^{-1}\bar\beta^p(T-1)$ satisifes (17a) and (17b) with p replaced by $p+1$, and also satisfies

$$\langle \bar\phi, \bar x^i\rangle_T = 0 \qquad \text{for } 1 \le i \le p. \quad (28)$$

In consideration of (28), condition (17c) will be satisfied for $p+1$, provided K is chosen such that

$$0 = \langle \bar\phi, \bar x^{p+1}\rangle_T = \langle \bar\alpha^p(T) + Kz^{-1}\bar\beta^p(T-1), \bar x^{p+1}\rangle_T,$$

i.e., if we choose $K = K_\alpha^p(T)$, where

$$K_\alpha^p(T) = -\frac{\langle \bar\alpha^p(T), \bar x^{p+1}\rangle_T}{\langle z^{-1}\bar\beta^p(T-1), \bar x^{p+1}\rangle_T}. \quad (29)$$

A similar analysis verifies (20b) where $K_\beta^p(T)$ is chosen to satisfy

$$K_\beta^p(T) = -\frac{\langle z^{-1}\bar\beta^p(T-1), \bar x^0\rangle_T}{\langle \bar\alpha^p(T), \bar x^0\rangle_T}. \quad (30)$$

The appearance of $T-1$ in the above equations should be noted. The z^{-1} was necessary in order to obtain a vector whose "$\bar x^{p+1}$ coefficient" was nonzero [see (29)], which in turn necessitated $T-1$ in order that (27) be valid [cf. remarks after (8)].

B. Simplification of K_α^p and K_β^p

Equations (29) and (30) may be rewritten in a form which provides a simple geometric interpretation. We first note that since $\bar{\beta}^p(T-1)$ is a proj_{T-1} of $\bar{x}^p$, we have $\langle \bar{\beta}^p(T-1), \bar{x}^p \rangle_{T-1} = \langle \bar{\beta}^p(T-1), \bar{\beta}^p(T-1) \rangle_{T-1} = \epsilon_\beta^p(T-1)$. Thus,

$$\langle z^{-1}\bar{\beta}^p(T-1), \bar{x}^{p+1} \rangle_T = \langle \bar{\beta}^p(T-1), \bar{x}^p \rangle_{T-1} = \epsilon_\beta^p(T-1). \tag{31}$$

Similarly,

$$\langle \bar{\alpha}^p(T), \bar{x}^0 \rangle_T = \epsilon_\alpha^p(T). \tag{32}$$

We now show that the numerators of (29) and (30) are both given by

$$k^p(T) \triangleq \langle \bar{\alpha}^p(T), z^{-1}\bar{\beta}^p(T-1) \rangle_T. \tag{33}$$

We have, using (27),

$$\langle \bar{\alpha}^p(T), z^{-1}\bar{\beta}^p(T-1) \rangle_T = \left\langle \bar{x}^0 - \sum_{i=1}^{p} a_i^p(T)\bar{x}^i, z^{-1}\bar{\beta}^p(T-1) \right\rangle_T = \langle \bar{x}^0, z^{-1}\bar{\beta}^p(T-1) \rangle_T \tag{34}$$

and from (17c)

$$\langle \bar{\alpha}^p(T), z^{-1}\bar{\beta}^p(T-1) \rangle_T = \left\langle \bar{\alpha}^p(T), z^{-1}(\bar{x}^p - \sum_{i=0}^{p-1} b_i^p(T-1)\bar{x}^i) \right\rangle_T = \langle \bar{\alpha}^p(T), z^{-1}\bar{x}^p \rangle_T = \langle \bar{\alpha}^p(T), \bar{x}^{p+1} \rangle_T. \tag{35}$$

It thus follows from (29), (30), and (33)–(35) that

$$K_\alpha^p(T) = -\frac{k^p(T)}{\epsilon_\beta^p(T-1)} \qquad K_\beta^p(T) = -\frac{k^p(T)}{\epsilon_\alpha^p(T)}. \tag{36}$$

For ease of notation, let $\bar{\alpha} = \bar{\alpha}^p(T)$ and $\bar{\beta} = z^{-1}\bar{\beta}^p(T-1)$. Then

$$K_\alpha = -\frac{\langle \bar{\alpha}, \bar{\beta} \rangle_T}{\| \bar{\beta} \|_T^2} \quad \text{and} \quad K_\beta = -\frac{\langle \bar{\alpha}, \bar{\beta} \rangle_T}{\| \bar{\alpha} \|_T^2}. \tag{37}$$

It follows from the Cauchy-Schwarz inequality that their product satisfies

$$0 < K_\alpha K_\beta < 1. \tag{38}$$

This condition is weaker than that of [3], where the gains are individually less than one. However, relations (37) are exact for arbitrary input $x(t)$, whereas [3] assumes a knowledge of the true correlation matrix, a condition rarely met in practice. For $w = 1$ and under reasonable conditions on x, $\langle \bar{x}^i, \bar{x}^j \rangle_T \triangleq R_{ij}$ approaches the true correlation matrix as $T \to \infty$, so that $\lim_{T\to\infty} |K_\alpha(T)| < 1$ and $\lim_{T\to\infty} |K_\beta(T)| < 1$. This is not true for $w < 1$, and either $|K_\alpha(T)|$ or $|K_\beta(T)|$ (but not both) may be greater than one. This may have serious stability consequences ([2], [3], [5], [6], [10]).

C. Order Updates for $\bar{\gamma}$, ϵ_α, ϵ_β

Equation (20c), the order update for $\bar{\gamma}^p$ (as well as the expression for K_γ^p), may be derived in exactly the same fashion as (20b). The order updates (21a) and (21b) for the squared vector magnitudes $\epsilon_\alpha^p = \|\bar{\alpha}^p\|^2$ and $\epsilon_\beta^p = \|\bar{\beta}^p\|^2$ are obtained by taking the scalar product of equations (20a) and (20b) with $\bar{\alpha}^p(T)$ and $\bar{\beta}^p(T)$, respectively.

D. Time Updates and $\bar{\xi}$

We introduce a set of auxiliary vectors $\bar{\xi}^p$ to aid in the time update defined as follows. Let $\bar{e}^T$ be the vector representing the Tth coordinate axis, i.e.,

$$[\bar{e}^T]_t = \begin{cases} 1 & t = T \\ 0 & t \neq T. \end{cases} \tag{39}$$

Then $\bar{\xi}^p(T)$ is defined as the proj_T of $\bar{e}^T$ onto S^p

$$\bar{\xi}^p(T) \in S^p \tag{40a}$$

$$\langle \bar{\xi}^p(T), \bar{y} \rangle_T = \langle \bar{e}^T, \bar{y} \rangle_T = [\bar{y}]_T \quad \text{for } \bar{y} \in S^p. \tag{40b}$$

(We could characterize $\bar{\xi}^p(T) = \Sigma_{i=0}^p c_i^p(T)\bar{x}^i$ as in (10) and (11), but this is not needed in the development.) Thus, $\bar{\xi}^p(T)$ picks out the Tth coordinate of all vectors $\bar{y}$ in S^p. It is worth observing that $\bar{\xi}^p$ is a measure of the influence of the most recent data point (coordinate axis $\bar{e}^T$) on the pth order model (subspace S^p).

We define the squared magnitude of $\bar{\xi}^p(T)$ to be $\sigma^p(T)$.

$$\sigma^p(T) = \| \bar{\xi}^p(T) \|_T^2. \tag{41}$$

A consistent notation for $\sigma^p(T)$ would be $\epsilon_\xi^p(T)$, but this variable plays a distinguished role, since it is the square of the cosine of the angle between the T-axis and S^p.

An important property of $\sigma^p(T)$ is that

$$0 \leq \sigma^p(T) \leq \sigma^q(T) \leq 1 \qquad \text{for } p \leq q. \tag{42}$$

Relation (42) follows from definition (41) and the nesting of the S^p's: $S^p \subset S^q$ so that the magnitude of the projections $\bar{\xi}^p$ must be an increasing function of p. Also, $[\bar{\xi}^p(T)]_T = \sigma^p(T) = \|\bar{\xi}^p(T)\|_T^2 \geq [\bar{\xi}^p(T)]_T^2$, which implies $[\bar{\xi}^p(T)]_T \leq 1$.

Since $S^{p+1} = S^p \oplus \bar{\beta}^{p+1}(T)$ and the set of vectors $\bar{\beta}^i(T)/\sqrt{\epsilon_\beta^i(T)}$ for $0 \leq i \leq p+1$ form an orthonormal_T basis of S^{p+1}, the proj_T of $\bar{e}^T$ onto S^{p+1} is equal to its projection onto S^p plus the vector $\langle \bar{e}^T, \bar{\beta}^{p+1}(T)/\sqrt{\epsilon_\beta^{p+1}(T)} \rangle_T \bar{\beta}^{p+1}(T)/\sqrt{\epsilon_\beta^{p+1}(T)}$. Thus,

$$\bar{\xi}^{p+1}(T) = \bar{\xi}^p(T) + \frac{\langle \bar{e}^T, \bar{\beta}^{p+1}(T) \rangle_T}{\epsilon_\beta^{p+1}(T)} \bar{\beta}^{p+1}(T) = \bar{\xi}^p(T) + \frac{[\bar{\beta}^{p+1}(T)]_T}{\epsilon_\beta^{p+1}(T)} \bar{\beta}^{p+1}(T) \tag{43a}$$

and

$$\sigma^{p+1}(T) = \sigma^p(T) + \frac{[\bar{\beta}^{p+1}(T)]_T}{\epsilon_\beta^{p+1}(T)} [\bar{\beta}^{p+1}(T)]_T. \tag{43b}$$

This gives order recursions for $\bar{\xi}^p(T)$ and $\sigma^p(T)$.

We now derive the time update recursion for $\bar{\beta}^p$. A useful relationship which follows immediately from definition (4) is

$$\langle \bar{x}, \bar{y} \rangle_T = w\langle \bar{x}, \bar{y} \rangle_{T-1} + [\bar{x}]_T[\bar{y}]_T. \tag{44}$$

Thus,

$$\begin{aligned} \langle \bar{\beta}^p(T-1), \bar{x}^i \rangle_T &= w\langle \bar{\beta}^p(T-1), \bar{x}^i \rangle_{T-1} + [\bar{\beta}^p(T-1)]_T[\bar{x}^i]_T \\ &= [\bar{\beta}^p(T-1)]_T[\bar{x}^i]_T \qquad 0 \leq i \leq p-1 \\ &= [\bar{\beta}^p(T-1)]_T \langle \bar{\xi}^{p-1}(T), \bar{x}^i \rangle_T. \end{aligned} \tag{45}$$

Consider $\bar{\phi} \triangleq \bar{\beta}^p(T-1) - [\bar{\beta}^p(T-1)]_T \bar{\xi}^{p-1}(T)$. Then from (45), $\langle\bar{\phi}, \bar{x}^i\rangle_T = 0$ for $0 \leq i \leq p-1$. Also, $\bar{\phi} - \bar{x}^p \in S^{p-1}$, and it follows that $\bar{\phi}$ satisfies (13a)–(13c), i.e., $\bar{\phi} = \bar{\beta}^p(T)$ and (22b) follows

$$\bar{\beta}^p(T) = \bar{\beta}^p(T-1) - [\bar{\beta}^p(T-1)]_T \bar{\xi}^{p-1}(T). \quad (46)$$

It can be seen from (10) that $\bar{\beta}^p(T-1) - \bar{\beta}^p(T)$ is the difference of the proj_{T-1} and proj_T of $\bar{x}^p$ onto the space S^{p-1}. In other words, $\bar{\beta}^p(T-1) - \bar{\beta}^p(T)$ is proportional to the projection of the T-axis onto S^{p-1}. This projection is precisely $\bar{\xi}^{p-1}(T)$. The coefficient $[\bar{\beta}^p(T-1)]_T$ is the projection of $\bar{\beta}^p(T-1)$ onto the T-axis, and thus acts as a measure of how large a modification to the prediction error is possible through the acquisition of the new data $x(T)$. Finally, we note that the entire term $[\bar{\beta}^p(T-1)]_T \bar{\xi}^{p-1}(T)$ is the projection of $\bar{\beta}^p(T-1)$ onto the T-axis followed by a projection onto S^{p-1}.

A parallel calculation (one must use (8) because of z^{-1}) yields (22a), the time update for $\bar{\alpha}^p$

$$\bar{\alpha}^p(T) = \bar{\alpha}^p(T-1) - [\bar{\alpha}^p(T-1)]_T z^{-1}\bar{\xi}^{p-1}(T-1) \quad (47)$$

and, similarly, (22c) follows.

The time update for k^p easily follows from (33), (35), and (47) by utilizing the identities (8) and (44)

$$\begin{aligned} k^p(T) &= \langle\bar{\alpha}^p(T), \bar{x}^{p+1}\rangle_T \\ &= \langle\bar{\alpha}^p(T-1), \bar{x}^{p+1}\rangle_T \\ &\quad - [\bar{\alpha}^p(T-1)]_T \langle z^{-1}\bar{\xi}^{p-1}(T-1), \bar{x}^{p+1}\rangle_T \\ &= wk^p(T-1) + [\bar{\alpha}^p(T-1)]_T [\bar{x}^{p+1}]_T \\ &\quad - [\bar{\alpha}^p(T-1)]_T \langle\bar{\xi}^{p-1}(T-1), \bar{x}^p\rangle_{T-1}. \end{aligned} \quad (48)$$

It follows from (48) (or simple geometric considerations) that

$$\begin{aligned} \langle\bar{\xi}^{p-1}(T-1), \bar{x}^p\rangle_{T-1} &= \langle\bar{\xi}^p(T-1), \bar{x}^p\rangle_{T-1} \\ &\quad - \frac{[\bar{\beta}^p(T-1)]_{T-1}}{\epsilon_\beta^p(T-1)} \langle\bar{\beta}^p(T-1), \bar{x}^p\rangle_{T-1} \\ &= [\bar{x}^p]_{T-1} - [\bar{\beta}^p(T-1)]_{T-1}. \end{aligned}$$

This, combined with (43), yields

$$k^p(T) = wk^p(T-1) + [\bar{\alpha}^p(T-1)]_T [\bar{\beta}^p(T-1)]_{T-1}. \quad (49)$$

The derivation of (25) for k_α^p is almost identical.

The time updates for $\bar{\xi}^p$, ϵ_α^p, and ϵ_β^p rely on the following two useful relations.

$$\begin{aligned} \langle z^{-1}\bar{\xi}^{p+1}(T-1), \bar{x}^0\rangle_T &= \langle z^{-1}\bar{\xi}^{p-1}(T-1), \bar{x}^0 - \bar{\alpha}^p(T)\rangle_T \\ &= \langle\bar{\xi}^{p-1}(T-1), z(\bar{x}^0 - \bar{\alpha}^p(T))\rangle_{T-1} \\ &= [\bar{x}^0 - \bar{\alpha}^p(T)]_T \\ &= [\bar{x}^0]_T - [\bar{\alpha}^p(T)]_T \end{aligned} \quad (50)$$

and, similarly,

$$\langle\bar{\xi}^{p-1}(T), \bar{x}^p\rangle_T = [\bar{x}^p]_T - [\bar{\beta}^p(T)]_T. \quad (51)$$

We can now calculate $\bar{\xi}^p(T)$ from $\bar{\xi}^p(T-1)$. Let $\bar{\phi} = z^{-1}\bar{\xi}^{p-1}(T-1) + K\bar{\alpha}^p(T)$. Then $\langle\bar{\phi}, \bar{x}^i\rangle_T = [x^i]_T$ for $p \geq i > 0$. For $i = 0$, we determine K by

$$\begin{aligned} [x^0]_T &= \langle\bar{\phi}, \bar{x}^0\rangle_T \\ &= \langle z^{-1}\bar{\xi}^{p-1}(T-1), \bar{x}^0\rangle_T + K\langle\bar{\alpha}^p(T), \bar{x}^0\rangle_T. \end{aligned}$$

Using (50), we have

$$K = \frac{[\bar{\alpha}^p(T)]_T}{\epsilon_\alpha^p(T)}.$$

Thus,

$$\begin{aligned} \bar{\xi}^p(T) &= z^{-1}\bar{\xi}^{p-1}(T-1) + \frac{[\bar{\alpha}^p(T)]_T}{\epsilon_\alpha^p(T)} \bar{\alpha}^p(T) \\ &= z^{-1}\bar{\xi}^p(T-1) - \frac{[\bar{\beta}^p(T-1)]_{T-1}}{\epsilon_\beta^p(T-1)} \\ &\quad \cdot \bar{\beta}^p(T-1) + \frac{[\bar{\alpha}^p(T)]_T}{\epsilon_\alpha^p(T)} \bar{\alpha}^p(T) \end{aligned} \quad (52)$$

where we have used (43).

Also, combining (50) with (47), we have

$$\begin{aligned} \epsilon_\alpha^p(T) &= \langle\bar{\alpha}^p(T), \bar{x}^0\rangle_T \\ &= \langle\bar{\alpha}^p(T-1), \bar{x}^0\rangle_T \\ &\quad - [\bar{\alpha}^p(T-1)]_T \langle z^{-1}\bar{\xi}^{p-1}(T-1), \bar{x}^0\rangle_T \\ &= w\langle\bar{\alpha}^p(T-1), \bar{x}^0\rangle_{T-1} \\ &\quad + [\bar{\alpha}^p(T-1)]_T ([\bar{x}^0]_T - \langle z^{-1}\bar{\xi}^{p-1}(T-1), \bar{x}^0\rangle_T) \\ &= w\epsilon_\alpha^p(T-1) + [\bar{\alpha}^p(T-1)]_T [\bar{\alpha}^p(T)]_T. \end{aligned} \quad (53)$$

Finally, it follows from (51) that

$$\epsilon_\beta^p(T) = w\epsilon_\beta^p(T-1) + [\bar{\beta}^p(T-1)]_T [\bar{\beta}^p(T)]_T. \quad (54)$$

E. A. Least Squares Lattice Algorithm

We preface our development with a very useful lemma, which may be used in deriving various forms of the least squares algorithm.

Lemma:

$$[\bar{\alpha}^p(T)]_T = [\bar{\alpha}^p(T-1)]_T (1 - \sigma^{p-1}(T-1)) \quad (55a)$$

$$[\bar{\beta}^p(T)]_T = [\bar{\beta}^p(T-1)]_T (1 - \sigma^{p-1}(T)) \quad (55b)$$

$$[\bar{\gamma}^p(T)]_T = [\bar{\gamma}^p(T-1)]_T (1 - \sigma^p(T)). \quad (55c)$$

Equations (55) are derived by taking the scalar products of (22a), (22b), and (22c) with $\bar{\xi}^p(T)$. The order update (43b) was also used in the derivation of (55b).

As mentioned in Section III, the output variables we seek are usually $[\bar{\gamma}^p(T-1)]_T$, as in noise canceling, or $[y_e^p(T-1)]_T \triangleq y(T) - [\bar{\gamma}^p(T-1)]_T$, as in equalization [12]. (We address the determination of $g_i^p(T)$ in the next section.) To achieve this, our algorithm need consider only a single component of the vectors $\bar{\alpha}^p$, $\bar{\beta}^p$, and $\bar{\gamma}^p$. More specifically, we choose

$$\alpha^p(T) = [\bar{\alpha}^p(T)]_T \quad (56a)$$

$$\beta^p(T) = [\bar{\beta}^p(T)]_T \quad (56b)$$

$$\gamma^p(T) = [\bar{\gamma}^p(T-1)]_T \quad (56c)$$

where removal of the bar indicates a particular vector component. We remark that relation (55) allows us to alternate between $[\cdot(T-1)]_T$ and $[\cdot(T)]_T$, but, for the applications mentioned, (56c) is the most prudent choice.

The algorithm is now obtained by taking the Tth component of (20a) and (20b) and the $T+1$st of (20c). These, combined with the order updates for ϵ_α, ϵ_β, and σ((21a), (21b), and (43b)), and the time updates for k and k_γ ((24) and (25)), yield the algorithm of [12]

$$\alpha^0(T) = \beta^0(T) = x(T) \quad (57a)$$

$$\epsilon_\alpha^0(T) = \epsilon_\beta^0(T) = w\epsilon_\alpha^0(T-1) + x^2(T) \quad (57b)$$

$$\sigma^{-1}(T) = 0 \quad (57c)$$

$$\gamma^{-1}(T) = y(T). \quad (57d)$$

For $p = 0, \cdots, p_0$

$$k^p(T) = wk^p(T-1) + \frac{\alpha^p(T)\beta^p(T-1)}{1-\sigma^{p-1}(T-1)} \tag{58a}$$

$$\alpha^{p+1}(T) = \alpha^p(T) - \frac{k^p(T)}{\epsilon_\beta^p(T-1)}\beta^p(T-1) \tag{58b}$$

$$\beta^{p+1}(T) = \beta^p(T-1) - \frac{k^p(T)}{\epsilon_\alpha^p(T)}\alpha^p(T) \tag{58c}$$

$$\epsilon_\alpha^{p+1}(T) = \epsilon_\alpha^p(T) - (k^p(T))^2/\epsilon_\beta^p(T-1) \tag{58d}$$

$$\epsilon_\beta^{p+1}(T) = \epsilon_\beta^p(T-1) - (k^p(T))^2/\epsilon_\alpha^p(T) \tag{58e}$$

$$\sigma^p(T) = \sigma^{p-1}(T) + (\beta^p(T))^2/\epsilon_\beta^p(T) \tag{58f}$$

$$y_e^p(T) = y_e^{p-1}(T) + \frac{k_\gamma^p(T-1)}{\epsilon_\beta^p(T-1)}\frac{\beta^p(T)}{1-\sigma^{p-1}(T)} \tag{58g}$$

$$\gamma^p(T) = y(T) - y_e^p(T) \tag{58h}$$

$$k_\gamma^p(T) = wk_\gamma^p(T-1) + \gamma^{p-1}(T)\beta^p(T). \tag{58i}$$

Note that (58g) follows from (19) and (20c). Equation (57b) is a consequence of the definitions of ϵ_α and ϵ_β and (44) and (57a). Aspects of the implementation of the above algorithm, such as initialization, may be found in [12].

V. THE NATURAL ISOMORPHISM

It is sometimes desirable to obtain the filter coefficients $g_i^p(T)$ (as in the fast Kalman algorithm [7], [8]). This may be achieved via a natural isomorphism which is the analog of that implicit in [3]. The only difference is that here, the inner product is defined with respect to the estimated correlation matrix, rather than the true correlation matrix.

More precisely, consider the set of polynomials in z^{-1} of the form

$$A(z^{-1}) = \sum_{i=0}^{\infty} a_i z^{-i} \tag{59}$$

where it is assumed that A has finite degree, i.e., $a_i = 0$ for $i > p$ for some $p < \infty$. Define a family of inner products $(\cdot, \cdot)_T$ on these polynomials by

$$(A(z^{-1}), B(z^{-1}))_T = \sum_{i,j} a_i R_{ij}(T) b_j \tag{60}$$

where

$$R_{ij}(T) = \sum_{t=0}^{T} w^{T-t} x(t-i)x(t-j). \tag{61}$$

Note that the conditions of the Appendix insure that $(\cdot, \cdot)_T$ is nonsingular on all polynomials of degree less than p_0.

$A(z)$ may be thought of as an operator on the vector space of (1), and the mapping $A(z^{-1}) \to A(z^{-1})\bar{x}$ maps the polynomials of degree p onto S^p. It is easy to see that this mapping is a metric space isomorphism (for $p \leq p_0$).

$$(A(z^{-1}), B(z^{-1}))_T = \langle A(z^{-1})\bar{x}, B(z^{-1})\bar{x}\rangle_T. \tag{62}$$

It is a simple matter to translate the recursions of the previous sections into recursions of polynomial coefficients via the inverse isometry. For example, let $\xi^p(T) = \Sigma_{i=0}^p c_i^p(T)\bar{x}^i$. Since the $\bar{x}^i$ form a basis for S^p, by matching the coefficients of $\bar{x}^i$ on both sides of (22b), we obtain

$$b_i^p(T) = b_i^p(T-1) - [\bar{\beta}^p(T-1)]_T c_i^{p-1}(T-1) \qquad 0 \leq i \leq p-1. \tag{63}$$

Similarly, since $\{\bar{y}, \bar{x}^i, i = 1, \cdots, p\}$ form a basis for $S^p \oplus y$, (22c) yields the filter coefficients

$$g_i^p(T) = g_i^p(T-1) - [\gamma^{p-1}(T-1)]_T c_i^p(T). \tag{64}$$

This approach, the development of the algorithm through the polynomial coefficients (a_i^p, b_i^p, c_i^p, g_i^p), is most practical if the order is fixed at some value $p = p_0$, and the time updates are used [8]. In this case, the computational burden is $0(p_0)$ per time update. If g_i^p for $0 \leq p \leq p_0$, and $0 \leq i \leq p$ is required, there are $p_0^2/2$ variables to evaluate, and the number of computations becomes $0(p_0^2)$. Note also that, for fixed $p = p_0$, the expressions such as $[\beta^{p_0}(T-1)]_T$ which appear throughout [as in (63)] may be computed directly from their definitions in $0(p_0)$ operations. For example, $[\beta^{p_0}(T-1)]_T = x(T-p_0) - \Sigma_{i=0}^{p_0-1} b_i^{p_0}(T-1)\,x(T-i)$.

VI. CONCLUDING REMARKS

In this paper we defined a class of metric spaces which resulted in a geometrical derivation of least squares lattice algorithms. This inner-product formalism is closely related to that found in [3], but involves the actual input data rather than its second-order statistical properties. This is reflected in the use of a set of metrics indexed by time to replace the expected value. Note that such an approach lends itself naturally to a time-recursive formulation. (We remark that, while this paper was under review, similar results, independently developed, have appeared in [13].)

An advantage of the above structure is that it provides a single framework which encompasses an entire family of fairly complex algorithms [7]-[14]. Its geometrical nature also provides a guide for the intuition, which should be of use in implementation and future investigations. The extension to multichannel inputs and ARMA modeling is straightforward. Such a treatment may be found in [14]. Finally, we wish to point out that there remain many questions which were beyond the scope of this paper. Among these are stability properties, the role of w for $w < 1$, the significance of the scalar gains K and σ, and the spectral properties of the lattice decomposition. Some of the present techniques may also generalize to the case treated in [15].

APPENDIX

INDEPENDENCE ASSUMPTION

The metric $\langle\rangle_T$ is singular on $S^p \Leftrightarrow \exists\, r_i$ not all zero and $\bar{y} = \Sigma_{i=0}^p r_i\bar{x}^i$ such that $\|\bar{y}\|_T^2 = 0 \Leftrightarrow \Sigma_{t=0}^T w^{T-t}[\bar{y}]_t^2 = 0 \Leftrightarrow [\bar{y}]_t = 0$ for $t \leq T \Leftrightarrow \Sigma_{i=0}^p r_i\bar{x}^{i'} = 0 \Leftrightarrow$ the $\bar{x}^{i'}$ are linearly dependent. Note that $\bar{x}^{i'}$ are defined in (5).

Note also $\langle\rangle_T$ is singular $\Leftrightarrow \exists\, r_i$ not all zero such that $\langle \Sigma_{i=0}^p r_i\bar{x}^i, x^j\rangle_T = 0$ for $0 \leq j \leq p \Leftrightarrow$ the matrix $R^p(T)$ defined by (60) with $i, j \leq p$ satisfies $\Sigma_{i=0}^p R_{ij}^p(T)\, r_i = 0 \Leftrightarrow R^p(T)$ is singular.

The independence condition on the vectors $\bar{x}^{i'}$ of (5) is thus a mixing condition. It says that by the time T all the modes of the estimated correlation matrix R^{p_0} must be excited, i.e., $R^{p_0}(T)$ is nonsingular.

ACKNOWLEDGMENT

The author wishes to thank E. Satorius for his always willing ear and numerous valuable suggestions.

REFERENCES

[1] K. Astrom and P. Eykhoff, "System identification—A survey," *Automatica*, vol. 7, pp. 123–162, 1971.

[2] F. Itakura and S. Saito, "Digital filtering techniques for speech analysis and systems," in *Proc. 7th Int. Conf.*, Budapest, 1971, paper 25-C-1, pp. 261–264.

[3] J. Markel and A. H. Gray, Jr., "On autocorrelation equations as applied to speech analysis," *IEEE Trans. Audio Electroacoust.*, vol. AU-21, pp. 69–79, Apr. 1973.

[4] M. Morf, "Fast algorithms for multivariable systems," Ph.D. dissertation, Stanford Univ., Stanford, CA, 1974.

[5] J. Markel and A. H. Gray, Jr., *Linear Prediction of Speech*. Berlin: Springer-Verlag, 1976.

[6] J. Makhoul, "Stable and efficient lattice methods for linear prediction," *IEEE Trans. Acoust., Speech, Signal Processing*, vol. ASSP-25, pp. 423–428, Oct. 1977.

[7] L. Ljung, M. Morf, and D. Falconer, "Fast calculation of gain matrices for recursive estimation schemes," *Int. J. Contr.*, vol. 27, no. 1, pp. 1–19, 1978.

[8] D. D. Falconer and L. Ljung, "Application of fast Kalman estimation to adaptive equalization," *IEEE Trans. Commun.*, vol. COM-26, pp. 1439–1446, Oct. 1978.

[9] M. Morf and D. Lee, "Fast algorithms for speech modeling," Defense Commun. Agency, Inform. Systems Lab., Stanford Univ., Stanford, CA, Final Rep., Contract DCA 100-77-C-0005, Nov. 1978.

[10] ——, "Recursive least squares ladder forms for fast parameter tracking," in *Proc. 1978 IEEE Conf. Decision Contr.*, San Diego, CA, Jan. 1979, pp. 1362–1367.

[11] E. H. Satorius and J. D. Pack, "Application of least squares lattice algorithms to adaptive equalization," *IEEE Trans. Commun.*, vol. COM-29, pp. 136–142, Feb. 1981.

[12] E. H. Satorius and M. J. Shensa, "On the application of recursive least squares methods to adaptive processing," presented at the Int. Workshop Applications of Adaptive Contr., Yale Univ., New Haven, CT, Aug. 1979.

[13] D. Lee and M. Morf, "Recursive square-root ladder estimation algorithms," presented at the IEEE Conf. Acoust., Speech, Signal Processing, Denver, CO, Apr. 1980.

[14] M. J. Shensa, "A least squares lattice decision feedback equalizer," presented at the IEEE Int. Conf. Commun., Seattle, WA, June 1980.

[15] B. Friedlander, M. Morf, T. Kailath, and L. Ljung, "New inversion formulas for matrices classified in terms of their distance from Toeplitz matrices," to be published.

Adaptive Tracking of Multiple Sinusoids Whose Power Levels are Widely Separated

WILLIAM S. HODGKISS, JR., MEMBER, IEEE, AND JOE A. PRESLEY, JR., STUDENT MEMBER, IEEE

***Abstract*—The behavior of the gradient transversal filter (LMS), the gradient lattice (GL), and the least squares lattice (LSL) when used to track multiple sinusoidal (or narrow-band) components whose power levels are widely separated is investigated. These approaches to the realization of the *p*th-order one-step linear predictor of the time series are recursive in time. The lags of the instantaneous frequency estimates from their actual underlying values are of particular interest. The frequency tracking characteristics of the LMS, GL, and LSL algorithms are illustrated in several situations. Included are simulations of dual sinusoids undergoing a variety of frequency versus time dynamics and a formant tracking example of real speech.**

I. INTRODUCTION

THE TRACKING of a single sinusoid whose frequency varies as a function of time has been treated traditionally in the context of phase-locked loops [1] or as a problem in extended Kalman filtering [2]. These approaches require an accurate initializing estimate of the sinusoid's frequency for the purpose of signal capture. Furthermore, the tracking of multiple sinusoids entails multiple implementations of the algorithms.

Numerous applications of frequency tracking inherently are characterized by imprecise initialization information due to abrupt onset, rapid transitions, uncertainty in the exact number of sinusoids to be tracked, and a wide disparity in their relative power levels. In the area of speech processing, for example, a great deal of emphasis has been placed on parameterizing the time-evolving spectral characteristics of the speech waveform. In particular, tracking of the multiple spectral peaks (formants) has been of interest. As opposed to the approaches mentioned above, the most common analysis techniques here have been the autocorrelation and covariance methods of linear prediction in which the observed signal is modeled as an autoregressive (all-pole) process [3]–[5]. As typically implemented, these are block data structured approaches which derive high resolution spectral estimates by creating a whitening or inverse filter for the available data block whose transfer function then is inverted [6].

Instead of block processing time series data in the method of linear prediction, an inverse filter can be implemented as a continuously updated adaptive transversal (all-zero) filter. Griffiths [7] used this approach to track the rapidly changing digital instantaneous frequency of various frequency modulated (FM) waveforms buried in noise. The resulting structure is appealing due to the unbroken flow of data through the filter and its natural exponentially decaying memory. The major drawback in the transversal filter steepest descent approach to adaptive filter synthesis is

Manuscript received April 17, 1980; revised November 13, 1980. This work was supported in part by the Naval Sea Systems Command, Code 03421, under subcontract to the Applied Research Laboratory, Pennsylvania State University, University Park, PA, and by the Fannie and John Hertz Foundation.

The authors are with the Marine Physical Laboratory, University of California, San Diego, CA 92152.

Reprinted from *IEEE Trans. Circuit Syst.*, vol. CAS-28, no. 6, pp. 550–561, June 1981.

that the convergence characteristics of the filter are determined by the eigenvalue spread of the input time series [8], [9]. In a frequency tracking application where the multiple sinusoids are widely separated in power, this convergence behavior will manifest itself in each sinusoid being tracked with a different lag from its actual underlying instantaneous frequency. The lag is inversely related to the sinusoid's power.

The method of least squares linear prediction leads to the requirement of solving a set of normal equations. An outgrowth from their computationally efficient, order recursive solution in the autocorrelation method has been the equivalent implementation of the inverse filter in the form of a lattice structure [4], [5], [10], [11]. This particular structure is attractive since it provides a Gram–Schmidt type of orthogonalization of the input time series as a natural by-product of the stage-by-stage creation of the filter. Such an inherently orthogonal form suggests that adaptive realizations of the lattice should have a high insensitivity to eigenvalue spread [11]–[18]. Experimental evidence in the context of adaptive channel equalization has supported this conjecture [19]. All noisy gradient implementations of the adaptive lattice are based on an instantaneous approximation to the local satisfaction of a minimum error power optimality criterion. Assuming stationarity of the input random process, local satisfaction of this optimality criterion implies global optimality of the filter as a linear predictor.

The lattice structures proposed by Morf *et al.* [20]–[23] have been derived in a significantly different manner than the adaptive forms previously cited in that they exactly satisfy a global least squares optimality criterion at every point in time. Performance results from an adaptive channel equalization experiment have indicated improved convergence characteristics over other gradient lattice implementations [24].

The intent of this paper is to focus on the behavior of the gradient transversal filter (LMS), the gradient lattice (GL), and the least squares lattice (LSL) when used to track multiple sinusoids whose power levels are widely separated. Of particular interest will be the lags of the instantaneous frequency estimates from their actual underlying values. The paper is organized as follows: Section II will lay the groundwork for spectral estimation via autoregressive (AR) modeling and will consider briefly the degradations encountered when the actual process is autoregressive-moving average (ARMA). Next, the tracking of time-evolving spectra will be introduced in Section III. The LMS, GL, and LSL algorithms whose tracking characteristics are to be compared in the sequel are summarized. Section IV illustrates their behavior in a variety of situations. Included are simulations of dual sinusoids whose power levels are significantly different. Three types of frequency versus time dynamics are considered: 1) coincident steps, 2) in-phase sinusoidal FM, and 3) crossing ramps. In addition, a formant tracking example of the phrase "we were away" is included and compared with a block data structured analysis [25]. Lastly, Section V concludes the paper with a summary of the major results.

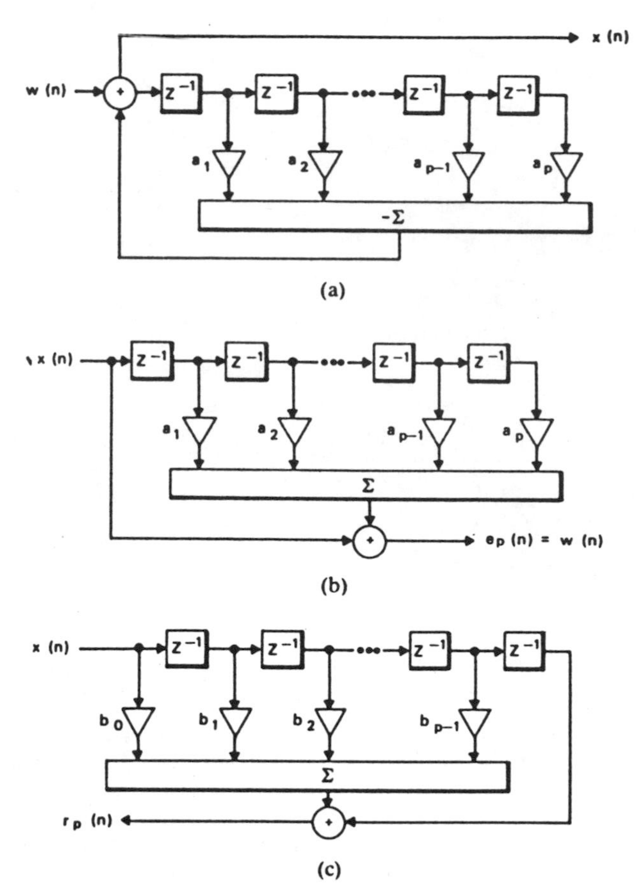

Fig. 1. (a) AR process generation model. (b) Corresponding whitening or inverse filter (one-step forward prediction error filter). (c) One-step backward prediction error filter.

II. Spectral Estimation

In recent years, a great deal of interest has been shown in high resolution spectral estimation techniques [6]. This interest typically has been motivated by the desire to resolve narrow-band frequency components in data records too short for adequate frequency separation via standard fast Fourier transform (FFT) techniques. By incorporating into the estimation problem assumptions about how the observed data was generated, rather remarkable results have been obtained.

For a number of physical reasons, modeling the observed data as an AR process has been accepted widely in speech processing and geophysical time series analysis. As portrayed in Fig. 1(a), an AR process of order p is obtained by passing the white noise sequence $w(n)$ having zero mean and variance σ_w^2 through a p pole filter. The corresponding input–output relationships are

$$x(n) = -a_1 x(n-1) - a_2 x(n-2) - \cdots - a_p x(n-p) + w(n)$$
$$= -\sum_{k=1}^{p} a_k x(n-k) + w(n) \quad (1)$$

where $w(n)$ is called the innovation of the process. By evaluating the all-pole filter's z-transform on the unit circle, the power spectrum of the AR process $x(n)$ is obtained as

$$S_x(\omega)=\left.\frac{\sigma_w^2}{|A(z)|^2}\right|_{z=e^{j\omega}}$$

$$=\frac{\sigma_w^2}{\left|1+\sum_{k=1}^{p} a_k \exp(-j\omega k)\right|^2}. \qquad (2)$$

Notice that the all-zero filter of order p which will recover $w(n)$ from $x(n)$ is $A(z)$. Appropriately, this filter has been called the whitening or inverse filter for the AR process and is illustrated in Fig. 1(b). A closer examination of Fig. 1(b) reveals that the structure is in the form of a one-step forward linear predictor. The most recent p samples of the AR process $\{x(n-1), x(n-2),\cdots,x(n-p)\}$ are linearly combined to form an estimate of $-x(n)$. Removing the predictable components from $x(n)$ (or, correspondingly, the coloring from $S_x(\omega)$) yields the white forward prediction error sequence $e_p(n)=w(n)$. A companion to Fig. 1(b) which will be needed for the discussion in Section III-C is the one-step backward linear predictor (coefficients $b_0, b_1,\cdots,b_{p-1}$) shown in Fig. 1(c) along with the backward prediction error sequence $r_p(n)$.

Now, given that the time series of interest is pth-order AR, the spectral estimation task becomes equivalently a problem of determining the pth-order linear predictor, $A(z)$, along with the power, σ_w^2, of its prediction error output sequence. This problem has been formulated both statistically (as a Wiener filtering problem) and deterministically using a minimum squared error optimality criterion [3], [4]. In the area of speech processing, the deterministic formulations are known as the autocorrelation and covariance methods of linear prediction. They are summarized below.

With reference to Fig. 1(b), the total squared error is expressed as

$$E_p^e=\sum_n e_p^2(n) \qquad (3)$$

where the interval over which the summation is carried out will be left unspecified for the moment. By substituting the expression for $e_p(n)$

$$e_p(n)=\sum_{k=0}^{p} a_k x(n-k), \qquad a_0=1 \qquad (4)$$

and defining

$$c_{ik}=\sum_n x(n-i)x(n-k) \qquad (5)$$

the total squared error can be written equivalently as

$$E_p^e=\sum_{i=0}^{p}\sum_{k=0}^{p} a_i c_{ik} a_k. \qquad (6)$$

The minimization of E_p^e with respect to the filter coefficients is carried out by setting $\partial E_p^e/\partial a_i=0$ ($i=1,2,\cdots,p$) thus yielding the normal equations

$$\sum_{k=1}^{p} c_{ki}a_k=-c_{0i}, \qquad i=1,2,\cdots,p. \qquad (7)$$

Substitution of (7) into (6) provides an expression for the total squared prediction error

$$E_p^e=c_{00}+\sum_{k=1}^{p} c_{k0}a_k. \qquad (8)$$

Notice that (7) and (8) can be combined in a single expression

$$\sum_{k=0}^{p} c_{ki}a_k=E_p^e\delta_{i0}, \qquad a_0=1 \text{ and } i=0,1,\cdots,p \qquad (9)$$

where $\delta_{i0}=1$ when $i=0$ and $\delta_{i0}=0$ for $i=1,\cdots,p$.

The distinction between the autocorrelation and covariance methods of linear prediction now will be made. Suppose the time series is observed only over the interval $n=0,1,\cdots,N-1$. In the autocorrelation method $x(n)$ is assumed identically zero outside this interval (i.e., equivalent to the application of an N-point rectangular window to the time series) and the lower and upper summation limits in (3) and (5) are set to $-\infty$ and ∞. In contrast, no assumption about the nature of $x(n)$ is made outside the observation interval in the covariance method, thus forcing the lower and upper summation limits in (3) and (5) to be p and $N-1$.

The impact of the assumed nature of $x(n)$ is seen in the structure of the matrix whose elements in (5) are c_{ik}. In the autocorrelation method, this matrix is both symmetric and Toeplitz. Thus $c_{ik}=c(|k-i|)$ and the $p+1$ unknowns in (9) ($a_1, a_2,\cdots,a_p$, and E_p^e) can be determined in a computationally efficient, order recursive fashion [26]–[28]:

Initialization

$$E_0^e=c(0). \qquad (10a)$$

Order update ($i=1,2,\cdots,p$)

$$K_i=\left\{-\sum_{k=0}^{i-1} a_k^{(i-1)}c(i-k)\right\}\Big/E_{i-1}^e, \qquad a_0=1 \quad (10b)$$

$$a_i^{(i)}=K_i \qquad (10c)$$

$$a_k^{(i)}=a_k^{(i-1)}+K_i a_{i-k}^{(i-1)}, \qquad 1\leqslant k\leqslant i-1 \qquad (10d)$$

$$E_i^e=(1-K_i^2)E_{i-1}^e. \qquad (10e)$$

In (10b)–(10d), the superscript (i) indicates the ith-order linear predictor.

Although still symmetric, the matrix is no longer Toeplitz in the covariance method and (9) typically has been solved using the more computationally expensive square root or Cholesky decomposition [29]. Recently, an algorithm whose structure is similar to that in (10) has been introduced for use in the covariance method [30].

An outgrowth from the order recursive solution of the normal equations in the autocorrelation method of linear prediction has been the implementation of the all-zero inverse filter $A(z)$ in the form of a lattice structure [4], [5],

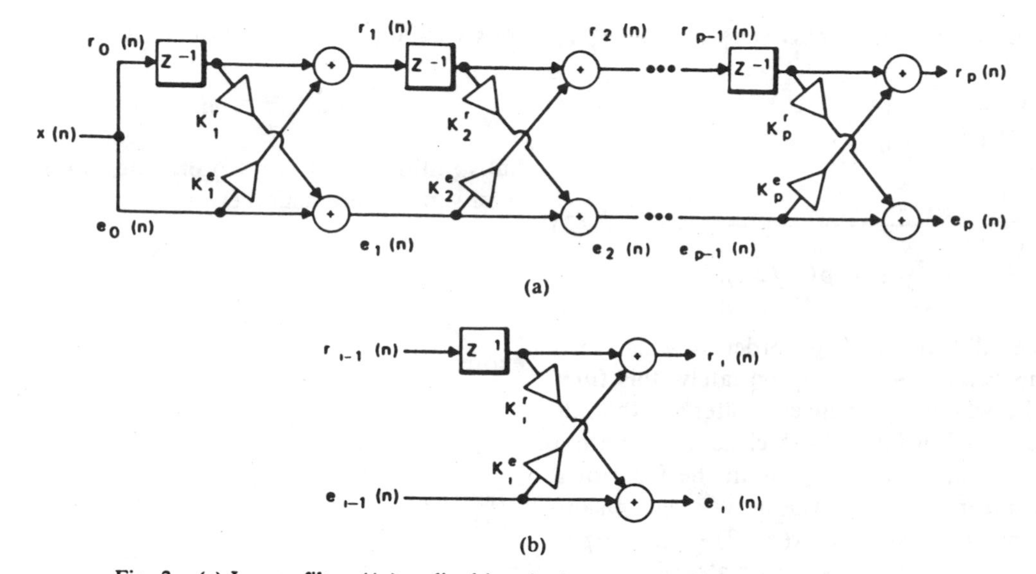

Fig. 2. (a) Inverse filter $A(z)$ realized in a lattice structure. (b) The ith stage of the lattice.

[10], [11]. Shown in Fig. 2, the lattice parameters $K_i^e = K_i^r = K_i$ are known either as reflection coefficients (based on their correspondence to physical parameters of the AR process generation model) or partial correlation coefficients (based on their statistical interpretation as correlation coefficients). They occur naturally in (10) as intermediate parameters in the solution algorithm. The $e_i(n)$ and $r_i(n)$ are the ith-order forward and backward prediction error sequences, respectively. Thus the pth-order linear predictor is created on a stage-by-stage basis a single order at a time. Of particular importance is that the backward prediction error sequences $r_i(n)$ are orthogonal.

The following relationships exist between the direct and lattice forms of the inverse filter $A(z)$ where i denotes filter order:

Lattice→Direct (iterate upward $i=1,2,\cdots,p$)

$$a_i^{(i)} = K_i \tag{10c}$$

$$a_k^{(i)} = a_k^{(i-1)} + K_i a_{i-k}^{(i-1)}, \qquad 1 \leq k \leq i-1. \tag{10d}$$

Direct→Lattice (iterate downward $i=p, p-1,\cdots,1$)

$$K_i = a_i^{(i)} \tag{11a}$$

$$a_k^{(i-1)} = \frac{a_k^{(i)} - K_i a_{i-k}^{(i)}}{1-K_i^2}, \qquad 1 \leq k \leq i-1. \tag{11b}$$

Up to this point, the tacit assumption has been made that the time series of interest in the frequency tracking problem are adequately modeled as AR of order p (at least over a short segment of time). The selection of a particular generation model for the observed process has a significant impact on the resulting spectral estimate [31]. For example, even if the process of interest is AR, when it is buried in an additive white background noise, the observed time series actually is ARMA [32]–[35]. For the limiting case of a single real sinusoid buried in white noise, the predominant effect on AR analysis techniques has been a nonradial trajectory of the zeros of the inverse filter from the unit circle inward as signal-to-noise ratio (SNR) is reduced [36]–[40]. This bias in the resulting frequency estimate has been shown experimentally to be substantially reduced or eliminated when the AR analysis is formulated for complex rather than real signals [41, [42]. The tracking behavior experiments which will be presented in Section IV were conducted on real signals, but at SNR's sufficiently large so that bias in the frequency estimates was negligible.

Central to the frequency tracking problem is a nonstationary observation process whose time-evolving spectral characteristics are of interest. As formulated in (3)–(9), the autocorrelation and covariance methods of linear prediction inherently are block data structure approaches. A more appealing solution to AR analysis would be one where the inverse filter in either direct (Fig. 1(b)) or lattice (Fig. 2) form was continuously updated with a memory which decays exponentially. The three algorithms summarized in Section III are such recursive in time approaches to the linear prediction problem.

III. Adaptive Modeling

Three continuously adapting approaches to the realization of the time varying all-zero inverse filter $A(z)$ are summarized below: 1) the gradient transversal filter (LMS), 2) the gradient lattice (GL), and 3) the least squares lattice (LSL). Each is based on a recursive in time solution to the linear prediction problem. Their behavior when used to track multiple sinusoids whose power levels are widely separated will be compared in Section IV.

A. Gradient Transversal Filter (LMS)

The statistical formulation of the linear prediction problem employing a minimum mean-square error optimality criterion leads to a set of equations identical in form to (9) where temporal averages are replaced by statistical expectations. A key result from Wiener filtering theory is that the error sequence $e_p(n)$ must be orthogonal to the data

$\{x(n-1), x(n-2), \cdots, x(n-p)\}$. The gradient transversal filter utilizes a coefficient up-date algorithm which attempts to implement this orthogonality condition directly.

The gradient transversal filter algorithm is summarized as follows (see Fig. 1(b)):

Initialization $(k=1,2,\cdots,p)$

$$a_k(-1)=0 \tag{12a}$$

$$x(-k)=x(-k-1)=0 \tag{12b}$$

$$e_p(-1)=0. \tag{12c}$$

Time update $(n\geqslant 0$ and $k=1,2,\cdots,p)$

$$a_k(n)=a_k(n-1)-2\mu e_p(n-1)x(n-k-1) \tag{12d}$$

$$e_p(n)=\sum_{k=0}^{p} a_k(n)x(n-k). \tag{12e}$$

In (12d), $2e_p(n-1)x(n-k-1)=\partial e_p^2(n-1)/\partial a_k$, an instantaneous estimate of the gradient of the minimum mean-square error performance surface along the a_k coordinate. The algorithm follows a steepest descent path as it iteratively approaches satisfaction of the optimality criterion. The gradient step size parameter μ, common to all filter coefficients, governs the rate at which the $a_k(n)$ adapt. Given that the time series, $x(n)$, is stationary, convergence in the mean of the $a_k(n)$ to their optimal values is guaranteed under relatively mild assumptions when

$$0<\mu<1/\lambda_{\max}. \tag{13}$$

In (13), $\lambda_{\max}$ is the maximum eigenvalue of the $p\times p$ data autocorrelation matrix whose components are given by (5) with the summation replaced by statistical expectation. Also, given stationarity, $c_{ik}=c(|k-i|)$ and the matrix is both symmetric and Toeplitz. The variances of the $a_k(n)$ about their means converge to values which are proportional to μ.

For the sequel, it will be convenient to introduce a new parameter α_{LMS} such that

$$\mu=\frac{\alpha_{\text{LMS}}}{p\cdot E_0^e} \tag{14}$$

where E_0^e is the power of the sequence, $x(n)$. Thus (13) becomes

$$0<\alpha_{\text{LMS}}<\frac{p\cdot E_0^e}{\lambda_{\max}}. \tag{15}$$

Since

$$p\cdot E_0^e=\sum_{i=1}^{p} c_{ii}=\sum_{i=1}^{p}\lambda_i\geqslant\lambda_{\max}$$

(15) will be satisfied when $0<\alpha_{\text{LMS}}<1$. More extensive discussions related to the derivation of (12) and its convergence properties can be found in [8], [9], [36], [37].

From a practical standpoint, E_0^e is replaced by the estimate, $\hat{E}_0^e(n)$, which is generated recursively by

$$\hat{E}_0^e(n)=(1-\alpha_{\text{LMS}})\hat{E}_0^e(n-1)+\alpha_{\text{LMS}}x^2(n) \tag{16}$$

with initialization $\hat{E}_0^e(-1)=\epsilon_{\text{LMS}}$, $\epsilon_{\text{LMS}}=E_0^e(0)$. Thus (14) becomes

$$\mu(n)=\frac{\alpha_{\text{LMS}}}{p\cdot\hat{E}_0^e(n)} \tag{17}$$

and the gradient transversal filter is provided the capability of adapting to power level fluctuations in the time series, $x(n)$. Note that the time constant of the recursive power estimator in (16) need not necessarily be tied to α_{LMS}.

The major drawback in the utilization of gradient transversal filters is that their convergence characteristics are determined by the eigenvalue spread of the input time series. As shown in (15), $\lambda_{\max}$ determines the largest permissible value of the gradient step size parameter. The system formed by the $a_k(n)$ can be broken into its natural modes by a decomposition based on the eigenvectors of the $p\times p$ data autocorrelation matrix. Each mode converges with a time constant inversely proportional to $\mu\lambda_i=\alpha_{\text{LMS}}\lambda_i/pE_0^e$. Thus the small eigenvalue modes govern the total time to convergence. By making a rough correspondence between the power of individual sinusoids in the input time series and eigenvalues of the $p\times p$ data autocorrelation matrix, it is seen that the gradient transversal filter will adapt to each sinusoidal component at a rate inversely proportional to its power [37].

Griffiths [7] was first to use the gradient transversal filter to track the rapidly changing digital instantaneous frequency of various waveforms buried in noise. The tracking of multiple sinusoids was demonstrated in that paper, but not for widely separated power levels. Subsequent work by a number of authors has dealt with potential applications [39], [43]–[46] and performance comparisons with other approaches to the frequency tracking problem (FM discriminator [47], MAP demodulation [48], and other AR-based spectral estimation techniques along with the conventional FFT [49]).

B. Gradient Lattice (GL)

The all-zero inverse filter $A(z)$ can be realized in direct form (Fig. 1(b)) or as a lattice structure (Fig. 2). When the input time series $x(n)$ is stationary, $K_i^e=K_i^r=K_i$ and the relationship between the K_i and the direct form filter coefficients a_k is summarized in (10c)–(10d) and (11). The $e_i(n)$ and $r_i(n)$ are the ith-order forward and backward prediction error sequences, respectively. Again drawing on a key result from Wiener filtering theory, the sequence pair $e_i(n)$ and $r_{i-1}(n-1)$ as well as the sequence pair $r_i(n)$ and $e_{i-1}(n)$ are orthogonal. Thus assuming stationarity, local satisfaction of the orthogonality principle on a stage-by-stage basis leads to global optimality of the entire structure as a pth-order linear predictor.

In the nonstationary situation actually of interest, the lattice structure in Fig. 2 with $K_i^e=K_i^r=K_i$ still will be assumed appropriate. The lattice coefficients are up-dated by a gradient steepest descent algorithm which attempts to locally satisfy the orthogonality condition between the two sequence pairs previously mentioned.

The GL algorithm is summarized as follows (see Fig. 2):

Initialization $(i=0,1,\cdots,p)$

$$K_i(-1)=0, \qquad i\neq 0 \tag{18a}$$

$$\hat{E}_i(-1)=\epsilon_{GL}, \qquad \epsilon_{GL}=2E_0^e(0) \text{ and } i\neq p \tag{18b}$$

$$e_i(-1)=r_i(-1)=0 \tag{18c}$$

$$r_i(-2)=0, \qquad i\neq p. \tag{18d}$$

Time update $(n\geq 0)$

$$e_0(n)=r_0(n)=x(n). \tag{18e}$$

Order update $(i=1,2,\cdots,p)$

$$K_i(n)=K_i(n-1)-\frac{2\alpha_{GL}}{\hat{E}_{i-1}(n-1)} \cdot\{e_i(n-1)r_{i-1}(n-2)+r_i(n-1)e_{i-1}(n-1)\} \tag{18f}$$

$$e_i(n)=e_{i-1}(n)+K_i(n)r_{i-1}(n-1) \tag{18g}$$

$$r_i(n)=r_{i-1}(n-1)+K_i(n)e_{i-1}(n) \tag{18h}$$

$$\hat{E}_{i-1}(n)=(1-\alpha_{GL})\hat{E}_{i-1}(n-1) +\alpha_{GL}\{e_{i-1}^2(n-1)+r_{i-1}^2(n-2)\}. \tag{18i}$$

The relationships in (10c)–(10d) are used to derive the direct form coefficients $a_k(n)$ from the lattice parameters $K_i(n)$ in (18f).

Notice that $\hat{E}_{i-1}(n)$ is simply an estimate of the total input power to the ith-stage of the lattice. Assuming this estimate is equal to the actual power $E_{i-1}(n)$, the coefficient $K_i(n)$ will converge in the mean approximately to its optimal value provided [14].

$$0<\alpha_{GL}<1. \tag{19}$$

The variance of $K_i(n)$ about its mean converges to a value which is proportional to α_{GL}.

The natural stage-by-stage decoupling which occurs in the lattice under stationary conditions suggests the adaptive realization in (16) should have a high insensitivity to eigenvalue spread [11]–[13], [15]–[18]. This can be understood intuitively as follows. Notice that the adaptive computation of the lattice coefficient at any particular stage does not depend on the computation of $K_i(n)$ at succeeding stages. Unlike the gradient transversal filter where the pth-order prediction error residual $e_p(n)$ is coupled back into the update expression (12d) for each coefficient $a_k(n)$, in the GL the ith-stage prediction error residuals required in (18f) are not functions of the $K_i(n)$ in succeeding stages as can be seen from Fig. (2). Essentially, convergence of the lattice occurs progressively on a stage-by-stage basis. The time constants at each stage are approximately the same which also is in contrast to the individual mode behavior in the LMS filter. Unlike (12d) where the gradient step size parameter is constant for each $a_k(n)$, (18f) and (18i) indicate that the gradient step size parameter is not the same at each stage of the adaptive lattice, but is inversely proportional to the estimate of power entering that stage. And, as discussed previously, the convergence rate of the steepest descent algorithm is determined by the product of a power quantity and the gradient step size parameter.

Only preliminary analytical results which investigate the detailed convergence behavior of the adaptive lattice are available [14]. However, experimental evidence in the context of adaptive channel equalization has supported the conjecture of a high insensitivity to eigenvalue spread [19].

C. *Least Squares Lattice (LSL)*

The adaptive lattice given in (18) presupposes a certain structure for the linear predictor ($K_i^e=K_i^r=K_i$) and utilizes a gradient descent approach in an attempt to locally satisfy a minimum power optimality criterion by a direct implementation of the orthogonality principle. Although intuitively appealing it is not clear that such an implementation of the inverse filter will be globally optimal in a least squares sense when operating in a nonstationary environment. A significantly different approach has been taken by Morf *et al.* [20]–[23]. Their structures have been allowed to evolve out of the mathematics of the temporal and order recursive solution to the globally optimal pth-order linear predictor. The lattice form of Fig. 2 still appears in a natural fashion, but, in general, $K_i^e\neq K_i^r$.

At every point in time N, the LSL solves for the pth-order linear predictor by minimizing the following modification of (3):

$$E_p^e(N)=\sum_{n=0}^{N}(1-\alpha_{LSL})^{N-n}e_p^2(n). \tag{20}$$

The fade factor, $(1-\alpha_{LSL})$, enables the linear predictor to adapt to a nonstationary environment by weighting recent errors more heavily than those which occurred in the distant past. Fig. 3 shows the values of $(1-\alpha_{LSL})$ for which the error of $n=0$ has 10 and 30 percent weighting relative to the error at $n=N$. The expression in (20) defines the prewindowed case where $x(n)$ is assumed identically zero for $n<0$ and no assumptions about the data are made for $n>N$.

The LSL algorithm is summarized as follows (see Fig. 2):

Initialization $(i=0,1,\cdots,p)$

$$r_i(-1)=0, \qquad i\neq p \tag{21a}$$

$$E_i^r(-1)=\epsilon_{LSL}, \qquad \epsilon_{LSL}=0.001 \text{ and } i\neq p \tag{21b}$$

$$\Delta_i(-1)=0, \qquad i\neq 0 \tag{21c}$$

$$b_k^{(i)}(-1)=0, \qquad 0\leq k\leq i-1, \qquad i\neq 0, \text{ and } i\neq p \tag{21d}$$

Time update $(N\geq 0)$

$$e_0(N)=r_0(N)=x(N) \tag{21e}$$

$$E_0^e(N)=E_0^r(N)=(1-\alpha_{LSL})E_0^r(N-1)+x^2(N) \tag{21f}$$

$$\gamma_{-1}(N-1)=0. \tag{21g}$$

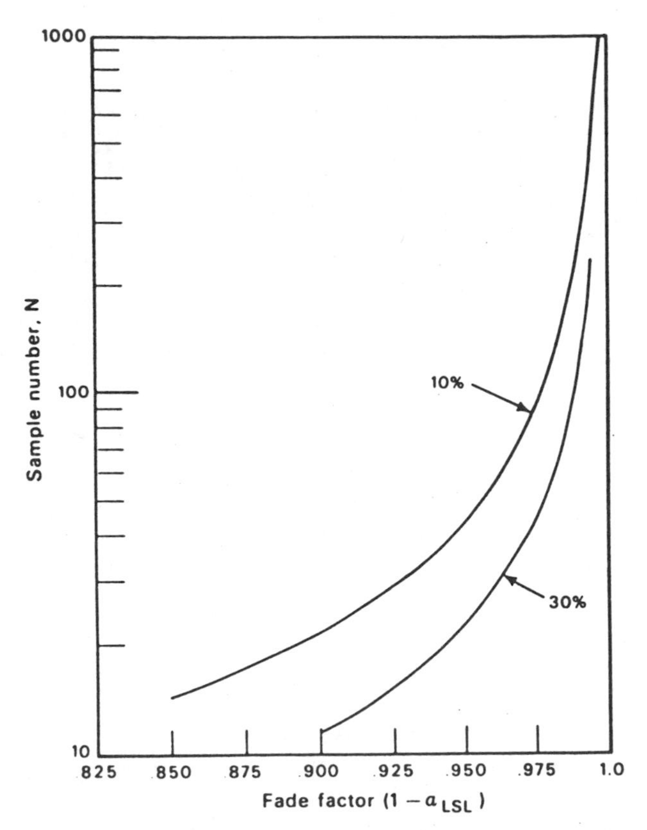

Fig. 3. Fade factor $(1-\alpha_{LSL})$ as a function of sample number N for 10 and 30 percent weighting of $e_p(n)$ relative to $e_p(N)$.

Order update $(i=1,2,\cdots,p)$

Lattice

$$\Delta_i(N)=(1-\alpha_{LSL})\Delta_i(N-1)-\frac{e_{i-1}(N)r_{i-1}(N-1)}{1-\gamma_{i-2}(N-1)} \quad (21h)$$

$$K_i^e(N)=\Delta_i(N)/E_{i-1}^e(N) \quad (21i)$$

$$K_i^r(N)=\Delta_i(N)/E_{i-1}^r(N-1) \quad (21j)$$

$$e_i(N)=e_{i-1}(N)+K_i^r(N)r_{i-1}(N-1) \quad (21k)$$

$$r_i(N)=r_{i-1}(N-1)+K_i^e(N)e_{i-1}(N) \quad (21l)$$

$$E_i^e(N)=E_{i-1}^e(N)-\Delta_i^2(N)/E_{i-1}^r(N-1) \quad (21m)$$

$$E_i^r(N)=E_{i-1}^r(N-1)-\Delta_i^2(N)/E_{i-1}^e(N) \quad (21n)$$

$$\gamma_{i-1}(N-1)=\gamma_{i-2}(N-1)+r_{i-1}^2(N-1)/E_{i-1}^r(N-1). \quad (21o)$$

Predictors

$$a_i^{(i)}(N)=K_i^r(N) \quad (21p)$$

$$b_0^{(i)}(N)=K_i^e(N) \quad (21q)$$

$$\left.\begin{aligned} a_k^{(i)}(N)&=a_k^{(i-1)}(N)+K_i^r(N)b_{k-1}^{(i-1)}(N-1), & (21r)\\ b_k^{(i)}(N)&=b_{k-1}^{(i-1)}(N-1)+K_i^e(N)a_k^{(i-1)}(N) & (21s)\end{aligned}\right\} \quad 1\leq k\leq i-1.$$

Although the LSL algorithm is fairly complex, it is interesting to note that the major difference between it and the GL (after substituting (18g) and (18h) into (18f)) is the presence of the gain factor $(1-\gamma_{i-2}(N-1))^{-1}$ in (21h). The variable $\gamma_{i-2}(N-1)$ takes on the values [0, 1] where the lower bound is reached for $\alpha_{LSL}=0$ and the upper bound is reached for $\alpha_{LSL}=1$. Under the assumption of zero mean Gaussian statistics, it has been interpreted as a measure of the likelihood that the $(i-1)$ samples, $\{x(N-1),\cdots,x(N-i+1)\}$, deviate from the $(i-1)$st-order multivariate Gaussian distribution parameterized by the sample covariance matrix whose elements are similar to those in (5) with the summation entrants weighted by $(1-\alpha_{LSL})^{N-n-1}$ [22], [23]. Thus $\gamma_{i-2}(N-1)$ enables the LSL to quickly modify the $K_i^e(n)$ and $K_i^r(n)$ when data of a significantly different character than that of the recent past is encountered.

Only preliminary work is available investigating the experimental behavior of the least LSL. However, performance results from an adaptive channel equalization experiment have indicated improved convergence characteristics over the adaptive transversal filter and GL [24]. Other work in the area of AR modeling has compared convergence properties of the LSL with those of the block data structured autocorrelation method of linear prediction outlined in (3)–(10) [22]. In addition, the same authors have investigated the fast adapting properties of the LSL in applications to speech processing.

IV. Tracking Behavior

The frequency tracking behavior of the three adaptive linear prediction algorithms summarized in Section III will be illustrated below in a variety of situations. Of particular interest is their response to sinusoidal (or narrow-band) components whose power levels are widely separated. Included are computer simulations of dual sinusoids as well as a format tracking example of real speech.

A. Dual Step

Fig. 4 illustrates the response of the LMS, GL, and LSL algorithms to dual constant frequency sinusoids which undergo an instantaneous step in frequency. The adaptation rate parameters, α, were chosen such that the response to the higher power sinusoid by all three algorithms was similar. The algorithms were allowed 1536 iterations to insure convergence prior to the beginning of the plots. Thereafter, frequency estimates were made every 4 iterations by a simple peak picking procedure on $1/|A(z)|^2$ evaluated at evenly spaced points around the unit circle via a 128 point FFT.

Notice the dramatic difference in the capability to track a step change in frequency of the lower power sinusoid between the LMS and the two lattice algorithms (GL and LSL). From the discussion in Section IIIA, the significant difference in adaptation rate for LMS is seen to be a manifestation of the approximate factor of ten spread in the eigenvalues corresponding to the two sinusoids.

It is interesting to observe the z-plane zero trajectories of $A(z)$ following the step change in frequency for the three algorithms. These are displayed in Fig. 5. Since the filter is

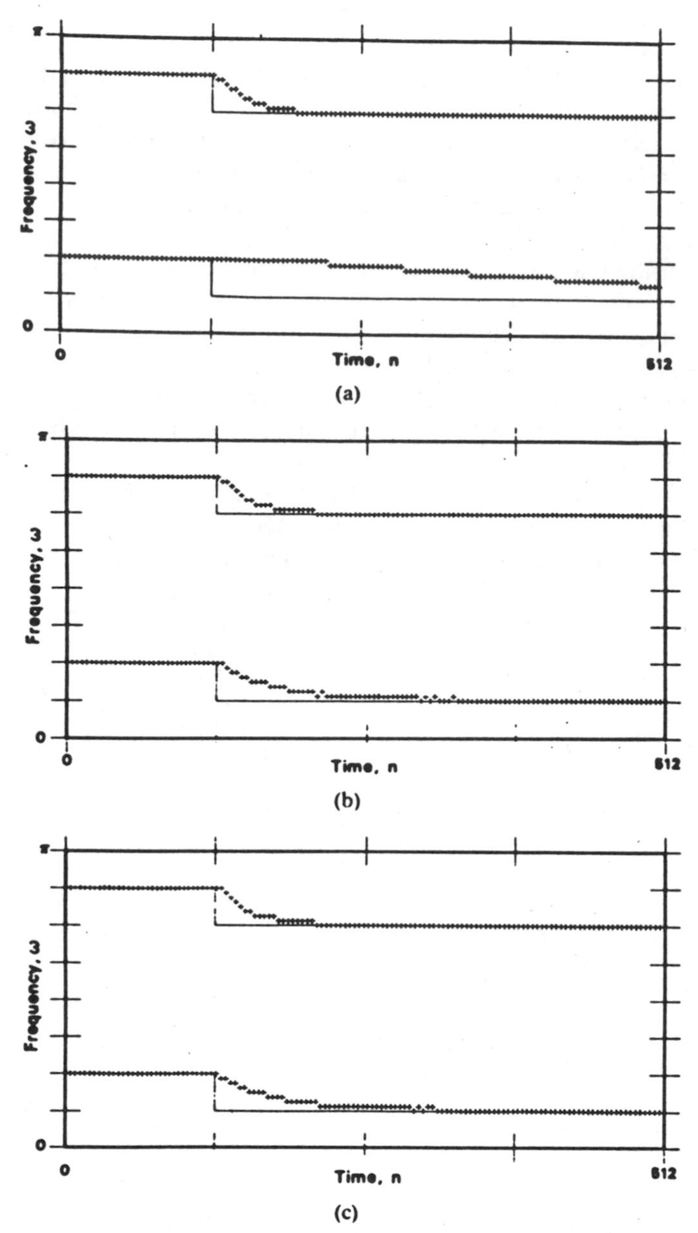

Fig. 4. Dual step, $SNR_1=20$ dB, $SNR_2=10$ dB, $p=6$. ($\omega_1=7\pi/8$, $\omega_2=\pi/4$, $\Delta\omega_1=\Delta\omega_2=-\pi/8$). (a) LMS($\alpha_{LMS}=0.04$). (b) GL($\alpha_{GL}=0.01$). (c) LSL($\alpha_{LSL}=0.02$).

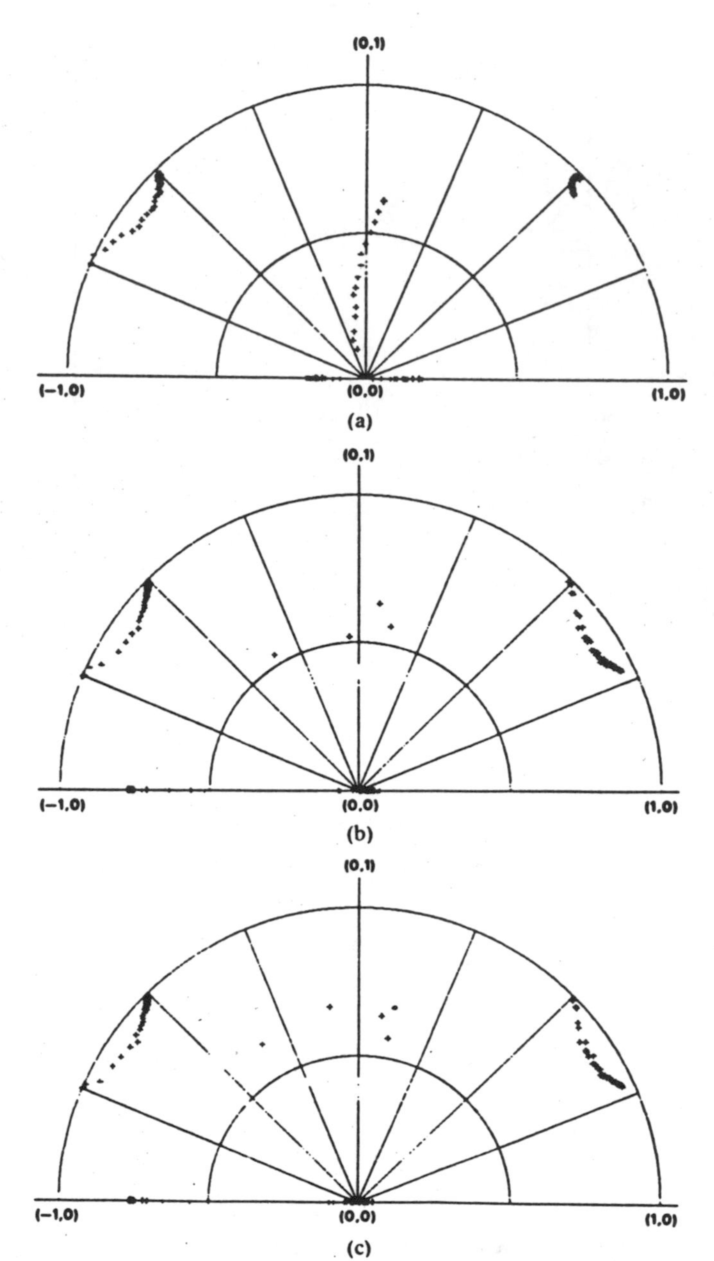

Fig. 5. Z-plane zero trajectories for dual step example. (a) LMS($\alpha_{LMS}=0.04$). (b) GL($\alpha_{GL}=0.01$). (c) LSL($\alpha_{LSL}=0.02$).

real, only the upper half of the z-plane is shown. Zeros off the real axis occur in complex-conjugate pairs. The zeros of $A(z)$ were determined at the same instants the frequency estimates were made in Fig. 4, starting with the estimate just prior to the step and continuing for a total of 32 zero sets. Since $p=6$, only four of the zeros (two complex-conjugate pairs) are captured by the two sinusoids, and their clockwise trajectories follow the underlying downward steps in frequency. Almost immediately, the two lattice algorithms send the excess pair of zeros to the real axis while their descent in the gradient transversal filter is more gradual.

B. Sinusoidal FM

Fig. 6 illustrates the response of the LMS, GL, and LSL algorithms to dual frequency components which undergo a sinusoidal FM with a period of 256 samples. As before, the adaptation rate parameters, α, were chosen such that the response to the higher power sinusoid by all three algorithms was similar. The plots start with the first sample of data and thus indicate an initial transient period. Frequency estimates were made every 4 iterations.

Once again, the LMS shows a significantly greater lag of the frequency estimate of the lower power component than is exhibited by the two lattice algorithms. For this particular example with a 10-dB separation in power and the tone modulation parameters chosen, tracking of the lower power component by the LMS borders on useless while that of the higher power component is quite good. Another point to note which will be discussed more fully in the next example is the relative noisiness of the frequency estimate derived from the GL algorithm.

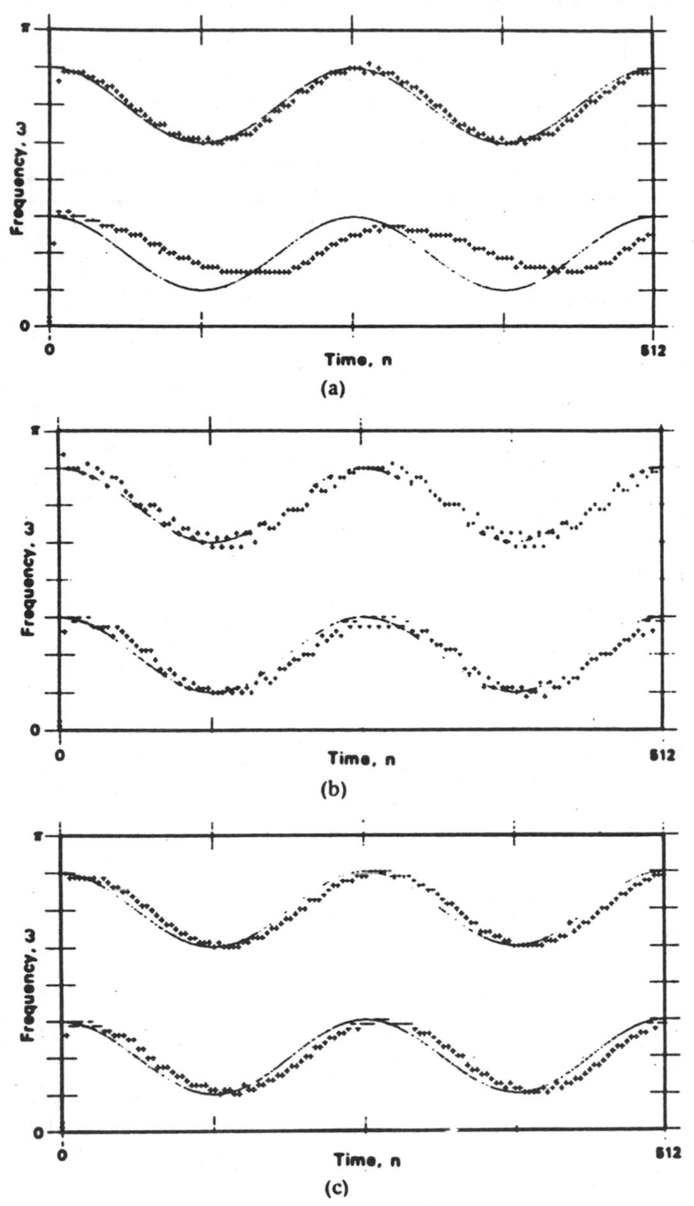

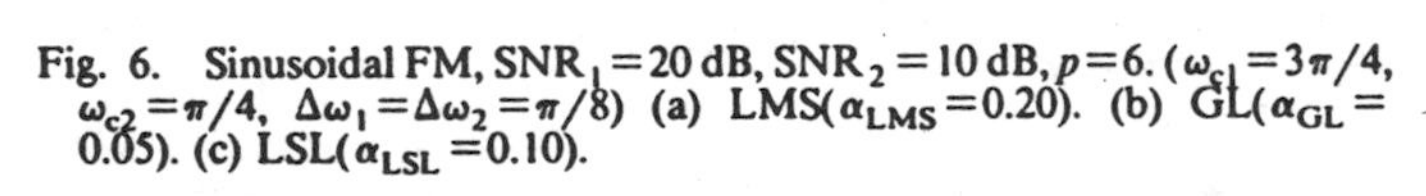

Fig. 6. Sinusoidal FM, $\mathrm{SNR}_1=20$ dB, $\mathrm{SNR}_2=10$ dB, $p=6$. ($\omega_{c1}=3\pi/4$, $\omega_{c2}=\pi/4$, $\Delta\omega_1=\Delta\omega_2=\pi/8$) (a) LMS($\alpha_{\mathrm{LMS}}=0.20$). (b) GL($\alpha_{\mathrm{GL}}=0.05$). (c) LSL($\alpha_{\mathrm{LSL}}=0.10$).

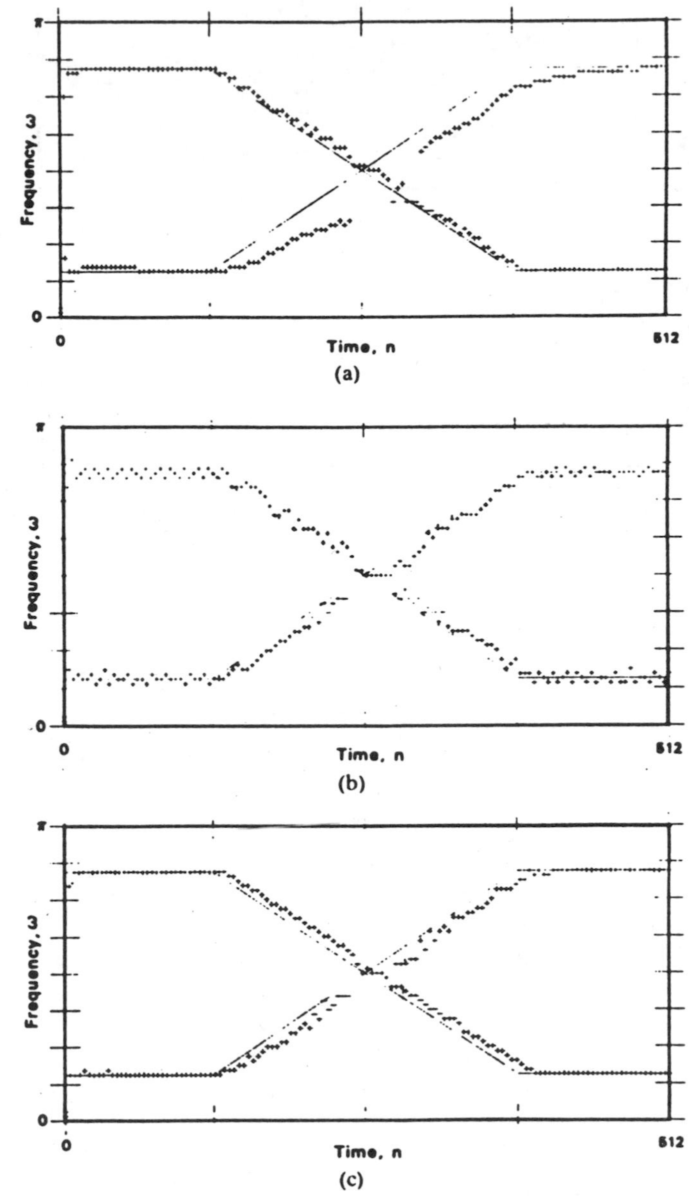

Fig. 7. Crossing Ramps. $\mathrm{SNR}_1=20$ dB, $\mathrm{SNR}_2=10$ dB, $p=6$. ($\omega_1=27\pi/32$, $\omega_2=9\pi/32$, $\Delta\omega_1=-\Delta\omega_2=-11\pi/16$). (a) LMS($\alpha_{\mathrm{LMS}}=0.20$). (b) GL($\alpha_{\mathrm{GL}}=0.05$). (c) LSL($\alpha_{\mathrm{LSL}}=0.10$).

C. Crossing Ramps

Fig. 7 illustrates the response of the LMS, GL, and LSL algorithms to dual frequency ramps (chirped FM) which cross each other midway through their frequency excursions. As in the previous two examples, the adaptation rate parameters α were chosen such that the response to the higher power sinusoid by all three algorithms was similar. The plots start with the first sample of data and frequency estimates were made every 4 iterations.

As seen previously, the LMS follows the frequency excursions of the lower power component with a considerably larger lag than is demonstrated by either of the two lattice algorithms. Note that the frequency estimate is particulary degraded in the region of crossover and a lengthy gap exists where the track of the weaker sinusoid disappears altogether. In contrast, the two lattice algorithms maintain a tight track on the weaker component and loose track only briefly during the exact time of frequency crossover.

Mentioned in connection with the sinusoidal FM results, the noisiness in the frequency estimates derived from the GL is clearly apparent in Fig. 7(b). This behavior is not shared by the LSL. The noisiness, nearly oscillatory in the constant frequency segments of Fig. 7(b), is a function of α_{GL} as can be seen by a comparison with Fig. 4(b) where an α_{GL} one-fifth as large was used. Essentially, this phenomena is a direct result of the approximation in (18f) of the true gradient by a noisy, instantaneous estimate. The resulting error is multiplied by α_{GL} prior to being coupled into the iterative update of $K_i(n)$.

D. Speech

The last example to be discussed considers the tracking of spectral peaks (formants) in speech. The phrase "we were away" spoken by a male was digitized at a sampling

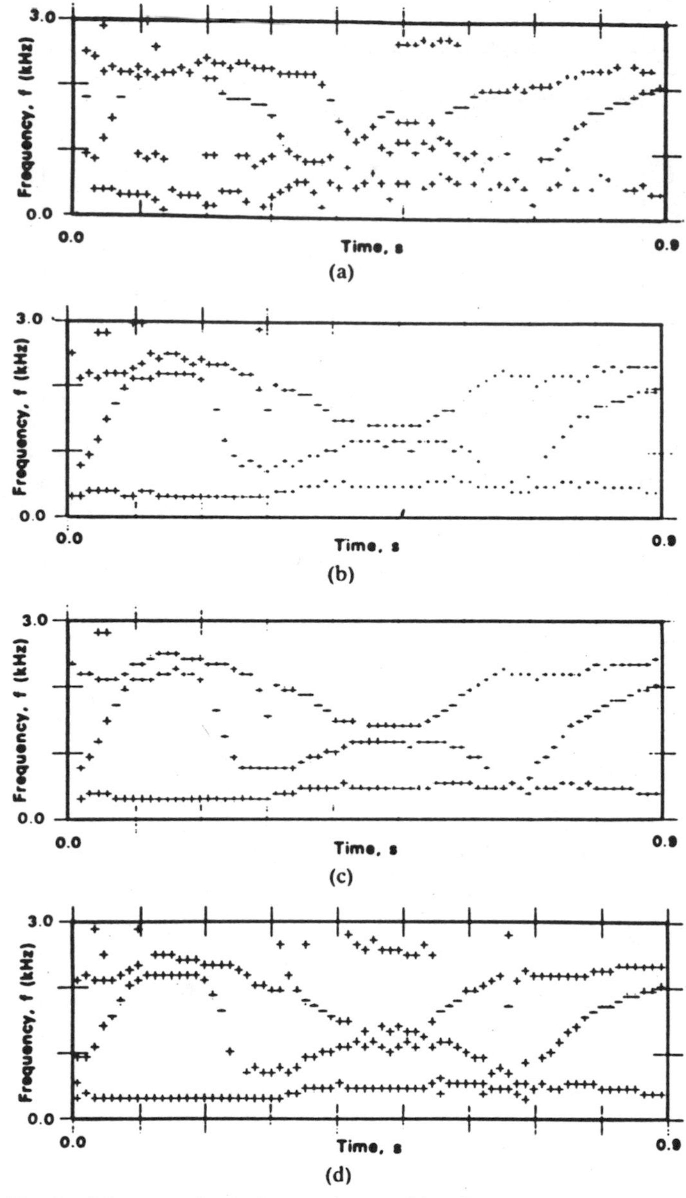

Fig. 8. Nonpreemphasized speech, $p=14$. ("we were away") (a) LMS($\alpha_{LMS}=0.20$). (b) GL($\alpha_{GL}=0.01$). (c) LSL($\alpha_{LSL}=0.02$). (d) DA (Hamming window, $N=256$).

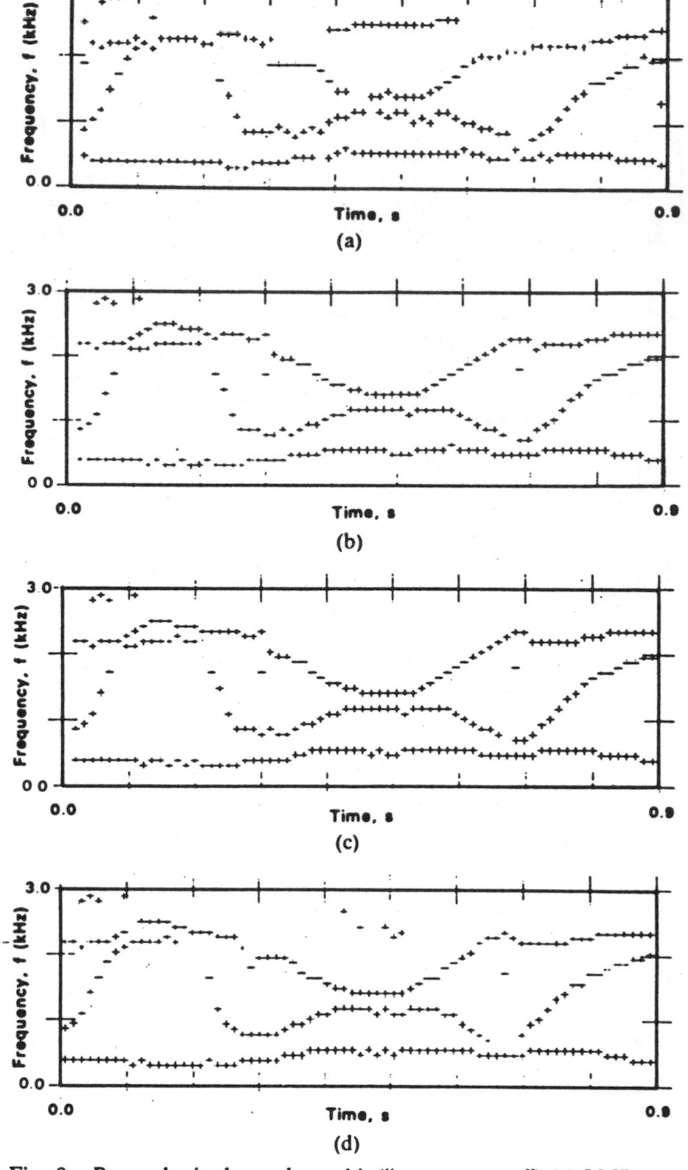

Fig. 9. Preemphasized speech, $p=14$. ("we were away") (a) LMS($\alpha_{LMS}=0.10$). (b) GL($\alpha_{GL}=0.005$). (c) LSL($\alpha_{LSL}=0.01$). (d) DA (Hamming window, $N=256$).

rate, $f_s=10$ kHz. The speech then was analyzed both without and with a 6-dB/octave preemphasis which was implemented by differencing the time series, $x(n)$. Since voiced speech typically exhibits a highly colored, negative sloping power spectrum with a dynamic range of 20 dB or greater over the range 0–5 kHz, this form of preemphasis tends to flatten out the gross spectral character of the speech time series [4], [5]. The adaptation rate parameters, α, for each algorithm were chosen to yield (qualitatively) the best overall tracking of the formants. Frequency estimates were made every 128 iterations and only those peaks in the range 0–3 kHz are plotted.

In order to provide a comparison with block data structured methods as well as with results previously published in the literature [25], a fourth approach to the analysis corresponding to the autocorrelation method of linear prediction is included. The algorithm implemented applied a Hamming window to a block of data prior to computing the data autocorrelation function as in (5). Taking advantage of the Toeplitz form of the matrix whose elements are the c_{ik}, Durbin's algorithm (DA) in (10) is used to compute the inverse filter coefficients a_k. A new data block consisting of the most recent $N=256$ samples of the speech time series is analyzed at the same instant the adaptive algorithms generate their frequency estimates. Thus a 50-percent overlap is achieved for successive analysis frames.

The tracking characteristics of the four algorithms are illustrated in Figs. 8 and 9. Of particular interest are the following regions: 1) the closely spaced second and third formants of the vowel |IY| (*bee*t) in "we" occurring at 0.15 s, 2) the rapid descent of the second formant following "we" in the vicinity of the 0.20 s, 3) the closely spaced second and third formants of the vowel |ER| (b*ir*d) in "were" occurring at 0.45 s, and 4) the closely spaced first and second formants of the semivowel |W| at 0.65 s which acts as a transition between the vowels |UH| (b*u*t) and |E| (h*a*te) in "away."

Consider first the case of nonpreemphasized speech in Fig. 8. The LMS tracks the first formant reasonably well, but follows the second and third formants with a significant lag except during the last half of "away." As seen previously in the context of the dual sinusoid experiments, this behavior is a direct result of the wide dynamic range present in nonpreemphasized speech. Note that in order to obtain even these marginal tracks on the second and third formants, such a large α_{LMS} was required that the first formant track has become extremely noisy due to gradient estimation error in (12d) [8], [9]. The two lattice algorithms (GL and LSL) perform quite well and nearly identically. Their only difficulty appears in following the second formant as it approaches the first formant in "away." The block data processing algorithm (DA) nicely follows all formants with a minor amount of difficulty separating the closely spaced formants in |ER| of "were" and |W| of "away." Both LMS and DA incorrectly suggest that spectral peaks occur near 2.5 kHz in |ER| of "were." Conventional FFT spectral analysis (Hamming window, $N=256$) indicates that during this vowel, the region 2–3 kHz actually contains a very low level of spectral energy (approximately 40 dB below the first formant while the first three formants are separated from each other by less than 10 dB).

Next, consider the case of preemphasized speech in Fig. 9. Notice the significantly improved performance of LMS. Essentially, the effect of the preemphasis has been to equalize the power levels of the first few formants. Note that LMS is still unable to separate the second and third formants in |IY| of "we." Both lattice algorithms also show improved performance in that they now are able to follow the second formant in the region where they lost track previously. Lastly, DA exhibits a more smooth tracking behavior than with the nonpreemphasized speech. A momentary loss of the second formant during |W| in "away" also is noticed.

V. Summary

The focus of this paper has been on the behavior of the LMS, the GL, and the LSL when used to track multiple sinusoidal (or narrow-band) components whose power levels are widely separated. All of these are recursive in time approaches to the realization of the pth-order, one-step linear predictor of the time series.

Numerous applications of frequency tracking inherently are characterized by the abrupt onset of a component, rapid transitions, uncertainty in the exact number of components to be tracked, and a wide disparity in their relative power levels. Particularly in the areas of speech processing and geophysical time series analysis, the modeling of $x(n)$ as an AR process has proven quite effective. Assuming that such a model for $x(n)$ is appropriate, the spectral estimation problem is equivalent to the derivation of a whitening or inverse filter in the form of a one-step linear predictor. This all-zero filter can be implemented either in direct form or as a lattice structure. Conventional approaches to its derivation are block data structured, however, rather than recursive.

The three algorithms chosen for this study continuously update their realizations of the inverse filter with an exponentially fading memory of past data. The form of each algorithm was summarized along with a brief discussion of its noteworthy characteristics. The first two were seen to be suboptimal in the sense that their structures (direct form for LMS and symmetrical lattice for GL) were assumed *a priori* and were not allowed to evolve freely out the mathematics of the derivation of the globally optimal linear predictor. No such constraints were applied in the derivation which led to the LSL whose structure is that of a nonsymmetrical lattice.

The tracking behaviors of LMS, GL, and LSL were illustrated in a variety of situations. Included were simulations of dual sinusoids whose power levels were significantly different. Three types of frequency versus time dynamics were considered: 1) coincident steps, 2) in-phase sinusoidal FM, and 3) crossing ramps. In all three cases, even though LMS nicely tracked the higher power sinusoid, it exhibited a substantial lag while attempting to follow the dynamics of the weaker sinusoid. Such behavior is a direct result of the inherent inability of the adaptive transversal filter to cope with data containing a wide eigenvalue spread. In contrast, the two lattice algorithms maintained a tight track on the dynamics of both components. The GL, however, exhibited a noisiness in its frequency estimates not seen with the LSL which arises from the particular mechanism by which the instantaneous gradient estimate is incorporated into the coefficient update procedure.

A formant tracking example of the phrase "we were away" also was included and was compared with a block data structured approach (Durbin's algorithm, DA). Tracking behaviors for both the nonpreemphasized and 6 dB/octave preemphasized speech waveform were illustrated. In the case of nonpreemphasized speech, LMS exhibited a significant lag while following the second and third formants. As in the dual sinusoid simulations, this behavior is due to the typically large power level spread between the first few formants of speech. Preemphasis significantly improved the ability of the LMS to track the second and third formants but some difficulty was noted in resolving closely spaced spectral peaks. In contrast, both lattice algorithms and DA performed quite well without preemphasis. Noticeable, but minor, improvements from GL, LSL, and DA were obtained when the preemphasized speech was analyzed.

References

[1] S. C. Gupta, "Phase-locked loops," *Proc. IEEE*, vol. 63, pp. 291–306, Feb. 1975.

[2] K. R. Brown *et al.*, "IQ-locked loop tracker," in *1977 Int. Conf. Acoust., Speech, Signal Processing*, (Hartford, CT), pp. 293–298, May 13–19, 1977.

[3] J. Makhoul, "Linear prediction: A tutorial review," *Proc. IEEE*, vol. 63, pp. 561–580, Apr. 1975.

[4] J. D. Markel and A. H. Gray, *Linear Prediction of Speech*. New York: Springer-Verlag, 1976.

[5] L. R. Rabiner and R. W. Schafer, *Digital Processing of Speech Signals*. Englewood Cliffs, NJ: Prentice-Hall, 1978.

[6] D. G. Childers (ed.), *Modern Spectrum Analysis.* New York: IEEE Press, 1978.
[7] L. J. Griffiths, "Rapid measurement of digital instantaneous frequency," *IEEE Trans. Acoust., Speech, Signal Processing*, vol. ASSP-23, pp. 207–222, Apr. 1975.
[8] B. Widrow *et al.*, "Adaptive noise cancelling: Principles and applications," *Proc. IEEE*, vol. 63, pp. 1692–1716, Dec. 1975.
[9] B. Widrow *et al.*, "Stationary and nonstationary learning characteristics of the LMS adaptive filter," *Proc. IEEE*, vol. 64, pp. 1151–1162, Aug. 1976.
[10] J. Makhoul, "Stable and efficient lattice methods for linear prediction," *IEEE Trans. Acoust., Speech, Signal Processing*, vol. ASSP-25, pp. 423–428, Oct. 1977.
[11] ____, "A class of all-zero lattice digital filters: Properties and applications," *IEEE Trans. Acoust., Speech, Signal Processing*, vol. ASSP-26, pp. 304–314, Aug. 1978.
[12] L. J. Griffiths, "A continuously adaptive filter implemented as a lattice structure," in *1977 Proc. IEEE Int. Conf. Acoust., Speech, Signal Processing*, (Hartford, CT), pp. 683–686, May 13–19, 1977.
[13] ____, "An adaptive lattice structure for noise cancelling applications," in *1978 Proc. IEEE Int. Conf. Acoust., Speech, Signal Processing*, (Tulsa, OK), pp. 87–90, Apr. 10–12, 1978.
[14] L. J. Griffiths and R. S. Medough, "Convergence properties of an adaptive noise cancelling lattice structure," in *1978 IEEE Conf. Decision and Control*, (San Diego, CA), pp. 1357–1361, Jan. 1979.
[15] L. J. Griffiths, "Adaptive structures for multiple-input noise cancelling applications," in *1979 Proc. IEEE Int. Conf. Acoust., Speech, Signal Processing*, (Washington, DC), pp. 925–928, Apr. 2–4, 1979.
[16] R. Viswanathan and J. Makhoul, "Sequential lattice methods for stable linear prediction," EASCON '76, pp. 155A–155H, Sept. 1976.
[17] J. Makhoul and R. Viswanathan, "Adaptive lattice methods for linear prediction," in *1978 Proc. IEEE Int. Conf. Acoust., Speech, Signal Processing*, (Tulsa, OK), pp. 83–86, Apr. 10–12, 1978.
[18] R. Viswanathan and J. Makhoul, "Adaptive lattice methods for linear predictive signal processing," in *Pattern Recognition and Signal Processing*, C. H. Chen (ed). The Netherlands: Sijthoff & Nordhoff, 1978, pp. 559–574.
[19] E. H. Satorius and S. T. Alexander, "Channel equalization using adaptive lattice algorithms," *IEEE Trans. Commun.*, vol. COM-27, pp. 899–905, June 1979.
[20] M. Morf *et al.*, "A classification of algorithms for ARMA models and ladder realizations," in *1977 Proc. Int. Conf. Acoust., Speech, Signal Processing*, (Hartford, CT) pp. 13–19, May 13–19, 1977.
[21] M. Morf, A. Vieira, and D. T. Lee, "Ladder forms for identification and speech processing," in *1977 IEEE Conf. Decision and Control*, pp. 1074–1078, Dec. 1977.
[22] M. Morf and D. Lee, "Fast algorithms for speech modeling," final rep. Defense Communication Agency, Contract DCA 100-77-C-0005, Information Systems Lab., Stanford Univ., Stanford CA, Nov. 1978.
[23] M. Morf and D. T. Lee, "Recursive least squares ladder forms for fast parameter tracking," in *1978 IEEE Conf. Decision and Control*, (San Diego, CA), pp. 1362–1367, Jan. 10–12, 1979.
[24] E. H. Satorius and J. D. Pack, "Applications of least squares lattice algorithms to adaptive equalization," IEEE Trans. *Commun.*, vol. COM-29, pp. 136–142 Feb. 1981.
[25] J. D. Markel, "Digital inverse filtering—A new tool for formant trajectory estimation," *IEEE Trans. Audio Electroacoust.*, vol. AU-20, pp. 129–137, June 1972.
[26] N. Levinson, "The Weiner RMS (root mean square) error criterion in filter design and prediction," *J. Math. Phys.*, vol. 25, pp. 261–278, 1947.
[27] J. Durbin, "The fitting of time-series models," *Rev. Inst. Int. Stat.*, vol. 28, p. 233–243, 1960.
[28] R. A. Wiggins and E. A. Robinson, "Recursive solution to the multichannel filtering problem," *J. Geophys. Res.*, vol. 70, pp. 1885–1891, 1965.
[29] G. H. Golub and C. Reinsch, "Singular value decomposition and least squares solutions," *Numer. Math.*, vol. 14, pp. 403–420, 1970.
[30] M. Morf et al., "Efficient solution of covariance equations for linear prediction," *IEEE Trans. Acoust., Speech, Signal Processing*, vol. ASSP-25, pp. 429–433, Oct. 1977.
[31] P. Gutowski, E. Robinson, and S. Treitel, "Spectral estimation: Fact or fiction," *IEEE Trans. Geosci. Electron.*, vol. GE-16, pp. 80–84, Apr. 1978.
[32] T. J. Ulrych and R. W. Clayton, "Time series modelling and maximum entropy," *Phys. Earth and Planetary Interiors*, vol. 12, pp. 188–200, 1976.
[33] S. T. Alexander, E. H. Satorius, and J. R. Zeidler, "Linear prediction and maximum entropy spectral analysis of finite bandwidth signals in noise," in *1978 Proc. IEEE Int. Conf. Acoust., Speech, Signal Processing*, (Tulsa, OK), pp. 188–191, Apr. 10–12, 1978.
[34] E. H. Satorius and J. R. Zeidler, "Maximum entropy spectral analysis of multiple sinusoids in noise," *Geophys.*, vol. 46, pp. 1111–1118, Oct. 1978.
[35] E. H. Satorius, J. R. Zeidler, and S. T. Alexander, "Linear predictive digital filtering of narrowband processes in additive broadband noise," NOSC TR-331, Naval Ocean Systems Center, San Diego, CA, Nov. 1, 1978.
[36] J. R. Treichler, "The spectral line enhancer—The concept, an implementation, and an application," Ph.D. dissertation, Stanford Univ., Stanford, CA, June 1977.
[37] ____, "Transient and convergent behavior of the adaptive line enhancer," *IEEE Trans. Acoust., Speech, Signal Processing*, vol. ASSP-27, pp. 53–62, Feb. 1979.
[38] J. R. Treichler, "γ-LMS and its use in a noise-compensating adaptive spectral analysis technique," in *1979 Proc. IEEE Int. Conf. Acoust., Speech, Signal Processing*, (Washington, DC) pp. 933–936, Apr. 2–4, 1979.
[39] R. J. Keeler and R. W. Lee, "Complex covariance/maximum entropy doppler estimates for pulsed CO_2 LIDAR," in *1978 Proc. IEEE Int. Conf. Acoust., Speech, Signal Processing*, (Tulsa, OK), pp. 365–368, Apr. 10–12, 1978.
[40] R. J. Keeler, "Uncertainties in adaptive maximum entropy frequency estimates," *IEEE Trans. Acoust., Speech, Signal Processing*, vol. ASSP-26, pp. 469–471, Oct. 1978.
[41] L. B. Jackson et al., "Frequency estimation by linear prediction," in *1978 Proc. IEEE Int. Conf. Acoust., Speech, Signal Processing*, (Tulsa, OK), pp. 352–356, April 10–12, 1978.
[42] S. M. Kay, "Maximum entropy spectral estimation using the analytical signal," *IEEE Trans. Acoust., Speech, Signal Processing*, vol. ASSP-26, pp. 467–469, Oct. 1978.
[43] D. R. Morgan and S. E. Craig, "Real-time adaptive linear prediction using the least square mean gradient algorithm," *IEEE Trans. Acoust., Speech, Signal Processing*, vol. ASSP-24, pp. 494–507, Dec. 1976.
[44] L. J. Griffiths and R. Prieto-Diaz, "Spectral analysis of natural seismic events using autoregressive techniques," *IEEE Trans. Geosci. Electron.*, vol. GE-15, pp. 13–25, Jan. 1977.
[45] R. J. Keeler and L. J. Griffiths, "Acoustic doppler extraction by adaptive linear-prediction filtering," *J. Acoust. Soc. Amer.*, vol. 61, p. 1218–1226, May 1977.
[46] J. R. Treichler, "Response of the adaptive line enhancer to chirped sinusoids," in *1978 IEEE Conf. Decision and Control*, pp. 1368–1373, Jan. 10–12, 1979.
[47] H. S. El-Ghoroury and S. C. Gupta, "Algorithmic measurement of digital instantaneous frequency," *IEEE Trans. Commun.*, vol. COM-24, pp. 1115–1122, Oct. 1976.
[48] D. W. Tufts and R. M. Rao, "Frequency tracking by MAP demodulation and by linear prediction techniques," *Proc. IEEE*, vol. 65, pp. 1120–1121, Aug. 1977.
[49] F. M. Hsu and A. A. Giordano, "Line tracking using autoregressive estimates," *IEEE Trans. Acoust., Speech, Signal Processing*, vol. ASSP-25, pp. 510–519, Dec. 1977.

Stochastic Convergence Properties of the Adaptive Gradient Lattice

GUY R. L. SOHIE, MEMBER, IEEE, AND LEON H. SIBUL, MEMBER, IEEE

Abstract—A stochastic fixed-point theorem is used as a basis for the study of stochastic convergence properties (in mean-squares sense) of the adaptive gradient lattice filter. Such properties include conditions on the stepsize in the adaptive algorithm and analytic expressions for the misadjustment and convergence rate.

Our results indicate that the limits on the stepsize are stricter than the ones obtained by considering convergence of the mean of the reflection coefficients and, therefore, only a slower convergence of the mean-square error can be obtained. It is shown that faster convergence is achieved for highly uncorrelated sequences than for almost deterministic sequences. The misadjustment is shown to be exponentially dependent on the number of stages in the lattice and is higher for uncorrelated sequences than for almost deterministic sequences.

I. Introduction

Research results in various areas such as speech processing [1], array processing [2], and adaptive tracking [3] have indicated that the gradient-lattice algorithms have superior behavior over classic methods such as Widrow's LMS approach applied to the transversal filter [4]. A variety of work has been devoted to the mean-square convergence of iterative algorithms (see, for instance, [24], [25]) or stochastic approximation schemes [19]. However, the application of these ideas to adaptive lattice filters has been very limited. Since the convergence of adaptive algorithms is primarily a stochastic problem, a mean-square criterion will be considered as opposed to the convergence of the mean of the reflection coefficients, which is mostly used in the literature [1], [3], [5]-[10].

A stochastic version of a fixed-point theorem, which was developed by Oza [11], [12], will provide the basis in establishing convergence conditions for the gradient-lattice algorithm. Such an approach has been taken in an earlier paper [13] to study convergence properties of the LMS algorithm and is based on the notion of a contraction mapping on a "stochastic" Hilbert space [13]. In the same framework, expressions for the convergence rate, conditions on the stepsize, and misadjustment are derived by considering a "distance" measure in the appropriate space. As such, our work differs from other studies on convergence of adaptive lattice filters [23], [26] in that convergence of the filter as an operator is considered, rather than convergence of some parameters which determine the filter.

Manuscript received December 27, 1982; revices August 11, 1983. This research was supported by the Naval Sea Systems Command and by the Office of Naval Research. Parts of this paper was presented at ICASSP '83 and an abreviated version of the paper was published in *Proc. Int. Conf. Acoust., Speech, Signal Processing*, 1983.

The authors are with the Applied Research Laboratory, Graduate Program in Acoustics, Pennsylvania State University, University Park, PA 16802.

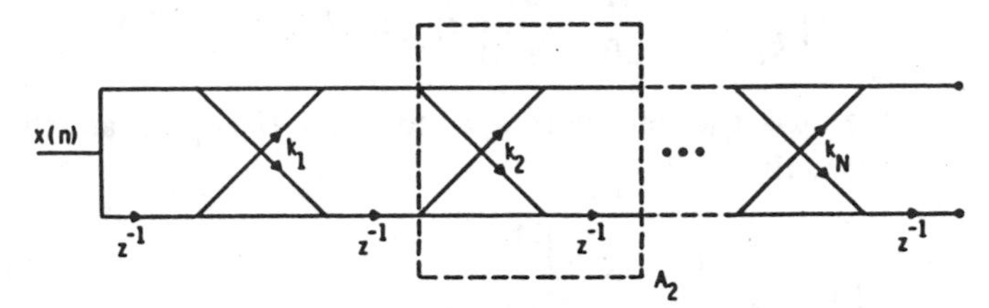

Fig. 1. Lattice filter as cascade of elementary operators A_i.

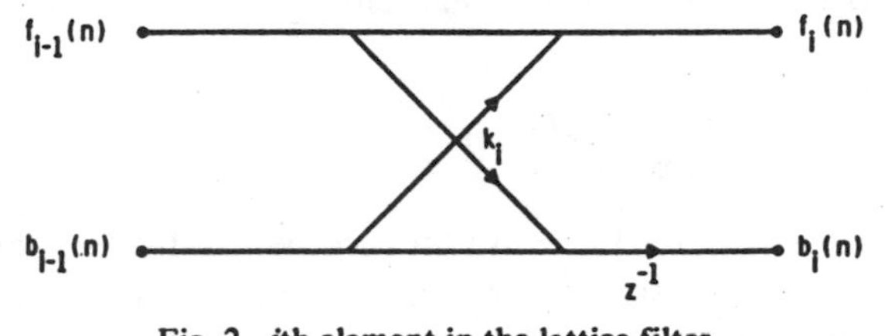

Fig. 2. ith element in the lattice filter.

II. Mathematical Background

A. The Lattice Filter

The lattice filter, as depicted in Fig. 1, will formally be considered as the cascade of elementary operators A_i (Fig. 2):

$$A = \prod_{i=1}^{N} A_i \tag{1}$$

where N is the order of the filter. Since the input of each element in the lattice can be represented by the "augmented" vector

$$\tilde{\mathbf{e}}_i(n) = \begin{bmatrix} \tilde{f}_{i-1}(n) \\ \tilde{b}_{i-1}(n) \end{bmatrix}, \tag{2}$$

A_i will be assumed to be an operator on $l_2(\Omega)$, the Hilbert space of zero-mean, wide sense stationary stochastic $(\Omega \times T)$ sequences with inner product

$$\langle \tilde{\mathbf{e}}_i(\cdot), \tilde{\mathbf{e}}_j(\cdot) \rangle \triangleq E\{\tilde{\mathbf{e}}_j^T(n)\, \tilde{\mathbf{e}}_i(n)\}. \tag{3}$$

It is also assumed that $\tilde{\mathbf{e}}_i(\cdot)$ has finite average power.[1] The correlation matrix at the input of each stage is given by

[1]The general theory developed in this paper is applicable to non-Gaussian stochastic processes. However, simple expressions for the appropriate norms can be obtained if Gaussian assumption is made.

Reprinted from *IEEE Trans. Acoust., Speech, Signal Processing,* vol. ASSP-32, no. 1, pp. 102–107, Feb. 1984.

$$E\{\tilde{\mathbf{e}}_i(n)\tilde{\mathbf{e}}_i^T(n)\} \triangleq \begin{bmatrix} E_i & C_i \\ C_i & E_i \end{bmatrix} \tag{4}$$

where, by symmetry, the average power of $\tilde{f}_i(n)$ and $\tilde{b}_i(n)$, are equal, with value denoted by E_i. The input-output relationship for the element A_i can be written in matrix form as

$$\mathbf{e}_i(n) = \begin{bmatrix} 1 & k_i z^{-1} \\ k_i & z^{-1} \end{bmatrix} \tilde{\mathbf{e}}_{i-1}(n) \triangleq A_i \tilde{\mathbf{e}}_{i-1}(n)$$

$$= \begin{bmatrix} 1 & k_i \\ k_i & 1 \end{bmatrix} \begin{bmatrix} 1 & 0 \\ 0 & z^{-1} \end{bmatrix} \tilde{\mathbf{e}}_{i-1}(n) \triangleq K_i U \tilde{\mathbf{e}}_{i-1}(n) \tag{5}$$

where z^{-1} is the unit shift operator, and U and K_i are the matrices defined by

$$U = \begin{bmatrix} 1 & 0 \\ 0 & z^{-1} \end{bmatrix} \quad K_i = \begin{bmatrix} 1 & k_i \\ k_i & 1 \end{bmatrix} \tag{6}$$

where the magnitudes of the "reflection coefficients" k_i are always taken to be smaller than one. The norm of A_i [14] is given by

$$\|A_i\| = \sup_{\|\tilde{\mathbf{e}}_{i-1}(\cdot)\|=1} \{\|UK_i\tilde{\mathbf{e}}_{i-1}(\cdot)\|\} \tag{7}$$

where $\|\cdot\|$ denotes the norm induced by the inner product in (3). Since it is straightforward to show that U is a unitary operator, (7) reduces to the simple matrix of K_i, which is given by its spectral radius [14]

$$\|A_i\| = |\lambda_{\max}| = 1 + |k_i| \tag{8a}$$

where $\lambda_{\max}$ is the largest eigenvalue of A_i. A lower bound on the output power can be obtained by replacing sup by inf and using the same procedure. This gives

$$(1 - |k_i|)\|\tilde{\mathbf{e}}_{i-1}(\cdot)\| \leq \|\tilde{\mathbf{e}}_{i-1}(\cdot)\| \leq (1 + |k_i|)\|\tilde{\mathbf{e}}_{i-1}(\cdot)\|. \tag{8b}$$

Note [10] that the lower bound is actually attained when $\tilde{\mathbf{e}}_{i-1}(\cdot)$ consists of the actual innovations processes (optimal linear predictor). Equation (8b) was previously obtained by Makhoul using different methods [10].

B. The Adaptive Lattice Filter

Thus far, we have considered the lattice filter as a fixed, deterministic operator on the space $l_2(\Omega)$. If the lattice is used in an adaptive algorithm, however, two major differences arise. First of all, if a time-recursive method is used, the operators A_i will become time-varying and thus the assumption of wide-sense stationarity of the input is no longer valid. In the sequel, however, we will assume that the stepsize in the adaptive algorithms is sufficiently small such that the inputs to successive stages are at least locally stationary. Secondly, the parameters determining the operators A_i will be based on measured data, thus resulting in a stochastic operator. Consequently, the assumption of Gaussian statistics of the input to each stage will be violated. Again, under the same assumption of a small stepsize, it can be assumed that the statistics of the sequences $\tilde{\mathbf{e}}_i(\cdot)$ will be sufficiently Gaussian.

In order to calculate the mean-square norm of the stochastic lattice element, we can again use (5) where now the reflection coefficients $\tilde{k}_i$ in the matrix $\tilde{K}_i$ are random variables:

$$\|\tilde{A}_i\|^2 = \sup_{\|\tilde{\mathbf{e}}_{i-1}(\cdot)\|=1} \{\|\tilde{K}_i\tilde{\mathbf{e}}_{i-1}(\cdot)\|^2\}$$

$$= \sup_{\|\tilde{\mathbf{e}}_{i-1}(\cdot)\|=1} E\{\tilde{\mathbf{e}}_{i-1}^T(n)\tilde{K}_i^2\tilde{\mathbf{e}}_{i-1}(n)\}. \tag{9}$$

Under the widely used assumption that the operator $\tilde{K}_i$ and the sequence $\tilde{\mathbf{e}}_{i-1}(\cdot)$ are uncorrelated, which corresponds to the case of uncorrelated data [4], (9) becomes

$$\|\tilde{A}_i\|^2 = \sup_{\|\tilde{\mathbf{e}}_{i-1}(\cdot)\|=1} E\{\tilde{\mathbf{e}}_{i-1}^T(n)E\{\tilde{K}_i^2\}\tilde{\mathbf{e}}_{i-1}(n)\}$$

$$\leq \|E\{\tilde{K}_i^2\}\| \tag{10}$$

(by the Schwarz inequality and the definition of matrix norm). Because the matrix norm is attained by an eigenvector, in which case the equality sign in the Schwarz inequality holds, we have

$$\|\tilde{A}_i\|^2 = \|E\{\tilde{K}_i^2\}\|. \tag{11}$$

Noting that

$$E\{\tilde{K}_i^2\} = E\begin{bmatrix} 1 + \tilde{k}_i^2 & 2\tilde{k}_i \\ 2\tilde{k}_i & 1 + \tilde{k}_i^2 \end{bmatrix}$$

$$= \begin{bmatrix} 1 + \sigma_i^2 + \bar{k}_i^2 & 2\bar{k}_i \\ 2\bar{k}_i & 1 + \sigma_i^2 + \bar{k}_i^2 \end{bmatrix} \tag{12}$$

where σ_i^2 and $\bar{k}_i$ are the variance and mean, respectively, of $\tilde{k}_i$, we can write bounds for $\|\tilde{\mathbf{e}}_i(\cdot)\|$ similarly to (8b):

$$[\sigma_i^2 + (1 - |\bar{k}_i|)^2]\,\|\tilde{\mathbf{e}}_{i-1}(\cdot)\|^2 \leq \|\tilde{\mathbf{e}}_i(\cdot)\|^2$$

$$\leq [\sigma_i^2 + (1 + |\bar{k}_i|)^2]\,\|\tilde{\mathbf{e}}_{i-1}(\cdot)\|^2. \tag{13}$$

C. The Stochastic Fixed-Point Theorem

The following theorem, which forms the basis for the discussions in the sequel, is a stochastic fixed-point theorem and was used by Oza [11], [12] in a system identification problem. For the proof, we refer to [11].

Theorem: Let $\{\tilde{T}_n\}_{n=0}^{\infty}$ be a sequence of (random) operators on a Hilbert space $\mathcal{H}$ and let $\tilde{T}_n \to \tilde{T}$ where $\tilde{T}$ is a contraction mapping, i.e.,

$$\lim_{n\to\infty} \|\tilde{T}_n y - \tilde{T}y\| = 0 \quad \forall\, y \in \mathcal{H} \tag{14}$$

and

$$\|\tilde{T}y_1 - \tilde{T}y_2\| < \|y_1 - y_2\| \quad \forall\, y_1 y_2 \in \mathcal{H}. \tag{15}$$

If it is assumed that $\tilde{T}$ has a fixed point, then the sequence generated by

$$\tilde{y}_{n+1} = \tilde{T}_n\tilde{y}_n$$

y_0 fixed, but arbitrary in $\mathcal{H}$, converges strongly to the fixed-point of $\tilde{T}$.

III. Convergence Properties

A. The Adaptive Time-Recursive Algorithm

It is well known [15] that the lattice filter is a natural implementation of the Levinson algorithm to solve the linear prediction problem. In the "optimal" case, the reflection coefficients k_i are determined in the Levinson recursion by

$$E_{i-1} k_i = -C_{i-1} \tag{16}$$

and the sequences $\tilde{f}_i(\cdot)$ and $\tilde{b}_i(\cdot)$ are the innovations processes of the ith-order forward and backward predictors [16]. Practically, however, neither the average power (E_i) nor the crosscorrelation (C_i) of the innovations processes are completely known, and an adaptive method then consists of calculating the reflection coefficients based on estimates of C_i and E_i which are obtained from actual data. A very common and simple form of adaptation is achieved when the average values in (4) are replaced by their instantaneous values and the solutions to (16) are obtained time-recursively. This leads to an adaptive algorithm given by [3]

$$\tilde{k}_{i,n+1} = \tilde{k}_{i,n} - \alpha[\tilde{E}_{i-1}(n)\tilde{k}_{i,n} + \tilde{C}_{i-1}(n)] \tag{17}$$

where

$$\tilde{E}_{i-1}(n) \triangleq \tfrac{1}{2}[\tilde{f}_{i-1}^2(n) + \tilde{b}_{i-1}^2(n)] \tag{18}$$

and

$$\tilde{C}_{i-1}(n) \triangleq \tilde{f}_{i-1}(n)\tilde{b}_{i-1}(n). \tag{19}$$

α is the so-called step size of the algorithm and $\tilde{k}_{i,n}$ denotes the adaptive estimate of the optimal reflection coefficient k_i at the nth recursion. Note that if the estimates are taken to be the exact values (i.e., $\tilde{E}_{i-1}(n) \equiv E_{i-1}$, $\tilde{C}_{i-1}(n) = C_{i-1}$), the sequence in (17) converges if $\alpha < (2/E_{i-1})$, the same condition as is obtained for convergence of the mean of the reflection coefficients [3].

B. Conditions on the Step Size (α)

Strictly speaking, an adaptive lattice element $\tilde{A}_{i,n}$ converges to a lattice element A_i (the "optimal" operator) if

$$\lim_{n\to\infty} \|\tilde{A}_{i,n}\tilde{\mathbf{e}}_{i-1}(\cdot) - A_i\tilde{\mathbf{e}}_{i-1}(\cdot)\| = 0. \tag{20}$$

Practically, however, most adaptive algorithms can only result in outputs which are within a certain "distance" of the optimal value, i.e.,

$$\lim_{n\to\infty} \|\tilde{A}_{i,n}\tilde{\mathbf{e}}_{i-1}(\cdot) - A_i\tilde{\mathbf{e}}_{i-1}(\cdot)\| = M_i < \infty \tag{21}$$

where M_i is the (unnormalized) misadjustment. Using (5), the norm in (21) can be written as

$$\|(\tilde{A}_{i,n} - A_i)\tilde{\mathbf{e}}_{i-1}(\cdot)\|^2 = \left\| \begin{bmatrix} 0 & \tilde{k}_{i,n} - k_i \\ \tilde{k}_{i,n} - k_i & 0 \end{bmatrix} \tilde{\mathbf{e}}_{i-1}(\cdot) \right\|^2 \tag{22}$$

$$= 2E\{|\tilde{k}_{i,n} - k_i|^2\} E_{i-1}. \tag{23}$$

Thus, convergence of the adaptive element is determined by convergence of the reflection coefficients.

In order to obtain limits for the step size α, (17) can be rewritten in terms of a stochastic fixed-point theorem as follows: write (17) as

$$\tilde{k}_{i,n+1} - k_i = (\tilde{k}_{i,n} - k_i)[1 - \alpha\tilde{E}_{i-1}(n)] - \alpha[\tilde{C}_{i-1}(n) + k_i\tilde{E}_{i-1}(n)] \tag{24}$$

and define the stochastic operator $\tilde{T}_{i,n}$ on the space of Gaussian random variables by

$$\tilde{T}_{i,n}k \triangleq [1 - \alpha\tilde{E}_{i-1}(n)]\,\tilde{k} - \alpha[\tilde{C}_{i-1}(n) + k_i\tilde{E}_{i-1}(n)] \tag{25}$$

for all $\tilde{k}$. Then (24) can be expressed as a fixed-point problem

$$(\tilde{k}_{i,n+1} - k_i) = \tilde{T}_{i,n}(\tilde{k}_{i,n} - k_i) \tag{26}$$

and using the contraction mapping principle of Section II-A, the recursion in (26) will converge if

$$\|[1 - \alpha\tilde{E}_{i-1}(n)]\,(y_1 - y_2)\| < \|y_1 - y_2\| \tag{27}$$

for y_1 and y_2 fixed but arbitrary.

Since the norm in (27) is given by

$$\|[1 - \alpha\tilde{E}_{i-1}(n)]\,(y_1 - y_2)\|^2 = E\{|1 - \alpha\tilde{E}_{i-1}(n)|^2\}\,\|y_1 - y_2\|^2 \tag{28}$$

this condition becomes (see Appendix A)

$$1 + \alpha^2[2E_{i-1}^2 + C_{i-1}^2] - 2\alpha E_{i-1} < 1 \tag{29}$$

or

$$0 < \alpha < \frac{2E_{i-1}}{2E_{i-1}^2 + C_{i-1}^2}. \tag{30}$$

Various interpretations can now be made. First of all, it is easy to see from (30) that the limits obtained in this fashion are stricter than the ones for convergence of the mean of $\tilde{k}_{i,n}$. Furthermore, while previous results showed a dependence of the upper limit on E_{i-1} only, and thus the upper limit for α could be made independent of the position in the lattice by normalizing it with respect to E_{i-1} [3], (30) shows that α is also dependent on the crosscovariance of $\tilde{f}_{i-1}(\cdot)$ and $\tilde{b}_{i-1}(\cdot)$.

Thus, in order to make the limits for the step size independent on the position in the ladder, normalizing with respect to the right-hand side of (30) requires the extra computation of C_{i-1}. As C_{i-1} is bounded by the Schwarz inequality as

$$0 \leq |C_{i-1}| \leq E_{i-1} \tag{31}$$

two extreme cases can be considered. If $|C_{i-1}|$ is very small (forward and backward innovations uncorrelated, which is the case if the input approaches "white" noise) the upper bound for α in (30) becomes approximately equal to $1/E_{i-1}$, while if the input sequence is almost deterministic (singular process), the limit for α approaches $\frac{2}{3}(1/E_{i-1})$. This indicates that faster convergence can be obtained for lower signal-to-noise ratios.

C. The Misadjustment

The misadjustment due to the ith stage in the lattice filter is defined by (21). Since (23) shows that

$$M_i^2 = 2 \lim_{n\to\infty} E\{|\tilde{k}_{i,n} - k_i|^2\} E_{i-1} \tag{32}$$

it is clear that the normalized misadjustment is nothing but the norm in the fixed-point problem, (26):

$$m_i^2 \triangleq \frac{M_i^2}{2E_{i-1}} \lim_{n\to\infty} \|\tilde{T}_{i,n}(\tilde{k}_{i,n} - k_i)\|. \tag{33}$$

The remark can be made that the fixed-point of T_i, the "ideal" contraction mapping given by

$$T_i k \triangleq [1 - \alpha E_{i-1}] k - \alpha[C_{i-1} + k_i E_{i-1}] \tag{34}$$

has as (only) fixed-point the zero element because

$$T_i 0 = 0 - \alpha[C_{i-1} + k_i E_{i-1}]$$
$$= 0 \tag{35}$$

by definition of k_i in (16). Thus, zero misadjustment can be obtained if the estimates in $\tilde{E}_{i-1}(\cdot)$ and $\tilde{C}_{i-1}(\cdot)$ converge to the actual values. This will be the case if a different estimate is used as

$$\tilde{E}_{i-1}(n) = \frac{1}{2n} \sum_{j=1}^{n} [\tilde{f}_{i-1}^2(j) + \tilde{b}_{i-1}(j)] \tag{36}$$

and

$$\tilde{C}_{i-1}(n) = \frac{1}{n} \sum_{j=1}^{n} [\tilde{f}_{i-1}(j) \tilde{b}_{i-1}(j)] \tag{37}$$

and if the process $\tilde{e}(\cdot)$ is correlation ergodic [18]. Note that in this case, α/n satisfies the established conditions for stochastic approximation [19].

In order to compute the misadjustment of the algorithm (17), the norm of (26) is needed.

$$\|\tilde{k}_{i,n+1} - k_i\|^2 = \|\tilde{T}_{i,n}(\tilde{k}_{i,n} - k_i)\|^2. \tag{38}$$

This norm is computed in Appendix B and the (normalized) misadjustment is found to be

$$m_i^2 = \alpha\left[\frac{E_{i-1}^2 - C_{i-1}^2}{E_{i-1}}\right]^2 \cdot \frac{1}{2E_{i-1} - \alpha[2E_{i-1}^2 + C_{i-1}^2]}. \tag{39}$$

This expression indicates that, while the limits on α can be made independent of the stage i by appropriate normalization of the stepsize, such normalization will not result in a stage-dependent misadjustment. Again, two extreme cases occur. If the input sequence approaches "white" noise, ($|C_{i-1}| \ll E_{i-1}$), (39) will equal approximately $[\alpha E_{i-1}/2(1 - \alpha E_{i-1})]$, while if the input sequence is almost deterministic, the misadjustment approaches zero. Therefore, M_i is bounded by

$$0 < M_i^2 < \frac{\alpha E_{i-1}^2}{2(1 - \alpha E_{i-1})}. \tag{40}$$

In this result, E_{i-1} is the average power in steady state conditions at the input of the ith stage. For the usual case where α is normalized with respect to E_{i-1}, the upper bound in (40) becomes

$$0 < M_i^2 < \frac{\alpha}{2(1 - \alpha)} E_{i-1}. \tag{41}$$

In order to compute the total misadjustment at the output of the cascade, (13) can be used. Essentially, (13) corresponds to the model shown in Fig. 3. The total output power in Fig. 3 is bounded by (13) where σ_i^2 and $\bar{k}_i$ are (since the lattice converged) equal to m_i^2 and k_i, respectively:

$$\prod_{i=1}^{N} [m_i^2 + (1 - |k_i|)^2] E_0 \leq E_N$$
$$\leq \prod_{i=1}^{N} [m_i^2 + (1 + |k_i|)^2] E_0. \tag{42}$$

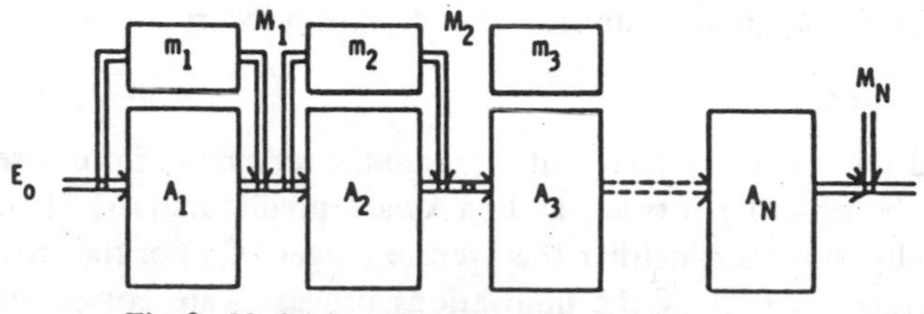

Fig. 3. Model for the total output misadjustment.

Since the optimal output power is given in (8b), the output misadjustment normalized with respect to the total output power is given by

$$\prod_{i=1}^{N} \left[\frac{m_i^2 + (1 - |k_i|)^2}{(1 - |k_i|)^2}\right] - 1 \leq m$$
$$\leq \prod_{i=1}^{N} \left[\frac{m_i^2 + (1 + |k_i|)^2}{(1 - |k_i|)^2}\right] - 1. \tag{43}$$

If the input sequence is very random, $|k_i| \ll 1$ and both bounds approach the same value. In this case, we will have approximately

$$m \cong \prod_{i=1}^{N} (1 + m_i^2) - 1. \tag{44}$$

If α is normalized with respect to E_{i-1}, which is the case in most applications, we have from (41)

$$m_i^2 = \frac{\alpha}{2(1 - \alpha)} \tag{45}$$

and finally

$$m = \left[\frac{2 - \alpha}{2(1 - \alpha)}\right]^N - 1. \tag{46}$$

Ungerboeck [21] and Widrow [22] showed that for the LMS algorithm, the misadjustment is linearly dependent on the order N. The exponential dependence on N for the lattice suggests that, in general, a higher misadjustment can be expected for the lattice filter.

D. The Convergence Rate

In order to compute the convergence rate of the ith adaptive element, the homogeneous part of the state equation (B8) has to be considered. This is denoted by

$$\mathbf{x}(n+1) = \begin{bmatrix} A & -2\alpha C \\ 0 & 1 - \alpha E_{i-1} \end{bmatrix} \mathbf{x}(n).$$
$$\equiv \mathfrak{A}\mathbf{x}(n). \tag{47}$$

The convergence rate is determined by the smallest eigenvalue of the matrix $\mathfrak{A}$ [20]. In the case of a small step size, it can be easily shown that this smallest eigenvalue is equal to A, the "slowest" mode, which determines convergence of the squared error, is given by

$$m_{i,n} = A^n. \tag{48}$$

Substituting for A using (A3) and assuming small α, an approximate expression for $m_{i,n}$ is obtained:

$$m_{i,n} = 1 - n\{2\alpha E_{i-1} - \alpha^2[2E_{i-1}^2 + C_{i-1}^2]\}. \quad (49)$$

Comparing (49) to the expression

$$e^{-t/\tau} = 1 - \frac{t}{\tau} + \cdots$$

the time-constant of the adaptation is

$$\tau_i = \frac{1}{2\alpha E_{i-1} - \alpha^2[2E_{i-1}^2 + C_{i-1}^2]}. \quad (50)$$

In the extreme case of almost deterministic inputs, this becomes

$$\tau_i \simeq \frac{1}{2\alpha E_{i-1}[1 - \frac{3}{2}\alpha E_{i-1}]} \quad (51)$$

and in case of highly uncorrelated input sequences, τ_i becomes

$$\tau_i \simeq \frac{1}{2\alpha E_{i-1}[1 - \alpha E_{i-1}]}. \quad (52)$$

Thus, a slower adaptation is obtained for deterministic sequences than for uncorrelated sequences.

Conclusion

Our results, which are based on a mean-square convergence criterion, suggest that the usual approach, which consists of the study of convergence properties of the mean of the reflection coefficients, may not be sufficient for a number of practical applications. The mean-square approach shows that convergence is only obtained under stricter conditions than previous studies suggested. The convergence rate of the adaptive gradient lattice is better for highly uncorrelated signals than for almost deterministic sequences, but a higher misadjustment can be expected if the input signal is extremely uncorrelated. The total output misadjustment varies exponentially with the filter order. Since the output misadjustment for the basic LMS algorithm varies linearly with the number of taps, the excess mean-square error for the lattice can be expected to be larger, especially for higher order filters.

Appendix A

Note that

$$\begin{aligned} E\{|1 - \alpha\tilde{E}_{i-1}(n)|^2\} &= E\{1 + \alpha^2|\tilde{E}_{i-1}(n)|^2 - 2\alpha\tilde{E}_{i-1}(n)\} \\ &= 1 + \alpha^2 E\{|\tilde{E}_{i-1}(n)|^2\} - 2\alpha E_{i-1}. \end{aligned} \quad (A1)$$

Now, using the well-known expression for 4th order Gaussian moments [17], this becomes (18)

$$\begin{aligned} E\{|\tilde{E}_{i-1}(n)|^2\} &= \tfrac{1}{4}E\{|\tilde{f}_{i-1}^2(n) + \tilde{b}_{i-1}^2(n)|^2\} \\ &= \tfrac{1}{4}[E\{\tilde{f}_{i-1}^4(n) + \tilde{b}_{i-1}^4(n)\} \\ &\quad + 2E\{\tilde{f}_{i-1}^2(n)\,\tilde{b}_{i-1}^2(n)\}] \\ &= \tfrac{1}{4}[3E_{i-1}^2 + 3E_{i-1}^2 + 2E_{i-1}^2 + 4C_{i-1}^2] \\ &= 2E_{i-1}^2 + C_{i-1}^2. \end{aligned} \quad (A2)$$

Substituting this in (A1), we finally obtain

$$\begin{aligned} E\{|1 - \alpha\tilde{E}_{i-1}(n)|^2\} &= \alpha^2[2E_{i-1}^2 + C_{i-1}^2] \\ &\quad + 1 - 2\alpha E_{i-1} \triangleq A. \end{aligned} \quad (A3)$$

Appendix B

Using (25), the following recursion can be written:

$$\begin{aligned} \|\tilde{k}_{i,n+1} - k_i\|^2 &= E\{|[1 - \alpha\tilde{E}_{i-1}(n)](\tilde{k}_{i,n} - k_i) \\ &\quad - \alpha[\tilde{C}_{i-1}(n) + k_i\tilde{E}_{i-1}(n)]|^2\} \\ &= E\{|[1 - \alpha\tilde{E}_{i-1}(n)]^2(\tilde{k}_{i,n} - k_i)|^2\} \\ &\quad + \alpha^2 E\{[\tilde{C}_{i-1}(n) + k_i\tilde{E}_{i-1}(n)]^2\} \\ &\quad - 2\alpha E\{[1 - \alpha\tilde{E}_{i-1}(n)][\tilde{C}_{i-1}(n) \\ &\quad + k_i\tilde{E}_{i-1}(n)](\tilde{k}_{i,n} - k_i)\}. \end{aligned} \quad (B1)$$

If we assume that the repetition rate of the algorithm is small enough such that $(\tilde{k}_{i,n} - k_i)$ is uncorrelated with $\tilde{C}_{i-1}(n)$ and $\tilde{E}_{i-1}(n)$, (B1) will become

$$\begin{aligned} \|\tilde{k}_{i,n+1} - k_i\|^2 &= E\{|1 - \alpha E_{i-1}(n)|^2\}\,\|\tilde{k}_{i,n} - k_i\|^2 \\ &\quad + \alpha^2 E\{[\tilde{C}_{i-1}(n) + k_i\tilde{E}_{i-1}(n)]^2\} \\ &\quad - 2\alpha E\{[1 - \alpha\tilde{E}_{i-1}(n)][\tilde{C}_{i-1}(n) \\ &\quad + k_i\tilde{E}_{i-1}(n)]\} \\ &\quad E\{\tilde{k}_{i,n} - k_i\} \\ &\triangleq A\|\tilde{k}_{i,n} - k_i\| + \alpha^2 B - 2\alpha C E\{\tilde{k}_{i,n} - k_i\}. \end{aligned} \quad (B2)$$

The coefficients in (B2) can again be computed using the expression for 4th order moments of Gaussian processes:

$$\begin{aligned} B &\triangleq E\{[\tilde{C}_{i-1}(n) + k_i\tilde{E}_{i-1}(n)]^2\} \\ &= E\{|\tilde{C}_{i-1}(n)|^2\} + k_i^2 E\{|\tilde{E}_{i-1}(n)|^2\} \\ &\quad + 2k_i E\{\tilde{C}_{i-1}(n)\,\tilde{E}_{i-1}(n)\}. \end{aligned} \quad (B3)$$

In (B3) we have

$$\begin{aligned} E\{|\tilde{C}_{i-1}(n)|^2\} &= E\{\tilde{f}_{i-1}^2(n)\,\tilde{b}_{i-1}^2(n)\} \\ &= E_{i-1}^2 + C_{i-1}^2 \end{aligned} \quad (B4)$$

$$\begin{aligned} E\{\tilde{C}_{i-1}(n)\,\tilde{E}_{i-1}(n)\} &= \tfrac{1}{2}E\{\tilde{f}_{i-1}(n)\,b_{i-1}(n)\,\tilde{f}_{i-1}^2(n)\} \\ &\quad + \tfrac{1}{2}E\{\tilde{f}_{i-1}(n)\,\tilde{b}_{i-1}(n)\,\tilde{b}_{i-1}(n)\} \\ &= 3C_{i-1}E_{i-1}. \end{aligned} \quad (B5)$$

Using (B4), (B5), and (A2), (B3) becomes

$$\begin{aligned} E\{[\tilde{C}_{i-1}(n) + k_i\tilde{E}_{i-1}(n)]^2\} &= E_{i-1}^2 + 2C_{i-1}^2 + 2k_i^2E_{i-1}^2 + k_i^2C_{i-1}^2 \\ &\quad + 6k_iC_{i-1}E_{i-1} \end{aligned} \quad (B6)$$

and with the value for the optimal reflection coefficient from (16), (B6) reduces to

$$\begin{aligned} B &\triangleq E\{[\tilde{C}_{i-1}(n) + k_i\tilde{E}_{i-1}(n)]^2\} = E_{i-1}^2 + 2C_{i-1}^2 + 2C_{i-1}^2 \\ &\quad + \frac{C_{i-1}^4}{E_{i-1}^2} - 6C_{i-1}^2 \\ &= E_{i-1}^2 - 2C_{i-1}^2 + \frac{C_{i-1}^4}{E_{i-1}^2} \\ &= \left[\frac{E_{i-1}^2 C_{i-1}^2}{E_{i-1}}\right]. \end{aligned} \quad (B7)$$

The recursion in (B2) can therefore be written as

$$\|\tilde{k}_{i,n+1} - k_i\|^2 = A\|\tilde{k}_{i,n} - k_i\|^2 + \alpha^2 B - 2\alpha C E\{\tilde{k}_{i,n} - k_i\}, \quad \text{(B8)}$$

where A is given by (A2), and B is given by (B7). Equation (B8) can now be considered as a difference equation in the variable

$$\mathbf{x}(n) \triangleq \begin{bmatrix} \|\tilde{k}_{i,n} - k_i\|^2 \\ E\{\tilde{k}_{i,n} - k_i\} \end{bmatrix}. \quad \text{(B9)}$$

Transfer analysis of this difference (20) and application of the final-value theorem of the Z-transform lead to an expression for the (normalized) misadjustment due to the ith stage.

$$m_i^2 = \frac{\alpha^2 B}{1 - A}. \quad \text{(B10)}$$

REFERENCES

[1] J. Makhoul, "Stable and efficient lattice methods for linear prediction," *IEEE Trans. Acoust., Speech, Signal Processing*, vol. ASSP-25, pp. 423-428, Oct. 1977.

[2] W. F. Gabriel, "Spectral analysis and adaptive array super-resolution techniques," *Proc. IEEE*, vol. 68, pp. 654-666, June 1980.

[3] W. S. Hodgkiss and J. A. Presley, Jr., "Adaptive trackings of multiple sinusoids whose power levels are widely separated," *IEEE Trans. Acoust., Speech, Signal Processing*, vol. ASSP-29, pp. 710-721, June 1981.

[4] B. Widrow, "Adaptive filters," in *Aspects of Network and System Theory*, R. Kalman and N. DeClaris, Eds. New York: Holt, Rinehart and Winston, 1971, pp. 563-587.

[5] J. Makhoul and R. Viswanathan, "Adaptive lattice methods for linear prediction," in *Proc. Int. Conf. Acoust., Speech, Signal Processing*, Apr. 1978, pp. 83-86.

[6] R. Viswanathan and J. Makhoul, "Sequential lattice methods for stable linear prediction," in *Proc. EASCON '76*, pp. 155A-155H.

[7] L. J. Griffiths, "Adaptive structures for multiple-input noise cancelling applications," in *Proc. Int. Conf. Acoust., Speech, Signal Processing*, Apr. 2-4, 1979, pp. 925-928.

[8] —, "An adaptive lattice structure for noise-cancelling applications," in *Proc. Int. Conf. Acoust., Speech, Signal Processing*, Apr. 10-12, 1978, pp. 87-90.

[9] —, "A continuously-adaptive filter as a lattice structure," in *Proc. Int. Conf. Acoust., Speech, Signal Processing*, May 13-19, 1977, pp. 683-686.

[10] J. Makhoul, "A class of all-zero lattice digital filters, properties and applications," *IEEE Trans. Acoust., Speech, Signal Processing*, vol. ASSP-26, pp. 304-314, Aug. 1978.

[11] K. G. Oza, "Identification problem and random contraction mappings," Ph.D. dissertation, Department of Electrical Engineering and Computer Sciences, University of California, Berkeley, 1967.

[12] K. G. Oza and E. I. Jury, "System identification and the principle of random contraction mapping," *SIAM J. Contr.*, vol. 6, pp. 249-257, 1968.

[13] G.R.L. Sohie and L. H. Sibul, "Application of Hilbert space theory to optimal and adaptive array processing," presented at 16th Annu. Conf. Inform. Sci. Syst., Princeton Univ., Mar. 17-19, 1982.

[14] F. Riesz and B. Sz-Nagy, *Functional Analysis*, Translated by F. Boron. New York: Ungar, 1955.

[15] J. D. Markel and A. H. Gray, *Linear Prediction of Speech*. New York: Springer-Verlag, 1979.

[16] T. Kailath, "A view of three decades of linear filtering theory," *IEEE Trans. Inform. Theory*, vol. IT-20, pp. 146-181, Mar. 1974.

[17] A. Papoulis, *Probability, Random Variables and Stochastic Processes*. New York: McGraw-Hill, 1965.

[18] A. Papoulis, *Signal Analysis*. New York: McGraw-Hill, 1977.

[19] J. Sakrison, "Stochastic approximation: A recursive method for solving regression problems," in *Advances in Communication Systems*, A. V. Balakrishnam, Ed. New York, London, England: Academic, 1966.

[20] H. Kwakernaak and R. Sivan, *Linear Optimal Control Systems*. New York: Wiley, 1972.

[21] G. Ungerboeck, "Theory on the speed of convergence in adaptive equalizers for digital communication," *IBM J. Res. Develop.*, Nov. 1972.

[22] B. Widrow *et al.*, "Stationary and nonstationary learning characteristics of the LMS adaptive filter," *Proc. IEEE*, vol. 64, pp. 1151-1162, Aug. 1976.

[23] C. J. Gibson and S. Haykin, "Learning characteristics of adaptive lattice filtering algorithms," *IEEE Trans. Acoust., Speech, Signal Processing*, vol. ASSP-28, pp. 681-691, Dec. 1980.

[24] O. Macchi and E. Eweda, "Second-order convergence analysis of stochastic adaptive linear filter," *IEEE Trans. Automat. Contr.*, vol. AC-28, pp. 76-85, Jan. 1983.

[25] L. Ljung, "Analysis of recursive stochastic algorithms," *IEEE Trans. Automat. Contr.*, vol. AC-22, pp. 551-575, Aug. 1977.

[26] M. L. Honig and D. G. Messerschmitt, "Convergence properties of an adaptive digital lattice filter," *IEEE Trans. Acoust., Speech, Signal Processing*, vol. ASSP-29, pp. 642-659, June 1981.

Bibliography

1. Acar, L., and Compton, R. T., Jr., "The Performance of an LMS Adaptive Array with Frequency Hopped Signals," *IEEE Trans. on Aerosp. Electron. Syst.*, vol. AES-21, no. 3, pp. 360–371, May 1985.
2. Adams, R. N., Horowitz, L. L., and Senne, K. D., "Adaptive Main-Beam Nulling for Narrow-Beam Antenna Arrays," *IEEE Trans. Aerosp. Electron. Syst.*, vol. AES-16, pp. 509-516, July 1980.
3. Agee, B. G., "The Least-Squares CMA: A New Technique for Rapid Correction of Constant Modulus Signals," in *Proc. IEEE Int. Conf. Acoust., Speech, Signal Processing,* Tokyo, Japan, pp. 953-956, 1986.
4. Ahmed, H. M., and Morf, M., "VLSI Array Architecture for Matrix Factorization," in *Proc. Workshop on Fast Algorithms for Linear Systems,* Aussois, France, 1981.
5. Ahmed, H. M., Ang, P. H., and Morf, M., "A VLSI Speech Analysis Chip Set Utilizing Coordinate Rotation Arithmetic," in *Proc. Conf. Circuits. Syst.,* Chicago, IL, 1981.
6. Ahmed, H. M., "VLSI Architectures for Real-Time Signal Processing," Ph.D. Dissertation, Department of Electrical Engineering, Stanford University, Stanford, CA, 1981.
7. Ahmed, H. M., Morf, M., Lee, D. T., and Ang, P. H., "A VLSI Speech Analysis Chip Set Based on Square Root Normalization Ladder Forms," in *Proc. Int. Conf. Acoust., Speech, Signal Processing,* Mar. 1981.
8. Ahmed, H. M., Delosme, J. M., and Morf, M., "Highly Concurrent Computing Structures for Matrix Arithmetic and Signal Processing," *IEEE Comput.,* Jan. 1982.
9. Ahmed, N. D., and Vijayendra, S., "An Adaptive Short-Term Correlator Algorithm," in *Proc. IEEE Int. Conf. Acoust., Speech, Signal Processing,* Boston, MA, 1983, pp. 25–28.
10. Ahmed, N. D., *et al,* "Detection of Multiple Sinusoids Using an Adaptive Cascaded Structure," in *Proc. IEEE Int. Conf. Acoust., Speech, Signal Processing,* San Diego, CA, 1984, pp. 21.3/1–4.
11. Alam, M. A., "Orthonormal Lattice Filter: A Multi-Stage Multi-Channel Estimation Technique," *Geophys.,* vol. 43, no. 7, pp. 1368–1383, Dec. 1978.
12. Alam, M. A., and Sage, A. P., "Sequential Estimation and Identification of Reflection Coefficients by Minimax Entropy Inverse Filtering," *Comput. Elec. Eng.,* vol. 2, pp. 315–338, 1975.
13. Albert, A. E., and Gardner, L. A. Jr., *Stochastic Approximation and Nonlinear Regression.* Cambridge, MA: Riverside Press, 1967.
14. Alexander, S. T., and Rajiala, S. A., "Adaptive Compression of Teleconference Sequences Using the LMS Algorithm," in *Proc. IEEE Int. Conf. Acoust., Speech, Signal Processing,* Boston, MA, 1983, pp. 1216–1219.
15. Alexander, S. T., and Miller T., "An Implementation of the LMS Adaptive Filter Using an SIMD Multiprocessor Ring Architecture," in *Proc. IEEE Int. Conf. Acoust., Speech, Signal Processing,* Tampa, FL, 1985, pp. 1605–1608.
16. Alexander, S. T., *Adaptive Signal Processing, Theory and Applications.* New York, NY: Springer Verlag, 1986.
17. Allen, J. L., "Array Antennas: New Applications for an Old Technique," *IEEE Spectrum,* vol. 35, pp. 115–130, Nov. 1964.
18. Amin, M., and Griffiths, L. J., "Time-Varying Spectral Estimation Using Symmetric Smoothing," in *Proc. IEEE Int. Conf. Acoust., Speech, Signal Processing,* Boston, MA, 1983, pp. 9–12.
19. Anderson, B. D. O., Hitz, K. L., and Diem, N. D., "Recursive Algorithm for Spectral Factorization," *IEEE Trans. Circuit Syst.,* vol. CAS-21, pp. 742–750, Nov. 1974.
20. Anderson, V. C., "Digital Array Phasing," *J. Acoust. Soc. Am.,* vol. 32, pp. 867–870, July 1960.
21. Anderson, V. C., "DICANNE, A Realizable Adaptive Process," *J. Acoust. Soc. Am.,* vol. 45, pp. 393–405, Mar. 1969.
22. Anderson, V. C., and Rudnick, P., "Rejection of a Coherent Arrival at an Array," *J. Acoust. Soc. Am.,* vol. 45, pp. 406–410, Apr. 1969.
23. Andresdottir, E., and Schafer, R. W., "Application of Adaptive Noise Cancelling in a Noisy Reverberant Environment," in *Proc. IEEE Int. Conf. Acoust., Speech, Signal Processing,* Boston, MA, 1983, pp. 57–60.
24. Aoki, M., *Optimization of Stochastic Systems.* New York, NY: Academic Press, 1967.
25. Appel, U., and Brandt, A. V., "Recursive Lattice Algorithms with Finite-Duration Windows," in *Proc. IEEE Int. Conf. Acoust., Speech, Signal Processing,* Paris, France, 1982, pp. 647-650.
26. Applebaum, S. P., "Adaptive Arrays," Syracuse University Research Corp. Rep. SPL-769, June 1964.
27. Applebaum, S. P., "Adaptive Arrays," *IEEE Trans. Antennas Propagat.,* vol. AP-24, no. 5, pp. 585–598, Sept. 1976.
28. Applebaum, S. P., and Chapman, D. J., "Adaptive Arrays with Mainbeam Constraints," *IEEE Trans. Antennas Propagat.,* vol. AP-24, pp. 650–662, Sept. 1976.
29. Ardalan, S. H., and Alexander, S. T., "Finite Wordlength Analysis of the Recursive Least Square Algorithm," presented at the Eighteenth Asilomer Conference on Circuits, Systems, and Computers, Santa Clara, CA, 1984.
30. Ardalan, S. H., "Derivation of the Fast Pole-Zero (ARMA) Recursive Least-Squares Algorithm Using Geometric Projections," in *Proc. IEEE Int. Conf. Acoust., Speech, Signal Processing,* Tokyo, Japan, 1986, pp. 1095–1098.
31. Ardalan, S. H., "Floating Point Error Analysis of Recursive Least-Squares and Least Mean Squares Adaptive Filters," in *Proc. IEEE Int. Conf. Acoust., Speech, Signal Processing,* Tokyo, Japan, 1986, pp. 513–516.
32. Åstrom̈, K. J., Ed., *Adaptive Systems in Control and Signal Processing.* Oxford, UK: Pergamon Press, 1986.
33. Atal, B. S., "On Determining Partial Correlation Coefficients by the Covariance Method of Linear Prediction," presented at 94th Meeting, Acoust. Soc. Amer., 1977.
34. Ayala, I. L., "On a New Adaptive Lattice Algorithm for Recursive Filters," *IEEE Trans. Acoust., Speech, Signal Processing,* vol. ASSP-30, no. 2, pp. 316–319, Apr. 1982.
35. Baird, C. A., "Recursive Processing for Adaptive Arrays," in *Proc. Adaptive Antenna Systems Workshop, Naval Research Laboratory,* Washington, DC, March 11–13, 1974, pp. 163–183.
36. Belfiore, C. A., and Park, J. H., "Decision Feedback Equalization," *Proc. IEEE,* vol. 67, no. 8, pp. 1143–1156, Aug. 1979.
37. Bellanger, M. and Evci, C. C., "An Efficient Step Size Adaption Technique for LMS Adaptive Filters," in *Proc. IEEE Int. Conf. Acoust., Speech, Signal Processing,* Tampa, FL, 1985, pp. 1153–1156.
38. Bellanger, M. and Evci, C. C., "Coefficient Wordlength Limitation in FLS Adaptive Filters," in *Proc. IEEE Int. Conf. Acoust., Speech, Signal Processing,* Tokyo, Japan, 1986.
39. Bellanger, M. and Evci, C. C., "On Computational Complexity in Adaptive Digital Filters," in *Proc. IEEE Int. Conf. Acoust., Speech, Signal Processing,* Boston, MA., 1983, pp. 45–48.
40. Benveniste, A., and Chaure, C., "Une Methode Rapide Pour Arriver a des Algorithms d'Estimation Rapides Pour des Fonctions de Transfert AR et ARMA Synthesisees en Treillis," IRISA Int. Pub., No. 116, Rennes, France, May 1979.
41. Benveniste, A., and Chaure, C., "AR and ARMA Identification Algorithms of Levinson Type: An Innovations Approach," *IEEE Trans. Automat. Contr.,* vol. AC-26, no. 6, pp. 1243–1260, Dec. 1981.
42. Bershad, N. J., and Qu, L. Z., "On the Probability Density Functions of the Weight for the Complex Scalar LMS Adaptive Algorithm," in *Proc. IEEE Int. Conf. Acoust., Speech, Signal Processing,* San Diego, CA, 1984, pp. 3.5.1–3.5.4.
43. Bershad, N. J., and Chang, Y. H., "Time Correlation Statistics of the LMS Adaptive Algorithm Weights," in *Proc. IEEE Int. Conf. Acoust., Speech, Signal Processing,* San Diego, CA, 1984, pp. 3.6.1–3.6.4.
44. Bershad, N. J., "Optimum Data Nonlinearity in LMS Adaptation," *IEEE Trans. Acoust. Speech, Signal Processing,* vol. ASSP-34, no. 1, pp. 69-76, Feb. 1986.
45. Bershad, N. J., and Feintuch, P. L., "Normalized Frequency-Domain LMS Adaptive Algorithm," *IEEE Trans. Acoust. Speech, Signal Processing,* vol. ASSP-34, no. 3, pp. 452–461, June 1986.
46. Bershad, N. J., "Normalized Least-Mean-Square Adaptive Filter

Algorithm with Gaussian Inputs; Analysis," *IEEE Trans. Acoust. Speech, Signal Processing,* vol. ASSP-34, no. 4, pp. 793–806, Aug. 1986.

47. Bierman, G. J., *Factorization Methods for Discrete Sequential Estimation.* New York, NY: Academic Press, 1977.
48. Bierman, G. J., *Modern Spectrum Estimation,* D. G. Childers, Ed. New York, NY: IEEE PRESS, 1978.
49. Blum, J. R., "Multidimensional Stochastic Approximation Methods," *Ann. Math. Stat.,* vol. 25, pp. 737–744, 1954.
50. Bode, H., and Shannon, C., "A Simplified Derivation of Linear Least Square Smoothing and Prediction Theory," *Proc. IRE,* vol. 38, no. 4, pp. 417– 425, April 1950.
51. Box, G. E. P., and Jenkins, G. M., *Time Series Analysis: Forecasting and Control.* San Francisco, CA: Holden-Day, 1970.
52. Brandt, A. V., "Detecting and Estimating Parameter Jumps Using Ladder Algorithms and Likelihood Ratio Tests," in *Proc. IEEE Int. Conf. Acoust., Speech, Signal Processing,* Boston, MA., 1983, pp. 1017–1020.
53. Brennan, L. E., and Reed, I. S., "Theory of Adaptive Radar," *IEEE Trans. Aerosp. Electron. Syst.,* vol. AES-9, pp. 237–252, Mar. 1973.
54. Bruton, L. T., "Low-Sensitivity Digital Ladder Filters," *IEEE Trans. Circuit Syst.,* vol. CAS-22, no. 3, pp. 168–176, Mar. 1975.
55. Bruton, L. T., and Vaughan-Pope, D. A., "Synthesis of Digital Ladder Filters from LC Filters," *IEEE Trans. Circuits Syst.,* vol. CAS-23, no. 6, pp. 395–402, June 1976.
56. Bryn, F., "Optimum Signal Processing of Three-Dimensional Arrays Operating on Gaussian Signals and Noise," *J. Acoust. Soc. Am.,* vol. 34, pp. 289–297, Mar. 1962.
57. Buckley, K. M., "Broadband Beamforming; Generalized Sidelobe Canceler," *IEEE Trans. Acoust., Speech, Signal Processing,* vol. ASSP-34, no. 5, pp. 1322–1323.
58. Bucy, R. S., "Stability and Positive Supermartingale," *J. Diff. Equations,* vol. 1, pp. 151–155, Apr. 1965.
59. Burg, J. P., "Three-Dimensional Filtering with an Array of Seismometers," *Geophysics,* vol. 29, pp. 693–713, Oct. 1964.
60. Burg, J. P., "Maximum Entropy Spectral Analysis," presented at the 37th Annual International Meeting, Society of Exploratory Geophysicists, Oklahoma City, OK, Oct. 31, 1967.
61. Burg, J. P., "Maximium Entropy Spectral Analysis," Ph.D. dissertation, Stanford University, Stanford, CA, 1975.
62. Cadzow, J. A., and Moses, R. L., "An Adaptive ARMA Spectral Estimator, Parts 1 and 2," in *Proc. 1st ASSP Workshop on Spectral Estimation,* McMaster Univ., Hamilton, Ont., Canada, Aug. 1982.
63. Capon, J., Greenfield, J. R., and Kolker, R. K., "Multidimensional Maximum-Likelihood Processing of a Large Aperture Seismic Array," *Proc. IEEE,* vol. 55, pp. 192–211, Feb. 1967.
64. Capon, J., "High-Resolution Frequency-Wave Number Spectrum Analysis," *Proc. IEEE,* vol. 57, no. 8, pp. 1408–1418, Aug. 1969.
65. Caraiscos, C., and Liu, B., "A Round-off Error Analysis of the LMS Adaptive Algorithm," in *Proc. IEEE Int. Conf. Acoust., Speech, Signal Processing,* Boston, MA, 1983, pp. 29–32.
66. Caraiscos, C., and Liu, B., "A Roundoff Error Analysis of the LMS Adaptive Algorithm," *IEEE Trans. on Acoust., Speech, Signal Processing,* vol. ASSP-32, no. 2, pp. 34-41, Feb. 1984.
67. Carayannis, G., Kalouptsidis, N., and Manolakis, D. G., "Fast Recursive Algorithms for a Class of Linear Equations," *IEEE Trans. Acoust., Speech, Signal Processing,* vol. ASSP-20, pp. 227–239, Apr. 1982.
68. Carayannis, G., Manolakis, D. G., and Kalouptsidis, N., "A Fast Sequential Algorithm for Least-Squares Filtering and Prediction," *IEEE Trans. Acoust., Speech, Signal Processing,* vol. ASSP-31, no. 6, pp. 1394–1402, Dec. 1983.
69. Carayannis, G., Manolakis, D. G. and Kalouptsidis, N., "A Unified View of Parametric Processing Algorithms for Prewindowed Signals," *Signal Processing,* pp. 335–368, Oct. 1986.
70. Carpenter, R. N., "Applications of Adaptive Signal Processing to the Reduction of Flow-Induced Noise in Small Underwater Vehicles," in *Proc. IEEE Int. Conf. Acoust., Speech, Signal Processing,* Boston, MA, 1983, pp. 969–972.
71. Carter, T. E., "Study of an Adaptive Lattice Structure for Linear Prediction Analysis of Speech," in *Proc. IEEE Int. Conf. Acoust., Speech, Signal Processing,* Tulsa OK, 1978, pp. 27–30.
72. Casar-Corredera, J., *et al,* "Data Echo Nonlinear Cancellation," in *Proc. IEEE Int. Conf. Acoust., Speech, Signal Processing,* Tampa, FL, 1985, pp. 1245–1248.
73. Chabries, D. M., *et al,* "Application of the LMS Adaptive Filter to Improve Speech Communication in the Presence of Noise," in *Proc. IEEE Int. Conf. Acoust., Speech, Signal Processing,* Paris, France, 1982, pp. 148–151.
74. Chang, R. W., "A New Equalizer Structure for Fast Start-Up Digital Communications," *Bell Syst. Tech. J.,* vol. 50, no. 6, pp. 1969–2014, July/Aug. 1971.
75. Chang, C. Y., "Multichannel Adaptive Filtering with a Feedback Convergence Function," in *Proc. IEEE Int. Conf. Acoust., Speech, Signal Processing,* Paris, France, 1982, pp. 667–670.
76. Chao-Huan, H., and Si-Zun, Y., "Adaptive Array Processing Using Predicted Coefficients as Constrained Conditions," in *Proc. IEEE Int. Conf. Acoust., Speech, Signal Processing,* Paris, France, 1982, pp. 1473–1476.
77. Chapman, D. J., "Partial Adaptivity for the Large Array," *IEEE Trans. Antennas Propagat.,* vol. AP-24, pp. 685–696, Sept. 1976.
78. Chen, H., *et al,* "Adaptive Spectral Estimation by the Conjugate Gradient Method," in *Proc. IEEE Int. Conf. Acoust., Speech, Signal Processing,* Tampa, FL, 1985, pp. 81–84.
79. Chen, Y. Y., "Stable Time- and Order-Recursive Algorithm for the Adaptive Lattice Filter," in *Proc. IEEE Int. Conf. Acoust., Speech, Signal Processing,* San Diego, CA, 1984.
80. Cheung, J. Y., "Convergence Conditions for an Optimal Solution in Adaptive Recursive Filters," in *Proc. IEEE Int. Conf. Acoust., Speech, Signal Processing,* San Diego, CA, 1984.
81. Chien, Y. T., and Fu, K. S., "Stochastic Learning of Time Varying Parameters in Random Environment," *IEEE Trans. Syst. Sci. Cybern.,* vol. SSC-5, pp. 237–246, Sept. 1969.
82. Childers, D. G., Ed., *Modern Spectrum Analysis.* New York, NY: IEEE PRESS, 1978.
83. Chu, P. L., and Messerschmitt, D. G., "Zero Sensitivity of the Digital Lattice Filter," in *Proc. IEEE Int. Conf. Acoust., Speech, Signal Processing,* 1980, pp. 89–93.
84. Chu, P. L., and Messerschmitt, D. G., "Zero Sensitivity Properties of the Digital Lattice Filter," *IEEE Trans. Acoust., Speech, Signal Processing,* vol. ASSP-31, no. 3, pp. 685–706, June 1983.
85. Chung, K. C., "Structure of Interference Extent on LMS Spatial Cancellation," in *Proc. IEEE Int. Conf. Acoust., Speech, Signal Processing,* San Diego, CA, 1984.
86. Cioffi, J. M., and Kailath, T., "Fast, Fixed-Order Least Squares Algorithms for Adaptive Filtering," in *Proc. IEEE Int. Conf. Acoust., Speech, Signal Processing,* Boston, MA, 1983, pp. 675–678.
87. Cioffi, J. M., and Kailath, T., "Fast, Recursive Least-Squares Transversal Filters for Adaptive Filtering," *IEEE Trans. Acoust., Speech, Signal Processing,* vol. ASSP-32, no. 2, pp. 304–337, Apr. 1984.
88. Cioffi, J. M., "Fast Transversal Filters for Communications Applications," Ph.D. dissertation, Stanford University, 1984.
89. Cioffi, J. M., "The Block-Processing FTF Adaptive Algorithm," in *Proc. IEEE Int. Conf. Acoust., Speech, Signal Processing,* Tampa, FL, 1985, pp. 1241–1244.
90. Cioffi, J. M., "Block Processing Fast-Transversal Filter Adaptive Algorithm," *IEEE Trans. Acoust. Speech, Signal Processing,* vol. ASSP-34, no. 1, pp. 77–90, Feb. 1986.
91. Cioffi, J. M., "A Covariance RLS Lattice for Adaptive Filtering," in *Proc. IEEE Int. Conf. Acoust., Speech, Signal Processing,* Tokyo, Japan, 1986.
92. Claasen, T. A. C. M., and Mecklenbrauker, W. F. G., "Comparison of the Convergence of Two Algorithms for Adaptive FIR Digital Filters," *IEEE Trans. Circuit Syst.,* vol. CAS-28, no. 6, pp. 510–518, June 1981.
93. Claerbout, J. F., "A Summary, by Illustrations, of Least Squares Filters with Constraints," *IEEE Trans. Inform. Theory,* vol. IT-14, pp. 269–272, Mar. 1968.
94. Clark, G. A., Mitra, S. K., and Parker, S. R., "Block Implementation of Adaptive Digital Filters," *IEEE Trans. Circuit Syst.,* vol. CAS-28, no. 6, pp. 584–592, June 1981.
95. Clark, G. A., Parker, S. R., and Mitra, S. K., "A Unified Approach to Time- and Frequency-Domain Realization of FIR Adaptive Digital Filters," *IEEE Trans. Acoust., Speech, Signal Processing,* vol. ASSP-31, no. 5, pp. 1073–1083, Oct. 1983.
96. Cochran, W. G., and Davis, M., "The Robbins-Monroe Method for Estimating the Median Lethal Dose," *J. R. Stat. Soc.,* Series B, vol. 27, pp. 28–44, 1965.
97. Comer, T. R., "Some Stochastic Approximation Procedures for Use in Process Control," *Math. Stat.,* vol. 35, pp. 86–98, 1964.
98. Compton, R. T., Jr., "An Adaptive Array in a Spread-Spectrum

Communication System,'' *Proc. IEEE,* vol. 66, p. 289, Mar. 1978.
99. Compton, R. T., Jr., Adaptive Antennas: Concepts and Applications. Englewood Cliffs, NJ: Prentice Hall, 1985.
100. Corl, D., ''A ACTD Adaptive Inverse Filter,'' *Electron. Lett.,* vol. 14, no. 3, pp. 60-62, Feb. 2, 1978.
101. Cowan, C. F. N., Mavor, J., and Arthur, J. W., ''Implementation of a 64-Point Adaptive Filter Using an Analogue CCD Programmable Filter,'' *Electron. Lett.,* vol. 14, no. 17, pp. 568–569, Aug. 17, 1978.
102. Cowan, C. F. N., Mavor, J., Arthur, J. W., and Denyer, P. B., ''An Evaluation of Analogue and Digital Adaptive Filter Realizations,'' International Specialists Seminar on Case Studies in Advanced Signal Processing, *IEEE Conference Proc. 180,* pp. 178–183, Sept. 1979.
103. Cowan, C. F. N., and Mavor, J., ''Miniature CCD-Based Analog Adaptive Filters,'' in *Proc. IEEE Inter. Conf. Acoust., Speech, Signal Processing,* Denver, CO, 1980, pp. 474–478.
104. Cowan, C. F. N., Arthur, J. W., Mavor, J., and Denyer, P. B., ''CCD-Based Adaptive Filters: Realization and Analysis,'' *IEEE Trans. Acoust., Speech, Signal Processing,* vol. ASSP-29, no. 2, pp. 220–229, Apr. 1981(1).
105. Cowan, C. F. N., and Mavor, J., ''New Digital Adaptive Filter Implementation Using Distributed-Arithmetic-Techniques,'' in *Proc. IEEE Acoust., Speech, Signal Processing,* vol. 128, pt. F, no. 4, pp. 225–230, Aug. 1981(2).
106. Cowan, C. F. N., Smith, S. G., and Elliott, J. H., ''A Digital Adaptive Filter Using a Memory-Accumulator Architecture: Theory and Realization,'' *IEEE Trans. Acoust., Speech, Signal Processing,* vol. ASSP-31, no. 3, pp. 541–549, June 1983.
107. Cowan, C. F. N., and Grant, P. M., Eds., *Adaptive Filters.* Englewood Cliffs, NJ: Prentice-Hall, 1985.
108. Cox, H., ''Optimum Arrays and the Schwarz Inequality,'' *J. Acoust. Soc. Am.,* vol. 45, pp. 228–232, 1969.
109. Cox, H., ''Resolving Power and Sensitivity to Mismatch of Optimum Array Processors,'' *J. Acoust. Soc. Am.,* vol. 54, no. 3, pp. 771–785, Sept. 1973.
110. Cox, J. R., Jr., and Medgyesi-Mitxchang, L. N., ''An Algorithmic Approach to Signal Estimation Useful in Fetal Electrocardiography,'' *IEEE Trans. Biomed. Eng.,* vol. BME-16, pp. 215–219, July 1969.
111. Crochiere, R. E., ''Digital Ladder Structures and Coefficient Sensitivity,'' *IEEE Trans. Audio Electroacoust.,* vol. AU-20, no. 4, pp. 240–246, Oct. 1972.
112. Cumani, A., ''On a Covariance-Lattice Algorithm for Linear Prediction,'' in *Proc. IEEE Int. Conf. Acoust., Speech, Signal Processing,* Paris, France, 1982, pp. 651–654.
113. Cybenko, G., ''The Numerical Stability of the Levinson-Durbin Algorithm for Toeplitz Systems of Equations,'' *SIAM J. Sci. Statist. Comput.,* vol. 1, pp. 303–320, 1980.
114. Cybenko, G., ''A General Orthogonalization Technique with Applications to Time Series Analysis and Signal Processing,'' *Math. Comput.,* vol. 40, no. 161, pp. 323–336, Jan. 1983.
115. Daniell, T., ''Adaptive Estimation with Mutually Correlated Training Samples,'' Ph.D. dissertation, Stanford Electronics Lab., Stanford Univ., Rep. SEL-68-083, Aug. 1968.
116. Daniell, T. P., ''Adaptive Estimation with Mutually Correlated Training Sequences,'' *IEEE Trans. Syst. Sci. Cybern.,* vol. SSC-6, pp. 12–19, Jan. 1970.
117. Darlington, S., ''Linear Least Squares Smoothing and Prediction, with Applications,'' *Bell Syst. Tech. J.,* vol. 37, pp. 1221–1294, Dec. 1958.
118. David, R. A., ''IIR Adaptive Algorithms Based on Gradient Search Techniques,'' Ph.D. dissertation, Department of Electrical Engineering, Stanford University, Stanford, CA, Aug. 1981.
119. David, R. A., ''Detection of Multiple Sinusoids Using a Parallel ALE,'' in *Proc. IEEE Int. Conf. Acoust., Speech, Signal Processing,* San Diego, CA, 1984.
120. Davisson, L. D., ''A Theory of Adaptive Filtering,'' *IEEE Trans. Inform. Theory.,* vol. IT-12, pp. 97–102, Apr. 1966.
121. Davisson, L. D., ''Steady State Error in Adaptive Mean Square Minimization,'' *IEEE Trans. Inform. Theory.,* vol. IT-16, pp. 382–385, July 1970.
122. Davisson, L. D., ''Convergence Probability Bounds for Stochastic Approximation,'' *IEEE Trans. Inform. Theory.,* vol. IT-16, pp. 680–685, Nov. 1970.
123. Delosme, J. M., ''Algorithms and Implementations for Linear Least Squares Estimation,'' Ph.D. dissertation, Stanford University, Stanford, CA, 1982.
124. Delosme, J. M., and Morf, M., in ''Mixed and Minimal Representations for Toeplitz and Related Systems,'' in *Proc. 14th Asilomar Conf. Circuits, Systems, and Computers,* Pacific Grove, CA, Nov. 1980, pp. 19–24.
125. Delosme, J. M., and Morf, M., ''A Tree Classification of Algorithms for Toeplitz and Related Equations Including Generalized Levinson and Doubling Type Algorithms,'' in *Proc. 19th IEEE Conf. Decision and Control,* Dec. 1980, pp. 42–46.
126. Delosme, J. M., and Morf, M., ''A Unified Stochastic Description of Efficient Algorithms for Second Order Processes,'' in *Proc. Workshop on Fast Algorithms for Linear Systems,* Aussois, France, 1981.
127. Delosme, J. M., and Morf, M., ''Constant-Gain Filters for Finite Shift-Rank Processes,'' in *Proc. IEEE Int. Conf. Acoust., Speech, Signal Processing,* Paris, France, May 1982, pp. 1732–1735.
128. Delosme, J. M., Genin, Y., Morf, M., and Van Dooren, P., ''Σ Contractive Embeddings and Interpretation of Some Algorithms for Recursive Estimation,'' in *Proc. 14th Asilomar Conf. Circuits Systems and Computers,* Pacific Grove, CA, Nov. 1980, pp. 25–28.
129. Demytko, N., and Mackechnie, L. K., ''A High Speed Digital Adaptive Echo Canceler,'' *Australian Telecommunication Rev.,* vol. 7, no. 1, pp. 20–28, 1973.
130. Dentino, N., McCool, J., and Widrow, B., ''Adaptive-Filtering in the Frequency Domain,'' *Proc. IEEE,* vol. 66, no. 12, pp. 1658–1659, Dec. 1978.
131. Deprettere, E., ''Orthogonal Digital Cascade Filters and Recursive Construction Algorithms for Stationary and Nonstationary Fitting or Orthogonal Models,'' Ph.D. dissertation, Delft University of Technology, Delft, The Netherlands, 1980.
132. Deprettere, E., and Dewilde, P., ''Orthogonal Cascade Realization of Real Multiport Digital Filter,'' Tech Rep., Dept. Elec. Eng., Delft University, Delft, The Netherlands, 1980.
133. Deprettere, E., and Dewilde, P., ''Generalized Orthogonal Filters for Stochastic Prediction and Modeling,'' in *Digital Signal Processing,* V. Cappellini and A. G. Constantinides, Eds. New York, NY: Academic Press, 1980, pp. 35–48.
134. Deprettere, E. F. A., ''Fast, Non-stationary Lattice Recursions for Adaptive Modeling and Estimation,'' in *Proc. IEEE Int. Conf. Acoust., Speech, Signal Processing,* Paris, France, 1982, pp. 1721–1726.
135. Derman, C., and Sacks, J., ''On Dvoretzky's Stochastic Approximation Theorem,'' *Ann. Math Stat.,* vol. 30, pp. 601–606, 1959.
136. Dewilde, P., Vieira, A. C., and Kailath, T., ''On a Generalized Szego-Levinson Realization Algorithm for Optimal Linear Predictors Based on a Network Synthesis Approach,'' *IEEE Trans. Circuit Syst.,* vol. CAS-25, no. 9, pp. 663–675, Sept. 1978.
137. Dickinson, B. W., ''An Approach to Stationary Autoregressive Estimation,'' in *Proc. Conf. Information Sciences and Systems,* Johns Hopkins University, Baltimore, MD, 1977, pp. 507–511.
138. Dickinson, B. W., ''Autoregressive Estimation Using Residual Energy Ratios,'' *IEEE Trans. Inform. Theory,* vol. IT-24, no. 4, pp. 503–506, July 1978.
139. Dickinson, B. W., ''Estimation of Partial Correlation Matrices Using Cholesky Decomposition,'' *IEEE Trans. Automat. Contr.,* vol. AC-24, no. 2, pp. 302–305, 1979.
140. Dickinson, B. W., and Turner, J. M., ''Reflection Coefficient Estimation Using Cholesky Decomposition,'' *IEEE Trans. Acoust., Speech, Signal Processing,* vol. ASSP-27, no. 2, pp. 146–149, Apr. 1979.
141. Dickinson, B. W., and Turner, J. M., ''Reflection Coefficient Estimates Based on a Markov Chain Model,'' in *Proc. IEEE Int. Conf. Acoust., Speech, Signal Processing,* Washington, DC, Apr. 1979, pp. 727–730.
142. Ding, H., and Yu, C., ''Adaptive Lattice Noise Canceller and Optimal Step Size,'' in *Proc. IEEE Int. Conf. Acoust., Speech, Signal Processing,* Tokyo, Japan, 1986, pp. 2939–2942.
143. Dupač, V., ''A Dynamic Stochastic Approximation Method,'' *Ann. Math. Stat.,* vol. 36, pp. 1695–1702, 1965.
144. Durbin, J., ''The Fitting of Time Series Models,'' *Rev. L'Institut Intl. de Statisque,* vol. 28, pp. 233–243, 1960.
145. Duttweller, D. L., J. E. Maxo, and D. G. Messerschmitt, ''An Upper Bound on the Error Probability in Decision-Feedback Equalization,'' *IEEE Trans. Inform. Theory,* vol. IT-20, no. 4, pp. 490–496, July 1974.
146. Duttweller, D. L., ''A Twelve Channel Digital Echo Canceler,'' *IEEE Trans. Commun.,* vol. COM-26, no. 5, pp. 647–653, May 1978.
147. Duttweller, D. L., and Chen, Y. S., ''A Single-Chip VLSI Echo Canceler,'' *Bell Syst. Tech. J.,* vol. 59, no. 2, pp. 149–160, Feb. 1980.

148. Duttweller, D. L., "Adaptive Filter Performance with Nonlinearities in the Correlation Multiplier," *IEEE Trans. Acoust., Speech, Signal Processing,* vol. ASSP-30, no. 4, pp. 578–586, Aug. 1982.
149. Dvoretzky, A., "On Stochastic Approximation," in *Proc. Third Berkeley Symp. Math. Stat. Probab.,* vol. I, 1956, pp. 39–55.
150. Earp, S. L., and Nolte, L. W., "Multichannel Adaptive Array Processing for Optimal Detection," in *Proc. IEEE Int. Conf. Acoust., Speech, Signal Processing,* San Diego, CA, 1984.
151. Edelblute, D. J., Fisk, J. M., and Kinnison G. L., "Criteria for Optimum Signal Detection Theory for Arrays," *J. Acoust. Soc. Am.,* vol. 41, pp. 199–205, Jan. 1967.
152. Egardt, B., and Morf, M., "Asymptotic analysis of a ladder algorithm for ARMA models," Preprint, July 1980.
153. Eleftheriou, E., and Falconer, D., "Steady-State Behavior of RLS Adaptive Algorithms," in *Proc. IEEE Int. Conf. Acoust., Speech, Signal Processing,* Tampa, FL, 1985, pp. 1145–1148.
154. Eleftheriou, E., and Falconer, D., "Tracking Properties and Steady-State Performance of RLS Adaptive Filter Algorithms," *IEEE Trans. Acoust., Speech, Signal Processing,* vol. ASSP-34, no. 5, pp. 1097–1110, Oct. 1986.
155. El-Sharkawy, M. A., "Recursive Adaptive Filters in Deterministic and Stochastic Environments," Ph.D. dissertation, Southern Methodist University, 1985.
156. El-Sharkawy, M. A., and Peikari, B., "An Adaptive Stochastic Filter with No Positive Real Condition," in *Proc. IEEE Int. Conf. Acoust., Speech, Signal Processing,* Tokyo, Japan, 1986, pp. 2111–2114.
157. El-Sherief, H., "Adaptive Least-Squares for Parametric Spectral Estimation and its Application to Pulse Estimation and Deconvolution of Seismic Data," *IEEE Trans. on Syst. Man, and Cybern.,* vol. SMC-16, vol. 2, pp. 299–303, Mar./Apr. 1986.
158. El-Sherief, H., "Adaptive Least-Squares for Parametric Spectral Estimation and its Application to Pulse Excitation and Deconvolution of Seismic Data," in *Proc. IEEE Int. Conf. Acoust., Speech, Signal Processing,* San Diego, CA, 1984, pp. 5.5.1–5.5.4.
159. Etter, D. M., *et al,* "Recursive Adaptive Filter Design Using an Adaptive Generic Algorithm," in *Proc. IEEE Int. Conf. Acoust., Speech, Signal Processing,* Paris, France, 1982, pp. 635–638.
160. Etter, D. M., *et al,* "IIR Algorithms for Adaptive Line Enhancement," in *Proc. IEEE Int. Conf. Acoust., Speech, Signal Processing,* Paris, France, 1982, pp. 17–20.
161. Etter, D. M., "Identification of Sparse Impulse Response Systems Using an Adaptive Delay Filter," in *Proc. IEEE Int. Conf. Acoust., Speech, Signal Processing,* Tampa, FL, 1985, pp. 1169–1172.
162. Etter, D. M., and Huang, C. T., "An Adaptive Delay Filter with Variable Gains and Variable Taps," in *Proc. IEEE Int. Conf. Acoust., Speech, Signal Processing,* Tokyo, Japan, 1986, pp. 2123–2126.
163. Evei, C. C., and Bellanger, M., "Characteristics of Adaptive Filters with Leakage," in *Proc. IEEE Int. Conf. Acoust., Speech, Signal Processing,* San Diego, CA, 1984.
164. Everly, F. J., "Application of Stochastic Approximation to Frequency Domain Adaptive Filters," Ph.D. thesis, Pennsylvania State University, 1976.
165. Fabian, V., "On Choice. of Design in Stochastic Approximation Methods," *Ann. Math. Stat.,* vol. 39, pp. 457–465, 1968.
166. Fabre, P., and Gueguen, C., "Fast Recursive Least-Squares Algorithms: Preventing Divergence," in *Proc. IEEE Int. Conf. on Acoust., Speech, Signal Processing,* Tampa, FL, 1985, pp. 1149–1152.
167. Falb, P. L., and Wolovich, W. A., "Decoupling in the Design and Synthesis of Multivariable Control Systems," *IEEE Trans. Automat. Contr.,* vol. AC-12, pp. 651–659, Dec. 1967.
168. Falconer, D. D., and Ljung, L., "Application of Fast Kalman Estimation to Adaptive Equalization," *IEEE Trans. Commun.,* vol. COM-26, no. 10, pp. 1439–1446, Oct. 1978.
169. Falconer, D. D., "Jointly Adaptive Equalization and Carrier Recovery in Two-Dimensional Digital Communication Systems," *Bell Syst. Tech. J.,* vol. 55, no. 3, pp. 317–334, Mar. 1976(1).
170. Falconer, D. D., "Application of Passband Decision Feedback Equalization in Two-Dimensional Data Communication Systems," *IEEE Trans. Commun.,* vol. COM-24, no. 10, pp. 1159–1166, Oct. 1976(2).
171. Falconer, D. D., and Mueller, K. H., "Adaptive Echo Cancellation/AGC Structures for Two-Wire, Full-Duplex Data Transmission," *Bell Syst. Tech. J.,* vol. 58, no. 7, pp. 1593–1616, Sept. 1979.
172. Farden, D. C., "Stochastic Approximation with Correlated Data," Ph.D. dissertation, Department of Electrical Engineering, Colorado State University, Fort Collins, CO.
173. Farden, D. C., "Stochastic Approximation with Correlated Data," *IEEE Trans. Inform. Theory,* vol. IT-27, no. 1, pp. 105–113, Jan. 1981.
174. Farden, D. C., "Tracking Properties of Adaptive Signal Processing Algorithms," *IEEE Trans. Acoust., Speech, Signal Processing,* vol. ASSP-29, no. 3, pp. 439–446, June 1981.
175. Farden, D. C. and Davis, R. M., "Orthogonal Weight Perturbation Algorithms in Partially-Adaptive Arrays," *IEEE Trans. Antennas Propagat.,* vol. AP-33, no. 1, pp. 56–63, Jan. 1985.
176. Feintuch, P. L., "An Adaptive Recursive LMS Filter," *Proc. IEEE,* vol. 64, no. 11, pp. 1622–1624, Nov. 1976.
177. Ferber, R., and Harris, H., "Adaptive Processing of Digital Broadband Seismic Data," *IEEE Trans. Geosci. and Remote Sensing,* vol. GES-23, vol. 6, pp. 789–796, Nov. 1985.
178. Fernandez-Gaucherand, E., and Cruz, J., "Equalization for Transmission Line Channels: A Discussion of Three IIR Adaptive Filtering Algorithms," in *Proc. IEEE Int. Conf. on Acoust., Speech, Signal Processing,* Tampa, FL, 1985, pp. 1249–1252.
179. Ferrara, E. R., "Fast Implementation of LMS Adaptive Filters," *IEEE Trans. Acoust., Speech, Signal Processing,* vol. ASSP-28, no. 4, pp. 474–475, Aug. 1980.
180. Ferrara, E. R., and Widrow, B., "Multichannel Adaptive Filtering for Signal Enhancement," *IEEE Trans. Circuit Syst.,* vol. CAS-28, no. 6, pp. 606–610, June 1981.
181. Fettweis, A., "Digital Filter Structures Related to Classical Filter Networks," *Arch. Elek. Ubertragung,* vol. 25, pp. 78–89, 1971.
182. Fettweis, A., "Some Principles of Designing Digital Filters Imitating Classical Filter Structures," *IEEE Trans. Circuit Theory,* vol. CT-18, pp. 314–316, Mar. 1971.
183. Fettweis, A., "Pseudopassivity, Sensitivity, and Stability of Wave Digital Filters," *IEEE Trans. Circuit Theory,* vol. CT-19, no. 6, pp. 668–673, Nov. 1972.
184. Fettweis, A., "Wave Digital Filters with Reduced Number of Delays," *Int. J. Circuit Theory Appl.,* vol. 2, no. 4, pp. 319–320, 1974.
185. Fettweis, A., "On Sensitivity and Roundoff Noise in Wave Digital Filters," *IEEE Trans. Acoust., Speech, Signal Processing,* vol. ASSP-22, no. 5, pp. 383–384, Oct. 1974.
186. Fettweis, A., Levin, H., and Sedlmeyer, A., "Wave Digital Lattice Filters," *Int. J. Circuit Theory Appl.,* vol. 2, pp. 203–211, 1974.
187. Fever, A., and Weinstein, E., "Convergence Analysis of LMS Filters with Uncorrelated Gaussian Data," *IEEE Trans. Acoust., Speech, Signal Processing,* vol. ASSP-33, vol. 1, pp. 222–230, Feb. 1985.
188. Fletcher, R., *Practical Methods of Optimization, Vol. 1: Unconstrained Optimization.* New York, NY: John Wiley, 1980.
189. Friedlander, B., Kailath, T., Morf, M., and Ljung, L., "Extended Levinson and Chandrasekhar Equations for General Discrete-Time Linear Estimation Problems," *IEEE Trans. Automat. Contr.,* vol. AC-23, no. 4, pp. 653–659, Aug. 1978.
190. Friedlander, B., Morf, M., Kailath, T., and Ljung, L., "New Inversion Formulas for Matrices Classified in Terms of Their Distance from Toeplitz Matrices," *Linear Algebra and Its Applications,* vol. 27, pp. 31–60, 1979.
191. Friedlander, B., "Recursive Lattice Forms for Adaptive Control," in *Proc. Joint Automatic Control Conf.,* San Francisco, CA, Aug. 1980, paper WP2-E.
192. Friedlander, B., and Morf, M., "Efficient Inversion Formulas for Sums of Products of Toeplitz and Hankel Matrices," in *Proc. 18th Annual Allerton Conf. Communication, Control and Computing,* Oct. 1980.
193. Friedlander, B., "A Pole-Zero Lattice Form for Adaptive Line Enhancement," in *Proc. 14th Asilomar Conf. Circuits, Systems, and Computers,* Pacific Grove, CA, Nov., 1980, pp. 380–384.
194. Friedlander, B., "Recursive Lattice Forms for Spectral Estimation and Adaptive Control," in *Proc. 19th IEEE Conf. Decision and Control,* Dec. 1980, pp. 466–471.
195. Friedlander, B., "A Modified Lattice Algorithm for Deconvolving Filtered Impulsive Processes," in *Proc. Int. Conf. Acoust., Speech, Signal Processing,* Atlanta, GA, Mar. 1981, pp. 865–868.
196. Friedlander, B., "Recursive Lattice Forms for Spectral Estimation," in *Proc. ASSP Workshop on Spectral Estimation,* McMaster Univ., Hamilton, Ont., Canada, Aug. 1981.
197. Friedlander, B., Ljung, L., and Morf, M., "Lattice Implementation of the Recursive Maximum Likelihood Algorithm," in *Proc. 20th IEEE Conf. Decision and Control,* Dec. 1981, pp. 1083–1084.

198. Friedlander, B., "A Lattice Algorithm for Factoring the Spectrum of a Moving Average Process," in *Proc. Conf. Information Sciences and Systems,* Princeton Univ., Princeton, NJ, Mar. 1982.
199. Friedlander, B., "Instrumental Variable Methods for ARMA Spectral Estimation," in *Proc. IEEE Int. Conf. Acoust., Speech, Signal Processing,* Paris, France, May 1982, pp. 248–251.
200. Friedlander, B., "Lattice Implementations of Some Recursive Parameter Estimation Algorithms," in *6th IFAC Symp. on Identification and System Parameter Estimation,* June 1982, pp. 481–486.
201. Friedlander, B., "A Passive Maximum Likelihood Algorithm for ARMA Spectral Estimation," *IEEE Trans. Inform. Theory,* vol. IT-28, no. 4, pp. 639–646, July 1982.
202. Friedlander, B., "A Recursive Maximum Likelihood Algorithm for ARMA Line Enhancement," *IEEE Trans. Acoust., Speech, Signal Processing,* vol. ASSP-30, no. 4, pp. 651–657, Aug. 1982.
203. Friedlander, B., "Lattice Filters for Adaptive Processing," *Proc. IEEE,* vol. 70, no. 8, pp. 829–867, Aug. 1982.
204. Friedlander, B., "Lattice Methods for Spectral Estimation," *Proc. IEEE,* vol. 70, no. 9, pp. 990–1017, Sept. 1982.
205. Friedlander, B., "System Identification Techniques for Adaptive Noise Cancelling," *IEEE Trans. Acoust., Speech, Signal Processing,* vol. ASSP-30, vol. 5, pp. 699–709, Oct. 1982.
206. Friedlander, B., "System Identification Techniques for Adaptive Signal Processing," *Circuits, Systems, Signal Processing,* vol. 1, no. 1, pp. 1–41, 1982.
207. Friedlander, B., "Efficient Computation of the Covariance Sequence of an Autoregressive Process," *IEEE Trans. Automat. Contr.,* vol. AC-28, no. 1, pp. 97–99, Jan. 1983.
208. Friedlander, B., "Instrumental Variable Methods for ARMA Spectral Estimation," *IEEE Trans. Acoust., Speech, Signal Processing,* vol. ASSP-31, no. 2, pp. 404–415, Apr. 1983.
209. Friedlander, B., "An Infinite Impulse Response Lattice Filter for Adaptive Line Enhancement," *Circuits, Systems, Signal Processing,* vol. 2, no. 4, pp. 391–420, 1983.
210. Friedman, S., "On Stochastic Approximation," *Ann. Math. Stat.,* vol. 34, pp. 343–364, 1963.
211. Frost, O. L., "An Algorithm for Linearly Constrained Adaptive Array Processing," *Proc. IEEE,* vol. 60, pp. 926–935, Aug. 1972.
212. Fu, K. S., *Sequential Methods in Pattern Recognition and Machine Learning.* New York, NY: Academic Press, 1968.
213. Fuhrman, D. R. and Liu, B., "Rotational Search Methods for Adaptive Pisarenko Harmonic Retrieval," *IEEE Trans. Acoust., Speech, Signal Processing,* vol. ASSP-34, no. 6, pp. 1550–1565.
214. Gabor, D., Wilby, W. P. L., and Woodcock, R., "A Universal Non-Linear Filter Predictor and Simulator Which Optimizes Itself by a Learning Process," *Proc. Inst. Elec. Eng.,* vol. 108B, July 1960.
215. Gabriel, W. F., "Spectral Analysis and Adaptive Array Super-Resolution Techniques," *Proc. IEEE,* vol. 68, pp. 654–666, June 1980.
216. Gabriel, W. F., "Adaptive Arrays—An Introduction," *Proc. IEEE,* vol. 64, no. 2, pp. 239–272, Feb. 1976.
217. Gardiner, T., *et al,* "Noise Cancellation Studies Using a Least-Squares Lattice Filter," in *Proc. IEEE Int. Conf. Acoust., Speech, Signal Processing,* Tampa, FL, 1985, pp. 1173–1176.
218. Gardner, W. A., "Learning Characteristics of Stochastic-Gradient Descent Algorithms: A General Study, Analysis, and Critique," *Signal Processing,* vol. 6, no. 2, pp. 113–133, Apr. 1984.
219. Gerlach, K., "Fast Orthogonalization Networks," *IEEE Trans. Antennas Propagat.* (Special issue on adaptive processing antenna systems), vol. AP-34, no. 3, pp. 458–462, Mar. 1986.
220. Gersho, A., "Adaptive Equalization of Highly Dispersive Channels," *Bell Syst. Tech. J.,* vol. 48, no. 1, pp. 55–70, Jan. 1969.
221. Gersho, A., and Lim, T. L., "Adaptive Cancellation of Intersymbol Interference for Data Transmission," *Bell Syst. Tech. J.,* vol. 60, no. 11, pp. 1997–2021, Nov. 1981.
222. Gibson, D., "On Reflection Coefficients and the Cholesky Decomposition," *IEEE Trans. Acoust., Speech, Signal Processing,* vol. ASSP-25, no. 1, pp. 93–96, 1977.
223. Gibson, C. J., and Haykin, S., "A Comparison of Algorithms for the Calculation of Adaptive Lattice Filter," in *Proc. IEEE Conf. on Acoust., Speech, Signal Processing,* 1979.
224. Gibson, C. J., and Haykin, S., "Learning Characteristics of Adaptive Lattice Filtering Algorithms," *IEEE Trans. Acoust., Speech, Signal Processing,* vol. ASSP-28, pp. 681–691, Dec. 1980.
225. Gibson, C. J., and Haykin, S., "Nonstationary Learning Characteristics of Adaptive Lattice Filters," in *Proc. IEEE Int. Conf. Acoust., Speech, Signal Processing,* Paris, France, 1982, pp. 671–674.
226. Gitlin, R. D., Mazo, J. E., and Taylor, M. G., "On the Design of Gradient Algorithms for Digitally Implemented Adaptive Filters," *IEEE Trans. Circuit Theory,* vol. CT-20, no. 2, pp. 125–136, Mar. 1973.
227. Gitlin, R. D., and Magee, F. R., "Self-Orthogonalization Algorithms," *IEEE Trans. Commun.,* vol. COM-25, no. 7, pp. 666–672, July 1977.
228. Gitlin, R. D., and Weinstein, S. B., "The Effects of Large Interference on the Tracking Capability of Digitally Implemented Echo Cancelers," *IEEE Trans. Commun.,* vol. COM-26, no. 6, pp. 833–839, June 1978.
229. Gitlin, R. D., and Weinstein, S. B., "On the Required Tap Weight Precision for Digitally Implemented Adaptive Equalizers," *Bell Syst. Tech. J.,* vol. 58, no. 2, pp. 301–321, Feb. 1979.
230. Gitlin, R. D., and Weinstein, S. B., "Fractionally-Spaced Equalization: An Improved Digital Transversal Equalizer," *Bell Syst. Tech. J.,* vol. 60, no. 2, pp. 275–296, Feb. 1981.
231. Gladyshev, E. G., "On Stochastic Approximation," *Theory Prob. Appl.,* vol. 10, pp. 275–278, Mar. 1965.
232. Glaser, E. M., "Signal Detection by Adaptive Filters," *IEEE Trans. Inform. Theory,* vol. IT-7, no. 2, pp. 87–98, Apr. 1961.
233. Glover, J., "Adaptive Noise Cancelling of Sinusoidal Interferences," Ph.D. dissertation, Stanford Univ., Stanford, CA, May 1975.
234. Godara, L. C., and Cantoni, A., "Analysis of the Performance of Adaptive Beamforming Using Perturbation Sequences," *IEEE Trans. Antennas Propagat.,* vol. AP-31, no. 2, pp. 268–279, Mar. 1983.
235. Godara, L. C., and Cantoni, A., "Analysis of Constrained LMS Algorithm with Application to Adaptive Beamforming Using Perturbation Sequence," in *Proc. IEEE Int. Conf. Acoust., Speech, Signal Processing,* Tampa, FL, 1985, pp. 1804–1807.
236. Godard, D., "Channel Equalization Using a Kalman Filter for Fast Data Transmission," *IBM J. Res. Develop.,* vol. 18, pp. 267–273, May 1974.
237. Golub, G. H. and C. F. Van Loan, *Matrix Computations.* Baltimore, MD: Johns Hopkins University Press, 1983.
238. Goode, B. B., "Adaptive Sensor Array Processing," Ph.D. dissertation, Stanford University, Stanford, CA, Nov. 1970.
239. Goodwin, G. C., and Payne, R. L., *Dynamic System Identification, Experimental Design and Data Analysis.* New York, NY: Academic Press, 1977.
240. Goodwin, G. C. and K. S. Sin, *Adaptive Filtering, Prediction and Control.* Englewood Cliffs, NJ: Prentice-Hall, 1984.
241. Grant, P. M., and Kino, G. S., "Adaptive Filter Based on SAW Monolithic Storage Correlators," *Electron. Lett.,* vol. 14, no. 7, pp. 562–564, Aug. 17, 1978.
242. Grant, P. M., and Morgul, A., "Frequency Domain Adaptive Filter Based on SAW Chirp Transform Processors," in *Proc. IEEE Ultrasonics Symposium,* 1982, pp. 186–189.
243. Gray, A. H., Jr., "Passive Cascade Lattice Digital Filters," *IEEE Trans. Circuit Syst.,* vol. CAS-27, no. 5, pp. 337–344, May 1980.
244. Gray, A. H., Jr., and Markel, J. D., "Digital Lattice and Ladder Filter Synthesis," *IEEE Trans. Audio Electroacoust.,* vol. AU-21, no. 6, pp. 491–500, 1973.
245. Gray, A. H., Jr., and Markel, J. D., "A Normalized Digital Filter Structure," *IEEE Trans. Acoust., Speech, Signal Processing,* vol. ASSP-23, no. 3, pp. 268–277, 1975.
246. Griffiths, L. J., "A Comparison of Multidimensional Wiener and Maximum-Likelihood Filters," *Proc. IEEE,* vol. 55, pp. 2045–2049, Nov. 1967.
247. Griffiths, L. J., "A Simple Adaptive Algorithm for Real-Time Processing in Antenna Arrays," *Proc. IEEE,* vol. 57, pp. 1969–1704, Oct. 1969.
248. Griffiths, L. J., "Rapid Measurement of Instantaneous Frequency," *IEEE Trans. Acoustics, Speech, Signal Processing,* vol. ASSP-23, pp. 209–222, Apr. 1975.
249. Griffiths, L. J., "A Continuously-Adaptive Filter as a Lattice Structure," in *Proc. Int. Conf. Acoust., Speech, Signal Processing,* 1977, pp. 683–686.
250. Griffiths, L. J., "Adaptive Lattice Structure for Noise Cancelling Applications," in *Proc. IEEE Int. Conf. Acoust., Speech, Signal Processing,* Tulsa, OK, Apr. 1978, pp. 87–90.
251. Griffiths, L. J., and Medaugh, R. S., "Convergence Properties of an Adaptive Noise Cancelling Lattice Structure," in *IEEE Conf. Decision and Control,* San Diego, CA, Jan. 1979, pp. 1357–1361.
252. Griffiths, L. J., "Adaptive Structures for Multiple-Input Noise Cancelling Applications," in *Proc. IEEE Int. Conf. Acoust., Speech, Signal Processing,* Washington, DC, Apr. 1979, pp. 925–

928.
253. Griffiths, L. J., "Adaptive Structures for Multiple-Input Noise Cancelling," in *Proc. Int. Conf. Acoust., Speech, Signal Processing,* 1979, pp. 925–928.
254. Griffiths, L. J., and Jim, C. W., "An Alternative Approach to Linearly Constrained Adaptive Beamforming," *IEEE Trans. Antennas Propagat.*, vol. AP-30, pp. 27–34, Jan. 1982.
255. Hague, Y. A., "An Adaptive Transversal Filter," in *Proc. Int. Conf. Acoust., Speech, Signal Processing,* Boston, MA, 1983, pp. 1208–1211.
256. Harris, R. W. *et al,* "Variable Step Filter Algorithms; Applications to FIR and IIR Filters," *IEEE Trans. Acoust., Speech, Signal Processing,* vol. ASSP-34, no. 2, pp 309–316, Apr. 1986.
257. Harrison, W. A. *et al,* "Adaptive Noise Cancellation Applied to Case Where Acoustic Barrier Exists Between Primary and Reference Microphones," *IEEE Trans. Acoust., Speech, Signal Processing,* vol. ASSP-34, no. 1, pp. 21–27, Feb. 1986.
258. Hartman, P., and Bynam, B., "Adaptive Equalization for Digital Microwave Radio Systems," in *Proc. IEEE Intern. Conf. on Communications (ICC),* paper 8.5, 1980.
259. Haykin, S., *Array Processing Application to Radar.* Stroudsburg, PA: Dowden, Hutchinson & Ross, Inc., 1980.
260. Haykin, S., *Introduction to Adaptive Filters.* New York, NY: MacMillan, 1984.
261. Haykin, S., Ed. *Adaptive Filter Theory.* Englewood Cliffs, NJ: Prentice-Hall, 1986.
262. Haykin, S., Ed. *Array Signal Processing.* Englewood Cliffs, NJ: Prentice-Hall, 1985.
263. Hodgkiss, W. S., and Alexandrou, D., "Power Normalization Sensitivity of Adaptive Frequency Tracking Architectures," in *Proc. 1st ASSP Workshop on Spectral Estimation,* McMaster Univ., Hamilton, Ont., Canada, Aug., 1981.
264. Hodgkiss, W. S., and Alexandrou, D., "Power Normalization Sensitivity of Adaptive Lattice Structures," *IEEE Trans. Acoust., Speech, Signal Processing,* vol. ASSP-32, no. 4, pp. 925–928, Aug., 1984.
265. Hodgkiss, W. S., and Presley, J. A., Jr., "Adaptive Trackings of Multiple Sinusoids Whose Power Levels are Widely Separated," *IEEE Trans. Acoust., Speech, Signal Processing,* vol. ASSP-29, pp. 710–721, June 1981.
266. Hodgkiss, W. S., and Presley, J. A., Jr., "The Complex Adaptive Least-Squares Lattice," *IEEE Trans. Acoust., Speech, Signal Processing,* vol. ASSP-30, no. 2, pp. 330–333, Apr. 1982.
267. Hodgkiss, W. S., and Alexandrou, A., "Application of Adaptive Linear Predictor Structures to the Prewhitening of Acoustic Reverberation Data," in *Proc. Int. Conf. Acoust., Speech and Signal Processing,* Boston, MA, 1983, pp. 599–602.
268. Holmes, J. K., "Two Stochastic Approximation Procedures for Identifying Linear Systems," *IEEE Trans. Automat. Contr.*, vol. AC-14, pp. 292–295, May 1969.
269. Hon, E., and Lee, S., "Noise Reduction in Fetal Electrocardiography," *Amer. J. Obst. and Gynecol.*, vol. 87, pp. 1987–1096, Dec. 15, 1963.
270. Honig, M. L., and Messerschmitt, D. G., "Convergence Models for Adaptive Gradient and Least-Squares Algorithms," in *Proc. Int. Conf. Acoust., Speech, Signal Processing,* Atlanta, GA, 1981, pp. 267–270.
271. Honig, M. L., and Messerschmitt, D. G., "Convergence Properties of an Adaptive Digital Lattice Filter," *IEEE Trans. Acoust., Speech, Signal Processing,* vol. ASSP-29, pp. 642–659, June 1981.
272. Honig, M. L., "Convergence Models for Lattice Joint Process Estimators and Least-Squares Algorithms," *IEEE Trans. Acoust., Speech, Signal Processing,* vol. ASSP-31, no. 2, pp. 415–425, Apr. 1983.
273. Honig, M. L., and Messerschmitt, D. G., *Adaptive Filters: Structures, Algorithms, and Applications.* Boston, MA: Kulwer Academic Publishers, 1984.
274. Horna, O. A., "Echo Canceler with Adaptive Transversal Filter Utilizing Pseudo-Logarithmic Coding," *COMSAT Tech. Rev.*, vol. 7, no. 2, pp. 393–428, Fall 1977.
275. Horna, O. A., "Cancellation of Acoustic Feedback," *COMSAT Tech. Rev.*, vol. 12, no. 2, pp. 319–333, Fall 1982.
276. Horowitz, L. L., and Senne, K. D., "Performance Advantage of Complex LMS for Controlling Narrow-Band Adaptive Arrays," *IEEE Trans. Acoust., Speech, Signal Processing,* vol. ASSP-29, no. 3, pp. 722–736, June 1981.
277. Horvath, S., "Lattice Form Adaptive Recursive Digital Filters: Algorithms and Applications," in *Proc. IEEE. International Symposium on Circuits and Systems (ISCAS),* 1980, pp. 128–133.
278. Howells, P. W., "Intermediate Frequency Side-Lobe Canceller," U.S. Patent 3,202,990, Aug. 24, 1965.
279. Hsia, T. C., *System Identification.* Lexington, MA: Lexington Books, 1978.
280. Hsia, T. C., "Convergence Analysis of LMS and NLMS Adaptive Algorithms," in *Proc. Int. Conf. Acoust., Speech, Signal Processing,* Boston, MA, 1983, pp. 667–670.
281. Hu, Y., "Adaptive Methods for Real-Time Pisarenko Spectrum Estimate," in *Proc. Int. Conf. Acoust., Speech, Signal Processing,* Tampa, FL, 1985, pp. 205–108.
282. Huang, Y. F., "Recursive Estimation Algorithm Using Selective Updating for Spectral Analysis and Adaptive Signal Processing," *IEEE Trans. Acoust., Speech, Signal Processing,* vol. ASSP-34, no. 5, pp. 1331–1334, Oct. 1986.
283. Hudson, J. E., *Adaptive Array Principles.* Stevenage, UK: P. Peregrinus, 1981.
284. Huhta, J. C., and Webster, J. G., "60-Hz Interference in Electrocardiography," *IEEE Trans. Biomed. Eng.*, vol. BME-20, pp. 91–101, Mar. 1973.
285. Hui, S. K., and Lim, Y. C., "A Block Adaptive Approach for Clutter Suppression," in *Proc. Int. Conf. Acoust., Speech, Signal Processing,* Tampa, FL, 1985, pp. 1368–1371.
286. Hui, S. K., and Lim, Y. C., "An Adaptive Recursive Algorithm for Array Processing of Coherent Signals," in *Proc. Int. Conf. Acoust., Speech, Signal Processing,* Tampa, FL, 1985, pp. 1845–1848.
287. Hush, D. R., and Ahmed, N., "Detection and Identification of Sinusoids in Broadband Noise via a Parallel Recursive ALE," in *Proc. Int. Conf. Acoust., Speech, Signal Processing,* Tampa, FL, 1985, pp. 1193–1196.
288. Hush, D. R., *et al,* "An Adaptive IIR Structure for Sinusoidal Enhancement, Frequency Estimation and Detection," *IEEE Trans. Acoust., Speech, Signal Processing,* vol. ASSP-34, no. 6, pp. 1380–1390, Dec. 1986.
289. Hwang, S. Y., "Realization of Canonical Digital Networks," *IEEE Trans. Acoust., Speech, Signal Processing,* vol. ASSP-22, no. 1, pp. 27–39, Feb. 1974.
290. Itakura, F., and Saito, S., "A Statistical Method for Estimation of Speech Spectral Density and Formant Frequencies," *Electron. Commun.*, vol. 53-A, pp. 36–43, 1970.
291. Itakura, F., and Saito, S., "Digital Filtering Techniques for Speech Analysis and Synthesis," in *Proc. 7th Int. Conf. Acoust.*, Budapest, Hungary, 1971, Paper 25-C-1, pp. 261–264.
292. Jakowatz, C. V., Shuey, R. L., and White, G. M., "Adaptive Waveform Recognition," in *Proc. 4th London Symposium on Information Theory,* Butterworth, London, Sept. 1960, pp. 317–326.
293. Janssen, A. J. E. M. *et al,* "Adaptive Interpolation of Discrete-Time Signals That Can Be Modeled as Autoregressive Processes," *IEEE Trans. Acoust., Speech, Signal Processing,* vol. ASSP-34, no. 2, pp. 317–330, Apr. 1986.
294. Jaynes, E. T., "On the Rationale of Maximum-Entropy Methods," *Proc. IEEE,* vol. 70, no. 9, pp. 939–952, Sept. 1982.
295. Jenkins, W. K., and Nayeri, M., "Adaptive Filters Realized with Second Order Sections," in *Proc. Int. Conf. Acoust., Speech, Signal Processing,* Tokyo, Japan, 1986, pp. 2103–2106.
296. Jiang, J., and Doraiswami, R., "A New Structure for Adaptive Signal Processing with Combined FIR and IIR Filtering Algorithms," in *Proc. Int. Conf. Acoust., Speech, Signal Processing,* Tokyo, Japan, 1986, pp. 3023–3027.
297. Johnson, C. R., Jr., and Larimore, M. G., "Comments On and Additions To "An Adaptive Recursive LMS Filter," *Proc. IEEE,* vol. 65, no. 9, pp. 1399–1401, Sept. 1977.
298. Johnson, C. R. Jr., Treichler, J. R. and Larimore, M. G., "Remarks on the Use of SHARF as an Output Error Identifier," in *Proc. 17th IEEE Conference on Decision and Control,* San Diego, CA, pp. 1094–1095, Jan. 1979(1).
299. Johnson, C. R. Jr., "A Convergence Proof for a Hyperstable Adaptive Recursive Filter," *IEEE Trans. Inform. Theory,* vol. IT-25, no. 6, pp. 745–749, Nov. 1979(2).
300. Johnson, C. R., Jr., Larimore, M. G., Treichler, J. R., and Anderson, B. D. O., "SHARF Convergence Properties," *IEEE Trans. Acoust. Speech, Signal Processing,* vol. ASSP-29, no. 3, pp. 659–670, June 1981.
301. Johnson, C. R., Jr., *et al,* "A New Adaptive Parameter Estimation Structure Applicable to ADPCM," in *Proc. Int. Conf. Acoust., Speech, Signal Processing,* Boston, MA, 1983, pp. 1–4.

302. Johnson, C. R., Jr., "Adaptive IIR Filtering: Current Results and Open Issues," *IEEE Trans. Inform. Theory,* vol. IT-30, no. 2, pp. 237–250, Mar. 1984.
303. Johnston, D. H., and De Graaf, S. R., "Improving the Resolution of Bearing in Passive Sonar Arrays by Eigenvalue Analysis," *IEEE Trans. Acoust. Speech, Signal Processing,* vol. ASSP-30, no. 4, pp. 638–647, Apr. 1982(1).
304. Johnston, D. H., "The Application of Spectral Estimation Methods to Bearing Estimation Problems," *Proc. IEEE,* vol. 70, no. 9, pp. 1018–1028, Sept. 1982(2).
305. Kadota, T., "Optimum Estimation of Nonstationary Gaussian Signals in Noise," *IEEE Trans. Inform. Theory,* vol. IT-15, pp. 201–221, Mar. 1969.
306. Kailath, T., "A View of Three Decades of Linear Filtering Theory," *IEEE Trans. Inform. Theory,* vol. IT-20, pp. 145–181, Mar. 1974.
307. Kailath, T., Ljung, L., and Morf, M., "Generalized Krein-Levinson Equations for Efficient Calculation of Fredholm Resolvents of Nondisplacement Kernels," in *Topics in Functional Analysis* (essays in honor of M. G. Krein, Advances in Mathematics Supplementary Studies, Vol. 3). New York, NY: Academic Press, 1978.
308. Kailath, T., Vieira, V., and Morf, M., "Inverses of Toeplitz Operators, Innovations, and Orthogonal Polynomials," *SIAM Rev.,* vol. 20, no. 1, pp. 1006–1019, 1978.
309. Kailath, T., Kung, S-Y, and Morf, M., "Displacement Ranks of Matrices and Linear Equations," *J. Math. Analy. Appl.,* vol. 68, no. 2, pp. 395–407, Apr. 1979.
310. Kailath, T., Kung, S-Y, and Morf, M., "Displacement Ranks of a Matrix," *Bull. Amer. Math. Soc.,* vol. 1, no. 5, pp. 769–773, Sept. 1979.
311. Kailath, T., Levy, B., Ljung, L., and Morf, M., "The Factorization and Representation of Operators in the Algebra Generated by Toeplitz Operators," *SIAM J. Appl. Math.,* vol. 37, no. 3, pp. 467–484, Dec. 1979.
312. Kailath, T., *Linear Systems.* Englewood Cliffs, NJ: Prentice-Hall, 1980.
313. Kailath, T., *Lectures on Wiener and Kalman Filtering.* New York, NY: Springer-Verlag, 1981.
314. Kailath, T., "Time-Variant and Time-Invariant Lattice Filters for Nonstationary Processes," *Mathematical Tools and Models for Control Systems Analysis and Signal Processing,* Vol. 2. Paris, France: CNRS Editions, 1982, pp. 417–464.
315. Kailath, T., Ed., *Modern Signal Processing.* New York, NY: Hemisphere Publishing, 1985.
316. Kalman, R. E., "A New Approach to Linear Filtering and Prediction Problems," *Trans. ASME,* vol. 82, Ser. D, pp. 35–43, Mar. 1960.
317. Kalman, R. E., "On the General Theory of Control," in *Proc. 1st IFAC Congress.* London: Butterworth, 1960.
318. Kalman, R. E., and Bucy, R. S., "New Results in Linear Filtering and Prediction Theory," *Trans. ASME,* vol. 83, Ser. D, pp. 95–108, Mar. 1961.
319. Kalson, S., "Recursive Least-Squares Filtering with Systolic Array Architectures: A Geometrical Approach," Ph.D. dissertation, Univ. of California (Los Angeles), 1986.
320. Kaneda, Y., and Ohga, J., "Adaptive Microphone-Array System for Noise Reduction," *IEEE Trans. Acoust., Speech, Signal Processing,* vol. ASSP-34, no. 6, pp. 1391–1400, Dec. 1986.
321. Kanemasa, A., and Sugiyama, A., "An Adaptive Filter Convergence Method for Echo Cancellation and Decision Feedback Equalization," in *Proc. IEEE Int. Conf. Acoust., Speech, Signal Processing,* Tokyo, Japan, 1986, pp. 957–960.
322. Karhunen, J., "Adaptive Algorithms for Estimating Eigenvectors of Correlation Type Matrices," in *Proc. IEEE Int. Conf. Acoust., Speech, Signal Processing,* San Diego, CA, 1984, pp. 14.6.1–14.6.4.
323. Kaunitz, J., "Adaptive Filtering of Broadband Signals as Applied to Noise Cancelling," Ph.D. dissertation, Stanford Electronics Lab., Stanford Univ., Stanford, CA, Rep. SU-SEL-72-038, August 1972.
324. Kay, S. M., and Marple, S. L., Jr., "Spectrum Analysis—A Modern Perspective," *Proc. IEEE,* vol. 69, no. 11, pp. 1380–1419, Nov. 1981.
325. Kay, S. M., "Recursive Maximum Likelihood Estimation of Autoregressive Processes," *IEEE Trans. Acoust., Speech, Signal Processing,* vol. ASSP-31, no. 1, pp. 56–65, Feb. 1983.
326. Kesler, S. B., Ed., *Modern Spectrum Analysis, II.* New York, NY: IEEE PRESS, 1986.
327. Kesten, H., "Accelerated Stochastic Approximation," *Ann. Math. Stat.,* vol. 29, pp. 41–59, 1958.
328. Kiefer, J., and Wolfowitz, J., "Stochastic Estimation of the Maximum of a Regression Function," *Ann. Math. Stat.,* vol. 23, pp. 462–466, 1952.
329. Kim, J. K., and Davisson, L. D., "Adaptive Linear Estimation for Stationary M-Dependent Processes," *IEEE Trans. Inform. Theory,* vol. IT-21, pp. 23–31, Jan. 1975.
330. Kirlin, R. L. and Moghaddamjoo, A., "Robust Adaptive Kalman Filtering for Systems with Unknown Step Inputs and Non-Gaussian Measurement Errors," *IEEE Trans. Acoust., Speech, Signal Processing,* vol. ASSP-34, no. 2, pp. 252–263, Apr. 1986.
331. Klein, J. D., and Dickinson, B. W., "A Normalized Ladder Form of the Residual Energy Ratio Algorithms for PARCOR Estimation Via Projections," *IEEE Trans. Automat. Contr.,* vol. AC-28, no. 10, pp. 943–952, Oct. 1983.
332. Kobayashi, H., "Iterative Synthesis Methods for a Seismic Array Processor," *IEEE Trans. Geosci. Electron.,* vol. GE-8, pp. 169–178, July 1970.
333. Koford, J., and Groner, G., "The Use of an Adaptive Threshold Element to Design a Linear Optimal Pattern Classifier," *IEEE Trans. Inform. Theory,* vol. IT-12, pp. 42–50, Jan. 1966.
334. Koh, T., and Powers, E. J., "An Adaptive Non-linear Digital Filter with Lattice Orthogonalization," in *Proc. IEEE Int. Conf. Acoust., Speech, Signal Processing,* Boston, MA, 1983, pp. 37–40.
335. Kretschimer, F. F., Jr. and Lewis, B. L., "A Digital Open Loop Adaptive Processor," *IEEE Trans. Aerosp. Electron. Syst.,* vol. AES-14, no. 1, pp. 165–171, Jan. 1978.
336. Krolik, J., "Application of the LMS Adaptive Line Enhancer in Time Delay Estimation," in *Proc. IEEE Int. Conf. Acoust., Speech, Signal Processing,* Tampa, FL, 1985, pp. 1766–1769.
337. Krolik, J., *et al,* "A Comparative Simulation Study of the LMS Adaptive Filter Versus Generalized Methods for Time Delay Estimation," in *Proc. IEEE Int. Conf. Acoust., Speech, Signal Processing,* San Diego, CA, 1984, pp. 15.11/1–4.
338. Ku, W. H., and Ng, S. M., "Floating-Point Coefficient Sensitivity and Roundoff Noise of Recursive Digital Filters Realized in Ladder Structures," *IEEE Trans. Circuits Syst.,* vol. CAS-22, no. 12, pp. 927–936, Dec. 1975.
339. Kumar, R. V. R., and Pal, R. N., "Separation of Sinusoids Using the Constrained Adaptive Line Enhancer," in *Proc. IEEE Int. Conf. Acoust., Speech, Signal Processing,* San Diego, CA, 1984, pp. 7.4/1–4.
340. Kung, S. Y., and Rao, D. V., "Analysis and Implementation of the Adaptive Notch Filter for Frequency Estimation," in *Proc. IEEE Int. Conf. Acoust., Speech, Signal Processing,* Paris, France, 1982, pp. 663–666.
341. Kuo, S. M., and Rodriquez, M. A., "An Adaptive Frequency-Sampling Line Enhancer," in *Eighteenth Asilomar Conf. on Circuits, Systems, and Computers,* Santa Clara, CA, 1984.
342. Kurosawa, K., and Tsujii, S., "A New IIR Type Adaptive Algorithm of Parallel Type Structure," in *Proc. IEEE Int. Conf. Acoust., Speech, Signal Processing,* Tokyo, Japan, 1986, pp. 2091–2094.
343. Kushner, H. J., "New Theorems: Examples in the Liapunov Theory of Stochastic Stability," in *Proc. Joint Auto. Control Conf.,* Troy, New York, 1965, pp. 613–619.
344. Kushner, H. J., "A Note on the Maximum Sample Excursions of Stochastic Approximation Processes," *Ann. Math. Stat.,* vol. 37, pp. 513–516, 1966.
345. Kushner, H. J., "Stochastic Approximation Algorithms for the Local Optimization of Functions with Nonunique Stationary Points," *IEEE Trans. Automat. Control,* vol. AC-17, pp. 646–654, Oct. 1972.
346. Kushner, H. J., and Gavin, T., "Extensions of Kesten's Adaptive Stochastic Approximation Method," *Ann. Math. Stat.,* vol. 1, pp. 851–862, 1973.
347. Kushner, H. J., and Gavin, T., "Stochastic Approximation Type Methods for Constrained Systems," *IEEE Trans. Automat. Control,* vol. AC-19, pp. 349–357, 1974.
348. Lacoss, R. T., "Adaptive Combining of Wideband Array Data for Optimal Reception," *IEEE Trans. Geosci. Electron.,* vol. GE-6, pp. 78–86, May 1968.
349. Lagunas-Hernandz, M. A., and Masgrau-Gomez, E., "What Does Parameter Mean in Adaptive Lattice Algorithms," in *Proc. IEEE Int. Conf. Acoust., Speech, Signal Processing,* Paris, France, 1982, pp. 643–646.
350. Lainiotis, D. G., and Sims, F. L., "Sensitivity Analysis of Discrete Kalman Filters," *Int. J. Contr.,* vol. 12, pp. 657–669, June 1970.
351. Landau, I. D., "A Survey of Model Reference Adaptive Techniques-Theory and Applications," *Automatica,* vol. 10, pp. 353–379, July 1974.

352. Landau, I. D., "Unbiased Recursive Identification Using Model Reference Adaptive Techniques," *IEEE Trans. Automat. Contr.*, vol. AC-21, no. 2, pp. 194–202, Apr. 1976.
353. Landau, I. D., *Adaptive Control — The Model Reference Approach.* New York, NY: Marcel Dekker, 1979.
354. Landau, I. D., *et al*, "Applications of Output Error Recursive Estimation Algorithms for Adaptive Signal Processing," in *Proc. IEEE Int. Conf. Acoust., Speech, Signal Processing,* Paris, France, 1982, pp. 639–642.
355. Landau, I. D., Tomizuka, M., and Auslander, D. M., Eds., *Adaptive Systems in Control and Signal Processing.* Oxford, UK: Pergamon Press, 1983.
356. Larimore, M. G., Treichler, J. R., and Johnson, C. R., Jr., "SHARF: An Algorithm for Adapting IIR Digital Filters," *IEEE Trans. Acoust. Speech, Signal Processing,* vol. ASSP-28, no. 4, pp. 428–440, Aug. 1980.
357. Larimore, M. G., and Treichler, J. R., "Convergence Behavior of the Constant Modulus Algorithm," in *Proc. IEEE Int. Conf. Acoust., Speech, Signal Processing,* Boston, MA, 1983, pp. 13–16.
358. Lawson, C. L., and Hanson, R. J., *Solving Least Squares Problems.* Englewood Cliffs, NJ: Prentice Hall, 1974.
359. Lee, D., "Canonical Ladder Form Realizations and Fast Estimation Algorithms," Ph.D. dissertation, Dept. Elec. Eng., Stanford Univ., Stanford, CA, Aug. 1980.
360. Lee, D., Friedlander, B., and Morf, M., "Recursive Ladder Algorithms for ARMA Modeling," in *Proc. 19th Conf. Decision and Control,* Dec. 1980, pp. 1225–1231.
361. Lee, D., and Morf, M., "Recursive Square Root Estimation Algorithms," in *Proc. IEEE Int. Conf. Acoust. Speech, Signal Processing,* Denver, CO, Apr. 1980, pp. 1005–1017.
362. Lee, D., Morf, M., and Friedlander, B., "Recursive Square-Root Ladder Estimation Algorithms," *IEEE Trans. Acoust., Speech, Signal Processing,* vol. ASSP-29, no. 3, pp. 627–641, June 1981.
363. Lee, D. T. L., Friedlander, B., and Morf, M., "Recursive Ladder Algorithms for ARMA Modeling," *IEEE Trans. Automat. Contr.*, vol. AC-27, no. 4, pp. 753–764, Aug., 1982.
364. Lee, D. T. L., and Morf, M., "A Novel Innovations Based Time-Domain Pitch Detector," in *Proc. IEEE Int. Conf. Acoust., Speech, Signal Processing,* Denver, CO, Apr. 1980, pp. 40–44.
365. Lee, D. T. L., and Morf, M., "Distortion Measures Via Ladder Forms," presented at the *IEEE Information Theory Symp.*, Santa Monica, CA, 1981.
366. Lee, D. T. L., and Morf, M., "Generalized CORDIC for Digital Signal Processing," in *Proc. IEEE Int. Conf. Acoust., Speech, Signal Processing,* Paris, France, May 1982.
367. Lee, J. C., and Un, C. K., "On the Convergence Behavior of Frequency Domain LMS Adaptive Filters," in *Proc. IEEE Int. Conf. Acoust., Speech, Signal Processing,* San Diego, CA, 1984.
368. Lee, J. C. and Un, C. K., "Transform Domain LMS Adaptive Digital Filters; Performance," *IEEE Trans. Acoust., Speech, Signal Processing,* vol. ASSP-34, no. 3, pp. 499–510, June 1986.
369. Lee, J. W., *et al*, "Adaptive Digital Filtering of Differentially Coded Signals," in *Proc. IEEE Int. Conf. Acoust., Speech, Signal Processing,* Tampa, FL, 1985, pp. 1257–1260.
370. Lee, Y. W., *Statistical Theory of Communication.* New York, NY: John Wiley & Sons, 1960.
371. Le Roux, J., and Gueguen, C., "A Fixed Point Computation of Partial Correlation Coefficients," *IEEE Trans. Acoust., Speech, Signal Processing,* vol. ASSP-25, pp. 257–259, 1977.
372. Lev-Ari, H., "Parameterization and Modeling of Nonstationary Processes," Ph.D. dissertation, Stanford Univ., Stanford, CA, 1982.
373. Lev-Ari, H., and Kailath, T., "Schur and Levinson Algorithms for Nonstationary Processes," in *Proc. IEEE Int. Conf. Acoust., Speech, Signal Processing,* Atlanta, GA, Mar. 1981, pp. 860–864.
374. Lev-Ari, H., and Kailath, T., "On Generalized Schur and Levinson Szego Algorithms for Quasi-Stationary Processes," in *Proc. 20th IEEE Conf. Decision and Control,* San Diego, CA, Dec. 1981, pp. 1077–1080.
375. Lev-Ari, H., and Kailath, T., "Ladder Form Filters for Nonstationary Processes," in *Proc. 19th IEEE Conf. Decision and Control,* Dec. 1980, pp. 960–961.
376. Lev-Ari, H., and Kailath, T., "Lattice Filter Parameterization and Modeling of Nonstationary Processes," *IEEE Trans. Inform. Theory,* vol. IT-30, no. 1, pp. 2–16, Jan. 1984.
377. Levinson, N., "The Wiener RMS (Root-Mean-Square) Error Criterion in Filter Design and Prediction," *J. Math. Phys.*, vol. 25, pp. 261–278, 1947.
378. Li, Q. H., "Signal Separation Theory by Using Adaptive Array," in *Proc. IEEE Int. Conf. Acoust., Speech, Signal Processing,* Boston, MA, 1983, pp. 363–366.
379. Lim, Y. C., and Parker, S. R., "On the Synthesis of Lattice Parameter Digital Filters," *IEEE Trans. Circuit Syst.*, vol. CAS-31, no. 7, pp. 593–601, July 1984.
380. Ling, F., and Proakis, J. G., "Generalized Least-Squares Lattice Algorithm and Its Application to Decision Feedback Equalization," in *Proc. IEEE Int. Conf. Acoust., Speech, Signal Processing,* Paris, France, 1982, pp. 1764–1769.
381. Ling, F., and Proakis, J. G., "A Generalized Multichannel Least-Squares Lattice Algorithm Based on Sequential Processing Stages," *IEEE Trans. Acoust., Speech, Signal Processing,* vol. ASSP-32, no. 2, pp. 381–389, Apr. 1984.
382. Ling, F., and Proakis, J. G., "Nonstationary Learning Characteristics of Least-Squares Adaptive Algorithms," in *Proc. IEEE Int. Conf. Acoust., Speech, Signal Processing,* San Diego, CA, 1984, pp. 3.7/1–4.
383. Ling, F., *et al*, "A Flexible, Numerically Robust Array Processing Algorithm and Its Relationship to the Given's Transformation," in *Proc. IEEE Int. Conf. Acoust., Speech, Signal Processing,* Tokyo, Japan, 1986, pp. 2127–2130.
384. Ling, F., *et al*, "A Family of Pseudo-Least Squares Estimation Algorithms Without Division," in *Proc. IEEE Int. Conf. Acoust., Speech, Signal Processing,* Tokyo, Japan, 1986, pp. 2943–2946.
385. Ljuang, S., and Ljuang, L., "Error Propagation Properties of Recursive Least-Squares Adaptation Algorithms," *Automatica,* vol. 2, no. 21, pp. 157–167, Mar. 1985.
386. Ljung, L., "Analysis of Recursive Stochastic Algorithms," *IEEE Trans. Automat. Contr.*, vol. AC-22, pp. 551–575, Aug. 1977.
387. Ljung, L., Morf, M., and Falconer, D., "Fast Calculation of Gain Matrices for Recursive Estimation Schemes," *Int. J. Cont.*, vol. 27, no. 1, pp. 1–19, 1978.
388. Ljung, L., "The ODE Approach to the Analysis of Adaptive Control Systems-Possibilities and Limitations," in *Proc. IEEE Joint Automatic Control Conference,* San Francisco, 1980, Paper WA2-C.
389. Ljung, L., "Analysis of a General Recursive Prediction Algorithm," *Automatica,* vol. 17, no. 1, pp. 89–99, Jan. 1981.
390. Ljung, L., "Recursive Identification Techniques," in *Proc. IEEE Int. Conf. Acoust., Speech, Signal Processing,* Paris, France, 1982, pp. 627–630.
391. Ljung, L., and Söderström, T., *Theory and Practice of Recursive Identification.* Cambridge, MA: MIT Press, 1983.
392. Lower, R. R., Stofer, R. C., and Shumway, N. E., "Homovital Transplantation of the Heart," *J. Thoracic and Cardiovascular Surgery,* vol. 41, p. 196, 1961.
393. Lucky, R. W., "Automatic Equalization for Digital Communication," *Bell Syst. Tech. J.*, vol. 44, pp. 547–588, Apr. 1965.
394. Lucky, R. W., *et al, Principles of Data Communication.* New York, NY: McGraw-Hill, 1968.
395. Lucky, R. W., "A Survey of the Communication Theory Literature: 1968–1973," *IEEE Trans. Inform. Theory,* vol. IT-19, no. 6, pp. 725–739, Nov. 1973.
396. Lucky, R. W., "Techniques for Adaptive Equalization of Digital Communication Systems," *Bell Syst. Tech. J.*, vol. 45, no. 2, pp. 255–286, Feb. 1986.
397. Macchi, O. and Eweda, E., "Second-Order Convergence Analysis of Stochastic Adaptive Linear Filter," *IEEE Trans. Automat. Contr.*, vol. AC-28, pp. 76–85, Jan. 1983.
398. Madan, B. B., and Kuriyan, C. V., "An Escalator Structure for Adaptive Beamforming," in *Proc. IEEE Int. Conf. Acoust., Speech, Signal Processing,* Boston, MA, 1983, pp. 356–358.
399. Madan, B., and Parker, S., "Adaptive Beam Forming in Correlated Interference Environment," in *Proc. IEEE Int. Conf. Acoust., Speech, Signal Processing,* Tampa, FL, 1985, pp. 1792–1795.
400. Makhoul, J., "Linear Prediction: A Tutorial Review," *Proc. IEEE,* vol. 63, no. 4, pp. 561–580, Apr. 1975.
401. Makhoul, J., "New Lattice Methods for Linear Prediction," in *IEEE Int. Conf. Acoust., Speech, Signal Processing,* Philadelphia, PA, 1976, pp. 462–465.
402. Makhoul, J., "Stable and Efficient Lattice Methods for Linear Prediction," *IEEE Trans. Acoust., Speech, Signal Processing,* vol. ASSP-25, pp. 423–428, Oct. 1977.
403. Makhoul, J., "A Class of All-Zero Lattice Digital Filters, Properties and Applications," *IEEE Trans. Acoust. Speech, Signal Processing,* vol. ASSP-26, pp. 304–314, Aug. 1978.
404. Makhoul, J., "Lattice Methods in Spectral Estimation," in *Proc.*

RADC Spectrum Estimation Workshop, pp. 159–174, May 1978.

405. Makhoul, J., and Viswanathan, R., "Adaptive Lattice Methods for Linear Prediction," in *Proc. Int. Conf. Acoust., Speech, Signal Processing,* pp. 83–86, Apr. 1978.
406. Makhoul, J. I., and Cosell, L. K., "Adaptive Lattice Analysis of Speech," *IEEE Trans. Acoust. Speech, Signal Processing,* vol. ASSP-29, no. 3, pp. 654–658, June 1981.
407. Manolakis, D., *et al,* "Efficient Least-Squares Algorithms for Finite Memory Adaptive Filtering," in *Proc. of the 1984 Conf. on Information Sciences and Systems,* Princeton, NJ, pp. 28–33.
408. Manolakis, D., *et al,* "Fast Algorithms for Direct and Ladder Wiener Filters with Linear Phase," in *Proc. of the 1984 Conf. on Information Sciences and Systems,* Princeton, NJ, pp. 34–39.
409. Mansour, D., and Gray, A. H., Jr., "Unconstrained Frequency-Domain Adaptive Filter," *IEEE Trans. Acoust. Speech, Signal Processing,* vol. ASSP-30, no. 5, pp. 726–734, Oct. 1982.
410. Mansour, D., "A Highly Parallel Architecture for Adaptive Multichannel Algorithms," in *Proc. IEEE Int. Conf. Acoust., Speech, Signal Processing,* Tokyo, Japan, 1986, pp. 2931–2934.
411. Marginedes, D., "Fast Frequency Tracking Using an Adaptive Lattice Filter for a Nortex Flowmeter Signal," in *Proc. IEEE Int. Conf. Acoust., Speech, Signal Processing,* San Diego, CA, 1984.
412. Markel, J. D., and Gray, A. H., Jr., "On Autocorrelation Equations as Applied to Speech Analysis," *IEEE Trans. Audio Electroacoust.,* vol. AU-21, no. 2, pp. 69–79, Apr. 1973.
413. Markel, J. D., and Gray, A. H., Jr., "Roundoff Noise Characteristics of a Class of Orthogonal Polynomial Structures," *IEEE Trans. Acoust., Speech, Signal Processing,* vol. ASSP-23, pp. 473–486, 1975.
414. Markel, J. D., and Gray, A. H., Jr., "Fixed-Point Implementation Algorithms for a Class of Orthogonal Polynomial Filter Structures," *IEEE Trans. Acoust., Speech, Signal Processing,* vol. ASSP-23, pp. 486–494, 1975.
415. Markel, J. D., and Gray, A.. H., *Linear Prediction of Speech.* New York, NY: Springer-Verlag, 1979.
416. Marple, S. L., "Efficient Least-Squares FIR System Identification," *IEEE Trans. Acoust., Speech, Signal Processing,* vol. ASSP-29, no. 1, pp. 62–73, Feb. 1981.
417. Marple, S. L., "A Fast Least-Squares Linear Phase Adaptive Filter," in *Proc. IEEE Int. Conf. Acoust., Speech, Signal Processing,* San Diego, CA, 1984, pp. 21.8.1–21.8.4.
418. Marucci, R., "Constrained Iterative Deconvolution Using a Conjugate Gradient Algorithm," *Proc. IEEE Int. Conf. Acoust., Speech, Signal Processing,* Paris, France, 1982, pp. 1845–1848.
419. Masenten, W. K., "Adaptive Signal Processing," International Specialist Seminar on Case Studies in Advanced Signal Processing, *IEE Conf. Proc. 180,* Sept. 1979, pp. 168–177.
420. Massey, N. R., Grant, P. M., and Mavor, J., "CCD Adaptive Filter Employing Parallel Coefficient Updating," *Electron. Lett.,* vol. 15, no. 18, pp. 573–574, Aug. 30, 1979.
421. Mayham, J. T., "Adaptive Nulling with Multiple-Beam Antennas," *IEEE Trans. Antennas Propagat.,* vol. AP-26, pp. 267–273, Mar. 1978.
422. Mayhan, J. T., "Some Techniques for Evaluating the Bandwidth Characteristics of Adaptive Nulling Systems," *IEEE Trans. Antennas Propagat.,* vol. AP-27, no. 3, pp. 363–373, May 1979.
423. Mazo, J. E., "Analysis of Decision Directed Convergence," *Bell Syst. Tech. J.,* vol. 59, no. 10, pp. 1858–1876, Dec. 1980.
424. McCool, J. M., "A Constrained Adaptive Beamformer Tolerant of Array Gain and Phase Errors," in *Aspects of Signal Processing,* Pt. 2, G. Tacconi, Ed. Dordrecht, Holland: D. Reidel Publishing Co., 1977, pp. 477–483.
425. McGarty, T. P., *Stochastic Systems and State Estimation.* New York, NY: John Wiley and Sons, 1974.
426. McShane, E. J., *Stochastic Calculus and Stochastic Models.* New York, NY: Academic Press, 1974.
427. McWhirter, J. G. *et al,* "A Digital Adaptive Noise Canceller Based on a Stabilized Version of the Widrow LMS Algorithm," in *Proc. IEEE Int. Conf. Acoust. Speech, Signal Processing,* Paris, France, 1982, pp. 1394–1397.
428. McWhirter, J. G., and Shepherd, T. J., "Least-Squares Lattice Algorithm and Adaptive Channel Equalization, A Simplified Derivation," *IEE Proc.,* vol. 130, part F, no. 6, pp. 532–542, Oct. 1983.
429. Medaugh, R. S., "A Comparison of Two Fast Linear Predictions," Ph.D. dissertation, Dept. Elec. Eng., Univ. Colorado, Boulder, CO, 1981.
430. Medaugh, R. S., and Griffiths, L. J., "Further Result of a Least-Squares and Gradient Adaptive Lattice Algorithm Comparison," in *Proc. IEEE Int. Conf. Acoust., Speech, Signal Processing,* Paris, France, 1982, pp. 1412–1415.
431. Medaugh, R. S., and Griffiths, L. J., "A Comparison of Two Fast Linear Predictors," in *Proc. IEEE Int. Conf. Acoust., Speech, Signal Processing,* pp. 293–296, 1981.
432. Mendel, J. M., and Fu, K. S., *Adaptive Learning and Pattern Recognition.* New York, NY: Academic Press, 1970.
433. Meng, T. H. Y., and Messerschmitt, D. G., "Implementations of Arbitrarily Fast Adaptive Lattice Filters with Multiple Slow Processing Elements," in *Proc. IEEE Int. Conf. Acoust., Speech, Signal Processing,* Tokyo, Japan, 1986, pp. 1153–1156.
434. Mermoz, H., "Adaptive Filtering and Optimal Utilization of an Antenna," Ph.D. thesis, Institute Polytechnique, Grenoble, France, Oct. 4, 1965.
435. Mermoz, H., "Optimal and Adaptive Receiving Antennas—A Review [Essai de Synthese sur les Antennes de Detection Optimales et Adaptives]," a translation from the French by J. M. Taylor, Jr., NRL Translation 1241, January 3, 1972, Original Source: *Ann. Telecommun.,* vol. 24, pp. 269–280, 1970.
436. Messerschmitt, D. G., "A Class of Generalized Lattice Filters," *IEEE Trans. Acoust., Speech, Signal Processing,* vol. ASSP-28, no. 2, pp. 198–204, Apr. 1980.
437. Messerschmitt, D. G., "Echo Cancellation in Speech and Data Transmission," *IEEE J. on Select. Areas Commun.,* vol. SAC-2, no. 2, pp. 283–297, Mar. 1984.
438. Michael, K., "The VLSI Implementation of the Adaptive Lattice Filter," M.S. thesis, Pennsylvania State University, 1984.
439. Michael, K., and Sibul, L. H., "VLSI Implementation of Adaptive Lattice Filter," *Proc. 1984 International Symposium on Circuits and Systems,* 1984, pp. 772–775.
440. Middleton, D., and Groginsky, H. L., "Detection of Random Acoustic Signals by Receivers with Distributed Elements: Optimum Receiver Structures for Normal Signal and Noise Fields," *J. Acoust. Soc. Am.,* vol. 38, pp. 727–737, Nov. 1965.
441. Mikhael, W. B., *et al,* "ARMA Modeling by Cascading a Linear-Zero Structure," *Eighteenth Asilomar Conf. Circuits, Systems, and Computers,* Santa Clara, CA, 1984.
442. Mitra, S. K., and Sherwood, R. J., "Canonical Realizations of Digital Filters Using the Continued Fraction Expansion," *IEEE Trans. Audio Electroacoust.,* vol. AU-20, no. 3, pp. 185–194, Aug. 1972.
443. Mitra, S. K., and Sherwood, R. J., "Digital Ladder Networks," *IEEE Trans. Audio Electroacoust.,* vol. AU-21, pp. 30–36, 1973.
444. Mitra, D., and Sondhi, M. M., "Adaptive Filtering with Non-ideal Multipliers-Applications to Echo Cancellation," in *Proc. IEEE Inter. Conf. on Communications (ICC),* 1975, pp. 30.11–30.15.
445. Mitra, S. K., Kamat, P. S., and Huey, D. C., "Cascaded Lattice Realization of Digital Filters," *Circuit Theory Appl.,* vol. 3, pp. 3–11, 1977.
446. Miyanaga, Y., *et al,* "Adaptive Identification of a Time-Varying ARMA Speech Model," *IEEE Trans. Acoust., Speech, Signal Processing,* vol. ASSP-34, no. 2, pp. 423–433, June 1986.
447. Mondal, K., and Mitra, S. K., "P-Normalized Digital Two-Pairs," *IEEE Trans. Acoust., Speech, Signal Processing,* vol. ASSP-26, no. 4, pp. 374–376, Aug. 1978.
448. Monsen, P., "Feedback Equalization for Fading Dispersive Channels," *IEEE Trans. on Inform. Theory,* vol. IT-17, no. 1, pp. 56–64, Jan. 1971.
449. Monsen, P., "Adaptive Equalization of the Slow Fading Channel," *IEEE Trans. Commun.,* vol. COM-22, no. 8, pp. 1064–1075, Aug. 1974.
450. Monsen, P., "Fading Channel Communications," *IEEE Commun. Soc. Mag.,* vol. 18, no. 1, pp. 27–36, Jan. 1980.
451. Montagna, R., and Nebbia, L., "Comparison of Some Algorithms for Tap Weight Evaluation in Adaptive Echo Cancellers," in *Proc. IEEE Int. Conf. Acoust., Speech, Signal Processing,* Paris, France, 1982, pp. 1404–1407.
452. Monzingo, R. A., and Miller, T. W., *Introduction to Adaptive Arrays.* New York, NY: John Wiley, 1980.
453. Morf, M., "Fast Algorithms for Multivariable Systems," Ph.D. dissertation, Dept. Elec. Eng., Stanford Univ., Stanford, CA, 1974.
454. Morf, M., "Ladder Forms in Estimation and System Identification," in *Proc. 11th Asilomar Conf. Circuits Systems and Computers,* Monterey, CA, Nov. 1977, pp. 424–429.
455. Morf, M., "Doubling Algorithms for Toeplitz and Related Equations," in *Proc. IEEE Int. Conf. Acoust., Speech, Signal Processing,* Apr. 1980, pp. 954–959.

456. Morf, M., and Delosme, J. M., "Matrix Decomposition and Inversions Via Elementary Signature-Orthogonal Transformation," presented at the Int. Symp. Mini and Micro-Computers in Control and Measurements, San Francisco, CA, May 1981.
457. Morf, M., Dickinson, B., Kailath, T., and Vieira, A., "Efficient Solutions of Covariance Equations for Linear Prediction," *IEEE Trans. Acoust., Speech, Signal Processing,* vol. ASSP-25, no. 5, pp. 429–435, 1977.
458. Morf, M., Kailath, T., and Dickinson, B. W., "General Speech Models and Linear Estimation," in *Speech Recognition,* R. Reddy, Ed. New York, NY: Academic Press, 1975.
459. Morf, M., Kailath, T., and Ljung, L., "Fast Algorithm for Recursive Identification," in *Proc. IEEE Conf. Decision and Control,* Clearwater Beach, FL, Dec. 1976, pp. 916–921.
460. Morf, M., and Lee, D., "Recursive Spectral Estimation of α-Stationary Processes," in *Proc. Rome Air Development Center Spectrum Estimation Workshop,* Griffiss Air Force Base, NY, May 1978, pp. 97–108.
461. Morf, M., and Lee, D., "Recursive Least Squares Ladder forms for Fast Parameter Tracking," in *Proc. IEEE Conf. Decision and Control,* San Diego, CA, Jan. 1979, pp. 1362–1367.
462. Morf, M., and Lee, D. T. L., "Fast Algorithms for Speech Modeling," Tech. Rep. M303-1, Information Systems Lab., Stanford Univ., Stanford, CA, Dec. 1978.
463. Morf, M., and Lee, D. T. L., "State-Space Structures of Ladder Canonical Forms," in *Proc. 19th IEEE Conf. Decision and Control,* Albuquerque, NM, Dec. 1980, pp. 1221–1224.
464. Morf, M., Lee, D. T., Nickolls, J. R., and Vieira, A., "A Classification of Algorithms for ARMA Models and Ladder Realizations," in *Proc. IEEE Conf. Acoust., Speech, Signal Processing,* Hartford, CT, Apr. 1977, pp. 13–19.
465. Morf, M., Lee, D. T., and Vieira, A., "Ladder Forms for Estimation and Detection," (abstracts of papers), in *IEEE Int. Symp. Information Theory,* Ithaca, NY, Oct. 1977, pp. 111–112.
466. Morf, M., Muravchik, C. H., Ang, P. H., and Delosme, J-M., "Fast Cholesky Algorithms and Adaptive Feedback Filters," in *Proc. IEEE Int. Conf. Acoust., Speech, Signal Processing,* Paris, France, 1982, pp. 1727–1731.
467. Morf, M., Muravchik, C. H., and Lee, D. T., "Hilbert Space Array Methods for Finite Rank Process Estimation and Ladder Realizations for Adaptive Signal Processing," in *Proc. IEEE Int. Conf. Acoust., Speech, Signal Processing,* Atlanta, GA, Mar. 1981, pp. 856–859.
468. Morf, M., Vieira, A., and Kailath T., "Covariance Characterization by Partial Autocorrelation Matrices," *Annals Stat.,* vol. 6, pp. 643–645, May 1978.
469. Morf, M., Vieira, A., Lee, D., and Kailath, T., "Recursive Multichannel Maximum Entropy Spectral Estimation," *IEEE Trans. Geosci. Electron.,* vol. GE-16, no. 2, pp. 85–94, Apr. 1978.
470. Morf, M., Vieira, A., and Lee, D., "Ladder Forms for Identification and Speech Processing," in *Proc. IEEE Conf. Decision and Control,* New Orleans, LA, Dec. 1977, pp. 1074–1078.
471. Morf, M., and Muravchik, C. H., "A New Stable Feedback Ladder Algorithm for the Identification of Moving Average Processes," in *Proc. IEEE Int. Conf. Acoust., Speech, Signal Processing,* Boston, MA, 1983, pp. 683–686.
472. Morgan, D. R., and Aridgides, A., "Adaptive Sidelobe Cancellation of Wideband Multipath Interference," *IEEE Trans. Antennas Propagat.,* vol. AP-33, no. 8, pp. 908–917, Aug. 1985.
473. Morgan, D. R., and Aridgides, A., "Adaptive Array Cancellation of Multipath Interference," *Proc. IEEE Int. Conf. Acoust., Speech, Signal Processing,* San Diego, CA, 1984, pp. 46.10.1–46.10.4.
474. Morgul, A., "Wideband Frequency Domain Adaptive Filter Module," *Proc. IEEE Int. Conf. Acoust., Speech, Signal Processing,* Boston, MA, 1983, pp. 5–8.
475. Moses, R. L., Cadzow, J. A., and Beex, A. A., "A Recursive Procedure for ARMA Modeling," *IEEE Trans. Acoust., Speech, Signal Processing,* vol. ASSP-33, no. 4, pp. 1188–1196, Oct. 1985.
476. Mueller, K. H., "A New Fast Converging Mean Square Algorithm for Adaptive Equalizers with Partial Response Signaling," *Bell Syst. Tech. J.,* vol. 54, pp. 143–153, Jan. 1975.
477. Mueller, K. H., and Spaulding, K. H., "Cyclic Equalization—A New Rapidly Converging Equalization Technique for Synchronous Data Communication," *Bell Syst. Tech. J.,* vol. 54, pp. 369–406, Feb. 1975.
478. Mueller, K. H., "Combined Echo Cancellation and Decision Feedback Equalization," *Bell Syst. Tech. J.,* vol. 58, no. 2, pp. 491–500, Feb. 1979.
479. Mueller, M. S., "Least-Squares Algorithms for Adaptive Equalizers," *Bell Syst. Tech. J.,* vol. 60, no. 8, pp. 1905–1925, Oct. 1981.
480. Mueller, M. S., "On the Rapid Initial Convergence of Least-Squares Equalizer Adjustment Algorithms," *Bell Syst. Tech. J.,* vol. 60, no. 10, pp. 2345–2358, Dec. 1981.
481. Mueller, M. S., and Werner, J. J., "Adaptive Echo Cancellation with Dispersion and Delay in the Adjustment Loop," in *Proc. IEEE Int. Conf. Acoust., Speech, Signal Processing,* Paris, France, p. 1384.
482. Mulgrew, B., and Cowan, C. F. N., "An Adaptive IIR Equalizer: A Kalman Filter Approach," in *Proc. IEEE Int. Conf. Acoust., Speech, Signal Processing,* Tokyo, Japan, 1986, pp. 2099–2102.
483. Mullis, C. T., and Roberts, R. A., "Roundoff Noise in Digital Filters: Frequency Transformations and Invariants," *IEEE Trans. Acoust., Speech, Signal Processing,* vol. ASSP-24, no. 6, pp. 538–550, Dec. 1976.
484. Mullis, C. T., and Roberts, R. A., "Synthesis of Minimum Roundoff Noise Fixed-Point Digital Filters," *IEEE Trans. Circuit Syst.,* vol. CAS-23, pp. 501–512, Sept. 1976.
485. Muravchik, C. H., *et al,* "Fast Cholesky Algorithms and Adaptive Feedback Filters," in *Proc. IEEE Int. Conf. Acoust., Speech, Signal Processing,* Paris, France, 1982, pp. 1727–1731.
486. Narayan, S. S., Peterson, A. M., and Narsimha, M. J., "Transform Domain LMS Algorithm," *IEEE Trans. Acoust., Speech, Signal Processing,* vol. ASSP-31, no. 3, pp. 609–615, June 1983.
487. Nehorai, A., "A Minimal Parameter Adaptive Notch Filter with Constrained Poles and Zeroes," in *Proc. IEEE Int. Conf. Acoust., Speech, Signal Processing,* Tampa, FL, 1985, pp. 1185–1188.
488. Nehorai, A. and Porat, B., "Adaptive Comb Filtering for Harmonic Signal Enhancement," *IEEE Trans. Acoust., Speech, Signal Processing,* vol. ASSP-34, no. 5, pp. 1124–1138, Oct. 1986.
489. Neissen, C. W., and Willim, D. K., "Adaptive Equalizer for Pulse Transmission," *IEEE Trans. Commun.,* vol. COM-18, no. 4, pp. 377– 395, Aug. 1970.
490. Nilsson, N., *Learning Machines.* New York, NY: McGraw-Hill, 1965.
491. Nitzberg, R., "Application of the Normalized LMS Algorithm to MSLC," *IEEE Trans. on Aerosp. Electron. Syst.,* vol. AES-21, no. 1, pp. 79–91, Jan. 1985.
492. Nouta, R., "Studies in Wave Digital Filter Theory and Design," Ph.D. dissertation, Delft University, Delft, The Netherlands, 1979.
493. Nuttal, A. H., "Direct Coherence Estimation Via a Constrained Least-Squares Linear-Predictive Fast Algorithm," in *Proc. IEEE Int. Conf. Acoust., Speech, Signal Processing,* Paris, France, 1982, pp. 1104–1107.
494. Ochiai, K., Araseki, T., and Oghara, T., "Echo Canceler with Two Echo Path Models," *IEEE Trans. Commun.,* vol. COM-25, no. 6, pp. 589–595, June 1977.
495. Ogawa, Y., Ohmiya, M., and Itoh, K., "An LMS Adaptive Array Using a Pilot Signal," *IEEE Trans. Aerosp. Electron. Syst.,* vol. AES-21, no. 6, pp. 777–782, Nov. 1985.
496. Olcer, S., "Convergence Analysis of Ladder Algorithms for AR and ARMA Models," in *23rd IEEE Conf. on Decision and Control,* Las Vegas, NV, 1984, pp. 440–445.
497. Oppenheim, A. V., and Schafer, R. W., *Digital Signal Processing.* Englewood Cliffs, NJ: Prentice Hall, 1975.
498. Orfanidis, S. J., and Vail, L. M., "Zero-Tracking Adaptive Filters," *IEEE Trans. Acoust., Speech, Signal Processing,* vol. ASSP-34, no. 6, pp. 1566–1572, Dec. 1986.
499. Orgren, A. C., *et al,* "Convergence of an Adaptive Echo Cancellation System with an Augmented Predictor," in *Proc. IEEE Int. Conf. Acoust., Speech, Signal Processing,* Tokyo, Japan, 1986, pp. 961–964.
500. Owsley, N. L., "Constrained Adaption," in *Array Processing Applications to Radar.* New York, NY: Academic Press, 1980.
501. Oza, K. G., "Identification Problem and Random Contraction Mappings," Ph.D. dissertation, Department of Electrical Engineering and Computer Sciences, University of California, Berkeley, CA, 1967.
502. Oza, K. G., and Jury, E. I., "System Identification and the Principle of Random Contraction Mapping," *SIAM J. Contr.,* vol. 6, pp. 244–257, Feb. 1968.
503. Pack, J. D., and Satorious, E. H., "Least Squares Adaptive Lattice Algorithms," NOSC Tech. Rep. TR 423, Apr. 1979.
504. Panda, G. *et al,* "A Self-Orthogonalizing Efficient Block Adaptive Filter," *IEEE Trans. Acoust., Speech, Signal Processing,* vol. ASSP-34, no. 6, pp. 1573–1582, Dec. 1986.
505. Parikh, D., Ahmed, N., and Stearns, S. D., "An Adaptive Lattice Algorithm for Recursive Filters," *IEEE Trans. Acoust., Speech,*

Signal Processing, vol. ASSP-28, no. 1, pp. 110–111, Feb. 1980.
506. Pasupathy, S., and Venetsanopoulos, A. N., "Optimum Active Array Processing Structure and Space-Time Factorability," *IEEE Trans. Aerosp. Electron. Syst.,* vol. AES-10, pp. 770–779, Nov. 1974.
507. Paul, I., and Woods, J. W., "Some Experimental Results in Adaptive Prediction DPCM Coding of Images," in *Proc. IEEE Int. Conf. Acoust., Speech, Signal Processing,* Boston, MA, 1983, pp. 1220–1223.
508. Perry, F. A., and Parker, S. R., "Recursive Solutions for Zero-Pole Modeling," in *Proc. 13th Asilomar Conf. Circuits, Systems, and Computers,* Pacific Grove, Monterey, CA, Nov. 1979, pp. 509–512.
509. Perry, F. A., and Parker, S. R., "Adaptive Solutions of Multichannel Lattice Models for Linear and Non-Linear Systems," in *Proc. IEEE Int. Symp. Circuits Syst.,* 1980, pp. 744–747.
510. Picchi, G., and Prati, G., "Self-Orthogonalizing Adaptive Algorithm for Channel Equalization in the Discrete Frequency Domain," *IEEE Trans. Commun.,* vol. COM-32, no. 4, pp. 371–379, Apr. 1984.
511. Pierre, D. A., *Optimization Theory with Applications.* New York, NY: John Wiley & Sons, Inc., 1969.
512. Porat, B., "Contributions to the Theory and Applications of Lattice Forms," Ph.D. dissertation, Stanford, Univ., Stanford, CA, 1982.
513. Porat, B., and Kailath, T., "Normalized Lattice Algorithms for Least-Squares FIR System Identification," *IEEE Trans. Acoust., Speech, Signal Processing,* vol. ASSP-31, no. 1, pp. 122–128, Feb. 1983.
514. Porat, B., Friedlander, B., and Morf, M., "Square-Root Covariance Ladder Algorithms," in *Proc. Conf. Acoust., Speech, Signal Processing,* Atlanta, GA, Mar. 1981, pp. 877–880.
515. Porat, B., Morf, M., and Morgan, D., "On the Relationship Among Several Square-Root Normalized Ladder Algorithms," in *Proc. Conf. Information Sciences and Systems,* John Hopkins Univ., Baltimore, MD, Mar. 1981.
516. Porat, B., *et al,* "Square-Root Covariance Ladder Algorithms," *IEEE Trans. Automat. Contr.,* vol. AC-27, no. 4, pp. 813–829, Aug. 1982.
517. Porat, B., and Friedlander, B., "Adaptive Detection of Transient Signals," in *Proc. IEEE Int. Conf. Acoust., Speech, Signal Processing,* Tampa, FL, 1985, pp. 1266–1269.
518. Porat, B., and Friedlander, B., "Adaptive Detection of Transient Signals," *IEEE Trans. Acoust., Speech, Signal Processing,* vol. ASSP-34, no. 6, pp. 1410–1418, Dec. 1986.
519. Pratt, W. K., "Generalized Wiener Filtering Computation Techniques," *IEEE Trans. Comput.,* vol. C-21, pp. 636–641, July 1972.
520. Proakis, J. G., "Advances in Equalization for Intersymbol Interference," in *Advances in Communication Systems,* Vol. 4. New York, NY: Academic Press, 1975.
521. Qureshi, S. U. H., "Fast Start-Up Equalization with Periodic Training Sequences," *IEEE Trans. Inform. Theory,* vol. IT-23, no. 5, pp. 553–563, Sept. 1977.
522. Qureshi, S. U. H., "Adaptive Equalization," *IEEE Commun. Society Magazine,* vol. 21, no. 2, pp. 9–16, Mar. 1982.
523. Qureshi, S. U. H., "Adaptive Equalization," *Proc. IEEE,* vol. 73, no. 9, pp. 1349–1387, Sept. 1985.
524. Rabiner, L. R., and Gold, B., *Theory and Application of Digital Signal Processing.* Englewood Cliffs, NJ: Prentice Hall, 1975.
525. Rabiner, L. R., Crochiere, R., and J. Allen, "FIR System Modeling and Identification in the Presence of Noise and with Band-Limited Inputs," *IEEE Trans. Acoust., Speech, Signal Processing,* vol. ASSP-26, pp. 319–333, Aug. 1978.
526. Raja, Kumar R. V. and Ranendra, N.D. "Recursive Center-Frequency Adaptive Filters for Enhancement of Bandpass Signals," *IEEE Trans. Acoust., Speech, Signal Processing,* vol. ASSP-34, no. 3, pp. 633–637, June 1986.
527. Rao, D. V. B., and Kung, S., "Adaptive Notch Filtering for the Retrieval of Sinusoids in Noise," *IEEE Trans. Acoust., Speech, Signal Processing,* vol. ASSP-32, no. 4, pp. 799–802, Aug. 1984.
528. Reddy, V. U., Egardt, B., and Kailath, T., "Optimized Lattice-Form Adaptive Line Enhancer for a Sinusoidal Signal in Broadband Noise," *IEEE Trans. Acoust., Speech, Signal Processing,* vol. ASSP-29, no. 3, pp. 702–710, June 1983.
529. Reddy, V. U., *et al,* "Application of Modified Least-Square Algorithm to Adaptive Echo Cancellation," in *Proc. IEEE Int. Conf. Acoust., Speech, Signal Processing,* Boston, MA, 1983, pp. 53–56.
530. Reed, F. A., and Feintuch, P. L., "A Comparison of LMS Adaptive Cancelers Implemented in the Frequency Domain and Time Domain," *IEEE Trans. Acoust., Speech, Signal Processing,* vol. ASSP-29, no. 3, pp. 770–775, June 1981.
531. Reed, F. A., in *et al,* "The Effects of Interference Extent on LMS Spatial Cancellation," in *Proc. IEEE Int. Conf. Acoust., Speech, Signal Processing,* San Diego, CA, 1984, pp. 33.9/1–4.
532. Reed, I. S., Mallett, J. D., and Brennan, L. E., "Rapid Convergences Rate in Adaptive Arrays," *IEEE Trans. Aerosp. Electron. Syst.,* vol. AES-10, pp. 853–863, Nov. 1974.
533. Renner, K., and Gupta, S. C., "On the Design of Wave Digital Filters with Low Sensitivity Properties," *IEEE Trans. Circuit Theory,* vol. CT-20, no. 5, pp. 555–567, Sept. 1973.
534. Rhodes, I. B., "A Tutorial Introduction to Estimation and Filtering," *IEEE Trans. Automat. Control,* vol. AC-16, pp. 688–706, Dec. 1971.
535. Riegler, R., and Compton, R., "An Adaptive Array for Interference Rejection," *Proc. IEEE,* vol. 61, pp. 748–758, June 1973.
536. Robbins, H., and Monroe, S., "A Stochastic Approximation Method," *Ann. Math. Stat.,* vol. 22, pp. 400–407, 1951.
537. Robinson, E. A., *Multichannel Time Series Analysis with Digital Computer Programs.* San Francisco, CA: Holden-Day, 1967.
538. Robinson, E. A., and Treitel, S., "Maximum Entropy and the Relationship of the Partial Autocorrelation to the Reflection Coefficients of a Layered System," *IEEE Trans. Acoust., Speech, Signal Processing,* vol. ASSP-28, no. 2, pp. 224–235, Apr. 1980.
539. Rohrs, C. E., *et al,* "A Stability Problem in Sign-Sign Adaptive Algorithms," in *Proc. IEEE Int. Conf. Acoust., Speech, Signal Processing,* Tokyo, Japan, 1986, pp. 2999–3001.
540. Rosen, J. B., "The Gradient Projection Method for Nonlinear Programming: Part I: Linear Constraints," *J. Soc. Ind. Appl. Math.,* pp. 181–217, 1960.
541. Rosenblatt, F., "The Perceptron: A Perceiving and Recognizing Automaton, Project PARA," Cornell Aeronaut. Lab., Rep. 85-460-1, Jan. 1957.
542. Rosenblatt, F., *Principles of Neurodynamics: Perceptrons and the Theory of Brain Mechanisms.* Washington, DC: Spartan Books, 1961.
543. Rosenberger, J., and Thomas, E., "Performance of an Adaptive Echo Canceller Operating in a Noisy, Linear, Time-Invariant Environment," *Bell Syst. Tech. J.,* vol. 50, pp. 785–813, Mar. 1971.
544. Rudnick, P., "Small Signal Detection in the DIMUS Array," *J. Acoust. Soc. Am.,* vol. 32, pp. 871–877, July 1960.
545. Rutter, M. J., "Theory Design and Application of Gradient Adaptive Lattice Filters," Ph.D. thesis, Department of Electrical Engineering, University of Edinburgh, Edinburgh, Sept. 1983(2).
546. Rutter, M. J., *et al,* "Design and Realization of Adaptive Lattice Filters," in *Proc. IEEE Int. Conf. Acoust., Speech, Signal Processing,* Boston, MA, 1983, pp. 21–24.
547. Sakai, H., "Circular Lattice Filtering Using Pagano's Method," *IEEE Trans. Acoust., Speech, Signal Processing,* vol. ASSP-30, no. 2, pp. 279–287, Apr. 1982.
548. Sakrison, D. J., "Iterative Design of Optimum Filters for Non-Mean Square Error Criteria," *IEEE Trans. Inform. Theory,* vol. IT-3, pp. 161–167, Feb. 1963.
549. Sakrison, D. J., "Efficient Recursive Estimation of the Parameters of Radio Astronomy Target," *IEEE Trans. Inform. Theory,* vol. IT-12, pp. 35–41, Jan. 1966.
550. Sakrison, D. J., "Stochastic Approximation: A Recursive Method for Solving Regression Problems," in *Advances in Communication Systems,* A. V. Balakrishnan, Ed. London, England: Academic Press, 1966.
551. Salz, J., "Optimum Mean-Square Decision Feedback Equalization," *Bell Syst. Tech. J.,* vol. 52, no. 8, pp. 1341–1373, Oct. 1973.
552. Samson, C., and Reddy, U. Y., "Fixed Point Error Analysis of the Normalized Ladder Algorithm," in *Proc. IEEE Int. Conf. Acoust., Speech, Signal Processing,* Paris, France, 1982, pp. 1385–1389.
553. Samson, C., and Reddy, V. U., "Fixed Point Error Analysis of the Normalized Ladder Algorithm," *IEEE Trans. Acoust., Speech, Signal Processing,* vol. ASSP-31, no. 5, pp. 1177–1191, Oct. 1983.
554. Saro, A., and Hashimoto, K., "Adaptive Recursive Scheme for Spectral Analysis of Sinusoids in Signals with Unknown Colored Spectrum," in *Proc. IEEE Int. Conf. Acoust., Speech, Signal Processing,* Tampa, FL, 1985, pp. 109–112.
555. Sari, H., "Performance Evaluation of Three Adaptive Equalization Algorithms," in *Proc. IEEE Int. Conf. Acoust., Speech, Signal Processing,* Paris, France, 1982, pp. 1385–1389.
556. Saridis, G. N., Nikolic, Z. Z., and Fu, K. S., "Stochastic Approximation Algorithms for System Identification, Estimation and Decomposition of Mixtures," *IEEE Trans. Syst., Sci., Cybern.,* vol. SSC-5, pp. 8–15, Jan. 1969.
557. Satorius, E. H., and Shensa, M. J., "On the Application of Recursive

Least Squares Methods to Adaptive Processing,'' presented at Int. Workshop on Applications of Adaptive Control, Yale University, New Haven, CT, Aug. 1977.

558. Satorius, E. H., Smith, J. D., and Reeves, P. M., ''Adaptive Noise Cancelling of a Sinusoidal Interference Using A Lattice Structure,'' in *Proc. IEEE Int. Conf. Acoust., Speech, Signal Processing,* Washington, DC, Apr. 1979, pp. 929–932.

559. Satorius, E. H., and Alexander, S. T., ''Channel Equalization Using Adaptive Lattice Algorithms,'' *IEEE Trans. Commun.,* vol. COM-27, no. 6, pp. 899–905, June 1979.

560. Satorius, E. H., and Pack, J., ''On the Application of Lattice Algorithms to Data Equalization,'' in *Proc. 13th Asilomar Conf. Circuits, Systems, and Computers,* Pacific Grove, CA, Nov. 1979, pp. 363–366.

561. Satorius, E. H., and Shensa, M. J., ''Recursive Lattice Filters—A Brief Overview,'' in *Proc. 19th IEEE Conf. Decision and Control,* Dec. 1980, pp. 955–959.

562. Satorius, E. H., and Pack, J. D., ''Application of Least-Squares Lattice Algorithms to Adaptive Equalization,'' *IEEE Trans. Commun.,* vol. COM-29, pp. 136–142, Feb. 1981.

563. Satorius, E. H., and Pack, J. D., ''A Least-Squares Adaptive Lattice Equalizer Algorithm,'' NOSC Tech. Rep. 575, Sept. 1982.

564. Satorius, E. H., *et al,* ''Fixed-Point Implementation of Adaptive Digital Filters,'' in *Proc. IEEE Int. Conf. Acoust., Speech, Signal Processing,* Boston, MA, 1983, pp. 33–36.

565. Savaji, M. H., ''A Variable Length Lattice Filter for Adaptive Noise Cancellation,'' in *Proc. IEEE Int. Conf. Acoust., Speech, Signal Processing,* Tokyo, Japan, 1986, pp. 2935–2938.

566. Schmidt, R., ''Multiple Emitter Location and Signal Parameter Estimation,'' in *Proc. RADC Spectral Estimation Workshop,* Griffiss AFB, Rome, NY, 1979, pp. 243–258.

567. Schreiber, R., ''Implementation of Adaptive Array Algorithms,'' *IEEE Trans. Acoust., Speech, Signal Processing,* vol. ASSP-34, no. 5, pp. 1038–1045, Oct. 1986.

568. Schultheiss, P. M., ''Passive Sonar Detection in the Presence of Interference,'' *J. Acoust. Soc. Am.,* vol. 43, pp. 418–425, Mar. 1968.

569. Schwartz, M., and Winkler, L. P., ''Adaptive Nonlinear Optimization of the Signal-to-Noise Ratio of an Array Subject to a Constraint,'' *J. Acoust. Soc. Am.,* vol. 52, pp. 39–51, Jan. 1972.

570. Schweppe, F. C., ''Sensor-Array Data Processing for Multiple-Signal Sources,'' *IEEE Trans. Inform. Theory,* vol. IT-14, pp. 294–305, Mar. 1968.

571. Sedlmeyer, A., and Fettweis, A., ''Digital Filters with True Ladder Configuration,'' *Int. J. Circuit Theory Appl.,* vol. 1, no. 1, pp. 5–10, Mar. 1973.

572. Senne, K., ''Adaptive Linear Discrete-Time Estimation,'' Ph.D. dissertation, Stanford Univ., Rep. SEL-68-090, June 1968.

573. Serra, J. C., and Esteves, N. L., ''A Blind Equalization Algorithm Without Decision,'' in *Proc. IEEE Int. Conf. Acoust., Speech, and Signal Processing,* San Diego, CA, 1984.

574. Sethares, W. A., *et al,* Parameter Drift in LMS Adaptive Filters,'' *IEEE Trans. Acoust., Speech, Signal Processing,* vol. ASSP-34, no. 4, pp. 868–879, Aug. 1986.

575. Shaffer, S., and Williams, C. S., ''Comparison of the LMS, α LMS, Algorithms,'' in *Seventeenth Asilomar Conf. on Circuits, Systems, and Computers,* Santa Clara, CA, 1983.

576. Shaffer, S., and Williams, C. S., ''The Filtered Error LMS Algorithm,'' in *Proc. IEEE Int. Conf. Acoust., Speech, Signal Processing,* Boston, MA, 1983, pp. 41–44.

577. Shan, T. J., and Kailath, T., ''New Adaptive Processor for Coherent Signals and Interference,'' *Proc. IEEE Int. Conf. Acoust., Speech, Signal Processing,* San Diego, CA, 1984, pp. 33.5.1–33.5.4.

578. Shan, T. J., and Kailath, T., ''Adaptive Beamforming for Coherent Signals and Interference,'' *IEEE Trans. Acoust., Speech, Signal Processing,* vol. ASSP-33, pp. 527–536, June 1985.

579. Sharman, K. C., and Durrani, T. S., ''A Triangular Adaptive Lattice Filter for Spatial Signal Filtering,'' in *Proc. IEEE Int. Conf. Acoust., Speech, Signal Processing,* Boston, MA, 1983, pp. 348–351.

580. Shensa, M. J., ''The Spectral Dynamics of Evolving LMS Adaptive Filters,'' in *Proc. IEEE Int. Conf. Acoust., Speech, Signal Processing,* 1979, pp. 950–953.

581. Shensa, M. J., ''A Least-Squares Lattice Decision Feedback Equalizer,'' in *Proc. IEEE Int. Commun. Conf.,* Seattle, WA, Aug. 1980, pp. 57.61–57.65.

582. Shensa, M. J., ''Recursive Least-Squares Lattice Algorithms: A Geometrical Approach,'' *IEEE Trans. Automat. Contr.,* vol. AC-26, no. 3, pp. 695–702, June 1981.

583. Sherwood, D. T. and Bershad, N. J., ''Nonlinear Quantization Effects in Frequency-Domain Complex Scalar LMS Adaptive Algorithm,'' *IEEE Trans. Acoust., Speech, Signal Processing,* vol. ASSP-34, no. 1, pp. 140–151, Feb. 1986.

584. Shichor, E., ''Fast Recursive Estimation Using the Lattice Structure,'' *Bell Syst. Tech. J.,* vol. 61, no. 1, pp. 97–115, Jan. 1982.

585. Shynk, J. J., and Gooch, R. P., ''Frequency Domain Adaptive Pole-Zero Filtering,'' *Proc. IEEE,* vol. 73, no. 10, pp. 1526–1528, Oct. 1985.

586. Shynk, J. J., and Widrow, B., ''Bandpass Adaptive Pole-Zero Filtering,'' in *Proc. IEEE Int. Conf. Acoust., Speech, Signal Processing,* Tokyo, Japan, 1986, pp. 2107–2110.

587. Shynk, J. J., ''A Complex Adaptive Algorithm for IIR Filtering,'' *IEEE Trans. Acoust., Speech, Signal Processing,* vol. ASSP-34, no. 5, pp. 1342–1343, Oct. 1986.

588. Sibul, L. H., and Sohie, G. R. L., ''Structure of a Multibeam Adaptive Space-Time Processor,'' *Proc. IEEE,* vol. 70, no. 3, pp. 303–304, 1982.

589. Sibul, L. H., and Sohie, G. R. L., ''Implementation of an Adaptive Space-Time Processor by an Unconstrained Multichannel Lattice,'' in *Proc. IEEE Int. Conf. on Acoust., Speech, and Signal Processing,* 1982, pp. 799–802.

590. Sibul, L. H., ''Application of Eigenvalue Preprocessors to Adaptive Beamforming and Signal Estimation,'' in *Proc. of the Princeton Conf. on Information Sciences and Systems,* 1984, pp. 532–536.

591. Sibul, L. H., ''Application of Singular Value Decomposition to Adaptive Beamforming,'' in *Proc. of the IEEE Int. Conf. on Acoustics, Speech, and Signal Processing,* vol. 3, pp. 33.11.1–33.11.4, 1984.

592. Sibul, L. H., and Fogelsanger, A. L., ''Application of Coordinate Rotation Algorithm to Singular Value Decomposition,'' in *Proc. of the 1984 Int. Symp. of Circuits and Systems,* vol. 2, pp. 821–824, 1984.

593. Sicuranza, G. L., *et al,* ''Adaptive Echo Cancellation with Nonlinear Digital Filters,'' in *Proc. IEEE Int. Conf. Acoust., Speech, and Signal Processing,* San Diego, CA, 1984, pp. 3.10.1–3.10.4.

594. Sicuranza, G. L., and Ramponi, G., ''Nonlinear Digital Filters Using Distributed Arithmetic,'' *IEEE Trans. Acoust. Speech, Signal Processing,* vol. ASSP-34, no. 3, pp. 518–526, June 1986.

595. Singer, R. A., and Frost, P. A., ''On the Relative Performance of the Kalman and Wiener Filters,'' *IEEE Trans. Automat. Contr.,* vol. AC-14, pp. 390–394, Aug. 1969.

596. Smith, J. O., and Friedlander, B., ''Adaptive Interpolated Time-Delay Estimation,'' in *Seventeenth Asilomar Conf. on Circuits, Systems, and Computers,* Santa Clara, CA, 1983.

597. Smith, S. G., *et al,* ''A New Structure for Adaptive Echo Cancellation,'' in *Proc. IEEE Int. Conf. Acoust., Speech, Signal Processing,* Boston, MA, 1983, pp. 49–52.

598. Smith, J. O., and Friedlander, B., ''Adaptive Multipath Delay Estimation,'' in *Proc. IEEE Int. Conf. Acoust., Speech, and Signal Processing,* San Diego, CA, 1984, pp. 15.9.1–15.9.4.

599. Snyder, D. L., *The State-Variable Approach to Continuous Estimation*. Cambridge, MA: MIT Press, 1969.

600. Soderstrand, M. A., Vernia, C., Paulson, D. W., and Vigil, M. C., ''Microprocessor Controlled Adaptive Digital Filters,'' in *Proc. IEEE Int. Symp. on Circuits and Systems,* Houston, TX, Apr. 1980, pp. 142–146.

601. Sohie, G. R. L., and Sibul, L. H., ''Application of Hilbert Space Theory to Optimal and Adaptive Array Processing,'' presented at *16th Annual. Conf. Inform. Sci. Syst.,* Princeton Univ., Mar. 17–19, 1982, pp. 17–22.

602. Sohie, G. R. L., and Sibul, L. H., ''Stochastic Convergence Properties of the Adaptive Gradient Lattice,'' in *Proc. IEEE Int. Conf. Acoust., Speech, Signal Processing,* Boston, MA, 1983, pp. 663–666.

603. Sohie, G. R. L., and Sibul, L. H., ''Stochastic Convergence Properties of the Adaptive Gradient Lattice,'' *IEEE Trans. Acoust., Speech, Signal Processing,* vol. ASSP-32, no. 1, pp. 102–107, Feb. 1984.

604. Sohie, G. R. L., ''Adaptive Systems as Optimal Processors,'' *IEEE Trans. Acoust., Speech, Signal Processing,* Tampa, FL, 1985, pp. 1261–1262.

605. Sohie, G. R. L., and Sibul, L. H., ''Stochastic Operator Norm for Two-Parameter Adaptive Lattice Filters,'' *Proc. IEEE Int. Conf. Acoust., Speech, Signal Processing,* vol. ASSP-34, no. 5, pp. 1162–1165, Oct. 1986.

606. Solomon, O. M., Jr., *et al,* ''A Parametric Method for Computing Magnitude Squared Coherency,'' in *Proc. IEEE Int. Conf. Acoust.,*

Speech, and Signal Processing, San Diego, CA, 1984, pp. 21.5.1–21.5.4.

607. Sondhi, M. M., "An Adaptive Echo Canceller," *Bell Syst. Tech. J.,* vol. 46, pp. 497–520, Mar. 1967.
608. Sondhi, M. M., and Berkley, D. A., "Silencing Echoes on the Telephone Network," *Proc. IEEE,* vol. 68, no. 8, pp. 948–963, Aug. 1980.
609. Soong, F. K., and Peterson, A. M., "Fast Least-Squares in the Voice Echo Cancellation Application," in *Proc. IEEE Int. Conf. Acoust., Speech, and Signal Processing,* Paris, France, 1982, pp. 1398–1403.
610. South, C. R., Hoppitt, C. E., and Lewis, A. V., "Adaptive Filters to Improve Loudspeaker Telephone," *Electron. Lett.,* vol. 15, no. 21, pp. 673–674, Oct. 1979.
611. South, C. R., and Lewis, A. V., "Extension Facilities and Performance of an LSI Adaptive Filter," in *Proc. IEEE Int. Conf. Acoust., Speech, and Signal Processing,* San Diego, CA, 1984, pp. 3.4.1–3.4.4.
612. Srinath, M. D., and Viswanathan, M. M., "Sequential Algorithm for Identification of Parameters of an Autoregressive Process," *IEEE Trans. Automat. Contr.,* vol. AC-20, no. 4, pp. 542–546, Aug. 1975.
613. Stao, S. M., and Lopresti, P. V., "On the Generalization of State Feedback Decoupling Theory," *IEEE Trans. Automat. Contr.,* vol. AC-16, pp. 133–140, Apr. 1971.
614. Stearns, S. D., and Elliott, G. R., "On Adaptive Recursive Filtering," in *Proc. 10th Asilomar Conf. on Circuits, Systems, and Computers,* Nov. 1976, pp. 5–11.
615. Stearns, S. D., "Error Surfaces of Adaptive Recursive Filters," *IEEE Trans. Acoust., Speech, Signal Processing,* vol. ASSP-28, no. 3, pp. 763–766, June 1981.
616. Steinberg, B. D., *Principles of Aperture and Array System Design, Including Random and Adaptive Arrays.* New York, NY: Wiley-Interscience, 1976.
617. Steinhardt, A., and Jones, W., "A Qualitative Instability Theory for Lattice Filters," in *Proc. IEEE Int. Conf. Acoust., Speech, and Signal Processing,* San Diego, CA, 1984, pp. 46.1.1–46.1.4.
618. Stewart, G. W., *Introduction to Matrix Computations.* New York, NY: Academic Press, 1973.
619. Strobach, P., "New Forms of Least-Squares Lattice Algorithms and a Comparison of Their Round-off Error Characteristics," in *Proc. IEEE Int. Conf. Acoust., Speech, Signal Processing,* Tokyo, Japan, 1986, pp. 573–576.
620. Su, Y., "A Complex Algorithm for Linearly Constrained Adaptive Arrays," *IEEE Trans. Antennas Propagat.,* vol. AP-31, no. 4, pp. 676–678, July 1983.
621. Swanson, D. C., and Symons, F. W., "Sources of Numerical Errors in Both the Square-Root Normalized and Unnormalized Least-Squares Lattice Algorithms," in *Proc. IEEE Int. Conf. Acoust., Speech, and Signal Processing,* San Diego, CA, 1984, pp. 45.4.1–45.4.4.
622. Swanson, D. C., and Symons, F. W., "The Unbiased Least-Squares Lattice," in *Proc. IEEE Int. Conf. Acoust., Speech, Signal Processing,* Tampa, FL, 1985, pp. 1189–1192.
623. Symons, F. W., Jr., "Spatial and Spectral Filtering as Applied to Array Processing," Ph.D. thesis, Pennsylvania State University, 1975.
624. Symons, F. W., Jr., and Sibul, L. H., "Dynamic Effects of Signal Normalizing on Adaptive Algorithms," *IEEE Trans. Aerosp. Electron. Syst.,* vol. AES-12, no. 6, pp. 728–735, Dec. 1976.
625. Takao, K., Fujita, M., and Nishi, T., "An Adaptive Antenna Array Under Directional Constraint," *IEEE Trans. Antennas Propagat.,* vol. AP-24, no. 5, pp. 662–669, Sept. 1976.
626. Takao, K., and Komiyama, K., "An Adaptive Antenna for Rejection of Wideband Interference," *IEEE Trans. Aerosp. Electron. Syst.,* vol. AES-16, no. 4, pp. 452–459, July 1980.
627. Talmon, J. L., *et al,* "Adaptive Gaussian Filtering in Routine ECG/VCG Analysis," *IEEE Trans. Acoust., Speech, Signal Processing,* vol. ASSP-34, no. 3, pp. 527–534, June 1986.
628. Taylor, N. G., Ed., Adaptive Antennas, Special Issue, *Proc. IEE,* vol. 130, pt. F, no. 1, pp. 1–151, Jan. 1983.
629. Tjahjadi, T., and Steenaart, W. J., "Adaptive Filter Realization with a Minimum Number of Multipliers," *IEEE Trans. Circuits Syst.,* vol. CAS-32, no. 3, pp. 209–216, Mar. 1985.
630. Treichler, J. R., "The Spectral Line Enhancer: The Concept, Implementation, and an Application," Ph.D. dissertation, Stanford Univ., Stanford, CA, June 1977.
631. Treichler, J. R., Larimore, M. G., and Johnston, C. R., Jr., "Simple Adaptive IIR Filtering," in *Proc. IEEE Int. Conf. Acoust., Speech, Signal Processing,* April 1978, pp. 118–122.
632. Treichler, J. R., "Transient and Convergent Behavior of the Adaptive Line Enhancer," *IEEE Trans. Acoust., Speech, Signal Processing,* vol. ASSP-27, no. 1, pp. 53–62, Feb. 1979.
633. Treichler, J. R., and Larimore, M. G., "Thinned Impulse Responses for Adaptive FIR Filters," in *Proc. IEEE Int. Conf. Acoust., Speech, Signal Processing,* Paris, France, 1982, pp. 631–634.
634. Treichler, J. R., and Agee, B. G., "A New Approach to Multipath Correction of Constant Modulus Signals," *IEEE Trans. Acoust., Speech, Signal Processing,* vol. ASSP-31, no. 2, pp. 459–472, Apr. 1983.
635. Treichler, J. R., and Larimore, M. G., "A Real-Arithmetic Implementation of the Constant Modulus Algorithm," in *Proc. IEEE Int. Conf. Acoust., Speech, Signal Processing,* San Diego, CA, 1984, pp. 3.2.1–3.2.4.
636. Treichler, J. R., "Adaptive Algorithms That Restore Signal Properties," in *Proc. IEEE Int. Conf. Acoust., Speech, Signal Processing,* San Diego, CA, 1984, pp. 21.4.1–21.4.4.
637. Treichler, J. R., and Larimore, M., "Convergence Rates for the Constant Modulus Algorithm with Sinusoidal Inputs," in *Proc. IEEE Int. Conf. Acoust., Speech, Signal Processing,* Tampa, FL, 1985, pp. 1157–1160.
638. Tsypkin, Y., *Adaptation and Learning in Automatic Systems.* New York, NY: Academic Press, 1971.
639. Turner, J. M., "The Lattice Structure and Its Use in Estimation and Filtering," in *Proc. Conf. Information Sciences and Systems,* Princeton Univ., Princeton, NJ, Mar. 1980.
640. Turner, J. M., "Application of Recursive Exact Least-Squares Ladder Estimation Algorithm for Speech Recognition," in *Proc. IEEE Int. Conf. Acoust., Speech, Signal Processing,* Paris, France, May 1982, pp. 543–545.
641. Turner, J. M., "Fast Approximate Whitening Ladder Filters," in *Proc. IEEE Int. Conf. Acoust., Speech, Signal Processing,* Paris, France, May 1982, pp. 655–658.
642. Turner, J., Dickinson, B., and Lai, D., "Characteristics of Reflection Coefficient Estimates Based on a Markov Chain Model," in *Proc. Int. Conf. Acoust., Speech, Signal Processing,* Denver, CO, 1980, pp. 131–134.
643. Tuteur, F. B., and Chang, J. H., "A New Class of Adaptive Array Processors," *J. Acoust. Soc. Am.,* vol. 49, pp. 639–649, Mar. 1971.
644. Ungerboeck, G., "Theory on the Speed of Convergence in Adaptive Equalizers for Digital Communication," *IBM J. Res. Develop.,* pp. 546–555, Nov. 1972.
645. Ungerboeck, G., "Adaptive Maximum-Likelihood Receiver for Carrier-Modulated Data Transmission Systems," *IEEE Trans. on Commun.,* vol. COM-22, no. 5, pp. 634–636, May 1974.
646. Ungerboeck, G., "Fractional Tap-Spacing Equalizer and Consequences for Clock Recovery in Data Modems," *IEEE Trans. Commun.,* vol. COM-24, no. 8, pp. 856–864, Aug. 1976.
647. Urkowitz, H., "On Detection and Estimation of Wave Fields for Surveillance," *IEEE Trans. Mil. Electron.,* vol. ME-12, pp. 44–56, Jan. 1970.
648. Vaccaro, R. J., "On Adaptive Implementations of Pisarenko's Harmonic-Retrieval Method," in *Proc. IEEE Int. Conf. Acoust., Speech, Signal Processing,* San Diego, CA, 1984, pp. 6.1.1–6.1.4.
649. Van Bemmel, J., "Detection of Weak Fetal Electrocardiograms by Autocorrelation and Crosscorrelation of Envelopes," *IEEE Trans. Biomed. Eng.,* vol. BME-15, pp. 17–23, Jan. 1968.
650. Van Trees, H. L., *Detection, Estimation and Linear Modulation Theory.* New York, NY: John Wiley & Sons, 1968.
651. Varga, R., *Matrix Iterative Analysis.* Englewood Cliffs, NJ: Prentice Hall, 1956.
652. Varner, L. W., *et al,* "A Simple Adaptive Filtering Technique for Speech Enhancement," in *Proc. IEEE Int. Conf. Acoust., Speech, Signal Processing,* Boston, MA, 1983, pp. 1126–1128.
653. Vaysboro, E. M., and Yudin, D. B., "Multiextremal Stochastic Approximation," *Eng. Cybern.,* vol. 6, pp. 1–11, Jan. 1968.
654. Ventner, J. H., "On Dvoretzky Stochastic Approximation Theorems," *Ann. Math. Stat.,* vol. 37, pp. 1534–1544, 1966.
655. Ventner, J. H., "An Extension of the Robbins-Monro Procedure," *Ann. Math. Stat.,* vol. 38, pp. 181–190, 1967.
656. Ventner, J. H., "On Convergence of the Keifer-Wolfowitz Approximation Procedures," *Ann. Math. Stat.,* vol. 38, pp. 1031–1036, 1967.
657. Vieira, A. C. G., "Matrix Orthogonal Polynomials, with Applications to Autoregressive Modeling and Ladder Forms," Ph.D. dissertation, Stanford University, Stanford, CA, Aug. 1977.
658. Vieira, A., and Kailath, T., "On Another Approach to the Schur-Cohn Criterion," *IEEE Trans. Circuit Syst.,* vol. CAS-24, pp. 218–220,

Apr. 1977.

659. Viswanathan, R., and Makhoul, J., "Sequential Lattice Methods for Stable Linear Prediction," in *Proc. EASCON '76,* pp. 155A–155H.

660. Vural, A. M., "Effects of Perturbations on the Performance of Optimum/Adaptive Arrays," *IEEE Trans. Aerosp. Electron. Syst.,* vol. AES-15, no. 1, pp. 76–87, Jan. 1979.

661. Wakita, H., "Direct Estimation of the Vocal Tract Shape by Inverse Filtering of Acoustic Speech Waveform," *IEEE Trans. Audio Electroacoust.,* vol. AU-12, pp. 417–427, Oct. 1973.

662. Wakita, H., and Gray, A. H., Jr., "Numerical Determination of the Lip Impedance and Vocal Tract Area Functions," *IEEE Trans. Acoust., Speech, Signal Processing,* vol. ASSP-23, pp. 574–558, 1975.

663. Walden, W., and Birnbaum, S., "Fetal Electrocardiography with Cancellation of Maternal Complexes," *Amer. J. Obst. and Gynecol.,* vol. 94, pp. 596–598, Feb. 1966.

664. Wasan, M. T., *Stochastic Approximation.* Cambridge, MA: Cambridge University Press, 1969.

665. Wegener, W., "On the Design of Wave Digital Lattice Filters With Short Coefficient Word Length and Optimal Dynamic Range," *IEEE Trans. Circuit Syst.,* vol. CAS-25, no. 12, p. 1091, Dec. 1978.

666. Weiss, A., and Mitra, D., "Digital Adaptive Filters: Conditions of Convergence, Rates of Convergence, Effects of Noise and Errors Arising from the Implementation," *IEEE Trans. Inform. Theory,* vol. IT-25, no. 6, pp. 637–652, Nov. 1969.

667. Whalen, A. D., *Detection of Signals in Noise.* New York, NY: Academic Press, 1971.

668. White, W. D., "Cascade Preprocessors for Adaptive Antennas," *IEEE Trans. Antennas Propagat.,* vol. AP-24, no. 5, pp. 670–684, Sept. 1976.
IEEE Trans. Antennas Propagat., vol. AP-24, no. 5, Sept. 1976.

669. Whittle, P., "On the Fitting of Multivariable Autoregressions and the Approximate Canonical Factorization of a Spectral Density Matrix," *Biometrika,* vol. 50, pp. 129–134, 1963.

670. Widrow, B., and Hoff, M., Jr., "Adaptive Switching Circuits," in *IRE WESCON Conf. Rec.,* pt. 4, pp. 96–104, 1960.

671. Widrow, B., "Adaptive Filters 1: Fundamentals," Stanford Electronics Lab., Stanford Univ., Rep. SU-SEL-66-126, Dec. 1966.

672. Widrow, B., *et al,* "Adaptive Antenna Systems," *Proc. IEEE,* vol. 55, no. 12, pp. 2143–2159, Dec. 1967.

673. Widrow, B., Mantey, P. E., Griffiths, L. J., and Goode, B. B., "Adaptive Antenna Systems," *Proc. IEEE,* vol. 55, pp. 2143–2159, Dec. 1967.

674. Widrow, B., "Adaptive Filters," in *Aspects of Network and System Theory,* R. Kalman and N. Declaris, Eds. New York, NY: Holt, Rinehart and Winston, 1971, pp. 563–587.

675. Widrow, B., and Glover, J. R., Jr., "Adaptive Noise Cancelling: Principles and Applications," *Proc. IEEE,* vol. 63, no. 12, pp. 1692–1716, Dec. 1975.

676. Widrow, B, *et al,* "Stationary and Nonstationary Learning Characteristics of the LMS Adaptive Filter," *Proc. IEEE,* vol. 64, no. 8, pp. 1151–1162, Aug. 1976.

677. Widrow, B., and McCool, J. M., "A Comparison of Adaptive Algorithms Based on the Methods of Steepest Descent and Random Search," *IEEE Trans. Antennas Propagat.,* vol. AP-24, no. 5, pp. 615–637, Sept. 1976.

678. Widrow, B., *et al,* "Signal Cancellation Phenomena in Adaptive Antennas: Causes and Cures," *IEEE Trans. Antennas Propagat.,* vol. AP-30, pp. 469–478, May 1982.

679. Widrow, B., and Walach, E., "Adaptive Signal Processing for Adaptive Control," in *Proc. IEEE Int. Conf. Acoust., Speech, Signal Processing,* San Diego, CA, 1984, pp. 21.1.1–21.1.4.

680. Widrow, B., and Stearns, S. D., *Adaptive Signal Processing.* Englewood Cliffs, NJ: Prentice-Hall, 1985.

681. Wiener, N., *Extrapolation, Interpolation and Smoothing of Stationary Time Series with Engineering Applications.* Cambridge, MA: MIT Press, and New York, NY: John Wiley, 1949.

682. Wiggins, R. A., and Robinson, E. A., "Recursive Solution to the Multichannel Filtering Problem," *J. Geophys. Res.,* vol. 70, pp. 1885–1891, Apr. 1966.

683. Wolfowitz, J., "On the Stochastic Approximation Method of Robbins and Monro," *Ann. Math. Stat.,* vol. 23, pp. 457–461, 1952.

684. Wolfowitz, J., "On the Stochastic Approximation Methods," *Ann. Math. Stat.,* vol. 27, pp. 1151–1156, 1956.

685. Wong, K. M., and Jan, Y. G., "Adaptive Walsh Equalizer for Data Transmission," *Proc. IEE,* vol. 130, pt. F, no. 2, pp. 153–160, Mar. 1983.

686. Xue, P. and Liu, B., "Adaptive Equalizer Using Finite-Bit Power-of-Two Quantizer," in *Proc. IEEE Int. Conf. Acoust., Speech, Signal Processing,* San Diego, CA, 1984, pp. 46.9.1–46.9.4.

687. Xue, P. and Liu, B., "Adaptive Equalizer Using Finite-Bit Power-of-Two Quantizer," *IEEE Trans. Acoust., Speech, Signal Processing,* vol. ASSP-34, no. 6, pp. 1603–1611, Dec. 1986.

688. Yaminysharif, M., and Durrani, T. S., "Adaptive Signal Processing Using a Modified Gradient Estimation Technique," in *Proc. IEEE Int. Conf. Acoust., Speech, Signal Processing,* Tokyo, Japan, 1986, pp. 2975–2978.

689. Yanagida, M., *et al,* "Least Squares Method for Multi-Dimensional Deconvolution," in *Proc. IEEE Int. Conf. Acoust., Speech, Signal Processing,* Paris, France, 1982.

690. Yassa, F., "A Generalized Filter Structure for IIR Adaptive Filters," *Proc. IEEE Int. Conf. Acoust., Speech, Signal Processing,* Tampa, FL, 1985, pp. 1177–1180.

691. Yeh, S., Betyar, L., and Hon, E., "Computer Diagnosis of Fetal Heart Rate Patterns," *Amer J. Obst. and Gynecol.,* vol. 114, pp. 890–897, Dec. 1972.

692. Youn, D. H., *et al,* "Estimation of Magnitude-Squared Coherence: An Adaptive Approach," in *Proc. IEEE Int. Conf. Acoust., Speech, Signal Processing,* Paris, France, 1982, pp. 1100–1103.

693. Youn, D. H., and Ahmed, N., "Comparison of Two Adaptive Methods for Time Delay Estimation," in *Proc. IEEE Int. Conf. Acoust., Speech, Signal Processing,* Boston, MA, 1983, pp. 883–886.

694. Youn, D. H., and Ahmed, N., "Time Delay Estimation Via Coherence: An Adaptive Approach," *J. Acoust. Soc. Amer.,* vol. 75, no. 2, pp. 505–514, Feb. 1984.

695. Youn, D. H., and Kim, J. H., "An Adaptive FIR Filter with an Augmented Predictor," in *Eighteenth Asilomar Conf. on Circuits, Systems, and Computers,* Santa Clara, CA, 1984.

696. Youn, D. H., and Prakash, S., "On Realizations and Related Algorithms for Adaptive Linear Phase Filtering," in *Proc. IEEE Int. Conf. Acoust., Speech, Signal Processing,* San Diego, CA, 1984, pp. 3.11.1–3.11.4.

697. Youn, D. H., *et al,* "Adaptive Realization of Phase Transform for Time Delay Estimation," in *Proc. IEEE Int. Conf. Acoust., Speech, Signal Processing,* San Diego, CA, 1984, pp. 15.10.1–15.10.4.

698. Youn, D., *et al,* "An Efficient Algorithm for Lattice Filter/Prediction," in *Proc. IEEE Int. Conf. Acoust., Speech, Signal Processing,* Tampa, FL, 1985, pp. 1181–1184.

699. Youn, D. H., and Chang, B., "Multichannel Lattice Filter for an Adaptive Array Processor with Linear Constraints," in *Proc. IEEE Int. Conf. Acoust., Speech, Signal Processing,* Tokyo, Japan, 1986, pp. 1829–1832.

700. Yule, G. U., "On a Method for Investigating Periodicities in Disturbed Series with Special Reference to Wolfer's Sunspot Numbers," *Philos. Trans. Roy. Soc. London, Ser. A,* vol. 226, pp. 267–298, 1927.

701. Zahm, C. L., "Application of Adaptive Arrays to Suppress Strong Jammers in the Presence of Weak Signals," *IEEE Trans. Aerosp. Electron. Syst.,* vol. AES-9, pp. 260–270, Mar. 1973.

702. Zeidler, J. R., Satorius, E. H., Chabries, D. M., and Wexler, H. T., "Adaptive Enhancement of Multiple Sinusoids in Uncorrelated Noise," *IEEE Trans. Acoust., Speech, Signal Processing,* vol. ASSP-26, no. 3, pp. 240–254, June 1978.

703. Zhang, Q. T., and Haykin, S., "Tracking Characteristics of the Kalman Filter in a Nonstationary Environment for Adaptive Filter Applications," *Proc. IEEE Int. Conf. Acoust., Speech, Signal Processing,* Boston, MA, 1983, pp. 671–674.

704. Zentner, C. R., "Frequency Domain Adaptive Decoupling in Multiple Output Array Processors," Ph.D. thesis, Pennsylvania State University, 1975.

705. Zinser, R., *et al,* "Some Experimental and Theoretical Results Using a New Adaptive Filter Structure for Noise Cancellation in the Presence of Crosstalk," *Proc. IEEE Int. Conf. Acoust., Speech, Signal Processing,* Tampa, FL, 1985, pp. 1253–1256.

706. *IEEE Trans. Antennas Propagat.,* Special Issue on Active and Adaptive Antennas, vol. AP-12, no. 2, Mar. 1964.

707. *IEEE Trans. Antennas Propagat.,* Special Issue on Adaptive Antennas, vol. AP-24, no. 5, Sept. 1976.

708. *IEEE Trans. Antennas Propagat.,* Special Issue on Adaptive Processing Antenna Systems, vol. AP-34, no. 3, Mar. 1986.

709. Joint Special Issue on Adaptive Signal Processing, *IEEE Trans. on Circuits and Systems,* vol. CAS-28, no. 6, and *IEEE Trans. on Acoust., Speech, and Signal Processing,* vol. ASSP-29, no. 3, June 1981.

710. Special Issue on Spectral Estimation, *Proc. IEEE,* vol. 70, no. 9, Sept. 1982.

711. *IEEE Trans. Inform. Theory,* Special Issue on Linear Adaptive Filtering, vol. IT-30, no. 2, Mar. 1984.

Subject Index to Bibliography

Author Index

Subject Index

Editor's Biography

Leon H. Sibul (S'52–A'53–M'60) was born in Võru, Estonia, on August 30, 1932. He received the B.E.E. degree from George Washington University, Washington, DC, in 1960, the M.E.E. degree from New York University, New York, in 1963, and the Ph.D. degree from The Pennsylvania State University, University Park, all in electrical engineering.

From 1960 to 1964, he was a member of the technical staff at Bell Telephone Laboratories working primarily on the electronic switching system. Since 1964, he has been with the Applied Research Laboratory, The Pennsylvania State University, engaged in various aspects of research in underwater systems. His primary research interests are in the areas of adaptive signal processing, array processing, stochastic system theory, and broadband signal ambiguity function theory. He directs a group that does research in these areas. He has developed and has taught graduate courses in adaptive signal processing. He is currently Senior Scientist at the Applied Research Laboratory and Professor of Acoustics.

Dr. Sibul is a member of Sigma Tau, Sigma Xi, and the Society for Industrial and Applied Mathematics. He is the Associate Editor for Sonar and Undersea Systems of IEEE TRANSACTIONS ON AEROSPACE AND ELECTRONIC SYSTEMS.